# Fundamental Theories of Physics

Volume 177

AF774199

*Series editors*

Henk van Beijeren, Utrecht, The Netherlands
Philippe Blanchard, Bielefeld, Germany
Paul Busch, York, UK
Bob Coecke, Oxford, UK
Dennis Dieks, Utrecht, The Netherlands
Detlef Dürr, München, Germany
Roman Frigg, London, UK
Christopher A. Fuchs, Waterloo, Canada
Giancarlo Ghirardi, Trieste, Italy
Domenico Giulini, Hannover, Germany
Gregg Jaeger, Boston, USA
Claus Kiefer, Cologne, Germany
Klaas Landsman, Nijmegen, The Netherlands
Christian Maes, Leuven, Belgium
Hermann Nicolai, Golm, Germany
Vesselin Petkov, Montreal, Canada
Alwyn van der Merwe, Denver, USA
Rainer Verch, Leipzig, Germany
Reinhard Werner, Hannover, Germany
Christian Wuthrich, La Jolla, USA

For further volumes:
http://www.springer.com/series/6001

Jiří Bičák · Tomáš Ledvinka
Editors

# General Relativity, Cosmology and Astrophysics

## Perspectives 100 years after Einstein's stay in Prague

Springer

*Editors*
Jiří Bičák
Faculty of Mathematics and Physics
Charles University
Praha 8
Czech Republic

Tomáš Ledvinka
Faculty of Mathematics and Physics
Charles University
Praha 8
Czech Republic

ISBN 978-3-319-38034-6 ISBN 978-3-319-06349-2 (eBook)
DOI 10.1007/978-3-319-06349-2
Springer Cham Heidelberg New York Dordrecht London

© Springer International Publishing Switzerland 2014
Softcover reprint of the hardcover 1st edition 2014
This work is subject to copyright. All rights are reserved by the Publisher, whether the whole or part of the material is concerned, specifically the rights of translation, reprinting, reuse of illustrations, recitation, broadcasting, reproduction on microfilms or in any other physical way, and transmission or information storage and retrieval, electronic adaptation, computer software, or by similar or dissimilar methodology now known or hereafter developed. Exempted from this legal reservation are brief excerpts in connection with reviews or scholarly analysis or material supplied specifically for the purpose of being entered and executed on a computer system, for exclusive use by the purchaser of the work. Duplication of this publication or parts thereof is permitted only under the provisions of the Copyright Law of the Publisher's location, in its current version, and permission for use must always be obtained from Springer. Permissions for use may be obtained through RightsLink at the Copyright Clearance Center. Violations are liable to prosecution under the respective Copyright Law.
The use of general descriptive names, registered names, trademarks, service marks, etc. in this publication does not imply, even in the absence of a specific statement, that such names are exempt from the relevant protective laws and regulations and therefore free for general use.
While the advice and information in this book are believed to be true and accurate at the date of publication, neither the authors nor the editors nor the publisher can accept any legal responsibility for any errors or omissions that may be made. The publisher makes no warranty, express or implied, with respect to the material contained herein.

Printed on acid-free paper

Springer is part of Springer Science+Business Media (www.springer.com)

# Preface

A few days before leaving Prague, after 15 months spent at the German part of Charles-Ferdinand University, Albert Einstein submitted a paper titled *Relativity and Gravitation. Reply to a Comment by M. Abraham*. It was received by Annalen der Physik on July 4, 1912. Here Einstein summarized the contemporary state of his relativistic theory of gravitation and, remarkably, anticipated what a future theory of gravity should look like.[1]

"Relativity and Gravitation: 100 years after Einstein in Prague," was the name of the conference held in Prague on June 25–29, 2012, inspired by the title, date, and significance of this last of Einstein's Prague papers. The aim of the conference was twofold. First, it was to review the present status of the general theory of relativity (both classical and quantum) and its applications in cosmology and astrophysics from a broad perspective. The second aim was to present the newest results in each of these fields. This volume is based on the invited plenary lectures at the conference. In another volume, "Relativity and Gravitation: 100 years after Einstein's stay in Prague," appearing in the "Springer Proceedings in Physics," articles based on contributed talks and posters are included; more on cultural and other events associated with the conference is recalled therein.

The articles included in this volume represent a broad and highly qualified view on the present state of general relativity, quantum gravity, and their cosmological and astrophysical implications. As such, it may serve as a valuable source of knowledge and inspiration for experts in these fields, as well as an advanced source of information for young researchers.

The contents is divided into four broad parts: (i) Gravity and Prague, (ii) Classical General Relativity, (iii) Cosmology and Quantum Gravity, and (iv) Numerical Relativity and Relativistic Astrophysics.

[1] See the contribution by J. Bičák in this Volume.

## Gravity and Prague

In the first contribution, **Julian Barbour** "honors" Kepler and Mach for their work and fundamental discoveries made in Prague. Barbour starts with Greek planetary astronomy, continues with Copernicus, and comes to a profound description of the work of Kepler, "the true discoverer of heliocentricity." After a short intermezzo about Doppler, he comes to Ernst Mach. Mach profoundly influenced several generations of experimentalists (not only) in Prague. However, within the context of the conference, the most important issue was Mach's influence on Einstein in the formulation of general relativity. At the same time, though, Einstein sometimes "distorted" Mach. A novel conception of Mach's principle within geometrodynamics, called by Barbour the "shape dynamics," concludes Barbour's excursion through almost 2,200 years.

**Jiří Bičák** describes how and why Einstein was invited to Prague, and what his influence was on Czech science and culture. The main themes that occupied him then were the principle of equivalence, the bending of light, gravitational lensing, gravitational redshift, and frame dragging effects. In the article, these topics are discussed from a present-day perspective. Perhaps most importantly, just before leaving Prague, Einstein summarized his views on what basic features a new theory of gravity should possess, including the invariance with respect to a larger group than the Lorentz group, the local significance of the equivalence principle, or the nonlinearity of the field equation for gravity.

## Classical General Relativity

The problem of measurement, of precise definition of observers and observables they employ, has been one of the most distinct features of both special and general relativity since their birth. **Donato Bini** summarizes the results obtained by his colleagues and him during the last two decades on the "measurement process." The process involves clearly defined geometrical and physical quantities arising from an identification of "space" and "time" relative to a given observer within a congruence of timelike worldlines.

A novel theme, unimaginable to be discussed 100 or even 20 years ago, is reviewed in **Gary Gibbons'** article: The Role of General Relativity in Other Parts of Physics. Gibbons concentrates on a number of specific problems in which geometrical ideas are employed: the description of shallow water waves is analogous to the behavior of the rays in the Schwarzschild metric or around straight cosmic strings. In optics, examples of a left-handed light moving in a medium with a negative refractive index can be described with the help of hyperbolic or Lobachevsky space. Zermelo's problem of minimizing the travel time of a boat moving with a fixed speed in a Riemannian metric under the presence of a "wind" is discussed using Finsler geometry. Other problems include: invisibility cloaks, hyperbolic metamaterials, gravitational kinks, Bloch walls, liquid crystal droplets, and last but not least the popular subject of graphene.

A very old theme in general relativity is the $N$-body problem. For over 30 years a leading figure in this field has been **Thibault Damour**. He first reviews ongoing post-Newtonian calculations, continuing to still higher orders in $v/c$, and combined with formalisms yielding gravitational waveforms. Next he describes the effective field theory approach employing diagrammatic methods of quantum field theory, numerical relativity simulations, and gravitational self-force theory. The main attention is then paid to the "Effective One Body" (EOB), formalism that Damour and his collaborators started to develop at the end of the 1990s. The goal of the EOB formalism is to obtain an analytical description of the motion and radiation of coalescing binaries during the *whole* process, from inspiral to the final black-hole ring-down. The results obtained by employing the EOB formalism are compared with those coming from numerical relativity and self-force computations. The ways in which EOB theory may progress are indicated and the conclusion is taken from Henri Poincaré: "There are no solved problems, there are only more-or- less solved problems... ."

The article by **Leor Barack** on the gravitational self-force follows. Originally, the self-force theory was developed for so-called "Extreme Mass Ratio Inspirals" (EMRIs), in which a test particle (say a $10M_{\odot}$ black hole) moves along a geodesic of a stationary background geometry of a large mass (say a $10^6M_{\odot}$ black hole). If the mass of the particle is taken into account, the effect of the perturbation of the background geometry on the particle gives rise to a gravitational self-force. To calculate the resulting deviation from the geodesic motion is a formidable task. Barack, a principal author of a powerful, practical method for doing this for EMRI orbits gives a brief but comprehensive review of both successes and difficulties. As an illustration, important cases of the self-force effects are discussed: the induced shift in the frequency of the innermost stable circular orbit or the correction to the periastron shift.

For a number of years, the problem of motion has been studied by **Gerhard Schäfer** and his collaborators concentrating on the use of the Hamiltonian treatments. The Hamiltonian formalisms of Arnowitt-Deser and Misner (ADM), Dirac and Schwinger are compared. The results based on the ADM approach are combined with the post-Newtonian/post-Minkowskian approximations. Lastly, Schäfer tells how the complicated problem of incorporating the spin of the particles into the formalism has been recently analyzed.

The article by **Marc Mars** focuses on geometric inequalities involving physical quantities like mass, charge, area, or angular momentum. This work is interwoven by definitions, propositions, lemmas, and theorems, though it also contains helpful intuitive remarks. The best known inequalities are the famous positive mass theorem and the Penrose inequality (the theorem under the presence of trapped surfaces). More recently, with new concepts of Marginally Outer Trapped Surfaces (MOTS) and dynamical horizons, a great interest arose in inequalities considering area and angular momentum. Other inequalities involving charge, the cosmological constant, and the topology of MOTS's, are concisely reviewed.

It has been a long-standing problem whether the spin–spin interaction between two black holes can balance their gravitational attraction. The article by **Gernot Neugebauer** and **Jörg Hennig**, based on Neugebauer's talk in Prague, provides a survey of the numerous papers addressing this problem in the past. It also describes the rigorous *non-existence* proof based on the formulation of a boundary value problem for the nonlinear Ernst equation under the assumption of the existence of two disconnected Killing horizons.

**Robert Wald's** article discusses a recently obtained dynamical stability criterion for black holes in $D \geq 4$ spacetime dimensions. This relates stability with respect to axisymmetric perturbations to the positivity of a certain canonical energy. This energy is determined by second variations in the mass, angular momentum, and horizon area like those appearing in the first law of black hole thermodynamics. One consequence is that black branes, corresponding to thermodynamically unstable black holes, are dynamically also unstable. Wald concludes that "The remarkable relationship between the laws of black hole physics and the laws of thermodynamics [...] extends to dynamical stability."

A brief contribution by **Piotr Bizoń** and **Andrzej Rostworowski**, based on the talk given by Rostworowski, summarizes their numerical and perturbative work on the instability of anti-de Sitter spacetime based on numerical and perturbative calculations. Results for spherically symmetric massless scalar fields strongly indicate that anti-de Sitter space is unstable to arbitrarily small perturbations—eventually forming black holes.

Higher dimensional black holes within classical general relativity in higher dimensions are discussed in the last two articles of this part. **Harvey Reall** analyzes stationary vacuum solutions within (i) Kaluza–Klein theory and (ii) general relativity. In addition to Myers-Perry black holes, solutions include the famous black rings and black Saturns found by Reall and Roberto Emparan. Generalizations of these are mentioned, as well as the problem of instabilities and the search for perturbative solutions.

The next article is based on the talk by **Valeri Frolov**. It concentrates on black holes with topologically spherical horizons but non-vanishing cosmological constant and NUT parameter. Crucial is the existence of the principal conformal Killing-Yano tensor which is admitted by Kerr-NUT-AdS metrics in four and higher dimensions. This object enabled Carter in 1968 to separate geodesic and wave equations in the four-dimensional Kerr geometry. The principal conformal Killing-Yano tensor allows similar results to be derived in higher dimensions as well.

## Cosmology and Quantum Gravity

**Lars Andersson's** article deals with cosmological models and their stability. The principle of equivalence and Mach's principle are mentioned and shown to have roles different from the "hierarchy of cosmological principles." A brief discussion follows of cosmological models and issues such as the coincidence problem and the role of inhomogeneities. The second part has a more technical character: The asymptotics and nonlinear stability of various models is surveyed, including de Sitter space, the deformed Milne model, and generalized anisotropic Kasner models. The present status of the Belinskii, Khalatnikov, Lifshitz proposal is also discussed.

The contribution by **Misao Sasaki** discusses inflation and the birth of cosmological perturbations. Inflationary universe, despite various objections, is being accepted by more and more active cosmologists not only because it helps to explain various problems of the standard big-bang theory, but also appears to be in agreement with new observational data on the anisotropy of the microwave background. Sasaki describes slow-roll inflation, the curvature perturbations arising from vacuum fluctuations of the inflation field, and tensor perturbations of the metric. Possible primordial non-Gaussianities, if observed, could significantly constrain cosmological models. The review concludes with a brief description of a powerful formalism which allows curvature perturbations to be calculated on superhorizon scales.

What preceded inflation? This question is asked and an answer is suggested in the article on Loop Quantum Cosmology (LQC) by **Abhay Ashtekar**. His discovery of new variables in 1986 initiated the birth of the Loop Quantum Gravity (LQG). Ashtekar suggests that "even though we are far from a complete theory [of quantum gravity], advances can occur by focusing on specific physical problems." One "grand" problem is the LQC based on a *truncated* LQG. It concentrates on (i) the resolution of the "initial" singularity, (ii) the formulation of effective LQC dynamics which leads to inflation, and (iii) the extension of cosmological perturbation theory to the Planck regime. Ashtekar reviews progress in these directions. He also indicates how LQC relates initial conditions (at a bounce) with observations.

One open aspect of inflation is the issue of how quantum fluctuations in the inflaton field transmute into observed classical inhomogeneities. This problem is addressed by **Daniel Sudarsky**. In the standard "philosophy" of quantum theory there exists a collapse of the wave function during a measuring process. How can this happen in the Universe? Some believe in Everett's many world interpretation. More radical ideas require "novel physics," such as gravity-induced collapse of the wave function. Such an approach is adopted by Sudarsky. He investigates it within a semiclassical treatment of gravity interacting with quantum fields during inflation.

A broad view on the state of quantum gravity (QG) from a particle physics perspective is given by **Hermann Nicolai**. It covers a large territory in a fairly nontechnical style. There are difficulties in both general relativity and quantum field theory associated with the use of a continuum at and below Planck-scale distances. The principle differences in the approaches to QG between two main candidates, string theory and LQG, are elucidated and considered to be a sign that "we are probably still very far from the correct answer!" A number of other issues that should be answered by a future theory of QG include the divergence problem of perturbative general relativity, the hierarchy problem (the smallness of $G$), and the question of whether the Standard Model of particle physics remains true up to the Planck scale. The incompleteness of the Standard Model is considered as "one of the strongest arguments in favor of quantizing gravity and searching for new concepts replacing classical notions of space and time."

## Numerical Relativity and Relativistic Astrophysics

Relativistic astrophysics celebrated its 50th birthday in December 2013. Seven articles included in this part demonstrate remarkably well the enormous progress this field experienced in the last half century.

**Luciano Rezzolla** illustrates the progress in numerical relativity in three examples: (1) numerical calculations of a binary merger of two neutron stars producing an extremely strong magnetic field, (2) numerical calculations leading to a black hole due to the collision of two self-gravitating fluids moving toward each other with ultra-relativistic velocities, (3) numerical study of the recoil dynamics of a black hole formed from head-on collision of two black holes with different masses. The presence of an "anti-kick" is studied from various points of view, including instructive figures, observational data, and issues of interest in mathematical relativity.

A comprehensive review on instabilities of relativistic stars is given by **John Friedman** and **Nikolaos Stergioulas**. It represents an extension of the talk given by John Friedman in Prague and is partially based on the book *Rotating Relativistic Stars* published by the authors in 2013. An action for perturbations leads to the canonical energy and momentum and to the criterion for stability. Various types of instabilities have been analyzed: convective instability, axisymmetric instability, nonaxisymmetric instabilities leading to the formation of bar modes, and instabilities driven by gravitational waves like the Chandrasekhar-Friedman-Schutz (CFS) instability. The CFS instability is primarily analyzed and questions of how it can be influenced by the complex physics in neutron stars are discussed.

In "Gravity talks: observing the universe with gravitational waves" (GW), **Bernard Schutz** indicates that, indeed, the detectors are like microphones, not pointed in some direction; and phases of the waves are more important than the amplitudes. The network of six GW detectors is portrayed eloquently. In this field, data analysis is most important. Its use at present is demonstrated by Einstein@Home platform which led to the discovery of new pulsars. The information we can obtain from detection (expected to occur by 2017) not only yields masses and spins of colliding black holes or neutron stars but also, for example, the distance to the sources. The properties of likely sources (neutron star or black hole binary coalescence, neutron stars interiors and pulsars) are discussed. The article concludes with information about LISA, substituted by somewhat restricted eLISA, which should be launched if the LISA Pathfinder is successful. The enormous advantages of detectors in space are highlighted.

**Gerhard Heinzel** (talking in Prague), and **Karsten Danzmann**, write about the LISA and eLISA missions in greater detail, focusing also on financial problems. It became clear in March 2011 that the originally designed joint ESA/NASA Mission cannot rely on support from NASA. The redesigned eLISA ("evolved" LISA) could be supported by ESA alone. Despite the cost reduction, thousands of compact white dwarfs binaries and hundreds of black hole binaries inspirals could still be observed. The authors conclude that "the technology is well developed, the team is strong and convinced that LISA must fly in the early 2020s... ."

A wide range of relativistic effects, including the strong-field regime, can be investigated by observing pulsars. **Michael Kramer** wrote a readable and comprehensive article on the current state-of-the art experiments involving these "cosmic lighthouses." Going from individual pulsars, he shows how about 10 % of about 2,000 known pulsars are in binary systems. These enable tests of general relativity and alternative theories with extraordinary precision. The effects include the precession of periastron, gravitational redshift, Shapiro delay, GW, and spin-orbit coupling; at the same time, the basic principles of the theory are tested. The detailed properties of the famous Hulse-Taylor pulsar are summarized and, in particular, of the double pulsar. It provides tests of the GW quadrupole formula far below the 0.1 % level. The masses and the orbital and relativistic spin-precession are measured with extremely high accuracy. There are good prospects of detecting GW by the "Pulsar Timing Array" method and a great challenge exists to test fundamental predictions like the "no-hair theorem" if a pulsar orbiting the black hole in the center of our Galaxy is discovered with the future Square Kilometer Array.

In the last two articles, based on talks by **Marek Abramowicz** and by **Ramesh Narayan** (with co-authors **Jeffrey McClintock** and **Alexander Tchekhovskoy**), regions of extreme gravity effects are discussed. Three instruments planned for missions in the near future (e.g. "The event horizon telescope") will provide angular and time resolution that will enable us to investigate the immediate neighborhoods of event horizons, ergospheres, innermost stable circular orbits (ISCO), and circular photon orbits around black holes. All these features of black holes are discussed in Abramowicz's review, the main attention being paid to ISCO because here the standard paradigm might have to be modified by the magneto-rotational instability leading to turbulence around ISCO.

In the last contribution, the authors deal with a long-standing, complex, and fascinating astrophysical problem: how jets from both stellar-mass and supermassive black holes form, and how are they powered. Results of recent computer simulations of black hole accretion and jets, in which magnetic fields twisted by the rotating black hole play an essential role, combined with recent observations, imply that (i) jets are powered from the black hole energy (rather than from a surrounding disk), (ii) "the first observational evidence for a correlation between jet power and black hole spin has finally been obtained."

The conference was organized[2] under the auspices of the Rector of Charles University, the oldest university north of the Alps (founded in 1348). It was sponsored by the Faculty of Mathematics and Physics of Charles University, and, in particular, by the Karel Janeček Foundation.[3] We also acknowledge a continuous support of the Czech Science Foundation, now under the grant No. 14-37086G (Albert Einstein Centre).

Prague, December 2013

Jiří Bičák
Tomáš Ledvinka

---

[2] The Scientific Organizing Committee included: M. Abramowicz, L. Andersson, A. Ashtekar, J. Barbour, J. Bičák, R. Blandford, B. Brügmann, P. Chruściel, T. Damour, K. Danzmann, F. de Felice, G. Ellis, J. Friedman, H. Friedrich, V. Frolov, G. Gibbons, G. Horowitz, J. Katz, K. Kuchař, J. Lewandowski, G. Neugebauer, H. Nicolai, I. Novikov, M. Rees, O. Reula, L. Rezzolla, M. Sasaki, G. Schäfer, B. Schmidt, A. Starobinsky, P. Tod, R. Wald, and C. Will.
The local organizing committee included: J. Bičák, M. Bursa, P. Hadrava, D. Heyrovský, V. Karas, D. Kofroň, P. Krtouš, J. Langer, T. Ledvinka, J. Podolský, V. Pravda, O. Semerák, Z. Stuchlík, O. Svítek, V. Špička, and M. Žofka, all of them associated (now or in the past) with the Institute of Theoretical Physics of the Faculty of Mathematics and Physics of the Charles University in Prague.

[3] We would also like to express our gratitude to Frank Schulz and Vera Oswald from the Living Reviews in Relativity, based in the Albert Einstein Institute in Golm, for their help with references.

# Contents

# Part I
# Gravity and Prague

# Kepler and Mach's Principle

**Julian Barbour**

**Abstract** The definitive ideas that led to the creation of general relativity crystallized in Einstein's thinking during 1912 while he was in Prague. At the centenary meeting held there to mark the breakthrough, I was asked to talk about earlier great work of relevance to dynamics done at Prague, above all by Kepler and Mach. The main topics covered in this chapter are: some little known but basic facts about the planetary motions; the conceptual framework and most important discoveries of Ptolemy and Copernicus; the complete change of concepts that Kepler introduced and their role in his discoveries; the significance of them in Newton's work; Mach's realization that Kepler's conceptual revolution needed further development to free Newton's conceptual world of the last vestiges of the purely geometrical Ptolemaic world view; and the precise formulation of Mach's principle required to place GR correctly in the line of conceptual and technical evolution that began with the ancient Greek astronomers.

## 1 Introduction

Some of the most important advances in science are associated with Prague. The meeting at which the talk on which this chapter is based celebrated Einstein's breakthrough to the key ideas of general relativity (GR) in 1912 near the end of his time in the Bohemian capital. In this chapter, I wish to honour Kepler and his discovery of the laws of planetary motion and Mach's critique of Newton's concepts of absolute space and time. The creation of GR is unthinkable without them. I also wish to give what I believe is the correct formulation of Mach's principle. I believe that misunderstanding about this, ironically due to Einstein, may well be holding back both cosmology and the discovery of the quantum law of the universe.

J. Barbour (✉)
Department of Physics, University of Oxford, Oxford, UK
e-mail: julian.barbour@physics.ox.ac.uk

J. Bičák and T. Ledvinka (eds.), *General Relativity, Cosmology and Astrophysics*,
Fundamental Theories of Physics 177, DOI: 10.1007/978-3-319-06349-2_1,

© Springer International Publishing Switzerland 2014

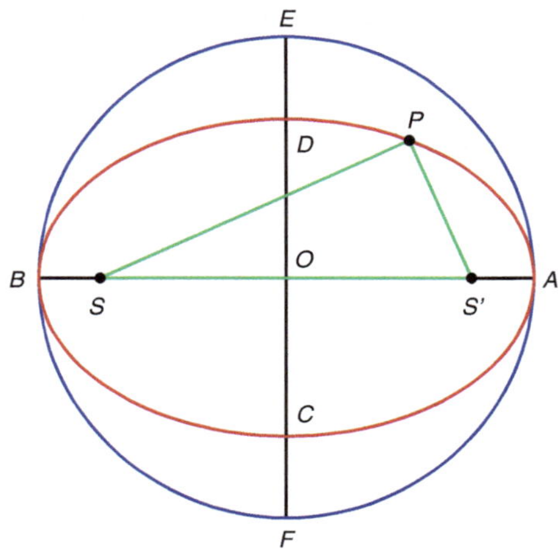


**Fig. 1** The eccentricity $e$ is OS/OB, the ellipticity $\varepsilon =$ DE/OE is $\varepsilon = e^2/2$ and very small for the naked-eye planets

I begin with some background to Kepler's discovery of the laws of planetary motion and then discuss the key intuitive ideas that enabled him to find them. We shall see that Kepler's reaction to the conceptual framework he inherited from Ptolemy, Copernicus and all previous astronomers was a clear anticipation of Mach's reaction to Newton's absolute space and time. In the broadest terms, one can see the creation of GR proceeding along a line of conceptual and technical development associated with six scientists: Ptolemy, Copernicus, Kepler, Newton, Mach, Einstein.

The main topics covered in this chapter are listed in the abstract, so I turn directly to their presentation.

## 2 Some Important Facts of Planetary Motion

Everyone knows Kepler's three laws: 1. The planets move in ellipses with the Sun at one focus. 2. The radius vector from the Sun to the planet sweeps out equal areas in equal times. 3. The period of each planet is proportional to $a^{3/2}$, where $a$ is the major axis of its ellipse. The first two laws were discovered in 1605, the third followed in 1618.

However, what really counts for understanding the history of planetary astronomy[1] up to Kepler's discovery of his first two laws is the form they take when the eccentricity of the ellipse is relatively small, as it is for all the planets. We must start with basic facts about ellipses (Fig. 1).

Because the ellipticity $\varepsilon$ is half the square of the eccentricity $e$, the magnitude of $\varepsilon$ is small for all the planets. The planet with the largest eccentricity, $e \approx 1/5$, is Mercury; then comes Mars with $e \approx 1/11$; Jupiter and Saturn have $e \approx 1/20$; the

[1] See [1] for a detailed discussion of the history.

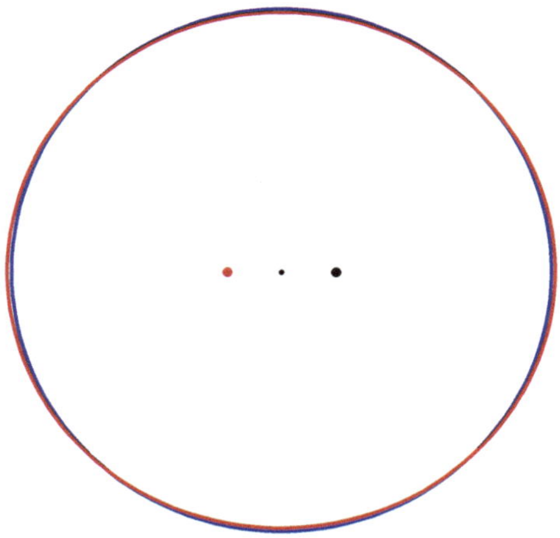

**Fig. 2** The orbit of Mercury is circular to one part in 50. The figure shows the foci and centre of the orbit and the *circle* that most closely approximates the *ellipse*. The *circle* is inside the *ellipse* along the line of the apsides joining the foci and outside at the quadrants. For Mars, the planet for which Kepler discovered ellipticity, the gap between the *circle* and *ellipse* is 4.5 times less. The figure shows clearly that the effects of eccentricity are far more readily observable than those of ellipticity

Earth ($e \approx 1/60$) and Venus ($e \approx 1/140$) have very small eccentricities and their orbits are wonderfully circular. This is crucial for the effects that even conscientious observers, who could use only the naked eye until 1610, were likely to find. The Sun and Moon subtend about 30 arc minutes on the sky. The accuracy of Ptolemy's observations was about a third of that, $10'$. Tycho Brahe's heroic observations, mostly in the period 1576–1597, pushed the accuracy to $2'$, as Brahe claimed, or $4'$ according to Kepler's more sober estimate.

What these facts about the accuracy of naked-eye astronomy mean is that the effects due to the orbit *eccentricity*, typically with a magnitude of degrees, were readily observable, while those due to the *ellipticity* were virtually undetectable. The only planet for which this is not strictly true is Mercury, but it is close to the Sun and seldom well seen, so it played no significant role in the discovery of Kepler's laws.[2] In one of several flukes in astronomy—the nearly equal apparent diameters of the Sun and Moon and the advance of the perihelion of Mercury included—it just so happens that among the remaining planets Mars was the most readily observable and has an ellipticity *just* large enough for Kepler's genius to espy it in the multitude of Brahe's observations.

To get an idea of Brahe and Kepler's achievement in the discovery of the ellipticity, Fig. 2 shows the orbit of Mercury. To the eye, it is a circle. One needs the

[2] It did help the belated recognition of Kepler's laws. His *Rudolphine Tables* (1627) led to the correct prediction and observation of the transit of Mercury across the Sun in 1631, a year after Kepler's death. The vastly superior accuracy of the *Tables* compared with the rivals, and the laws on which they were based, could no longer be denied.

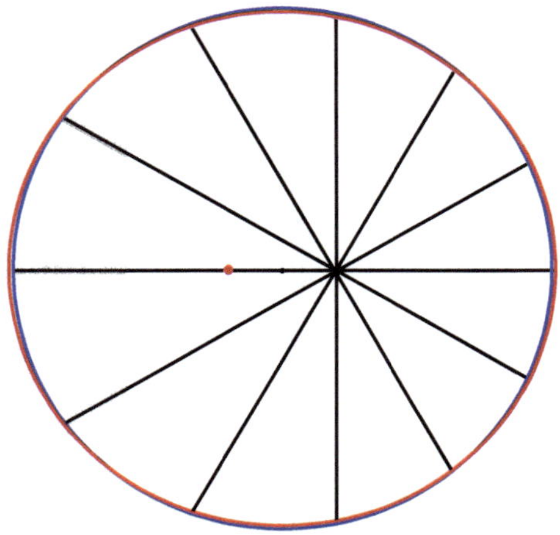

**Fig. 3** Illustration of the empty-focus effect for Mercury. The Sun is *left* of the centre of the orbit. An observer at the empty focus on the *right* sees the planet move round a great *circle* with near perfect uniformity

circumscribing circle to see the difference. Fapp—for all practical purposes, to use John Bell's acronym—the planetary orbits are circles as far as naked-eye astronomy is concerned.

If the near circularity of the orbits is little known even among many astronomers, another remarkable fact is virtually unknown. It relates to a property of the *empty* focus of the planet's ellipse. If you could hover in a spacecraft just above the Sun's surface and watch a planet on the celestial sphere, you would see it move in a great circle with a decidedly non-uniform motion: first because, in accordance with Kepler's 2nd law, its physical speed in space does change, and, second, because the Sun is displaced from the centre of the orbit. This geometrical effect *doubles* the nonuniformity of the observed angular speed in the small-eccentricity approximation appropriate for the planets. If you then fly to the centre of the orbit and hover there, the geometrical distortion is eliminated, and the observed angular speed reflects the true variable speed. But a miracle happens if you journey on to the empty focus: the geometrical effect that enhanced the non-uniformity above the Sun is now *reversed*: you see the planet move round its great circle with near perfect uniformity. Figure 3 illustrates the combined effect of the circularity and empty-focus effect.

The way to understand the actual process of discovery of the laws of planetary motions is through approximations to Kepler's laws. If the eccentricity $e$ is zero, the orbits are circles and the speed uniform; if $e$ is small, the orbits are still effectively circles but eccentric, and the speed on the circle is nonuniform though seen from the empty focus it is amazingly uniform. The major advances in the early history of planetary astronomy were largely due to these two effects. They are, respectively, very good approximations to Kepler's first two laws. It is no exaggeration to say that without them celestial dynamics could not have begun. However, they later became a source of great confusion: Copernicus and Brahe were literally going round in circles trying to make sense of the circles they imagined really were there in the sky. That's the story to which we now turn, beginning with the Greek astronomy.

## 3 Greek Planetary Astronomy

Let's start with what is readily observable: the Sun. Aristotle already knew that its motion around the ecliptic is non-uniform. Sometime around 150 BC (or 150 BCE as we should now say), Hipparchus, the first great astronomer of antiquity, made accurate measurements and a model to *explain* the non-uniformity. He supposed that the Sun moves on a perfect circle and with perfect uniformity and that the observed non-uniformity of its motion was due to a displacement $D$ of the centre of the circle from the centre of the Earth, assumed to be the centre of the Universe. The ratio of $D$ to $R$, the radius of the circle, was its *eccentricity*, the origin of both the technical and non-technical meanings of the word.

Hipparchus needed only two observations, the times taken by the Sun from the vernal equinox to the summer solstice and from there to the autumnal equinox, 94.5 and 92.5 days, respectively. These two data were enough to fix the two unknown parameters of his model: the magnitude of $e = D/R$ and the direction of the eccentric centre. This defined the *line of the apsides*, which joins *apogee* and *perigee* (or aphelion and perihelion in heliocentric astronomy). Hipparchus found $e = 1/24$, at that epoch more than twice the eccentricity of the Earth's heliocentric, or Sun's geocentric, orbit, which was then $\approx$1/57.

Part of the inaccuracy was due to observational error, but the major contribution was a flaw in the theory. Having not the remotest reason to suspect non-uniformity of the motion, Hipparchus had inadvertently doubled the eccentricity. What was truly remarkable was that, when the observational accuracy had been pushed to its naked-eye limit by Islamic astronomers and Brahe, Hipparchus's incorrect model proved to be amazingly accurate in its predictions. In fact, it gives deviations from the true positions never greater than 3/4 of an arc minute, way below the detection level of naked-eye observations.[3] For this reason, Hipparchus's model, converted appropriately to heliocentric motion of the Earth, passed unscathed through the Copernican revolution. One of the gems in Kepler's work was, as we shall see, the dethronement of the Hipparchan model. It had reigned supreme for over 1700 years.

We now move on 300 years to the next great astronomer of antiquity: Claudius Ptolemy. He worked in Alexandria and had access there to, among much else, Babylonian observations made nearly a millennium earlier. He embarked on what was surely the first rationally and comprehensively planned scientific-research project in human history—the theoretical explanation through uniform circular motions of all

---

[3] This remarkable accuracy is due to the very small eccentricity of the Earth's orbit, currently about 1/60. A solar theory developed by a Martian Hipparchus would not have survived for long because Mars has eccentricity $\approx 1/11$ (crucial for Kepler's discoveries). It is worth mentioning that the theory-independent quantity most immediately observable is always *twice* the eccentricity. For the Sun observed from the Earth, this is currently 1/30 of a radian or about 2° (four apparent solar diameters), half of which comes from the relative geometrical displacement of 1/60 and half from the physical non-uniformity described by Kepler's 2nd law. These equal contributions to observable effects are very important for understanding the history of ancient astronomy. By comparison, the observable ellipticity effects in the solar motion are about 120 times smaller at 1/30 of the apparent solar diameter, i.e., $1'$.

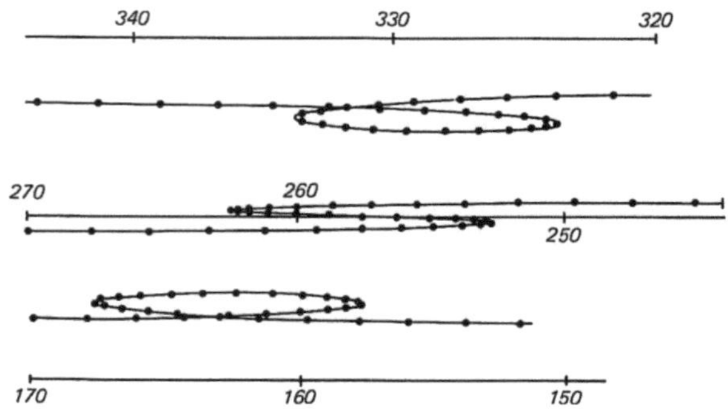


**Fig. 4** The retrograde motions of Jupiter. Against the backdrop of the fixed stars, Jupiter (like all the planets) generally moves eastwards, never straying far from the ecliptic (the great *circle* on which the Sun moves). But about once every 13 months it executes its retrograde motions, in which the eastward motion comes to a stop and reverses into a westward motion until that halts and the eastward motion is resumed. Any attentive observer who keeps reasonably good records is likely to try to describe the complicated observed motion as a superposition of two simpler motions

the seven planets. Meaning *wanderers*, these included the Sun and Moon as well as Mercury, Venus, Mars, Jupiter and Saturn. Ptolemy's astronomical compendium, known through its Arabic title as *The Almagest* and written around 150 CE, was the handbook of astronomy for close on 1500 years. His epicycles tend to be mocked today as the paradigm of poor *ad hoc* science, often by scientists who should know better,[4] but they were one of the great contributions to the advance of science. On top of that, Ptolemy made what is arguably the first great discovery in the history of dynamics. To that I now turn.

Compared with the observed motion of a planet, that of the Sun is simplicity itself, being just the mirror image of the purely periodic motion of a single planet, Mother Earth, around the Sun. But the motion of a planet seen from the Earth is a compound of two incommensurate periodic motions. This compounding leads to the famous observed retrograde motions (Fig. 4).

Following earlier proposals, which were probably only qualitative and may have been suggested by the great mathematician Apollonius (*circa* 255–170 BCE), Ptolemy attempted to describe them with the epicycle–deferent model (Fig. 5), which I describe in the caption only for the simpler case of the three outer planets.

In heliocentric terms, the motion of the guide point D is the *actual* motion of the planet, while the epicyclic motion is the reflection of the Earth's motion seen through the relative motion of the planet against the stars. The specific eccentricities of the various planetary orbits played a crucial role in the details of Ptolemy's theory. The key thing to understand is that the Earth's eccentricity, $e_E$, is significantly smaller

[4] I suspect some of them trust the scientifically very ignorant account in Koestler's *Sleepwalkers*, in which it is stated that "There is something profoundly distasteful about Ptolemy's universe; it is the work of a pedant with much patience and little originality, doggedly piling 'orb in orb'." The *Almagest* use the bare minimum of epicycles.

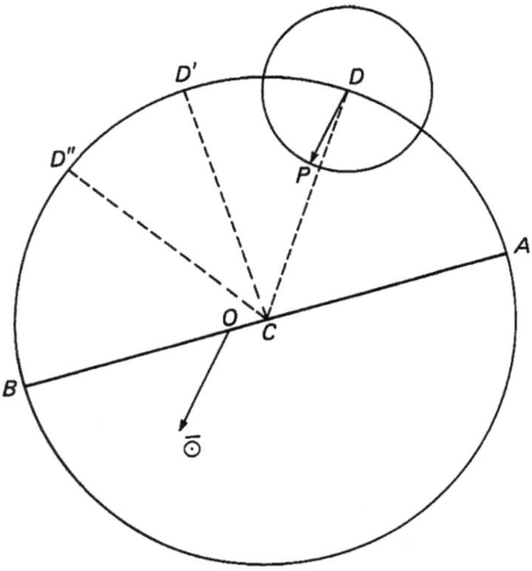


**Fig. 5** Ptolemy sought to explain the regular eastward motion of, say, Jupiter, through uniform eastward motion of an invisible guide point $D$ on a circle called the deferent. Around $D$ a spoke of length less than the deferent radius rotated on an epicycle with perfect uniformity carrying the planet $P$ at its tip. The centre $C$ of the deferent is displaced from the position of the terrestrial observer $O$ in order to explain the fact that even without the epicyclic motion the general eastward motion of Jupiter is manifestly non-uniform like that of the Sun as described by Hipparchus's model

that that of the three outer planets, $e_{\mathrm{P}}$, which are moreover further from the Sun. On the sky, the observable effect $O$ due to the orbital eccentricities is, to a first approximation,

$$O = \frac{e_{\mathrm{P}}}{e_{\mathrm{E}}}\frac{a_{\mathrm{E}}}{a_{\mathrm{P}}},$$

where $a_{\mathrm{E}}$ and $a_{\mathrm{P}}$ are the Earth's and the planet's semi-major axes, respectively. For Saturn, Jupiter and Mars $O \approx 1/30, 1/15, 1/8$, respectively. This meant that the nonuniformity in the planet's motion, represented in Ptolemy's theoretical model by the motion of the invisible guide point D, was readily observable and could not be ignored. In a first attempt to describe it, Ptolemy copied the Hipparchan solar model exactly by a simple displacement of the centre of the deferent from the terrestrial observer. For the epicycle motion, he assumed perfect uniformity around D. In heliocentric terms, this corresponds to an exactly zero-eccentricity circular orbit of the Earth. Note that the error due to circularity is tiny, around one part in 3,600 and unobservable; the error due to the zero eccentricity is only 1/60. However, both of these are reduced by the ratio $a_{\mathrm{E}}/a_{\mathrm{P}}$ and escaped Ptolemy. They would in any case have been very difficult for him, with his rudimentary mathematics, to model.

What is extremely interesting is the way Ptolemy fixed the parameters of the deferent. The epicyclic motion being simply the reflection of the Earth's motion, the epicycle always points in the direction of the Sun. Ptolemy knew this; it did not prompt him to heliocentricity, but it did help him to fix the deferent parameters and to make that great discovery I mentioned.

His task was to fix the position of the invisible guide point: mission impossible you might think. But no; when the Earth is exactly between the planet and the Sun (so that the planet, in opposition, is due south at midnight), the epicycle, with the planet on its tip, points simultaneously towards the Sun and the Earth. Moreover, seen from the Earth, the guide point D is exactly behind the planet. Using such observations, spread necessarily over many years, Ptolemy could mimic what Hipparchus had done for the Sun. He was able to determine the eccentricity and line of the apsides of the deferent. But this was simultaneously the *heliocentric* orbit of the planet! However, because Ptolemy directly copied Hipparchus and did not suspect physical nonuniformity in the motion of D, he too found double the actual eccentricity.

Now in the solar motion there was no possibility of detecting the error. But Ptolemy's model was not yet complete. He had to fix the ratio of the epicycle and deferent radii. For this he needed just one more observation, of necessity made when the planet is not in opposition. There is an almost poetic touch worth mentioning here. When the planet is in opposition and due south a midnight, it rises at sunset, *acronychal* in Greek.[5] Ptolemy needed just one non-acronychal observation to determine the length of the epicycle.

With the model complete, Ptolemy—good scientist that he was—tested it using further non-acronychal observations. Dismay: the model failed to predict them correctly. After a long period of trial and error that, as Ptolemy admitted, had no principled basis except fidelity to observation—and hence truth—he found a deferent model that worked very well.

He discovered that it was necessary to *halve* his previous deferent eccentricity and introduce an 'equalizing point', or *equant* as it is now called. It lay on the other side of the centre of the orbit from the observer along the line of the apsides, which remained unchanged. Around the equant one had to imagine a spoke that rotated with perfectly uniform (hence equalizing) angular velocity and cut the deferent circle in its new position at the point when the guide point D must be. As before, the planet-carrying epicycle rotated with perfect uniformity about D.

With this model (somewhat modified for Venus and Mercury), Ptolemy found he could describe and predict the motion of all the planets with surprisingly good accuracy. What, in the long run, was truly significant for astronomy and dynamics, was that he had found a wonderfully good approximation to what Kepler's second law predicts. For, in heliocentric terms, Ptolemy's equant is none other than the empty focus of the planet's orbit—and I have already explained what a superb approximation that is. Because he was also working with eccentric, perfectly circular orbits, he also had an excellent approximation to Kepler's first law.

If we discount the barely observable and hence 'thankless' Mercury, Ptolemy's theory was correct for all the other planets to excellent accuracy. For Mars, with the largest eccentricity, the maximal deviation from Kepler's laws was only one part in 225. That is the measure of his achievement.

---

[5] Acronychal and non-acronychal observations were still vital in Kepler's work.

Final comments before we move on to Copernicus and Kepler. First, all of Ptolemy's work was based on measurement of *angles* between objects *that could be seen.* Second, none of his work or anything really accurate in astronomy could have been done up to the invention of truly accurate clocks in the 20th century without the diurnal revolution of the stars, aka the Earth's rotation. It was the one and only clock that could be used. Time was also read off it by measurement of angles between visible objects.

## 4 Copernicus

It is ironic that Copernicus stumbled on his revolutionary idea by trying to undo Ptolemy's greatest discovery: the equant. Despite great admiration for Ptolemy's technical skill and achievements, Copernicus strongly disliked the equant's violation of the literally sacred principle that all the divine objects in the heavens must move with perfect uniformity in perfect circles. Ptolemy had maintained the circles (with good reason—they worked) but had discarded uniformity for the sake of truth. Copernicus, like at least one Islamic astronomer before him, sought to replace the equant device by a combination of uniform circular motions that, of necessity, was more complicated than Ptolemy's solution if fidelity to observational facts was to be maintained. While working on this project, he realized that all the retrograde motions of the planets could be understood as effects of *relative motion* if one assumes that the Earth is not at rest but moves in a circle.

There is an important point here that needs to be emphasized: Copernicus proposed a theory of *terrestrial mobility*, not heliocentricity. This was still the main point for Galileo, as shown by his famous retort "Eppur si muove." All that Copernicus needed, and said, was that the Sun must be near the centre of the circle in which the Earth moved.[6] For Copernicus, the Sun and its precise position had no physical significance. He said the Sun had a worthy place in the heavens, placed to illuminate the dance of the planets. As we shall see, the true discoverer of heliocentricity was Kepler.

Copernicus made four great contributions: first, he unavoidably, though without having an inkling of its significance, drew attention to the Sun, which was very important for Kepler; second, he explained the retrograde motions; third, as an under-appreciated consequence of that, he brought to planetary astronomy a unity entirely lacking in Ptolemy's universe. Fourth, his arrangement of the solar system and the absence of observed parallax of any of the stars required the stars to be immensely farther away than Saturn.

The third contribution needs a little elaboration. Since Ptolemy worked with angles, he had no way of determining any distances. He therefore set all deferent radii equal to the nominal value unity and found the epicycle radius as a ratio to unity. Copernicus realized that this ratio, different for each planet, simply reflected

[6] Copernicus actually thought that the Sun might have some slow motion of its own.

the ratio of the radii of his various circles: the Earth's to that of the planets. He could use the radius of the Earth's orbit as a trigonometric base line to determine the distances to the planets. This immediately gave him a very good overall picture of the solar system. He worked out the correct order, distances and average speeds of all the planets, measured of course in terms of the radius of the Earth's orbit (now, of course, the astronomical unit) and the terrestrial day. He obtained a qualitative form of Kepler's third law and facts that Kepler subsequently found highly suggestive. It is often said that the Copernican and Ptolemaic arrangements are kinematically identical. This is not strictly true and does not do justice to Copernicus. Geometrical dispositions are part of kinematics. Ptolemy's *Almagest* did not have them or the lower bound on the distance to the stars.

A nice way to compare the respective achievements is this: when Ptolemy died, he could predict what the sky would look like—where the planets would be—as seen from Alexandria centuries after his death, but he had no idea what it would look like from Mars. When Copernicus died in 1543, he did know or, at least, knew how to calculate the positions of the Sun and planets as seen from Mars. In fact, the possibility was only literally confirmed in the space age.

Copernicus did great things, but, from a modern point of view, he bequeathed a most odd solar system to posterity. I have already mentioned the Sun's role as a mere lantern. Really strange was the location of the 'centre of the Copernican universe'. Ptolemy had discovered an equant in the deferents of all the planets, essentially because the Earth, unbeknown to him, was a spaceship that allowed him to look at the planets' positions from a whole circle in the solar system and not just from the Sun's position (as in the acronychal observations). But because the Sun's motion is merely the Earth's 180° out of phase, there was no way observation could force an equant on the solar motion. Ptolemy left the Hipparchan model unchanged.

And so did Copernicus. Even though he made Mother Earth a planet like the others and contrived makeshift substitutes for their equants, it never occurred to him that the Earth should get anything equivalent, so he simply inverted the Hipparchan model and gave the Earth an eccentricity twice what it should have. That simultaneously singled out the empty focus of the Earth's orbit as a special point. Moreover, because of the fluke of the Earth's eccentricity being so small Copernicus was misled into thinking that the lines of the apsides of the planets all converged, not on the Sun, but at the very same point that Kepler was later to identify as the Earth's empty focus. Figure 6 shows how small the mismatch was—but also that Copernicus did not propose a truly heliocentric system.

In fact, the clearest evidence of that is in the diagrams which Copernicus drew to show the orbits of the planets. *They do not show the Sun.* It was in no way an integral part of his scheme.

There were many other oddities, some very bizarre, in the Copernican cosmos, most of which arose because Copernicus simply inverted the Ptolemaic models. When *De Revolutionibus* was published in the same year 1543 that he died, Copernicus knew he had made a monumental discovery, but, like Ptolemy, his insights and methods were purely geometrical and kinematical. Kepler commented "he was unaware of his riches".

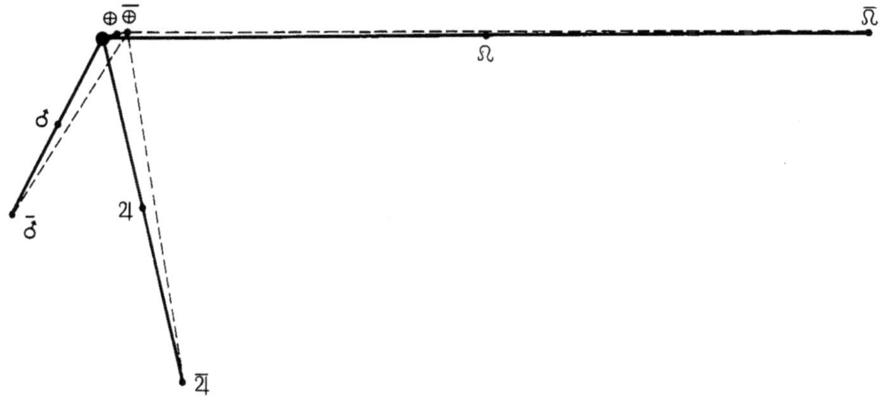


**Fig. 6** Copernicus believed that the lines of the apsides of the three outer planets converged on a void point near the Sun that was actually the empty focus of the Earth's heliocentric orbit. The true Sun is the *black disc* on the *left*, where the *solid lines* of the apsides converge. The *dashed lines* show Copernicus's belief. The mismatch is small but shows that Copernicus thought solely in geometrical terms, for which such an arrangement with the Sun playing no physical role is perfectly acceptable

## 5 Kepler

In this section, I want to concentrate on the huge conceptual change that Kepler (1571–1630), for whom a magnificent portrait (Fig. 7) survives, introduced and how he anticipated Mach's attitude to dynamics. I am firmly of the belief that more is still to come of it. I can only pick out the highlights. The details, which are absorbing, can be found in [1].

The best place to start is Brahe's observations of the comet of 1577, which established its distance as interplanetary. For Kepler, the supreme importance of the observations were that they 'destroyed' the crystal spheres widely believed to carry the planets. In his fascinating account of how he mastered the motion of Mars, the *Astronomia Nova* published in 1609, he repeatedly pointed out that Brahe's observations proved that the planets were not carried by spheres. The comet had passed clean through the solar system without crashing into them. They could not be there. In one of the great intuitive insights in the history of science, he proclaimed: "Henceforth the planets must find their way through the void like the birds through the air. We must philosophize about these things differently."

Crucial questions then arose. What moves the planets? How do they find their way? What if anything is directing them? Let us start with the second and third questions, which reveal Kepler's affinity with Mach—or better Mach's with Kepler. Birds find their way around the world by reference to features in the terrain and sky. But, according to the astronomy Kepler inherited from Ptolemy, Copernicus and all previous astronomers including Tycho Brahe, literally everything was controlled and directed by void points, above all the equants that Ptolemy had discovered, in empty

**Fig. 7** Johannes Kepler. The epitaph that he composed for himself read "I used to measure the Heavens, now I measure the shadows of Earth. The mind belonged to Heaven, the body's shadow lies here." His grave and tombstone in Regensburg, where he died in 1630, have been lost

featureless space. The difference from birds was blatant. Like them, the planets must use *visible objects*. Motion is relative to things you can observe. The only significant visible things that the planets could be guided by were the Sun and the distant stars. The guiding and determining role of visible matter is exactly what Mach was insisting on two and a half centuries later and led him to argue so persuasively against Newton's absolute space.

Moreover, if crystals spheres do not carry the planets, whence comes their motion? The planets must either have inherent motive force or be subject to it. This was a veritable change of mindset. A prominent part of the immensely long subtitle to the *Astronomia Nova* proclaimed it to be Celestial Physics. Kepler introduced forces into the heavens. True, they were Aristotelian, with the force assumed to determine the velocity it imparts and not the acceleration as in Newtonian dynamics. The astronomical data could give Kepler no hints in that direction, ironically for the same reason that Einstein three centuries later was able to subsume gravitational forces and inertia into a single geodesic law. Indeed, until very late in his work, Kepler believed the planets moved in circles. What more perfect and self-contained motion exists than that?

What was really important about Kepler's forces was not their mode of action but their conjectured *source*: the Sun and the planets themselves. In this key respect, Kepler's forces correctly prefigured Newton's. Their sources and controlling power resided in *physical* bodies, not void points in empty space. Here too, in identifying motion-controlling power with bodies and not space, Kepler anticipated Mach, who insisted that apparently force-free inertial motion was nothing of the sort but the

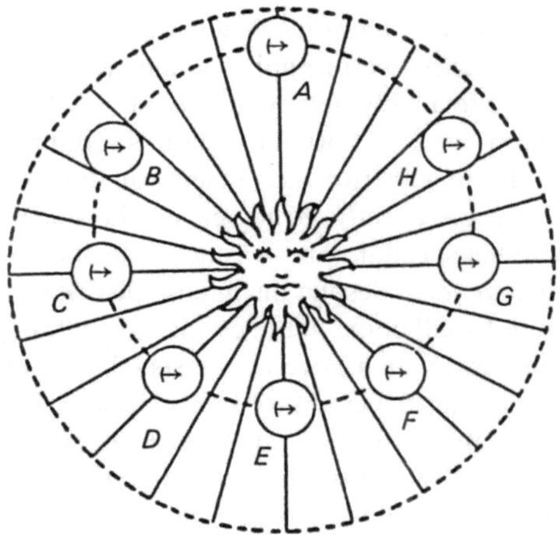


**Fig. 8** This diagram encapsulates the difference between Kepler and all his predecessors. The arrival of a new mind on the astronomical scene is demonstrated nowhere more clearly than in the comparison of the diagrams in *De Revolutionibus* and the *Astronomia Nova.* The Sun is prominent by its absence in Copernicus's; in Kepler's, as here, it takes pride of place, controlling the motion of the planet through physical forces

outcome of an as yet unknown physical effect of all the matter in the Universe. Let me here quote Mach [2, p. 296]: "The natural investigator must feel the need of further insight—of knowledge of the *immediate* connections, say, of the masses of the Universe. There will hover before him as an ideal an insight into the principles of the whole matter, from which accelerated and inertial motion result in the *same* way. The progress from Kepler's discovery to Newton's law of gravitation, and the impetus given by this to the finding of a physical understanding of the attraction in the manner in which electrical actions at a distance have been treated, may here serve as a model." That Mach sensed an affinity between himself and Kepler comes through in this quotation.

Let us return to details. Kepler had to explain two different kinds of motion: the eccentric circular motion around the Sun and the motion towards and away from the Sun during its course. To explain the circular motion, Kepler conjectured (nearly a decade before Galileo observed it!) that the Sun rotates about an axis perpendicular to the ecliptic and that what one might called ethereal 'spokes', rotating with the Sun, protruded from its equator. These, he assumed, swept the planets along in their circular motion, their strength diminishing with increasing distance from the Sun in order to explain why the more distant planets moved slower. As for the motion towards and away from the Sun, he conjectured that that it housed a powerful magnet and each planet a lesser one. The alignment of the magnetic poles would pull the planet towards the Sun on one side of the orbit and repel it on the other side.[7] These ideas are illustrated in Kepler's diagram shown in Fig. 8.

By modern standards, Kepler's forces were rather primitive and could not have survived detailed quantitative testing. What was decisive was that they focussed all of Kepler's interest on the Sun and its preciselocation. He was firmly convinced that

[7] William Gilbert's influential book on magnetism, published in 1600, strongly influenced Kepler's thinking.

the centre of the solar system did not lie at that mysterious void point that, post his discoveries, we recognize as the empty focus of the Earth's orbit, but at the entre of the relatively nearby mighty physical Sun (the distance between the Sun and the empty focus is 1/30 of the Earth's semimajor axis).

There were two ways to confirm this: first, to show that the lines of the apsides all converged *exactly* on the Sun, not the void point relatively close to it. Second, to show that the speed in orbit was not controlled by the void equant, as it was in Ptolemy's and, *de facto* despite his intense dislike of it, in Copernicus's astronomy, but by the Sun. I shall come to the crucial steps through which Kepler eventually came to his area law, which governs the speed, in a moment. First, I want to make clear why Kepler has the credit for heliocentricity in a way that Copernicus does not. Modest as the move from the empty to the occupied focus as centre of the solar system may appear, it was a small step that anticipated and made possible Newton's giant leap of understanding in the workings of the world. It identified the turning point.

Kepler's conceptual ideas drove all the technical work done at Brahe's behest in Prague—to establish the precise motion of Mars to the same accuracy that the Dane's incomparable observations allowed. His primary tool was trigonometry, which he put to use like no one before him. Kepler was the first man who could roam truly freely in imagination through the solar system. Appropriately, he also wrote almost the first work of science fiction: a dream of a journey to the Moon. The entire thrust of his trigonometric work was to establish heliocentricity beyond gainsaying, above all to show that the knitting-needle lines of the apsides all converged bang in the middle of the Sun and not as in the Copernican scheme (Fig. 6) at the nearby void point.[8]

There is even a sense in which Kepler anticipated gauge theory: he knew perfectly well that, kinematically, all of his precise geometrical results could be expressed just as adequately in the Tychonic[9] or geocentric Ptolemaic schemes as in the Copernican arrangement, but triumphantly pointed out that *in all three* the lines of the apsides meet at one point in the Sun. That was the 'gauge-invariant' content of his discoveries. The area law, the discovery of which we have still to discuss, had the same status.

Let us go through the most important technical advances to which Kepler was led by his intuition. I said that Ptolemy?s acronychal observations were made at opposition, when the observed planet is due south at midnight and the Sun is directly behind the terrestrial observer. This is not quite true; to facilitate computations and very likely because he did not realize the importance of the difference, Ptolemy's actual acronychal observations were made not when the true Sun was behind the observer but a substitute, a mathematically defined 'mean Sun' that moved around the ecliptic with perfectly uniform speed, coinciding with the true Sun only at the equinoxes. The angular distance between the true and mean Sun could be as much

[8] Kepler's work was actually a much more logically consistent and definitive proof of the Copernican cosmology than Galileo was able to muster. In fact, magnificent as his many achievements were, the Tuscan completely failed to recognize or begin to comprehend Kepler's achievement.

[9] Tycho Brahe could not believe in the immensity of the Universe that, given the absence of any observable stellar parallax, followed from Copernicus's proposal and therefore proposed that the Sun goes round the Earth while Mercury and Venus orbit it.

as those 2° corresponding to twice the Earth's eccentricity. Brahe had continued the Ptolemaic practice of using the mean Sun; Kepler meticulously corrected his observations by interpolation to make them correspond to his beloved true Sun. This was a first useful sharpening of accuracy.

Kepler was also the first person to understand how to take into account correctly the fact that the planets do not all move in the ecliptic. This gives rise to significant effects, causing the retrogression loops to have an out-of-ecliptic component (Fig. 4); without proper understanding of and correction for the effect Kepler could never have found the ellipticity of Mars's orbit. In fact, Kepler's first major result is what one might call his zeroth law: each planet moves in plane that passes through the Sun and is fixed in the frame defined by the stars.

But the real gem in Kepler's work that prepared the ground for his greatest discoveries was his finding of the *true location of the Earth*. All motion is relative. If you are trying to determine the position and motion of a distant object, you will surely make errors if you are mistaken about your own position and motion.[10] This is what Kepler understood perfectly—and he had good reason to be concerned. According to Copernicus and Brahe, the Earth's orbit was *sui generis*: unlike those of the other planets, it had no equant. Kepler's sense of the uniformity of nature told him that could not be true. The Earth had to have an equant. If so, that would mean it would have only *half* the eccentricity attributed to it by Copernicus and Brahe. Existing theory must be putting the Earth in the wrong place and thereby distorting the interpretation of the observations of all the extraterrestrial bodies.

In an article written in 1930 to mark the 300th anniversary of Kepler's death, Einstein described—with good reason—Kepler's halving of the Earth's eccentricity as one of the most beautiful things in all of science. Kepler knew that too. He had the finest diagram in his book engraved to show the way it was done (Fig. 9).

His stroke of genius was to use a trigonometric base line formed by the Sun and Mars at times when he knew Mars was at exactly the same point in its orbit. One of the great clarifications due to Copernicus's insights was that the planets traced out invariable orbits in the space defined by the Sun and fixed stars, returning to the same orbital position after completion of one orbit. Now among all data, heliocentric periods were the easiest to determine accurately; that of Mars was known to be 685 terrestrial days. Kepler searched among Brahe's 21-year treasury for Martian observations that by chance were separated by multiples of 685 days. At them, Mars must be at the same point in space. Kepler found three such observations. Acronychal observations of Mars and the theory of them, which he could trust, told Kepler the direction of Mars as seen from the Sun at all times. The direction to the Earth was also known, so Kepler could determine the angle between Mars and the Earth seen from the Sun. Brahe's observations gave him the angle between the Sun and Mars as seen from the Earth. Kepler had the one fixed Sun–Mars side of the triangle and two angles of the Sun–Mars–Earth triangle. Three such observations gave three positions

[10] I was told some years ago that the largest uncertainty in many high-precision tests of GR was the uncertainty in the Earth's position that results from the perturbing influence of the asteroids, whose size is known but not, to sufficient accuracy, their densities.

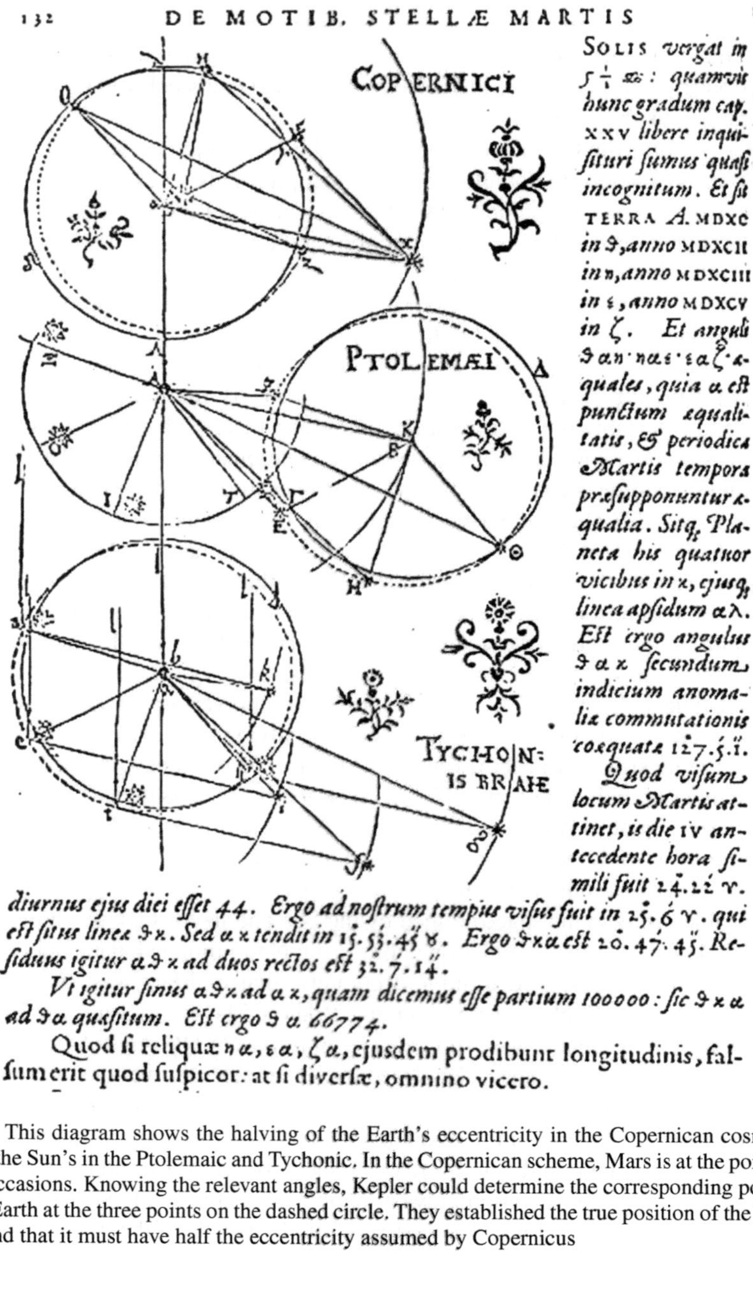
132 DE MOTIB. STELLÆ MARTIS



SOLIS vergat in 5 ½ ♒: quamvis hunc gradum cap. XXV libere inquisituri simus quasi incognitum. Et sit TERRA A. MDXC in ϑ, anno MDXCII in η, anno MDXCIII in ε, anno MDXCV in ζ. Et anguli ϑ α η · η α ε · ε α ζ · æquales, quia α est punctum æqualitatis, & periodica Martis tempora præsupponuntur æqualia. Sitque Planeta his quatuor vicibus in κ, ejusque linea apsidum α λ. Est ergo angulus ϑ α κ secundum indicium anomaliæ commutationis coæquatæ 127.5.1.

Quod visum locum Martis attinet, is die IV antecedente hora simili fuit 24.22 ♈. diurnus ejus diei esset 44. Ergo ad nostrum tempus visus fuit in 25.6 ♈. qui est situs lineæ ϑ κ. Sed α κ tendit in 15.53.45 ♉. Ergo ϑ κ α est 20.47.45. Residuus igitur α ϑ κ ad duos rectos est 32.7.14.

Vt igitur sinus α ϑ κ ad α κ, quam dicemus esse partium 100000: sic ϑ κ α ad ϑ α quæsitum. Est ergo ϑ α 66774.

Quod si reliquæ η α, ε α, ζ α, ejusdem prodibunt longitudinis, falsum erit quod suspicor: at si diversæ, omnino vicero.

**Fig. 9** This diagram shows the halving of the Earth's eccentricity in the Copernican cosmology and of the Sun's in the Ptolemaic and Tychonic. In the Copernican scheme, Mars is at the point $x$ on three occasions. Knowing the relevant angles, Kepler could determine the corresponding positions of the Earth at the three points on the dashed circle. They established the true position of the Earth's orbit and that it must have half the eccentricity assumed by Copernicus of the Earth. But one only needs three known points to fix the position and size of a circle.[11] Kepler's determination of the position of the Earth's orbit finally revealed the error in Hipparchus's solar theory. The halving of the Earth's eccentricity created a firm foundation for astronomy. It was hugely important.

Having properly located the Earth, and still believing in circular orbits, Kepler set about fitting the parameters of Mars's orbit to Brahe's observations. He still made use of the Ptolemaic equant even though convinced the speed in orbit must somehow be determined relative to the Sun and not the void equant. It was a wonderful piece of work and a posthumous triumph for Ptolemy's circles and equant. But the merry-go-round just would not get everything right. Kepler tweaked here, he tweaked there, but whatever he did an occasional error of up to just 8 min of arc in the position of Mars would show up. He could never have done it without Brahe's observations, which were both as accurate as they could be and also, most importantly, comprehensive.[12] This is the place to quote Kepler:

"We, whom God in his goodness has given such a careful observer in Tycho Brahe, and whose observations reveal the $8'$ error of Ptolemy's calculations, should thankfully recognize the goodness of God and make use of it. That is, we should make the effort (supported by the arguments for the falsity of our assumptions) to find at last the true form of the celestial motions... These $8'$ alone reveal the need for reformulation of the whole of astronomy; they become the material of a great part of my work."

Now it is time to talk about the area law. Along with establishing where the lines of the apsides meet all in one place, this was the other great bonus of the 'Machian' shifting from a void point to the Sun. For the apside adjustment, the shift was from the Earth's empty focus. In the case of the area law, it was from Mars's empty focus.

Kepler was keenly aware of the value of the equant phenomenon: mathematically, in pre-calculus days, anything that involved uniform motion (angular velocity about the equant in this case) was a significant plus. Non-uniform motion was barely tractable. For this reason astronomers were still using the equant to calculate ephemerides in Newton's time three quarters of a century after Kepler, absolutely insistent on a physical interpretation of celestial motions, had done away with it.

The area law was one of the most serendipitous discoveries—of which there are so many—in science. Kepler was looking for a law, governed by the Sun, that would determine the speed in orbit of each planet. His physical intuition told him the Sun

---

[11] Keplers' work on Mars began under the assumption that the Earth has an exactly circular orbit. Because the Earth's eccentricity is $\approx 1/60$, this assumption is accurate to better than one part in 7000. Even when Kepler knew the Earth's orbit could not be a perfect circle, he could assume it to be so for his work on Mars with its far larger eccentricity $\approx 1/11$. In the story of the discovery of the laws of their motion, the planets were like the characters in a good novel. Each had an individual personality determined by its eccentricity and semimajor axis. The interaction of these personalities, reflected in the observational data, is what makes the discovery of the laws of the planets' motion such an absorbing story.

[12] Ptolemy had made and used relatively few observations obtained at times and orbital positions he expected to be especially valuable for construction and testing of his observations. Brahe believed in blanket coverage of the orbits: who could know what would be relevant and revealing? This was truly prescient and of immense value to Kepler.

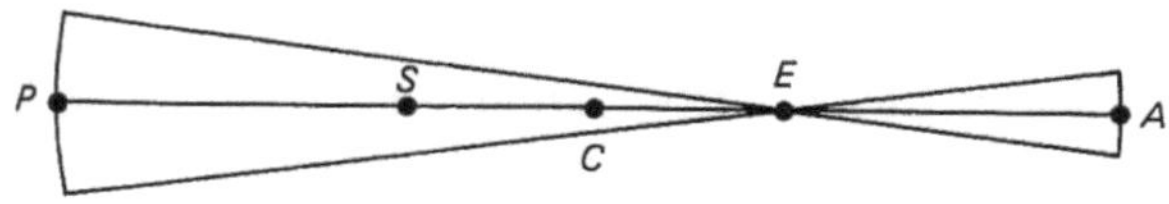


**Fig. 10** The exact area law and the speed law that follows from the empty-focus effect both predict that the orbital speeds at aphelion and perihelion are inversely proportional to the distance from the Sun

must exert a force on the planet that would be stronger, the closer the planet was to the Sun. In fact, that was clearly indicated by observations, which showed that the planets moved fastest when closest to the Sun. Plausibly enough, Kepler guessed that the speed would be inversely proportional to the distance from the Sun. Strong support for this came from the fact that the equant law showed the orbital speeds at aphelion and perihelion to be exactly in inverse proportion (Fig. 10).

But the mathematics of the putative law, applied to the eccentric orbit of Mars, proved to be beyond Kepler's abilities. He therefore decided to replace what he regarded as the exact law by an approximation in the form of the area law! He was encouraged to this by his recollection of the way Archimedes had estimated the area of a circle by dividing it into ever smaller segments. As his work progressed and he gained an increasing number of accurate locations of Mars through his application of trigonometry to Brahe's observations, always under the key assumption that the Sun was the centre of the solar system and the controller of planetary motions, he came to realize that the area law did actually govern the speed in orbit.

Kepler's final, very tortuous breakthrough to the joint discovery of ellipticity of the orbit and the area law was in fact somewhat delayed by his enthusiasm for theory. The moment he found unambiguous evidence that Mars's orbit could not be circular, he started to speculate and initially guessed an egg-shaped orbit, i.e., fatter at one end than the other. Slowly, as he accumulated more and more accurate locations, the egg was abandoned. A chance glance at a table of logarithmic tables was what finally led him to the ellipse—more serendipity. The full truth at last came to him around Easter 1605.

## 6 Kepler's Significance

It would be a futile counterfactual exercise to ask how science would have developed without Brahe and Kepler's extraordinary efforts. However, it is entirely possible that, even without the discovery of the telescope and the possibility that gave for more accurate observations, decades could have passed before the discovery of the laws of planetary motion. What is absolutely certain is that Newton's *Principia* is inconceivable without Kepler's discoveries. All three of Kepler's laws were important: from the third, Newton deduced the $1/r^2$ force law for gravity; from the first, that the planet's elliptical motions could be understood as the outcome of two competing tendencies—rectilinear inertial motion and gravitationally induced deflection from it along the direction to the Sun. In many ways, Kepler's second law was actually the most important. Newton recognized this by making it the subject of his very first proposition in the *Principia*. It will be worth saying something about this.

By 1670 at the latest, Newton had most of the elements of a rudimentary dynamics, above all the notion of inertial motion that would persist forever were it not changed by the action of other bodies. He understood elastic collisions and the nature of centrifugal force. What hindered a full blossoming of dynamics was the prevailing mechanistic conception of the world due above all to Descartes. According to this view, all mechanical action took place through direct contact: collisions. The Cartesian cosmos was a terribly crowded world crammed full of pieces of matter in continual collision. Newton basically subscribed to this view. Although he had laws to describe collisions, there was little he could do with them.

The real advance almost certainly came in 1679, when Robert Hooke, newly appointed as secretary of the Royal Society, pressed Newton hard to confirm his (Hooke's) proposal "of compounding the celestial motions of the planets of a direct motion by the tangent & an attractive motion towards the central body." Calculations of Newton that he may well have made as a result of Hooke's letters of 1679 have survived and include the key result that became Proposition 1 in the *Principia*:

**Proposition 1** *The areas which revolving bodies describe by radii drawn to an immovable centre of force do lie in the same immovable planes, and are proportional to the times in which they are described.*

This is the theoretical explanation of Kepler's area law. It had far-reaching implications, for it told Newton that nature should be described, mathematically at least, by forces that act over distances. Huygens had coined the expression *centrifugal force*; in explicit imitation, Newton called his new forces *centripetal*. He was well aware of the revolutionary nature of what he was doing; he was proposing to give universally despised *occult* forces a decisive role in physics. He was very cautious about this and emphasized, in *hypotheses non fingo*, that he was not making any assumptions about the physical mode of action of the forces he introduced. What he did stress was that such forces, introduced mathematically, could explain at a stroke a vast number of diverse phenomena. They opened up a whole field for exploration that is still ongoing. Kepler's $8'$ led to more than the reformulation of astronomy.

## 7 Intermezzo: Christian Doppler

Before we move on to Mach and Einstein, brief mention should be made of Christian Doppler (Fig. 11) and the important eponymous effect that he predicted in 1842 while a professor at the Czech Polytechnic in Prague. Ironically, Doppler was seeking an explanation of the different colours of binary stars; the effect he proposed to explain the difference was physically correct but completely wrong in his application. Binary

**Fig. 11** Christian Doppler (1803–1853)

stars have different colours because, in the first place, their temperatures (and to some extent their chemical compositions) are different. Doppler suggested, without at that time any experimental support, that the observed frequency of the light emitted by the stars depended on their orbital speeds, which are different (and epoch dependent). This is true, but the effect was far too small to explain the colour differences.

The Dutch meteorologist Buys Ballot (1817–1890) made the first experimental confirmation of the Doppler effect in 1845 by getting a group of musicians to play a calibrated note on a train on the line between Utrecht and Amsterdam. Of course, in those days there were no police sirens that make the effect so evident today. Despite this confirmation of the effect for sound, Doppler's proposal of a dependence of the observed frequency of light on the speed of the source remained controversial for a surprisingly long time—decades. One person who helped to establish it was Ernst Mach.

## 8 Mach and Kinematic Residues in Dynamics

Mach (Fig. 12) was one of the great experimentalists of all times and a man of wide interests. His name is associated with three very diverse things in science: Mach bands in psychology, the Mach number in aerodynamics, and Mach's principle in the theory of gravity and inertia. Mach was twice nominated for the Nobel Prize for his discovery of shock waves (Fig. 13), but so many exciting discoveries were being made in the early 20th century that he missed the honour he deserved.

**Fig. 12** Ernst Mach (1838–1916)

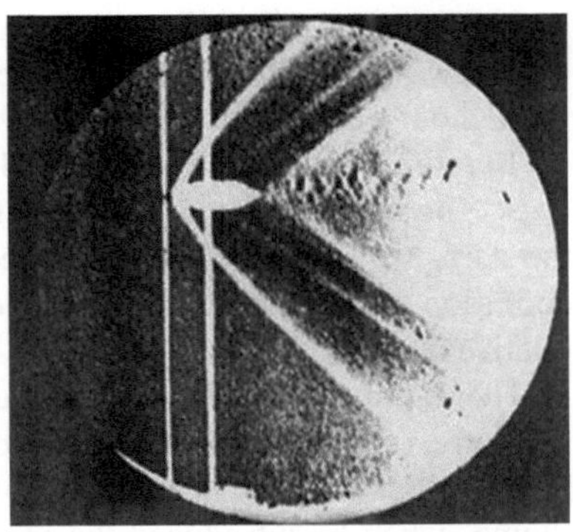

**Fig. 13** Mach's flash photograph of a supersonic bullet and the shock wave generated by it

Before we discuss Mach's critique of Newton's absolute space and time and Einstein's reaction to it, it is worth reproducing the diagram (Fig. 14) in Blackmore's valuable informative biography of Mach [3]. This shows the instrument Mach devised soon after completing his doctorate to demonstrate the Doppler effect for sound. A vertical tube AA rotates in the plane perpendicular to the page. Air forced through the column creates sound in the whistle at G. A person standing in the plane of the rotating tube hears a clearly modulated pitch of the whistle as it rotates at the end of the tube, while someone standing some distance way at right angles to the plane of

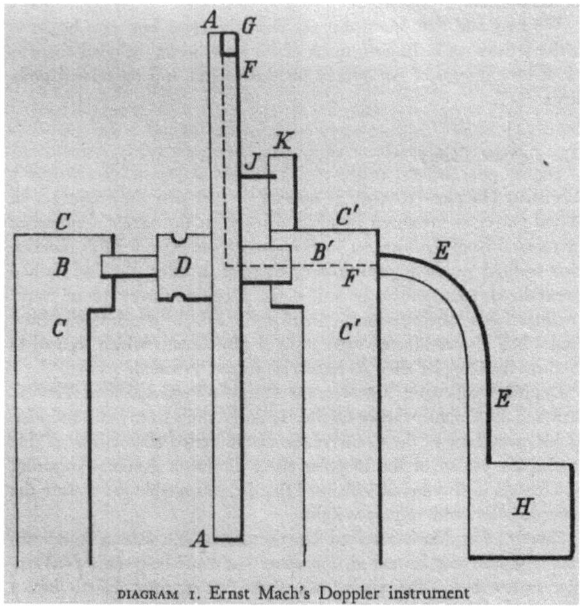


**Fig. 14** Mach's device to confirm the Doppler effect for sound, reproduced from [3]

rotation hears a constant pitch. The Doppler effect for sound is demonstrated in this simple way. The apparatus "became a frequently used class demonstration device throughout Central Europe for many years" [3, p. 19].

To return to our topic, visible and physical markers were key to Kepler's discoveries. The Sun had a dual role: it defined and created motion of the planets. As for the stars, Kepler, like Copernicus, declared them the ultimate frame of reference, by definition at rest. His laws did not contradict this, and the stars did not exhibit any relative motion.

The star-studded shell of this closed world retained the Sun's warmth generated at its centre, or *focus*. Kepler introduced the Latin word for hearth into scientific usage, first in optics and then in astronomy. Descartes (1596–1650) shattered Kepler's cosy 'home' when he introduced the mechanical philosophy. He, above all, marks the transition from the closed world to the infinite universe. In it, all bodies, including the stars and their constituents, move relative to each other.

Descartes actually had two diametrically opposed concepts of motion: absolute and relative. The origins of both are worth retelling (for more details, see [1]). Let us start with the first. One day, lying on his bed, he is said to have spotted a fly on the ceiling and saw he could fix its position by its two distances from the walls. The story is *ben trovato*, apposite even if invented. Cartesian coordinates, so convenient for defining straight lines, were born. The idea of rectilinear inertial motion (not yet named so) was already in circulation; in a book, *Le Monde*, ready for publication in 1632, Descartes made it the foundation of mechanics long before Newton.

Implicit here is an unchanging reified space like the ceiling on which the fly crawled: Newton's absolute frame in all but name. No longer do the Sun and stars define motion. Space does. Whereas pre-Kepler void points governed the planets' motions, now invisible space controls all motion. This is occult forces in spades!

Descartes was just about to publish *Le Monde*, which assumed correctness of the Copernican cosmology, when he heard about Galileo's condemnation by the Inquisition. In alarm—piety he claimed—Descartes hurriedly withdrew his book and thought long and hard how he could save his mechanical philosophy. Eventually he introduced a quite different definition of place and motion in his *Principles of Philosophy* (1644). He declared all motion to be relative. Now any one body has infinitely many positions and motions according to which bodies are used to define them. However, he did grant the existence of 'one true philosophical definition of position', according to which the position of any one body is defined by its envelope, i.e., the immediately adjacent matter that surrounds it. The reason for this definition, actually a throwback to the Aristotelian notion of *topos*, is to be found in Descartes' contention that the Earth is carried around the Sun by a vortex, which is thus its *immediate envelope*. The point then is that the Earth does not move relative to the vortex and therefore does not move in accordance with the true definition. Since terrestrial mobility (and not heliocentricity) was the Inquisition's objection to Copernicus, Descartes felt he had secured his position and explicitly stated that in accordance with his proposal the Earth does not move.

But after this avowal of pure relationalism, Descartes, failing to note the contradiction, reverted to uniform rectilinear motion as the first principle of mechanics. This made no sense in a world with position and motion defined relatively in either way. It required an implicit absolute space.

Newton studied Descartes' book closely and did see the contradiction. Knowing what could be done with the law of inertia, he recoiled from the virtual impossibility of expressing it rigourously in Descartes' shifting cosmos. The prominence given to absolute space and time in the Scholium at the start of the *Principia* are a covert dismissal of Descartes, even though Newton does grant the great difficulty of distinguishing "the true motions of particular bodies from the apparent; because the parts of that immovable space in which those motions are performed do by no means come under the observation of our senses."

In fact, as Mach was later to remark, Newton's laws were never verified relative to absolute space and time but to exactly the same referents that Kepler had used: the effectively fixed stars and the time-measuring clock supplied by the diurnal revolution of the stars.

Descartes' absolute and relative are the origin of the reductionistic–holistic dichotomy. Mach the holist reacted to Newton the reductionist when he spoke of '*immediate* connections' and the ideal that hovers before the natural investigator as 'an insight into the principles of the whole matter'. The essence of reductionism is threefold: simple objects, atoms, that move in accordance with simple laws, primarily the law of inertia, in a simple background: absolute space. But if position is relative, only the totality of separations between objects is real: the world is held together by an indissoluble network of relations, and history is nothing but their evolution. As

Mach said: "The universe is not twice given, with an earth at rest and an earth in motion; but only once, with its relative motions alone determinable" (Mach, p.284).

But Mach went much further than this epistemological verity. Kepler had given the Sun a dynamical role. Mach extended it to the stars or, rather, the totality of masses of the Universe. They should not only define but also control motion. That Mach envisaged this done by some as yet unknown physical mechanism is confirmed by his famous refutation of Newton's bucket argument [13] for absolute motion:

"Newton's experiment with the rotating vessel of water simply informs us that the relative rotation of the water with respect to the sides of the vessel produces no noticeable centrifugal forces, but that such forces are produced by its relative rotation with respect to the mass of the earth and the other celestial bodies. No one is competent to say how the experiment would turn out if the sides of the vessel increased in thickness and mass till they were ultimately several leagues thick."

It is well known that Mach's critique made a powerful impression on the late teenage Einstein. It was the main stimulus to his attempt to eliminate all trace of Newton's absolute space through the creation of general relativity. The idea that the totality of the masses of the universe 'work together' to create the local inertial frames in which force-free bodies move rectilinearly and uniformly is what Einstein called *Mach's principle*. For a variety of reasons, this has had a tangled history, for which several factors are responsible.

## 9 Einstein's Reaction to Mach

Let me start with Einstein's strange confusion of two distinct meanings of inertia: there is inertial *motion*, as defined in Newton's first law, and inertial *mass*. That they are distinct is evident: the concept of inertial mass does not enter into the statement of the first law. Mach gave a much admired operational definition of inertial mass, which he defined through the accelerations bodies impart to each other when they interact. These are inversely proportional to their *intrinsic* inertial masses. Mach's disagreement with Newton on this score was not about substance but proper formulation. What really concerned Mach was the origin of inertial motion: Newton believed absolute space governed it, Mach the totality of masses in the universe.

Reading Einstein's various comments about Mach and inertia I am forced to conclude he was the victim of semantic confusion. He does not seem to have seen any difference between the two meanings of the word inertia. His most egregious distortion of Mach is in the 1917 paper in which he laid the foundation of modern relativistic cosmology. He claims:

---

[13] Newton introduced the bucket to make a serious scientific argument but simultaneously a fool of Descartes, whose mechanical philosophy relied heavily on centrifugal force. Many people writing on the absolute–relative debate and unaware of the background to the bucket argument have been misled into thinking the issue is about the difference between linear and circular motion, which is not true. In fact, I increasingly think Newton confused himself.

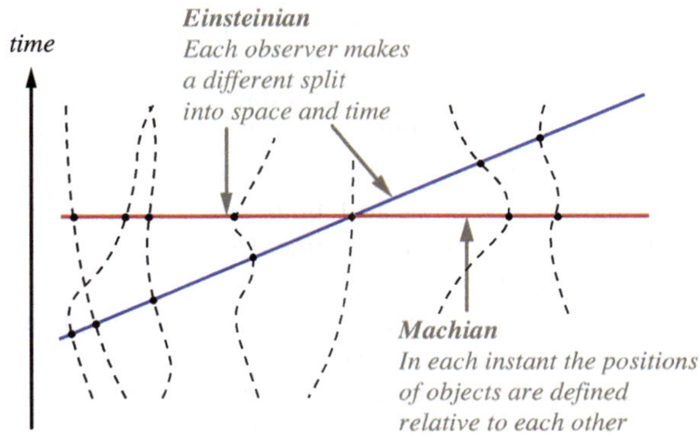


**Fig. 15** The two quite different meanings of relativity. Relativity as defined by Minkowski and Einstein refers to the ambiguity in the splitting of four-dimensional spacetime into time and space. Relativity as defined by Mach means that the position of any one body is defined at a given instant by its distances to all the other bodies in the Universe at that instant. The conflict of concepts is evident: Machian relativity makes complete sense if there exists a distinguished notion of simultaneity, but that is denied as the first principle of Einsteinian relativity

"In a consistent theory of relativity there can be no inertia *relatively to 'space'*, but only an inertia of masses *relatively to one another*. If, therefore, I have a mass at a sufficient distance from all other masses in the universe, its inertia must fall to zero."

Mach would have dismissed this comment as a gross distortion of his ideas; Einstein is clearly substituting a bogus issue about inertial mass for Mach's proper concern with inertial motion. Unfortunately, Einstein's 1917 comment led to several misguided attempts to implemented a Mach's principle along inappropriate lines.

However, the complexity of the Machian issue has a much more solid basis and raises a real dilemma, which is illustrated in Fig. 15. What is at stake is the very meaning of the word *relativity*. Einstein and Minkowski had in mind the observer dependence of the split of spacetime into space and time and, more generally, to the complete freedom to lay down coordinates on spacetime in any suitably continuous way. Einstein spoke of general covariance; today one speaks of four-dimensional diffeomorphism invariance. That may be called Einsteinian relativity. The most important aspect of it is the denial of simultaneity as a physically significant concept.

In contrast, relativity as originally formulated by Mach makes no sense without an underlying notion of simultaneity: it asserts that the position of any given object at a given instant is defined by its distance to *all* the other objects in the universe *in that instant*. Most relativists today would say that this is a hopelessly obsolete notion because Einstein and Minkowski showed that our intuitive notion of simultaneity has no counterpart in the physical world. Does this mean that Mach's principle is a dead duck?

Not necessarily. In his only article not devoted to quantum mechanics, John Bell wrote on special relativity and sought "to drive home the lesson that the laws of physics in any *one* reference frame account for all physical phenomena, including the observations of moving observers" [4, p. 77]. Attempts to find a generically distinguished frame in Minkowski space are doomed to fail on account of its high symmetry, but the case is altered in GR: gravity brings structure into spacetime. I want to use the remainder of this chapter to give what I believe is the correct definition of Mach's principle and to show how Einsteinian gravity is much more compatible with Machian relativity than one might imagine. I shall even suggest that Machian relativity is the deeper principle.

However, I must first briefly recount how Einstein set out to implement the idea that Mach had espoused: that inertial motion should not be governed by absolute space but the totality of masses in the Universe (a more detailed account is given in [5]).

What was decisive for Einstein was his discovery of special relativity in 1905. This arose from his successful reconciliation of Maxwell's electrodynamics, with its only apparent need for an ether, and Galilean relativity applied to all physical phenomena. The lesson Einstein drew from his success was that uniform motion through Newton's absolute space could not be defined—it was impossible to associate any actual speed with it. Although he said nothing explicit at the time, this result already suggested to Einstein the way to implement Mach's idea: to show that the alleged absolute space had no observable effects at all, for then one could argue that it does not exist. The impossibility of determining a speed of uniform motion through space was the first step in that direction.

The decisive idea that set in motion Einstein's long search for a new theory of gravity was the equivalence principle, that 'happiest thought' of his life which occurred to Einstein in 1907. Its importance for Einstein was not so much that gravity and inertia are identical in essence but the possibility "that the principle of relativity is also satisfied for systems moving relatively to each other with acceleration" [6]. The equivalence principle suggested that this could be done for at least uniform accelerations.

Einstein's strategy from then on was clear and settled. He would attempt to extend the relativity principle ever further. The next step, clearly suggested by Mach's retort to Newton's bucket argument, called for extension to uniform circular motion, for which the magnitude of the acceleration, as in the equivalence-principle elevator, is constant in magnitude but its direction changes. From there, the logical step to complete relativity of motion was not too difficult. Einstein advanced the principle of general covariance as the physical foundation of the new theory of gravity he was seeking.

Two aspects of Einstein's approach should be noted. His principle of relativity did not in any way directly address the way in which the Universe itself behaved. It merely said that the description of its behaviour should be the same in whatever coordinate system one cared to describe it. Einstein's immediate acceptance of Kretschmann's objection that general covariance in itself had no physical content but was merely a requirement of mathematical consistency was a remarkable volte-face that has generated much argument and confusion about the foundations of general relativity and, in particular, Mach's principle. The only conclusion I wish to draw from this brief discussion is that Einstein did not attempt a direct implementation of Mach's ideas but attacked the problem indirectly. This comes out especially clearly in a comment he made in 1918 [7]:

"We want to distinguish more clearly between quantities that belong to a physical system as such ... and quantities that depend on the coordinate system. Ones initial reaction would be to require that physics should introduce in its laws only quantities of the first kind. However, the scientific development has not confirmed this conjecture. It cannot dispense with coordinate systems."

There is a clear anticipation here of the distinction, now commonplace due to developments in gauge theory, between so-called true degrees of freedom and redundant degrees of freedom. What I want to question is whether Einstein had correctly identified what are the "quantities that belong to a physical system as such". In my final section before brief conclusions, I wish to suggest that he may have made the incorrect identification.

## 10 The Machian Approach: Shape Dynamics

There is no doubt what Mach regarded as the true physical quantities: bodies that possess intrinsic mass and distances (in Euclidean space) between them. He most certainly did not think time had any ontological reality [2, p. 273]: "It is utterly beyond our power to measure the changes of things by time. Quite the contrary, time is an abstraction at which we arrive from the changes of things." But Minkowski, followed by Einstein, had given time the same ontological status as space. This led Einstein to identify the "quantities that belong to a physical system as such" with four-dimensional spacetime intervals, whereas Mach had identified them with exclusively three-dimensional spatial entities. Let us see where such a standpoint takes us.

Let us start with one thing on which we can be sure Mach and Einstein would have agreed: if the local frames in which force-free particles move inertially are determined by the universe, there must be a sense in which the universe is a closed dynamical system, for otherwise one could never close the circle and say the whole determines the parts: local inertial frames. This underlying sense is implicit in Mach[14] and explicit in Einstein's 1917 cosmological model.

Let us then allow the notion of simultaneity and assume that the universe is a closed dynamical system. We can consider two models: an island universe of $N$ point particles, which matches the ontology of Mach's original proposal, and a three-dimensional Riemannian geometry closed up on itself,which corresponds to

[14] See his comment (Mach, p. 287) "Nature does not begin with elements, as we are obliged to begin with them. It is certainly fortunate for us that we can, from time to time, turn aside our eyes from the overpowering unity of the All, and allow them to rest on individual details. But we should not omit, ultimately to complete and correct our views by a thorough consideration of the things which for the time being we left out of account." How can completion come without a definite sense in which the universe is closed?

closed-space vacuum GR. In the point-particle case, the difference between Newton and Mach is easily expressed in terms of configuration spaces. Newton's is $R^{3N}$, three coordinates for each particles. But Mach said only the inter-particle separations are real. We need to quotient $R^{3N}$ by the Euclidean translations and rotations to obtain the $3N$-6-dimensional Machian *relative configuration space* (RCS). Absolute position and orientation are removed from the RCS. In fact, although Mach did not recognize the need, one must go a step further since *distance* presupposes an absolute scale. We also need to quotient by dilatations; this takes us to the $3N$-7-dimensional *shape space* $\mathscr{S}$. I would say that an instantaneous shape of the universe matches Mach's requirement that we grasp the 'immediate connections'.

However, that does not yet mean that we have gained 'an insight into the principles of the whole matter'. Machian histories of the universe will be curves in $\mathscr{S}$. The issue now is this: what determines these curves? The fact is that any Newtonian history can be represented as a curve in $\mathscr{S}$: one simply plots the representative points of the successive shapes. In what way would a Machian history be distinguished from an arbitrary Newtonian one plotted in $\mathscr{S}$?

A problem with Mach is that he tended to speak in general intuitive terms. It is here that a penetrating analysis by Poincaré [8], who analyzed the problem in much more precise terms, provides the guide. Poincaré asked: what defect, if any, arises from Newton's use of absolute space? His answer was that a true believer in relationalism, convinced that only inter-particle separations $r_{ab}$ have physical significance, would pose the initial-value problem of particle dynamics in these terms: $r_{ab}, \dot{r}_{ab}$ should determine the evolution $r_{ab}(t), a, b = 1, 2, ..., N$, uniquely. This matches the formulation in terms of the particle coordinates and velocities; in accordance with Laplacian determinism, $\boldsymbol{r}_a, \dot{\boldsymbol{r}}_a, a = 1, 2, ..., N$, determine the evolution.

Poincaré pointed out that the rather natural transfer of this requirement from $R^{3N}$ to the RCS fails. The reason is that the data $r_{ab}, \dot{r}_{ab}$ contain no information about the angular momentum $\boldsymbol{L}$ in the system, whereas this information is encoded in $\boldsymbol{r}_a, \dot{\boldsymbol{r}}_a$ (under the assumption that the masses are known). Although the presence or absence of $\boldsymbol{L}$ is undetectable in $r_{ab}, \dot{r}_{ab}$ initial data, the curves that result do encode information about $\boldsymbol{L}$. Poincaré said that this fact, reflected in the manifest presence of angular momentum in the solar system, was the true evidence for the existence of a dynamically active agent in addition to the separations $r_{ab}$ and their rates of change $\dot{r}_{ab}$. As a convinced believer that only relative motions should have dynamical effect, Poincaré said he found this state of affairs repugnant but that it was necessary to accept the empirical evidence.

It is strange that Poincaré did not consider a Machian resolution to the problem, namely that for *for a dynamically closed universe as a whole* the relative data do determine the future uniquely. One can then attribute the failure of this requirement in subsystems of the universe precisely to the fact that the masses of the universe do determine local inertial frames of reference. Poincaré formulated his ideas in the context of the RCS, but they can be directly extended to and made more stringent in shape space. This leads me to the formulation of the Mach-Poincaré principle for particle dynamics in these terms.

*Mach–Poincarè Principle*. Specification of a point and direction (strong form) or point and tangent vector (weak form) in shape space $\mathscr{S}$ should determine the evolution in $\mathscr{S}$ uniquely.[15]

It is necessary to allow for the weaker form if one is to model expansion of the Universe. For discussion of this delicate issue, see my introduction to shape dynamics [9]. The extension to dynamical geometry is relatively obvious; the shape space in this case is conformal superspace, which provides the natural framework for describing the dynamics of three-dimensional conformal geometries. I cannot describe in detail this work; the most important chapters are [10–12] (see also Koslowski's contribution to this conference proceedings).

What one can say is that, if the Universe is spatially closed, there is a well-defined sense in which GR implements Mach–Poincaré principle in the weak form very well, indeed perfectly if there is no cosmological constant. For all the details I must refer the reader to the references already cited, but the key conclusions do need to be stated, at least for vacuum gravity. First, by virtue of its clearly formulated first principles shape dynamics introduces of necessity a notion of simultaneity into GR and insists that the physical entity which is evolving is the conformal three-geometry on successive leaves of a foliation of spacetime by hypersurfaces of constant-mean-(extrinsic)-curvature (CMC). Second, the spacetime in which these hypersurfaces are embedded is completely determined by specification of a point and tangent vector in conformal superspace. This fact was first demonstrated in [13].

If the ideas of shape dynamics, which do follow very naturally from Mach's ideas, are vindicated, it will be incorrect to view shape dynamics as a rule to select certain special solutions—those that are globally hyperbolic and CMC foliable—from among the full set allowed by GR. Rather GR might have to be seen as an extension of shape dynamics beyond its physical domain. I suspect we shall have to await the quantum theory of gravity to see if this view is justified.

## 11 Conclusions

Galileo said "He that attempts natural philosophy without geometry is lost." He meant of course *three-dimensional* geometry, which was still Euclidean in his day, though I am sure he would have greeted Riemann's generalization with enthusiasm.

---

[15] In my mind, the great virtue of Poincaré's analysis is that he formulates requirements on the form of a dynamical theory of the Universe in terms of the *initial data* that one regards as belonging "to the physical system as such". This allows a much more precise formulation than Einstein's requirement that all coordinate systems should be on an equal footing, which is actually void of content, or that the action should satisfy certain symmetry requirements, which is also amenable to adjustment, as one sees with the passage from standard Newtonian dynamics to parametrized particle dynamics, which adds reparametrization invariance as a symmetry without changing the physical content of the theory. In contrast, implementation of Mach's ideas boils down (in the case of the weak Mach–Poincaré principle) to identification of the true (configurational) degrees of freedom and construction of a theory in which they and their velocities wrt an independent variable uniquely determine the evolution of the true degrees of freedom.

The first step in the still ongoing creation of the dynamical theory of the Universe was Hipparchus's theory of the Sun's motion. It is important that all the great work in astronomy reviewed in this chapter studied the evolution of *intrinsic shapes*, of which the fixed stars formed part. Absolute position, orientation and size played role at all; I have already emphasized that every conclusion drawn in astronomy was based on measurement of angles between observed physical objects. These included measurements of what was called time but was actually the diurnal rotation of the stars. Even now, with geometry curved and made dynamical, the irreducible epistemological basis of science is observed angles. We now see the Universe as almost infinitely flexible, but we cannot do without angles.

A conformal geometry supplies the 'immediate connections' that Mach exhorted us to grasp. As regards "the principles of the whole matter", I would say that as far as classical physics is concerned they are encapsulated in the weak Mach–Poincaré principle applied to a closed Universe whose possible spatial configurations are defined by conformal geometry.

Let me end with my sincere thanks to the organizers, above all Jiří Bičák, for the invitation to speak at the wonderful conference in Prague.

## References

1. Barbour, J.B.: Absolute or Relative Motion?: The Discovery of Dynamics. Absolute or Relative Motion?: A Study from a Machian Point of View of the Discovery and the Structure of Dynamical Theories. The Discovery of Dynamics. Cambridge University Press, Cambridge (1989)
2. Mach, E.: The Science of Mechanics. Open Court, LaSalle (1960)
3. Blackmore, J.T.: Ernst Mach. His Life, Work, and Influence. University of California Press, Berkeley (1972)
4. Bell, J.S.: Speakable and Unspeakable in Quantum Mechanic. Cambridge University Press, Cambridge (1987)
5. Barbour, J.B.: The Part Played by Mach's Principle in the Genesis of Relativistic Cosmology, p. 47. Cambridge University Press, Cambridge (1990)
6. Einstein, A.: Berichtigungen zu der Arbeit: "Über das Relativitätsprinzip und die aus demselben gezogenen Folgerungen". Jahrbuch der Radioaktivität und Elektronik **5**, 98 (1908)
7. Einstein, A.: Prinzipielles zur allgemeinen Relativitätstheorie. Annalen der Physik **360**, 241 (1918). doi:10.1002/andp.19183600402
8. Poincaré, H.: Science and Hypothesis. Walter Scott, London (1905)
9. Barbour, J.B.: Shape Dynamics: An Introduction. Birkäuser, Berlin (2012)
10. Anderson, E., Barbour, J., Foster, B.Z., Kelleher, B., Murchadha, N.Ó.: The physical gravitational degrees of freedom. Class. Quantum Gravity **22**, 1795 (2005). doi:10.1088/0264-9381/22/9/020
11. Gomes, H., Gryb, S., Koslowski, T.: Einstein gravity as a 3D conformally invariant theory. Class. Quantum Gravity **28**(4), 045005 (2011). doi:10.1088/0264-9381/28/4/045005
12. Gomes, H., Koslowski, T.: The link between general relativity and shape dynamics. Class. Quantum Gravity **29**(7), 075009 (2012). doi:10.1088/0264-9381/29/7/075009
13. Barbour, J., Murchadha, N.Ó.: Conformal superspace: the configuration space of general relativity. ArXiv e-prints arxiv:1009.3559 [gr-qc] (2010)

# Einstein in Prague: Relativity Then and Now

**Jiří Bičák**

**Abstract** It was during his stay in Prague that Einstein started in earnest to develop his ideas about general relativity. I will recall those days in 1911 and 1912, discuss Einstein's papers on gravitation from that period and emphasize which new concepts and ideas he introduced. I also want to indicate how the main themes that preoccupied him then, the principle of equivalence, bending of light, gravitational redshift and frame dragging effects, are alive in contemporary relativity.

## 1 Introduction

I would like to start as I did in my talk at the conference, quoting what Einstein wrote soon after his arrival in Prague on April 3, 1911: "The city of Prague is very fine, so beautiful that it is worth a long journey for itself" (from the letter to his friend M. Besso on May 13, 1911); or, "I have a magnificent institute here in which I work very comfortably... By the way, Czechs are much more harmless than one thinks" (from the letter to M. Grossmann on April 27, 1911). I hope the conference participants, 100 years after Einstein in Prague, had a similar impression.

The quote from Hesiod's "Works and Days" from the seventh century BC—*The price of achievement is toil* ['Schinderei']; *and the gods have ruled that you must pay in advance*—enables me to give a brief summary: *Einstein paid much in Prague*. The days: April 1911–July 1912. The works: principle of equivalence, light bending, dragging of inertial frames; features of a future theory of gravity.

It was not until 1911, only after his arrival in Prague, that Einstein's interest in quantum theory started to diminish and his systematic concentration on the problems of a new theory of gravity began. There were specific issues which Einstein analyzed

J. Bičák (✉)
Institute of Theoretical Physics, Charles University in Prague, V Holešovičkách 2, 180 00 Prague, Czech Republic
e-mail: bicak.troja@gmail.com

J. Bičák and T. Ledvinka (eds.), *General Relativity, Cosmology and Astrophysics*, Fundamental Theories of Physics 177, DOI: 10.1007/978-3-319-06349-2_2,

© Springer International Publishing Switzerland 2014

in Prague, for example that the bending of light due to the Sun's gravitational field is observable. However, as we shall see later, just before leaving Prague, on July 4, 1912, Einstein submitted a short paper to Annalen der Physik in which a number of fundamental features of the final general relativity were anticipated. Concerning new ideas and liberating oneself from old views, especially in this respect, "Einstein paid much in Prague". Before we turn to Einstein's work on gravity in Prague we shall recall his days here. We cannot describe the atmosphere in Prague of those days, with tensions between three main groups of its inhabitants, Czechs, Germans and Jews, on one hand, and with many intellectually inspiring aspects and interactions between them on the other hand, in any detail, but literature of a special interest will be mentioned.

## 2 Why and How He was Invited to Prague

Charles University in Prague, founded in 1348 as the first university "beyond the Alps", was originally one educational center for Czechs, Germans, Poles and various south-European nations. Nationalism led to its division in 1882 into the German and Czech parts.[1] In 1911 the Czech part had 4432 students, the German part 1844 students but, for example, Mach's lectures in the German part were visited by a number of Czech students. In the German part, the head of mathematical physics, Professor F. Lippich, was due to retire and the German university decided to have the Institute of Theoretical Physics that existed in the Czech University already. The chief advocate of the proposal to appoint Albert Einstein was Anton Lampa, professor of experimental physics, a great "Machian", a Czech by origin but an ardent supporter of Germanization who hoped that Einstein would develop further Mach's ideas.[2] Another member of the commission for choosing a candidate for theoretical physics was Georg Pick, a mathematician of broad interests, with whom Einstein made later, during his stay in Prague, a close friendship. In his recommendation letter from 1910, Max Planck wrote that if Einstein's theories were to be confirmed, Einstein would be considered the Copernicus of the twentieth century.

---

[1] In 1879/80 and 1883/84 Ernst Mach was elected Rector of the University. However, during his second Rectorship he resigned since he disapproved of the division of Prague University into Czech and German parts.

[2] Anton Lampa (1968–1938) was an experimental physicist, interested also in philosophy, history and cultural aspects of physics. Interestingly, he was the first to publish a paper on the appearance of a moving rod according to special relativity (Zeits. f. Phys., 1924). For the life and work of Lampa, see [1].

## 3 Days in Prague

No view of Einstein in Prague can be more original and immediate than that of Philipp Frank,[3] Einstein's successor as head of the Institute of Theoretical Physics in the German university in Prague until 1938. In 1929, Frank organized the famous first meeting on the Epistemology of the Exact Sciences in Prague. From 1939 till 1941 he wrote the biography of Einstein [3], which Einstein endorsed. Its whole Chapter IV, "Einstein at Prague", besides information on Einstein and the theory of relativity, paints amusingly also the atmosphere in Prague at the beginning of the twentieth century. Frank, when writing Einstein's biography, did not have manuscripts and other documents at his disposal, so for an earnest Einstein biographer, other sources are important. Nevertheless, Frank tells various entertaining stories which are probably not far from the truth. So, for example, it was the custom for a newly appointed professor to pay a visit to his colleagues. Einstein started to pay these visits and decided to go first to romantic old parts of Prague. But later he found that these calls, which numbered around forty, were really a waste of time and he stopped his visits. The professors whom he had not visited felt puzzled but in fact the main reason was that they did not live in interesting parts of the city or their name was far back in the alphabet. Another nice story supports our point that it was in Prague where Einstein for the first time started to concentrate on gravity rather than on quantum physics. Frank recalls how during his first visit of Einstein in Prague, Einstein took him to the window of his office from which they could overlook a large garden behind the wall (see Fig. 1, with the wall on the right). Einstein told Frank that he saw people there in deep meditation but also groups very vividly discussing. Only quite later he learned that the park belonged to the insane asylum of Bohemia. Einstein pointed to people walking there, turned to Frank and said: "Those are the madmen who do *not* occupy themselves with the quantum theory". This is not true today. The park and gardens still belong to the mental hospital, however, it is open to the public during the day. And physicists from our Faculty nearby quite often walk across it to get to the center of the city.

In Prague, Einstein associated himself with a group of Jewish intellectuals who gathered in the evenings at Berta Fanta's salon in the house "At the Unicorn" in the Old Town Square. There philosophy was discussed and music played. Einstein met an ardent Zionist, Hugo Bergmann, the son-in-law of Berta Fanta, but he was then not able to arouse Einstein's interest in his ideas. In the letter to Hedwig Born in 1916, Einstein wrote that Zionists in Prague are "a small troop of unrealistic people, harking back to the Middle Ages". Still later, after Bergmann became a Hebrew University professor in Jerusalem, Einstein called him "the serious saint from Prague". During

[3] P. Frank (1884, Vienna—1966, Cambridge, Mass.) was a theoretical physicist and logical positivist, a member of the Vienna Circle. His work in relativity is summarized in detail in the comprehensive article by Havas [2]. With R. von Mises, Frank published a book on differential and integral equations in physics; later he wrote several books on the philosophy of science. His booklet *Relativity–a richer truth*, with a foreword by Albert Einstein on the "Laws of science and the laws of ethics", published in 1951, is less known. It touches on a number of philosophical and ethical issues.

**Fig. 1** The building of the former Faculty of Philosophy of the German University on Viničná 7 wherein Einstein had "an excellent Institute with a beautiful library"

the evenings Einstein perhaps met Franz Kafka, although Kafka, in fact, did not like to go there, but he certainly was acquainted with Max Brod, a writer and journalist who later, after Czechoslovakia was founded in 1918, played a significant role in promoting the Czech culture. Almost all of Einstein biographers, inspired by Philipp Frank, make the point that Max Brod drew on Einstein's character for his portrait of Kepler in his novel *Tycho Brahe's Way to God* and some even suggest that in this way certain egocentric features of Einstein's personality were disclosed. However, when reading Brod's autobiography *Streitbares Leben*, we discover that he was quite unhappy with Frank's interpretation and even wrote a letter to Einstein to explain that he never noticed any egocentric features in his behavior and, in any case, that it was rather the poet and writer Franz Werfel, Brod's friend, who contributed to the portrait of Kepler.

Einstein visited Prague once again in 1921 when he accepted an invitation from Urania, Prague's German Society, to give a lecture on the theory of relativity. Accompanied by Frank, he visited the Physical Institute of the Czech part of Charles University. "By this visit Einstein wanted to express his sympathy for the new Czechoslovak Republic and its democratic policy under Masaryk's leadership", writes Frank. Abraham Pais in his celebrated biography of Einstein [4] gives the list of people Einstein suggested for the Nobel Prize. Masaryk is among them, proposed by Einstein for the Peace Prize. Einstein was later in correspondence with Masaryk regarding the fate of a pacifist Přemysl Pitter.[4]

[4] Copies of their letters are available in "Einstein Archives Online"—see http://www.alberteinstein.info/

# 4 The Czech Culture and Science Responding to Einstein's Work

Karel Čapek (1890–1938), a humanist with encyclopedic knowledge, one of the best known writers and journalists of the Masaryk era, who was a close friend of Masaryk, wrote "philosophical" novels, charming detective stories, anti-war dramas, and also was a pioneer of the Czech science fiction. Alan J. Friedman and Carol C. Donley in their book *Einstein as myth and muse* (Cambridge University Press 1985) start their section "Approaches to relativity in fiction" emphasizing that from the use of the profoundly wrong aphorism "everything is relative", various authors explore intricate possibilities of Einstein and his theories. And they continue: "A remarkable early exposition of the possibilities appeared in 1924, with Karel Čapek's novel *Krakatit*. Čapek's awareness of science and technology was indicated by mentions... of the leading scientists of the day, including Einstein, Rutherford, Planck, Bohr and Millikan. The plot concerns an inventor who has discovered a way to release atomic energy... The technical details are as accurate as they could be in the early 1920s, and atomic energy is correctly seen as a possibility emerging from the radioactivity work of Becquerel and Rutherford, and not from Einstein's theories... The inventor, Prokop, is torn in the traditional struggle between God and the devil... Prokop's bewilderment, in the literal form of a fever, is described by the first metaphor from relativity: ... It appeared he was moving with velocity approaching velocity of light; in some way his heart was compressed. But that was only Fitzgerald-Lorentz contraction... " Prokop then finds himself in the closed Einstein universe... Curiously, Einstein's closed universe attracted also one of the most sophisticated Czech art critics and writers, F. X. Šalda. In 1928 already, he makes analogies between Einstein's conception of the finite, closed Universe and the conception of space in paintings by Cézanne.

We cannot continue here with more examples of the inspiring role which Einstein's theories exerted on the Czech culture. Let us just look at Fig. 2 where the portrait of Einstein by well-known Czech oil painter and graphic artist Max Švabinský (1873–1962) is shown. In a recent interview in the University magazine *Babylon*, Švabinský's son-in-law, originally a mathematician, declared that he sent one of the copies of this lithography to Robert Oppenheimer and it was hanging on the wall of his office in the Institute of Advanced Studies in Princeton.

## *4.1 Impact on Czech Physics and Astronomy*

Relativity theory was popularized and even taught quite soon by the Czech physicists and astronomers. The first papers were written by A. Dittrich and A. Žáček in 1912. One of the main protagonists of Einstein's theories was professor of theoretical physics at the Czech part of the Charles University, František Záviška (1879–1945).

**Fig. 2** Einstein's portrait from 1955 by Max Švabinský

In 1925, his semi-popular book *Einstein's principle of relativity and theory of gravity* including basic principles of general relativity appeared. He was in good relations with Philipp Frank and translated Frank's book *Das Ende der mechanischen Physik* (The end of mechanistic Physics), published in 1935, into Czech. It is a criticism of the totalitarian (Nazi) philosophy from the point of view of the theory of knowledge and philosophy of science. An interesting impact of Einstein's prediction of light bending on a Czech astronomer, F. Link, will be discussed below. At present there is a rather extensive literature on Einstein's influence on culture and science in the Czech lands between World Wars I and II available, and on Einstein's work done during the Prague stay.[5]

[5] To give some examples, we quote a booklet [5] published on the occasion of Einstein's centenary in 1979 (in Czech and partially in German), the 3rd number of the Czechoslovak Journal of Physics from the same year dedicated to Einstein and the comprehensive article on his route to general relativity, concentrated primarily on the Prague period [6]. Two articles in English [7, 8] are texts of talks about Einstein's Prague papers on gravity given at the Conference of the European Physical Society in Prague in 1984 and at the Marcel Grossmann meeting in Perth in 1988. And a very recent detailed work [9] by Těšínská, a historian of science (containing 57 references) concentrates

After World War II Miroslav Brdička (1912–2007) wrote one of the first original papers on general relativity ("On gravitational waves", Proc. Roy. Irish Acad., 1951) after his stay 1948–1949 as a scholar in the Dublin Institute for Advanced Studies. It has been curious to see this work quoted and Brdička's name observed in the title of the very recent paper by G. Gibbons and C. Rugina.[6] The communist upheaval in Czechoslovakia in 1948 led to a large wave of emigration. Within this there was also a geometer Václav Hlavatý who became professor at Indiana University and started to work extensively in relativity, in particular on Einstein's unified field theories. In another wave of emigration, after August 1968, Karel Kuchař came first to Princeton, following the invitation of John Wheeler, and then became professor in Salt Lake City. His influential work in the quantum theory of covariant systems, canonical quantum gravity and the issue of time is well known to contemporary relativists, including some of participants of this conference.

Finally, I am glad to say that at present there are several groups active in relativity, relativistic astrophysics and cosmology in the Czech Republic—at the Faculty of Mathematics and Physics (Theoretical Physics, Astronomy), in the Academy of Sciences (Institute of Astronomy, Mathematical Institute) in Prague, and at the Silesia University in Opava and Masaryk University in Brno. A volume *Gravitation: following the Prague inspiration* [10] contains comprehensive essays by 14 Czechoslovak relativists and astrophysicists about their work. Some most recent results are included in the Proceedings [11] of this conference, containing contributions based on oral and poster presentations.

Before I finish this "Czech intermezzo" let me present the Fig. 3 from a two-day celebratory meeting on the occasion of Einstein's centenary which took place in the Carolinum–the same place as our conference–on February 26 and 27, 1979. Among more than 200 participants there were several distinguished guests from abroad, including two associated directly with Einstein—P. G. Bergmann and J. A. Wheeler. Peter Bergmann met his wife Margot when they both were students of Physics at Prague German University.[7]

## 5 Lectures, Seminars and Papers of Albert Einstein in Prague

Below in Table 1 we see that the lectures Einstein gave in Prague were on classical subjects like mechanics, thermodynamics and molecular theory of heat. It is not known whether in the seminars he organized more modern topics were included.

(Footnote 5 continued)
on Einstein's call to Prague and on development of theoretical physics at the Czech university in Prague in relation to Einstein's work.

[6] See Coryacher-Chaplygin, Kovalevskaya, and Brdička-Eardley-Nappi-Witten pp-waves spacetimes with higher rank Stäckel-Killing tensor, J. Math. Phys **52**, 122901 (2011).

[7] Unfortunately they did not recall where Einstein's or even Frank's office was. However, their faces turned into a fine smile when we went through the big door to see the Einstein memorial tablet in the entrance hall inside because they could clearly identify the door into the building in Fig. 1.

**Fig. 3** During Einstein's centenary celebrations outside Carolinum in February 1979: John Archibald Wheeler, Andrzej Trautman, Mrs. Melcher, Ernst Schmutzer, Jiří Langer, Margot Bergmann, Peter Bergmann, and Horst Melcher (from *left* to *right*). The present conference took place in the same place... (Photograph taken by the author)

**Table 1** Einstein's lectures and seminars in Prague

| Period | Title | No. of students |
|---|---|---|
| 20.4.1911–31.7.1911 | Mechanik diskreter Massenpunkte (3h) | 13 |
| | Thermodynamik (2h) | 12 |
| | Seminar | 6 |
| 19.10.1911–27.3.1912 | Mechanik diskreter Massenpunkte (3h) | 12 |
| | Wärmelehre (2h) | 13 |
| | Seminar | 7 |
| 12.4.1912–31.7.1912 | Mechanik der Kontinua (2h) | 10 |
| | Molekulartheorie der Wärme (3h) | 11 |
| | Seminar | 7 |

During the first year, the lectures were given in Klementinum in the centre of the historical part of Prague, then they moved to Viničná 7.

Concerning the Prague papers of Albert Einstein, five were on thermodynamics, radiation theory and quantum theory, among them also just brief notes. An exception was the review on the problem of specific heats which Einstein presented at the first Solvay Congress in November 1911 where he and Friedrich Hasenöhrl from Vienna were the only representatives of the Austro-Hungarian Empire. Since we are concerned primarily with gravity, we shall not analyze those papers, but we give now the *complete list of the works of Albert Einstein in Prague on the theory of relativity and gravitation.*[8] The first three papers appeared in 1911, and the following four in 1912. The titles are here given in English, the exact German title and bibliographical details are given in the References.

1. The Theory of Relativity [12].
2. On the Ehrenfest Paradox. Comment on Varičak's Paper [13].
3. On the Influence of Gravitation on the Propagation of Light [14].
4. The Speed of Light and the Statics of the Gravitational Field [15].
5. On the Theory of the Static Gravitational Field and Note Added in Proof [16].
6. Is There a Gravitational Effect Which Is Analogous to Electrodynamic Induction? [17].
7. Relativity and Gravitation. Reply to a Comment by M. Abraham [18].

The first paper, assigning Prague as Einstein's address already, is based on a lecture given in Zurich in January 16, 1911, before Einstein's arrival in Prague. The lecture, in which Einstein used the term "Relativity Theory" in a title for the first time, was followed by the discussion published in April 1912 (see "The collected papers of Albert Einstein", Vol. 3). The second paper is a short note on a paradox involving moving measuring rigid rods in special relativity. Our main attention will be paid to five papers on the development of a new theory of gravity. Before we turn to them in detail, we shall first present a document, valuable in connection with Einstein's stay in Prague, though largely unknown. As a document of historical importance it was first published and commented upon in 1979 in [5], then quoted in the biography of Pais [4], and later also elsewhere. The document concerns the 1923 Czech translation of Einstein's little book *About the Special and General Theory of Relativity in Plain Terms.*[9] Einstein wrote a special foreword to the Czech edition; this appeared in both the German original and the Czech translation. In the foreword he recalls what he did during his Prague stay: "*I am pleased that this small book, in which the main ideas of the theory of relativity are explained without mathematical elaboration, should now appear in the native language of the country in which I found the necessary concentration for developing the basic idea of the general theory of relativity which I had already conceived in 1908. In the quiet rooms of the Institute of Theoretical Physics of Prague's German University in Viničná Street, I discovered that the principle of equivalence implies the deflection of light rays near the Sun by an observable amount, without at that time knowing that a similar result*

[8] Einstein's long review on the Special Theory of Relativity for the *Handbuch der Radiologie* was started in Prague at the beginning of 1912, and continued after Einstein's move to Zurich (the quality of ink and paper improved after the move). The First World War interrupted the publication. Einstein was later unwilling to add material on general relativity or even revise the existing manuscript. Still, it is an extraordinarily precious document since it is the earliest and most significant of the surviving scientific manuscripts written by Einstein before World War I. A fine facsimile was published in 1996 by George Braziller, Inc., in association with J. Safra Foundation and the Israel Museum, Jerusalem. For more details on the manuscript, see the Collected papers of Albert Einstein, Volume 4.

[9] The book was translated by an excellent physics teacher at a highly regarded Czech Gymnasium in Prague, V. Štíbr. Incidentally, after World War II Štíbr was the physics teacher of Karel Kuchař.

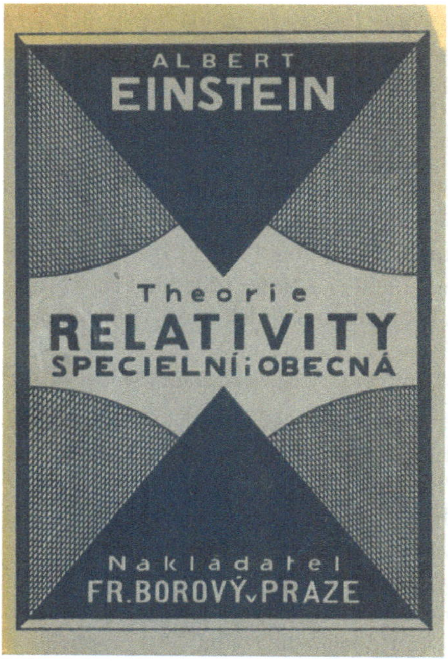


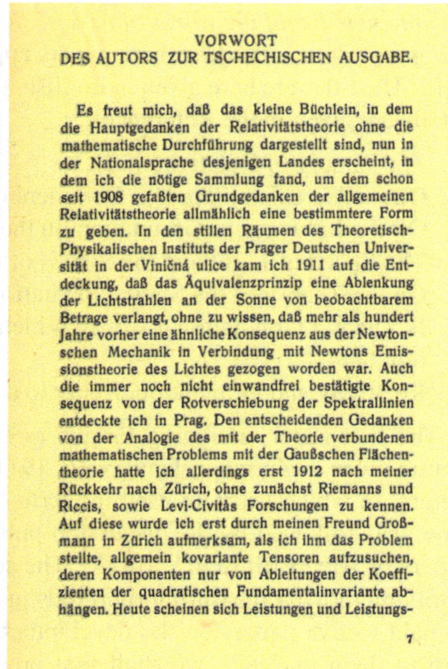
VORWORT
DES AUTORS ZUR TSCHECHISCHEN AUSGABE.

Es freut mich, daß das kleine Büchlein, in dem die Hauptgedanken der Relativitätstheorie ohne die mathematische Durchführung dargestellt sind, nun in der Nationalsprache desjenigen Landes erscheint, in dem ich die nötige Sammlung fand, um dem schon seit 1908 gefaßten Grundgedanken der allgemeinen Relativitätstheorie allmählich eine bestimmtere Form zu geben. In den stillen Räumen des Theoretisch-Physikalischen Instituts der Prager Deutschen Universität in der Viničná ulice kam ich 1911 auf die Entdeckung, daß das Äquivalenzprinzip eine Ablenkung der Lichtstrahlen an der Sonne von beobachtbarem Betrage verlangt, ohne zu wissen, daß mehr als hundert Jahre vorher eine ähnliche Konsequenz aus der Newtonschen Mechanik in Verbindung mit Newtons Emissionstheorie des Lichtes gezogen worden war. Auch die immer noch nicht einwandfrei bestätigte Konsequenz von der Rotverschiebung der Spektrallinien entdeckte ich in Prag. Den entscheidenden Gedanken von der Analogie des mit der Theorie verbundenen mathematischen Problems mit der Gaußschen Flächentheorie hatte ich allerdings erst 1912 nach meiner Rückkehr nach Zürich, ohne zunächst Riemanns und Riccis, sowie Levi-Civitàs Forschungen zu kennen. Auf diese wurde ich erst durch meinen Freund Großmann in Zürich aufmerksam, als ich ihm das Problem stellte, allgemein kovariante Tensoren aufzusuchen, deren Komponenten nur von Ableitungen der Koeffizienten der quadratischen Fundamentalinvariante abhängen. Heute scheinen sich Leistungen und Leistungs-

7

**Fig. 4** The cover of the Czech edition of Einstein's popular book "About the Special and General Theory of Relativity in Plain Terms" from 1923 and the foreword in original German which Einstein wrote for the Czech edition

*had been derived from Newton's mechanics and his corpuscular theory of light. In Prague I also discovered the shift of spectral lines towards the red which is not yet completely confirmed. However, the decisive idea of the analogy between the mathematical formulation of the theory and the Gaussian theory of surfaces came to me only in 1912 after my return to Zürich, without being aware at that time of the work of Riemann, Ricci, and Levi-Civita. This was first brought to my attention by my friend Grossmann when I posed to him the problem of looking for generally covariant tensors whose components depend only on derivatives of the coefficients of the quadratic fundamental invariant. It now appears that it is already possible to evaluate the achievements and limitations of the whole theory. It gives a deep knowledge of the physical nature of space, time, matter and gravity; however, it does not provide sufficient means for solving the problems of quanta and of the atomic constitution of the elementary electric units of which matter is composed.*"

In Fig. 4, the cover and the German foreword are displayed.

During talks about Einstein's Prague period a question often arises whether the idea of using Riemannian geometry in building a new theory of gravity emerged during the Prague stay already, or only after his return to Zurich when he started to collaborate with his friend Marcel Grossmann. In the foreword above, Einstein mentions that the work of Riemann, Ricci, and Levi-Civita was brought to his attention

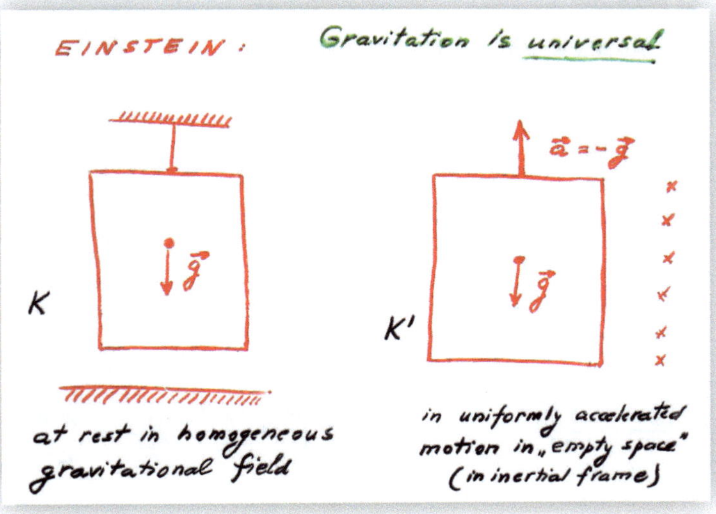


**Fig. 5** All physical phenomena will proceed in the same way in the systems $K$ and $K'$—principle of equivalence

first by Grossmann. Nevertheless, Einstein did not care much about giving precise historical statements; for example, he writes that in Prague he discovered the shift of spectral lines towards red although this effect is contained in his paper from 1907 already in which he tackled the problem of gravity for the first time and then left it until the beginning of the Prague period. That the idea of using the Riemannian geometry was conceived in Prague already as a result of discussions with his friend mathematician Georg Pick is stated in the biography by Frank [3], as well as in the "Personal Reminiscence" by Reinhold Fürth who studied at the German University of Prague during 1912–1916 and became then an important member of the Faculty.[10]

## 6 The Principle of Equivalence

At the beginning of his first paper on gravity [14] the equivalence of the systems $K$ and $K'$ indicated in Fig. 5 is discussed. To paraphrase Einstein's words in English

[10] Fürth says that "Pick, professor of mathematics at Prague and a fellow violin player, drew his [Einstein's] attention to the Italian, Levi-Civita, and his absolute differential calculus... (see *Einstein—the first hundred years*, ed. by M. Goldsmith et al. Pergamon Press 1980). Fürth became a member of Max Born's group in Edinburgh in 1939 but still in October 1938, after the Munich agreement, he was in Prague and published the text "Der Streit um die Deutung der Relativitaetstheorie" (The struggle about the meaning of the relativity theory) in which he defends relativity against philosophers like Prof. Kraus in Prague who wanted to find inconsistencies of the theory by "metaphysical" arguments.

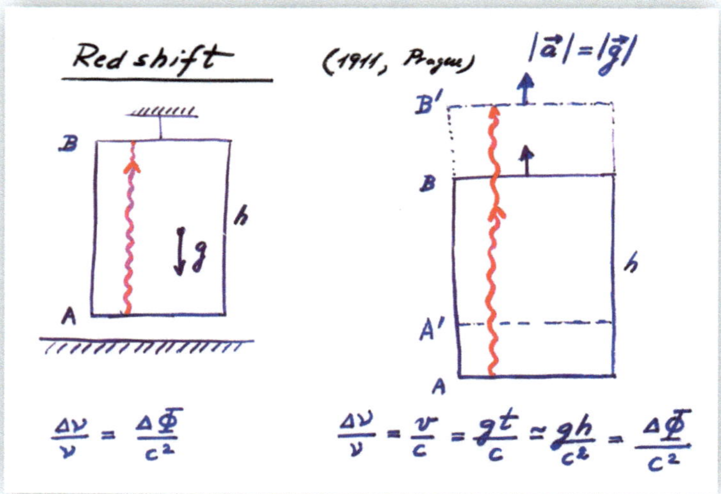


**Fig. 6** The principle of equivalence implies the gravitational redshift

he clearly states that the equivalence of both systems is certain as far as we restrict ourselves to purely mechanical phenomena, but it will acquire a deeper meaning if we extend the equivalence to *all laws of nature*. Then we have a principle which has a great heuristic meaning. In his biography *A. Einstein: Creator and Rebel* Banesh Hoffmann, Einstein's direct collaborator, writes: "In the paper of 1907... Einstein had already begun his attack on the problem of acceleration, and he returned to it in his Prague paper of 1911. His arguments, particularly in its 1911 form, must rank as one of the most remarkable in the history of science." The principle of equivalence is then applied to derive the gravitational redshift—for the first time in a beautifully pedagogical way (see Fig. 6). As Mark Twain writes: "The nice thing about Science is that one gets such wholesale returns of conjecture from such a trifling investment of fact"...

What is the present-day formulation of the (weak) equivalence principle? Employing Cliff Will's formulation from his Living Reviews article [19]:

- Test bodies fall with the same acceleration independently of their structure or composition.
- The outcome of any local non-gravitational experiment is independent of: (a) the velocity of the local inertial frame in which it is performed, (b) where and when in the universe it is performed.

From the time of Newton and Eötvös it has been a continuing effort to measure a possible violation of the first item. After the 1950s, it is connected with the names of Dicke in Princeton, Braginskij in Moscow and, most recently, with the group at the University of Washington which used a torsion balance tray to study the accelerations

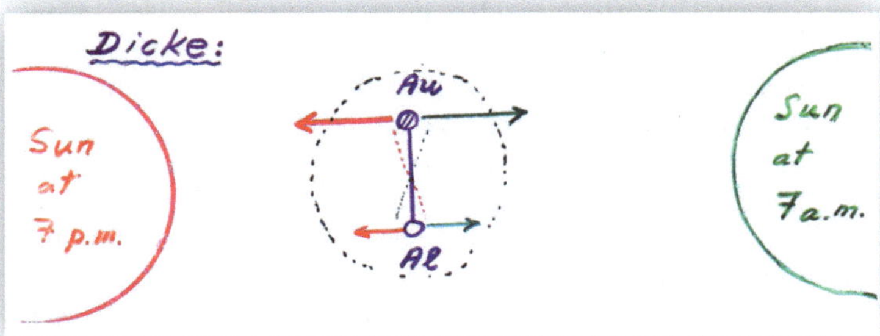


**Fig. 7** Dicke used the field of the Sun to put the limits on difference in acceleration of various materials. (The suspension of the torsion balance is perpendicular to the picture)

of various materials toward local masses, towards the Sun (as Dicke did—see Fig. 7) and even towards the Galaxy.

The present best limits on the fractional difference in acceleration of various materials/bodies $A$ and $B$ are

$$\eta = \frac{a_A - a_B}{\frac{1}{2}(a_A + a_B)} \quad \begin{cases} (0.3 \pm 1.8) \times 10^{-13}, & \text{Eöt-Wash} \\ (-1.0 \pm 1.4) \times 10^{-13}, & \text{LLR} \end{cases} \tag{1}$$

where the first result is given by experiments at the University of Washington, whereas the second comes from Lunar laser ranging. See [19] for more details.

An endeavor to improve the limit on $\eta$ will undoubtedly continue since some theories, inspired primarily by string theory, predict the violation of the weak equivalence principle due to the presence of dilatons. That is why a *Satellite Test of Equivalence Principle* (*"STEP"*) has been conceived to improve the limit on $\eta$ by about five orders of magnitude to $\eta = 10^{-17} - 10^{-18}$. It is a drag-free satellite consisting of an outer shell around an inner test mass [20].

## 6.1 Gravitational Redshift Today

Concerning the gravitational redshift, the first reliable experiment was the Pound-Rebka-Snider experiment in 1960 using the Mössbauer effect to measure precisely the frequency shift of $\gamma$-rays in a 22.6 m tower at Harvard. These experiments, yielding an accuracy of the order of 10 %, were followed by clocks placed in an aircraft and a rocket; the clock rates were compared with the same clocks on the ground. An accuracy of $7 \times 10^{-5}$ in determining $\Delta\nu$ was obtained by using a hydrogen maser clock in the rocket by Vessot et al. in 1980. Only 30 years later this accuracy was claimed to have been improved upon by laboratory experiments based on quantum interference of atoms [21] which yielded an accuracy of $7 \times 10^{-9}$. However, the interpretation was criticized in the work [22] which considers the experiment as a test of the universality of free fall but not the redshift effect. An intriguing idea for producing and measuring a "force-free gravitational redshift" (a gravitational analogue of the Aharonov-Bohm effect) imagines two atomic matter waves which serve as clocks at different gravitational potentials and the redshift can be caused just by a potential difference although a force vanishes. This suggestion was published 100 years after Einstein was going to leave Prague, just two weeks before our conference in [23]. Therein the authors address also the criticism raised in [22]. Anyway, the best known "common" proof of gravitational redshift we currently have available is, of course, the daily successful operation of the Global Positioning System (GPS).

What is the present situation concerning the gravitational shift of the spectral lines from the surface of the Sun relative to the corresponding laboratory lines which, in his Prague paper [12], Einstein predicted to be $\Delta\lambda = 2.1 \times 10^{-6}\,\lambda$, which in the velocity scale is approximately 600 $\mathrm{ms}^{-1}$? The measurements are complicated because there are various sources of wavelength shifts, such as radial currents different at different levels, convection, inner asymmetries of spectral lines etc. The first convincing measurements appeared only in 1991 by using the infrared oxygen triplet [24]. Converted to velocities, the three chromospheric oxygen lines yielded $627 \pm 10\ \mathrm{ms}^{-1}$ which is $0.99 \pm 0.02$ of the value predicted by the principle of equivalence. Precise measurements of the solar redshift are planned [25] which should reach an accuracy of $10^{-6}$ and so test the second order relativistic effects. For the most recent pioneering work on the gravitational redshift of galaxies in clusters, see [26].

## 7 Bending of Light

After the formulation of the principle of equivalence and its application in deriving the formula for redshift, Einstein turned in the same paper [14] to the problem of light propagation in a gravitational field. The equivalence of the systems $K'$ and $K$ in Fig. 6 implies the frequency shift, i.e., different rates of clocks located at different gravitational potentials $\Phi$ in the case of a *homogeneous* gravitational field. However, typically for his grasp of the laws of nature, Einstein assumes that the same effect takes place in a *general* static field.[11] As a consequence, the velocity of light, measured by clocks influenced by a gravitational potential, depends on the value of the potential. Denoting the velocity of light at origin of coordinates by $c_0$, then at a place in which gravitational potential with respect to the origin is $\Phi$, the velocity of light is given by the relation $c = c_0(1 + \Phi/c^2)$. Employing a simple picture based on Huygens

[11] Einstein notices the effect of gravity on the light propagation in his paper from 1907 already in which, for the first time, the gravitational field and equivalence principle are mentioned. However, only a homogeneous field is analyzed and the effect considered in the field of the Earth is found to be unobservable.

principle, Einstein then derives the formula for an angle $\alpha$ by which a light ray will be deflected in the direction $n'$ along any trajectory $s$:

$$\alpha = -\frac{1}{c^2}\int \frac{\partial \Phi}{\partial n'}\, ds. \tag{2}$$

According to this formula a ray propagating around a celestial body with mass $M$ deflects towards the body by the value given in Fig. 8—the picture of the final page of Einstein's best known paper from the Prague period. Under the formula the resulting value of 0.83 arcseconds for a light ray passing close to the Sun is given, half of the value following from the final form of general relativity. Einstein promptly contacted various observatories in the world and tried to encourage observation of the effect during a total eclipse.

The only response came from Erwin Freundlich (1885–1965), an assistant in the Royal Observatory in Berlin.[12] Freundlich took part in a solar eclipse expedition in 1914 in Crimea but World War I broke and he was interned in Russia. It was only just after the war, in 1919, when the effect was measured for the first time by the British expeditions led by Eddington and Dyson. There is much literature on these observations, some of which attributed the confirmation of the relativistic value to Eddington's high regard of general relativity and also his wish, of a pacifist, to show how despite the war English astronomers confirm a theory by a German physicist. It is true, difficulties during the expeditions appeared, but in a recent article, D. Kennefick carefully re-analyzed the case and showed that Eddington et al. had good reasons for the claim that general relativity was confirmed, and Newtonian theory was not (see Fig. 9).

The light deflection is one of the finest examples of the continuing success of general relativity as the best gravity theory available. When one allows a more general form of a static spherically symmetric vacuum metric as it can follow from an alternative theory of gravity by inserting parameters $\beta$ and $\gamma$ into the Schwarzschild metric (see Fig. 10), one finds that it is $\gamma$ which enters the formula for the deflection of waves.

If general relativity is correct, $\gamma = 1$. Let us have a look at Fig. 11 from Cliff Will's Living Reviews in Relativity article [19] on "The Confrontation between General Relativity and Experiment".[13] A 2004 analysis of $\approx 2$ million VLBI observations of 541 radio sources at 87 VLBI sites imply $\gamma - 1 = (-1.7 \pm 4.5) \times 10^{-4}$!

---

[12] For a detailed account on Freundlich and tests of relativity theory, see [27]. In fact, the first attempt to measure light deflection was already made during an eclipse in October 1912 by C. D. Perrine from Cordoba, Argentina, who was inspired by Freundlich. However, rain frustrated all efforts. (I thank Jorge Pullin for pointing this out in the discussion after my lecture in the conference.) Freundlich later wrote one of the first books on Einstein's gravitation theory (Springer 1916, Cambridge 1920). From the "Prague perspective" it is interesting to notice that in 1937 he was appointed professor of astronomy at the Charles University in Prague, however, was forced to leave because of Hitler's policies towards Czechoslovakia in January 1939.

[13] Will's reviews on the verification of Einstein's theory are well known. Perhaps less recognized is his "Resource Letter" on the tests of gravity [28] which gives many references on both current literature as well on some of historical papers.

908 *A. Einstein. Einfluß der Schwerkraft usw.*

Nach Gleichung (4) erleidet ein an einem Himmelskörper vorbeigehender Lichtstrahl eine Ablenkung nach der Seite sinkenden Gravitationspotentials, also nach der dem Himmelskörper zugewandten Seite von der Größe

$$\alpha = \frac{1}{c^2} \int_{\vartheta = -\frac{\pi}{2}}^{\vartheta = +\frac{\pi}{2}} \frac{k\,M}{r^2} \cos \vartheta \cdot ds = \frac{2\,k\,M}{c^2\,\Delta},$$

wobei $k$ die Gravitationskonstante, $M$ die Masse des Himmelskörpers, $\Delta$ den Abstand des Lichtstrahles vom Mittelpunkt des Himmelskörpers bedeutet. *Ein an der Sonne vorbeigehender Lichtstrahl erlitte demnach eine Ablenkung vom Betrage $4 \cdot 10^{-6}$*
[11] *= 0,83 Bogensekunden.* Um diesen Betrag erscheint die Winkeldistanz des Sternes vom Sonnenmittelpunkt durch die Krümmung des Strahles vergrößert. Da die Fixsterne der der Sonne zugewandten Himmelspartien bei totalen Sonnenfinsternissen sichtbar werden, ist diese Konsequenz der Theorie mit der Erfahrung vergleichbar.
[12] Beim Planeten Jupiter erreicht die zu erwartende Verschiebung etwa $^1/_{100}$ des angegebenen Betrages. Es wäre dringend zu wünschen, daß sich Astronomen der hier aufgerollten Frage annähmen, auch wenn die im vorigen gegebenen Überlegungen ungenügend fundiert oder gar abenteuerlich erscheinen sollten. Denn abgesehen von jeder Theorie muß man sich fragen, ob mit den heutigen Mitteln ein Einfluß der Gravitationsfelder auf die Ausbreitung des Lichtes sich konstatieren läßt.

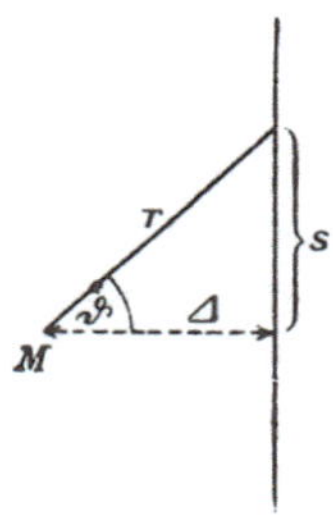


Fig. 3.

Prag, Juni 1911.

(Eingegangen 21. Juni 1911.)

**Fig. 8** The last page of Einstein's most famous Prague paper on deflection of light

For example, scalar-tensor theories must have parameter $\omega > 40000$ (which their proponents like R. Dicke considered to be around 6), to be compatible with the light deflection observations.

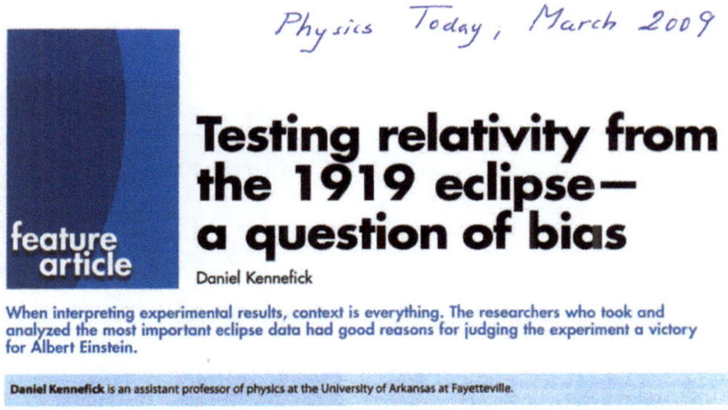

Physics Today, March 2009

feature article

**Testing relativity from the 1919 eclipse— a question of bias**

Daniel Kennefick

When interpreting experimental results, context is everything. The researchers who took and analyzed the most important eclipse data had good reasons for judging the experiment a victory for Albert Einstein.

**Daniel Kennefick** is an assistant professor of physics at the University of Arkansas at Fayetteville.

**Fig. 9** Portrait of Erwin Freundlich by Max Pechstein from 1918. The heading of a recent article which showed undoubtedly that Eddington's and Dyson's expeditions of 1919 to measure light bending, despite difficulties they encountered, could state safely that their observations ruled out Newton theory and confirmed general relativity

A fundamental consequence of the influence of gravity on the light propagation is the existence of black holes formed by gravitational collapse. I dare to state this commonly known fact only because I recall how in Erice in 1974 Roger Penrose drew a picture similar to Fig. 12. He only "forgot" to write "Prague 1911" beneath the left picture.

## 8 Gravitational Lensing

In contemporary astronomy, nowhere is light deflection so extensively and profitably used as in the phenomenon of gravitational lensing. Going now from the present to the past we may look at just two well-known examples: the images of two galaxy clusters after collision (Fig. 13) indicating the existence of dark matter in the clusters; it moves further after the collision (its distribution, shown by green counters, being known thanks to gravitational weak-lensing) than ionized gas (yellow-red in the right part observed in X-rays); and a Horseshoe Einstein Ring from Hubble telescope (Fig. 14).

Here I wish to make just a few remarks on the history of discovering the effect, in which Prague played a significant role, unrecognized generally yet. It was long thought that the effect was first described by Einstein in a paper published in *Science* on December 4, 1936, after his interaction with a Czech amateur scientist Rudi Mandl. Later Einstein's Scratch Notebook from 1912 (when he was still in Prague) was found and it became evident that the effect was found by him then already. The complete history of this finding and its relation to the 1936 paper is comprehensively described in several publications by Renn and Sauer; see, for example, [30].

It is well-known that the effect was also discussed by Eddington in 1920 and by Chwolson in 1924. However, the role of a Czech astronomer, an expert in the eclipse phenomena of the Moon and planets, František Link (1906–1984), remains

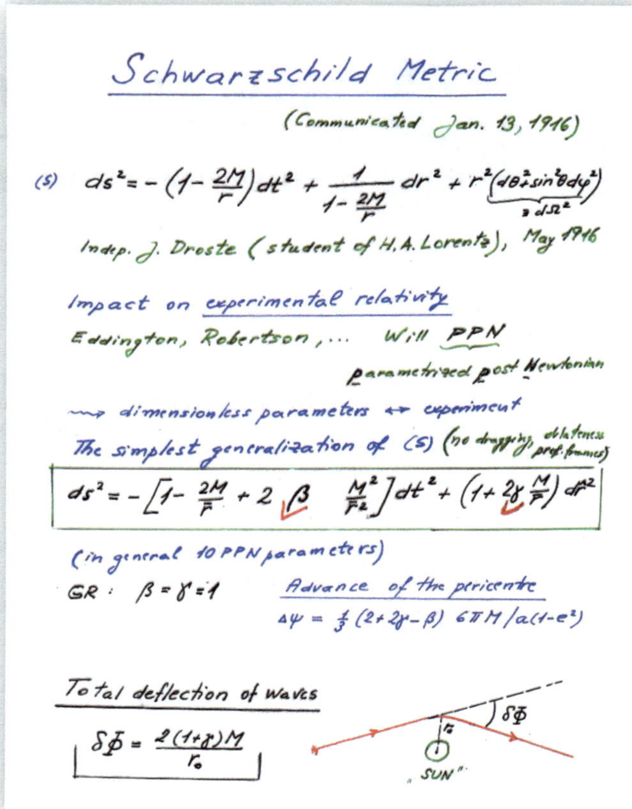


**Fig. 10** The generalization of the Schwarzschild metric to include alternative theories of gravity

apparently unknown outside his country.[14] In Fig. 15 it is seen that Link published his first paper on using Einstein's bending in photometry in Comptes Rendus in March 1936, more than eight months before Einstein's paper appeared in *Science*. In this paper already, his expression for $\rho_\infty = \sqrt{Kk\alpha_1}$ corresponds to the angular size of the Einstein ring, usually denoted as $\Theta_E$. A much more detailed work on the "Photometric consequences of the Einstein deflection" by Link appeared in Bulletin Astronomique in 1937. In that time Link did not notice Einstein's paper in *Science*

[14] On the occasion of the centenary of Link's birth there was a seminar organized by astronomers and historians of science at Charles University on November 29, 2006, where gravitational lensing was extensively discussed and the role of Link, the first director of the Astronomical Institute of the Academy, recalled. Incidentally, after my talk was prepared, including the part on lensing and Link, a preprint appeared in the arXiv:1206.1165v1 [physics.hist-ph] by D. Valls-Gabaud in which the pioneering role of Link in the origins of gravitational lensing is emphasized.

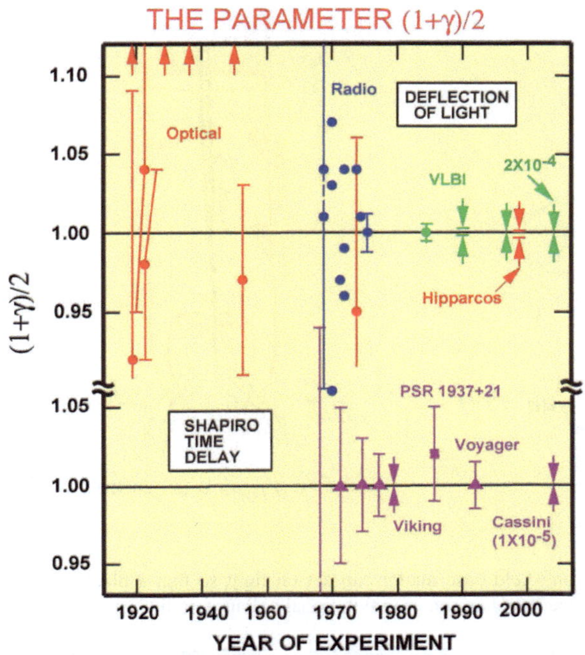


**Fig. 11** Limits on parameter $\gamma$ from measurements of the light deflection. $\gamma = 1$ in general relativity. Taken from [19]

but he gave a very detailed account of the history of lensing and many mathematical details later in his monograph [31].

## 9 Prague Works on Gravitation from 1912

These are four papers: two on the static gravitational field, [15, 16], including non-linear field equations and the motion of test particles in a given field, the first work on a new effect–on linear gravitational dragging [17], and last but not least, the reply to a comment by Abraham [18] in which Einstein outlines the main features a future theory of gravity should possess.

In the first paper, an inertial frame $\mathscr{I}_0$ is first considered and a uniformly accelerated frame $\mathscr{K}$ with an acceleration $a$ with respect to $\mathscr{I}_0$ (Fig. 16). Denoting coordinates in these frames by $\mathscr{I}_0(\tau, \xi, \eta, \zeta)$ and $\mathscr{K}(t, x, y, z)$, Einstein assumed the following approximate form of the transformation between the frames: $\xi = \lambda(x) + \alpha(x)t^2 + \mathscr{O}(t^3)$, $\tau = \beta(x) + \gamma(x)t + \delta(x)t^2 + \mathscr{O}(t^3)$, $\eta = y$, $\zeta = z$, where functions $\lambda(x)$, $\alpha(x)$, ... are to be determined as follows. First assume that at $t = 0$ the origins coincide, $\xi = 0$, $x = 0$. Consider the line elements in $\mathscr{I}_0$ and $\mathscr{K}$ to have the form

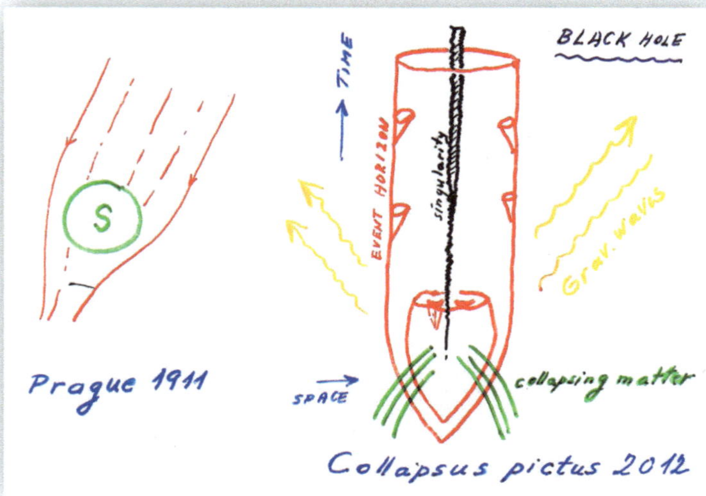


**Fig. 12** In the strong-field case matter can act on light so that a black hole is formed. This is indicated in the spacetime diagram of gravitational collapse on the right

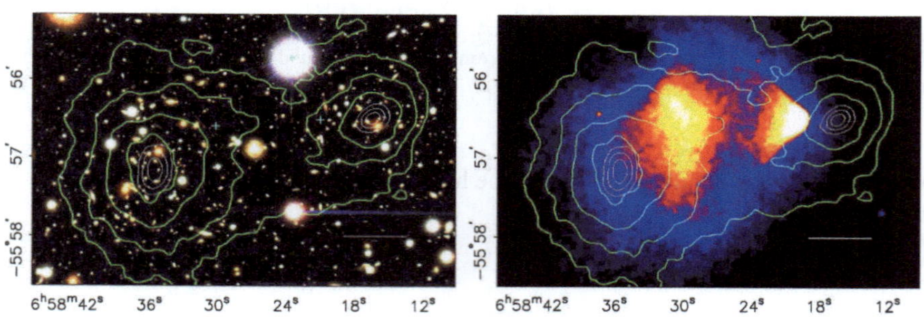


**Fig. 13** "The bullet cluster"—the collision of two galaxy clusters (1E0657–558) provides the best current evidence for the nature of dark matter (taken from [29]). See the text for details

$$
\begin{aligned}
\mathrm{d}s^2_{\mathscr{I}} &= -c_*^2 \mathrm{d}\tau^2 + \mathrm{d}\xi^2 + \mathrm{d}\eta^2 + \mathrm{d}\zeta^2, \quad c_* = 1 \\
\mathrm{d}s^2_{\mathscr{K}} &= -c^2(x)\mathrm{d}t^2 + \mathrm{d}x^2 + \mathrm{d}y^2 + \mathrm{d}z^2.
\end{aligned}
$$

Requiring then that $\mathrm{d}s^2_{\mathscr{I}_0} = 0 \Leftrightarrow \mathrm{d}s^2_{\mathscr{K}} = 0$, one finds functions $\lambda(x), \alpha(x), \ldots$ in the transformation and arrives at the following final form of the transformation between the inertial and uniformly accelerated frames:

$$
\xi = x + \tfrac{1}{2} act^2, \ \tau = ct, \ c = c_0 + ax. \tag{3}
$$

**Fig. 14** A Horseshoe Einstein Ring (Credit: ESA / Hubble & NASA)

ASTROPHYSIQUE. — *Sur les conséquences photométriques de la déviation d'Einstein.* Note de M. F. Link, présentée par M. Charles Fabry.

Dès le début de la théorie de la relativité on a cherché de vérifier la déviation des rayons lumineux passant normalement au champ de gravitation d'un corps céleste. La déviation $\omega$ est

**Fig. 15** The first publication on gravitational lensing: Link F., Comptes Rendus **202** (16 Mar 1936), 917–919. Notice that the paper was communicated by C. Fabry

## *9.1 Intermezzo*

Today we know that the transformation to the rigid uniformly accelerated ("Rindler") frame reads

$$\xi = \frac{1}{a}(\cosh at - 1) + x \cosh at, \tag{4}$$

$$\tau = \frac{1}{a}\sinh at + x \sinh at. \tag{5}$$

For small $t$ (neglecting $\mathcal{O}\left(t^3\right)$) Einstein's "Prague transformation" above immediately follows from these two relations. The spacetime orbits of uniformly accelerated particles are hyperbolas; hyperbolas (timelike at $z^2 > t^2$, spacelike at $z^2 < t^2$) are

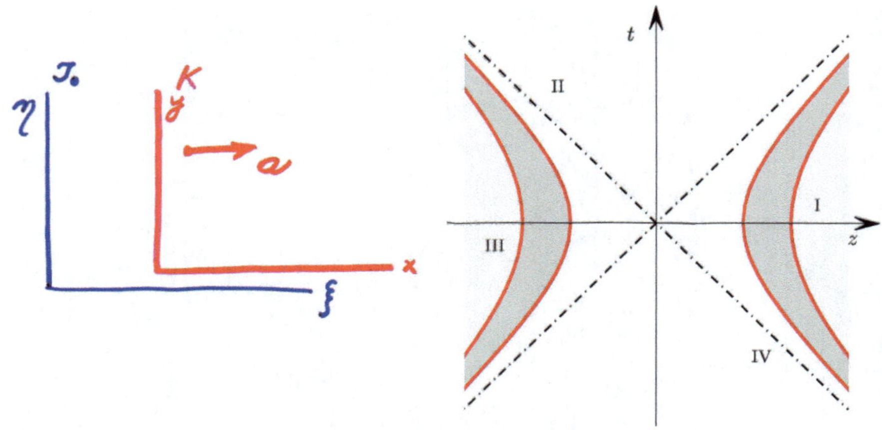


**Fig. 16** The system $\mathscr{K}$, considered by Einstein in [15], is "uniformly accelerated" (in a relativistic sense) with respect to inertial frame $\mathscr{I}_0$; and a schematic spacetime diagram of the C-metric representing two *black* holes uniformly accelerated in opposite directions

orbits of the boost Killing vector. Spacetimes with axial/rotational symmetry and boost symmetry, so called boost-rotation symmetric spacetimes, play an important role in general relativity: they are radiative, with a non-vanishing news function, have a plausible Newtonian limit, admit global null infinity; they have been used as test-beds in numerical relativity. One of the best known examples is the C-metric, representing two uniformly accelerated black holes in opposite directions—a schematic diagram is on the right in Fig. 16. In the Rindler-type coordinates (which can be introduced in the quadrants in which sources occur and the boost Killing vector is timelike) the spacetime is static but in the other two quadrants it is dynamical (locally of the Einstein-Rosen type).[15]

Returning back to 1912, we notice that the velocity of light in a uniformly accelerated frame, Eq. 3, satisfies Laplace's equation. Now invoking the equivalence principle, Einstein postulates that the field equation even for a *general* static gravitational field will read

$$\Delta c = 0 \text{ (in vacuum)}, \qquad \Delta c = kc\rho \text{ (in matter)}.$$

However, he soon realized that one gets contradictions with conservation of energy and momentum ($\int \mathbf{f}\,\mathrm{d}V \neq 0$, with $\mathbf{f} = -\rho \operatorname{grad} c$). Hence, in a shortly published following paper [16] he avoided the inconsistency by modifying the field equation into

[15] I dared to make this intermezzo since our group in Prague has been devoting quite an effort to understand these spacetimes (see, e.g., contribution by Bičák and Kofroň in the Proceedings [11] and the literature quoted therein).

$$\Delta c = k\left[c\rho + \frac{1}{2k}\frac{\text{grad}^2 c}{c}\right].$$

The field equation becomes *nonlinear*; and the additional term on the right-hand side describes the energy density of the gravitational field which thus itself contributes as a source to the field. (An interesting discussion of this theory of gravity from a purely Newtonian point of view was presented in this conference by Domenico Giulini—see [11].) In the same paper, a local ("pocket") temperature is introduced and shown to depend on $c$, i.e., on a local gravitational potential; and in the Appendix the equations of motion for a test particle in a given gravitational field are derived from the variational/Hamilton principle–the foreshadow of a variational principle for geodesics. Still more remarkably, Einstein introduces, for the first time, a "local view" on the equivalence principle since the nonlinear field equation above is not satisfied by $c = c_0 + ax$. The local transformation between the two frames in a small neighborhood is generalized to

$$\xi = x + \frac{1}{2}c\frac{\mathrm{d}c}{\mathrm{d}x}t^2, \quad c(x) \text{ arbitrary}.$$

## *9.2 Dragging of Inertial Frames*

The following paper from the Prague period is not directly connected to a construction of a new theory of gravity. However, it fits well to the Prague physics since it is related to Ernst Mach–"the fact is that Mach exercised a great influence upon our generation through his historical-critical writings... ", wrote Einstein in 1916. By using equations of motion derived in the preceding paper, Einstein considers a shell of matter and its influence on a mass point placed in its center; as the shell starts to accelerate, a force on the point mass appears and its inertial mass increases (see Fig. 17).

In the "Machian spirit" Einstein puts forward an idea that perhaps the inertia of a point mass is *fully* determined by the action of all other masses. We know today that this does *not* arise in general relativity, however, it is so that *inertial frames* are, under suitable conditions, determined by averages over matter distributions. However, Einstein's considerations from Prague show, for the first time, that an inertial frame will be dragged by moving masses. The effect was soon exhibited also for rotating masses by Einstein and Thirring and has been studied extensively until the present, including recently the dragging effects due to gravitational waves (see contributions by Pfister, by Lynden-Bell and Katz, and by Bičák, Katz, Ledvinka and Lynden-Bell in the Proceedings [11] of this conference).

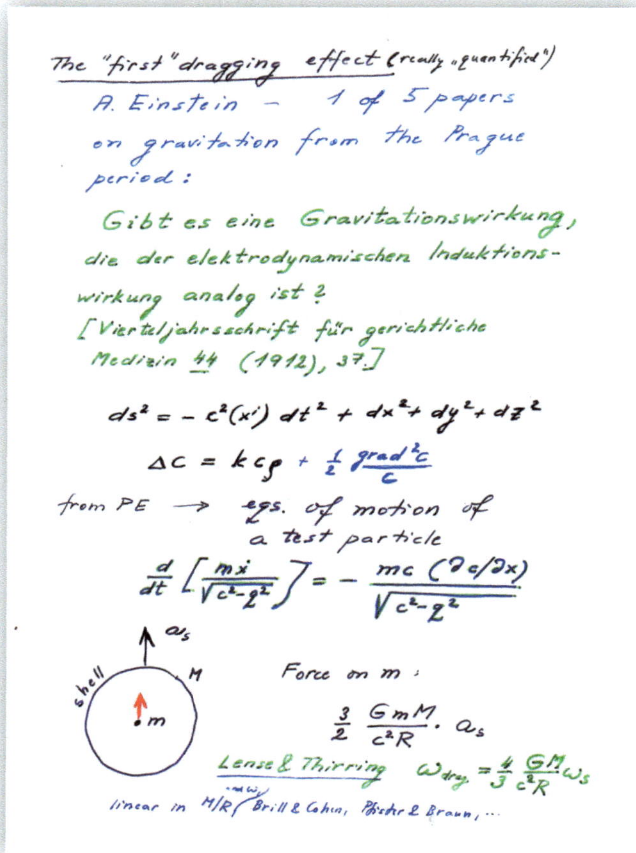


**Fig. 17** From his "Prague theory of gravity" Einstein discovered a (linear) dragging effect

The planning and eventually launching of the Stanford gyroscopic experiment, later called *Gravity Probe B*, started in 1959 and ended in 2011.[16] The spacecraft with gyros were orbiting the Earth during a mission that ran from April 2004 to September 2005; however, many simulations had to be performed after the measurements finished. There were eventually four superconducting gyros in the satellite and both the geodetic effect (cf. the "left" gyroscope in Fig. 18) and (much smaller) dragging effect due to the rotation of the Earth were measured. In an article published in Physical Review Letters from May 31, 2011, "the authors reported analysis

[16] It started thus contemporaneously with my studies at the Faculty of Mathematics and Physics in Prague and closed close to my retirement; but I enjoyed this long time span because the experiment is interesting and well explainable to a general public.

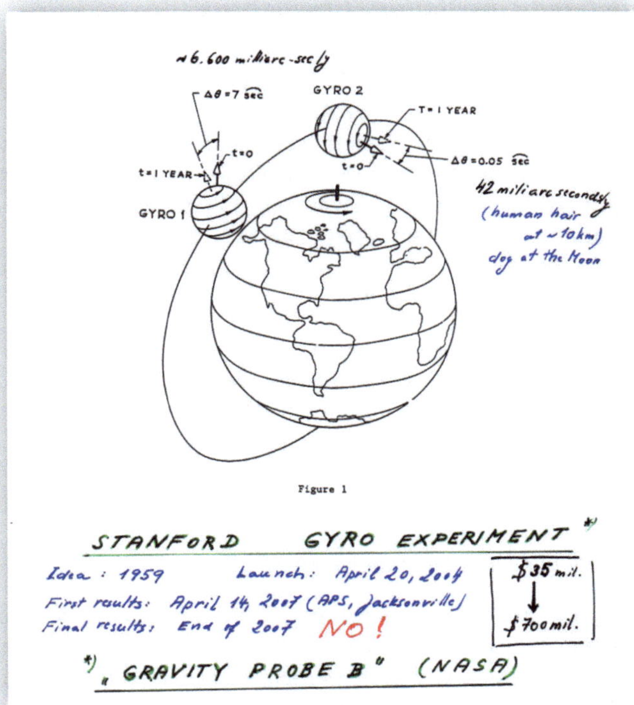


**Fig. 18** Stanford gyro experiment–"Gravity Probe B"–measured both a geodetic precession due to curvature around Earth (gyro 1) and a dragging effect due to the rotation of Earth (gyro 2)

of the data from all four gyroscopes results in a geodetic drift rate of $-6601.8 \pm 18.3$ milliarcsecond/year (≡ mas/yr; 1 mas = $4.84 \times 10^{-9}$ rad) and a frame-dragging drift rate of $-37.2 \pm 7.2$ mas/yr, to be compared with the GR predictions of $-6606.1$ mas/yr and $-39.2$ mas/yr, respectively". An extensive literature on science, technology, sociology and politics associated with the experiment is available and interesting to read (see, e.g., http://physics.aps.org/articles/v4/43). Principally, however, despite the doubts about the initial results (burdened by some unexpected phenomena like, for example, random patches of electrostatic potentials on both gyros and their housing) it is now firmly established that dragging effects, or as many relativists like to say—gravitomagnetic effects, were measured and so general relativity is confirmed by using stationary rather than just static gravitational fields. Another confirmation of the dragging (Lense-Thirring) effect came from the laser tracking of two Earth-orbiting LAGEOS satellites, analyzed in a number of papers by I. Ciufolini and his collaborators.

### 9.3 *Relativität and Gravitation. Erwiderung...*

In his last paper from the Prague period "Relativity and Gravitation. Reply to a Comment by M. Abraham" received by *Annalen der Physik* on July 4, 1912, just a few days before he left Prague, Einstein was inspired by Abraham's criticism to summarize the views on the contemporary state of the theory of relativity and its description of gravity. From Einstein's summary it is evident that he learned many fundamental issues during his Prague stay. This was also the reason we decided to give the name to the conference following this work. The main points which Einstein expressed in his summary are as follows:

- Local significance of equivalence principle.
- Equations of motion for point masses (variational principle).
- Equations of the electromagnetic field when gravity is present.
- Nonlinear field equation for gravity (energy density of gravitational field itself as a source).
- All equations must be form invariant with respect to a larger group than the Lorentz group.
- *Spacetime coordinates lose their simple physical meaning.*

Let us quote the last point in more detail, in Einstein's own words: "Man sieht schon aus dem bisher behandelten, daß die Raum-Zeit-Koordinaten ihre einfache physikalische Deutung einbüßen werden, und es ist noch nicht abzusehen, welche Form die allgemeinen raumzeitlichen Transformationsgleichungen haben könnten..."

In Prague it was thus for the first time that Einstein realized that spacetime coordinates need not determine the distances between spacetime points. Later, in his *Autobiographical Notes*,[17] he comments "Why were another seven years required for the construction of the general theory of relativity? The main reason lies in the fact that it is not so easy to free oneself from the idea that coordinates must have an immediate metrical meaning". At the end of his Prague stay, Einstein was on the right track towards general relativity. Important features of the new theory were understood, but gravitation was described wholly by one function only, the variable velocity of light.

As we discussed above, Einstein might have learned about Riemannian geometry in Prague already from Georg Pick, however, this does not seem to be of great importance since, in any case, a few months after he left Prague for Zurich, he started to collaborate with Marcel Grossmann and, with some exaggeration perhaps, one may say that they had general relativity almost in hand:

[17] In Albert Einstein-Philospher and Scientist, The Library of the Living Philosphers Vol. VII, ed. P. A. Schild, Open Court Publ. Co. 1949.

- Gravitation described by ten components of the metric tensor $g_{\mu\nu}$.
- Correct variational principle for the geodesic equation.
- Maxwell equations in general gravitational field covariant under general coordinate transformations.
- The source of gravity $T^{\mu\nu}$ covariantly conserved, $\nabla_\nu T^{\mu\nu} = 0$.
- The field equations $\Gamma_{\mu\nu} = \kappa T_{\mu\nu}$ where $\Gamma_{\mu\nu}$ is formed from the metric tensor and its derivatives.

Although in 1913 they first considered the Ricci tensor at the left-hand side of the field equations (on the right-hand side of which was a symmetrical energy-momentum tensor of matter $T^{\mu\nu}$ ), they wrongly concluded that they do not get a correct Newtonian limit, i.e., in vacuum the Laplace equation for the Newtonian potential.[18] And Einstein went astray from his route to a correct theory by assuming that the field equations, in order to give a meaningful Newtonian limit, cannot be generally covariant... The world had to wait another three years to see the equations $R_{\mu\nu} - \frac{1}{2}Rg_{\mu\nu} = \kappa T_{\mu\nu}$. About their Genesis, see the monumental survey quoted in [32].

## *9.4 Coda*

On Saturday, May 15, 2004, the Prague daily newspaper "Lidové Noviny" published an article "Einstein through the eyes of Johanna from Czechia"; see Fig. 19. It was just in those days when Princeton University published the diary of Johanna Fanta, the daughter-in-law of Berta Fanta whose salon Einstein liked to visit during his stay in Prague in 1911–1912. Einstein and Johanna knew each other from Einstein's Berlin period already, but became quite close friends, spending often time on the Lake Carnegie in a boat, when Einstein was in Princeton and Johanna was employed in the library of the University. During the last two years Johanna made regular notes of what Einstein said or wrote. This is most interesting, often touching and sad reading. Let us quote the notes from just three days. April 13, 1954: "Expresses annoyance at Oppenheimer for letting the McCarthy and Atomic Energy Commission affairs bother him so much. Already told the press that he has great respect for Oppenheimer, both as a human being and as a scientist." October 24, 1954: "He calculated like crazy again today but accomplished nothing." April 10, 1955, 8 days before Einstein's death: "He tried all day to compose a radio message on behalf of Israel and did not succeed in finishing it. He claims he is totally stupid—that he has always thought so, and that only once in a while was he able to accomplish something."

In a rather poor, hilly part of south-western Moravia, close to the place where Gustav Mahler was born, there lived various "country-type" sages interested in literature, art, music, religion, and also science. In the left side of the last Fig. 20 there is the

[18] For a detailed analysis of the "Zurich notebook" which Einstein began to write soon after leaving Prague, see [32].

20 Sobota 15. května 2004 LIDOVÉ NOVINY

Zajímavosti týdne

# Einstein očima Johanny z Čech

V těchto dnech by Princetonská univerzita měla začít plnit to, co před měsícem slíbila: odstartovat sérii článků o vztahu mezi Albertem Einsteinem a Johannou Fantovou. Jedním z podkladů jsou i nedávno objevené zápisky, které si fyzikova přítelkyně vedla v posledních dvou letech jeho života.

Marie Homolová

Ty zápisky se našly šťastnou náhodou. Princetonská univerzita se rozhodla připravit seriál o slavných párech, které se vyskytovaly v její historii. A protože se vědělo, že Einstein, který za amerického pobytu v Princetonu působil, a Fantová, která tam pracovala v knihovně, měli k sobě blízko, otevřeli autoři seriálu letos v únoru její dokumentaci. Kromě jiného tu ke svému překvapení našli i deník, obsahující na 62 stránkách více než dvě stovky záznamů.

„V 90. letech se mě řada lidí dotazovala, zda existuje," říká Don Skemer, kurátor zdejší sbírky rukopisů. „Odkazoval jsem je na experty a sbírky po celých Stá-

UDÁLOST TÝDNE

„Chtěla jsem k tomu, jak Einsteina chápeme, přidat nový pohled," napsala Fantová v úvodu. „Nezachytit ho jako žijící legendu ani jako věhlasného vědce, ale jako člověka."

**Z Prahy do Princetonu**

Johanna Fantová se narodila roku 1901 v severních Čechách jako Johanna Bobaschová. Její muž Otto byl synem manželů Fantových, kteří před první světovou válkou vedli v pražském domě U bílého jednorožce slavný salon, kde se scházel pražský kulturní svět. Mezi hosty se objevoval i Franz Kafka, Max Brod a... Albert Einstein, který tehdy v letech 1911 a 1912 přednášel v Praze na německé univerzitě fyziku.

Johanna, které tehdy bylo deset a běhala na zcela jiné adrese, se s Einsteinem sešla až v roce 1929 v Berlíně, kde jí nabídl, aby mu uspořádala jeho rozsáhlou, ale chaotickou knihovnu. Ačkoliv je dělilo dvaadvacet let věku, velmi se spřátelili, a nakonec jí přizval i ke své nejmilejší zábavě, jízdě na plachetnici. Když v roce 1939 odjela do Spojených států, znovu se k Einsteinovi, který tam už několik let žil v emigraci, přihlásila.

Einstein už měl za sebou dvě manželství. V roce 1902 se oženil se srbskou studentkou fyziky Milevou, měli spolu dva syny, ale časem se rozvedli. Druhá manželka, sestřenice Elsa, ho doprovázela i do Ameriky, kam emigroval

Albert Einstein a Johanna Fantová spolu v Americe rádi jachtařili na Lake Carnegie Repro LN

týdně spolu dlouze telefonovali, ona mu stříhala vlasy, on jí psal básně, obnovili i příjemné chvíle na pla-

„Unavený z dlouhého ticha chci ti tímto jasně ukázat, jak silné myšlenky na tebe budou vždy sídlit v jedné

že všichni jsme malé Země, ale každý si myslí, že je jejich středem...

A tak dále.

mera, který vedl americký program vývoje atomové bomby. Když se Oppenheimer, kterého si Einstein velmi považoval, ocitl před McCarthyovým Výborem pro vyšetřování neamerické činnosti, Fantová si zaznamenala, že se to u nich stalo častým předmětem konverzace. „Politická perzekuce jeho přítele byla pro něj hořkým vystřízlivěním," zapsala si.

Soucítil s Oppenheimerovým zklamáním, ale o sobě říkal, že se „narodil se sloní kůží" a nikdo mu nemůže ublížit. „Každá kritika po mně steče jako voda."

Přesto si nikdy nepřestal vyčítat, že je svým způsobem odpovědný za tragédie způsobené atomovou bombou. Nejen proto, že k bombě dal jistý podnět svými vědeckými závěry, ale především za druhé světové války napsal prezidentu Rooseweltovi dopis, ve kterém ho varoval, že Němci jsou blízko k sestrojení a je důležité předběhnout je. Ten dopis byl jedním z impulsů, které způsobily, že se americký projekt opravdu rozjel na plné obrátky. Odpovědnost za to, co následovalo, „ho velmi stresovala", zapsala si Fantová.

Zápisy dosvědčují, jak se trápil vývojem světa i tím, že už ho nemůže ovlivnit. Sovětům vrátil mírovou cenu, nechtěl nést nálepku bolševika. A zároveň odmítl varovat před preventivní válkou s Čínou, protože věděl, že by jeho slova padla do prázdna. Zlobil se, že jméno Einstein „zna-

**Fig. 19** An article about Einstein and his friend Johanna Fanta in Lidové noviny, May 2004

**Fig. 20** Covers of two books: Einstein's course published in Moravian village Stará Říše in September 1926, containing the translation of parts of Eddington's popular book *Space, Time and Gravitation* and the correspondence of Otokar Březina, poet and essayist, who created almost all of his works in a nearby village Nová Říše

cover of the "Einsteinian course" published in September 1926 by Josef Florian in the village Stará Říše. It contains a very fine translation of Eddington's famous book with conversation between an experimental physicist, a pure mathematician. and a relativist who advocates the newer conceptions of time and space in physics. On the cover one can see symbols of Noah, Moses, Jesus, etc but, at the left low corner, there is also the correct expression for the variational principle for geodesics! On the right side there is the cover of the second volume of the correspondence of Otokar Březina, the top poet of Czech symbolism who lived and taught in a neighboring village Nová Říše (cf. Wikipedia). In his correspondence he mentions Planck's quantum theory, Bohr, Minkowski and Einstein in the 1920s. In a letter to the professor of classical literature, František Novotný in Brno, who sent him his monumental work about Plato, Březina wrote encouraging words which I used both at the end of my talk in June and in [5] on the occasion of our Einstein centenary celebrations in 1979 in Karolinum as the present conference took place:

"There will always be minds who, by the united power of knowledge and dreams, science and poetry, will strive for a unified picture of all phenomena in the Universe, an image that in equal measure corresponds both to the eternal longing of the human spirit for harmony and beauty and to the thirst of the heart for justice." Is not then the example of Albert Einstein a reason for an enduring optimism?

**Acknowledgments** I am grateful to David Kofroň for the help with the manuscript, especially with the figures, and for his suggestions improving the text. I also thank the Albert Einstein Institute, Golm, where most of this text was written after the conference, for the kind hospitality, Lars Andersson for comments and Ian Hinder for correcting some of my Czechisms. The partial support from the Czech Science Foundation No 14-37086G (A. Einstein Center) is also acknowledged.

## References

1. Kleinert, A.: Anton Lampa und Albert Einstein-Die Neubesetzung der physikalischen Lehrstühle an der deutschen Univesrität Prag 1909 und 1910. Gesnerus **32**, 285 (1975)
2. Havas, P.: Einstein, relativity and gravitation research in Vienna before 1938. In: Goenner, H., Renn, J., Ritter, J., Sauer, T. (eds.) The Expanding Worlds of General Relativity, pp. 161–206. Birkhauser, Boston-Basel-Berlin (1999)
3. Frank, P.: Einstein, his Life and Times. Alfred A. Knopf, New York (1947)
4. Pais, A.: 'Subtle is the Lord..' The Science and the Life of Albert Einstein. Clarendon Press, Oxford (1982)
5. Bičák, J.: Einstein a Praha (in Czech and German). Prometeus, Prague (1979)
6. Bičák, J.: Einstein's route to the general theory of relativity (in Czech). Československý Časopis pro Fyziku **A29**, 222 (1979)
7. Bičák, J.: Einstein's Prague articles on gravitation. In: Blair, D.G., Buckingham, M.J. (eds.) Proceedings of the 5th M. Grossmann Meeting on General Relativity, pp. 1325–1334. World Scientific, Singapore (1989)

8. Bičák, J.: Einstein's Prague ideas on gravitation: the history and the present. In: Janta, J., Pantoflíček, J. (eds.) Trends in Physics, pp. 65–75. Union of Czechoslovak Mathematics and Physics, Prometheus, Prague (1985)
9. Těšínská, E.: Theoretical physics at the Prague University and its connections with Prague stay of A. Einstein 100 years ago (in Czech), Pokroky Matematiky, Fyziky a Astronomie **57**, 146–168 (2012)
10. Semerák, O., Podolský, J., Žofka, M. (eds.): Gravitation: Following the Prague Inspiration. World Scientific, New Jersey (2002)
11. Bičák, J., Ledvinka, T. (eds.): Relativity and Gravitation: 100 Years After Einstein in Prague. Springer, Switzerland (2014)
12. Einstein, A.: Die Relativitäts-Theorie. Vierteljahrsschrift der Naturforschenden Gesellschaft in Zürich **56**, 1 (1911)
13. Einstein, A.: Zum Ehrenfestschen Paradoxon. Bemerkung zu V. Varičaks Aufsatz. Phys. Z. **12**, 509 (1911)
14. Einstein, A.: Über den Einfluss der Schwerkraft auf die Ausbreitung des Lichtes. Ann. d. Phys. **35**, 898 (1911)
15. Einstein, A.: Lichtgeschwindigkeit und Statik des Gravitationsfeldes. Ann. d. Phys. **38**, 355 (1912)
16. Einstein, A.: Zur theorie des statischen Gravitationsfeldes und Nachtrag zur Korrektur. Ann. d. Phys. **38**, 443 (1912)
17. Einstein, A.: Gibt es eine Gravitationswirkung, die der elektrodynamischen Induktionswirkung analog ist? Vierteljahrsschrift für gerichtliche Medizin und öffentliches Sanitätswesen **44**, 37 (1912)
18. Einstein, A.: Relativität und Gravitation: Erwiderung auf eine Bemerkung von M. Abraham. Ann. d. Phys. **38**, 1059 (1912)
19. Will, C.M.: The confrontation between general relativity and experiment. Living Rev. Relativ. **9**(3), (2006). http://www.livingreviews.org/lrr-2006-3
20. Sumner, T.J., et al.: Satellite test of equivalence principle. Adv. Space Res. **39**, 254 (2007). doi:10.1016/j.asr.2006.09.019
21. Müller, H., Peters, A., Chu, S.: A precision measurement of the gravitational redshift by the interference of matter waves. Nature **463**, 926 (2010). doi:10.1038/nature08776
22. Wolf, P., et al.: Does an atom intorferometer test the gravitational redshift at the Compton frequency? Class. Quantum Grav. **28**, 145017 (2011). doi:10.1088/0264-9381/28/14/145017
23. Hohensee, M.A., Estey, B., Hamilton, P., Zeilinger, A., Müller, H.: Force-free gravitational redshift: proposed gravitational Aharonov-Bohm experiment. Phys. Rev. Lett. **108**, 230404 (5 pages) (2012). doi:10.1103/PhysRevLett.108.230404
24. LoPresto, J.C., Schrader, C., Pierce, A.K.: Solar gravitational redshift from the infrared oxygen triplet. Astrophys. J. **376**, 757 (1991). doi:10.1086/170323
25. Cacciani, A., Briguglio, R., Massa, F., Rapex, P.: Precise measurement of the solar gravitational redshift. Celest. Mech. and Dyn. Astrometry **95**, 425 (2006). doi:10.1007/s10569-006-9014-0
26. Wojtak, R., Hansen, S.H., Hjorth, J.: Gravitational redshift of galaxies in clusters as predicted by general relativity. Nature **477**, 567 (2011). doi:10.1038/nature10445
27. Hentschel, K.: Erwin Finlay Freundlich and testing Einstein's theory of relativity. Arch. Hist. Exact Sci. **47**, 143 (1994)
28. Will, C.M.: Resource letter PTG-1: precision tests of gravity. Amer. J. Phys. **78**, 1240 (2010). doi:10.1119/1.3481700
29. Clowe, D., et al.: A direct empirical proof of the existence of dark matter. Astrophys. J. Lett. **648**, L109 (2006)
30. Renn, J., Sauer, T.: Eclipses of the stars-Mandl, Einstein, and the early history of gravitational lensing. In: Ashtekar, A., Cohen, R., Howard, D., et al. (eds.) Revisiting the Foundations of Relativistic Physics—Festschrift in Honour of John Stachel, pp. 69–92. Kluwer Academic Publishers, Dordrecht (2003)

31. Link, F.: Eclipse Phenomena in Astronomy. Springer, New York (1969)
32. Renn, J., et al.: Einstein's Zurich notebook: gravitation in the twilight of classical physics. In: Renn J. et al. (eds.) The Genesis of General Relativity, Boston Studies in the Philosophy of Science, vol. 1–4. Springer, Dordrecht (2007)

# Part II
# Classical General Relativity

# Observers, Observables and Measurements in General Relativity

**Donato Bini**

**Abstract** To perform any physical measurement it is necessary to identify in a non ambiguous way both the observer and the observable. A given observable can be then the target of different observers: a suitable algorithm to compare among their measurements should necessarily be developed, either formally or operationally. This is the task of what we call "theory of measurement," which we discuss here in the framework of general relativity.

## 1 Introduction

The spacetime (or absolute) point of view constitutes a unified scenario for quantities which, in the pre-relativistic physics, were associated with distinct notions: time and space themselves, energy and momentum, mechanical power and force, electric and magnetic fields and so on. In every day experience, however, our intuition is still compatible with the perception of a three-dimensional space and a one-dimensional time and therefore any physical measurement requires a local recovery of the pre-relativistic type of separation between space and time. To this purpose we need some prescription in order to perform the required splitting, and hence identifying a "space" and a "time" relative to any given observer. Any such prescription requires a congruence of timelike world lines with a future-pointing unit tangent vector field $u$ (i.e., the local time direction) which we interpret as the world lines of a family of (test) observers with associated 4-velocity $u$.

The splitting of the tangent space at each point of the congruence into the local time direction $u$ and the local rest space spanned by vectors orthogonal to $u$ (hereafter $LRS_u$), allows one to decompose all spacetime tensors, including tensorial operators, and tensor equations into their spatial and temporal components. One may ask then

D. Bini (✉)
Istituto per le Applicazioni del Calcolo "M. Picone," CNR, 00185 Rome, Italy
e-mail: binid@icra.it

J. Bičák and T. Ledvinka (eds.), *General Relativity, Cosmology and Astrophysics*, Fundamental Theories of Physics 177, DOI: 10.1007/978-3-319-06349-2_3,

© Springer International Publishing Switzerland 2014

if there exist natural or special observer families in a given spacetime. This is clearly the case of a stationary spacetime where a special observer (timelike) congruence is associated with the timelike Killing vector field. Also, any spacetime admitting a spacelike foliation has naturally associated with it a timelike congruence, namely that of the normal directions to the slicing itself. It is known that any spacetime admitting separable geodesics (e.g., as a consequence of the existence of a Killing tensor of rank 2) also admits a foliation. For example, Kerr spacetime with the metric written in standard Boyer-Lindquist coordinates summarizes these three conditions: it has the family of static observers (whose world lines are aligned with the coordinate time lines, i.e., with unit tangent vector parallel to the timelike Killing vector $\partial_t$), the family of locally nonrotating observers or ZAMOs (whose world lines are orthogonal to the $t = \text{constant}$ hypersurfaces) and finally, the families of Painlevé-Gullstrand observers who follow geodesic timelike lines since the latter have a separable dependence from the coordinates. A review of the essential splitting formalism follows below. For more details one can refer for instance to Refs. [1–6] (and references therein).

## 2 Orthogonal Decompositions

Let g be the four-dimensional spacetime metric with signature +2 and components $g_{\alpha\beta}$ ($\alpha, \beta = 0, 1, 2, 3$), $\nabla$ its associated covariant derivative operator and $\eta$ the unit volume 4-form which assures spacetime orientation. Let $u$ be a future-pointing unit timelike vector field which identifies an observer, $u \cdot u = -1$. The local splitting of the tangent space into orthogonal sub-spaces uniquely related to the given observer $u$, is accomplished by a temporal projection operator $T(u)$ (along $u$) and a spatial projection operator $P(u)$ (generating the $LRS_u$). These operators, in mixed form, are defined as follows

$$T(u) = -u^\sharp \otimes u^\flat \quad P(u) = I + u^\sharp \otimes u^\flat \tag{1}$$

where $I \equiv \delta^\alpha{}_\beta$ is the identity on the tangent spaces of the manifold and the symbols $\sharp$ and $\flat$ identify the fully contravariant and covariant representation of tensors, respectively. In terms of components the above relations write

$$T(u)^\alpha{}_\beta = -u^\alpha u_\beta, \quad P(u)^\alpha{}_\beta = \delta^\alpha{}_\beta + u^\alpha u_\beta\,. \tag{2}$$

Given a $\binom{p}{q}$-tensor $S$, let us denote as $[P(u)S]$ its fully spatial projection obtained by acting with the operator $P(u)$ on all of its indices,

$$[P(u)S]^{\alpha\dots}{}_{\beta\dots} = P(u)^\alpha{}_\gamma \cdots P(u)^\delta{}_\beta \cdots S^{\gamma\dots}{}_{\delta\dots} \tag{3}$$

The splitting of $S$ relative to a given observer is the set of tensors which arise from the spatial and temporal projection of each of its indices as we are going to discuss.

This observer-dependent set of tensors represent $S$ and it is termed as its (geometrical) measurement by the observer $u$.

1. Splitting of a vector
   If $S$ is a vector field then its splitting gives rise to a scalar field and a spatial vector field

$$S \quad \leftrightarrow \quad \{u \cdot S, [P(u)S]\}. \tag{4}$$

   In terms of components they read

$$S^\alpha \quad \leftrightarrow \quad \{u_\gamma S^\gamma, P(u)^\alpha{}_\gamma S^\gamma\}. \tag{5}$$

   In fact, with respect to the observer $u$, the vector $S$ admits then the following representation

$$S^\alpha = [T(u)S]^\alpha + [P(u)S]^\alpha = -(u_\gamma S^\gamma)u^\alpha + P(u)^\alpha{}_\gamma S^\gamma. \tag{6}$$

2. Splitting of a $\binom{1}{1}$-tensor
   If $S$ is a mixed $\binom{1}{1}$-tensor field, then its splitting consists of a scalar field, a spatial vector field, a spatial 1-form and a spatial $\binom{1}{1}$-tensor field, namely

$$S^\alpha{}_\beta \leftrightarrow \{u^\delta u_\gamma S^\gamma{}_\delta,\ P(u)^\alpha{}_\gamma u^\delta S^\gamma{}_\delta,\ P(u)^\delta{}_\alpha u_\gamma S^\gamma{}_\delta,\ P(u)^\alpha{}_\gamma P(u)^\delta{}_\beta S^\gamma{}_\delta\}.$$

   In terms of these fields, the tensor $S$ admits the following representation with respect to the observer $u$

$$\begin{aligned} S^\alpha{}_\beta &= [T(u)^\alpha{}_\gamma + P(u)^\alpha{}_\gamma][T(u)^\delta{}_\beta + P(u)^\delta{}_\beta]S^\gamma{}_\delta \\ &= (u^\delta u_\gamma S^\gamma{}_\delta)u^\alpha u_\beta - u^\alpha u_\gamma P(u)^\delta{}_\beta S^\gamma{}_\delta \\ &\quad -u^\delta u_\beta P(u)^\alpha{}_\gamma S^\gamma{}_\delta + [P(u)S]^\alpha{}_\beta. \end{aligned} \tag{7}$$

   The local spatial and temporal projections of a $\binom{p}{q}$-tensor is easily generalized. For example, the metric tensor $g_{\alpha\beta}$ has the (trivial) representation

$$g_{\alpha\beta} = P(u)_{\alpha\beta} + T(u)_{\alpha\beta}.$$

3. Splitting of $p$-forms
   Given a p-form

$$S = S_{[\alpha_1 \ldots \alpha_p]}\omega^{\alpha_1} \otimes \cdots \otimes \omega^{\alpha_p} \equiv \frac{1}{p!} S_{\alpha_1 \ldots \alpha_p}\omega^{\alpha_1} \wedge \cdots \wedge \omega^{\alpha_p} \tag{8}$$

we define the electric part of $S$ relative to the observer $u$ the quantity

$$\left[S^{(\mathrm{E})}(u)\right]_{\alpha_1\ldots\alpha_{p-1}} = -u^\sigma S_{\sigma\alpha_1\ldots\alpha_{p-1}} \tag{9}$$

or in a more compact form $S^{(\mathrm{E})}(u) = -u \,\lrcorner\, S$. Similarly we define as the magnetic part of $S$ the quantity

$$\left[S^{(\mathrm{M})}(u)\right]_{\alpha_1\ldots\alpha_p} = P(u)^{\beta_1}{}_{\alpha_1}\ldots P(u)^{\beta_p}{}_{\alpha_p} S_{\beta_1\ldots\beta_p} \tag{10}$$

or, in a compact form, $S^{(\mathrm{M})}(u) = P(u)S$. From the above definitions we deduce the following representation of $S$

$$S = u^\flat \wedge S^{(\mathrm{E})}(u) + S^{(\mathrm{M})}(u), \tag{11}$$

or in components

$$S_{\alpha_1\ldots\alpha_p} = p!u_{[\alpha_1}[S^{(\mathrm{E})}(u)]_{\alpha_2\ldots\alpha_p]} + [S^{(\mathrm{M})}(u)]_{\alpha_1\ldots\alpha_p}. \tag{12}$$

For example, the splitting of the unit volume 4-form $\eta$ gives rise to the following representation

$$\eta = -u^\flat \wedge \eta(u), \tag{13}$$

that is $[u^\flat \wedge \eta(u)]_{\alpha\beta\gamma\delta} = \left[2u_{[\alpha}\eta(u)_{\beta]\gamma\delta} + 2u_{[\gamma}\eta(u)_{\delta]\alpha\beta}\right]$, where the spatial unit volume 3-form

$$\eta(u)_{\alpha\beta\gamma} = u^\delta \eta_{\delta\alpha\beta\gamma} \tag{14}$$

is the only nontrivial spatial field which arises from the splitting of the volume 4-form. Using the spacetime (Hodge) duality operation (*), one can associate with any $p$-form $S$ (with $0 \le p \le 4$) a $(4-p)$-form. Similarly a spatial duality operation ($^{*(u)}$) is defined for a spatial $p$-form $S$ ($u \,\lrcorner\, S = 0$) replacing $\eta$ with $\eta(u)$, namely

$$^{*(u)}S_{\alpha_1\ldots\alpha_{3-p}} = \frac{1}{p!} S_{\beta_1\ldots\beta_p}\eta(u)^{\beta_1\ldots\beta_p}{}_{\alpha_1\ldots\alpha_{3-p}}. \tag{15}$$

For example, given a spatial 2-form $S$, its spatial dual is

$$[^{*(u)}S]^\alpha = \frac{1}{2}\eta(u)^{\alpha\beta\gamma} S_{\beta\gamma}. \tag{16}$$

This operation satisfies the property ${}^{*(u)*(u)}S = S$. Let us now consider the splitting of ${}^{*}S$ where $S$ is given by (11). We have

$$\begin{aligned} {}^{*}S &= u^{\flat} \wedge [{}^{*}S]^{(\mathrm{E})}(u) + [{}^{*}S]^{(\mathrm{M})}(u) \\ &= {}^{*}[u^{\flat} \wedge S^{(\mathrm{E})}(u) + S^{(\mathrm{M})}(u)] \\ &= {}^{*(u)}S^{(\mathrm{E})}(u) + {}^{*}[{}^{*(u)}[{}^{*(u)}[S^{(\mathrm{M})}(u)]]] \\ &= {}^{*(u)}S^{(\mathrm{E})}(u) + (-1)^{p-1}u^{\flat} \wedge {}^{*(u)}[S^{(\mathrm{M})}(u)]. \end{aligned} \tag{17}$$

Comparing the first and the last line we have

$$[{}^{*}S]^{(\mathrm{E})}(u) = (-1)^{p-1\,*(u)}[S^{(\mathrm{M})}(u)], \quad [{}^{*}S]^{(\mathrm{M})}(u) = {}^{*(u)}S^{(\mathrm{E})}(u). \tag{18}$$

4. Splitting of differential operators
In general relativity one has several spacetime tensorial differential operators which act on tensor fields. Let us recall them: if $T$ is a tensor field of any rank, we have

   a. The Lie derivative of $T$ along the direction of a given vector field $X$: $[\pounds_X T]$.
   b. The covariant derivative of $T$: $\nabla T$.
   c. The absolute derivative of $T$ along a curve with unit tangent vector $X$ and parameterized by $s$: $\nabla_X T \equiv DT/ds$.
   d. The Fermi–Walker derivative of $T$ along a non-null curve with unit tangent vector $X$ and parameterized by $s$ defined by

$$\frac{D_{(\mathrm{fw},X)}T^{\alpha}{}_{\beta}}{ds} = \frac{DT^{\alpha}{}_{\beta}}{ds} \pm \left([a(X) \wedge X]^{\alpha}{}_{\gamma}T^{\gamma}{}_{\beta} - [a(X) \wedge X]^{\gamma}{}_{\beta}T^{\alpha}_{\gamma}\right),$$

   where $\pm$ refer to transport along timelike or spacelike curves, respectively. Finally if $S$ is a $p$-form, one has

5. The exterior derivative of $S$: $dS$.
Application of the spatial projection into the $LRS_u$ of a family of observers $u$ to the spacetime derivatives (i)–(v), yields new operators which can be more easily confronted with those defined in a three-dimensional Euclidean space. Given a tensor field $T$ of components $T^{\alpha\cdots}{}_{\beta\ldots}$ we have in fact

   a. The spatially projected Lie derivative along a vector field $X$

$$[\pounds(u)_X T]^{\alpha\cdots}{}_{\beta\ldots} \equiv P(u)^{\alpha}{}_{\sigma}\ldots P(u)^{\rho}{}_{\beta}\ldots[\pounds_X T]^{\sigma\cdots}{}_{\rho\ldots}; \tag{19}$$

   when $X = u$ we use also the notation

$$\nabla(u)_{(\mathrm{lie})}T \equiv \pounds(u)_u T, \tag{20}$$

   and this operation will be termed "spatial-Lie temporal derivative".
   b. The spatially projected covariant derivative along any $e_{\gamma}$ frame direction

$$\nabla(u)_\gamma T \equiv P(u)\nabla_\gamma T, \tag{21}$$

namely

$$[\nabla(u)_\gamma T]^{\alpha\dots}{}_{\beta\dots} = P(u)^\alpha{}_{\alpha_1}\dots P(u)^{\beta_1}{}_\beta \dots P(u)^\sigma{}_\gamma \nabla_\sigma T^{\alpha_1\dots}{}_{\beta_1\dots}.$$

c. The spatially projected absolute derivative along a curve with unit tangent vector $X$

$$[P(u)\nabla_X T]^{\alpha\dots}{}_{\beta\dots} = P(u)^\alpha{}_{\alpha_1}\dots P(u)^{\beta_1}{}_\beta \dots [\nabla_X T]^{\alpha_1\dots}{}_{\beta_1\dots} \tag{22}$$

d. The spatially projected "Fermi–Walker derivative" along a curve with unit tangent vector $X$ and parameterized by $s$

$$\left[P(u)\frac{D_{(\mathrm{fw},X)}T}{ds}\right]^{\alpha\dots}{}_{\beta\dots} = P(u)^\alpha{}_\sigma \dots P(u)^\rho{}_\beta \dots \left[\frac{D_{(\mathrm{fw},X)}T}{ds}\right]^{\sigma\dots}{}_{\rho\dots}.$$

e. the spatially projected exterior derivative of a $p$-form $S$

$$d(u)S \equiv P(u)dS, \tag{23}$$

namely $[d(u)S]_{\alpha_1\dots\alpha_p\beta} = P(u)^{\beta_1}{}_{\alpha_1}\dots P(u)^\sigma{}_\beta [dS]_{\beta_1\dots\sigma}$.

Note that all these spatial differential operators are well defined since they arise from the spatial projection of spacetime differential operators. From their definitions it is clear that both the Fermi–Walker and the Lie derivatives of the vector field $u$ along itself vanish identically (and so do the projections orthogonal to $u$ of these derivatives). The only derivative of $u$ along itself which is meaningful being different than zero is the covariant derivative

$$P(u)\nabla_u u = \nabla_u u = a(u). \tag{24}$$

## 3 Three-Dimensional Notation

Let $u$ be a given a family of observers and $X$ a spatial vector with respect to $u$. It is then convenient to introduce the 3-dimensional vector notation for the spatial inner product and the spatial cross product of two spatial vector fields $X$ and $Y$. The spatial inner product is defined as

$$X \cdot_u Y = P(u)_{\alpha\beta} X^\alpha Y^\beta \tag{25}$$

while the spatial cross product is

$$[X \times_u Y]^\alpha = \eta(u)^\alpha{}_{\beta\gamma} X^\beta Y^\gamma, \tag{26}$$

where $\eta(u)^\alpha{}_{\beta\gamma} = u_\sigma \eta^{\sigma\alpha}{}_{\beta\gamma}$ as stated.

In terms of the above definitions we can define spatial gradient, curl and divergence operators of functions $f$ and spatial vector fields $X$ as

$$\text{grad}_u f = \nabla(u) f , \quad \text{curl}_u X = \nabla(u) \times_u X , \quad \text{div}_u X = \nabla(u) \cdot_u X. \tag{27}$$

In components these relations read

$$\begin{aligned} [\text{grad}_u f]_\alpha &= \nabla(u)_\alpha f = P(u)^{\alpha\beta} e_\beta(f), \\ [\text{curl}_u X]^\alpha &= \eta(u)^{\alpha\beta\gamma} \nabla(u)_\beta X_\gamma = u^\sigma \eta_\sigma{}^{\alpha\beta\gamma} \nabla_\beta X_\gamma, \\ [\text{div}_u X] &= \nabla(u)_\alpha X^\alpha = P(u)^{\alpha\beta} \nabla_\alpha X_\beta. \end{aligned} \tag{28}$$

It is useful to extend the above definitions to

1. the spatial cross product of a vector $X$ by a symmetric tensor $A$,

$$[X \times_u A]^{\alpha\beta} = \eta(u)^{\gamma\delta(\alpha} X_\gamma A^{\beta)}{}_\delta, \tag{29}$$

2. the spatial cross product of two symmetric spatial tensors $A$ and $B$,

$$[A \times_u B]_\alpha = \eta(u)_{\alpha\beta\gamma} A^\beta{}_\delta \, B^{\delta\gamma}, \tag{30}$$

3. the spatial inner product of two symmetric spatial tensors $A$ and $B$,

$$[A \cdot_u B]_\alpha{}^\beta = A_{\alpha\gamma} B^{\gamma\beta}, \tag{31}$$

4. the trace of the above tensor product

$$\text{Tr}[A \cdot_u B] A_{\alpha\beta} B^{\alpha\beta}, \tag{32}$$

5. the spatial divergence of a spatial tensor

$$[\text{div}_u X]^{\alpha\ldots\beta} = \nabla(u)_\sigma X^{\sigma\alpha\ldots\beta} , \quad X = P(u)X. \tag{33}$$

## 4 Kinematics of the Observer's Congruence

Let us consider now the splitting of the covariant derivative $\nabla_\beta u^\alpha$. This operation generates two spatial fields namely the acceleration vector field $a(u)$ and the kinematical tensor field $k(u)$defined as

$$a(u) = P(u)\nabla_u u, \quad k(u) = -\nabla(u) u = \omega(u) - \theta(u). \tag{34}$$

where

$$[\omega(u)]_{\alpha\beta} = -P(u)^{\mu}_{\alpha} P(u)^{\nu}_{\beta} \nabla_{[\mu} u_{\nu]},$$
$$[\theta(u)]_{\alpha\beta} = P(u)^{\mu}_{\alpha} P(u)^{\nu}_{\beta} \nabla_{(\mu} u_{\nu)} = \frac{1}{2} [\pounds(u)_u P(u)]_{\alpha\beta}\,, \tag{35}$$

are the components of tensor fields $\omega(u)$ and $\theta(u)$ having the meaning respectively of vorticity (whose sign depends on convention[1]) and expansion. From the above definitions, the tensor field $\nabla_\beta u^\alpha$ can be written as

$$\nabla_\beta u^\alpha = -a(u)^\alpha u_\beta - k(u)^\alpha{}_\beta. \tag{36}$$

The expansion tensor field $\theta(u)$ may itself be decomposed into its trace-free and pure trace parts

$$\theta(u) = \sigma(u) + \frac{1}{3}\Theta(u) P(u), \tag{37}$$

where the trace-free tensor field $\sigma(u)$ ($\sigma(u)^\alpha{}_\alpha = 0$) is termed shear and the scalar

$$\Theta(u) = \nabla_\alpha u^\alpha \tag{38}$$

is termed volumetric (or isotropic) scalar expansion.

Define also the vorticity vector field $\omega(u) = 1/2\ \mathrm{curl}_u\, u$ as the spatial dual of the spatial rotation tensor, and given by

$$\omega(u)^\alpha = \frac{1}{2}\eta(u)^{\alpha\beta\gamma}\omega(u)_{\beta\gamma} = \frac{1}{2}\eta^{\sigma\alpha\beta\gamma} u_\sigma \nabla_\beta u_\gamma. \tag{39}$$

Although we use the same symbol for the vorticity tensor and the associated vector they can be easily distinguished by the context.

## 5 Adapted Frames

Given a field of observers $u$, a frame $\{e_\alpha\}$ with $\alpha = 0, 1, 2, 3$ (with dual $\omega^\alpha$) is termed adapted to $u$ if $e_0 = u$ and $e_a$ with $a = 1, 2, 3$ are orthogonal to $u$, namely $u \cdot e_a = 0$. From this it follows that $\omega^0 = -u^\flat$. In this section all indices denote components relative to the frame $\{e_\alpha\}$. The evolution of the frame vectors along the world lines of $u$ is governed by the relations $\nabla_u e_\alpha = e_\sigma \Gamma^\sigma{}_{\alpha 0}$ and one can express the connection coefficients in terms of the kinematical quantities of the observer congruence $u$. The result is the following

[1] We have adopted the $\nabla$-convention differently from a ;-convention also widely used.

$$\Gamma^a{}_{00} = a(u)^a, \qquad \Gamma^0{}_{a0} = a(u)_a, \qquad \Gamma^b{}_{a0} = C_{(\mathrm{fw})}{}^b{}_a, \\ \Gamma^b{}_{0a} = -k(u)^b{}_a, \quad \Gamma^0{}_{ba} = -k(u)_{ba}, \tag{40}$$

where the Fermi–Walker structure functions $C_{(\mathrm{fw})ba}$ are introduced so that

$$P(u)\nabla_u e_a = C_{(\mathrm{fw})}{}^b{}_a e_b. \tag{41}$$

They can also we written as

$$C_{(\mathrm{fw})}{}^b{}_a = C_{(\mathrm{lie})}{}^b{}_a - k(u)^b{}_a, \quad C_{(\mathrm{lie})}{}^b{}_a \equiv \omega^b(\pounds(u)_u e_a), \tag{42}$$

implying

$$P(u)\pounds_u e_a = \pounds(u)_u e_a = C_{(\mathrm{lie})}{}^b{}_a e_b. \tag{43}$$

Similarly, it is straightforward to express the structure functions in terms of kinematical quantities. In fact, from the definition

$$e_\alpha C^\alpha{}_{\beta\gamma} = [e_\beta, e_\gamma] = \nabla_{e_\beta} e_\gamma - \nabla_{e_\gamma} e_\beta \tag{44}$$

we have

$$\begin{aligned} e_\alpha C^\alpha{}_{0b} &= \nabla_u e_b - \nabla_{e_b} u = a(u)_b u + [C_{(\mathrm{fw})}{}^c{}_b + k(u)^c{}_b] e_c \\ &= a(u)_b u + C_{(\mathrm{lie})}{}^c{}_b e_c, \end{aligned} \tag{45}$$

so that $C^0{}_{0b} = a(u)_b$ and $C^c{}_{0b} = C_{(\mathrm{lie})}{}^c{}_b$. Similarly

$$e_\alpha C^\alpha{}_{bc} = \nabla_{e_b} e_c - \nabla_{e_c} e_b = 2\omega(u)_{bc} u + 2\Gamma^d{}_{[cb]} e_d, \tag{46}$$

so that $C^0{}_{bc} = 2\omega(u)_{bc}$ and $C^d{}_{bc} = 2\Gamma^d{}_{[cb]}$.

Finally, the structure functions satisfy the Jacobi identities which can also be given a $3+1$ form.

## *5.1 Spatial-Fermi–Walker and Spatial-Lie Temporal Derivatives*

We have introduced the Fermi–Walker structure functions $C_{(\mathrm{fw})}{}^b{}_a$,

$$P(u)\nabla_u e_a = C_{(\mathrm{fw})}{}^b{}_a e_b, \tag{47}$$

as well as the Lie structure functions $C_{(\mathrm{lie})}{}^b{}_a$ entering the projected Lie derivative along $u$ (which we also termed as "spatial-Lie temporal derivative", see Eq. (20))

$$\pounds(u)_u e_a = P(u)\pounds_u e_a = C_{(\text{lie})}{}^b{}_a e_b = C^b{}_{0a} e_b. \tag{48}$$

It is then useful to handle both these operations with a unified notation [1]

$$\nabla(u)_{(\text{tem})} e_a = C_{(\text{tem})}{}^b{}_a e_b, \quad \text{tem} = \text{fw, lie}, \tag{49}$$

where we define

$$\nabla(u)_{(\text{fw})} e_a \equiv P(u)\nabla_u e_a, \qquad \nabla(u)_{(\text{lie})} e_a \equiv P(u)\pounds_u e_a = \pounds(u)_u e_a. \tag{50}$$

Therefore if $X$ is a vector field orthogonal to $u$, i.e. $X \cdot u = 0$, we have

$$\begin{aligned} \nabla(u)_{(\text{tem})} X &= \nabla(u)_{(\text{tem})}(X^a e_a) = \frac{dX^a}{d\tau_u} e_a + X^a C_{(\text{tem})}{}^b{}_a e_b \\ &= \left(\frac{dX^b}{d\tau_u} + X^a C_{(\text{tem})}{}^b{}_a\right) e_b = (\nabla(u)_{(\text{tem})} X^b) e_b. \end{aligned} \tag{51}$$

The operation $\nabla(u)_{(\text{fw})} = P(u)\nabla_u$ is termed "spatial-Fermi–Walker temporal derivative." It can be extended to non-spatial fields. If we apply this operation to the vector field $u$ itself we have

$$\nabla(u)_{(\text{fw})} u = P(u)\nabla_u u = a(u). \tag{52}$$

Hence the temporal derivatives so defined through their action on purely spatial and purely temporal fields can now act on any spacetime field.

## 5.2 Frame Components of the Riemann Tensor

From the definition

$$e_\alpha R^\alpha{}_{\beta\gamma\delta} = [\nabla_{e_\gamma}, \nabla_{e_\delta}] e_\beta - C^\sigma{}_{\gamma\delta} \nabla_{e_\sigma} e_\beta, \tag{53}$$

we have

$$\begin{aligned} e_\alpha R^\alpha{}_{0b0} &= [\nabla_{e_b}, \nabla_u] u - C^\sigma{}_{b0} \nabla_{e_\sigma} u \\ &= \left\{ [\nabla(u)_b + a(u)_b] a(u)^c + \nabla(u)_{(\text{fw})} k(u)^c{}_b - [k(u)^2]^c{}_b \right\} e_c, \end{aligned} \tag{54}$$

so that

$$R^c{}_{0b0} = [\nabla(u)_b + a(u)_b] a(u)^c + \nabla(u)_{(\text{fw})} k(u)^c{}_b - [k(u)^2]^c{}_b, \tag{55}$$

where

$$\nabla(u)_{(\mathrm{fw})}k(u)^c{}_b = \nabla_u k(u)^c{}_b + C_{(\mathrm{fw})}{}^c{}_f k(u)^f{}_b - C_{(\mathrm{fw})}{}^f{}_b k(u)^c{}_f. \tag{56}$$

Similarly one obtains

$$R^0{}_{bcd} = -2[\nabla(u)_{[c}k(u)_{|b|d]} + \omega(u)_{cd}a(u)_b] \tag{57}$$

and

$$R^f{}_{bcd} = R_{(\mathrm{fw})}{}^f{}_{bcd} - 2k(u)_{b[c}k(u)^f{}_{d]}, \tag{58}$$

where

$$\begin{aligned} R_{(\mathrm{fw})}{}^f{}_{bcd} = & \; 2e_{[c}\left(\Gamma^f{}_{|b|d]}\right) + 2\Gamma^s{}_{b[c}\Gamma^f{}_{|s|d]} - C^s{}_{cd}\Gamma^f{}_{bs} \\ & -2\omega(u)_{cd}C_{(\mathrm{fw})}{}^f{}_b. \end{aligned} \tag{59}$$

This tensor is termed "Fermi–Walker spatial Riemann tensor"[2]; it can be written in invariant form as follows

$$\begin{aligned} R_{(\mathrm{fw})}(u)(X,Y)Z = & \; \{[\nabla(u)_X, \nabla(u)_Y]Z - \nabla(u)_{[X,Y]}\}Z \\ & -2\omega(u)(X,Y)\nabla(u)_{(\mathrm{fw})}Z, \end{aligned} \tag{60}$$

where $X$, $Y$ and $Z$ are spatial fields with respect to $u$ and we note that

$$[X,Y] = P(u)[X,Y] - 2\omega^\flat(X,Y)\,u. \tag{61}$$

The Fermi–Walker spatial Riemann tensor has not all the symmetries of a three dimensional Riemann tensor. For instance it does not satisfy the Ricci identities. In fact we have

$$0 = R^f{}_{[bcd]} = R_{(\mathrm{fw})}{}^f{}_{[bcd]} - 2k(u)_{[bc}k(u)^f{}_{d]}, \tag{62}$$

and hence

$$R_{(\mathrm{fw})}{}^f{}_{[bcd]} = 2k(u)^f{}_{[b}\omega(u)_{cd]}. \tag{63}$$

From the latter one can construct a new Riemann tensor with all the necessary symmetries. Ferrarese [2] has shown that the symmetry-obeying Riemann tensor, denoted as $R_{(\mathrm{sym})}{}^{ab}{}_{cd}$, is related to the Fermi–Walker Riemann tensor (60) by

[2] A Lie spatial Riemann tensor can be defined similarly, replacing the Fermi–Walker structure functions $C_{(\mathrm{fw})}{}^f{}_b$ with the corresponding Lie structure functions $C_{(\mathrm{lie})}{}^f{}_b$ according to Eq. (42).

$$\begin{aligned} R_{(\mathrm{sym})}{}^{ab}{}_{cd} &= R_{(\mathrm{fw})}{}^{ab}{}_{cd} - 2\omega(u)^{ab}\omega(u)_{cd} - 4\theta(u)^{[a}{}_{[c}\omega(u)^{b]}{}_{d]} \\ &= R^{ab}{}_{cd} + 2k(u)^{b}{}_{[c}k(u)^{a}{}_{d]} - 2\omega(u)^{ab}\omega(u)_{cd} \\ &\quad -4\theta(u)^{[a}{}_{[c}\omega(u)^{b]}{}_{d]}. \end{aligned} \qquad (64)$$

Together with the spatial symmetric Riemann tensor $R_{(\mathrm{sym})}{}^{ab}{}_{cd}$ we also introduce the spatial symmetric Ricci tensor, $R_{(\mathrm{sym})}{}^{a}{}_{b} = R_{(\mathrm{sym})}{}^{ca}{}_{cb}$ as well as the the associated scalar $R_{(\mathrm{sym})} = R_{(\mathrm{sym})}{}^{a}{}_{a}$.

## 6 Comparing Families of Observers

Let $u$ and $U$ be two unitary timelike vector fields. Define the relative spatial velocity of $U$ with respect to $u$ from the splitting relations

$$U = \gamma(U,u)[u + \nu(U,u)] = \gamma(U,u)[u + ||\nu(U,u)||\hat{\nu}(U,u)], \qquad (65)$$

and

$$u = \gamma(u,U)[U + \nu(u,U)] = \gamma(u,U)[U + ||\nu(u,U)||\hat{\nu}(u,U)]. \qquad (66)$$

where $\hat{\nu}(U,u)$ is the unitary vector giving the direction of $\nu(U,u)$ in the rest frame of $u$. Both the spatial relative velocity vectors have the same magnitude

$$||\nu(U,u)|| = [\nu(U,u)_\alpha \nu(U,u)^\alpha]^{1/2} = [\nu(u,U)_\alpha \nu(u,U)^\alpha]^{1/2}.$$

The common gamma factor is related to that magnitude by

$$\gamma(U,u) = \gamma(u,U) = [1 - ||\nu(U,u)||^2]^{-1/2} = -U_\alpha u^\alpha \qquad (67)$$

hence we recognize it as the relative Lorentz factor. It is convenient to abbreviate $\gamma(U,u)$ by $\gamma$ and $||\nu(U,u||$ by $\nu$ when their meaning is clear from the context and there are no more than two observers involved.

Let us notice here that by substituting Eq. (65) into Eq. (66) we obtain the following relation

$$-\hat{\nu}(u,U) = \gamma[\hat{\nu}(U,u) + \nu u], \qquad (68)$$

which together with $U = \gamma[u + \nu\hat{\nu}(U,u)]$ yields the relative boost $B(U,u)$ from $u$ to $U$, namely

$$\begin{aligned} B(U,u)u &= U = \gamma[u + \nu\hat{\nu}(U,u)] \\ B(U,u)\hat{\nu}(U,u) &= -\hat{\nu}(u,U) = \gamma[\hat{\nu}(U,u) + \nu u]. \end{aligned} \qquad (69)$$

The inverse relations hold by interchanging $U$ with $u$. The boost acts as the identity on the intersection of their local rest spaces $LRS_u \cap LRS_U$.

1. Maps between LRSs
   The spatial measurements of two observers in relative motion can be compared only relating their respective LRSs. Let $U$ and $u$ be two such observers and $LRS_U$ and $LRS_u$ their LRSs. There exists several maps between these LRSs; for example, by combining the projection operators $P(U)$ and $P(u)$ one can form the following "mixed projection" maps:

   a. $P(U, u)$ from the $LRS_u$ into $LRS_U$, defined as

$$P(U, u) = P(U)P(u) : LRS_u \to LRS_U \,, \tag{70}$$

   with inverse: $P(U, u)^{-1} : LRS_U \to LRS_u$;

   b. $P(u, U)$ from the $LRS_U$ into $LRS_u$, defined as

$$P(u, U) = P(u)P(U) : LRS_U \to LRS_u \,, \tag{71}$$

   with inverse: $P(u, U)^{-1} : LRS_u \to LRS_U$.

   Note that $P(U, u) \neq P(u, U)^{-1}$ as it follows from their representations

$$\begin{aligned} P(U, u) &= P(u) + \gamma \nu U \otimes \hat{\nu}(U, u), \\ P(U, u)^{-1} &= P(U) + \nu U \otimes \hat{\nu}(u, U), \\ P(u, U) &= P(U) + \gamma \nu u \otimes \hat{\nu}(u, U), \\ P(u, U)^{-1} &= P(u) + \nu u \otimes \hat{\nu}(U, u). \end{aligned}$$

   One can then show that

$$\begin{aligned} P(U, u)\hat{\nu}(U, u) &= -\gamma \hat{\nu}(u, U), \\ P(u, U)^{-1}\hat{\nu}(U, u) &= -\frac{1}{\gamma}\hat{\nu}(u, U). \end{aligned} \tag{72}$$

   Note that

$$P(U, u) = P(U) \llcorner P(u) = P(U) \llcorner P(U, u) = P(U, u) \llcorner P(u). \tag{73}$$

   Moreover the following relations hold

$$P(U) = P(U, u) \llcorner P(U, u)^{-1}, \quad P(u) = P(U, u)^{-1} \llcorner P(U, u). \tag{74}$$

2. The boost maps

Similarly to what we have done combining the projection maps, also the boost $B(U,u)$ induces an invertible map between the local rest spaces of the given observers defined as

$$B_{(\mathrm{lrs})}(U,u) \equiv P(U)B(U,u)P(u) : LRS_u \to LRS_U. \tag{75}$$

It acts as the identity on the intersection of their subspaces $LRS_U \cap LSR_u$. Being the boost an isometry, exchanging the role of $U$ and $u$ in (75) leads to the inverse boost

$$B_{(\mathrm{lrs})}(U,u)^{-1} \equiv B_{(\mathrm{lrs})}(u,U) : LRS_U \to LRS_u. \tag{76}$$

The representations of the boost and its inverse can be given in terms of the associated tensors

$$B_{(\mathrm{lrs})u}(U,u), \quad B_{(\mathrm{lrs})U}(U,u), \quad B_{(\mathrm{lrs})u}(u,U), \quad B_{(\mathrm{lrs})U}(u,U), \tag{77}$$

defined by:

$$\begin{aligned} B_{(\mathrm{lrs})u}(U,u) &= P(U,u)^{-1} \llcorner B_{(\mathrm{lrs})}(U,u), \\ B_{(\mathrm{lrs})U}(U,u) &= B_{(\mathrm{lrs})}(U,u) \llcorner P(U,u)^{-1}, \end{aligned} \tag{78}$$

with the corresponding expressions for the inverse boost obtained simply by exchanging the role of $U$ and $u$ and with

$$B_{(\mathrm{lrs})}(U,u) = B_{(\mathrm{lrs})U}(U,u) \llcorner P(U,u) = P(U,u) \llcorner B_{(\mathrm{lrs})u}(U,u). \tag{79}$$

The explicit expression of $B_{(\mathrm{lrs})u}(U,u)$, for example, is given by

$$B_{(\mathrm{lrs})u}(U,u) = P(u) + \frac{1-\gamma}{\gamma}\hat{\nu}(U,u) \otimes \hat{\nu}(U,u). \tag{80}$$

This can be shown as follows. Let $X \in LRS_u$ then

$$\begin{aligned} B_{(\mathrm{lrs})u}(U,u)X &= P(U,u)^{-1}[B_{(\mathrm{lrs})}(U,u)X] \\ &= P(U,u)^{-1}[B_{(\mathrm{lrs})}(U,u)X^{||}\hat{\nu}(U,u) + X^{\perp}], \end{aligned} \tag{81}$$

where $X^{||} = X \cdot \hat{\nu}(U,u)$ and $X^{\perp} = X - X^{||}\,\hat{\nu}(U,u)$, that is

$$X = X^{||}\,\hat{\nu}(U,u) + X^{\perp}, \tag{82}$$

and we have used the fact that the boost reduces to the identity for vectors not belonging to the boost plane, as $X^{\perp}$. In this case the boost plane is spanned by the vectors $u$ and $\hat{\nu}(U,u)$. Taking into account (69), namely

$$B_{(\mathrm{lrs})}(U,u)\hat{\nu}(U,u) = -\hat{\nu}(u,U), \tag{83}$$

as well as the linearity of the boost map, we have

$$\begin{aligned} B_{(\mathrm{lrs})u}(U,u)X &= P(U,u)^{-1}[X - X^{||}\hat{\nu}(u,U) - X^{||}\hat{\nu}(U,u)] \\ &= X + \frac{1-\gamma}{\gamma}X^{||}\hat{\nu}(U,u), \end{aligned} \tag{84}$$

where $P(U,u)^{-1}X = X$ because $X \in LRS_u$:

$$\begin{aligned} P(U,u)^{-1}X &= P(U,u)^{-1}P(u)X = P(U,u)^{-1}P(U)P(u)X \\ &= P(U,u)^{-1}P(U,u)X = P(u)X = X; \end{aligned} \tag{85}$$

hence $P(U,u)^{-1}\hat{\nu}(U,u) = \hat{\nu}(U,u)$ because $\nu(U,u)$ belongs to $LRS_u$. Moreover, from (72), by exchanging the roles of $U$ and $u$, we find

$$\begin{aligned} P(U,u)^{-1}\hat{\nu}(u,U) &= -\frac{1}{\gamma}\hat{\nu}(U,u) = \frac{1}{\gamma}B(u,U)\hat{\nu}(u,U) \\ &= \frac{1}{\gamma}B(U,u)^{-1}\hat{\nu}(u,U). \end{aligned} \tag{86}$$

Therefore:

$$\begin{aligned} B_{(\mathrm{lrs})u}(U,u)X &= X - X^{||}\left(-\frac{1}{\gamma}+1\right)\hat{\nu}(U,u) \\ &= \left[P(u) - \frac{\gamma-1}{\gamma}\hat{\nu}(U,u)\otimes\hat{\nu}(U,u)^{\flat}\right]\llcorner X \end{aligned} \tag{87}$$

which is equivalent to (80).
Similarly, for the inverse boost $B_{(\mathrm{lrs})}(u,U)$ one has

$$\begin{aligned} B_{(\mathrm{lrs})u}(u,U) &= P(u) - \frac{\gamma-1}{\gamma}\hat{\nu}(U,u)\otimes\hat{\nu}(U,u)^{\flat}, \\ B_{(\mathrm{lrs})U}(u,U) &= P(U) - \frac{\gamma-1}{\gamma}\hat{\nu}(u,U)\otimes\hat{\nu}(u,U)^{\flat}. \end{aligned} \tag{88}$$

Thus, if $S$ is a vector field such that $S \in LRS_U$, then its inverse boost is the vector belonging to $LRS_u$

$$B_{(\mathrm{lrs})}(u,U)S = [P(u) - \gamma(\gamma+1)^{-1}\nu(U,u)\otimes\nu(U,u)^{\flat}]\llcorner P(u,U)S. \tag{89}$$

# 7 Splitting of Derivatives Along a Timelike Curve

Consider a congruence of curves $\mathscr{C}_U$ with tangent vector field $U$ and proper time $\tau_U$ as parameter. We know, at this stage, that the evolution along $\mathscr{C}_U$ of any tensor field can be specified by one of the following spacetime derivatives:

1. the absolute derivative along $\mathscr{C}_U$: $D/d\tau_U = \nabla_U$,
2. the Fermi–Walker derivative along $\mathscr{C}_U$: $D_{(\text{fw},U)}/d\tau_U$,
3. the spacetime Lie derivative along $\mathscr{C}_U$: $£_U$, for which we use also the notation $D_{(\text{lie},U)}/d\tau_U = £_U$.

The action of the Fermi–Walker and Lie derivatives on a vector field $X$ is related to the absolute derivative as follows

$$
\begin{aligned}
\frac{D_{(\text{fw},U)}X}{d\tau_U} &= \nabla_U X + a(U)(U\cdot X) - U(a(U)\cdot X)\\
&= P(U)\nabla_U X - U\nabla_U(X\cdot U) + a(U)(X\cdot U)\,,\\
\frac{D_{(\text{lie},U)}X}{d\tau_U} &= [U,X] = \nabla_U X + a(U)(U\cdot X) - k(U)\llcorner X\,,
\end{aligned}
\tag{90}
$$

where $k(U) = \omega(U) - \theta(U)$ is the kinematical tensor of the congruence $\mathscr{C}_U$ defined in (34). For $X = U$ we have

$$
\frac{D_{(\text{fw},U)}U}{d\tau_U} = 0, \qquad \frac{D_{(\text{lie},U)}U}{d\tau_U} = 0,
\tag{91}
$$

whereas $DU/d\tau_U = a(U)$.

If $X$ is spatial with respect to $U$, namely $X\cdot U = 0$, we have instead

$$
\begin{aligned}
\frac{D_{(\text{fw},U)}X}{d\tau_U} &= P(U)\nabla_U X\\
\frac{D_{(\text{lie},U)}X}{d\tau_U} &= \nabla_U X - k(U)\llcorner X\\
&= P(U)\nabla_U X - U(a(U)\cdot X) - k(U)\llcorner X.
\end{aligned}
\tag{92}
$$

The projection orthogonal to $U$ of $D_{(\text{lie},U)}X/d\tau_U$ as in (92) gives

$$
P(U)\frac{D_{(\text{lie},U)}X}{d\tau_U} = \frac{D_{(\text{fw},U)}}{d\tau_U}X - k(U)\llcorner X.
\tag{93}
$$

Let $u$ be another family of observers whose world lines have as parameter the proper time $\tau_u$. One can introduce on the congruence $\mathscr{C}_U$ whose unit tangent vector field can be written as

$$
U = \gamma(U,u)[u + \nu(U,u)],
\tag{94}
$$

two new parametrizations $\tau_{(U,u)}$ and $\ell_{(U,u)}$ as follows

$$\frac{d\tau_{(U,u)}}{d\tau_U} = \gamma(U,u), \qquad \frac{d\ell_{(U,u)}}{d\tau_U} = \gamma(U,u)||\nu(U,u)||, \tag{95}$$

where $\tau_{(U,u)}$ corresponds to the proper times of the observers $u$ when their curves are crossed by a given curve of $\mathscr{C}_U$ and $\ell_{(U,u)}$ corresponds to the proper length on $\mathscr{C}_U$. The projection orthogonal to $u$ of the absolute derivative along $U$ is expressed by

$$\begin{aligned} P(u)\frac{D}{d\tau_U} &= P(u)\nabla_U = \gamma[P(u)\nabla_u + P(u)\nabla_{\nu(U,u)}] \\ &= \gamma[P(u)\nabla_u + \nabla(u)_{\nu(U,u)}]. \end{aligned} \tag{96}$$

We note that in the above equation the derivative operation $P(u)\nabla_u$ is just what we have termed spatial-Fermi–Walker temporal derivative, i.e. $\nabla(u)_{(\mathrm{fw})}$, in (50). For a vector field $X$ we can write then

$$\begin{aligned} P(u)\frac{D_{(\mathrm{fw},U)}X}{d\tau_U} &\equiv \frac{D_{(\mathrm{fw},U,u)}X}{d\tau_U} \\ &= P(u)\frac{DX}{d\tau_U} + P(u,U)a(U)(U\cdot X) \\ &- \gamma\nu(U,u)(a(U)\cdot X), \end{aligned} \tag{97}$$

$$\begin{aligned} P(u)\frac{D_{(\mathrm{lie},U)}X}{d\tau_U} &\equiv \frac{D_{(\mathrm{lie},U,u)}X}{d\tau_U} \\ &= P(u)\frac{DX}{d\tau_U} + P(u,U)a(U)(U\cdot X) \\ &- P(u)[k(U) \llcorner X]. \end{aligned} \tag{98}$$

We shall now examine the projected absolute derivative in detail.

## 7.1 Projected Absolute Derivative

Consider the absolute derivative of $u$ along $U$, namely $\nabla_U u$. Since $u$ is unitary, then $u \cdot Du/d\tau_U = 0$ and we can write

$$\begin{aligned} \frac{Du}{d\tau_U} &= P(u)\frac{Du}{d\tau_U} = \gamma[P(u)\nabla_u u + P(u)\nabla_{\nu(U,u)}u] \\ &= \gamma[\nabla(u)_{(\mathrm{fw})}u + P(u)\nabla_{\nu(U,u)}u] \\ &= \gamma[a(u) + \omega(u)\times_u \nu(U,u) + \theta(u)\llcorner \nu(U,u)]. \end{aligned} \tag{99}$$

Let us denote the above quantity as (minus) Fermi–Walker gravitational force, namely

$$F^{(G)}_{(\mathrm{fw},U,u)} = -\frac{Du}{d\tau_U} = -\gamma[a(u) + \omega(u) \times_u \nu(U,u) + \theta(u) \llcorner \nu(U,u)]\,.$$

It should be stressed here that, although $F^{(G)}_{(\mathrm{fw},U,u)}$ is referred to as a gravitational force, it contains contributions by true gravity and by inertial forces.

Consider now the case of $X$ orthogonal to $u$, i.e. $X \cdot u = 0$. The projection onto $LRS_u$ of the absolute derivative of $X$ along $U$ gives

$$P(u)\frac{DX}{d\tau_U} = \gamma[P(u)\nabla_u X + \nabla(u)_{\nu(U,u)}X] \equiv \frac{D_{(\mathrm{fw},U,u)}X}{d\tau_U}. \tag{100}$$

This differential operator plays an important role since both Fermi–Walker and Lie derivatives along $U$ can be expressed in terms of it. In terms of (adapted) frame components the above expression reads

$$P(u)\frac{DX}{d\tau_U} = \left\{\frac{dX^b}{d\tau_U} + \gamma\left[X^a\left(C_{(\mathrm{fw})}{}^b{}_a + \nu(U,u)^c\,\Gamma^b{}_{ac}\right)\right]\right\} e_b, \tag{101}$$

where we set $X = X^a e_a$. Introducing the relative standard time parametrization $\tau(U,u)$ defined in (95), we have

$$P(u)\frac{DX}{d\tau_{(U,u)}} = \frac{D_{(\mathrm{fw},U,u)}X}{d\tau_{(U,u)}} = \left(\frac{D_{(\mathrm{fw},U,u)}X}{d\tau_{(U,u)}}\right)^a e_a \tag{102}$$

or in components

$$\left(\frac{D_{(\mathrm{fw},U,u)}X}{d\tau_{(U,u)}}\right)^b = \frac{dX^b}{d\tau_{(U,u)}} + X^a\left(C_{(\mathrm{fw})}{}^b{}_a + \nu(U,u)^c\Gamma(u)^b{}_{ac}\right). \tag{103}$$

A particular vector field which is orthogonal to $u$ and is defined all along $\mathscr{C}_U$ is the field of relative velocities, $\nu(U,u)$. We introduce the acceleration of $U$ relative to $u$ by

$$a_{(\mathrm{fw},U,u)} = \frac{D_{(\mathrm{fw},U,u)}}{d\tau_{(U,u)}}\nu(U,u) = \gamma P(u)\frac{D}{d\tau_U}\nu(U,u). \tag{104}$$

Considering instead the unit vector $\hat{\nu}(U,u)$, this quantity can be written as

$$a_{(\mathrm{fw},U,u)} = P(u)\frac{D}{d\tau_{(U,u)}}[\nu\hat{\nu}(U,u)], \tag{105}$$

where $\nu = ||\nu(U,u)||$. Finally we have

$$a_{(\mathrm{fw},U,u)} = \hat{\nu}(U,u)\frac{d\nu}{d\tau_{(U,u)}} + \nu P(u)\frac{D}{d\tau_{(U,u)}}\hat{\nu}(U,u). \quad (106)$$

It is therefore quite natural to denote the first term as a tangential Fermi–Walker acceleration, $a^{(T)}_{(\mathrm{fw},U,u)}$ of $U$ relative to $u$ and the second as centripetal Fermi–Walker acceleration $a^{(C)}_{(\mathrm{fw},U,u)}$ of $U$ relative to $u$:

$$a_{(\mathrm{fw},U,u)} = a^{(T)}_{(\mathrm{fw},U,u)} + a^{(C)}_{(\mathrm{fw},U,u)}, \quad (107)$$

where

$$\begin{aligned} a^{(T)}_{(\mathrm{fw},U,u)} &= \hat{\nu}(U,u)\,\frac{d\nu}{d\tau_{(U,u)}}, \\ a^{(C)}_{(\mathrm{fw},U,u)} &= \nu\, P(u)\frac{D}{d\tau_{(U,u)}}\hat{\nu}(U,u) = \nu\frac{D_{(\mathrm{fw},U,u)}}{d\tau_{(U,u)}}\hat{\nu}(U,u). \end{aligned} \quad (108)$$

To generalize the classical mechanics notion of centripetal acceleration we need to convert the relative standard time parametrization into an analogous relative standard length parametrization[3]:

$$d\ell_{(U,u)} = \nu d\tau_{(U,u)}. \quad (109)$$

With this parametrization we have

$$\begin{aligned} a^{(C)}_{(\mathrm{fw},U,u)} &= \nu^2 P(u)\frac{D}{d\ell_{(U,u)}}\hat{\nu}(U,u) = \nu\frac{D_{(\mathrm{fw},U,u)}}{d\tau_{(U,u)}}\hat{\nu}(U,u) \\ &= \frac{\nu^2}{\mathcal{R}_{(\mathrm{fw},U,u)}}\hat{\eta}_{(\mathrm{fw},U,u)} = \nu^2 k_{(\mathrm{fw},U,u)}\hat{\eta}_{(\mathrm{fw},U,u)}, \end{aligned} \quad (110)$$

where $\hat{\eta}_{(\mathrm{fw},U,u)}$ is a unit spacelike vector orthogonal to $\hat{\nu}(U,u)$, $k_{(\mathrm{fw},U,u)}$ is the Fermi–Walker relative curvature and $\mathcal{R}_{(\mathrm{fw},U,u)}$ is the curvature radius of the curve such that

$$k_{(\mathrm{fw},U,u)}\hat{\eta}_{(\mathrm{fw},U,u)} = \frac{\hat{\eta}_{(\mathrm{fw},U,u)}}{\mathcal{R}_{(\mathrm{fw},U,u)}} = P(u)\frac{D}{d\ell_{(U,u)}}\hat{\nu}(U,u). \quad (111)$$

Clearly, if geometrically or physically motivated, one can replace the Spatial-Fermi–Walker temporal derivative with the Spatial-Lie temporal derivative defining the corresponding quantities. Doing this one really understands the power of the notation used. For example and for later use one can define Lie relative curvature of a curve and the associated curvature radius

[3] In fact the Euclidean space definition involves spatial orbits parameterized by the (spatial) curvilinear abscissa.

$$k_{(\mathrm{lie},U,u)}\hat{\eta}_{(\mathrm{lie},U,u)} = \frac{\hat{\eta}_{(\mathrm{lie},U,u)}}{\mathcal{R}_{(\mathrm{lie},U,u)}} = \frac{D_{(\mathrm{lie},U,u)}}{d\ell_{(U,u)}}\hat{\nu}(U,u). \tag{112}$$

Difficult to think of other efficient relativistic generalizations of the classical concepts of inertial forces besides this one.

## 8 Preferred Slicing in Spacetimes Admitting Separable Geodesics

After providing the general framework of spacetime splitting techniques let us briefly review now some recent results concerning the existence of preferred slicing in those spacetimes admitting separable geodesics (see [7] and references therein).

Let the coordinates $x^\alpha$ ($\alpha = 0\ldots3$, with $x^0 = t$) be such that the geodesic equations are separable in the metric $ds^2 = g_{\alpha\beta}dx^\alpha dx^\beta$. Using the Hamilton–Jacobi formalism we can write the tangent vector $U^\alpha = dx^\alpha(\lambda)/d\lambda$ to the affinely parametrized timelike geodesics as the gradient of the fundamental action function $S = S(x^\alpha, \lambda)$, $U_\alpha = \partial_\alpha S$, satisfying the Hamilton–Jacobi equation

$$-\frac{\partial S}{\partial \lambda} = H(x^\alpha, \partial_\alpha S), \tag{113}$$

with $\lambda$ an affine parameter for the integral curves of $U$ and the Hamiltonian

$$H = \frac{1}{2}g^{\alpha\beta}\partial_\alpha S\partial_\beta S = -\frac{1}{2}\mu^2 = const, \tag{114}$$

the latter identity following from the normalization condition $U^\alpha U_\alpha = -\mu^2$ for timelike geodesics. Assume that $S$ can be separated in its dependence on the variables $x^\alpha$ and $\lambda$, namely

$$S = \frac{1}{2}\mu^2\lambda + S_t(t) + S_1(x^1) + S_2(x^2) + S_3(x^3). \tag{115}$$

Thus we have for the 1-form $U^\flat \equiv U_\alpha dx^\alpha = \partial_\alpha S dx^\alpha = d(S - \frac{1}{2}\mu^2\lambda)$, where here $d$ stands for the spacetime differential only. Moreover, since in this case $U$ is a gradient it is also necessarily vorticity-free: $dU^\flat = 0$, and there exists a distribution of constant action hypersurfaces $T \equiv -S + \frac{1}{2}\mu^2\lambda = const$ with

$$-dT = U_\alpha dx^\alpha, \tag{116}$$

such that $U^\alpha$ is the associated unit normal vector field. When one sets $\mu = 1$, then the time function $T$ measures the proper time along the geodesics and the corresponding lapse function has the fixed value $N = 1$. For a stationary spacetime in which $t$ is taken to be a Killing time coordinate, then $U_t = -E$ is a constant interpreted as a conserved energy, with $S_t(t) = -Et$, and the metric is independent of $t$. One then has

$$-dT = -Edt + U_a dx^a. \tag{117}$$

## 8.1 Static Spherically Symmetric Spacetimes

Consider the case of static spherically symmetric spacetimes. The metric written in standard spherical-like coordinates is

$$ds^2 = -e^{\nu} dt^2 + e^{\lambda} dr^2 + r^2(d\theta^2 + \sin^2\theta d\phi^2), \tag{118}$$

where the functions $\nu$ and $\lambda$ depend only on the radial coordinate. Then $L = U_\phi$ is an additional Killing constant associated with the conserved angular momentum so

$$U_\alpha dx^\alpha = -Edt + (\partial_r S_r)dr + (\partial_\theta S_\theta)d\theta + Ld\phi, \tag{119}$$

and the corresponding Hamilton–Jacobi equation

$$-e^{-\nu}E^2 + e^{-\lambda}(\partial_r S_r)^2 + \frac{1}{r^2}\left[(\partial_\theta S_\theta)^2 + \frac{L^2}{\sin^2\theta}\right] = -\mu^2 \tag{120}$$

can be easily separated in its dependence on the coordinates by setting the square bracket expression to a separation constant $\mathscr{K}$, leading to

$$\frac{dS_r}{dr} = \varepsilon_r e^{\lambda/2}\sqrt{E^2 e^{-\nu} - \frac{\mathscr{K} + \mu^2 r^2}{r^2}}, \qquad \frac{dS_\theta}{d\theta} = \varepsilon_\theta\sqrt{\mathscr{K} - \frac{L^2}{\sin^2\theta}}, \tag{121}$$

where $|\epsilon_r| = 1 = |\varepsilon_\theta|$.

As stated above, we can set $\mu = 1$ to characterize a new foliation by a new temporal coordinate $T$ measuring proper time along the orthogonal geodesics. We are left to specify $E$, $L$, and $\mathscr{K}$ to obtain a specific family of timelike geodesics covering the spacetime. The simplest choice would be a spherically symmetric 4-velocity field involving only radial motion of the geodesics relative to the original coordinates. We can achieve this in two steps. First we can require that this family of geodesics be tangent to the equatorial plane $\theta = \pi/2$, which requires $\mathscr{K} = L^2$ to make $U_\theta = 0$, resulting in

$$U_r^2 = e^{\lambda-\nu}\left[E^2 - e^{\nu}\left(1 + \frac{L^2}{r^2}\right)\right]. \tag{122}$$

We then impose the radial condition $L = 0$, so that

$$U_r = \varepsilon_r e^{(\lambda-\nu)/2}\sqrt{E^2 - e^\nu}, \tag{123}$$

leaving finally the choice of the energy constant $E$. For spatially asymptotically flat spacetimes where $e^\nu < 1$ approaches 1 as $r \to \infty$, to have a choice which works even at spatial infinity, we must have $E \geq 1$, in which case the value may be interpreted as the energy of the radially moving geodesics at spatial infinity. Of course one could choose $E < 1$ but this would limit the slicing to the interior of a cylinder in spacetime inside the radial turning point of the geodesic motion.

The new time differential is then

$$dT = Edt - U_r\,dr. \tag{124}$$

A new global coordinate system for static spacetimes is given by $(X^\alpha) = (T, R, \theta, \phi)$ with $R = r$ and $\theta$ and $\phi$ unchanged and $T = Et + f(r)$ given by integrating the differential equation $f'(r) = -U_r$. This leads to

$$\partial_T = E^{-1}\partial_t, \quad \partial_R = \partial_r + \frac{U_r}{E}\partial_t, \tag{125}$$

and the transformed metric is

$$ds^2 = -dT^2 + \gamma_{ab}(dX^a + N^a dT)(dX^b + N^b dT), \tag{126}$$

with unit lapse function and the shift vector field aligned with the new radial direction, i.e.,

$$N^a = -\delta^a_R U^r = -\delta^a_R e^{-\lambda} U_r = -\delta^a_R \epsilon_r e^{-(\lambda+\nu)/2}\sqrt{E^2 - e^\nu}\,. \tag{127}$$

The 3-metric induced on the $T = const$ hypersurfaces is then given by

$${}^{(3)}ds^2 = \frac{e^{\lambda+\nu}}{E^2}dr^2 + r^2(d\theta^2 + \sin^2\theta d\phi^2). \tag{128}$$

In the case of vacuum as well as in the presence of a nonzero cosmological constant one has $\lambda + \nu = 0$, so that the induced metric is then

$${}^{(3)}ds^2 = \frac{dr^2}{E^2} + r^2(d\theta^2 + \sin^2\theta d\phi^2), \tag{129}$$

whose only nonvanishing component of the spatial Riemann curvature tensor and the spatial curvature scalar are

$${}^{(3)}R^{\theta\phi}{}_{\theta\phi} = \frac{1 - E^2}{r^2} = \frac{1}{2}{}^{(3)}R, \tag{130}$$

with positive or negative curvature respectively for $0 < E < 1$ (bound geodesics) or $E > 1$ (unbound geodesics). The choice $E = 1$ leads to a flat 3-geometry. The additional sign choice $\epsilon_r = -1$ corresponds to the radially infalling geodesics which start at rest at spatial infinity. This is the case for the Schwarzschild spacetime where the Painlevé-Gullstrand coordinates were originally found [8, 9].

# 9 Discussion

The results summarized above have been developed over a period of about 20 years (starting from the 1990s), initially motivated by the necessity to correctly define inertial forces in general relativity. As a consequence, the whole "measurement process" was reformulated, paying much attention to involve only quantities with a clear geometrical and physical meaning.

Actually, the relativistic generalization of well known classical quantities necessitated the introduction of the so called observer's viewpoint and, formally, the systematic use of "$1+3$" spacetime splitting techniques. Relative Frenet–Serret frames, for instance, were perhaps the most suited tools to explain how inertial forces could enter the general relativistic dynamics of test particles, in full similarity with the classical situation.

A lot of progress in this field (which was not at all a newborn field) was possible because of the international competition started in analyzing explicit applications of formalism to test particle motion in black hole spacetimes. The original enthusiasm was swept away when a satisfactory understanding of the problem was obtained. Nevertheless, many aspects of the formalism developed in this field may be exported in different contexts and hence one should wait for another wave of splitting formalism when new applications will be taken into account.

**Acknowledgments** I'm indebted to my "teachers," Profs. R.T. Jantzen and F. de Felice, for a more than 20 years of collaboration and friendship. I also acknowledge the numerous useful discussions with Dr. A. Geralico. Finally, I warmly thank the organizers of this wonderful meeting in Prague for all their work.

# References

1. R.T. Jantzen, P. Carini, D. Bini, The many faces of gravitoelectromagnetism, Ann. Physics 215, 1 (1992). doi:10.1016/0003-4916(92)90297-Y
2. Ferrarese, G.: Proprietà di secondo ordine di un generico riferimento fisico in Relatività generale. Rend. Mat. Roma **24**, 57 (1965)
3. D. Bini, P. Carini, R.T. Jantzen, The intrinsic derivative and centrifugal forces. I: Theoretical foundations, Int. J. Mod. Phys. D 6, 1 (1997). doi:10.1142/S0218271897000029
4. D. Bini, P. Carini, R.T. Jantzen, The intrinsic derivative and centrifugal forces. II: Applications to some familiar stationary axisymmetric spacetimes, Int. J. Mod. Phys. D 6, 143 (1997). doi:10.1142/S021827189700011X

5. D. Bini, F. de Felice, R.T. Jantzen, Absolute and relative Frenet-Serret frames and Fermi-Walker transport, Class. Quantum Grav. 16, 2105 (1999). doi:10.1088/0264-9381/16/6/333
6. de Felice, F., Bini, D.: Classical Measurements in Curved Space-Times. Cambridge Monographs on Mathematical Physics (Cambridge University Press, Cambridge; New York (2010)
7. D. Bini, A. Geralico, R.T. Jantzen, Separable geodesic action slicing in stationary spacetimes, Gen. Rel. Grav. 44, 603 (2012). doi:10.1007/s10714-011-1295-2
8. Painlevé, P.: La mécanique classique et la theorie de la relativité. C. R. Acad. Sci. **173**, 677 (1921)
9. Gullstrand, A.: Allgemeine Lösung des statischen Einkörperproblems in der Einsteinschen Gravitationstheorie. Ark. Mat. Astron. Fys. **16**(8), 1 (1922)

# Some Links Between General Relativity and Other Parts of Physics

**Gary W. Gibbons**

**Abstract** Now that General Relativity has become such a central part of modern physics, its geometrical formalism being taught as part of almost all undergraduate physics courses, it is natural to ask: how can its basic concepts and techniques be used to illuminate areas of physics which have no connection with gravity? Another way of asking this question is: are the analogues situations to those occurring in General Relativity? The search for such analogues is of course an old one, but recently, because of advances in technology, these questions have become more topical. In this talk I will illustrate this theme by examples drawn from optics, acoustics, liquid crystals, graphene and the currently popular topic of cloaking.

## 1 Introduction

General Relativity, its mathematical techniques and conceptual framework are by now part of the tool-kit of (almost) all theoretical physicists and at least some pure mathematicians. They have become part of the natural language of physics. Indeed, parts of the subject are passing into mathematics departments. It is natural therefore to ask to what extent can they illuminate other (non-relativistic) areas of physics. It is also the case that the relentless onward progress of technology makes possible analogue experiments illustrating basic ideas in General Relativity. In this talk I will illustrate this ongoing process of *unification*.

G. W. Gibbons(✉)
DAMTP, Centre for Mathematical Sciences, Wilberforce Rd, Cambridge CB3 0WA, UK
e-mail: gwg1@damtp.cam.ac.uk

J. Bičák and T. Ledvinka (eds.), *General Relativity, Cosmology and Astrophysics*, Fundamental Theories of Physics 177, DOI: 10.1007/978-3-319-06349-2_4,
© Springer International Publishing Switzerland 2014

### 1.1 Dynamical Casimir Effect

As a topical example of the relentless progress of technology November 2011 saw the demonstration in the laboratory [1] some 40 years after the original prediction [2, 3] of a very basic mechanism in semi-classical General Relativity: *amplification of vacuum fluctuations in a time-dependent environment*. This is the basis of all we believe about inflationary perturbations, Hawking evaporation, Black Hole information "Paradox" and much of AdS/CFT correspondence etc.

### 1.2 Some Previous Work

The idea of finding analogue models for general relativistic effects is not new, but the pace has quickened of late. Some important early work was done on cosmic strings modelled by *vortices in superfluid helium 4* and by Volovik [4], who noted that the order parameter of some phases of superfluid helium 3 is a triad $\mathbf{e}_i$ such that $\mathbf{e}_i \cdot \mathbf{e}_j = \delta_{ij}$. More recently, the emphasis has shifted to the optics of metamaterials and most recently to graphene. There are also interesting analogies in liquid crystals.

## 2 Shallow Water Waves

Let's start with a very simple example which will illustrate some basic ideas. If $\eta = \eta(t, x, y)$ is the height of the water above its level when no waves are present and $h = h(x, y)$ the depth of the water, then shallow water waves satisfy the non-dispersive wave equation (this is the analogue of the Einstein Equivalence Principle)

$$(a_g h \eta_x)_x + (a_g h \eta_y)_y = \eta_{tt},$$

where $a_g$ is the acceleration due to gravity. From now on we adopt units in which $a_g = 1$. The wave operator coincides with the covariant d'Alembertian

$$\frac{1}{\sqrt{-g}} \partial_\mu (\sqrt{-g}\, g^{\mu\nu} \partial_\nu \eta) = 0,$$

with respect to the $2 + 1$ dimensional spacetime metric

$$ds^2 = -h^2 dt^2 + h(dx^2 + dy^2).$$

Applying ray theory and geometrical optics, one writes

$$\eta = A e^{-i\omega(t - W(x,y))},$$

where $A(x, y)$ is slowly varying. To lowest order $W$ satisfies the Hamilton-Jacobi equation

$$\left(\frac{\partial W}{\partial x}\right)^2 + \left(\frac{\partial W}{\partial y}\right)^2 = \frac{1}{h},$$

and the rays are solutions of

$$\frac{\mathrm{d}x}{\mathrm{d}t} = h\,\frac{\partial W}{\partial x}.$$

Given any static spacetime metric

$$\mathrm{d}s^2 = -V^2\,\mathrm{d}t^2 + g_{ij}\,\mathrm{d}x^i\mathrm{d}x^j,$$

the projection $x^i = x^i(t)$ of light rays, that is characteristic curves of the covariant wave equation or the Maxwell or the Dirac equations, onto the spatial sections are geodesics of the Fermat or optical metric given by

$$\mathrm{d}s_o^2 = \frac{g_{ij}}{V^2}\,\mathrm{d}x^i\mathrm{d}x^j.$$

In the special case of shallow water waves, the rays are easily seen to be geodesics of the metric

$$\mathrm{d}s_o^2 = \frac{\mathrm{d}x^2 + \mathrm{d}y^2}{h}.$$

For a linearly shelving beach,

$$h \propto y \qquad y > 0,$$

the rays are cycloids, and all ray's strike the shore, i.e. $y = 0$, orthogonally. For a quadratically shelving beach,

$$h \propto y^2 \qquad y > 0,$$

the rays are circles centred on the shore at $y = 0$, and again every ray intersects the shore at right angle. In fact the optical metric in this case is

$$\mathrm{d}s_o^2 = \frac{\mathrm{d}x^2 + \mathrm{d}y^2}{y^2},$$

which is Poincaré's metric of constant curvature on the upper half plane. If $x$ is periodically identified, one obtains the metric induced on a tractrix of revolution in $\mathbb{E}^3$, sometimes called the Beltrami trumpet (i.e. $H^2/\mathbb{Z}$).

For an embedded surface of revolution the induced metric is

$$h_{ij}\,\mathrm{d}x^i\mathrm{d}x^j = \mathrm{d}\rho^2 + C^2(\rho)\,\mathrm{d}\phi^2, \qquad 0 \le \phi < 2\pi,$$
$$C^2(\rho) = x^2 + y^2 = R^2, \qquad \mathrm{d}\rho^2 = \mathrm{d}R^2 + \mathrm{d}z^2,$$

with Gauss curvature $K = -\frac{C''}{C}$.

For Beltrami's trumpet we have $\rho \ge 0$ and thus

$$C(\rho) = a\exp\left(-\frac{\rho}{a}\right), \qquad K = -\frac{1}{a^2},$$

and if we denote

$$w = a\phi + ia\exp\left(\frac{\rho}{a}\right),$$

then the metric is

$$h_{ij}\,\mathrm{d}x^i\mathrm{d}x^j = \frac{a^2|\mathrm{d}w|^2}{(\Im w)^2}.$$

Note that $(\mathrm{d}z/\mathrm{d}\rho)^2 > 0$ and thus the embedding can never reach the conformal boundary at $y = 0$. This will be significant later.

The optical time for rays to reach the shore in the second example above is infinite. This reminds one of the behaviour of event horizons. In fact there is a rather precise correspondence. The Droste-Schwarzschild metric in isotropic coordinates (setting $G = 1 = c$) is

$$\mathrm{d}s^2 = -\frac{\left(1 - \frac{m}{2r}\right)^2}{\left(1 + \frac{m}{2r}\right)^2}\mathrm{d}t^2 + \left(1 + \frac{m}{2r}\right)^4 (\mathrm{d}x^2 + \mathrm{d}y^2 + \mathrm{d}z^2),$$

with $r = \sqrt{x^2 + y^2 + z^2}$. The isotropic radial coordinate $r$ is related to the Schwarzschild radial coordinate $R$ by

$$R = r\left(1 + \frac{m}{2r}\right)^2.$$

The event horizon is at $R = 2m$, $r = \frac{m}{2}$. If we restrict the Schwarzschild metric to the equatorial plane $z = 0$ we obtain

$$\mathrm{d}s^2 = -\frac{\left(1 - \frac{m}{2r}\right)^2}{\left(1 + \frac{m}{2r}\right)^2}\mathrm{d}t^2 + \left(1 + \frac{m}{2r}\right)^4 (\mathrm{d}x^2 + \mathrm{d}y^2).$$

The optical metric is

$$ds_o^2 = \frac{(1+\frac{m}{2r})^6}{(1-\frac{m}{2r})^2}\,(dx^2+dy^2),$$

and

$$h = \frac{(r-\frac{m}{2})^2}{(r+\frac{m}{2})^6}\,r^4.$$

We get the analogue of a black hole: a circularly symmetric island whose edge is at $r = \frac{m}{2}$ and away from which the beach shelves initially in a quadratic fashion and ultimately levels out as $r \to \infty$. Since

$$\frac{1}{h}\frac{dh}{dr} = \frac{2}{r-\frac{m}{2}} + \frac{4}{r} - \frac{6}{r+\frac{m}{2}} > 0,$$

the beach shelves monotonically.

To obtain a cosmic strings for which the optical metric is a flat cone with deficit angle $\delta = \frac{2\pi p}{p+1}$ one needs a submerged mountain with

$$h \propto (x^2+y^2)^{\frac{p}{p+1}}.$$

As $p \to \infty$, we get a parabola of revolution and the optical metric approaches that of an infinitely long cylinder. If $p = 1$ the mountain is conical, like a submerged volcano. In physical coordinates $x$, $y$, the rays are bent, but one may introduce coordinates in which they are flat:

$$ds^2 = d\tilde{r}^2 + \tilde{r}^2\,d\tilde{\phi}^2, \qquad 0 \le \tilde{\phi} \le \frac{2\pi}{p+1}.$$

In these coordinates the rays are straight lines. One could multiply these examples to cover such things as cosmic strings, moving water and vortices. To take into account the fact that the Earth is round we replace $\mathbb{E}^2$ by $S^2$

$$dx^2 + dy^2 \to d\theta^2 + \sin^2\theta\,d\phi^2,$$

which gives Einstein's static universe in $2+1$ dimensions. To take into account that it is rotating, we replace the static, i.e. time-reversal invariant metric, by a stationary metric

$$d\theta^2 + \sin^2\theta\,d\phi^2 \to d\theta^2 + \sin^2\theta\,(d\phi - \Omega\,dt)^2\,.$$

## 3 Optics and Maxwell's Equations

Maxwell's source-free equations in a medium are

$$\operatorname{curl}\mathbf{E} = -\frac{\partial \mathbf{B}}{\partial t}, \qquad \operatorname{div}\mathbf{B} = 0,$$
$$\operatorname{curl}\mathbf{H} = +\frac{\partial \mathbf{D}}{\partial t}, \qquad \operatorname{div}\mathbf{D} = 0,$$

or if $(\varepsilon_{ijk} = \pm 1, 0)$

$$F = -E_i\,\mathrm{d}t \wedge \mathrm{d}x^i + \frac{1}{2}\,\varepsilon_{ijk} B_i\,\mathrm{d}x^j \wedge \mathrm{d}x^k,$$
$$G = \quad H_i\,\mathrm{d}t \wedge \mathrm{d}x^i + \frac{1}{2}\,\varepsilon_{ijk} D_i\,\mathrm{d}x^j \wedge \mathrm{d}x^k,$$
$$\mathrm{d}F = 0 = \mathrm{d}G.$$

*In what follows it will be important to realise that these equations hold in any coordinate system and they do not require the introduction of a spacetime metric.*

However to "close the system", one must relate $F$ to $G$ by means of a "constitutive equation". If the medium is assumed to be *static* and *linear*, then

$$D_i = \varepsilon_{ij} E_j, \qquad B_i = \mu_{ij} H_j,$$

where $\varepsilon_{ij}$ is the dielectric permittivity tensor and $\mu_{ij}$ the magnetic permeability tensor. If they are assumed symmetric: $\varepsilon_{ij} = \varepsilon_{ji}$, $\mu_{ij} = \mu_{ji}$, then $\mathscr{E} = \frac{1}{2}\big(E_i D_i + H_i B_i\big)$ may be regarded as the energy density and $\mathbf{S} = \mathbf{E} \times \mathbf{H}$ the energy current or Poynting vector since Maxwell's equations imply

$$\operatorname{div}\mathbf{S} + \frac{\partial \mathscr{E}}{\partial t} = 0.$$

"In olden days a glimpse of stocking was thought of as something shocking" and certainly $\mu_{ij}$ and $\varepsilon_{ij}$ were assume positive definite "but now", with the advent of nanotechnology and the construction of metamaterials "anything goes".

### *3.1 Left-handed Light*

As long ago as 1964, V.G. Vestilago pointed out that isotropic substances with $\mu_{ij} = \mu\delta_{ij}$, $\varepsilon_{ij} = \varepsilon\delta_{ij}$ and for which

$$\mu < 0, \quad \varepsilon < 0,$$

give rise to left-handed light moving in a medium with a negative refractive index. In 2001, Shelby et al. [5] produced this effect for microwave frequencies. In 2002, Smith et al. [6] appeared to have produced this effect in the laboratory.

Assuming a spacetime dependence proportional to an arbitrary function of $\mathbf{k} \cdot \mathbf{x} - \omega t$, with $\omega > 0$, one finds

$$\mathbf{k} \times \mathbf{E} = \omega \mathbf{B}, \qquad \mathbf{k} \times \mathbf{H} = -\omega \mathbf{D},$$
$$\mathbf{k} \times \mathbf{E} = \mu\omega \mathbf{H}, \qquad \mathbf{k} \times \mathbf{H} = -\varepsilon\, \omega \mathbf{E}.$$

It is always the case that $(\mathbf{E}, \mathbf{H}, \mathbf{S})$ form a right-handed orthogonal triad but if both $\mu$ and $\varepsilon$ are negative then $(\mathbf{E}, \mathbf{H}, \mathbf{k})$ form a left-handed orthogonal triad and so $\mathbf{S}$ and $\mathbf{k}$ are anti-parallel rather than parallel as is usually the case. Since the wave vector $\mathbf{k}$ must be continuous across a junction between a conventional medium and an exotic medium with $\mu < 0$, $\varepsilon < 0$, this gives rise to *backward bending light*.

The speed of propagation $v = \frac{1}{n}$, where $n$ is the refractive index, is given by

$$v^2 = \frac{\omega^2}{\mathbf{k}^2} = \frac{1}{\mu\varepsilon};$$

it is natural to take the negative square root to get the refractive index

$$n = -\frac{1}{\sqrt{\mu\varepsilon}}.$$

Given a spacetime metric $g_{\mu\nu}$ one has a natural way of specifying a constitutive relation:

$$G = \star_g F,$$

where $\star_g$ denotes the Hodge dual with respect to the spacetime metric $g$ such that $\star_g \star_g = -1$. If

$$\mathrm{d}s^2 = -V^2(x^k)\,\mathrm{d}t^2 + g_{ij}(x^k)\,\mathrm{d}x^i \mathrm{d}x^j,$$

Tamm [7], Skrotskii, [8] and Plebański [9] showed that

$$\mu_{ij} = \varepsilon_{ij} = \sqrt{\frac{\det g_{lm}}{V^2}}\, g^{ij}.$$

A medium with $\mu_{ij} = \varepsilon_{ij}$ is said to be *impedance matched*. A similar result holds for resistivity problems such as that Calderon [10] encountered oil prospecting

$$\nabla \cdot \mathbf{j} = 0, \qquad \mathbf{E} = -\nabla\phi, \qquad j_i = \sigma_{ij} E_j,$$

$$\partial_i \left(\sigma_{ij}\, \partial_j \phi\right) = 0 \quad \Rightarrow \quad \nabla_g^2 \phi = \frac{1}{\sqrt{g}}\, \phi_i \left(\sqrt{g}\, g^{ij}\, \partial_j \phi\right) = 0,$$

with

$$\sigma_{ij} = \sqrt{g}\, g^{ij}, \qquad g_{ij} = (\det \sigma_{ij}) \rho_{ij}.$$

If

$$\sigma_{ij} = \frac{1}{z}\, \delta_{ij},$$

we get Poincaré metric on upper half space model of hyperbolic or Lobachevsky space $H^2$:

$$\mathrm{d}s^2 = \frac{\mathrm{d}x^2 + \mathrm{d}y^2 + dz^2}{z^2}.$$

*The conformal boundary is a perfect conductor.*

In physics we may choose either the West Coast signature convention $(-, +, +, +)$, so that $g_{tt} < 0$ and $g_{ij}$ is positive definite or the East Coast convention $(+, -, -, -)$ for which $g_{tt} > 0$ and $g_{ij}$ is negative definite. By Sylvester's law of inertia the signature is locally constant, however running between the East Coast and the West coast there must be a curve on which the spacetime signature flips (as originally suggested in a different context by Arthur Eddington in 1922). Clearly, light passing from Coast to Coast will get bent back.

By *Fermat's principle* electromagnetic waves move along geodesics of the optical metric

$$\mathrm{d}s_o^2 = V^{-2} g_{ij}\, \mathrm{d}x^i \mathrm{d}x^j,$$

but this is invariant under signature change.

## 3.2 Zermelo-Randers-Finsler Geometry

If time reversal symmetry is broken a *stationary metric may be cast in three different forms* [11]:

$$\begin{aligned} \mathrm{d}s^2 &= -U(\mathrm{d}t + \omega_i \mathrm{d}x^i)^2 + \gamma_{ij} \mathrm{d}x^i \mathrm{d}x^j \\ &= U\left[-(\mathrm{d}t - b_i \mathrm{d}x^i)^2 + a_{ij} \mathrm{d}x^i \mathrm{d}x^j\right] \\ &= \frac{U}{1 - h_{ij} W^i W^j}\left[-\mathrm{d}t^2 + h_{ij}(\mathrm{d}x^i - W^i \mathrm{d}t)(\mathrm{d}x^j - W^j \mathrm{d}t)\right]. \end{aligned}$$

Fermat's principle for light rays now generalises to Zermelo's problem: minimize the travel time of a boat moving with fixed speed with respect to a Riemannian metric $h_{ij}$ in the presence of a "wind" $W^i$.

One may also think of the problem as one of a particular type of *Finsler geometry* considered first by *Randers* with a Finsler function of homogeneous degree one in velocity $v^i = \frac{dx^i}{d\lambda}$ defining a line element $ds = F d\lambda$, given by

$$F = \sqrt{a_{ij} v^i v^j} + b_i v^i .$$

Alternatively, one may think of a *charged particle of unit mass and unit charge*, moving on a Riemannian manifold with metric $a_{ij}$ and magnetic field $B_{ij} = \partial_i b_j - \partial_i b_j$. In General Relativity, this is *gravito-magnetism* verified recently by the GP-B satellite experiment.

In the absence of time reversal symmetry there is a *magneto-electric effect* first predicted by L. Landau and E.M. Lifshitz in 1956 and exhibited for instance by $Cr_2O_3$:

$$B_i = \mu_{ij} H_j + \alpha_{ji} E_j, \qquad D_i = \varepsilon_{ij} E_j + \alpha_{ij} H_j,$$

$$\mathscr{E} = \frac{1}{2} \mu_{ij} H_i H_j + \alpha E_i H_j + \frac{1}{2} \varepsilon_{ij} E_i E_j .$$

If we take as constitutive relation $G = \star_g F$, then $\mu_{ij}, \varepsilon_{ij}$ and $\alpha_{ij}$ may be read off from the spacetime metric.

In a *moving medium*, a typical sound or light wave satisfies

$$\left[ (\partial_t - W^i \partial_i)^2 - h^{ij} \, \partial_i \partial_j \right] u = 0.$$

The *rays* solve the Zermelo problem with wind $W^i$. For sound waves this is known to explain the curious (and irritating) propagation of traffic noise. The rays behave like charged magnetic particles, the magnetic field being given by the vertical gradient of the horizontal wind. Of course a vertical gradient in temperature and hence refractive index will also provide an anti-mirage effect. This produces a curved metric $h_{ij}$. Claude Warnick and I have recently modelled this by a charged particle moving in a magnetic field on the upper half plane [12].

For black or white holes Zermelo picture is equivalent to the use of *Painlevé-Gullstrand* coordinates. Here is a low-tech example involving just a kitchen sink [13]. The ripples are surface tension ripples (Fig. 1).

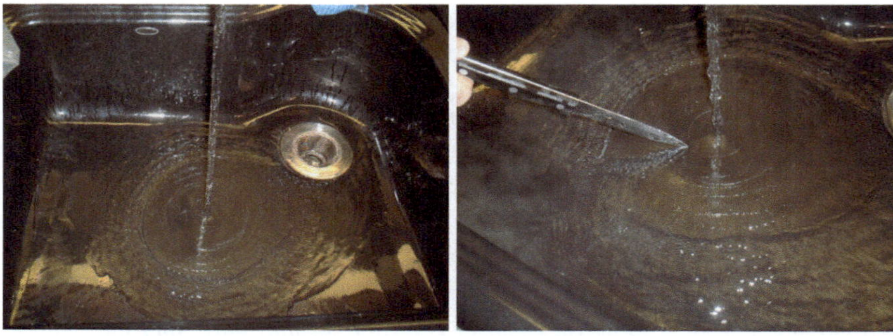

**Fig. 1** Here is the Mach cone

## *3.3 Invisibility Cloaks*

Designing *invisibility cloaks*, analogue black holes, etc. using metamaterials and *transformation optics*. The basic idea is to start with a metric and read off $\varepsilon_{ij}$ and $\mu_{ij}$. The metric could even be flat and obtained by a local diffeomorphism from the flat metric by which a beam or pencil of parallel straight lines in Cartesian coordinates are taken to the desired set of light rays in an *impedance matched* metamaterial medium. This technique has been much exploited by Pendry, Leonhardt and their collaborators and followers recently.

As pointed out by Uhlmann and others, similar problems arise in Calderon's inverse problem: given a measurement of $\mathbf{E}$ and $\phi$ on the boundary of some domain, can you determine uniquely the conductivity in the interior or can a reservoir of oil be invisible to the prospector?

In general one needs anisotropic materials.

To obtain an *isotropic* metamaterial medium the local diffeomorphisms should be *conformal*. The oldest and best known example of this is *Maxwell's fish eye lens* which makes use use of Hipparchus's *stereographic projection*. This is the basis of the Luneburg lens [14].

A variant due to Minano [15] pulls back the round metric on $S^2(\theta, \phi)$ to $R^2(x, y)$ using

$$x = \left(\frac{1-\sin\theta}{\cos\theta}\right)^{\frac{1}{p}} \cos\left(\frac{\phi}{p}\right), \quad y = \left(\frac{1-\sin\theta}{\cos\theta}\right)^{\frac{1}{p}} \sin\left(\frac{\phi}{p}\right),$$

to get

$$\mathrm{d}s_o^2 = \mathrm{d}\theta^2 + \cos^2\theta\, \mathrm{d}\phi^2 = n^2(\mathrm{d}x^2 + \mathrm{d}y^2), \qquad n = 2p^2 \frac{r^{p-1}}{r^{2p}+1}.$$

To get a *black hole* start again with the Droste-Schwarzschild metric in isotropic coordinates

$$ds^2 = -\frac{\left(1-\frac{m}{2r}\right)^2}{\left(1+\frac{m}{2r}\right)^2}\,dt^2 + \left(1+\frac{m}{2r}\right)^4\left(dx^2+dy^2+dz^2\right),$$

$$n = \mu = \varepsilon = \left(1+\frac{m}{2r}\right)^3\left(1-\frac{m}{2r}\right)^{-1}.$$

The *original cloak construction* by Uhlmann works like this. We consider a spherical shell or solid annulus $a < r < 2a$ in $r, \theta, \phi$ space and map it onto the punctured disc $0 < \tilde{r} < 2a$ by

$$\tilde{r} = 2(r-a), \qquad \tilde{\theta} = \theta, \qquad \tilde{\phi} = \phi.$$

The map is the identity: $r = \tilde{r}$ for $r > 2a, \tilde{r} > 2a$. Now pull back the flat metric $d\tilde{r}^2 + \tilde{r}^2(d\tilde{\theta}^2 + \sin^2\tilde{\theta}\,d\tilde{\phi}^2)$ and straight lines in $\tilde{r}, \tilde{\theta}, \tilde{\phi}$ space:

$$ds^2 = 4\,dr^2 + 4(r-a)^2(d\theta^2 + \sin^2\theta\,d\phi^2),$$

$$\varepsilon = \mu = \text{diag}\left(2(r-a)^2\sin\theta,\ 2\sin\theta,\ \frac{2}{\sin\theta}\right).$$

*No light ray (or electric current) enters the solid ball $r < a$.*

The construction just given is strikingly similar to that used in the *hole argument* which played a big part in Einstein's understanding of the concept of general covariance and his search for covariant field equations in the years from 1913 to 1915 [16].

It is remarkable that what hitherto has been of interest almost exclusively to philosophers and historians of science is now at the centre of a new technology!

## 3.4 Hyperbolic Metamaterials and Two-Time Physics

Another possibility are *hyperbolic metamaterials* for which $\varepsilon_{ij}$ is an indefinite matrix. The dispersion relation for a bi-refringent medium with $\mu_{ij} = \delta_{ij}$ is a quartic cone of two sheets:

$$\left(\frac{k_x^2}{n_o^2} + \frac{k_y^2}{n_o^2} + \frac{k_z^2}{n_o^2} - \frac{\omega^2}{c^2}\right)\left(\frac{k_x^2}{n_e^2} + \frac{k_y^2}{n_e^2} + \frac{k_z^2}{n_o^2} - \frac{\omega^2}{c^2}\right) = 0,$$

with $n_o^2 = \varepsilon_z, n_e^2 = \varepsilon_x = \varepsilon_y$. Exceptional electromagnetic waves in a *uniaxial* medium thus obey

$$\frac{1}{c^2}\frac{\partial^2 E}{\partial t^2} = \frac{1}{\varepsilon_1}\frac{\partial^2 E}{\partial z^2} + \frac{1}{\varepsilon_2}\left(\frac{\partial^2 E}{\partial x^2} + \frac{\partial^2 E}{\partial y^2}\right).$$

The idea is [17] that dipole moments in some crystals such as $\alpha$ *quartz* interact with lattice vibrations to form *phonon-polariton modes* called *restrahlen bands* in the mid infra-red region for which both $\varepsilon_1$ and $\varepsilon_2$ can become negative. Moreover because of crystal anisotropy $\varepsilon_1$ and $\varepsilon_2$ change sign at slightly different temperatures. This would allow an effective *two-time physics.*

In a model in a layered composite dielectric material

$$\varepsilon_2 = n_m + (1 - n_m)\varepsilon_d, \qquad \varepsilon_1 = \frac{\varepsilon_m \varepsilon_d}{(1 - n_m)\varepsilon_m + n_m \varepsilon_d},$$

where the subscripts $d$ and $m$ stand for dielectric and metal respectively and $\varepsilon_m$ is frequency dependent and can be negative; $n_m$ is the volume fraction of metal. In a simple Drude model

$$\varepsilon_m = 1 - \frac{\omega_p^2}{\omega^2 + i\omega\gamma},$$

with $\frac{\gamma}{\omega_p}$ being small. If $n_m \ll 1$ we have

$$\varepsilon_2 \approx \varepsilon_d - \frac{n_m \omega_p^2}{\omega^2 + i\omega\gamma}, \qquad \varepsilon_1 \approx \varepsilon_d.$$

## 4 Chiral Nematics

Rather than consider artificial impedance matched or hyperbolic metamaterials, we may consider realistic substances such as chiral nematics in their helical phase. Up to a divergence the *Frank-Oseen free energy* is

$$F = \frac{1}{2}\int \Big(|\nabla^q \mathbf{n}|^2 - \lambda(\mathbf{n}\cdot\mathbf{n} - 1)\Big)\, \mathrm{d}^3x,$$

$$\nabla_i^q n_j = \partial_j n_j + q\, \varepsilon_{ijk}\, n_k,$$

where $\nabla^q$ is Euclidean metric preserving connection with torsion. The free energy density would vanish if $\mathbf{n}$ were covariantly constant with respect to $\nabla^q$, i.e. $\nabla_i^q n_j = 0$. But rather like an anti-ferromagnet it is frustrated since

$$(\nabla_i^q \nabla_j^q - \nabla_j^q \nabla_i^q) n_k \neq 0.$$

The substance may adopt a compromise configuration called the *helical phase* which satisfies the second order equations but not the first order Bogomolnyi type equation

$$\mathbf{n} = (\cos(pz),\ \sin(pz),\ 0).$$

Optics in a nematic liquid crystal is governed by Fermat's principle using the *Joets-Ribotta metric*

$$ds_o^2 = n_e^2\, d\mathbf{x}^2 + (n_o^2 - n_e^2)(\mathbf{n}\cdot d\mathbf{x})^2,$$

where $n_o$ is the refractive index of the ordinary ray and $n_e$ that of the extra-ordinary ray.

Introducing tree one-forms with Maurer-Cartan relations

$$\begin{aligned} \lambda^1 &= \cos(pz)\, dx + \sin(pz)\, dy, & d\lambda^1 &= \lambda^3 \wedge \lambda^2, \\ \lambda^2 &= \cos(pz)\, dx - \sin(pz)\, dy, & d\lambda^2 &= \lambda^3 \wedge \lambda^1, \\ \lambda^3 &= p\, dz, & d\lambda^3 &= 0, \end{aligned}$$

we find the Joets-Ribotta metric to be

$$ds_o^2 = n_o^2(\lambda^1)^2 + n_e^2(\lambda^2)^2 + \frac{n_e^2}{p^2}(\lambda^3)^2.$$

This is a left-invariant metric on $\tilde{E}(2)$, the universal cover of the two-dimensional Euclidean group $E(2)$ whose Lie algebra $e(2)$ is of Type $VII_0$ in Bianchi's classification.

Thus the helical phase of chiral nematic crystals gives rise to a static Bianchi $VII_0$ cosmology:

$$ds^2 = -dt^2 + n_o^2(\lambda^1)^2 + n_e^2(\lambda^2)^2 + \frac{n_e^2}{p^2}(\lambda^3)^2,$$

and one may, and we did, use all the standard tools of general relativistic cosmology to describe its optical and electromagnetic properties, including solving Maxwell's equations, applying the Floquet-Bloch theorem and the associated Mathieu-Hill equation.

## 5 Gravitational Kinks

The topology of a Lorentzian metric may be (partially) captured by a direction field $n^i$. Given a Riemannian metric $g^R_{ij}$, and a unit direction field $n^i$ such that $g^R_{ij}\, n^i n^j = 1$, we may construct a Lorentzian metric $g^L_{ij}$ via

$$g^L_{ij} = g^R_{ij} - \frac{1}{\sin^2\alpha} n_i n_j, \qquad g^{ij}_L = g^{ij}_R - \frac{1}{\cos^2\alpha} n^i n^i, \qquad n_i = g^R_{ij} n^j.$$

Conversely, given $g^L_{ij}$ and $g^R_{ij}$ we may reconstruct $n_i$ up to a sign. Fixing the sign amounts to fixing a time orientation. In what follows we will choose $g^R_{ij}$ to be the usual flat Euclidean metric:

$$\mathrm{d}s^2_L = g^{ij}_L\, \mathrm{d}x^i \mathrm{d}x^j = \mathrm{d}\mathbf{x}^2 - \frac{1}{\cos^2\alpha}(\mathbf{n}\cdot \mathrm{d}\mathbf{x})^2.$$

Given a closed surface enclosing a domain $D$, Finkelstein and Misner quantified the notion of tumbling light cones– the light cone tips over on $\Sigma = \partial D$ – by introducing a *kink number* which counts how many times the light cone tips over on $\Sigma$. The outward unit normal $\boldsymbol{\nu}$ gives a 2-dimensional cross section of the four-dimensional bundle $S(\Sigma)$ of unit 3-vectors over $\Sigma$. In the orientable case, the director field gives another 2-dimensional cross section of $S(\Sigma)$. The kink number, $\mathrm{kink}(\Sigma, g^L)$, is number of intersections of these two sections with attention paid to signs. In the non-orientable case, one considers the bundle of directions. If the Lorentzian metric is non-singular, we have

$$\chi(D) = \mathrm{kink}(\partial D, g^L).$$

For planar domains $\mathrm{kink}(\partial D, g^L)$ is the obvious winding number.

Disclination line is defined by

$$\mathbf{n} = (\cos(s\phi),\ \sin(s\phi),\ 0), \qquad \phi = \arctan\left(\frac{y}{x}\right),$$

where $s \in \mathbb{Z} \cup \mathbb{Z} + \frac{1}{2}$. If $s$ is half integral, then we just have a direction field, not a vector field.

$$\mathbf{n}\cdot \mathrm{d}\mathbf{x} = \cos((s-1)\phi)\, \mathrm{d}r + \sin((s-1)\phi) r\, \mathrm{d}\phi.$$

For $\alpha = \frac{\pi}{2}$ we get

$$\mathrm{d}s^2_L = g^L_{ij} \mathrm{d}x^i \mathrm{d}x^j = -\cos(2(s-1)\phi)\left(\mathrm{d}r^2 - r^2\, \mathrm{d}\phi^2\right) - 2\sin(2(s-1)\phi) r\, \mathrm{d}r \mathrm{d}\phi.$$

Moving around a circle $r$ = constant, the radial coordinate is timelike and the angular coordinate spacelike or *vice versa* depending upon the sign of $\cos(2(s-1)\phi)$ (tumbling light cones). The metric components $g^L_{ij}$ are finite and $\det g^L_{ij} = -r^2 \Rightarrow$ metric non-singular if $r > 0$.

## 5.1 Bloch Walls

If parity symmetry holds then a typical free energy functional takes the form

$$F[\mathbf{M}] = \frac{1}{2}\int \mathrm{d}x \left(\alpha_{ij}\partial_i\mathbf{M}\cdot\partial_j\mathbf{M} + \beta_{ij}M_iM_j\right).$$

In the uniaxial case with the easy direction along the third direction $\alpha_{ij} = \mathrm{diag}(\alpha_1, \alpha_1, \alpha_2)$, $\beta_{ij} = \mathrm{diag}(\beta, \beta, 0)$. For a domain wall separating a region $x \ll -1$ and with $\mathbf{M}$ pointing along the positive 3rd direction, from the region $x \gg +1$ where it points along the negative 3rd direction, we have

$$\mathbf{M} = M\,(0,\ \sin\theta(x),\ \cos\theta(x)), \qquad M = \text{constant},$$

and find that $\theta$ must satisfy the quadrantal pendulum equation, $l = \sqrt{\frac{\alpha_1}{\beta}}$,

$$\theta^2 - \frac{1}{l^2}\sin^2\theta = \text{constant}'.$$

If we impose the boundary condition that $\theta \to 0$ as $x \to -\infty$ and $\theta \to \pi$ as $x \to +\infty$, then constant$' = 0$ and

$$\cos\theta = -\tanh\left(\frac{x}{l}\right).$$

The Lorentzian metric (if $\alpha = \frac{\pi}{2}$) is

$$\mathrm{d}s^2 = g^L_{ij}\,\mathrm{d}x^i\mathrm{d}x^j = \mathrm{d}x^2 + \cos(2\theta)\left(\mathrm{d}y^2 - \mathrm{d}z^2\right) - 2\sin 2\theta\,\mathrm{d}z\mathrm{d}y.$$

This closely resembles our previous examples and clearly exhibits the phenomenon of tumbling light cones. We note, *en passant*, that in principle the tensor $\alpha_{ij}$ could itself vary with position. If so, we might interpret it in terms of an effective metric $g_{ij}$ with inverse $g^{ij}$ and $g = \det g_{ij}$ obeying

$$\alpha_{ij} = \sqrt{g}\,g^{ij}.$$

## 5.2 Liquid Crystal Droplets

The normal $\nu_i = \partial_i S$ to the surface $S =$ constant of a droplet of anisotropic nematic phase inside a domain $D$ with unit outward normal $\boldsymbol{\nu}$ surrounded by an isotropic phase satisfies the constant angle condition

$$\mathbf{n} \cdot \boldsymbol{\nu} = \cos\alpha = \text{constant}.$$

That is

$$\boldsymbol{\nu} \cdot \boldsymbol{\nu} - \frac{1}{\cos^2\alpha}(\boldsymbol{\nu} \cdot \mathbf{n})(\boldsymbol{\nu} \cdot \mathbf{n}) = 0 = g_L^{ij}\, \nu_i \nu_j = g_L^{ij}\, \partial_i S\, \partial_j S.$$

The surface $\partial D$ of the droplet $\partial D$ is a *null-hypersurface* or *wave surface* (a solution of the zero-rest-mass *Hamilton-Jacobi* equation).

Taking the $z$-coordinate as time so time runs vertically upwards and making the ansatz

$$S = \frac{z}{\sin\alpha} + W(x, y), \qquad \nabla W \cdot \nabla W = 1.$$

Simple solutions of this Eikonal equation are given by sandpiles with $\frac{\pi}{2} - \alpha$ the angle of repose.

These describe *Bitter domains* in a ferromagnetic film with $\mathbf{n} = \frac{\mathbf{M}}{|\mathbf{M}|}$ with normal $\boldsymbol{\nu}$ and boundary condition $\mathbf{M} \cdot \boldsymbol{\nu} = 0$:

$$\nabla \cdot \mathbf{M} = 0, \qquad |\mathbf{M}| = \text{constant};$$

$$\nabla \cdot \mathbf{n} = 0 \quad \Rightarrow \quad n_x = \partial_y \psi, \qquad n_y = -\partial_x \psi \qquad |\nabla \psi| = 1.$$

The axisymmetric solution is the spiral wave surface swept out by the involute of a circle, a helical developable:

$$S = \pm \frac{z}{\sin\alpha} \pm a \left( \sqrt{\frac{r^2}{a^2} - 1} - \arctan\left( \sqrt{\frac{r^2}{a^2} - 1} \right) \right) \pm a\phi.$$

## 5.3 Helical Phase

We make the ansatz

$$S = F(z) + x\cos\theta + y\sin\theta,$$

where $F(z)$ solves the quadrantal pendulum equation

$$\cos^2(\theta - pz) - \cos^2\alpha = \left( \cos\alpha \frac{\mathrm{d}F}{\mathrm{d}z} \right)^2.$$

Thus

$$F = \frac{1}{\cos\alpha}\int \mathrm{d}z\sqrt{\cos^2(\theta - pz) - \cos^2\alpha}.$$

The surface is ruled by horizontal straight lines making a constant angle $\theta$ with the $x$-axis and is bounded by $|pz - (\theta + n\pi)| < \alpha$, $n \in \mathbb{Z}$. In other words it is horizontal cylinder or tube. The angle that the director $\mathbf{n}$ makes with the fixed direction $(\cos\theta, \sin\theta, 0)$ cannot be less than $\alpha$.

# 6 Graphene

The hexagonal Graphene "lattice" in $\mathbf{x}$-space has a hexagonal Brillouin zone in the dual $\mathbf{p}$-space and is the sum of two triangular (true) lattices, A and B in $\mathbf{x}$-space. Each lattice has a Fermi surface in $\mathbf{p}$-space and these two Fermi surfaces, governing the conduction and valence bands, touch in two conical Dirac points inside a Brillouin zone. Thus the dispersion relation for small $\mathbf{p}$ is

$$E = \pm|\mathbf{p}|.$$

Low energy excitations are governed by

$$E\Psi = \boldsymbol{\sigma}\cdot\mathbf{p}\,\Psi,$$

where the two-component $\Psi$ has two pseudo-spin states.

But this is the massless Dirac equation! (*cf.* [18])

On a curved graphene sheet it becomes the Dirac equation on a curved surface $\Sigma \subset \mathbb{E}^3$ in Euclidean 3-space with metric

$$\mathrm{d}s^2 = -\mathrm{d}t^2 + h_{ij}\,\mathrm{d}x^i\mathrm{d}x^j, \qquad i, j = 1, 2,$$

where $h_{ij}$ is the induced metric.

Since the massless Dirac equation is conformally invariant we may think of this metric on $R \times \Sigma$ as the optical metric of a static metric with $g_{tt} \neq$ constant.

If $\Sigma$ is a Beltrami trumpet with metric of constant negative curvature, we have the optical metric of identified Rindler spacetime. This is also near horizon optical geometry of a general 2-dimensional black hole. Unfortunately, we cannot find an isometric embedding of $H^2/\mathbb{Z}$ into $\mathbf{E}^3$ all the way down to $y = 0$, the horizon.

This is a general problem: a global theorem of Hilbert forbids isometric embeddings of complete surface of constant negative curvature into Euclidean space $\mathbb{E}^3$.

More generally, we may consider a BTZ black hole [19]:

$$\mathrm{d}s^2_{BTZ} = -\Delta\,\mathrm{d}t^2 + \frac{dr^2}{\Delta} + r^2\left(\mathrm{d}\phi - \frac{J}{2r^2}\,\mathrm{d}t\right)^2,$$
$$\Delta(r) = \frac{r^2}{l^2} - M + \frac{J^2}{4r^2}.$$

Zermelo metric is

$$h_{ij}\,\mathrm{d}x^i\mathrm{d}x^j = \frac{\mathrm{d}r^2}{\Delta^2} + \frac{r^2}{\Delta}\,\mathrm{d}\phi^2,$$

Wind

$$W^i\partial_i = \frac{J}{2r^2}\,\partial_\phi.$$

In the near horizon limit $h_{ij}$ is of Beltrami trumpet form.

The massless Dirac equation in the Zermelo frame is

$$\left[\gamma^1\left(\partial_\rho + \frac{1}{2}\frac{C'}{C}\right) + \gamma^2\frac{1}{C}\partial_\phi + \gamma^0(\partial_t + W\partial_\phi) + \frac{1}{4}\gamma^0\gamma^1\gamma^2 CW'\right]\Psi = 0,$$
$$\gamma^0 = i\sigma_2, \qquad \gamma^1 = \sigma_1, \qquad \gamma^2 = \sigma_3. \qquad \gamma^0\gamma^1\gamma^2 = 1,$$

and we get a position dependent "mass-like" term and a connection term. If $\Psi \propto e^{-i\omega t + im\phi}$ we have that

$$-ieA_0 = imW, \qquad \Rightarrow \qquad eA_0 = -mW.$$

A stationary Zermelo metric induces in the Dirac equation an effective, position dependent radial electric field.

We could have done this calculation in the Randers frame. The detailed form of the Randers metric is considerably more complicated. The embedding is qualitatively similar, but different.

More interestingly, because now the roles of $t$ and $\phi$ have essentially been interchanged, we now find that there is an effective *magnetic* vector potential in the Dirac equation. Therefore, the magnetic vector potential in the Randers frame appears as an electric potential in the Zermelo one.

Since the Zermelo and Randers frames are in relative motion, this is just a manifestation of the fact that under boosts magnetic and electric fields transform into themselves. In either case, these effects could be mimicked by applying external electric (Zermelo) or magnetic (Randers) fields to the two different graphene sheets.

### *6.1 Cold Atoms*

Instead of graphene one may consider, and people have discussed, metrics in the context of cold atoms in Bose-Einstein condensates [20].

## 7 Conclusion and Propects

In this talk I have described on some areas of non-gravitational physics where analogues of basic ideas in general relativity come into play. They include

1. Dynamic Casimir effect
2. Water and sound waves
3. Cloaking and other devices using metamaterials
4. Nematic liquid crystals
5. Graphene

Other areas not covered include

1. Bose-Einstein condensate
2. Dirac metals
3. Smectic and blue phases in liquid crystals

## References

1. Wilson, C.M., Johansson, G., Pourkąbirian, A., et al.: Observation of the dynamical Casimir effect in a superconducting circuit. Nature **479**, 376 (2011). doi:10.1038/nature10561
2. Moore, G.T.: Quantum theory of the electromagnetic field in a variable-length one-dimensional cavity. J. Math. Phys. **11**, 2679 (1970). doi:10.1063/1.1665432
3. Fulling, S.A., Davies, P.C.W.: Radiation from a moving mirror in two dimensional space-time—Conformal anomaly. Proc. R. Soc. London Ser. A **348**, 393 (1976). doi:10.1098/rspa.1976.0045
4. Volovik, G.E.: The Universe in a Helium Droplet, International Series of Monographs on Physics, vol. 117. Oxford University Press, Oxford, New York (2003)
5. Shelby, R.A., Smith, D.R., Schutz, S.: Experimental verification of a negative index of refraction. Science **292**, 77 (2001). doi:10.1126/science.1058847
6. Smith, D.R., Schurig, D., Pendry, J.B.: Negative refraction of modulated electromagnetic waves. Appl. Phys. Lett. **81**, 2713 (2002). doi:10.1063/1.1512828
7. Tamm, I.E.: Èlektrodinamika anizotropnoj sredy v special'noj teorii otnositel'nosti. Zh. Rus. Fiz.-Khim. Obshchestva, Otd. Fiz 56, 248 (1924).
8. Skrotskii, G.V.: On the influence of gravity on the light propagation. Dokl. Akad. Nauk. SSSR 114, 73 (1957). Soviet Physics Doklady 2, 226 (1957).
9. Plebański, J.: Electromagnetic waves in gravitational fields. Phys. Rev. **118**, 1396 (1960). doi:10.1103/PhysRev.118.1396
10. Calderón, A.P.: On an inverse boundary value problem. In: Meyer, W.H., Raupp, M.A. (eds.) Seminar in Numerical Analysis and its Applications to Continuum Physics, pp. 65–73. Sociedade Brasileira de Matemática, Rio de Janeiro (1980)

11. Gibbons, G.W., Herdeiro, C.A.R., Warnick, C.M., Werner, M.C.: Stationary metrics and optical Zermelo-Randers-Finsler geometry. Phys. Rev. D **79**, 044022 (2009). doi:10.1103/PhysRevD.79.044022
12. Gibbons, G.W., Warnick, C.M.: The geometry of sound rays in a wind. Contemp. Phys. **52**, 197 (2011). doi:10.1080/00107514.2011.563515
13. Jannes, G., Rousseaux, G.: The circular jump as a hydrodynamic white hole. ArXiv e-prints. arXiv:1203.6505 [gr-qc] (2012)
14. Luneberg, R.K.: Mathematical Theory of Optics. Brown University, Providence, RI (1944)
15. Miñano, J.C.: Perfect imaging in a homogeneous threedimensional region. Opt. Exp. **14**, 9627 (2006). doi:10.1364/OE.14.009627
16. Norton, J.D.: The hole argument. In: Zalta, E.N. (ed.) The Stanford Encyclopedia of Philosophy, Fall 2011 edn. Stanford University, Stanford (2011) http://plato.stanford.edu/archives/fall2011/entries/spacetime
17. Smolyaninov, I.I.: Critical opalescence in hyperbolic metamaterials. J. Opt. **13**, 125101 (2011). doi:10.1088/2040-8978/13/12/125101
18. Semenoff, G.W.: Condensed-matter simulation of a three-dimensional anomaly. Phys. Rev. Lett. **53**, 2449 (1984). doi:10.1103/PhysRevLett.53.2449
19. Cvetič, M., Gibbons, G.W.: Graphene and the zermelo optical metric of the BTZ black hole. Ann. Phys. (N.Y.) **327**, 2617 (2012). doi:10.1016/j.aop.2012.05.013
20. Boada, O., Celi, A., Latorre, J.I., Lewenstein, M.: Dirac equation for cold atoms in artificial curved spacetimes. New J. Phys. **13**, 035002 (2011). doi:10.1088/1367-2630/13/3/035002

# The General Relativistic Two Body Problem and the Effective One Body Formalism

**Thibault Damour**

**Abstract** A new analytical approach to the motion and radiation of (comparable mass) binary systems has been introduced in 1999 under the name of Effective One Body (EOB) formalism. We review the basic elements of this formalism, and discuss some of its recent developments. Several recent comparisons between EOB predictions and Numerical Relativity (NR) simulations have shown the aptitude of the EOB formalism to provide accurate descriptions of the dynamics and radiation of various binary systems (comprising black holes or neutron stars) in regimes that are inaccessible to other analytical approaches (such as the last orbits and the merger of comparable mass black holes). In synergy with NR simulations, post-Newtonian (PN) theory and Gravitational Self-Force (GSF) computations, the EOB formalism is likely to provide an efficient way of computing the very many accurate template waveforms that are needed for Gravitational Wave (GW) data analysis purposes.

## 1 Introduction

The general relativistic $N$-body problem has been investigated from the early days of Einstein's gravitation theory (and even earlier, because it was already tackled by Johannes Droste within the framework of the 1913 Einstein-Grossmann "Entwurf" theory). Here, we shall focus on the general relativistic two-body problem. This problem has been the subject of many investigations within the post-Newtonian (PN) formalism, since the pioneering works of Einstein (1915; when $m_1 \ll m_2$), Lorentz and Droste (1917), Levi-Civita (1937) and Einstein, Infeld and Hoffmann (1938). [See, *e.g.*, [1] for a review and references to the early literature.] For many years, the first post-Newtonian (1PN) approximation (*i.e.* the inclusion of the leading-order relativistic corrections, proportional to $(v/c)^2$ or $GM/c^2r$, to the Newtonian

T. Damour (✉)
Institut des Hautes Études Scientifiques, 35 route de Chartres, 91440 Bures-sur-Yvette, France
e-mail: damour@ihes.fr

J. Bičák and T. Ledvinka (eds.), *General Relativity, Cosmology and Astrophysics*, Fundamental Theories of Physics 177, DOI: 10.1007/978-3-319-06349-2_5, 
© Springer International Publishing Switzerland 2014

equations of motion) appeared as being accurate enough for applying Einstein's theory to known binary systems. The situation changed in the mid 1970s with the discovery of the Hulse-Taylor binary pulsar PSR 1913 + 16. The need to compare the accurate observations of this system (by Taylor and collaborators) to the predictions of Einstein's theory motivated the development of improved relativistic theories of binary systems, applicable to strongly self-gravitating bodies, and including terms up to the 2.5PN approximation (*i.e.* $O[(v/c)^5]$ beyond Newton). [See [2] and references therein.] The situation has again changed recently with the development of interferometric gravitational wave (GW) detectors, and the prospect of detecting the GW's emitted during the last orbits and the coalescence of binary systems made of black holes or neutron stars. The latter prospect motivated the development (or improvement) of several different methods of computing the motion and radiation of binary systems.

First, this motivated pushing PN calculations of the dynamics of binary systems to the 3PN level [3–7], with inclusion of 3.5PN radiation-reaction terms [8–10]. Second, this motivated the development of new, accurate GW generation formalisms, notably the Blanchet-Damour-Iyer (matched) "multipolar post-Minkowskian" formalism [11–16] and the "direct integration of the relaxed Einstein's equations" formalism of Will and collaborators [17–19], which extended previous work by Epstein and Wagoner [20] and Thorne [21]. These GW generation formalisms allowed one to compute emitted gravitational waveforms with an unprecedented PN accuracy.[1] After the 1PN correction to the waveform [12, 20, 22], there is a 1.5PN "tail" (*i.e.* hereditary) correction [15, 23, 24], then a "direct" 2PN term [17, 25, 26], followed by higher-order corrections [27–35]. [See [36] for a detailed account and more references.] Parallely to these improved PN computations of the GW emission of comparable-mass systems (with $m_1 \sim m_2$), other authors developed the analytical theory of the GW emission of extreme mass-ratio systems (with $m_1 \ll m_2$); see Refs. [37–40] and the review of Sasaki and Tagoshi [41].

Some of the PN calculations of the dynamics, and/or GW emission, of comparable-mass systems have been recently (re)done (e.g. the 3PN dynamics [42]) by using a somewhat different formalism, dubbed "effective field theory" [43]. Let us, however, note that most of the technical aspects of the effective-field-theory approach had already been introduced and used before. For instance: (i) Ref. [44] discussed the (Fokker) two-body effective action due to the exchange of a linear field (of spin $s = 0$, 1 and 2) ; (ii) Ref. [45] explicitly discussed the representation (and computation) of the (nonlinear) effective two-body action in terms of Feynman-like diagrams (made of concatenated propagators and vertices); (iii) The appendix A of Ref. [46] discussed finite-size effects in terms of nonminimal worldline couplings in the effective action; (iv) The (quantum field theory) technique of dimensional regularization (together with a diagrammatic analysis of ultraviolet divergences) was crucially used

---

[1] For gravitational waveforms, one conventionally defines the PN accuracy as the *fractional* PN accuracy with respect to the leading-order, $O(c^{-5})$, quadrupolar emission. *E.g.*, a 1PN-accurate waveform retains next-to-leading order terms, *i.e.* terms smaller than the quadrupolar waveform by a factor $O(c^{-2})$.

to derive the 3PN dynamics in Refs. [5, 6], and 3PN radiation in Ref. [31]; and (v) The exponential parametrization of the metric (which suppresses the leading-order gravitational cubic vertex) had been introduced in Ref. [12] and then standardly used in many PN works. It is, however, possible that the more systematic (and automatized) implementation of such diagrammatic methods, together with the tapping of standard techniques for computing Feynman graphs (as exemplified in [42]) may allow one to be more efficient in computing higher-order processes, or, at least, to open new ways of understanding them (see, in this respect, Ref. [47]).

Separately from these purely analytical approaches to the motion and radiation of binary systems, which have been developed since the early days of Einstein's theory, Numerical Relativity (NR) simulations of Einstein's equations have relatively recently (2005) succeeded (after more than thirty years of developmental progress) to stably evolve binary systems made of comparable mass black holes [48–51]. This has led to an explosion of works exploring many different aspects of strong-field dynamics in General Relativity, such as spin effects, recoil, relaxation of the deformed horizon formed during the coalescence of two black holes to a stationary Kerr black hole, high-velocity encounters, etc.; see [52] for a review. In addition, recently developed codes now allow one to accurately study the orbital dynamics, and the coalescence of binary neutron stars. Much physics remains to be explored in these systems, especially during and after the merger of the neutron stars (which involves a much more complex physics than the pure-gravity merger of two black holes).

Recently, a new source of information on the general relativistic two-body problem has opened: gravitational self-force (GSF) theory. This approach goes one step beyond the test-particle approximation (already used by Einstein in 1915) by taking into account self-field effects that modify the leading-order geodetic motion of a small mass $m_1$ moving in the background geometry generated by a large mass $m_2$. After some ground work (notably by DeWitt and Brehme) in the 1960s, GSF theory has recently undergone rapid developments (mixing theoretical and numerical methods) and can now yield numerical results that yield access to new information on strong-field dynamics in the extreme mass-ratio limit $m_1 \ll m_2$. See Ref. [53] for a review (see also the chapter by L. Barack in this volume).

Each of the approaches to the two-body problem mentioned so far, PN theory, NR simulations and GSF theory, have their advantages and their drawbacks. It has become recently clear that the best way to meet the challenge of accurately computing the gravitational waveforms (depending on several continuous parameters) that are needed for a successful detection and data analysis of GW signals in the upcoming LIGO/Virgo/GEO/... network of GW detectors is to combine knowledge from all the available approximation methods: PN, NR and GSF. Several ways of doing so are a priori possible. For instance, one could try to directly combine PN-computed waveforms (approximately valid for large enough separations, say $r \gtrsim 10\,G(m_1 + m_2)/c^2$) with NR waveforms (computed with initial separations $r_0 > 10\,G(m_1 + m_2)/c^2$ and evolved up to merger and ringdown). However, this method still requires too much computational time, and is likely to lead to waveforms of rather poor accuracy, see, *e.g.*, [54–56].

On the other hand, five years before NR succeeded in simulating the late inspiral and the coalescence of binary black holes, a new approach to the two-body problem was proposed: the Effective One Body (EOB) formalism [57–60]. The basic aim of the EOB formalism is to provide an analytical description of both the motion and the radiation of coalescing binary systems over the entire merger process, from the early inspiral, right through the plunge, merger and final ringdown. As early as in 2000 [58], this method made several quantitative and qualitative predictions concerning the dynamics of the coalescence, and the corresponding GW radiation, notably: (i) a blurred transition from inspiral to a 'plunge' that is just a smooth continuation of the inspiral, (ii) a sharp transition, around the merger of the black holes, between a continued inspiral and a ring-down signal, and (iii) estimates of the radiated energy and of the spin of the final black hole. In addition, the effects of the individual spins of the black holes were investigated within the EOB [60, 61] and were shown to lead to a larger energy release for spins parallel to the orbital angular momentum, and to a dimensionless rotation parameter $J/E^2$ always smaller than unity at the end of the inspiral (so that a Kerr black hole can form right after the inspiral phase). All those predictions have been broadly confirmed by the results of the recent numerical simulations performed by several independent groups (for a review of numerical relativity results and references see [52]). Note that, in spite of the high computer power used in NR simulations, the calculation, checking and processing of one sufficiently long waveform (corresponding to specific values of the many continuous parameters describing the two arbitrary masses, the initial spin vectors, and other initial data) takes on the order of one month. This is a very strong argument for developing analytical models of waveforms.

## 2 EOB Description of the Conservative Dynamics of Two Body Systems

Before reviewing some of the technical aspects of the EOB method, let us indicate the historical roots of this method. First, we note that the EOB approach comprises three, rather separate, ingredients:

1. a description of the conservative (Hamiltonian) part of the dynamics of two bodies;
2. an expression for the radiation-reaction part of the dynamics;
3. a description of the GW waveform emitted by a coalescing binary system.

For each one of these ingredients, the essential inputs that are used in EOB works are high-order post-Newtonian (PN) expanded results which have been obtained by many years of work, by many researchers (see references above). However, one of the key ideas in the EOB philosophy is to avoid using PN results in their original "Taylor-expanded" form (*i.e.* $c_0 + c_1\, v/c + c_2\, v^2/c^2 + c_3\, v^3/c^3 + \cdots + c_n\, v^n/c^n$), but to use them instead in some *resummed* form (*i.e.* some non-polynomial function of $v/c$, defined so as to incorporate some of the expected non-perturbative features of the exact result). The basic ideas and techniques for resumming each ingredient of the EOB are different and have different historical roots.

Concerning the first ingredient, *i.e.* the EOB Hamiltonian, it was inspired by an approach to electromagnetically interacting quantum two-body systems introduced by Brézin et al. [62].

The resummation of the second ingredient, *i.e.* the EOB radiation-reaction force $\mathcal{F}$, was initially inspired by the Padé resummation of the flux function introduced by Damour et al. [63]. More recently, a new and more sophisticated resummation technique for the radiation reaction force $\mathcal{F}$ has been introduced by Damour and Nagar [64, 65].

As for the third ingredient, *i.e.* the EOB description of the waveform emitted by a coalescing black hole binary, it was mainly inspired by the work of Davis et al. [66] which discovered the transition between the plunge signal and a ringing tail when a particle falls into a black hole. [Additional motivation for the EOB treatment of the transition from plunge to ring-down came from work on the, so-called, "close limit approximation" [67].] In addition, a very efficient resummation of the waveform has been introduced by Damour et al. [64, 68, 69]. It will be discussed in detail below.

Within the usual PN formalism, the conservative dynamics of a two-body system is currently fully known up to the 3PN level [3–7, 42] (see below for the partial knowledge beyond the 3PN level). Going to the center of mass of the system ($\boldsymbol{p}_1 + \boldsymbol{p}_2 = 0$), the 3PN-accurate Hamiltonian (in Arnowitt-Deser-Misner-type coordinates) describing the relative motion, $\boldsymbol{q} = \boldsymbol{q}_1 - \boldsymbol{q}_2$, $\boldsymbol{p} = \boldsymbol{p}_1 = -\boldsymbol{p}_2$, has the structure

$$H_{\rm 3PN}^{\rm relative}(\boldsymbol{q}, \boldsymbol{p}) = H_0(\boldsymbol{q}, \boldsymbol{p}) + \frac{1}{c^2}\, H_2(\boldsymbol{q}, \boldsymbol{p}) + \frac{1}{c^4}\, H_4(\boldsymbol{q}, \boldsymbol{p}) + \frac{1}{c^6}\, H_6(\boldsymbol{q}, \boldsymbol{p}) \,, \quad (1)$$

where

$$H_0(\boldsymbol{q}, \boldsymbol{p}) = \frac{1}{2\mu}\, \boldsymbol{p}^2 - \frac{GM\mu}{|\boldsymbol{q}|} \,, \quad (2)$$

with

$$M \equiv m_1 + m_2 \quad \text{and} \quad \mu \equiv m_1 m_2/M \,, \quad (3)$$

corresponds to the Newtonian approximation to the relative motion, while $H_2$ describes 1PN corrections, $H_4$ 2PN ones and $H_6$ 3PN ones. In terms of the rescaled variables $\boldsymbol{q}' \equiv \boldsymbol{q}/GM$, $\boldsymbol{p}' \equiv \boldsymbol{p}/\mu$, the explicit form (after dropping the primes for readability) of the 3PN-accurate rescaled Hamiltonian $\widehat{H} \equiv H/\mu$ reads [5, 70, 71]

$$\widehat{H}_N(\boldsymbol{q},\boldsymbol{p}) = \frac{\boldsymbol{p}^2}{2} - \frac{1}{q}\,, \tag{4}$$

$$\widehat{H}_{1\text{PN}}(\boldsymbol{q},\boldsymbol{p}) = \frac{1}{8}(3\nu-1)(\boldsymbol{p}^2)^2 - \frac{1}{2}\left[(3+\nu)\boldsymbol{p}^2 + \nu(\boldsymbol{n}\cdot\boldsymbol{p})^2\right]\frac{1}{q} + \frac{1}{2q^2}\,, \tag{5}$$

$$\begin{aligned}\widehat{H}_{2\text{PN}}(\boldsymbol{q},\boldsymbol{p}) = {}& \frac{1}{16}(1-5\nu+5\nu^2)(\boldsymbol{p}^2)^3 \\ & + \frac{1}{8}\left[(5-20\nu-3\nu^2)(\boldsymbol{p}^2)^2 - 2\nu^2(\boldsymbol{n}\cdot\boldsymbol{p})^2\boldsymbol{p}^2 - 3\nu^2(\boldsymbol{n}\cdot\boldsymbol{p})^4\right]\frac{1}{q} \\ & + \frac{1}{2}\left[(5+8\nu)\boldsymbol{p}^2 + 3\nu(\boldsymbol{n}\cdot\boldsymbol{p})^2\right]\frac{1}{q^2} - \frac{1}{4}(1+3\nu)\frac{1}{q^3}\,,\end{aligned} \tag{6}$$

$$\begin{aligned}\widehat{H}_{3\text{PN}}(\boldsymbol{q},\boldsymbol{p}) = {}& \frac{1}{128}(-5+35\nu-70\nu^2+35\nu^3)(\boldsymbol{p}^2)^4 \\ & + \frac{1}{16}\Big[(-7+42\nu-53\nu^2-5\nu^3)(\boldsymbol{p}^2)^3 + (2-3\nu)\nu^2(\boldsymbol{n}\cdot\boldsymbol{p})^2(\boldsymbol{p}^2)^2 \\ & \qquad + 3(1-\nu)\nu^2(\boldsymbol{n}\cdot\boldsymbol{p})^4\boldsymbol{p}^2 - 5\nu^3(\boldsymbol{n}\cdot\boldsymbol{p})^6\Big]\frac{1}{q} \\ & + \left[\frac{1}{16}(-27+136\nu+109\nu^2)(\boldsymbol{p}^2)^2 + \frac{1}{16}(17+30\nu)\nu(\boldsymbol{n}\cdot\boldsymbol{p})^2\boldsymbol{p}^2 \right. \\ & \qquad \left. + \frac{1}{12}(5+43\nu)\nu(\boldsymbol{n}\cdot\boldsymbol{p})^4\right]\frac{1}{q^2} \\ & + \left\{\left[-\frac{25}{8} + \left(\frac{1}{64}\pi^2 - \frac{335}{48}\right)\nu - \frac{23}{8}\nu^2\right]\boldsymbol{p}^2 \right. \\ & \qquad \left. + \left(-\frac{85}{16} - \frac{3}{64}\pi^2 - \frac{7}{4}\nu\right)\nu(\boldsymbol{n}\cdot\boldsymbol{p})^2\right\}\frac{1}{q^3} \\ & + \left[\frac{1}{8} + \left(\frac{109}{12} - \frac{21}{32}\pi^2\right)\nu\right]\frac{1}{q^4}\,.\end{aligned} \tag{7}$$

In these formulas $\nu$ denotes the symmetric mass ratio:

$$\nu \equiv \frac{\mu}{M} \equiv \frac{m_1 m_2}{(m_1+m_2)^2}\,. \tag{8}$$

The dimensionless parameter $\nu$ varies between 0 (extreme mass ratio case) and $\frac{1}{4}$ (equal mass case) and plays the rôle of a deformation parameter away from the test-mass limit.

It is well known that, at the Newtonian approximation, $H_0(\boldsymbol{q},\boldsymbol{p})$ can be thought of as describing a 'test particle' of mass $\mu$ orbiting around an 'external mass' $GM$. The EOB approach is a *general relativistic generalization* of this fact. It consists in looking for an 'effective external spacetime geometry' $g^{\text{eff}}_{\mu\nu}(x^\lambda; GM, \nu)$ such that the geodesic dynamics of a 'test particle' of mass $\mu$ within $g^{\text{eff}}_{\mu\nu}(x^\lambda; GM, \nu)$ is *equivalent* (when expanded in powers of $1/c^2$) to the original, relative PN-expanded dynamics (1).

Let us explain the idea, proposed in [57], for establishing a 'dictionary' between the real relative-motion dynamics, (1), and the dynamics of an 'effective' particle of mass $\mu$ moving in $g^{\rm eff}_{\mu\nu}(x^\lambda, GM, \nu)$. The idea consists in 'thinking quantum mechanically'.[2] Instead of thinking in terms of a classical Hamiltonian, $H(\boldsymbol{q}, \boldsymbol{p})$ (such as $H^{\rm relative}_{\rm 3PN}$, Eq. (1)), and of its classical bound orbits, we can think in terms of the quantized energy levels $E(n, \ell)$ of the quantum bound states of the Hamiltonian operator $H(\hat{\boldsymbol{q}}, \hat{\boldsymbol{p}})$. These energy levels will depend on two (integer valued) quantum numbers $n$ and $\ell$. Here (for a spherically symmetric interaction, as appropriate to $H^{\rm relative}$), $\ell$ parametrizes the total orbital angular momentum ($\boldsymbol{L}^2 = \ell(\ell+1)\,\hbar^2$), while $n$ represents the 'principal quantum number' $n = \ell + n_r + 1$, where $n_r$ (the 'radial quantum number') denotes the number of nodes in the radial wave function. The third 'magnetic quantum number' $m$ (with $-\ell \leq m \leq \ell$) does not enter the energy levels because of the spherical symmetry of the two-body interaction (in the center of mass frame). For instance, the non-relativistic Newton interaction, Eq. (2), gives rise to the well-known result

$$E_0(n, \ell) = -\frac{1}{2}\,\mu \left(\frac{GM\mu}{n\,\hbar}\right)^2, \tag{9}$$

which depends only on $n$ (this is the famous Coulomb degeneracy). When considering the PN corrections to $H_0$, as in Eq. (1), one gets a more complicated expression of the form

$$\begin{aligned} E^{\rm relative}_{\rm 3PN}(n, \ell) = -\frac{1}{2}\mu\frac{\alpha^2}{n^2}\biggl[1 &+ \frac{\alpha^2}{c^2}\left(\frac{c_{11}}{n\ell} + \frac{c_{20}}{n^2}\right) + \frac{\alpha^4}{c^4}\left(\frac{c_{13}}{n\ell^3} + \frac{c_{22}}{n^2\ell^2} + \frac{c_{31}}{n^3\ell} + \frac{c_{40}}{n^4}\right) \\ &+ \frac{\alpha^6}{c^6}\left(\frac{c_{15}}{n\ell^5} + \cdots + \frac{c_{60}}{n^6}\right)\biggr], \end{aligned} \tag{10}$$

where we have set $\alpha \equiv GM\mu/\hbar = G\,m_1\,m_2/\hbar$, and where we consider, for simplicity, the (quasi-classical) limit where $n$ and $\ell$ are large numbers. The 2PN-accurate version of Eq. (10) had been derived by Damour and Schäfer [72] as early as 1988 while its 3PN-accurate version was derived by Damour, Jaranowski and Schäfer in 1999 [70]. The dimensionless coefficients $c_{pq}$ are functions of the symmetric mass ratio $\nu \equiv \mu/M$, for instance $c_{40} = \frac{1}{8}(145 - 15\nu + \nu^2)$. In classical mechanics (*i.e.* for large $n$ and $\ell$), it is called the 'Delaunay Hamiltonian', *i.e.* the Hamiltonian expressed in terms of the *action variables*[3] $J = \ell\hbar = \frac{1}{2\pi}\oint p_\varphi\, d\varphi$, and $N = n\hbar = I_r + J$, with $I_r = \frac{1}{2\pi}\oint p_r\, dr$.

The energy levels (10) encode, in a *gauge-invariant* way, the 3PN-accurate relative dynamics of a 'real' binary. Let us now consider an auxiliary problem: the 'effective' dynamics of one body, of mass $\mu$, following (modulo the $Q$ term discussed below) a

[2] This is related to an idea emphasized many times by John Archibald Wheeler: quantum mechanics can often help us in going to the essence of classical mechanics.

[3] We consider, for simplicity, 'equatorial' motions with $m = \ell$, *i.e.*, classically, $\theta = \frac{\pi}{2}$.

geodesic in some $\nu$-dependent 'effective external' (spherically symmetric) metric[4]

$$g_{\mu\nu}^{\rm eff}\, dx^\mu\, dx^\nu = -A(R;\,\nu)\, c^2\, dT^2 + B(R;\,\nu)\, dR^2 + R^2(d\theta^2 + \sin^2\theta\, d\varphi^2)\,. \quad (11)$$

Here, the *a priori unknown* metric functions $A(R;\,\nu)$ and $B(R;\,\nu)$ will be constructed in the form of expansions in $GM/c^2R$:

$$A(R;\,\nu) = 1 + \widetilde{a}_1\, \frac{GM}{c^2R} + \widetilde{a}_2 \left(\frac{GM}{c^2R}\right)^2 + \widetilde{a}_3 \left(\frac{GM}{c^2R}\right)^3 + \widetilde{a}_4 \left(\frac{GM}{c^2R}\right)^4 + \cdots\,;$$
$$B(R;\,\nu) = 1 + \widetilde{b}_1\, \frac{GM}{c^2R} + \widetilde{b}_2 \left(\frac{GM}{c^2R}\right)^2 + b_3 \left(\frac{GM}{c^2R}\right)^3 + \cdots\,, \quad (12)$$

where the dimensionless coefficients $\widetilde{a}_n, \widetilde{b}_n$ depend on $\nu$. From the Newtonian limit, it is clear that we should set $\widetilde{a}_1 = -2$. In addition, as $\nu$ can be viewed as a deformation parameter away from the test-mass limit, we require that the effective metric (11) tend to the Schwarzschild metric (of mass $M$) as $\nu \to 0$, *i.e.* that

$$A(R;\,\nu = 0) = 1 - 2GM/c^2R = B^{-1}(R;\,\nu = 0)\,.$$

Let us now require that the dynamics of the "one body" $\mu$ within the effective metric $g_{\mu\nu}^{\rm eff}$ be described by an "effective" mass-shell condition of the form

$$g_{\rm eff}^{\mu\nu}\, p_\mu^{\rm eff}\, p_\nu^{\rm eff} + \mu^2\, c^2 + Q(p_\mu^{\rm eff}) = 0\,,$$

where $Q(p)$ is (at least) *quartic* in $p$. Then by solving (by separation of variables) the corresponding 'effective' Hamilton-Jacobi equation

$$g_{\rm eff}^{\mu\nu}\, \frac{\partial S_{\rm eff}}{\partial x^\mu}\, \frac{\partial S_{\rm eff}}{\partial x^\nu} + \mu^2 c^2 + Q\left(\frac{\partial S}{\partial x^\mu}\right) = 0\,,$$

$$S_{\rm eff} = -\mathcal{E}_{\rm eff}\, t + J_{\rm eff}\, \varphi + S_{\rm eff}(R)\,, \quad (13)$$

one can straightforwardly compute (in the quasi-classical, large quantum numbers limit) the effective Delaunay Hamiltonian $\mathcal{E}_{\rm eff}(N_{\rm eff}, J_{\rm eff})$, with $N_{\rm eff} = n_{\rm eff}\, \hbar$, $J_{\rm eff} = \ell_{\rm eff}\, \hbar$ (where $N_{\rm eff} = J_{\rm eff} + I_R^{\rm eff}$, with $I_R^{\rm eff} = \frac{1}{2\pi} \oint p_R^{\rm eff} dR$, $P_R^{\rm eff} = {\rm d}S_{\rm eff}(R)/dR$). This yields a result of the form

[4] It is convenient to write the 'effective metric' in Schwarzschild-like coordinates. Note that the effective radial coordinate $R$ differs from the two-body ADM-coordinate relative distance $R^{\rm ADM} = |\boldsymbol{q}|$. The transformation between the two coordinate systems has been determined in Refs. [57, 59].

$$\mathscr{E}_{\rm eff}(n_{\rm eff},\ell_{\rm eff}) = \mu c^2 - \frac{1}{2}\mu\frac{\alpha^2}{n_{\rm eff}^2}\Bigg[1+\frac{\alpha^2}{c^2}\left(\frac{c_{11}^{\rm eff}}{n_{\rm eff}\ell_{\rm eff}}+\frac{c_{20}^{\rm eff}}{n_{\rm eff}^2}\right)$$
$$+\frac{\alpha^4}{c^4}\left(\frac{c_{13}^{\rm eff}}{n_{\rm eff}\ell_{\rm eff}^3}+\frac{c_{22}^{\rm eff}}{n_{\rm eff}^2\ell_{\rm eff}^2}+\frac{c_{31}^{\rm eff}}{n_{\rm eff}^3\ell_{\rm eff}}+\frac{c_{40}^{\rm eff}}{n_{\rm eff}^4}\right)$$
$$+\frac{\alpha^6}{c^6}\left(\frac{c_{15}^{\rm eff}}{n_{\rm eff}\ell_{\rm eff}^5}+\dots+\frac{c_{60}^{\rm eff}}{n_{\rm eff}^6}\right)\Bigg], \qquad (14)$$

where the dimensionless coefficients $c_{pq}^{\rm eff}$ are now functions of the unknown coefficients $\widetilde{a}_n, \widetilde{b}_n$ entering the looked for 'external' metric coefficients (12).

At this stage, one needs to define a 'dictionary' between the real (relative) two-body dynamics, summarized in Eq. (10), and the effective one-body one, summarized in Eq. (14). As, on both sides, quantum mechanics tells us that the action variables are quantized in integers ($N_{\rm real} = n\hbar$, $N_{\rm eff} = n_{\rm eff}\hbar$, etc.), it is most natural to identify $n = n_{\rm eff}$ and $\ell = \ell_{\rm eff}$. One then still needs a rule for relating the two different energies $E_{\rm real}^{\rm relative}$ and $\mathscr{E}_{\rm eff}$. Buonanno and Damour [57] proposed to look for a general map between the real energy levels and the effective ones (which, as seen when comparing (10) and (14), cannot be directly identified because they do not include the same rest-mass contribution[5]), namely

$$\frac{\mathscr{E}_{\rm eff}}{\mu c^2} - 1 = f\left(\frac{E_{\rm real}^{\rm relative}}{\mu c^2}\right) = \frac{E_{\rm real}^{\rm relative}}{\mu c^2}\left(1+\alpha_1\frac{E_{\rm real}^{\rm relative}}{\mu c^2}+\alpha_2\left(\frac{E_{\rm real}^{\rm relative}}{\mu c^2}\right)^2\right.$$
$$\left.+\alpha_3\left(\frac{E_{\rm real}^{\rm relative}}{\mu c^2}\right)^3+\dots\right). \qquad (15)$$

The 'correspondence' between the real and effective energy levels is illustrated in Fig. 1.

Finally, identifying $\mathscr{E}_{\rm eff}(n,\ell)/\mu c^2$ to $1 + f(E_{\rm real}^{\rm relative}(n,\ell)/\mu c^2)$ yields a system of equations for determining the unknown EOB coefficients $\widetilde{a}_n, \widetilde{b}_n, \alpha_n$, as well as the three coefficients $z_1, z_2, z_3$ parametrizing a general 3PN-level quartic mass-shell deformation:

$$Q_{\rm 3PN}(p) = \frac{1}{c^6}\frac{1}{\mu^2}\left(\frac{GM}{R}\right)^2\left[z_1\,\boldsymbol{p}^4 + z_2\,\boldsymbol{p}^2(\boldsymbol{n}\cdot\boldsymbol{p})^2 + z_3(\boldsymbol{n}\cdot\boldsymbol{p})^4\right].$$

[The need for introducing a quartic mass-shell deformation $Q$ only arises at the 3PN level.]

The above system of equations for $\widetilde{a}_n, \widetilde{b}_n, \alpha_n$ (and $z_i$ at 3PN) was studied at the 2PN level in Ref. [57], and at the 3PN level in Ref. [59]. At the 2PN level it

[5] Indeed $E_{\rm real}^{\rm total} = Mc^2 + E_{\rm real}^{\rm relative} = Mc^2 + \text{Newtonian terms} + 1\text{PN}/c^2 + \cdots$, while $\mathscr{E}_{\rm effective} = \mu c^2 + N + 1\text{PN}/c^2 + \cdots$.

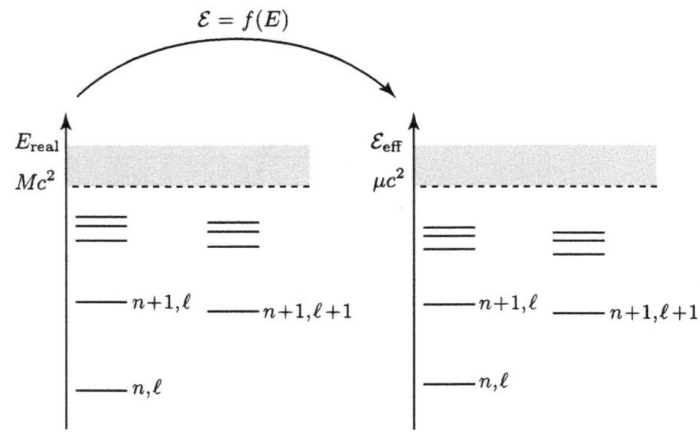


**Fig. 1** Sketch of the correspondence between the quantized energy levels of the real and effective conservative dynamics. $n$ denotes the 'principal quantum number' ($n = n_r + \ell + 1$, with $n_r = 0, 1, \ldots$ denoting the number of nodes in the radial function), while $\ell$ denotes the (relative) orbital angular momentum ($\boldsymbol{L}^2 = \ell(\ell+1)\,\hbar^2$). Though the EOB method is purely classical, it is conceptually useful to think in terms of the underlying (Bohr-Sommerfeld) quantization conditions of the action variables $I_R$ and $J$ to motivate the identification between $n$ and $\ell$ in the two dynamics

was found that, if one further imposes the natural condition $\widetilde{b}_1 = +2$ (so that the linearized effective metric coincides with the linearized Schwarzschild metric with mass $M = m_1 + m_2$), there exists a *unique* solution for the remaining five unknown coefficients $\widetilde{a}_2, \widetilde{a}_3, \widetilde{b}_2, \alpha_1$ and $\alpha_2$. This solution is very simple:

$$\widetilde{a}_2 = 0, \quad \widetilde{a}_3 = 2\nu, \quad \widetilde{b}_2 = 4 - 6\nu, \quad \alpha_1 = \frac{\nu}{2}, \quad \alpha_2 = 0\,. \tag{16}$$

At the 3PN level, it was found that the system of equations is consistent, and underdetermined in that the general solution can be parametrized by the arbitrary values of $z_1$ and $z_2$. It was then argued that it is natural to impose the simplifying requirements $z_1 = 0 = z_2$, so that $Q$ is proportional to the fourth power of the (effective) radial momentum $p_r$. With these conditions, the solution is unique at the 3PN level, and is still remarkably simple, namely

$$\widetilde{a}_4 = a_4\,\nu, \quad \widetilde{d}_3 = 2(3\nu - 26)\nu, \quad \alpha_3 = 0, \ z_3 = 2(4 - 3\nu)\nu.$$

Here, $a_4$ denotes the number

$$a_4 = \frac{94}{3} - \frac{41}{32}\pi^2 \simeq 18.6879027, \tag{17}$$

while $\widetilde{d}_3$ denotes the coefficient of $(GM/c^2R)^3$ in the PN expansion of the combined metric coefficient

$$D(R) \equiv A(R)\,B(R).$$

Replacing $B(R)$ by $D(R)$ is convenient because (as was mentioned above), in the test-mass limit $\nu \to 0$, the effective metric must reduce to the Schwarzschild metric, namely

$$A(R; \nu = 0) = B^{-1}(R; \nu = 0) = 1 - 2\left(\frac{GM}{c^2 R}\right),$$

so that

$$D(R; \nu = 0) = 1.$$

The final result is that the three EOB potentials $A, D, Q$ describing the 3PN two-body dynamics are given by the following very simple results. In terms of the EOB "gravitational potential"

$$u \equiv \frac{GM}{c^2 R},$$

$$A_{3PN}(R) = 1 - 2u + 2\,\nu\,u^3 + a_4\,\nu\,u^4, \tag{18}$$

$$D_{3PN}(R) \equiv (A(R)B(R))_{3PN} = 1 - 6\nu u^2 + 2(3\nu - 26)\nu u^3, \tag{19}$$

$$Q_{3PN}(\boldsymbol{q}, \boldsymbol{p}) = \frac{1}{c^2}\,2(4 - 3\nu)\nu\,u^2\,\frac{p_r^4}{\mu^2}. \tag{20}$$

In addition, the map between the (real) center-of-mass energy of the binary system $E_{\rm real}^{\rm relative} = H^{\rm relative} = \mathscr{E}_{\rm relative}^{\rm tot} - Mc^2$ and the effective one $\mathscr{E}_{\rm eff}$ is found to have the very simple (but non trivial) form

$$\frac{\mathscr{E}_{\rm eff}}{\mu c^2} = 1 + \frac{E_{\rm real}^{\rm relative}}{\mu c^2}\left(1 + \frac{\nu}{2}\,\frac{E_{\rm real}^{\rm relative}}{\mu c^2}\right) = \frac{s - m_1^2\,c^4 - m_2^2\,c^4}{2\,m_1\,m_2\,c^4}, \tag{21}$$

where $s = (\mathscr{E}_{\rm real}^{\rm tot})^2 \equiv (Mc^2 + E_{\rm real}^{\rm relative})^2$ is Mandelstam's invariant $s = -(p_1 + p_2)^2$.

It is truly remarkable that the EOB formalism succeeds in *condensing* the complicated, original 3PN Hamiltonian, Eqs. (4)–(2), into the very simple potentials $A, D$ and $Q$ displayed above, together with the simple energy map Eq. (21). For instance, at the 1PN level, the already somewhat involved Lorentz-Droste-Einstein-Infeld-Hoffmann 1PN dynamics (Eqs. (4) and (5)) is simply described, within the EOB formalism, as a test particle of mass $\mu$ moving in an external Schwarzschild background of mass $M = m_1 + m_2$, together with the (crucial but quite simple) energy transformation (21). [Indeed, the $\nu$-dependent corrections to $A$ and $D$ start only at the 2PN level.] At the 2PN level, the seven rather complicated $\nu$-dependent coefficients of $\widehat{H}_{2PN}(\boldsymbol{q}, \boldsymbol{p})$, Eq. (6), get condensed into the two very simple additional contributions $+\,2\nu u^3$ in $A(u)$, and $-\,6\nu u^2$ in $D(u)$. At the 3PN level, the eleven quite complicated $\nu$-dependent coefficients of $\widehat{H}_{3PN}$, Eq. (2), get condensed into only three simple contributions: $+a_4\nu u^4$ in $A(u)$, $+2(3\nu - 26)\nu u^3$ in $D(u)$, and $Q_{3PN}$ given by Eq. (20). This simplicity of the EOB results is not only due to the reformulation of the PN-expanded Hamiltonian into an effective dynamics. Indeed, the $A$-potential happens to be much simpler than it could a priori have been: (i) as already noted it is not modified at the 1PN level, while one would a priori expect to have found a 1PN potential $A_{1PN}(u) = 1 - 2u + \nu a_2 u^2$ with some non zero $a_2$; and (ii) there are striking cancellations taking place in the calculation of the 2PN and 3PN coefficients $\widetilde{a}_2(\nu)$ and $\widetilde{a}_3(\nu)$, which were a priori of the form $\widetilde{a}_2(\nu) = a_2\nu + a_2'\nu^2$, and $\widetilde{a}_3(\nu) = a_3\nu + a_3'\nu^2 + a_3''\nu^3$, but for which the $\nu$-nonlinear contributions $a_2'\nu^2$, $a_3'\nu^2$ and $a_3''\nu^3$ precisely cancelled out.

The fact that the 3PN coefficient $a_4$ in the crucial 'effective radial potential' $A_{3PN}(R)$, Eq. (18), is rather large and positive indicates that the $\nu$-dependent nonlinear gravitational effects lead, for comparable masses ($\nu \sim \frac{1}{4}$), to a last stable (circular) orbit (LSO) which has a higher frequency and a larger binding energy than what a naive scaling from the test-particle limit ($\nu \to 0$) would suggest. Actually, the PN-expanded form (18) of $A_{3PN}(R)$ does not seem to be a good representation of the (unknown) exact function $A_{EOB}(R)$ when the (Schwarzschild-like) relative coordinate $R$ becomes smaller than about $6GM/c^2$ (which is the radius of the LSO in the test-mass limit). By continuity with the test-mass case, one a priori expects that $A_{3PN}(R)$ always exhibits a simple zero defining an EOB "effective horizon" that is smoothly connected to the Schwarzschild event horizon at $R = 2GM/c^2$ when $\nu \to 0$. However, the large value of the $a_4$ coefficient does actually prevent $A_{3PN}$ to have this property when $\nu$ is too large, and in particular when $\nu = 1/4$. It was therefore suggested [59] to further resum[6] $A_{3PN}(R)$ by replacing it by a suitable Padé ($P$) approximant. For instance, the replacement of $A_{3PN}(R)$ by[7]

$$A_3^1(R) \equiv P_3^1[A_{3PN}(R)] = \frac{1 + n_1 u}{1 + d_1 u + d_2 u^2 + d_3 u^3} \tag{22}$$

ensures that the $\nu = \frac{1}{4}$ case is smoothly connected with the $\nu = 0$ limit.

The same kind of $\nu$-continuity argument, discussed so far for the $A$ function, needs to be applied also to the $D_{3PN}(R)$ function defined in Eq. (19). A straightforward way to ensure that the $D$ function stays positive when $R$ decreases (since it is $D = 1$ when $\nu \to 0$) is to replace $D_{3PN}(R)$ by $D_3^0(R) \equiv P_3^0\,[D_{3PN}(R)]$, where $P_3^0$ indicates the (0, 3) Padé approximant and explicitly reads

$$D_3^0(R) = \frac{1}{1 + 6\nu u^2 - 2(3\nu - 26)\nu u^3}. \tag{23}$$

[6] The PN-expanded EOB building blocks $A_{3PN}(R)$, $B_{3PN}(R)$, ... already represent a *resummation* of the PN dynamics in the sense that they have "condensed" the many terms of the original PN-expanded Hamiltonian within a very concise format. But one should not refrain to further resum the EOB building blocks themselves, if this is physically motivated.

[7] We recall that the coefficients $n_1$ and $(d_1, d_2, d_3)$ of the (1, 3) Padé approximant $P_3^1[A_{3PN}(u)]$ are determined by the condition that the first four terms of the Taylor expansion of $A_3^1$ in powers of $u = GM/(c^2 R)$ coincide with $A_{3PN}$.

## 3 EOB Description of Radiation Reaction and of the Emitted Waveform During Inspiral

In the previous Section we have described how the EOB method encodes the conservative part of the relative orbital dynamics into the dynamics of an 'effective' particle. Let us now briefly discuss how to complete the EOB dynamics by defining some *resummed* expressions describing radiation reaction effects, and the corresponding waveform emitted at infinity. One is interested in circularized binaries, which have lost their initial eccentricity under the influence of radiation reaction. For such systems, it is enough (in first approximation [58]; see, however, the recent results of Bini and Damour [73]) to include a radiation reaction force in the $P_\varphi$ equation of motion only. More precisely, we are using phase space variables $R$, $P_R$, $\varphi$, $P_\varphi$ associated to polar coordinates (in the equatorial plane $\theta = \frac{\pi}{2}$). Actually it is convenient to replace the radial momentum $P_R$ by the momentum conjugate to the 'tortoise' radial coordinate $R_* = \int dR(B/A)^{1/2}$, *i.e.* $P_{R_*} = (A/B)^{1/2} P_R$. The real EOB Hamiltonian is obtained by first solving Eq. (21) to get $H_{\rm real}^{\rm total} = \sqrt{s}$ in terms of $\mathcal{E}_{\rm eff}$, and then by solving the effective Hamilton-Jacobi equation to get $\mathcal{E}_{\rm eff}$ in terms of the effective phase space coordinates $\boldsymbol{q}_{\rm eff}$ and $\boldsymbol{p}_{\rm eff}$. The result is given by two nested square roots (we henceforth set $c = 1$):

$$\hat{H}_{\rm EOB}(r, p_{r_*}, \varphi) = \frac{H_{\rm EOB}^{\rm real}}{\mu} = \frac{1}{\nu}\sqrt{1 + 2\nu\,(\hat{H}_{\rm eff} - 1)}\,, \tag{24}$$

where

$$\hat{H}_{\rm eff} = \sqrt{p_{r_*}^2 + A(r)\left(1 + \frac{p_\varphi^2}{r^2} + z_3\,\frac{p_{r_*}^4}{r^2}\right)}\,, \tag{25}$$

with $z_3 = 2\nu\,(4 - 3\nu)$. Here, we are using suitably rescaled dimensionless (effective) variables: $r = R/GM$, $p_{r_*} = P_{R_*}/\mu$, $p_\varphi = P_\varphi/\mu\,GM$, as well as a rescaled time $t = T/GM$. This leads to equations of motion for $(r, \varphi, p_{r_*}, p_\varphi)$ of the form

$$\frac{d\varphi}{dt} = \frac{\partial\,\hat{H}_{\rm EOB}}{\partial\,p_\varphi} \equiv \Omega\,, \tag{26}$$

$$\frac{dr}{dt} = \left(\frac{A}{B}\right)^{1/2} \frac{\partial\,\hat{H}_{\rm EOB}}{\partial\,p_{r_*}}\,, \tag{27}$$

$$\frac{dp_\varphi}{dt} = \hat{\mathcal{F}}_\varphi\,, \tag{28}$$

$$\frac{dp_{r_*}}{dt} = -\left(\frac{A}{B}\right)^{1/2} \frac{\partial\,\hat{H}_{\rm EOB}}{\partial\,r}\,, \tag{29}$$

which explicitly read

$$\frac{d\varphi}{dt} = \frac{A p_\varphi}{\nu r^2 \hat{H} \hat{H}_{\rm eff}} \equiv \Omega\,, \qquad (30)$$

$$\frac{dr}{dt} = \left(\frac{A}{B}\right)^{1/2} \frac{1}{\nu \hat{H} \hat{H}_{\rm eff}} \left(p_{r_*} + z_3 \frac{2A}{r^2} p_{r_*}^3\right)\,, \qquad (31)$$

$$\frac{dp_\varphi}{dt} = \hat{\mathscr{F}}_\varphi\,, \qquad (32)$$

$$\frac{dp_{r_*}}{dt} = -\left(\frac{A}{B}\right)^{1/2} \frac{1}{2\nu \hat{H} \hat{H}_{\rm eff}} \left\{A' + \frac{p_\varphi^2}{r^2}\left(A' - \frac{2A}{r}\right) + z_3 \left(\frac{A'}{r^2} - \frac{2A}{r^3}\right) p_{r_*}^4\right\}\,, \qquad (33)$$

where $A' = dA/dr$. As explained above the EOB metric function $A(r)$ is defined by Padé resumming the Taylor-expanded result (12) obtained from the matching between the real and effective energy levels (as we were mentioning, one uses a similar Padé resumming for $D(r) \equiv A(r)\,B(r)$). One similarly needs to resum $\hat{\mathscr{F}}_\varphi$, *i.e.*, the $\varphi$ component of the radiation reaction which has been introduced on the r.h.s. of Eq. (28).

Several methods have been tried during the development of the EOB formalism to resum the radiation reaction $\hat{\mathscr{F}}_\varphi$ (starting from the high-order PN-expanded results that have been obtained in the literature; see references in the Introduction above). Here, we shall briefly explain the new, *parameter-free* resummation technique for the multipolar waveform (and thus for the energy flux) introduced in Ref. [68, 69] and perfected in [64]. To be precise, the new results discussed in Ref. [64] are twofold: on the one hand, that work generalized the $\ell = m = 2$ *resummed factorized waveform* of [68, 69] to higher multipoles by using the most accurate currently known PN-expanded results [33–35] as well as the higher PN terms which are known in the test-mass limit [39, 40]; on the other hand, it introduced a *further resummation procedure* which consists in considering a new theoretical quantity, denoted as $\rho_{\ell m}(x)$, which enters the $(\ell, m)$ waveform (together with other building blocks, see below) only through its $\ell$-th power: $h_{\ell m} \propto (\rho_{\ell m}(x))^\ell$. Here, and below, $x$ denotes the invariant PN-ordering parameter given during inspiral by $x \equiv (GM\Omega/c^3)^{2/3}$.

The main novelty introduced by Refs. [64, 68, 69] is to write the $(\ell, m)$ multipolar waveform emitted by a circular nonspinning compact binary as the *product* of several factors, namely

$$h_{\ell m}^{(\varepsilon)} = \frac{GM\nu}{c^2 R} n_{\ell m}^{(\varepsilon)} c_{l+\varepsilon}(\nu) x^{(\ell+\varepsilon)/2} Y^{\ell-\varepsilon,-m}\left(\frac{\pi}{2}, \Phi\right) \hat{S}_{\rm eff}^{(\varepsilon)} T_{\ell m} e^{{\rm i}\delta_{\ell m}} \rho_{\ell m}^\ell. \qquad (34)$$

Here $\varepsilon$ denotes the parity of $\ell + m$ ($\varepsilon = \pi(\ell + m)$), i.e. $\varepsilon = 0$ for "even-parity" (mass-generated) multipoles ($\ell + m$ even), and $\varepsilon = 1$ for "odd-parity" (current-generated) ones ($\ell + m$ odd); $n_{\ell m}^{(\varepsilon)}$ and $c_{l+\varepsilon}(\nu)$ are numerical coefficients; $\hat{S}_{\rm eff}^{(\varepsilon)}$ is a $\mu$-normalized effective source (whose definition comes from the EOB formalism); $T_{\ell m}$ is a resummed version [68, 69] of an infinite number of "leading logarithms" entering the *tail effects* [15, 27]; $\delta_{\ell m}$ is a supplementary phase (which corrects the phase effects not included in the *complex* tail factor $T_{\ell m}$), and, finally, $(\rho_{\ell m})^\ell$ denotes the $\ell$-th power of the quantity $\rho_{\ell m}$ which is the new building block introduced in [64]. Note that in previous papers [68, 69] the quantity $(\rho_{\ell m})^{\ell}$ was denoted as $f_{\ell m}$ and we will often use this notation below. Before introducing explicitly the various elements entering the waveform (34) it is convenient to decompose $h_{\ell m}$ as

$$h_{\ell m}^{(\varepsilon)} = h_{\ell m}^{(N,\varepsilon)} \hat{h}_{\ell m}^{(\varepsilon)}, \tag{35}$$

where $h_{\ell m}^{(N,\varepsilon)}$ is the Newtonian contribution (*i.e.* the product of the first five factors in Eq. (34)) and

$$\hat{h}_{\ell m}^{(\varepsilon)} \equiv \hat{S}_{\text{eff}}^{(\varepsilon)} T_{\ell m} e^{\mathrm{i}\delta_{\ell m}} f_{\ell m} \tag{36}$$

represents a resummed version of all the PN corrections. The PN correcting factor $\hat{h}_{\ell m}^{(\varepsilon)}$, as well as all its building blocks, has the structure $\hat{h}_{\ell m}^{(\varepsilon)} = 1 + \mathscr{O}(x)$.

The reader will find in Ref. [64] the definitions of the quantities entering the "Newtonian" waveform $h_{\ell m}^{(N,\varepsilon)}$, as well as the precise definition of the effective source factor $\hat{S}_{\text{eff}}^{(\varepsilon)}$, which constitutes the first factor in the PN-correcting factor $\hat{h}_{\ell m}^{(\varepsilon)}$. Let us only note here that the definition of $\hat{S}_{\text{eff}}^{(\varepsilon)}$ makes use of EOB-defined quantities. For instance, for even-parity waves ($\varepsilon = 0$) $\hat{S}_{\text{eff}}^{(0)}$ is defined as the $\mu$-scaled *effective* energy $\mathscr{E}_{\text{eff}}/\mu c^2$. [We use the "$J$-factorization" definition of $\hat{S}_{\text{eff}}^{(\varepsilon)}$ when $\varepsilon = 1$, *i.e.* for odd parity waves.]

The second building block in the factorized decomposition is the "tail factor" $T_{\ell m}$ (introduced in Refs. [68, 69]). As mentioned above, $T_{\ell m}$ is a resummed version of an infinite number of "leading logarithms" entering the transfer function between the near-zone multipolar wave and the far-zone one, due to *tail effects* linked to its propagation in a Schwarzschild background of mass $M_{\text{ADM}} = H_{\text{EOB}}^{\text{real}}$. Its explicit expression reads

$$T_{\ell m} = \frac{\Gamma(\ell + 1 - 2\mathrm{i}\hat{\hat{k}})}{\Gamma(\ell + 1)} e^{\pi \hat{\hat{k}}} e^{2\mathrm{i}\hat{\hat{k}} \log(2kr_0)}, \tag{37}$$

where $r_0 = 2GM/\sqrt{e}$ and $\hat{\hat{k}} \equiv G H_{\text{EOB}}^{\text{real}} m\Omega$ and $k \equiv m\Omega$. Note that $\hat{\hat{k}}$ differs from $k$ by a rescaling involving the *real* (rather than the *effective*) EOB Hamiltonian, computed at this stage along the sequence of circular orbits.

The tail factor $T_{\ell m}$ is a complex number which already takes into account some of the dephasing of the partial waves as they propagate out from the near zone to infinity. However, as the tail factor only takes into account the leading logarithms, one needs to correct it by a complementary dephasing term, $e^{\mathrm{i}\delta_{\ell m}}$, linked to subleading logarithms and other effects. This subleading phase correction can be computed as being the phase $\delta_{\ell m}$ of the complex ratio between the PN-expanded $\hat{h}_{\ell m}^{(\varepsilon)}$ and the above defined source and tail factors. In the comparable-mass case ($\nu \neq 0$), the 3PN $\delta_{22}$ phase correction to the leading quadrupolar wave was originally computed in Ref. [69] (see also Ref. [68] for the $\nu = 0$ limit). Full results for the subleading partial waves to the highest possible PN-accuracy by starting from the currently known 3PN-accurate $\nu$-dependent waveform [35] have been obtained in [64]. For higher-order test-mass ($\nu \to 0$) contributions, see [74, 75]. For extensions of the (non spinning) factorized waveform of [64], see [76–78].

The last factor in the multiplicative decomposition of the multipolar waveform can be computed as being the modulus $f_{\ell m}$ of the complex ratio between the PN-expanded $\hat{h}^{(\varepsilon)}_{\ell m}$ and the above defined source and tail factors. In the comparable mass case ($\nu \neq 0$), the $f_{22}$ modulus correction to the leading quadrupolar wave was computed in Ref. [69] (see also Ref. [68] for the $\nu = 0$ limit). For the subleading partial waves, Ref. [64] explicitly computed the other $f_{\ell m}$'s to the highest possible PN-accuracy by starting from the currently known 3PN-accurate $\nu$-dependent waveform [35]. In addition, as originally proposed in Ref. [69], to reach greater accuracy the $f_{\ell m}(x;\nu)$'s extracted from the 3PN-accurate $\nu \neq 0$ results are completed by adding higher order contributions coming from the $\nu = 0$ results [39, 40]. In the particular $f_{22}$ case discussed in [69], this amounted to adding 4PN and 5PN $\nu = 0$ terms. This "hybridization" procedure was then systematically pursued for all the other multipoles, using the 5.5PN accurate calculation of the multipolar decomposition of the gravitational wave energy flux of Refs. [39, 40].

The decomposition of the total PN-correction factor $\hat{h}^{(\varepsilon)}_{\ell m}$ into several factors is in itself a resummation procedure which already improves the convergence of the PN series one has to deal with: indeed, one can see that the coefficients entering increasing powers of $x$ in the PN expansion of the $f_{\ell m}$'s tend to be systematically smaller than the coefficients appearing in the usual PN expansion of $\hat{h}^{(\varepsilon)}_{\ell m}$. The reason for this is essentially twofold: (i) the factorization of $T_{\ell m}$ has absorbed powers of $m\pi$ which contributed to make large coefficients in $\hat{h}^{(\varepsilon)}_{\ell m}$, and (ii) the factorization of either $\hat{H}_{\rm eff}$ or $\hat{j}$ has (in the $\nu = 0$ case) removed the presence of an inverse square-root singularity located at $x = 1/3$ which caused the coefficient of $x^n$ in any PN-expanded quantity to grow as $3^n$ as $n \to \infty$.

To further improve the convergence of the waveform several resummations of the factor $f_{\ell m}(x) = 1 + c_1^{\ell m} x + c_2^{\ell m} x^2 + \cdots$ have been suggested. First, Refs. [68, 69] proposed to further resum the $f_{22}(x)$ function via a Padé (3,2) approximant, $P^3_2\{f_{22}(x;\nu)\}$, so as to improve its behavior in the strong-field-fast-motion regime. Such a resummation gave an excellent agreement with numerically computed waveforms, near the end of the inspiral and during the beginning of the plunge, for different mass ratios [68, 79, 80]. As we were mentioning above, a new route for resumming $f_{\ell m}$ was explored in Ref. [64]. It is based on replacing $f_{\ell m}$ by its $\ell$-th root, say

$$\rho_{\ell m}(x;\nu) = [f_{\ell m}(x;\nu)]^{1/\ell}. \tag{38}$$

The basic motivation for replacing $f_{\ell m}$ by $\rho_{\ell m}$ is the following: the leading "Newtonian-level" contribution to the waveform $h^{(\varepsilon)}_{\ell m}$ contains a factor $\omega^\ell r^\ell_{\rm harm} v^\varepsilon$, where $r_{\rm harm}$ is the harmonic radial coordinate used in the MPM formalism [12, 14]. When computing the PN expansion of this factor one has to insert the PN expansion of the (dimensionless) harmonic radial coordinate $r_{\rm harm}$, $r_{\rm harm} = x^{-1}(1 + c_1 x + \mathcal{O}(x^2))$, as a function of the gauge-independent frequency parameter $x$. The PN re-expansion of $[r_{\rm harm}(x)]^{\ell}$ then generates terms of the type $x^{-\ell}(1+\ell c_1 x+\cdots)$. This is one (though not the only one) of the origins of 1PN corrections in $h_{\ell m}$ and $f_{\ell m}$ whose coefficients grow linearly with $\ell$. The study of [64] has pointed out that these $\ell$-growing terms are problematic for the accuracy of the PN-expansions. The replacement of $f_{\ell m}$ by $\rho_{\ell m}$ is a cure for this problem.

Several studies, both in the test-mass limit, $\nu \to 0$ (see Fig. 1 in [64]) and in the comparable-mass case (see notably Fig. 4 in [65]), have shown that the resummed factorized (inspiral) EOB waveforms defined above provided remarkably accurate analytical approximations to the "exact" inspiral waveforms computed by numerical simulations. These resummed multipolar EOB waveforms are much closer (especially during late inspiral) to the exact ones than the standard PN-expanded waveforms given by Eq. (35) with a PN-correction factor of the usual "Taylor-expanded" form

$$\widehat{h}^{(\varepsilon)\rm PN}_{\ell m} = 1 + c_1^{\ell m} x + c_{3/2}^{\ell m} x^{3/2} + c_2^{\ell m} x^2 + \cdots$$

See Fig. 1 in [64], and slide 29 in my (June 2012) Prague presentation.

Finally, one uses the newly resummed multipolar waveforms (34) to define a resummation of the *radiation reaction force* $\mathscr{F}_{\varphi}$ as

$$\mathscr{F}_{\varphi} \equiv -\frac{1}{\Omega} F^{(\ell_{\max})}, \tag{39}$$

where the (instantaneous, circular) GW flux $F^{(\ell_{\max})}$ is defined as

$$F^{(\ell_{\max})} \equiv \frac{2}{16\pi G} \sum_{\ell=2}^{\ell_{\max}} \sum_{m=1}^{\ell} (m\Omega)^2 \, |R h_{\ell m}|^2 \, . \tag{40}$$

Summarizing: Eqs. (34) and (39), (40) define resummed EOB versions of the waveform $h_{\ell m}$, and of the radiation reaction $\hat{\mathscr{F}}_{\varphi}$, during inspiral. A crucial point is that these resummed expressions are *parameter-free*. Given some current approximation to the conservative EOB dynamics (*i.e.* some expressions for the $A$, $D$, $Q$ potentials), they *complete* the EOB formalism by giving explicit predictions for the radiation reaction (thereby completing the dynamics, see Eqs. (26)–(29)), and for the emitted inspiral waveform.

## 4 EOB Description of the Merger of Binary Black Holes and of the Ringdown of the Final Black Hole

Up to now we have reviewed how the EOB formalism, starting only from *analytical* information obtained from PN theory, and adding extra resummation requirements (both for the EOB conservative potentials $A$, Eq. (22), and $D$, Eq. (23), and for the waveform, Eq. (34), and its associated radiation reaction force, Eqs. (39), (40)) makes specific predictions, both for the motion and the radiation of binary black holes. The analytical calculations underlying such an EOB description are essentially based on skeletonizing the two black holes as two, sufficiently separated point masses, and therefore seem unable to describe the merger of the two black holes, and the subsequent ringdown of the final, single black hole formed during the merger. However, as early as 2000 [58], the EOB formalism went one step further and proposed a specific strategy for describing the *complete* waveform emitted during the entire coalescence process, covering inspiral, merger and ringdown. This EOB proposal is somewhat crude. However, the predictions it has made (years before NR simulations could accurately describe the late inspiral and merger of binary black holes) have been broadly confirmed by subsequent NR simulations. [See the Introduction for a list of EOB predictions.] The original EOB proposal (which was motivated partly by the closeness between the 2PN-accurate effective metric $g_{\mu\nu}^{\rm eff}$ [57] and the Schwarzschild metric, and by the results of Refs. [66] and [67]) consists of:

(i) defining, within EOB theory, the instant of (effective) "*merger*" of the two black holes as the (dynamical) EOB time $t_m$ where the orbital frequency $\Omega(t)$ reaches its *maximum*;

(ii) describing (for $t \leq t_m$) the inspiral-plus-plunge (or simply *insplunge*) waveform, $h^{\rm insplunge}(t)$, by using the inspiral EOB dynamics and waveform reviewed in the previous Section; and

(iii) describing (for $t \geq t_m$) the merger-plus-ringdown waveform as a superposition of several quasi-normal-mode (QNM) complex frequencies of a final Kerr black hole (of mass $M_f$ and spin parameter $a_f$, self-consistency estimated within the EOB formalism), say

$$\left(\frac{Rc^2}{GM}\right) h_{\ell m}^{\rm ringdown}(t) = \sum_N C_N^+ \, e^{-\sigma_N^+ (t-t_m)}, \tag{41}$$

with $\sigma_N^+ = \alpha_N + i\,\omega_N$, and where the label $N$ refers to indices $(\ell, \ell', m, n)$, with $(\ell, m)$ being the Schwarzschild-background multipolarity of the considered (metric) waveform $h_{\ell m}$, with $n = 0, 1, 2\ldots$ being the 'overtone number' of the considered Kerr-background Quasi-Normal-Mode, and $\ell'$ the degree of its associated spheroidal harmonics $S_{\ell' m}(a\sigma, \theta)$;

(iv) determining the excitation coefficients $C_N^+$ of the QNM's in Eq. (41) by using a simplified representation of the transition between plunge and ring-down obtained by smoothly *matching* (following Ref. [68]), on a $(2p+1)$-toothed "comb" $(t_m - p\delta, \ldots, t_m - \delta, t_m, t_m + \delta, \ldots, t_m + p\delta)$ centered around the merger (and matching) time $t_m$, the inspiral-plus-plunge waveform to the above ring-down waveform.

Finally, one defines a complete, quasi-analytical EOB waveform (covering the full process from inspiral to ring-down) as:

$$h_{\ell m}^{\rm EOB}(t) = \theta(t_m - t)\, h_{\ell m}^{\rm insplunge}(t) + \theta(t - t_m)\, h_{\ell m}^{\rm ringdown}(t)\,, \tag{42}$$

where $\theta(t)$ denotes Heaviside's step function. The final result is a waveform that essentially depends only on the choice of a resummed EOB $A(u)$ potential, and, less importantly, on the choice of resummation of the main waveform amplitude factor $f_{22} = (\rho_{22})^2$.

We have emphasized here that the EOB formalism is able, in principle, starting only from the best currently known analytical information, to predict the full waveform emitted by coalescing binary black holes. The early comparisons between 3PN-accurate EOB predicted waveforms[8] and NR-computed waveforms showed a satisfactory agreement between the two, within the (then relatively large) NR uncertainties [81, 82]. Moreover, as we shall discuss below, it has been recently shown that the currently known Padé-resummed 3PN-accurate $A(u)$ potential is able, as is, to describe with remarkable accuracy several aspects of the dynamics of coalescing binary black holes [83, 84].

On the other hand, when NR started delivering high-accuracy waveforms, it became clear that the 3PN-level analytical knowledge incorporated in EOB theory was not accurate enough for providing waveforms agreeing with NR ones within the high-accuracy needed for detection, and data analysis of upcoming GW signals. [See, *e.g.*, the discussion in Sect. II of Ref. [77].] At that point, one made use of the *natural flexibility* of the EOB formalism. Indeed, as already emphasized in early EOB work [60, 85], we know from the analytical point of view that there are (yet uncalculated) further terms in the $u$-expansions of the EOB potentials $A(u)$, $D(u)$, ... (and in the $x$-expansion of the waveform), so that these terms can be introduced either as "free parameter(s) in constructing a bank of templates, and [one should] wait until" GW observations determine their value(s) [60], or as "*fitting parameters* and adjusted so as to reproduce other information one has about the exact results" (to quote Ref. [85]). For instance, modulo logarithmic corrections that will be further discussed below, the Taylor expansion in powers of $u$ of the main EOB potential $A(u)$ reads

$$A^{\rm Taylor}(u;\nu) = 1 - 2u + \widetilde{a}_3(\nu)u^3 + \widetilde{a}_4(\nu)u^4 + \widetilde{a}_5(\nu)u^5 + \widetilde{a}_6(\nu)u^6 + \cdots\,,$$

where the 2PN and 3PN coefficients $\widetilde{a}_3(\nu) = 2\nu$ and $\widetilde{a}_4(\nu) = a_4\nu$ are known, but where the 4PN, 5PN, ... coefficients, $\widetilde{a}_5(\nu)$, $\widetilde{a}_6(\nu)$, ... have not yet been calculated (see, however, below). A first attempt was made in [85] to use numerical data (on circular orbits of corotating black holes) to fit for the value of a (single, effective) 4PN parameter of the simple form $\widetilde{a}_5(\nu) = a_5\nu$ entering a Padé-resummed 4PN-level $A$ potential, *i.e.*

[8] The new, resummed EOB waveform discussed above was not available at the time, so that these comparisons employed the coarser "Newtonian-level" EOB waveform $h_{22}^{(N,\varepsilon)}(x)$.

$$A_4^1(u; a_5, \nu) = P_4^1\left[A_{3\text{PN}}(u) + \nu a_5 u^5\right]. \tag{43}$$

This strategy was pursued in Refs. [69, 86] and many subsequent works. It was pointed out in Ref. [65] that the introduction of a further 5PN coefficient $\widetilde{a}_6(\nu) = a_6\nu$, entering a Padé-resummed 5PN-level $A$ potential, *i.e.*

$$A_5^1(u; a_5, a_6, \nu) = P_5^1\left[A_{3\text{PN}}(u) + \nu a_5 u^5 + \nu a_6 u^6\right], \tag{44}$$

helped in having a closer agreement with accurate NR waveforms.

In addition, Refs. [68, 69] introduced another type of flexibility parameters of the EOB formalism: the non quasi-circular (NQC) parameters accounting for uncalculated modifications of the quasi-circular inspiral waveform presented above, linked to deviations from an adiabatic quasi-circular motion. These NQC parameters are of various types, and subsequent works [65, 77, 79, 80, 87–89] have explored several ways of introducing them. They enter the EOB waveform in two separate ways. First, through an explicit, additional complex factor multiplying $h_{\ell m}$, *e.g.*

$$f_{\ell m}^{\text{NQC}} = (1 + a_1^{\ell m} n_1 + a_2^{\ell m} n_2)\exp[i(a_3^{\ell m} n_3 + a_4^{\ell m} n_4)],$$

where the $n_i$'s are dynamical functions that vanish in the quasi-circular limit (with $n_1, n_2$ being time-even, and $n_3, n_4$ time-odd). For instance, one usually takes $n_1 = (p_{r_*}/r\Omega)^2$. Second, through the (discrete) choice of the argument used during the plunge to replace the variable $x$ of the quasi-circular inspiral argument: *e.g.* either $x_\Omega \equiv (GM\Omega)^{2/3}$, or (following [90]) $x_\varphi \equiv v_\varphi^2 = (r_\omega\Omega)^2$, where $v_\varphi \equiv \Omega\, r_\omega$, and $r_\omega \equiv r[\psi(r, p_\varphi)]^{1/3}$ is a modified EOB radius, with $\psi$ being defined as

$$\psi(r, p_\varphi) = \frac{2}{r^2}\left(\frac{dA(r)}{dr}\right)^{-1}\left[1 + 2\nu\left(\sqrt{A(r)\left(1 + \frac{p_\varphi^2}{r^2}\right)} - 1\right)\right]. \tag{45}$$

For a given value of the symmetric mass ratio, and given values of the $A$-flexibility parameters $\widetilde{a}_5(\nu), \widetilde{a}_6(\nu)$, one can determine the values of the NQC parameters $a_i^{\ell m}$'s from accurate NR simulations of binary black hole coalescence (with mass ratio $\nu$) by imposing, say, that the complex EOB waveform $h_{\ell m}^{\text{EOB}}(t^{\text{EOB}}; \widetilde{a}_5, \widetilde{a}_6; a_i^{\ell m})$ *osculates* the corresponding NR one $h_{\ell m}^{\text{NR}}(t^{\text{NR}})$ at their respective instants of "merger", where $t_{\text{merger}}^{\text{EOB}} \equiv t_m^{\text{EOB}}$ was defined above (maximum of $\Omega^{\text{EOB}}(t)$), while $t_{\text{merger}}^{\text{NR}}$ is defined, say, as the (retarded) NR time where the modulus $|h_{22}^{\text{NR}}(t)|$ of the quadrupolar waveform reaches its maximum. The order of osculation that one requires between $h_{\ell m}^{\text{EOB}}(t)$ and $h_{\ell m}^{\text{NR}}(t)$ (or, separately, between their moduli and their phases or frequencies) depends on the number of NQC parameters $a_i^{\ell m}$. For instance, $a_1^{\ell m}$ and $a_2^{\ell m}$ affect only the modulus of $h_{\ell m}^{\text{EOB}}$ and allow one to match both $|h_{\ell m}^{\text{EOB}}|$ and its first time derivative, at merger, to their NR counterparts, while $a_3^{\ell m}, a_4^{\ell m}$ affect only the phase of the EOB waveform, and allow one to match the GW frequency $\omega_{\ell m}^{\text{EOB}}(t)$ and its first time derivative, at merger, to their NR counterparts. The above EOB/NR matching scheme has been developed and declined in various versions in Refs. [65, 77, 79, 80, 87–89, 91, 92]. One has also extracted the needed matching data from accurate NR simulations, and provided explicit, analytical $\nu$-dependent fitting formulas for them [65, 77, 92].

Having so "calibrated" the values of the NQC parameters by extracting non-perturbative information from a sample of NR simulations, one can then, for any choice of the $A$-flexibility parameters, compute a full EOB waveform (from early inspiral to late ringdown). The comparison of the latter NQC-completed EOB waveform to the results of NR simulations is discussed in the next Section.

## 5 EOB Versus NR

There have been several different types of comparison between EOB and NR. For instance, the early work [81] pioneered the comparison between a purely analytical EOB waveform (uncalibrated to any NR information) and a NR waveform, while the early work [93] compared the predictions for the final spin of a coalescing black hole binary made by EOB, *completed* by the knowledge of the energy and angular momentum lost during ringdown by an extreme mass ratio binary (computed by the test-mass NR code of [94]), to comparable-mass NR simulations [95]. Since then, many other EOB/NR comparisons have been performed, both in the comparable-mass case [65, 69, 79, 80, 82, 86, 87], and in the small-mass-ratio case [68, 88, 89, 96, 97]. Note in this respect that the numerical simulations of the GW emission by extreme mass-ratio binaries have provided (and still provide) a very useful "laboratory" for learning about the motion and radiation of binary systems, and their description within the EOB formalism.

Here we shall discuss only two recent examples of EOB/NR comparisons, which illustrate different facets of this comparison.

### *5.1 EOB[NR] Waveforms Versus NR Ones*

We explained above how one could complete the EOB formalism by calibrating some of the natural EOB flexibility parameters against NR data. First, for any given mass ratio $\nu$ and any given values of the $A$-flexibility parameters $\widetilde{a}_5(\nu)$, $\widetilde{a}_6(\nu)$, one can use NR data to uniquely determine the NQC flexibility parameters $a_i$'s. In other words, we have (for a given $\nu$)

$$a_i = a_i[\text{NR data}; a_5, a_6] \,,$$

where we defined $a_5$ and $a_6$ so that $\widetilde{a}_5(\nu) = a_5\nu, \widetilde{a}_6(\nu) = a_6\nu$. [We allow for some residual $\nu$-dependence in $a_5$ and $a_6$.] Inserting these values in the (analytical) EOB waveform then defines an NR-completed EOB waveform which still depends on the two unknown flexibility parameters $a_5$ and $a_6$.

In Ref. [65] the $(a_5, a_6)$-dependent predictions made by such a NR-completed EOB formalism were compared to the high-accuracy waveform from an equal-mass binary black hole ($\nu = 1/4$) computed by the Caltech-Cornell-CITA group [98], (and then made available on the web). It was found that there is a strong degeneracy between $a_5$ and $a_6$ in the sense that there is an excellent EOB-NR agreement for an extended region in the $(a_5, a_6)$-plane. More precisely, the phase difference between the EOB (metric) waveform and the Caltech-Cornell-CITA one, considered between GW frequencies $M\omega_{\rm L} = 0.047$ and $M\omega_{\rm R} = 0.31$ (i.e., the last 16 GW cycles before merger), stays smaller than 0.02 radians within a long and thin banana-like region in the $(a_5, a_6)$-plane. This "good region" approximately extends between the points $(a_5, a_6) = (0, -20)$ and $(a_5, a_6) = (-36, +520)$. As an example (which actually lies on the boundary of the "good region"), we shall consider here (following Ref. [99]) the specific values $a_5 = 0$, $a_6 = -20$ (to which correspond, when $\nu = 1/4$, $a_1 = -0.036347$, $a_2 = 1.2468$). [Damour and Nagar [65] did not make use of the NQC phase flexibility; *i.e.* it took $a_3 = a_4 = 0$. In addition, it used $n_2 = \ddot{r}/r\Omega^2$ and introduced a (real) modulus NQC factor $f_{\ell m}^{\rm NQC}$ only for the dominant quadrupolar wave $\ell = 2 = m$.] We henceforth use $M$ as time unit. This result relies on the proper comparison between NR and EOB time series, which is a delicate subject. In fact, to compare the NR and EOB phase time-series $\phi_{22}^{\rm NR}(t_{\rm NR})$ and $\phi_{22}^{\rm EOB}(t_{\rm EOB})$ one needs to shift, by additive constants, both one of the time variables, and one of the phases. In other words, we need to determine $\tau$ and $\alpha$ such that the "shifted" EOB quantities

$$t'_{\rm EOB} = t_{\rm EOB} + \tau \ , \quad \phi_{22}^{'\rm EOB} = \phi_{22}^{\rm EOB} + \alpha \tag{46}$$

"best fit" the NR ones. One convenient way to do so is first to "pinch" (*i.e.* constrain to vanish) the EOB/NR phase difference at two different instants (corresponding to two different frequencies $\omega_1$ and $\omega_2$). Having so related the EOB time and phase variables to the NR ones we can straigthforwardly compare the EOB time series to its NR correspondant. In particular, we can compute the (shifted) EOB–NR phase difference

$$\Delta^{\omega_1,\omega_2}\phi_{22}^{\rm EOBNR}(t_{\rm NR}) \equiv \phi_{22}^{'\rm EOB}(t^{'\rm EOB}) - \phi_{22}^{\rm NR}(t^{\rm NR}). \tag{47}$$

Figure 2 compares[9] (the real part of) the analytical EOB *metric* quadrupolar waveform $\Psi_{22}^{\rm EOB}/\nu$ to the corresponding (Caltech-Cornell-CITA) NR *metric* waveform $\Psi_{22}^{\rm NR}/\nu$. [Here, $\Psi_{22}$ denotes the Zerilli-normalized asymptotic quadrupolar waveform, *i.e.* $\Psi_{22} \equiv \widehat{R}h_{22}/\sqrt{24}$ with $\widehat{R} = Rc^2/GM$.] This NR metric waveform has been obtained by a double time-integration (following the procedure of Ref. [80]) from the original, publicly available, *curvature* waveform $\psi_4^{22}$ [98]. Such a curvature waveform has been extrapolated *both* in resolution and in extraction radius. The agreement between the analytical prediction and the NR result is striking, even around

[9] The two "pinching" frequencies used for this comparison are $M\omega_1 = 0.047$ and $M\omega_2 = 0.31$.

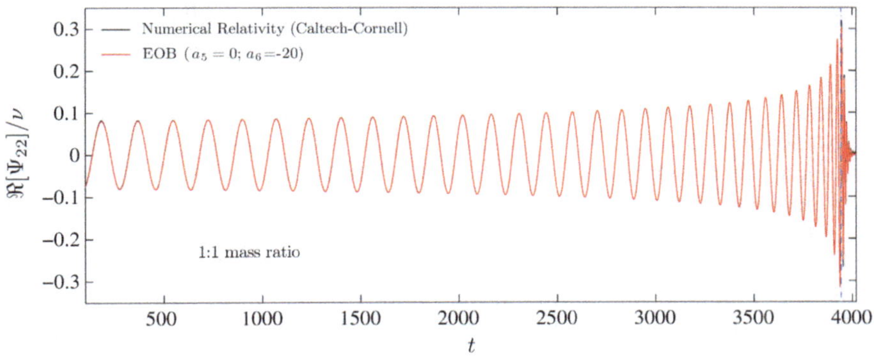


**Fig. 2** This figure illustrates the comparison (made in Refs. [65, 99]) between the (NR-completed) EOB waveform (Zerilli-normalized quadrupolar ($\ell = m = 2$) metric waveform (42) with parameter-free radiation reaction (39) and with $a_5 = 0$, $a_6 = -20$) and one of the most accurate numerical relativity waveform (equal-mass case) nowadays available [98]. The phase difference between the two is $\Delta\phi \leq \pm 0.01$ radians during the entire inspiral and plunge, which is at the level of the numerical error

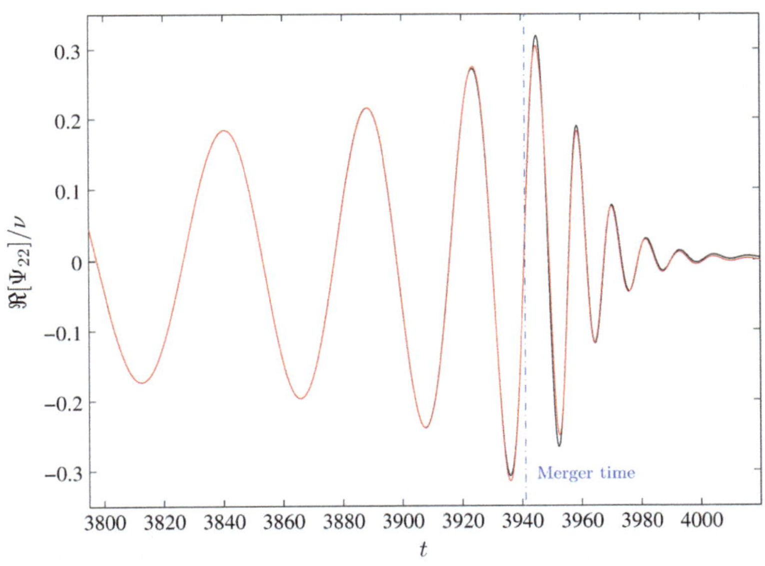


**Fig. 3** Close up around merger of the waveforms of Fig. 2. Note the excellent agreement between *both* modulus and phasing also during the ringdown phase

the merger. See Fig. 3 which closes up on the merger. The vertical line indicates the location of the EOB-merger time, i.e., the location of the maximum of the orbital frequency.

The phasing agreement between the waveforms is excellent over the full time span of the simulation (which covers 32 cycles of inspiral and about 6 cycles of ringdown), while the modulus agreement is excellent over the full span, apart from two cycles after merger where one can notice a difference. More precisely, the phase difference, $\Delta\phi = \phi^{\rm EOB}_{\rm metric} - \phi^{\rm NR}_{\rm metric}$, remains remarkably small ($\sim \pm 0.02$ radians) during the entire inspiral and plunge ($\omega_2 = 0.31$ being quite near the merger). By comparison, the root-sum of the various numerical errors on the phase (numerical truncation, outer boundary, extrapolation to infinity) is about 0.023 radians during the inspiral [98]. At the merger, and during the ringdown, $\Delta\phi$ takes somewhat larger values ($\sim \pm 0.1$ radians), but it oscillates around zero, so that, on average, it stays very well in phase with the NR waveform whose error rises to $\pm 0.05$ radians during ringdown. In addition, Ref. [65] compared the EOB waveform to accurate numerical relativity data (obtained by the Jena group [80]) on the coalescence of *unequal mass-ratio* black-hole binaries. Again, the agreement was good, and within the numerical error bars.

This type of high-accuracy comparison between NR waveforms and EOB[NR] ones (where EOB[NR] denotes a EOB formalism which has been completed by fitting some EOB-flexibility parameters to NR data) has been pursued and extended in Ref. [77]. The latter reference used the "improved" EOB formalism of Ref. [65] with some variations (*e.g.* a third modulus NQC coefficient $a_i$, two phase NQC coefficients, the argument $x_\Omega = (M\Omega)^{2/3}$ in $(\rho^{\rm Taylor}_{\ell m}(x))^\ell$, eight QNM modes) and calibrated it to NR simulations of mass ratios $q = m_2/m_1 = 1, 2, 3, 4$ and 6 performed by the Caltech-Cornell-CITA group [56, 100]. They considered not only the leading $(\ell, m) = (2, 2)$ GW mode, but the subleading ones $(2, 1)$, $(3, 3)$, $(4, 4)$ and $(5, 5)$. They found that, for this large range of mass ratios, EOB[NR] (with suitably fitted, $\nu$-dependent values of $a_5$ and $a_6$) was able to describe the NR waveforms essentially within the NR errors. This confirms the usefulness of the EOB formalism in helping the detection and analysis of upcoming GW signals.

Here, having in view GW observations from ground-based interferometric detectors, we focussed on comparable-mass systems. The EOB formalism has also been compared to NR results in the extreme mass-ratio limit $\nu \ll 1$. In particular, Ref. [88] found an excellent agreement between the analytical and numerical results.

## 5.2 EOB[3PN] Dynamics Versus NR One

Let us also mention other types of EOB/NR comparisons. Recently, two examples of EOB/NR comparisons have been performed directly at the level of the *dynamics* of a binary black hole, rather than at the level of the waveform. Moreover, contrary to the waveform comparisons of the previous subsection which involved an NR-completed EOB formalism ("EOB[NR]"), the dynamical comparisons we are going to discuss involve the purely analytical 3PN-accurate EOB formalism ("EOB[3PN]"), without any NR-based improvement.

First, Le Tiec et al. [83] have extracted from accurate NR simulations of slightly eccentric binary black-hole systems (for several mass ratios $q = m_1/m_2$ between

1/8 and 1) the function relating the periastron-advance parameter

$$K = 1 + \frac{\Delta\Phi}{2\pi},$$

(where $\Delta\Phi$ is the periastron advance per radial period) to the dimensionless averaged angular frequency $M\Omega_\varphi$ (with $M = m_1 + m_2$ as above). Then they compared the NR-estimate of the mass-ratio dependent functional relation

$$K = K(M\Omega_\varphi; \nu),$$

where $\nu = q/(1+q)^2$, to the predictions of various analytic approximation schemes: PN theory, EOB theory and two different ways of using GSF theory. Let us only mention here that the prediction from the purely analytical EOB[3PN] formalism for $K(M\Omega_\varphi; \nu)$ [101] agreed remarkably well (essentially within numerical errors) with its NR estimate for all mass ratios, while, by contrast, the PN-expanded prediction for $K(M\Omega_\varphi; \nu)$ [70] showed a much poorer agreement, especially as $q$ moved away from 1.

Second, Damour et al. [84] have recently extracted from accurate NR simulations of black-hole binaries (with mass ratios $q = m_2/m_1 = 1, 2$ and 3) the gauge-invariant relation between the (reduced) binding energy $E = (\mathscr{E}^{\text{tot}} - M)/\mu$ and the (reduced) angular momentum $j = J/(G\mu M)$ of the system. Then they compared the NR-estimate of the mass-ratio dependent functional relation

$$E = E(j; \nu)$$

to the predictions of various analytic approximation schemes: PN theory and various versions of EOB theory (some of these versions were NR-completed). Let us only mention here that the prediction from the purely analytical, 3PN-accurate EOB[3PN] for $E(j; \nu)$ agreed remarkably well with its NR estimate (for all mass ratios) essentially *down to the merger*. This is illustrated in Fig. 4 for the $q = 1$ case. By contrast, the 3PN expansion in (powers of $1/c^2$) of the function $E(j; \nu)$ showed a much poorer agreement (for all mass ratios).

# 6 Other Developments

## 6.1 EOB with Spinning Bodies

We lack space here for discussing the extension of the EOB formalism to binary systems made of spinning bodies. Let us start by mentioning that the spin-extension of the EOB formalism was initiated in Ref. [60], that the first EOB-based analytical calculation of a complete waveform from a spinning binary was performed in Ref. [61], and that the first attempt at calibrating a spinning EOB model to accu-

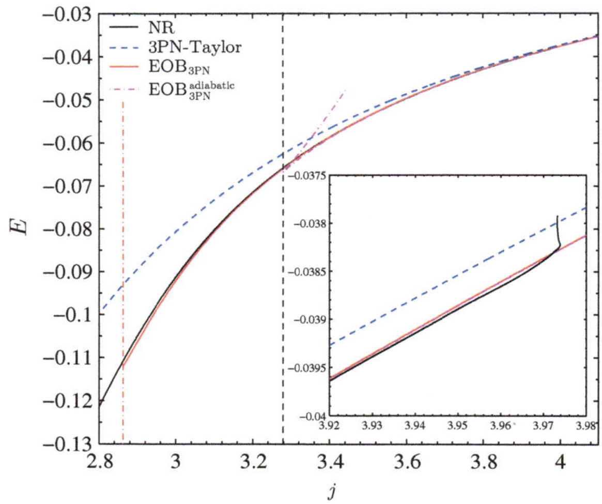


**Fig. 4** Comparison (made in [84]) between various analytical estimates of the energy-angular momentum functional relation and its numerical-relativity estimate (equal-mass case). The standard "Taylor-expanded" 3PN $E(j)$ curve shows the largest deviation from NR results, especially at low $j$'s, while the two (adiabatic and nonadiabatic) 3PN-accurate, *non-NR-calibrated* EOB $E(j)$ curves agree remarkably well with the NR one

rate NR simulations of spinning (non precessing) black-hole binaries was presented in [102]. In addition, several formal aspects related to the inclusion of spins in the EOB formalism have been discussed in Refs. [103–107] (see references within these papers for PN works dealing with spin effects) and a generalization of the factorized multipolar waveform of Ref. [64] to spinning, non-precessing binaries has been constructed in Refs. [76, 78].

## *6.2 EOB with Tidally Deformed Bodies*

In binary systems comprising *neutron stars*, rather than black holes, the tidal deformation of the neutron star(s) will significantly modify the phasing of the emitted gravitational waveform during the late inspiral. As GW's from binary neutron stars are expected sources for upcoming ground-based GW detectors, it is important to extend the EOB formalism by including tidal effects (see [108] and references therein). This extension has been defined in Refs. [109, 110]. The comparison between this tidal-extended EOB and state-of-the-art NR simulations of neutron-star binaries has been discussed in Refs. [111, 112]. It appears from these comparisons that the tidal-extended EOB formalism is able to describe the motion and radiation of neutron-star binaries within NR errors. More accurate simulations will be needed to ascertain whether one needs to calibrate some higher-order flexibility parameters of the tidal-EOB formalism, or whether the currently known analytic accuracy is sufficient.

## *6.3 EOB and GSF*

We mentioned in the Introduction that GSF theory has recently opened a new source of information on the general relativistic two-body problem. Let us briefly mention here that there has been, recently, a quite useful transfer of information from GSF theory to EOB theory. The program of using GSF theory to improve EOB theory was first highlighted in Ref. [101]. That work pointed to several concrete gauge-invariant calculations (within GSF theory) that would provide accurate information about the $O(\nu)$ contributions to several EOB potentials. More precisely, let us define the functions $a(u)$ and $\bar{d}(u)$ as the $\nu$-linear contributions to the EOB potentials $A(u;\nu)$ and $\overline{D}(u;\nu) \equiv D^{-1}(u;\nu)$:

$$A(u;\nu) = 1 - 2u + \nu\, a(u) + O(\nu^2)\,,$$

$$\overline{D}(u;\nu) = (AB)^{-1} = 1 + \nu\, \bar{d}(u) + O(\nu^2)\,.$$

Reference [101] has shown that a computation of the GSF-induced correction to the periastron advance of slightly eccentric orbits would allow one to compute the following combination of EOB functions

$$\bar{\rho}(u) = a(u) + u\, a'(u) + \frac{1}{2}\, u(1-2u)\, a''(u) + (1-6u)\, \bar{d}(u)\,.$$

The GSF-calculation of the EOB function $\bar{\rho}(u)$ was then performed in Ref. [113] (in the range $0 \le u \le \frac{1}{6}$).

More recently, a series of works by Le Tiec and collaborators [114–116] have (through an indirect route) shown how GSF calculations could be used to compute the EOB $\nu$-linear $a(u)$ function separately from the $\bar{d}(u)$ one. Reference [116] then gave a fitting formula for $a(u)$ over the interval $0 \le u \le \frac{1}{5}$ as well as accurate estimates of the coefficients of the Taylor expansion of $a(u)$ around $u = 0$ (corresponding to the knowledge of the PN expansion of $a(u)$ to a very high PN order). Very recently, Ackay et al. [117] succeeded in accurately computing (through GSF theory) the EOB $a(u)$ function over the larger interval $0 \le u \le \frac{1}{3}$. It was (surprisingly) found that $a(u)$ *diverges* like $a(u) \approx 0.25(1-3u)^{-1/2}$ at the light-ring limit $u \to \left(\frac{1}{3}\right)^{-}$. The meaning for EOB theory of this singular behavior of $a(u)$ at the light-ring is discussed in detail in Ref. [117].

## 6.4 Toward Further Improvements to EOB

Let us finally mention some avenues for further progress in EOB theory.

Logarithmic contributions to the $A(u)$ and $\overline{D}(u)$ functions have been recently computed at the 4PN level [101, 118] and even the 5PN one [116, 119].[10] They have been incorporated in a recent, improved implementation of the EOB formalism [92].

Two groups have embarked on a calculation of the (full) conservative dynamics at the 4PN level [120, 121]. If they succeed, it will be important to translate their *gauge-dependent* results in the *gauge-invariant* form used in EOB theory. [Remember that EOB theory is essentially based on the gauge-invariant Delaunay Hamiltonian $H(I_a)$.]

More generally, let us emphasize that the EOB formalism provides a convenient, gauge-invariant way of packaging both the conservative dynamics and the multipolar waveform. This EOB packaging has often turned out to be very *economical*. We recommend that authors computing high-order PN corrections to either the dynamics or the waveform reexpress their results in terms of the EOB building blocks.

For instance, Jaranowski and Schaëfer [121] have recently given a partial result at 4PN, expressed in terms of the (gauge-invariant) function $E(M\Omega_\varphi; \nu)$. In terms of this function, the 4PN contribution is a polynomial of the *fourth degree* in $\nu$, namely, with $x \equiv (M\Omega_\varphi)^{2/3}$ and

$$E(x;\nu) = -\frac{1}{2}\mu c^2 x(1 + e_{1PN}(\nu)x + e_{2PN}(\nu)x^2 + e_{3PN}(\nu)x^3 + e_{4PN}(\nu; \ln x)x^4 + O(x^5 \ln x)),$$

they found

$$e_{4PN}(\nu; \ln x) = -\frac{3969}{128} + c_1\nu + c_2\nu^2 + \frac{301}{1728}\nu^3 + \frac{77}{31104}\nu^4 + \frac{448}{15}\nu \ln x, \quad (48)$$

where they could not compute the values of the coefficients $c_1$ and $c_2$ of the terms linear and quadratic in $\nu$, but only the contributions cubic and quartic in $\nu$. We wish to point out that their result is re-expressed in a more economical (and more informative) way in terms of the basic EOB potential $A(u;\nu)$. Indeed, in terms of the PN expansion, of $A(u;\nu)$,

$$A^{\text{Taylor}}(u;\nu) = 1 - 2u + \tilde{a}_3(\nu)u^3 + \tilde{a}_4(\nu)u^4 + \tilde{a}_5(\nu; \ln u)u^5 + \tilde{a}_6(\nu; \ln u)u^6 + \cdots$$

the information contained in the above result can be entirely re-expressed in terms of the 4PN-level coefficient $\tilde{a}_5(\nu; \ln u)$. When doing this re-expression, one then finds that the information content of Eq. (48) is that the 4PN-level EOB coefficient $\tilde{a}_5(\nu; \ln u)$ is *no more than quadratic* in $\nu$, i.e.

[10] See Ref. [113], which quoted and used some combinations of the logarithmic contributions to $a(u)$ and $\bar{d}(u)$. given in Ref. [119].

$$\widetilde{a}_5(\nu; \ln u) = (a_5 + \frac{64}{5} \ln u)\nu + a_5' \nu^2 ,$$

*without* contributions of degree $\nu^3$ and $\nu^4$. We recall that similar cancellations of higher $\nu^n$ terms were found at lower PN orders in the EOB $A(u; \nu)$ function. Namely, they were found to contain only terms *linear* in $\nu$, while $\widetilde{a}_3(\nu)$ could a priori have been quadratic in $\nu$, and $\widetilde{a}_4(\nu)$ could a priori have been cubic in $\nu$. The fact that similar remarkable cancellations still hold, according to the result of [121], at the 4PN level, is a clear indication that the EOB packaging of information of the dynamics in the $A(u; nu)$ potential is quite compact. Indeed, it says that the two complicated terms $\frac{301}{1728}\nu^3 + \frac{77}{31104}\nu^4$ in the energy function are already encoded in the structure of the EOB formalism. Finally, note that the full gauge-invariant content of a 4PN computation of the dynamics, when interpreted within the EOB formalism, is described by only three EOB terms: the coefficient $\widetilde{a}_5(\nu; \ln u)$ in $A(u; \nu)$, an analogous coefficient $\tilde{\bar{d}}_4(\nu; \ln u)$ in $\bar{D}(u; \nu)$, and an additional contribution to $Q(p)$.

Regarding the waveform, let us mention another recent example where it would have been useful and clarifying to use the EOB packaging. Namely, when re-expressing it in terms of the factorized EOB waveform, the new content of the recent 3.5PN level computation by Faye et al. [122] of the PN-expanded quadrupolar waveform $h_{22}$, is entirely contained in an additional 3.5PN-level contribution to the supplementary phase, namely $\delta_{22} = (30995/1134\,\nu + 962/135\,\nu^2)\,x^{7/2}$. Indeed, the 3.5PN-level contributions to the modulus computed in [122] were already included in the factorized EOB waveform of Ref. [65].

## 7 Conclusions

We hope that this brief review has made it clear that:

1. There is a *complementarity* between the various current approaches to the general relativistic two-body problem: post-Newtonian, Effective One Body, gravitational self-force and numerical relativity simulations (of both comparable-mass and extreme-mass-ratio systems).
2. The effective one body formalism offers a convenient framework for combining, in a synergetic manner, information coming from the other approaches. This formalism seems to constitute an efficient way to analytically describe the motion and radiation of circularized[11] binaries, and to provide accurate gravitational wave templates for detection and data analysis.
3. The general relativistic two-body problem is more lively than ever. It provides an example of Poincaré's sentence: "Il n'y a pas de problèmes résolus, il y a seulement des problèmes plus ou moins résolus." ["There are no (definitely) solved problems, there are only more-or-less solved problems."]

[11] See [73] for a recent extension of the EOB formalism to non-circular (ellipticlike or hyperboliclike) motions.

# References


1. Damour, T.: The problem of motion in Newtonian and Einsteinian gravity. In: Hawking, S.W., Israel, W. (eds.) Three Hundred Years of Gravitation, pp. 128–198. Cambridge University Press, Cambridge (1987)
2. Damour, T.: Gravitational radiation and the motion of compact bodies. In: Deruelle, N., Piran, T. (eds.) Gravitational Radiation, pp. 59–144. North-Holland, Amsterdam (1983)
3. Jaranowski, P., Schäfer, G.: Third post-Newtonian higher order ADM Hamilton dynamics for two-body point-mass systems. Phys. Rev. D **57**, 7224 (1998). doi:10.1103/PhysRevD.57.7274 [Erratum: ibid. **63**, 029902 (2001)]
4. Blanchet, L., Faye, G.: General relativistic dynamics of compact binaries at the third post-Newtonian order. Phys. Rev. D **63**, 062005 (2001). doi:10.1103/PhysRevD.63.062005
5. Damour, T., Jaranowski, P., Schäfer, G.: Dimensional regularization of the gravitational interaction of point masses. Phys. Lett. B **513**, 147 (2001)
6. Blanchet, L., Damour, T., Esposito-Farèse, G.: Dimensional regularization of the third post-Newtonian dynamics of point particles in harmonic coordinates. Phys. Rev. D **69**, 124007 (2004). doi:10.1103/PhysRevD.69.124007
7. Itoh, Y., Futamase, T.: New derivation of a third post-Newtonian equation of motion for relativistic compact binaries without ambiguity. Phys. Rev. D **68**, 121501 (2003). doi:10.1103/PhysRevD.68.121501
8. Pati, M.E., Will, C.M.: Post-Newtonian gravitational radiation and equations of motion via direct integration of the relaxed Einstein equations. II: Two-body equations of motion to second post-Newtonian order, and radiation-reaction to 3.5 post-Newton. Phys. Rev. D **65**, 104008 (2002). doi:10.1103/PhysRevD.65.104008
9. Konigsdorffer, C., Faye, G., Schäfer, G.: The binary black-hole dynamics at the third-and-a-half post-Newtonian order in the ADM-formalism. Phys. Rev. D **68**, 044004 (2003). doi:10.1103/PhysRevD.68.044004
10. Nissanke, S., Blanchet, L.: Gravitational radiation reaction in the equations of motion of compact binaries to 3.5 post-Newtonian order. Class. Quantum Gravity **22**, 1007 (2005). doi:10.1088/0264-9381/22/6/008
11. Blanchet, L., Damour, T.: Radiative gravitational fields in general relativity I. General structure of the field outside the source. Philos. Trans. R. Soc. Lond. Ser. A **320**, 379 (1986)
12. Blanchet, L., Damour, T.: Post-Newtonian generation of gravitational waves. Ann. Inst. Henri Poincare A **50**, 377 (1989)
13. Damour, T., Iyer, B.R.: Multipole analysis for electromagnetism and linearized gravity with irreducible cartesian tensors. Phys. Rev. D **43**, 3259 (1991). doi:10.1103/PhysRevD.43.3259
14. T. Damour, B.R. Iyer, Post-Newtonian generation of gravitational waves. II. The spin moments. Ann. Inst. Henri Poincare A **54**, 115 (1991)
15. Blanchet, L., Damour, T.: Hereditary effects in gravitational radiation. Phys. Rev. D **46**, 4304 (1992). doi:10.1103/PhysRevD.46.4304
16. Blanchet, L.: Second-post-Newtonian generation of gravitational radiation. Phys. Rev. D **51**, 2559 (1995). doi:10.1103/PhysRevD.51.2559
17. Will, C.M., Wiseman, A.G.: Gravitational radiation from compact binary systems: gravitational waveforms and energy loss to second post-Newtonian order. Phys. Rev. D **54**, 4813 (1996). doi:10.1103/PhysRevD.54.4813
18. Will, C.M.: Generation of post-Newtonian gravitational radiation via direct integration of the relaxed Einstein equations. Prog. Theor. Phys. Suppl. **136**, 158 (1999). doi:10.1143/PTPS.136.158
19. Pati, M.E., Will, C.M.: Post-Newtonian gravitational radiation and equations of motion via direct integration of the relaxed Einstein equations. I: Foundations. Phys. Rev. D **62**, 124015 (2000). doi:10.1103/PhysRevD.62.124015
20. Epstein, R., Wagoner, R.V.: Post-Newtonian generation of gravitational waves. Astrophys. J. **197**, 717 (1975). doi:10.1086/153561

21. Thorne, K.S.: Multipole expansions of gravitational radiation. Rev. Mod. Phys. **52**, 299 (1980). doi:10.1103/RevModPhys.52.299
22. Wagoner, R.V., Will, C.M.: Post-Newtonian gravitational radiation from orbiting point masses. Astrophys. J. **210**, 764 (1976). doi:10.1086/154886
23. Wiseman, A.G.: Coalescing binary systems of compact objects to (post)$^{\frac{5}{2}}$-Newtonian order. IV. The gravitational wave tail. Phys. Rev. D **48**, 4757 (1993). doi:10.1103/PhysRevD.48.4757
24. Blanchet, L., Schäfer, G.: Gravitational wave tails and binary star systems. Class. Quantum Gravity **10**, 2699 (1993). doi:10.1088/0264-9381/10/12/026
25. Blanchet, L., Damour, T., Iyer, B.R., Will, C.M., Wiseman, A.G.: Gravitational-radiation damping of compact binary systems to second post-Newtonian order. Phys. Rev. Lett. **74**, 3515 (1995). doi:10.1103/PhysRevLett.74.3515
26. Blanchet, L., Damour, T., Iyer, B.R.: Gravitational waves from inspiralling compact binaries: energy loss and waveform to second-post-Newtonian order. Phys. Rev. D **51**, 5360 (1995). doi:10.1103/PhysRevD.51.5360 [Erratum: ibid. **54**, 1860 (1996)]
27. Blanchet, L.: Gravitational-wave tails of tails, Class. Quantum Gravity **15**, 113 (1998). [Erratum ibid. **22**, 3381 (2005)]
28. Blanchet, L., Iyer, B.R., Joguet, B.: Gravitational waves from inspiralling compact binaries: Energy flux to third post-Newtonian order. Phys. Rev. D **65**, 064005 (2002). [Erratum: ibid. **71**, 129903 (2005)]
29. Blanchet, L., Iyer, B.R.: Hadamard regularization of the third post-Newtonian gravitational wave generation of two point masses. Phys. Rev. D **71**, 024004 (2005)
30. Blanchet, L., Damour, T., Esposito-Farèse, G., Iyer, B.R.: Gravitational radiation from inspiralling compact binaries completed at the third post-Newtonian order. Phys. Rev. Lett. **93**, 091101 (2004). doi:10.1103/PhysRevLett.93.091101
31. Blanchet, L., Damour, T., Esposito-Farèse, G., Iyer, B.R.: Dimensional regularization of the third post-Newtonian gravitational wave generation from two point masses. Phys. Rev. D **71**, 124004 (2005)
32. Blanchet, L.: Quadrupole-quadrupole gravitational waves. Class. Quantum Gravity **15**, 89 (1998). doi:10.1088/0264-9381/15/1/008
33. Berti, E., Cardoso, V., Gonzalez, J.A., et al.: Inspiral, merger and ringdown of unequal mass black hole binaries: a multipolar analysis. Phys. Rev. D **76**, 064034 (2007). doi:10.1103/PhysRevD.76.064034
34. Kidder, L.E.: Using full information when computing modes of post-Newtonian waveforms from inspiralling compact binaries in circular orbit. Phys. Rev. D **77**, 044016 (2008). doi:10.1103/PhysRevD.77.044016
35. Blanchet, L., Faye, G., Iyer, B.R., Sinha, S.: The third post-Newtonian gravitational wave polarisations and associated spherical harmonic modes for inspiralling compact binaries in quasi-circular orbits. Class. Quantum Gravity **25**, 165003 (2008). doi:10.1088/0264-9381/25/16/165003
36. Blanchet, L.: Gravitational radiation from post-Newtonian sources and inspiralling compact binaries. Living Rev. Relativ. **5**, lrr-2002-3 (2002). http://www.livingreviews.org/lrr-2002-3
37. Poisson, E.: Gravitational radiation from a particle in circular orbit around a black hole. I. Analytic results for the nonrotating case. Phys. Rev. D **47**, 1497 (1993)
38. Sasaki, M.: Post-Newtonian expansion of the ingoing-wave Regge-Wheeler function. Progress Theoret. Phys. **92**, 17 (1994). doi:10.1143/PTP.92.17
39. Tagoshi, H., Sasaki, M.: Post-Newtonian expansion of gravitational waves from a particle in circular orbit around a Schwarzschild black hole. Progress Theoret. Phys. **92**, 745 (1994). doi:10.1143/PTP.92.745
40. Tanaka, T., Tagoshi, H., Sasaki, M.: Gravitational waves by a particle in circular orbit around a Schwarzschild black hole. Progress Theoret. Phys. **96**, 1087 (1996). doi:10.1143/PTP.96.1087
41. Sasaki, M., Tagoshi, H.: Analytic black hole perturbation approach to gravitational radiation. Living Rev. Relativ. **6**, lrr-2003-6 (2003). http://www.livingreviews.org/lrr-2003-6

42. Foffa, S., Sturani, R.: Effective field theory calculation of conservative binary dynamics at third post-Newtonian order. Phys. Rev. D **84**, 044031 (2011). doi:10.1103/PhysRevD.84.044031
43. Goldberger, W.D., Rothstein, I.Z.: An effective field theory of gravity for extended objects. Phys. Rev. D **73**, 104029 (2006). doi:10.1103/PhysRevD.73.104029
44. Damour, T., Esposito-Farèse, G.: Tensor multiscalar theories of gravitation. Class. Quantum Gravity **9**, 2093 (1992). doi:10.1088/0264-9381/9/9/015
45. Damour, T., Esposito-Farèse, G.: Testing gravity to second post-Newtonian order: a field theory approach. Phys. Rev. D **53**, 5541 (1996). doi:10.1103/PhysRevD.53.5541
46. Damour, T., Esposito-Farèse, G.: Gravitational-wave versus binary-pulsar tests of strong-field gravity. Phys. Rev. D **58**, 042001 (1998). doi:10.1103/PhysRevD.58.042001
47. Goldberger, W.D., Ross, A.: Gravitational radiative corrections from effective field theory. Phys. Rev. D **81**, 124015 (2010). doi:10.1103/PhysRevD.81.124015
48. Pretorius, F.: Evolution of binary black hole spacetimes. Phys. Rev. Lett. **95**, 121101 (2005). doi:10.1103/PhysRevLett.95.121101
49. Campanelli, M., Lousto, C.O., Marronetti, P., Zlochower, Y.: Accurate evolutions of orbiting black-hole binaries without excision. Phys. Rev. Lett. **96**, 111101 (2006). doi:10.1103/PhysRevLett.96.111101
50. Baker, J.G., Centrella, J., Choi, D.I., Koppitz, M., van Meter, J.: Gravitational wave extraction from an inspiraling configuration of merging black holes. Phys. Rev. Lett. **96**, 111102 (2006). doi:10.1103/PhysRevLett.96.111102
51. Boyle, M., Brown, D.A., Kidder, L.E., et al.: High-accuracy comparison of numerical relativity simulations with post-Newtonian expansions. Phys. Rev. D **76**, 124038 (2007). doi:10.1103/PhysRevD.76.124038
52. Pretorius, F.: Binary black hole coalescence. In: Colpi, M., et al. (eds.) Physics of Relativistic Objects in Compact Binaries: from Birth to Coalescense, Astrophysics and Space Science Library, vol. 359, pp. 305–370. Springer/Canopus, Dordrecht (2009)
53. Barack, L.: Gravitational self force in extreme mass-ratio inspirals. Class. Quantum Gravity **26**, 213001 (2009). doi:10.1088/0264-9381/26/21/213001
54. Hannam, M., Husa, S., Sperhake, U., Bruegmann, B., Gonzalez, J.A.: Where post-Newtonian and numerical-relativity waveforms meet. Phys. Rev. D **77**, 044020 (2008)
55. Hannam, M., Husa, S., Bruegmann, B., Gopakumar, A.: Comparison between numerical-relativity and post-Newtonian waveforms from spinning binaries: the orbital hang-up case. Phys. Rev. D **78**, 104007 (2008)
56. MacDonald, I., Mroué, A.H., Pfeiffer, H.P., et al.: Suitability of hybrid gravitational waveforms for unequal-mass binaries. Phys. Rev. D **87**, 024009 (2013). doi:10.1103/PhysRevD.87.024009
57. Buonanno, A., Damour, T.: Effective one-body approach to general relativistic two-body dynamics. Phys. Rev. D **59**, 084006 (1999). doi:10.1103/PhysRevD.59.084006
58. Buonanno, A., Damour, T.: Transition from inspiral to plunge in binary black hole coalescences. Phys. Rev. D **62**, 064015 (2000). doi:10.1103/PhysRevD.62.064015
59. Damour, T., Jaranowski, P., Schäfer, G.: On the determination of the last stable orbit for circular general relativistic binaries at the third post-Newtonian approximation. Phys. Rev. D **62**, 084011 (2000)
60. Damour, T.: Coalescence of two spinning black holes: an effective one-body approach. Phys. Rev. D **64**, 124013 (2001). doi:10.1103/PhysRevD.64.124013
61. Buonanno, A., Chen, Y., Damour, T.: Transition from inspiral to plunge in precessing binaries of spinning black holes. Phys. Rev. D **74**, 104005 (2006). doi:10.1103/PhysRevD.74.104005
62. Brézin, E., Itzykson, C., Zinn-Justin, J.: Relativistic balmer formula including recoil effects. Phys. Rev. D **1**, 2349 (1970). doi:10.1103/PhysRevD.1.2349
63. Damour, T., Iyer, B.R., Sathyaprakash, B.S.: Improved filters for gravitational waves from inspiralling compact binaries. Phys. Rev. D **57**, 885 (1998). doi:10.1103/PhysRevD.57.885
64. Damour, T., Iyer, B.R., Nagar, A.: Improved resummation of post-Newtonian multipolar waveforms from circularized compact binaries. Phys. Rev. D **79**, 064004 (2009). doi:10.1103/PhysRevD.79.064004

65. Damour, T., Nagar, A.: An improved analytical description of inspiralling and coalescing black-hole binaries. Phys. Rev. D **79**, 081503 (2009). doi:10.1103/PhysRevD.79.081503
66. Davis, M., Ruffini, R., Tiomno, J.: Pulses of gravitational radiation of a particle falling radially into a Schwarzschild black hole. Phys. Rev. D **5**, 2932 (1972). doi:10.1103/PhysRevD.5.2932
67. Price, R.H., Pullin, J.: Colliding black holes: the close limit. Phys. Rev. Lett. **72**, 3297 (1994). doi:10.1103/PhysRevLett.72.3297
68. Damour, T., Nagar, A.: Faithful effective-one-body waveforms of small-mass-ratio coalescing black-hole binaries. Phys. Rev. D **76**, 064028 (2007)
69. Damour, T., Nagar, A.: Comparing effective-one-body gravitational waveforms to accurate numerical data. Phys. Rev. D **77**, 024043 (2008)
70. Damour, T., Jaranowski, P., Schäfer, G.: Dynamical invariants for general relativistic two-body systems at the third post-Newtonian approximation. Phys. Rev. D **62**, 044024 (2000). doi:10.1103/PhysRevD.62.044024
71. Damour, T., Jaranowski, P., Schäfer, G.: Poincaré invariance in the ADM Hamiltonian approach to the general relativistic two-body problem. Phys. Rev. D **62**, 021501 (2000). doi:10.1103/PhysRevD.62.021501 [Erratum: ibid. D **63**, 029903 (2000)]
72. Damour, T., Schäfer, G.: Higher order relativistic periastron advances and binary pulsars. Nuovo Cimento B **101**, 127 (1988). doi:10.1007/BF02828697
73. Bini, D., Damour, T.: Gravitational radiation reaction along general orbits in the effective one-body formalism. Phys. Rev. D **86**, 124012 (2012). doi:10.1103/PhysRevD.86.124012
74. Fujita, R., Iyer, B.R.: Spherical harmonic modes of 5.5 post-Newtonian gravitational wave polarisations and associated factorised resummed waveforms for a particle in circular orbit around a Schwarzschild black hole. Phys. Rev. D **82**, 044051 (2010). doi:10.1103/PhysRevD.82.044051
75. Fujita, R.: Gravitational radiation for extreme mass ratio inspirals to the 14th post-Newtonian order. Progress Theoret. Phys. **127**, 583 (2012). doi:10.1143/PTP.127.583
76. Pan, Y., Buonanno, A., Fujita, R., Racine, E., Tagoshi, H.: Post-Newtonian factorized multipolar waveforms for spinning, non-precessing black-hole binaries. Phys. Rev. D **83**, 064003 (2011). doi:10.1103/PhysRevD.83.064003
77. Pan, Y., Buonanno, A., Boyle, M., et al.: Inspiral-merger-ringdown multipolar waveforms of nonspinning black-hole binaries using the effective-one-body formalism. Phys. Rev. D **84**, 124052 (2011). doi:10.1103/PhysRevD.84.124052
78. Taracchini, A., Pan, Y., Buonanno, A., et al.: Prototype effective-one-body model for nonprecessing spinning inspiral-merger-ringdown waveforms. Phys. Rev. D **86**, 024011 (2012). doi:10.1103/PhysRevD.86.024011
79. Damour, T., Nagar, A., Dorband, E.N., Pollney, D., Rezzolla, L.: Faithful effective-one-body waveforms of equal-mass coalescing black-hole binaries. Phys. Rev. D **77**, 084017 (2008). doi:10.1103/PhysRevD.77.084017
80. Damour, T., Nagar, A., Hannam, M., Husa, S., Bruegmann, B.: Accurate effective-one-body waveforms of inspiralling and coalescing black-hole binaries. Phys. Rev. D **78**, 044039 (2008)
81. Buonanno, A., Cook, G.B., Pretorius, F.: Inspiral, merger and ring-down of equal-mass black-hole binaries. Phys. Rev. D **75**, 124018 (2007). doi:10.1103/PhysRevD.75.124018
82. Pan, Y., Buonanno, A., Baker, J.G., et al.: A data-analysis driven comparison of analytic and numerical coalescing binary waveforms: nonspinning case. Phys. Rev. D **77**, 024014 (2008). doi:10.1103/PhysRevD.77.024014
83. Le Tiec, A., Mroue, A.H., Barack, L., et al.: Periastron advance in black hole binaries. Phys. Rev. Lett. **107**, 141101 (2011). doi:10.1103/PhysRevLett.107.141101
84. Damour, T., Nagar, A., Pollney, D., Reisswig, C.: Energy versus angular momentum in black hole binaries. Phys. Rev. Lett. **108**, 131101 (2012). doi:10.1103/PhysRevLett.108.131101
85. Damour, T., Gourgoulhon, E., Grandclément, P.: Circular orbits of corotating binary black holes: comparison between analytical and numerical results. Phys. Rev. D **66**, 024007 (2002). doi:10.1103/PhysRevD.66.024007
86. Buonanno, A., Pan, Y., Baker, J.G., et al.: Toward faithful templates for non-spinning binary black holes using the effective-one-body approach. Phys. Rev. D **76**, 104049 (2007). doi:10.1103/PhysRevD.76.104049

87. Buonanno, A., Pan, Y., Pfeiffer, H.P., et al.: Effective-one-body waveforms calibrated to numerical relativity simulations: coalescence of non-spinning, equal-mass black holes. Phys. Rev. D **79**, 124028 (2009). doi:10.1103/PhysRevD.79.124028
88. Bernuzzi, S., Nagar, A., Zenginoglu, A.: Binary black hole coalescence in the extreme-mass-ratio limit: testing and improving the effective-one-body multipolar waveform. Phys. Rev. D **83**, 064010 (2011). doi:10.1103/PhysRevD.83.064010
89. Barausse, E., Buonanno, A., Hughes, S.A., et al.: Modeling multipolar gravitational-wave emission from small mass-ratio mergers. Phys. Rev. D **85**, 024046 (2012). doi:10.1103/PhysRevD.85.024046
90. Damour, T., Gopakumar, A.: Gravitational recoil during binary black hole coalescence using the effective one body approach. Phys. Rev. D **73**, 124006 (2006). doi:10.1103/PhysRevD.73.124006
91. Bernuzzi, S., Nagar, A., Zenginoglu, A.: Binary black hole coalescence in the large-mass-ratio limit: the hyperboloidal layer method and waveforms at null infinity. Phys. Rev. D **84**, 084026 (2011). doi:10.1103/PhysRevD.84.084026
92. Damour, T., Nagar, A., Bernuzzi, S.: Improved effective-one-body description of coalescing nonspinning black-hole binaries and its numerical-relativity completion. ArXiv e-prints arxiv:1212.4357 [gr-qc] (2012)
93. Damour, T., Nagar, A.: Final spin of a coalescing black-hole binary: an effective-one-body approach. Phys. Rev. D **76**, 044003 (2007)
94. Damour, T., Nagar, A., Tartaglia, A.: Binary black hole merger in the extreme mass ratio limit. Class. Quantum Gravity **24**, S109 (2007)
95. Gonzalez, J.A., Sperhake, U., Bruegmann, B., Hannam, M., Husa, S.: Total recoil: the maximum kick from nonspinning black-hole binary inspiral. Phys. Rev. Lett. **98**, 091101 (2007). doi:10.1103/PhysRevLett.98.091101
96. Yunes, N., Buonanno, A., Hughes, S.A., Miller, M.C., Pan, Y.: Modeling extreme mass ratio inspirals within the effective-one-body approach. Phys. Rev. Lett. **104**, 091102 (2010). doi:10.1103/PhysRevLett.104.091102
97. Yunes, N., Buonanno, A., Hughes, S.A., et al.: Extreme mass-ratio inspirals in the effective-one-body approach: Quasi-circular, equatorial orbits around a spinning black hole. Phys. Rev. D **83**, 044044 (2011). doi:10.1103/PhysRevD.83.044044
98. Scheel, M.A., Boyle, M., Chu, T., et al.: High-accuracy waveforms for binary black hole inspiral, merger, and ringdown. Phys. Rev. D **79**, 024003 (2009). doi:10.1103/PhysRevD.79.024003
99. Damour, T., Nagar, A.: The effective one-body description of the two-body problem. In: Blanchet, L., Spallicci, A., Whiting, B. (eds.) Mass and Motion in General Relativity, Fundamental Theories of Physics, vol. 162, pp. 211–252. Springer, Berlin (2011). doi:10.1007/978-90-481-3015-3_7
100. Buchman, L.T., Pfeiffer, H.P., Scheel, M.A., Szilágyi, B.: Simulations of non-equal mass black hole binaries with spectral methods. Phys. Rev. D **86**, 084033 (2012). doi:10.1103/PhysRevD.86.084033
101. Damour, T.: Gravitational self force in a Schwarzschild background and the effective one body formalism. Phys. Rev. D **81**, 024017 (2010). doi:10.1103/PhysRevD.81.024017
102. Pan, Y., Buonanno, A., Buchman, L.T., et al.: Effective-one-body waveforms calibrated to numerical relativity simulations: coalescence of non-precessing, spinning, equal-mass black holes. Phys. Rev. D **81**, 084041 (2010). doi:10.1103/PhysRevD.81.084041
103. Damour, T., Jaranowski, P., Schäfer, G.: Effective one body approach to the dynamics of two spinning black holes with next-to-leading order spin-orbit coupling. Phys. Rev. D **78**, 024009 (2008)
104. Barausse, E., Racine, E., Buonanno, A.: Hamiltonian of a spinning test-particle in curved spacetime. Phys. Rev. D **80**, 104025 (2009). doi:10.1103/PhysRevD.80.104025 [Erratum: ibid. D **85**, 069904 (2012)]
105. Barausse, E., Buonanno, A.: An improved effective-one-body Hamiltonian for spinning black-hole binaries. Phys. Rev. D **81**, 084024 (2010). doi:10.1103/PhysRevD.81.084024

106. Nagar, A.: Effective one body Hamiltonian of two spinning black-holes with next-to-next-to-leading order spin-orbit coupling. Phys. Rev. D **84**, 084028 (2011). doi:10.1103/PhysRevD.84.084028
107. Barausse, E., Buonanno, A.: Extending the effective-one-body Hamiltonian of black-hole binaries to include next-to-next-to-leading spin-orbit couplings. Phys. Rev. D **84**, 104027 (2011). doi:10.1103/PhysRevD.84.104027
108. Damour, T., Nagar, A., Villain, L.: Measurability of the tidal polarizability of neutron stars in late-inspiral gravitational-wave signals. Phys. Rev. D **85**, 123007 (2012). doi:10.1103/PhysRevD.85.123007
109. Damour, T., Nagar, A.: Effective one body description of tidal effects in inspiralling compact binaries. Phys. Rev. D **81**, 084016 (2010). doi:10.1103/PhysRevD.81.084016
110. Bini, D., Damour, T., Faye, G.: Effective action approach to higher-order relativistic tidal interactions in binary systems and their effective one body description. Phys. Rev. D **85**, 124034 (2012). doi:10.1103/PhysRevD.85.124034
111. Baiotti, L., Damour, T., Giacomazzo, B., Nagar, A., Rezzolla, L.: Analytic modelling of tidal effects in the relativistic inspiral of binary neutron stars. Phys. Rev. Lett. **105**, 261101 (2010). doi:10.1103/PhysRevLett.105.261101
112. Baiotti, L., Damour, T., Giacomazzo, B., Nagar, A., Rezzolla, L.: Accurate numerical simulations of inspiralling binary neutron stars and their comparison with effective-one-body analytical models. Phys. Rev. D **84**, 024017 (2011). doi:10.1103/PhysRevD.84.024017
113. Barack, L., Damour, T., Sago, N.: Precession effect of the gravitational self-force in a Schwarzschild spacetime and the effective one-body formalism. Phys. Rev. D **82**, 084036 (2010). doi:10.1103/PhysRevD.82.084036
114. Le Tiec, A., Blanchet, L., Whiting, B.F.: The first law of binary black hole mechanics in general relativity and post-Newtonian theory. Phys. Rev. D **85**, 064039 (2012). doi:10.1103/PhysRevD.85.064039
115. Le Tiec, A., Barausse, E., Buonanno, A.: Gravitational self-force correction to the binding energy of compact binary systems. Phys. Rev. Lett. **108**, 131103 (2012). doi:10.1103/PhysRevLett.108.131103
116. Barausse, E., Buonanno, A., Le Tiec, A.: The complete non-spinning effective-one-body metric at linear order in the mass ratio. Phys. Rev. D **85**, 064010 (2012). doi:10.1103/PhysRevD.85.064010
117. Akcay, S., Barack, L., Damour, T., Sago, N.: Gravitational self-force and the effective-one-body formalism between the innermost stable circular orbit and the light ring. Phys. Rev. D **86**, 104041 (2012). doi:10.1103/PhysRevD.86.104041
118. Blanchet, L., Detweiler, S.L., Le Tiec, A., Whiting, B.F.: High-order post-Newtonian fit of the gravitational self-force for circular orbits in the Schwarzschild geometry. Phys. Rev. D **81**, 084033 (2010). doi:10.1103/PhysRevD.81.084033
119. Damour, T.: The 5PN contributions to EOB potentials. (unpublished) (2010)
120. Foffa, S., Sturani, R.: The dynamics of the gravitational two-body problem at fourth post-Newtonian order and at quadratic order in the Newton constant. ArXiv e-prints 1206.7087 [gr-qc] (2012)
121. Jaranowski, P., Schäfer, G.: Towards the 4th post-Newtonian Hamiltonian for two-point-mass systems. Phys. Rev. D **86**, 061503 (2012). doi:10.1103/PhysRevD.86.061503
122. Faye, G., Marsat, S., Blanchet, L., Iyer, B.R.: The third and a half post-Newtonian gravitational wave quadrupole mode for quasi-circular inspiralling compact binaries. Class. Quantum Gravity **29**, 175004 (2012). doi:10.1088/0264-9381/29/17/175004

# Gravitational Self-Force: Orbital Mechanics Beyond Geodesic Motion

**Leor Barack**

**Abstract** The question of motion in a gravitationally bound two-body system is a longstanding open problem of General Relativity. When the mass ratio $\eta$ is small, the problem lends itself to a perturbative treatment, wherein corrections to the geodesic motion of the smaller object (due to radiation reaction, internal structure, etc.) are accounted for order by order in $\eta$, using the language of an effective *gravitational self-force*. The prospect for observing gravitational waves from compact objects inspiralling into massive black holes in the foreseeable future has in the past 15 years motivated a program to obtain a rigorous formulation of the self-force and compute it for astrophysically interesting systems. I will give a brief survey of this activity and its achievements so far, and will identify the challenges that lie ahead. As concrete examples, I will discuss recent calculations of certain conservative post-geodesic effects of the self-force, including the $O(\eta)$ correction to the precession rate of the periastron. I will highlight the way in which such calculations allow us to make a fruitful contact with other approaches to the two-body problem.

## 1 Background: The Self-Force Domain of the Two-Body Problem and Astrophysical Relevance

In Newtonian gravity, the dynamics of a two-body system is extremely simple: an isolated system of two gravitationally-bound point masses admits two conserved integrals—the energy and angular momentum—and the resulting motion is precisely periodic. The corresponding general-relativistic problem is radically more difficult. In General Relativity (GR), the orbits in a bound binary are *never* periodic: gravitational radiation removes energy and angular momentum from the system, and the radiation back-reaction gradually drives the two objects tighter together until they

L. Barack (✉)
School of Mathematics, University of Southampton, Southampton SO17 1BJ, UK
e-mail: leor@soton.ac.uk

J. Bičák and T. Ledvinka (eds.), *General Relativity, Cosmology and Astrophysics*, Fundamental Theories of Physics 177, DOI: 10.1007/978-3-319-06349-2_6, 
© Springer International Publishing Switzerland 2014

eventually merge. It is testimony to the intricacy of the problem, that this qualitative picture of radiative damping has been fiercely debated in the literature until well into the 1960s (with some results famously predicting an energy *gain* in binary systems emitting waves!). Work by Hermann Bondi and others in the early 1960s eventually settled the dispute, and the prevailing view was later fully vindicated with the 1975 Nobel-Prize observation by Hulse and Taylor of radiative decay in the binary pulsar PSR 1913+16 [1], consistent with Einstein's quadrupole formula.

The dynamics in the Hulse-Taylor binary is very nearly Newtonian, due to the relatively large separation between the two components. We now know that the universe around us abounds with much more strongly gravitating systems, from pairs of inspiralling white dwarfs in nearby galaxies (likely progenitors of most Type Ia supernovae) to the dramatic coalescence of supermassive black holes at cosmological distances—by far the brightest events in the universe by energy output. In the coming years, direct observations of gravitational waves (GWs) will open a new window on the universe and allow us to peer deeply into these strong-field processes, which are largely invisible in electromagnetic spectrum. This is an exciting prospect, because GWs from coalescing compact binaries and black holes will probe the hitherto unexplored strong-field, highly-dynamical sector of Einstein's theory, where a variety of exotic nonlinear effects manifest themselves. With this prospect comes the need to have an accurate theoretical model of the two-body dynamics in the strong-field regime, and a prediction of the emitted gravitational waveforms. These waveforms are needed not only to allow interpretation of the signals and facilitate precision tests of GR theory, but also to enable the very extraction of some of the weaker signals from the noisy detector output.

In general, the description of the nonlinear strong-field dynamics in the binary system entails a full Numerical-Relativistic (NR) treatment, whereby the Einstein field equations are formulated as an initial/boundary problem and solved numerically. Efforts to obtain numerical solutions for (in particular) black hole binaries date back to work in the 1960s and 1970s [2, 3], but it was not until 2005 that first successful simulations were performed [4–6]. Today NR codes are capable of tracking the complicated nonlinear evolution of a spacetime containing two (spinning) black holes during the final stages of the merger. But NR methods have a limited utility in situations where the two black holes are far apart, or when one of the components is much heavier than the other. Each of these two regimes of the two-body problem (see Fig. 1) is characterized by two greatly separate lengthscales (the distance between the objects vs. their individual radii of curvature in the former case; the radius of curvature of the larger black hole vs. that of the smaller one in the latter case), which is difficult to accommodate in an NR framework due to the high resolution requirements and long evolution time.

Fortunately, the presence of two separate lengthscales also means that the problem becomes amenable to a simpler, perturbative treatment. In the first of the above regimes—at sufficiently large separations—the dynamics is best analyzed using the tools of *post-Newtonian* (PN) theory [7], whose roots go as far back as 1938, to the classical paper by Einstein, Infeld and Hoffman [8]. In PN theory, corrections to the Newtonian dynamics are incorporated into the equations of motion (essentially)

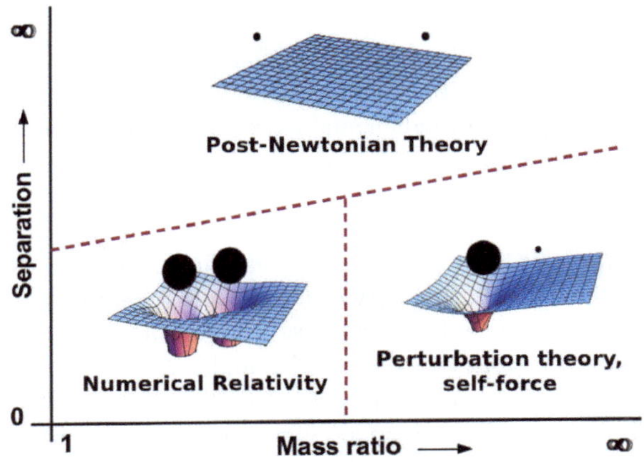


**Fig. 1** Schematic representation of the three domains of the binary black-hole problem, with the corresponding natural computational frameworks. Numerical solutions of the full Einstein equations are effectual for very close binaries of comparable masses. Widely separated binaries are treated most efficiently using the tools of post-Newtonian theory (less so when the mass ratio is extreme). Strongly gravitating binaries of large mass ratios are the realm of self-force theory. Much current study focuses on comparing the predictions of the different approaches where they interface (*dashed lines*)

order by order in the binary separation. PN results are useful (a priori) only when the binary separation is not too small, but no assumption is made about the masses of the two objects, which can be kept arbitrary.

The second, so-called *extreme mass-ratio* regime, is most naturally explored within the framework of black hole perturbation theory. Here the "zeroth-order" configuration is that of a test particle moving along a geodesic of the fixed, stationary background spacetime of the larger object (say, a black hole). This can then serve as a basis for a perturbative scheme, whereby corrections due to the finite mass of the small object (and due also, ultimately, to its internal structure) are included order by order in the small mass-ratio $\eta$. At $O(\eta)$, the gravitational field of the small object is a linear perturbation of the background geometry. The back-reaction from this perturbation gives rise to an effective *gravitational self-force* (GSF) that gradually diverts the small object from its geodesic motion. In this picture, it is the GSF that is responsible for the radiative decay of the orbit. It is also responsible for a variety of conservative effects arising from finite-$\eta$ corrections to the background gravitational potential. The GSF description is useful (a priori) when $\eta$ is sufficiently small, but the separation between the two objects need not be large. Indeed, GSF theory covers precisely the domain accessible to neither NR nor PN: strongly gravitating binaries of small $\eta$.

The basic notion of a "self-force" (aka *back-reaction* force or *braking* force) in a radiating system is an old one, dating many decades back, in the context of electrodynamics, to the classical works by Lorentz [9] and Dirac [10] on the electron's equation of motion. In 1960 DeWitt and Brehme generalized this idea to GR by deriving an equation of motion for an electric charge moving in curved space [11]. Their method involved a careful application of Gauss's theorem on a thin worldtube surrounding the particle's worldline, with an imposition of local energy-momentum conservation. This work has a fundamental importance also in that it develops the essential mathematical toolkit underpinning contemporary GSF theory: the calculus of bi-tensors in curved spacetime.

The failure of the Huygens principle in 3 + 1D curved spacetime means that the self-interaction effect in GR is fundamentally *nonlocal*: the self-acceleration formally depends on the entire past history of the particle in question. This represents a significant departure from the flat-space case, and it brings about significant complications, technical as well as conceptual. The application of the self-force idea to the motion of a *mass* particle in GR presents several more complications. Not least among these is the fact that the very notion of a point mass is ill-defined in GR [12]. A rigorous derivation of the GSF equations of motion cannot, therefore, rely on a point-mass assumption, as conveniently done in a linear theory like electrodynamics. For many years, a robust formulation of the GSF remained an open problem in GR theory.

Interest in the problem grew in the mid 1990s, when it was proposed that the planned space-based GW detector Laser Interferometer Space Antenna (LISA [13]) could observe signals from the inspiral of compact objects into massive black holes in galactic nuclei. (The inspiraling objects need to be compact—white dwarfs, neutron stars of stellar-mass black holes—because main-sequence stars will be tidally destroyed before they can produce an interesting GW signal.) Later work confirmed that a LISA-like mission should be able to detect hundreds of events, out to cosmological distances ($z \sim 1$) [14]. Dubbed Extreme Mass Ratio Inspirals (*EMRI*s), these sources have a unique facility as precision probes of strong-field gravity. In a typical LISA-band EMRI (a $\sim 10M_\odot/10^6 M_\odot$ system), the inspiralling object spends the last few years of inspiral in a very tight orbit around the massive hole, emitting some $10^5$–$10^6$ gravitational wave cycles. The inspiral trajectories show extreme versions of periastron precession, Lense-Thirring precession of the orbital plane, and other strong-field effects. This complex dynamics is encoded in the GWs, which then carry a detailed map of the spacetime geometry around the massive hole. It was calculated, for example, that LISA will be able to measure fractional deviations as small as 1:1000 in the quadrupole moment of the black hole metric [15], allowing precision tests of GR and setting tight bounds on the parameters of alternative theories of gravity.

It is a crucial prerequisite for realizing this science potential that accurate theoretical templates of the inspiral waveforms are at hand. This, in turn, requires a detailed understanding of the radiative evolution. In a typical LISA-band EMRI, the GSF drives the orbital decay over a timescale of months, and it *dephases* the orbit over mere hours. A useful model of the long-term orbital phase evolution therefore ought to account properly for GSF effects, which one must be able to calculate for generic (eccentric, arbitrarily inclined) strong-field orbits around a Kerr black hole of arbitrary spin.

## 2 The GSF Program: Foundation

The EMRI problem provided an important impetus to rapid progress in GSF research. In 1997 Mino et al. [16] derived (what has since been "canonized" as) the formal expression for the $O(m^2)$ GSF [that is, the $O(m)$ self-acceleration] acting on a particle of mass $m$ moving in an arbitrary vacuum spacetime of characteristic radius of curvature $\mathscr{R} \gg m$. The expression was derived using two methods (a third, axiomatic method was presented around the same time by Quinn and Wald [17], leading to the same result). The first method was a direct application of DeWitt and Brehme's analysis to the gravitational case, assuming from the onset that the particle can be represented as a point source to the perturbed Einstein equations. The second method removed this assumption: the small object was taken to be a (Schwarzschild) black hole, and its representative "worldline" on the background spacetime was defined and derived using the procedure of *matched asymptotic expansions*.

In its general form, matched asymptotic expansions is a common tool in physics for studying problem involving multiple scales (most relevantly, it has been applied in PN theory [18]). In the particular implementation of [16], the equation of motion is obtained by matching together two series representations of the metric: a "far field" expansion in $m/r$ (where $r$ is a suitable measure of spatial distance from the small black hole), and a "near" field expansion in $r/\mathscr{R}$. The first expansion treats the field of the small black hole at $r \gg m$ as a small perturbation on the external geometry, and the second expansion accounts for the background curvature at $r \ll \mathscr{R}$ via the small tidal deformations it induces on the metric of the small black hole. The assumption $m \ll \mathscr{R}$ means there is a "buffer zone" $m \ll r \ll \mathscr{R}$ where both descriptions apply, and demanding that the descriptions agree in this zone constrains the motion of the small black hole. Matching at leading order shows that the $O(m^0)$ motion is a geodesic on the background geometry. Matching at the next order gives the $O(m)$ acceleration of the small black hole on the background geometry, interpreted as a GSF effect. The accelerated "worldline" is defined from a far-field point of view, via a suitable limiting process. In subsequent work [19–22], this procedure was generalized and put on a more mathematically firm footing. For example, the small object was allowed to possess spin and consist of any form of matter (not necessarily a black hole). The most elegant and rigorous derivation was presented by Gralla and Wald [20], whose analysis derives both near and far zones as certain limits of a *single* global metric. For a thorough and self-contained review of these theoretical developments, see Ref. [19].

At first post-geodesic order in the GSF approximation, the equation of motion has the form

$$mu^{\alpha}\nabla_{\alpha}u^{\beta} = F^{\beta}_{\rm self}, \tag{1}$$

where $u^{\alpha}$ is the particle's four-velocity on the background spacetime, $\nabla_{\alpha}$ is a covariant derivative on the background spacetime, and $F^{\beta}_{\rm self}(\propto m^2)$ is the GSF. The above foundational work gives an expression for $F^{\beta}_{\rm self}$ in terms of the "tail" part of the physical metric perturbation associated with the point particle—the part arising from the piece of the Green's function supported *inside* (rather than *on*) the past light-cone of the source. Roughly speaking, it establishes that it is the back-reaction from the *tail* part of the self-field (which is finite) that is responsible for the self-acceleration.

Two comments are in order. First, as is easily seen, the GSF itself is a gauge-dependent notion, and so is the accelerated trajectory in the background geometry: an $O(m)$ gauge transformation in the perturbed geometry results in a *physically distinct* accelerated trajectory. Thus, a meaningful information about the post-geodesic motion is contained only in the *combination* of the GSF and the metric perturbation (in a particular gauge). The above foundational derivations of the GSF all involve a specific gauge choice—the so called *Lorenz gauge* (in which the divergence of the trace-reversed metric perturbation is set to zero). This is a convenient choice because (i) it preserves the local isotropic nature of the particle singularity, and (ii) the perturbation equations in the Lorenz gauge are fully hyperbolic and admit a well-posed initial-value formulation. It should not be assumed without a careful examination that the GSF is meaningful or well defined in any other given gauge (this has been a source of much confusion and debate in the GSF literature). A gauge transformation formula for the GSF was derived in Ref. [23], which also proposed some criteria for admissible GSF gauges. The topic is further developed in Ref. [24].

A second comment is that Eq. (1) is only guaranteed to hold momentarily at each point along the trajectory. It is quite a separate task to formulate a scheme that faithfully accounts for the long-term evolution of the orbit. A subtlety is that the Lorenz gauge condition cannot be imposed consistently when the source's worldline is accelerating. Ref. [20] suggested a scheme where the Lorenz-gauge perturbation equations and the equation of motion (1) are solved as a coupled set in a self-consistent manner, without actively imposing the Lorenz gauge conditions (a similar scheme of "gauge relaxation" has been used in PN theory); the gauge violations which would then occur at $O(m^2)$ will presumably be accounted for within a consistent second-order GSF formulation once this becomes available. In a more recent work, Pound [25] has used techniques from singular and multiple-scale perturbation theories in attempt to put the idea of gauge relaxation on sound mathematical footing, but it seems the issue remains somewhat controversial for now. Stronger consensus is likely to be reached soon, with the advent of the second-order GSF formulations (see Sect. 6). In any case, we note that a computation of the local GSF $F^{\beta}_{\rm self}$ will constitute a necessary input for any ultimate scheme for the long-term evolution of the orbit.

In a 2003 paper [26] Detweiler and Whiting proposed an appealing reinterpretation of Eq. (1) in terms of geodesic motion in a smooth perturbed spacetime. They showed that the GSF $F^{\beta}_{\rm self}$ can be interpreted as the back-reaction force from a certain smooth metric perturbation $h^R_{\alpha\beta}$, which, unlike the "tail" field mentioned above, is a *vacuum* solution of the linearized Einstein equations. The particle can be thought to be moving along a geodesic of $g_{\alpha\beta}+h^R_{\alpha\beta}$, where $g_{\alpha\beta}$ is the background metric. The equation of motion is reformulated as

$$m\tilde{u}^{\alpha}\tilde{\nabla}_{\alpha}\tilde{u}^{\beta}=0, \tag{2}$$

where $\tilde{u}^\alpha$ and $\tilde{\nabla}$ are the four-velocity and covariant derivative in $g_{\alpha\beta} + h^R_{\alpha\beta}$. A pedagogical review of this construction is given in Ref. [19]. It should be emphasized that the perturbation $h^R_{\alpha\beta}$ is *not* the particle's physical metric perturbation (for example, it is not causal), but rather a mathematical construct that serves as an *effective* potential for the motion. The two alternative formulations of the post-geodesic motion, Eqs. (1) and (2), are equivalent and both are useful; workers in the field often flip between the two points of view as necessary to highlight different aspects of the problem.

# 3 The GSF Program: Computation

## *3.1 Mode-Sum Regularization*

The above formulation is directly applicable to the EMRI problem, where the "large" scale $\mathscr{R}$ is provided by the mass $M$ of the large black hole. A practical method for calculating the GSF for EMRI orbits, known as *mode-sum regularization* was introduced in 2000 [27], and subsequently became the main working framework for GSF calculations in black hole spacetimes. The method is an implementation of the robust formulation discussed above (no extra regularization is introduced), and we shall give a schematic description of it here. A detailed review can be found in Ref. [28].

As mentioned above, the GSF can be interpreted as the effective force due to the Detweiler-Whiting R-field: $F^\alpha_{\rm self} = m\nabla^{\alpha\beta\gamma} h^R_{\beta\gamma}$, where $\nabla^{\alpha\beta\gamma}$ is a suitable derivative operator defined along the particle's worldline in the background metric. (The original mode-sum scheme was formulated in terms of the tail field but we shall use here the equivalent R-field formulation for simplicity.) The R-field itself can be obtained from the subtraction $h^R_{\beta\gamma} = h^{\rm full}_{\beta\gamma} - h^S_{\beta\gamma}$, where the "full" field is the physical (retarded) solution of the linearized Einstein equation sourced by the particle's energy-momentum, and the "S"-field (for *singular* field) is a particular solution prescribed by Detweiler and Whiting [26]. The fields $h^{\rm full}_{\beta\gamma}$ and $h^S_{\beta\gamma}$ have the same singular structure near the moving particle, so that their difference, $h^R_{\beta\gamma}$, is a smooth ($C^\infty$) function.

In the mode-sum scheme one essentially performs the above subtraction mode-by-mode in a multipole expansion, and the GSF is then reconstructed from a sum over multipole contributions:

$$F^\alpha_{\rm self} = m\nabla^{\alpha\beta\gamma} \sum_{l=0}^{\infty} \left[ (h^{\rm full}_{\beta\gamma})^l - (h^S_{\beta\gamma})^l \right]. \tag{3}$$

Here a superscript '$l$' denotes the $l$-multipole of the corresponding field (defined, as usual in black hole perturbation theory, via integrals over two-spheres surrounding the *large* black hole), summed over azimuthal ('$m$') modes. The advantage of the multipole decomposition is twofold: First, numerical methods in black-hole perturbation theory are usually based on multipole expansions, so that numerical calculations normally output individual modal contributions anyway. Second, and more crucially, each of the individual modal contributions $(h^{\rm full}_{\beta\gamma})^l$ [or $(h^S_{\beta\gamma})^l$] is finite and (piecewise) differentiable at the particle's location, which makes the subtraction more manageable in practice.

Now, as first suggested in Ref. [27], Eq. (3) can be put into a more useful form using some analytic input. One can analytically study the large-$l$ behavior of the S-field modes $(h^S_{\beta\gamma})^l$ and their derivatives at the particle, and it turns out that (generically) the derivatives admit a large-$l$ expansion in $1/l$, whose leading term is of $O(l)$. (The last statement depends somewhat on the gauge, but here we shall ignore this subtlety for simplicity.) If the first few terms in this expansion are known, one can rewrite Eq. (3) in the form

$$F^\alpha_{\rm self} = m\sum_{l=0}^{\infty}\left[\nabla^{\alpha\beta\gamma}(h^{\rm full}_{\beta\gamma})^l - A^\alpha l - B^\alpha - C^\alpha/l\right], \tag{4}$$

where $A^\alpha$, $B^\alpha$ and $C^\alpha$ ("regularization parameters") depend on the particle's location and velocity (and on the background spacetime) but not on $l$; importantly, it was shown [28, 29] that the residue $\sum_{l=0}^{\infty}\left[\nabla^{\alpha\beta\gamma}(h^S_{\beta\gamma})^l - A^\alpha l - B^\alpha - C^\alpha/l\right]$ vanishes along any geodesic orbit in Kerr spacetime. The regularization parameters were calculated analytically for generic orbits in Schwarzschild [30] and later for generic orbits in Kerr [28, 29]. With the regularization parameters given analytically, the task of computing the GSF along a given (pre-specified) orbit reduces to that of obtaining the full modes $(h^{\rm full}_{\beta\gamma})^l$ to serve as input in the mode-sum formula (4). This is usually done numerically, by solving the suitable set of mode-decomposed perturbation equations with retarded boundary conditions, sourced by the particle orbit in question.

Numerical implementations of the mode-sum formula (4) are reviewed in Ref. [28]. Typically, the particle is taken to move on a fixed geodesic orbit, and the perturbation equations are solved for the corresponding source (the back-reaction effect of the GSF on the orbit has only recently been accounted for in a numerical simulation—see Sect. 4 below). The most advanced implementations of mode-sum regularization are capable of computing the GSF along any (bound) geodesic in Schwarzschild spacetime—these codes where presented in Refs. [31, 32] (time-domain version) and [33, 34] (frequency-domain version). There are also calculations in Kerr spacetime [35, 36], but for now they are restricted to the toy model of a scalar-field self-force. (Shah et al. recently used mode-sum regularization to compute a certain GSF-related effect on circular equatorial orbit in Kerr spacetime [37], but they have not computed the GSF itself.) Further advance in mode-sum calculations is represented by the recent analytic derivation of higher-order regularization parameters [terms of $O(l^{-2})$ and higher in the large-$l$ decomposition of the singular field at the particle] by Heffernan et al. [38, 39]. This now helps to accelerate the convergence of the mode-sum in numerical implementations, leading to much improved precision in GSF calculations.

## 3.2 Puncture Method

This alternative computation method has been in development since 2007 [40–42]. The idea here is to "regularize" the field equations themselves, rather than (as in the mode-mode method) their solutions. This method works best with time-domain numerical implementations in $2+1$ or $3+1$ dimensional, and can benefit from recent advances in numerical method for time-domain evolution of hyperbolic equations in GR. Other advantages: the method offers a direct route to the Kerr problem (still a challenge for mode-sum regularization), and it offers a convenient framework for studies of the orbital evolution under the GSF effect. Following is a schematic description of the method as applied to a scalar-field analogue model; a fuller review can be found in Ref. [28].

Let us write the scalar-field equation in the schematic form $\Box\phi^{\rm full} = S$, where $\Box$ is a suitable wave operator (depending on the scalar-field theory), $S$ is a source term corresponding to a point particle of scalar charge $q$, and $\phi^{\rm full}$ is the sought-for retarded solution. Let $\phi^R$ and $\phi^S$ be the scalar-field analogues of Detweiler–Whiting's R and S fields, respectively, so that $F^\alpha_{\rm self} = q\tilde{\nabla}^\alpha\phi^R$ is the scalar-field self-force, with $\tilde{\nabla}^\alpha$ a suitable gradient operator. The implementation of the puncture scheme begins with finding an analytic approximation to $\phi^S$, denoted $\phi^P$ (the "puncture"), with the property that $\phi^P - \phi^S$ and $\tilde{\nabla}^\alpha(\phi^P - \phi^S)$ both vanish along the particle's worldline (the field $\phi^S$ can be extended globally as convenient). Then the self-force can be computed via $F^\alpha_{\rm self} = q\tilde{\nabla}^\alpha\phi^{\rm Res}$ (evaluated at the particle), where the "residual" field is $\phi^{\rm Res} := \phi^{\rm full} - \phi^P$. The latter satisfies the "punctured" equation

$$\Box\phi^{\rm Res} = S - \Box\phi^P := S_{\rm eff}, \tag{5}$$

where the "effective source" $S_{\rm eff}$ no longer contains a delta function. The residual non-smoothness of $S_{\rm eff}$ arises from the fact that $\phi^P$ is only a finite approximation to $\phi^S$ (a full expression for $\phi^S$ is not known in explicit form); one can improve the smoothness of $S_{\rm eff}$ by designing a "higher-order" input puncture $\phi^P$, for which higher-order derivatives of $(\phi^P - \phi^S)$ also vanish at the particle. High order punctures for the GSF, and corresponding effective sources, were derived in Refs. [43, 44].

The task of computing the self-force now reduces to solving the field equation $\Box\phi^{\rm Res} = S_{\rm eff}$ with suitable boundary conditions. One usually truncates (or otherwise attenuates) the support of $\phi^P$ far from the particle, so that the necessary boundary conditions are the usual "retarded" ones. Several groups are now engaged in code development for self-force calculations in the puncture approach. Most work so far has been confined to the toy model of a scalar field (as a platform for code development and testing) [43–46], but a first implementation of the GSF has very recently been presented [47] for circular orbits around a Schwarzschild black hole. Existing puncture codes use a 2+1 (or 3+1)-dimensional grid, which avoids the issue of separability of the field equations in Kerr spacetime. This makes the extension of Schwarzschild codes to Kerr rather straightforward. Indeed, following on from [47], a first implementation of the GSF in Kerr will be presented shortly in a forthcoming paper [48].

As mentioned, time-domain codes based on the puncture method provide a natural platform for studying the orbital evolution under the self-force: rather than pre-specifying the orbit, one can compute the self-acceleration at any time slice, then modify the source accordingly "in real time" so as to compute the evolving orbit in a self-consistent manner (say, using the idea of Lorenz-gauge relaxation). Unfortunately, current codes are not sufficiently efficient computationally to track the evolution in an EMRI-relevant system over many orbits. (A first self-consistent evolution simulation was presented recently for the scalar-field self-force [49], but computational cost restricts the ability of the code to compute more than a handful of orbital cycles.) In the past few years, this computational challenge has attracted some interest from experts in Numerical Relativity, leading to several programs to develop custom-built advanced numerical techniques for integrating the perturbation equations with pointlike sources. These include methods based on finite elements [50], adaptive mesh refinement [51], and hyperboloidal slicing [52]. It is hoped that this activity will lead to a dramatic improvement in the computational performance of time-domain GSF codes.

## 4 Orbital Evolution Under the GSF Effect

An important milestone in the GSF program was reached last year, with a first computation of the long-term orbital evolution under the full (first order) GSF [34]. This computation was based on a frequency-domain implementation of the mode-sum method in Schwarzschild spacetime, developed in [33, 35, 36]. Rather than evolving the orbit in a fully self-consistent manner as described above (which is not easily achievable in a frequency-domain framework), an approximation was used, in which the value of the GSF at each point along the evolving orbit is taken to be that computed along a fixed geodesic tangent to the orbit at that point. This approximation is a good one in situations where the timescale on which the orbit evolves is much longer than the effective "memory" time associated with the tail field that produces the GSF. In an EMRI-relevant strong-field system with $\eta \ll 1$, the former is expected to be larger than the latter by an amount of order $1/\eta \gg 1$ over much of the inspiral; the approximation will cease to be useful only very near the last stable orbit, where the adiabatic inspiral transits to a direct plunge and the orbit evolves quickly. Let us briefly review the method of [34], then present some results for illustration.

The calculation in [34] is a general-relativistic adaptation of the standard method of "variation of parameters", or "osculating orbits", used in celestial mechanics. In this *osculating geodesics* approach, the inspiral motion is reconstructed from a smooth sequence of tangent geodesics. In practice, this amounts to solving evolution equations for all the orbital elements that characterize the geodesic motion (principal as well as positional), with the driving force provided by the GSF. The necessary GSF information is prepared in advance, in the form of a global interpolation formula based on a dense data grid over the relevant phase space.

Let us give some more detail. Bound geodesics of the Schwarzschild geometry can be parametrized by their semilatus rectum $pM$ and eccentricity $e$, defined via $r_\pm = pM/(1 \mp e)$, where $r = r_+$ and $r = r_-$ are the apastron and periastron radii, respectively [hereafter $(t, r, \theta, \varphi)$ are standard Schwarzschild coordinates on the background spacetime]. The geodesic motion of a test particle is then described by [53]

$$r = r_g(t; p, e, \chi_0) = \frac{pM}{1 + e\cos[\chi(t) - \chi_0]}, \tag{6}$$

$$\varphi = \varphi_g(t; p, e, \chi_0) = \int_{\chi(0)}^{\chi(t)} \frac{p^{1/2}\, d\chi'}{\sqrt{p - 6 - 2e\cos(\chi' - \chi_0)}}, \tag{7}$$

where $\chi(t)$ is a monotonically increasing parameter along the orbit (a relativistic generalization of mean anomaly), related to $t$ via $d\chi/dt = (p - 2 - 2e\cos v)(1 + e\cos v)^2(p-6-2e\cos v)^{1/2}[(p-2)^2-4e^2]^{-1/2}/(Mp^2)$, with $v := \chi - \chi_0$. Without loss of generality we have assumed that the motion takes place in the equatorial plane ($\theta = \pi/2$), and took $t(\chi_0) = \varphi(\chi_0) = 0$ (i.e., at $t = 0$ the particle is at periastron at $\varphi = 0$); $p$ and $e$ are *principal* elements, which determine the "shape" of the orbit, $\chi_0$ is a *positional* element, which describes the orientation of the major axis. Both principal and positional elements evolve secularly under the effect of the GSF; the secular evolution of $p$ and $e$ is *dissipative*, while that of $\chi_0$ is *conservative*—it describes the precession effect of the GSF. Both principal and positional elements also exhibit quasi-periodic oscillations.

In the osculating geodesics approach, the inspiral motion is described by $r = r_g(t; p(t), e(t), \chi_0(t))$ and $\varphi = \varphi_g(t; p(t), e(t), \chi_0(t))$, where $p(t), e(t), \chi_0(t)$ are called *osculating elements*. The rate of change of these elements is determined from the local self-acceleration (i.e., $F^\alpha_{\rm self}$ per unit $m$) of the tangent geodesic. Evolution formulas for the osculating elements, given the GSF, where obtained in Refs. [54] (Schwarzschild case) and [55] (Kerr case). These formulas require as input the function $F^\alpha_{\rm self}(\chi - \chi_0; p, e)$. In the implementation of Ref. [34] this function was obtained from numerical GSF data computed along a sample of 1,100 geodesics covering the $p, e$ parameter space. A suitable interpolation model was derived, based on a Fourier representation of the $\chi$-dependence and a power-law series ansatz for the $p, e$ dependence. With this GSF input at hand, the evolution equations for $\{p(t), e(t), \chi_0(t)\}$ were then solved numerically starting from some initial values. An example is shown in Fig. 2.

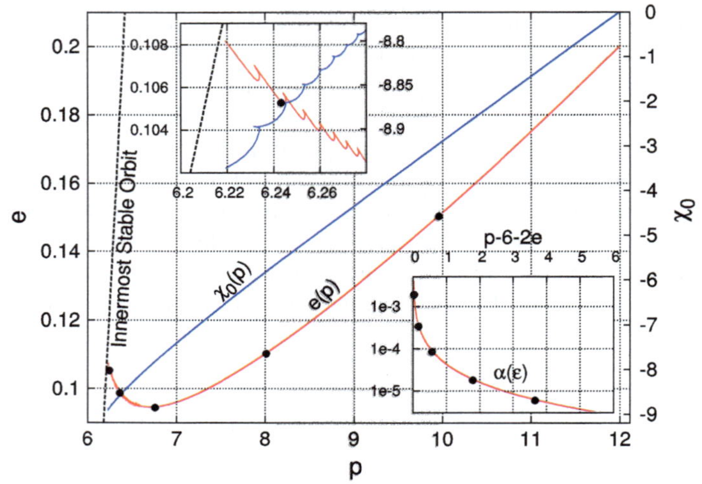


**Fig. 2** Evolution of the osculating elements in a sample case with $m = 10M_\odot$ and $M = 10^6 M_\odot$. We show the eccentricity $e$ (*red*, *left axis*) and periastron phase $\chi_0$ (*blue*, *right axis*) as functions of semi-latus rectum $p$, as the binary inspirals from $(p, e) = (12, 0.2)$ down to the innermost stable orbit (*dashed curve*). Marks along the *curves* count down (*from right to left*) 500 days, 100 days, 10 days, 1 day and 1 h to the onset of plunge. Note the orbit initially circularizes, but near the plunge the eccentricity begins to increase. Note also the phase $\chi_0$ decreases monotonically, implying that the conservative GSF acts to *reduce* the rate of relativistic precession. The *upper* inset is an enlargement of the near-plunge region; the manifest oscillatory behavior is due to the variation of the GSF with the radial phase. The *lower* inset shows the magnitude of the adiabaticity parameter $\alpha := \langle|\dot{p}/p|\rangle T$ (the average is over a radial period $T$) versus the distance $\varepsilon = p - 6 - 2e$ to the innermost stable orbit, confirming that the evolution is strongly adiabatic until very near the end. (Graphics from Ref. [34].)

# 5 Gauge Invariant Conservative Effects and Comparison with Other Methods

In the last few years GSF results have been used in a variety of applications going beyond the original EMRI program. GSF data can be used to compute gauge-invariant "observables" that describe post-geodesic corrections to the gravitational potential in the two-body system. These can then be utilized as reference points for comparison with the predications of PN theory and with results from full NR simulations. Our current knowledge allows us to go only one order beyond the geodesic approximation [i.e., to $O(\eta)$], but at this order the computed corrections are *exact*, and they give us a direct and hitherto unavailable handle on the very-strong-field conservative dynamics.

What do we mean by "conservative" dynamics? The precession effect already mentioned is an example. More generally, the GSF can be split in a unique way into dissipative and conservative bits. The dissipative bit is obtained, for example, by replacing $h^{\rm full}_{\beta\gamma}$ (the retarded metric perturbation) in Eq. (4) with the "half retarded minus half advanced" metric perturbation, and the conservative piece is similarly obtained from the "half retarded *plus* half advanced" perturbation. (This decomposition bears on time-symmetry and not on secularity: in general, both dissipative and conservative pieces of the GSF would have secular effects on the orbit.) In the simple case of (quasi)circular motion in the equatorial plane of a Kerr black hole, the dissipative piece is given by the coordinate components $F_t^{\rm self}$ and $F_\varphi^{\rm self}$ (related to dissipation rate of energy and angular momentum, respectively), and the conservative piece is given by $F^r_{\rm self}$. In more general cases it is still straightforward to construct the two pieces of the GSF separately in practice, either by obtaining both retarded and advanced solutions of the perturbation equations, or (more economically, as described in [28]) by exploiting the time-symmetry of bound geodesics in Kerr spacetime. Indeed, what makes the communication between GSF and PN so natural is the fact that in both approaches (and unlike in NR) the conservative and dissipative aspects of the dynamics can each be studied easily in isolation (this is true at least at first post-geodesic order in the GSF approximation, and through several orders in the PN expansion). In any case, to return to our question, what we mean by *conservative dynamics* is described by solutions to the equation of motion (1), with the GSF on the right-hand side replaced by its conservative piece (or, equivalently, with the dissipative piece "turned off").

A first gauge-invariant "observable" was proposed by Detweiler in 2005 [56]. Considering the effect of the conservative GSF on *circular* orbits in Schwarzschild spacetime, it is easy to see that both components $u^t$ and $u^\varphi$ of the particle's four-velocity $u^\alpha$ (on the background spacetime) are invariant through $O(m)$ under $O(m)$ gauge transformations that respect the helical symmetry of the perturbed spacetime. The combination $\Omega := u^\varphi/u^t$, which is the "observable" $t$-frequency of the perturbed orbit, is obviously also invariant. Detweiler proposed to utilize the $O(m)$ piece of the function $u^t(\Omega)$—let us denote it $u_1^t(\Omega)$—as a concrete gauge-invariant measure of the post-geodesic effect. [A simple calculation shows that $u_1^t(\Omega)$ does not actually involve the GSF itself; rather, it is constructed from the scalar contraction $h^R_{\alpha\beta}u^\alpha u^\beta$, where $h^R_{\alpha\beta}$, recall, is the Detweiler-Whiting R field.] Detweiler also suggested an interpretation of $u_1^t(\Omega)$ as a measure of the GSF correction to the gravitational redshift along a line of sight perpendicular to the orbital plane (but note this interpretation is subtle: the actual redshift from the point particle obviously diverges; rather, it is the redshift defined in the nonphysical effective metric $g_{\alpha\beta} + h^R_{\alpha\beta}$ that this interpretation alludes to). Detweiler used $u_1^t(\Omega)$ as a first contact point with PN theory, showing that the predictions from perturbation theory agree with PN formulas in the weak field [57]. This impressive comparison was pushed to higher-order in the PN expansion in subsequent work [58], also showing how by fitting to GSF data one can derive numerical values for [the $O(m)$ pieces of] higher-order, yet unknown PN coefficients. The quantity $u_1^t(\Omega)$ also served a reference point in a first comparison of GSF calculations carried out in different gauges [59]. Finally, in [60] the notion of redshift variable was generalized to eccentric orbits (using certain orbital averages); preliminary comparison with PN calculations in the eccentric case show a very good agreement [61].

In 2009 Barack and Sago computed the conservative GSF-induced shift in the frequency of the ISCO of a Schwarzschild black hole, using GSF analysis of slightly perturbed circular orbits [62]. They found, in fractional terms,

$$\left(\frac{\Delta\Omega}{\Omega}\right)_{\rm isco} = (0.25101546 \pm 0.00000005) \times \eta, \tag{8}$$

where the uncertainty is due to the finite numerical accuracy of the GSF computation. (We cite here the higher precision value obtained more recently in [63], and we account for a certain gauge correction introduced by Damour in [64], which "regularizes" the Lorenz-gauge time coordinate, known to be otherwise non-asymptotically-flat [65].) This was arguably a first concrete physical result, with a clear physical interpretation, to have emerged from the GSF program. Its importance was in that it provided a long-sought benchmark in the strong field. The ISCO shift result was immediately used as an accurate reference point in an exhaustive study of the performance of various PN methods [66]. It was also used to inform an "empirical" formula (based also on results from NR and PN) for the remnant masses and spins in binary black hole mergers [67], and to constrain some of the analytical parameters of the Effective One Body (EOB) potential [64].

The latter work, especially, highlighted the promise of a synergy between the GSF and other approaches. EOB was introduced by Buonnano and Damour in 1999 [68] as an analytical framework for modelling the two-body dynamics across all mass ratios (see T. Damour's contribution in this volume). At the heart of EOB is an effective one-body Hamiltonian, whose form is chosen to reproduce the known results at the test-particle limit, as well as all known PN results. The EOB Hamiltonian includes a number of "calibration" functions that can be adjusted by available NR data—and now also based of GSF information as it becomes available. In [64] Damour made the point that GSF results are particularly useful for calibrating EOB theory (even more so than NR data) given their accuracy, cleanness, and the fact that conservative effects can be computed separately. In this way, GSF calculations, whose validity is a priori restricted to the extreme mass ratio regime, can indirectly contribute to the development of a universal model of the two-body dynamics across all mass ratios.

In more recent work, Barack and Sago computed the GSF correction to the periastron precession of eccentric orbits around a Schwarzschild black hole [60]. In the limit of zero eccentricity the result is gauge invariant, and can be used to test the GSF prediction against that of PN theory in the weak field regime. This indeed was done in Ref. [69], where the precession results where also used to improve the calibration of EOB. In a recent culmination of this effort, a four-way collaboration between groups working on NR, PN, EOB and GSF presented a complete comparison between the predictions of all these methods, using the relativistic precession as a reference point [70]. The results of this study suggest, remarkably, that GSF theory may be applicable well beyond its natural extreme-mass-ratio domain. See Figs. 3 and 4 for an illustration.

In parallel, there has been progress in utilizing Detweiler's redshift variable $u_1^\alpha$ for EOB studies. This followed from a new formulation by Le Tiec and collaborators

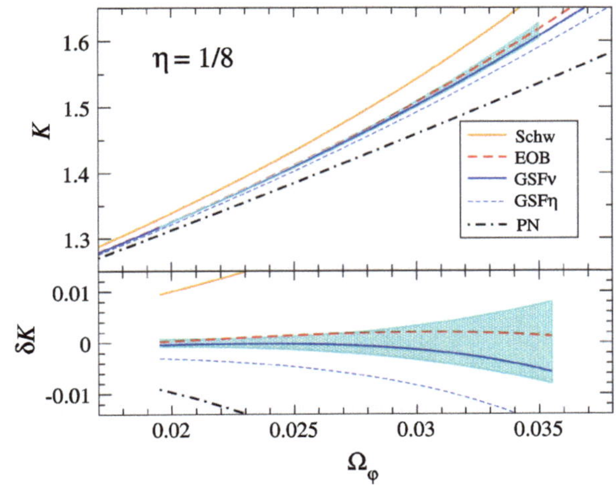


**Fig. 3** The relativistic periastron advance per orbit, $K$, for a close binary of mass ratio $\eta = m{:}M = 1{:}8$. As independent variable we use the invariant frequency $\Omega_{\varphi}$, given here in units of inverse-total-mass. (Note the periapsis advances by as much as *half a cycle per radial period* in this extreme regime of GR, corresponding to separations of just a few Schwarzschild radii.) The results, extracted from Ref. [70], show a comparison between the predictions of all methods available today. The *shaded region* comes from full NR simulations (with error margins), and the lower panel shows the relative difference between the predictions of each approximation method and the NR data. The *curve* labelled 'Schw' is the test-particle (geodetic) result, given for reference. 'GSF$\eta$' refers to the standard GSF prediction, whereas in 'GSF$\nu$' the mass ratio $\eta$ has been replaced with the *symmetric* mass ratio $\nu \equiv mM/(m+M)^2$ [this replacement is "allowed" since $\nu = \eta$ through $O(\eta)$]. 'PN' is the best available (3PN) PN result, and 'EOB' is a certain EOB model (see [70] for details). The GSF approximation, with the replacement $\eta \to \nu$, seems to perform remarkably well even though the mass ratio is not very extreme

[71] of a general "first law of binary black hole mechanics", relating infinitesimal variations of the total (invariant) energy and angular momentum of the binary system to variations of the individual rest masses—a relation which turned out to involve the redshift variable. Further work [72] then related the redshift variable to the EOB potentials, in a way that established a new useful link between GSF data and the EOB functions, leading to a complete determination of two of the main EOB functions at $O(m)$. Most recently [63], this analysis was extended to the very strong-field regime below the ISCO and down to the "light-ring" at $r = 3M$, revealing interesting new features of the EOB potentials.

## 6 Outlook

The primary ambition of the EMRI program (in the "experimental" context of low-frequency gravitational-wave astronomy) is to obtain a faithful model of the long-term orbital evolution in—and emitted gravitational waves from—compact-object

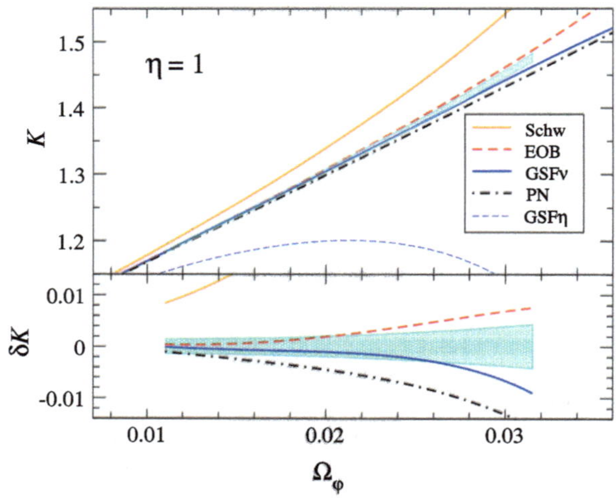


**Fig. 4** Same as in Fig. 3, now for $\eta = 1$ (equal masses). Even though this system is a priori well outside the reach of perturbation theory, the GSF$\nu$ prediction does extremely well, outperforming all other approximation methods. It remains to be seen whether this remarkable agreement is merely fortuitous or representative of a more general behavior

binaries of small mass ratios ($\eta \sim 10^{-4}$–$10^{-9}$), allowing both objects to spin. We are still quite far from achieving this goal. We have accomplished the important prerequisite of being able to compute the leading-order self-acceleration of orbits which are not evolving, in the Schwarzschild background case. But we are yet to learn how to extend this to orbits in Kerr, how to consistently evolve the orbit under the GSF effect, and how to incorporate the second-order GSF and the small object's spin in our calculations (both will be necessary in order to achieve the phase accuracy needed for LISA applications). At the same time, we have learned to appreciate how GSF calculations provide us with new, "high-fidelity" quantitative information about the dynamics in two-body systems, and we are beginning to learn how this can be used to test the faithfulness of PN calculations and inform the development of a universal EOB model. In that, GSF calculations are proving useful far beyond their original motivation. As the field matures, a wider range of applications become apparent. There has been recent work to explore the role of GSF in high-energy black hole scattering (of hypothetical relevance in experimental particle physics) [73, 74], and other applications are foreseeable.

Below we give a brief summary of (what we consider to be) the main challenges that lie ahead in the GSF field.

**Foundational issues:**—The first-order GSF is well understood at the foundational level, but the situation is less clear at second order. There currently exist at least two independent formulations of the second-order equation of motion, one by Pound [75, 76] and another by Gralla [77]. Both use (variants of) matched asymptotic expansions but each chooses to represent the motion in a rather different way, making a direct comparison nontrivial. It is important to understand the relation between the two formulations, and whether they are compatible. It is also important to translate (either or both) of the second-order formulations into a practical computational scheme for the GSF in Kerr geometry—perhaps akin to mode-sum regularization or to the puncture scheme. There is some initial work in this direction, but this problem is likely to remain an important research front for the coming years.

Tightly coupled to the question of a valid second-order formulation is the issue of long-term orbital evolution. Two-timescale expansion methods have been invoked [25, 78] to suggest an evolution scheme and control its error, but it is not yet clear how they would preform in practice (for example, how "runaway" self-acceleration terms would behave in a self-consistent evolution).

Generic inspirals in Kerr experience resonant episodes, where the ratio of the radial and longitudinal orbital frequencies crosses a low-order rational value, and the otherwise ergodic orbit becomes (quasi-)periodic. During resonant epochs radiation reaction acts on quite a different timescale (because the usual "averaging" effect of ergodicity is lost), leading effectively to a sudden jump in the values of the principal orbital elements, and a "resetting" of the orbital phases. If the goal is to obtain accurate phase-coherent waveforms for EMRI systems, it is important to derive an accurate model of the resonant crossing. Some recent work began to address this problem [79–81].

**Computational issues:—**A high priority task for the community is to extend existing methods and working codes for GSF calculations in Schwarzschild to a Kerr background. There are two main avenues of approach to the Kerr problem. In the *time-domain* approach one computes the metric perturbation due to the particle by evolving the linearized Einstein equations in the Lorenz gauge on a time-spatial grid using a finite-difference scheme. The 1 + 1D (time-radial) treatment that works so well in the Schwarzschild case is no longer useful in Kerr, because the Lorenz-gauge perturbation equations in Kerr are not separable (in any known way) into individual multipole modes in the time-domain. Instead, one has to work in 2 + 1D or 3 + 1D. This can be done using the puncture method described above, but so far implementations have been restricted to Schwarzschild (refraining from a 1+1D decomposition for the sake of preparing the ground for a Kerr implementation), or to a scalar field. One of the major technical obstacles in moving on to Kerr is the treatment of the "non-radiative" piece of the metric perturbation (the piece which reduces to the monopole and dipole modes in the Schwarzschild case). This piece has a numerically important contribution to the GSF, but so far attempts to compute it via time-domain evolution have been futile due to numerical instabilities. A few ad-hoc solutions to this problem were suggested recently in Ref. [47] (including the use of a judiciously chosen generalized Lorenz-gauge in an intermediate step), but they are yet to be tested in the Kerr case.

The alternative approach is based on a *frequency domain* treatment, in which the metric perturbation is solved for mode-by-mode in a multipole-Fourier decomposition. The advantage is that one now only deals with *ordinary* differential equations, but the method is only applicable for bound orbits of sufficiently small eccentricities. The problem of seperability in Kerr remains if one insists on working in the

Lorenz gauge, but there are suggestions to overcome this using a specially designed gauge in which (1) the metric perturbation is reconstructable from curvature scalars, which obey fully separable equations in Kerr (Teukolsky's formalism); and (2) the GSF is still mathematically well-defined and physically meaningful, as it is in the Lorenz gauge. A proposal of this kind was put forward long ago in [23], and a variant thereof is being under active development by Shah and collaborators [37, 82, 83]. This approach offers a computationally efficient route to the GSF, but much further development is needed.

**Synergy with other methods:—**The initial work described in Sect. 5 bears the promise of much further fruitful exchange between GSF and other computational approaches to the two-body problem, exploiting new GSF data as they become available. One of the challenges is to devise computable gauge-invariant quantities to facilitate a common language between the various approaches. Second-order GSF results, when at hand, will allow us to refine our comparisons and constrain the EOB potentials with exquisite accuracy. It is important to understand if and why the "GSF$\nu$" always provides a good (how good?) approximation even when the mass ratio is not extreme. What are the aspects of the dynamics in equal-mass binaries that can be modelled faithfully using purely perturbative methods?

GSF/PN comparison so far has been limited to circular or slightly eccentric orbits. There is merit in extending this comparison to fully eccentric orbits, which can be done using the available GSF results in Schwarzschild. Here are some specific invariant quantities that should be accessible (at least in principle) to existing Schwarzschild GSF codes, and could serve at contact points with—and strong-field benchmarks for—other approaches: (1) the GSF correction to the angular momentum and azimuthal frequency of a marginally-bound orbit on the capture threshold [64]; (2) the GSF correction to the Lyapunov exponent of unstable circular orbits below the ISCO; (3) the GSF correction to the function relating the azimuthal and radial frequencies on the "singular curve" identified in Appendix A of Ref. [60], where the transformation to orbital frequencies as system parameters becomes singular. Once GSF results in Kerr are available, one could compute the GSF correction to the Lense–Thirring precession of the orbital plane, and other spin-related effects of the GSF.

**Acknowledgments** This work was supported by the European Research Council under grant No. 304978; and by STFC in the UK through grant number PP/E001025/1.

## References

1. Hulse, R.A., Taylor, J.H.: Discovery of a pulsar in a binary system. Astrophys. J. **195**, L51 (1975). doi:10.1086/181708
2. Hahn, S.G., Lindquist, R.W.: The two-body problem in geometrodynamics. Ann. Phys. (N.Y.) 29, 304 (1964). doi:10.1016/0003-4916(64)90223-4
3. Smarr, L.: Space-time generated by computers: black holes with gravitational radiation. Ann. N.Y. Acad. Sci. **302**, 569 (1977). doi:10.1111/j.1749-6632.1977.tb37076.x

4. Pretorius, F.: Evolution of binary black-hole spacetimes. Phys. Rev. Lett. **95**, 121101 (2005). doi:10.1103/PhysRevLett.95.121101
5. Campanelli, M., Lousto, C.O., Marronetti, P., Zlochower, Y.: Accurate evolutions of orbiting black-hole binaries without excision. Phys. Rev. Lett. **96**, 111101 (2006). doi:10.1103/PhysRevLett.96.111101
6. Baker, J.G., Centrella, J., Choi, D.I., Koppitz, M., van Meter, J.: Gravitational-wave extraction from an inspiraling configuration of merging black holes. Phys. Rev. Lett. **96**, 111102 (2006). doi:10.1103/PhysRevLett.96.111102
7. Blanchet, L., Gravitational radiation from post-Newtonian sources and inspiralling compact binaries. Living Rev. Relativ. **9**, lrr-2006-4 (2006). http://www.livingreviews.org/lrr-2006-4
8. Einstein, A., Infeld, L., Hoffmann, B.: The gravitational equations and the problem of motion. Ann. Math. **39**, 65 (1938). doi:10.2307/1968714
9. Lorentz, H.A.: Theory of Electrons (1915) 2nd ed. Dover, New York (1952)
10. Dirac, P.A.M.: Classical theory of radiating electrons. Proc. R. Soc. Lond. Ser. A **167**, 148 (1938). doi:10.1098/rspa.1938.0124
11. DeWitt, B.S., Brehme, R.W.: Radiation damping in a gravitational field. Anna. Phys. **9**, 220 (1960). doi:10.1016/0003-4916(60)90030-0
12. Geroch, R., Traschen, J.: Strings and other distributional sources in general relativity. Phys. Rev. D **36**, 1017 (1987). doi:10.1103/PhysRevD.36.1017
13. LISA Project Office. http://lisa.nasa.gov
14. Gair, J.R., Barack, L., Creighton, T., et al.: Event rate estimates for lisa extreme mass ratio capture sources. Class. Quantum Grav. **21**, S1595 (2004). doi:10.1088/0264-9381/21/20/003
15. Barack, L., Cutler, C.: Using lisa extreme-mass-ratio inspiral sources to test off-kerr deviations in the geometry of massive black holes. Phys. Rev. D **75**, 042003 (2007). doi:10.1103/PhysRevD.75.042003
16. Mino, Y., Sasaki, M., Tanaka, T.: Gravitational radiation reaction to a particle motion. Phys. Rev. D **55**, 3457 (1997). doi:10.1103/PhysRevD.55.3457
17. Quinn, T.C., Wald, R.M.: Axiomatic approach to electromagnetic and gravitational radiation reaction of particles in curved spacetime. Phys. Rev. D **56**, 3381 (1997). doi:10.1103/PhysRevD.56.3381
18. Blanchet, L., Damour, T.: Hereditary effects in gravitational radiation. Phys. Rev. D **46**, 4304 (1992). doi:10.1103/PhysRevD.46.4304
19. Poisson, E., Pound, A., Vega, I.: The motion of point particles in curved spacetime. Living Rev. Relativ. **14**(7), lrr-2011-7 (2011). http://www.livingreviews.org/lrr-2011-7
20. Gralla, E., Wald, R.M.: A rigorous derivation of gravitational self-force. Class. Quantum Grav. **25**, 205009 (2008). doi:10.1088/0264-9381/25/20/205009
21. Pound, A.: Self-consistent gravitational self-force. Phys. Rev. D **81**, 024023 (2010). doi:10.1103/PhysRevD.81.024023
22. Harte, A.I.: Mechanics of extended masses in general relativity. Class. Quantum Grav. **29**, 055012 (2012). doi:10.1088/0264-9381/29/5/055012
23. Barack, L., Ori, A.: Gravitational self-force and gauge transformations. Phys. Rev. D **64**, 124003 (2001). doi:10.1103/PhysRevD.64.124003
24. Gralla, S.E.: Gauge and averaging in gravitational self-force. Phys. Rev. D **84**, 084050 (2011). doi:10.1103/PhysRevD.84.084050
25. Pound, A.: Singular perturbation techniques in the gravitational self-force problem. Phys. Rev. D **81**, 124009 (2010). doi:10.1103/PhysRevD.81.124009
26. Detweiler, S., Whiting, B.F.: Self-force via a green's function decomposition. Phys. Rev. D **67**, 024025 (2003). doi:10.1103/PhysRevD.67.024025
27. Barack, L., Ori, A.: Mode sum regularization approach for the self-force in black hole spacetime. Phys. Rev. D **61**, 061502 (2000). doi:10.1103/PhysRevD.61.061502
28. Barack, L.: Gravitational self-force in extreme mass-ratio inspirals. Class. Quantum Grav. **26**, 213001 (2009). doi:10.1088/0264-9381/26/21/213001
29. Barack, L., Ori, A.: Gravitational self-force on a particle orbiting a Kerr black hole. Phys. Rev. Lett. **90**, 111101 (2003). doi:10.1103/PhysRevLett.90.111101

30. Barack, L., Mino, Y., Nakano, H., Ori, A., Sasaki, M.: Calculating the gravitational self-force in Schwarzschild spacetime. Phys. Rev. Lett. **88**, 091101 (2002). doi:10.1103/PhysRevLett.88.091101
31. Barack, L., Sago, N.: Gravitational self-force on a particle in circular orbit around a Schwarzschild black hole. Phys. Rev. D **75**, 064021 (2007). doi:10.1103/PhysRevD.75.064021
32. Barack, L., Sago, N.: Gravitational self-force on a particle in eccentric orbit around a Schwarzschild black hole. Phys. Rev. D **81**, 084021 (2010). doi:10.1103/PhysRevD.81.084021
33. Akcay, S.: Fast frequency-domain algorithm for gravitational self-force: circular orbits in Schwarzschild spacetime. Phys. Rev. D **83**, 124026 (2011). doi:10.1103/PhysRevD.83.124026
34. Warburton, N., Akcay, S., Barack, L., Gair, J.R., Sago, N.: Evolution of inspiral orbits around a Schwarzschild black hole. Phys. Rev. D **85**, 061501 (2012). doi:10.1103/PhysRevD.85.061501
35. Warburton, N., Barack, L.: Self-force on a scalar charge in Kerr spacetime: circular equatorial orbits. Phys. Rev. D **81**, 084039 (2010). doi:10.1103/PhysRevD.81.084039
36. Warburton, N., Barack, L.: Self-force on a scalar charge in Kerr spacetime: eccentric equatorial orbits. Phys. Rev. D **83**, 124038 (2011). doi:10.1103/PhysRevD.83.124038
37. Shah, A.G., Friedman, J.L., Keidl, T.S.: Extreme-mass-ratio inspiral corrections to the angular velocity and redshift factor of a mass in circular orbit about a Kerr black hole. Phys. Rev. D **86**, 084059 (2012). doi:10.1103/PhysRevD.86.084059
38. Heffernan, A., Ottewill, A., Wardell, B.: High-order expansions of the Detweiler-Whiting singular field in Schwarzschild spacetime. Phys. Rev. D **86**, 104023 (2012). doi:10.1103/PhysRevD.86.104023
39. Heffernan, A., Ottewill, A., Wardell, B.: High-order expansions of the Detweiler-Whiting singular field in Kerr spacetime. Phys. Rev. D **89**, 024030 (2014). doi:10.1103/PhysRevD.89.024030
40. Barack, L., Golbourn, D.A.: Scalar-field perturbations from a particle orbiting a black hole using numerical evolution in $2+1$ dimensions. Phys. Rev. D **76**, 044020 (2007). doi:10.1103/PhysRevD.76.044020
41. Barack, L., Golbourn, D.A., Sago, N.: m-mode regularization scheme for the self-force in Kerr spacetime. Phys. Rev. D **76**, 124036 (2007). doi:10.1103/PhysRevD.76.124036
42. Vega, I., Detweiler, S.: Regularization of fields for self-force problems in curved spacetime: foundations and a time-domain application. Phys. Rev. D **77**, 084008 (2008). doi:10.1103/PhysRevD.77.084008
43. Dolan, S.R., Barack, L., Wardell, B.: Self-force via m-mode regularization and 2+1D evolution. II. Scalar-field implementation on Kerr spacetime. Phys. Rev. D 84, 084001 (2011). doi:10.1103/PhysRevD.84.084001
44. Wardell, B., Vega, I., Thornburg, J., Diener, P.: Generic effective source for scalar self-force calculations. Phys. Rev. D **85**, 104044 (2012). doi:10.1103/PhysRevD.85.104044
45. Vega, I., Diener, P., Tichy, W., Detweiler, S.: Self-force with (3+1) codes: a primer for numerical relativists. Phys. Rev. D **80**, 084021 (2009). doi:10.1103/PhysRevD.80.084021
46. Dolan, S.R., Barack, L.: Self-force via m -mode regularization and $2+1$D evolution: Foundations and a scalar-field implementation on Schwarzschild spacetime. Phys. Rev. D **83**, 024019 (2011). doi:10.1103/PhysRevD.83.024019
47. Dolan, S.R., Barack, L.: Self-force via m-mode regularization and 2+1D evolution: III. Gravitational field on Schwarzschild spacetime, ArXiv e-prints arXiv:1211.4586 [gr-qc] (2012)
48. Dolan, S., Barack, L.: Self-force via m-mode regularization and 2+1D evolution: IV. Gravitational field on Kerr spacetime. (In preparation)
49. Diener, P., Vega, I., Wardell, B., Detweiler, S.: Self-consistent orbital evolution of a particle around a Schwarzschild black hole. Phys. Rev. Lett. **108**, 191102 (2012). doi:10.1103/PhysRevLett.108.191102
50. Cañizares, P., Sopuerta, C.F.: Efficient pseudospectral method for the computation of the self-force on a charged particle: circular geodesics around a Schwarzschild black hole. Phys. Rev. D **79**, 084020 (2009). doi:10.1103/PhysRevD.79.084020
51. Thornburg, J.: Adaptive mesh refinement for characteristic grids. Gen. Rel. Grav. **43**, 1211 (2011). doi:10.1007/s10714-010-1096-z

52. Zenginoglu, A.: Hyperboloidal layers for hyperbolic equations on unbounded domains. J. Comput. Phys. **230**, 2286 (2011). doi:10.1016/j.jcp.2010.12.016
53. Darwin, C.: The gravity field of a particle. Proc. R. Soc. Lond. Ser. A **249**, 180 (1959). doi:10.1098/rspa.1959.0015
54. Pound, A., Poisson, E.: Multiscale analysis of the electromagnetic self-force in a weak gravitational field. Phys. Rev. D **77**, 044012 (2008). doi:10.1103/PhysRevD.77.044012
55. Gair, J.R., Flanagan, É., Drasco, S., Hinderer, T., Babak, S.: Forced motion near black holes. Phys. Rev. D **83**, 044037 (2011). doi:10.1103/PhysRevD.83.044037
56. Detweiler, S.L.: Perspective on gravitational self-force analyses. Class. Quantum Grav. **22**, S681 (2005). doi:10.1088/0264-9381/22/15/006
57. Detweiler, S.: Consequence of the gravitational self-force for circular orbits of the Schwarzschild geometry. Phys. Rev. D **77**, 124026 (2008). doi:10.1103/PhysRevD.77.124026
58. Blanchet, L., Detweiler, S., Le Tiec, A., Whiting, B.F.: Post-newtonian and numerical calculations of the gravitational self-force for circular orbits in the Schwarzschild geometry. Phys. Rev. D **81**(064004), 2010 (2010). doi:10.1103/PhysRevD.81.064004.Erratum:ibid.81,084033
59. Sago, N., Barack, L., Detweiler, S.: Two approaches for the gravitational self-force in black hole spacetime: comparison of numerical results. Phys. Rev. D **78**, 124024 (2008). doi:10.1103/PhysRevD.78.124024
60. Barack, L., Sago, N.: Beyond the geodesic approximation: conservative effects of the gravitational self-force in eccentric orbits around a Schwarzschild black hole. Phys. Rev. D **83**, 084023 (2011). doi:10.1103/PhysRevD.83.084023
61. Barack, L., Le Tiec, A., Sago, N.: In progress
62. Barack, L., Sago, N.: Gravitational self-force correction to the innermost stable circular orbit of a Schwarzschild black hole. Phys. Rev. Lett. **102**, 191101 (2009). doi:10.1103/PhysRevLett.102.191101
63. Akcay, S., Barack, L., Damour, T., Sago, N.: Gravitational self-force and the effective-one-body formalism between the innermost stable circular orbit and the light ring. Phys. Rev. D **86**, 104041 (2012). doi:10.1103/PhysRevD.86.104041
64. Damour, T.: Gravitational self-force in a Schwarzschild background and the effective one-body formalism. Phys. Rev. D **81**, 024017 (2010). doi:10.1103/PhysRevD.81.024017
65. Barack, L., Lousto, C.O.: Perturbations of Schwarzschild black holes in the Lorenz gauge: formulation and numerical implementation. Phys. Rev. D **72**, 104026 (2005). doi:10.1103/PhysRevD.72.104026
66. Favata, M.: Conservative self-force correction to the innermost stable circular orbit: comparison with multiple post-Newtonian-based methods. Phys. Rev. D **83**, 024027 (2011). doi:10.1103/PhysRevD.83.024027
67. Lousto, C.O., Nakano, H., Zlochower, Y., Campanelli, M.: Statistical studies of spinning black-hole binaries. Phys. Rev. D **81**, 084023 (2010). doi:10.1103/PhysRevD.81.084023
68. Buonanno, A., Damour, T.: Effective one-body approach to general relativistic two-body dynamics. Phys. Rev. D **59**, 084006 (1999). doi:10.1103/PhysRevD.59.084006
69. Barack, L., Damour, T., Sago, N.: Precession effect of the gravitational self-force in a Schwarzschild spacetime and the effective one-body formalism. Phys. Rev. D **82**, 084036 (2010). doi:10.1103/PhysRevD.82.084036
70. Le Tiec, A., Mroué, A.H., Barack, L., et al.: Periastron advance in black-hole binaries. Phys. Rev. Lett. **107**, 141101 (2011). doi:10.1103/PhysRevLett.107.141101
71. Le Tiec, A., Blanchet, L., Whiting, B.F.: First law of binary black hole mechanics in general relativity and post-Newtonian theory. Phys. Rev. D **85**, 064039 (2012). doi:10.1103/PhysRevD.85.064039
72. Barausse, E., Buonanno, A., Le Tiec, A.: Complete nonspinning effective-one-body metric at linear order in the mass ratio. Phys. Rev. D **85**, 064010 (2012). doi:10.1103/PhysRevD.85.064010
73. Barausse, E., Cardoso, V., Khanna, G.: Testing the cosmic censorship conjecture with point particles: the effect of radiation reaction and the self-force. Phys. Rev. D **84**, 104006 (2011). doi:10.1103/PhysRevD.84.104006

74. Gundlach, C., Akcay, S., Barack, L., Nagar, A.: Critical phenomena at the threshold of immediate merger in binary black hole systems: the extreme mass ratio case. Phys. Rev. D **86**, 084022 (2012). doi:10.1103/PhysRevD.86.084022
75. Pound, A.: Second-order gravitational self-force. Phys. Rev. Lett. **109**, 051101 (2012). doi:10.1103/PhysRevLett.109.051101
76. Pound, A.: Nonlinear gravitational self-force: field outside a small body. Phys. Rev. D **86**, 084019 (2012). doi:10.1103/PhysRevD.86.084019
77. Gralla, S.E.: Second-order gravitational self-force. Phys. Rev. D **85**, 124011 (2012). doi:10.1103/PhysRevD.85.124011
78. Hinderer, T., Flanagan, É.: Two-timescale analysis of extreme mass ratio inspirals in Kerr spacetime: orbital motion. Phys. Rev. D **78**, 064028 (2008). doi:10.1103/PhysRevD.78.064028
79. Gair, J., Yunes, N., Bender, C.M.: Resonances in extreme mass-ratio inspirals: asymptotic and hyperasymptotic analysis. J. Math. Phys. **53**, 032503 (2012). doi:10.1063/1.3691226
80. Flanagan, É., Hinderer, T.: Transient resonances in the inspirals of point particles into black holes. Phys. Rev. Lett. **109**, 071102 (2012). doi:10.1103/PhysRevLett.109.071102
81. Flanagan, É., Hughes, S.A., Ruangsri, U.: Resonantly enhanced and diminished strong-field gravitational-wave fluxes. ArXiv e-prints arXiv:1208.3906 [gr-qc] (2012)
82. Keidl, T.S., Shah, A.G., Friedman, J.L., Kim, D.H., Price, L.R.: Gravitational self-force in a radiation gauge. Phys. Rev. D **82**, 124012 (2010). doi:10.1103/PhysRevD.82.124012
83. Shah, A.G., Keidl, T.S., Friedman, J.L., Kim, D.H., Price, L.R.: Conservative, gravitational self-force for a particle in circular orbit around a Schwarzschild black hole in a radiation gauge. Phys. Rev. D **83**, 064018 (2011). doi:10.1103/PhysRevD.83.064018

# Hamiltonian Formalism for Spinning Black Holes in General Relativity

**Gerhard Schäfer**

**Abstract** A Hamiltonian treatment of gravitationally interacting spinning black holes is presented based on a tetrad generalization of the Arnowitt-Deser-Misner (ADM) canonical formalism of general relativity. The formalism is valid through linear order in the single spins. For binary systems, higher-order post-Newtonian Hamiltonians are given in explicit analytic forms. A next-to-leading order in spin generalization is presented, others are mentioned. Comparisons between the Hamiltonian formalisms by ADM, Dirac, and Schwinger are made.

## 1 Introduction

About half a century after Einstein's invention of general relativity, Hamiltonian formulations of the theory became available. Dirac developed his formalism in the years 1958–1959 [1–3], Arnowitt, Deser, and Misner (ADM) in the years 1959–1960 [4–6], for a summary see [7], and Schwinger in 1963 [8]. The three formalisms came up from quite different action functionals: Dirac used the Einstein action, ADM the Einstein-Hilbert action in Palatini form, and Schwinger the tetrad-generalized Palatini-based Einstein-Hilbert action. In the following, the three approaches are summarized and compared. Often in the article, the units $16\pi G = 1$ and $c = 1$ are applied.

In the context of his constraint dynamics formalism with various types of constraints, weakly and strongly vanishing ones, Dirac gave the full Hamiltonian, i.e. before applying the constraint equations and coordinate conditions, in the form, with $\partial_i$ denoting partial space-coordinate derivative and $i^0$ spatial infinity,

G. Schäfer (✉)
Friedrich-Schiller-Universität Jena, Theoretisch-Physikalisches Institut, Max-Wien-Platz 1, 07743 Jena, Germany
e-mail: gos@tpi.uni-jena.de

J. Bičák and T. Ledvinka (eds.), *General Relativity, Cosmology and Astrophysics*, Fundamental Theories of Physics 177, DOI: 10.1007/978-3-319-06349-2_7,

© Springer International Publishing Switzerland 2014

$$H_{\rm D} = -\int d^3x \partial_i (g^{-1/2}\partial_j(g\gamma^{ij})) + \int d^3x (N\mathscr{H} - N^i\mathscr{H}_i)$$
$$= -\oint_{i^0} dS_i g^{-1/2}\partial_j(g\gamma^{ij}) + \int d^3x(N\mathscr{H} - N^i\mathscr{H}_i) \quad (1)$$

(correctness of this form within the full non-linear theory was shown only in 1974, by Regge and Teitelboim [9]), whereas ADM and Schwinger gave the Hamiltonian in fully reduced forms, i.e. after applying constraint equations and appropriate coordinate conditions,

$$H_{\rm ADM} = \int d^3x \partial_i\partial_j(g_{ij} - \delta_{ij}g_{kk}) = \oint_{i^0} dS_i\partial_j(g_{ij} - \delta_{ij}g_{kk}), \quad (2)$$

$$H_{\rm S} = -\int d^3x \partial_i\partial_j(g\gamma^{ij}) = -\oint_{i^0} dS_i\partial_j(g\gamma^{ij}). \quad (3)$$

Though derived under specific coordinate conditions, here the form of the three-dimensional metric $g_{ij}$, $(i, j = 1, 2, 3)$, with inverse metric $\gamma^{ij}$, is still left general to better illuminate the general expressions behind. All three surface integrals coincide under the assumptions made: asymptotic flat spacetimes with coordinates of the form $N = 1 + O(1/r)$, $g_{ij} = \delta_{ij} + O(1/r)$, and $N^i = O(1/r)$ at spacelike infinity $(r \to \infty)$ including the first derivatives of the metric functions to decay as $1/r^2$. The Lagrangian multipliers $N \equiv (-g^{00})^{-1/2}$ and $N^i \equiv \gamma^{ij}g_{0i}$ are respectively coined "lapse" and "shift" functions by Wheeler [10]. The clear identification of $H_{\rm ADM}$ with the total energy of the system is one of the merits of ADM in the Hamiltonian approach to general relativity. The constraint equations read

$$\mathscr{H} = 0 \text{ and } \mathscr{H}_i = 0, \quad (4)$$

with the Hamilton density of weight one, used by ADM and Dirac,

$$\mathscr{H} \equiv -g^{1/2}{\rm R} + \frac{1}{g^{1/2}}\left(g_{ik}g_{jl}\pi^{ij}\pi^{kl} - \frac{1}{2}(g_{ij}\pi^{ij})^2\right) + \mathscr{H}_M \quad \text{(ADM)}$$
$$= B + \partial_i(g^{-1/2}\partial_j(g\gamma^{ij})) + \frac{1}{g^{1/2}}\left(g_{ik}g_{jl}\pi^{ij}\pi^{kl} - \frac{1}{2}(g_{ij}\pi^{ij})^2\right) + \mathscr{H}_M \quad \text{(D)}, \quad (5)$$

where

$$B \equiv \frac{1}{4}g^{1/2}g_{ij,k}g_{lm,n}[(\gamma^{il}\gamma^{jm} - \gamma^{ij}\gamma^{lm})\gamma^{kn} + 2(\gamma^{ik}\gamma^{lm} - \gamma^{il}\gamma^{mk})\gamma^{jn}], \quad (6)$$

or with Schwinger's weight-two density,

$$g^{1/2}\mathscr{H} = Q + \partial_i\partial_j q^{ij} + q^{ik}q^{jl}\Pi_{ij}\Pi_{kl} - (q^{ij}\Pi_{ij})^2 + g^{1/2}\mathscr{H}_M \quad \text{(S)}, \qquad (7)$$

where

$$Q \equiv -\frac{1}{4}q^{mn}\partial_m q^{kl}\partial_n q^{kl} - \frac{1}{2}q_{ln}\partial_m q^{kl}\partial_k q^{mn} - \frac{1}{2}q^{kl}\partial_k \ln(q^{1/2})\partial_l \ln(q^{1/2}), \qquad (8)$$

with $\Pi_{ij} = -g^{-1}(\pi_{ij} - \frac{1}{2}\pi g_{ij})$ and $q^{ij} = g\gamma^{ij}$, $q = g^2$, where $g$ denotes the determinant of $g_{ij}$ and lowering of indices is with $g_{ij}$, $\pi = g_{ij}\pi^{ij}$. The Dirac and ADM canonical field momentum is given by $\pi_{ij} = -g^{1/2}(K_{ij} - Kg_{ij})$, with $K = \gamma^{ij}K_{ij}$, where $K_{ij} = -N\Gamma^0_{ij}$ is the extrinsic curvature of a spacelike hypersurface defined through constant-in-time slice, $t = x^0 = \text{const}$. $\Gamma^0_{ij}$ denote Christoffel symbols. Schwinger's canonical field momentum $\Pi_{ij}$ is just $g^{-1/2}K_{ij}$. The intrinsic curvature scalar reads $R$. The expressions $g_{ij}dx^i dx^j$ and $K_{ij}dx^i dx^j$ are respectively called first and second fundamental form of a hypersurface. Both tensors $g_{ij}$ and $K_{ij}$ are symmetric.

The momentum density of weight one takes the forms

$$\begin{aligned}\mathscr{H}_i &\equiv 2g_{ij}D_k\pi^{jk} + \mathscr{H}_{Mi} \quad \text{(ADM and D)}\\ &= -\Pi_{lm}\partial_i q^{lm} + \partial_i(2\Pi_{lm}q^{lm}) - \partial_l(2\Pi_{im}q^{lm}) + \mathscr{H}_{Mi} \quad \text{(S)},\end{aligned} \qquad (9)$$

where $D_i$ denotes the three-dimensional covariant space derivative. The given densities are densities with respect to three-dimensional coordinate transformations. $\mathscr{H}_M$ and $\mathscr{H}_{Mi}$ are matter densities.

In 1967, DeWitt [11] and more refined later on in 1974, Regge and Teitelboim [9] explicitly showed that in asymptotically flat spacetimes the Hamiltonian, before applying constraint equations and coordinate conditions, takes the form

$$H = \oint_{i^0} dS_i\partial_j(g_{ij} - \delta_{ij}g_{kk}) + \int d^3x(N\mathscr{H} - N^i\mathscr{H}_i). \qquad (10)$$

This general Hamiltonian—not yet to be identified with the energy of the system—delivers all field equations also including those for the lapse and shift functions after imposing appropriate coordinate conditions.

Undoubtedly, originating from the very useful coordinate choice made by ADM—one may call it "maximal isotropic" —,

$$\pi^{ii} = 0, \quad 3\partial_j g_{ij} - \partial_i g_{jj} = 0 \text{ or } g_{ij} = \psi\delta_{ij} + h^{\text{TT}}_{ij}, \qquad (11)$$

the ADM formalism became the most often applied canonical formalism. The independent field variables therein are $\pi^{ij}_{\text{TT}}$ and $h^{\text{TT}}_{ij}$. Both variables are traceless and tracefree (TT). Already in 1961 Kimura [12] used this formalism for applications. The Poisson bracket reads

$$\{F, G\} = \delta_{ij}^{\mathrm{TT}kl}\left(\frac{\delta F}{\delta h_{ij}^{\mathrm{TT}}}\frac{\delta G}{\delta \pi_{\mathrm{TT}}^{kl}} - \frac{\delta G}{\delta h_{ij}^{\mathrm{TT}}}\frac{\delta F}{\delta \pi_{\mathrm{TT}}^{kl}}\right), \tag{12}$$

with

$$\delta_{ij}^{\mathrm{TT}kl} = \frac{1}{2}(P_{il}P_{jk} + P_{ik}P_{jl} - P_{kl}P_{ij}), \quad P_{ij} = \delta_{ij} - \frac{\partial_i \partial_j}{\nabla^2}, \tag{13}$$

where $1/\nabla^2$ denotes the inverse Laplacian. The nonlocality of the TT-operator $\delta_{ij}^{\mathrm{TT}kl}$ is just the gravitational analogue of the well-known nonlocality of the Coulomb gauge in the electrodynamics. If Schwinger would have chosen coordinate conditions corresponding to those introduced above (ADM also introduced another set of coordinate conditions to which Schwinger adjusted),

$$\Pi_{ii} = 0, \quad q^{ij} = \varphi\delta_{ij} + f_{\mathrm{TT}}^{ij}, \tag{14}$$

a similar simple technical formalism for detailed calculations would have resulted with the independent field variables $\Pi_{ij}^{\mathrm{TT}}$ and $f_{\mathrm{TT}}^{ij}$. To our best knowledge, only the paper by Kibble [13] delivers an application of Schwinger's formalism, apart from Schwinger himself, namely a Hamilton formulation of the Dirac spinor field in gravity. Much later in 1978, Nelson and Teitelboim [14] completed the same task within the tetrad-generalized Dirac formalism [15]. The Poisson bracket in the Schwinger formalism resembles very much the ADM one.

In terms of the ADM variables, Schwinger's fundamental field components take the quite simple form, [16],

$$q^{ij} = (\psi^2 - \frac{1}{2}h_{kl}^{\mathrm{TT}}h_{kl}^{\mathrm{TT}})\delta_{ij} - \psi h_{ij}^{\mathrm{TT}} + h_{ik}^{\mathrm{TT}}h_{kj}^{\mathrm{TT}}. \tag{15}$$

Dirac on the other side had chosen the following coordinate system or gauge, called "maximal slicing" because of the field momentum condition,

$$\pi = 0, \quad \partial_j(g^{1/3}\gamma^{ij}) = 0. \tag{16}$$

The corresponding independent field variables are

$$\tilde{\pi}^{ij} = (\pi^{ij} - \frac{1}{3}\gamma^{ij}\pi)g^{1/3}, \quad \tilde{g}_{ij} = g^{-1/3}g_{ij}. \tag{17}$$

To leading order linear in the metric functions, the Dirac gauge coincides with the ADM gauge. The full reduction of the Dirac-form dynamics to the independent degrees of freedom has been performed by Regge and Teitelboim [9]. The Poisson bracket reads

$$\{F, G\} = \tilde{\delta}^{kl}_{ij} \left( \frac{\delta F}{\delta \tilde{g}_{ij}} \frac{\delta G}{\delta \tilde{\pi}^{kl}} - \frac{\delta G}{\delta \tilde{g}_{ij}} \frac{\delta F}{\delta \tilde{\pi}^{kl}} \right) + \frac{1}{3} (\tilde{\pi}^{ij} \tilde{g}^{kl} - \tilde{\pi}^{kl} \tilde{g}^{ij}) \frac{\delta F}{\delta \tilde{\pi}^{ij}} \frac{\delta G}{\delta \tilde{\pi}^{kl}}, \tag{18}$$

with

$$\tilde{\delta}^{kl}_{ij} = \frac{1}{2}(\delta^k_i \delta^l_j + \delta^l_i \delta^k_j) - \frac{1}{3} \tilde{g}_{ij} \tilde{g}^{kl}, \quad \tilde{g}_{ij} \tilde{g}^{jl} = \delta^l_i. \tag{19}$$

The following quoted results, apart from the last one, are based on the ADM approach. The first post-Newtonian (1PN) equations of motion, usually called Einstein-Infeld-Hoffmann equations of motion, were rederived by Kimura in 1961 [12]. In 1974, Ohta et al. [17] derived the 2PN binary equations with shortcomings corrected by Damour and Schäfer [18] only much later. About the same time, in 1985, Schäfer succeeded with the dissipative 2.5PN level, [16]. Using for the first time in post-Newtonian calculations the technique of dimensional regularization, the 3PN binary Hamiltonian was obtained in 2001 by Damour et al. [19]. Much simpler to derive was the dissipative 3.5PN level [20]. Based on a post-linear paper by Schäfer [21], the first post-Minkowskian (1PM) n-body Hamiltonian was achieved in closed form by Ledvinka, Schäfer, and Bičák in 2008 [22]. The general relativistic Hamilton dynamics of compact objects with spin (proper rotation) has found several explicit results. Counting the order of the spin as $1/c$ (notice the spin of a miximally rotating black hole reading $GM^2/c$), the leading order spin-orbit coupling is decribed by an 1.5PN Hamiltonian [23], the next to leading order one by a 2.5PN Hamiltonian, [24, 25], and the next-to-next to leading order one by a 3.5PN Hamiltonain [26]. The leading order radiation damping from spin-orbit coupling is decribed by an 4PN Hamiltonian [27]. In case of spin(1)–spin(2) coupling, the conservative 2PN Hamiltonian is given in [23], the 3PN Hamiltonian by [25, 28], and the 4PN one by [29]. The 4.5PN Hamiltonian of the leading order radiation damping from spin(1)–spin(2) coupling is given in [30]. Results on the spin(1)–spin(1) coupling are the 2PN and 3PN Hamiltonians for black holes by respectively [23] and [31]. Several leading higher-order-in-spin Hamiltonians were obtained from Kerr metric considerations, [32]. Spinning test-particles in the Kerr metric have been treated by Barausse, Racine, and Buonanno in 2009, [33] using Dirac's constraint dynamics formalism.

## 2 Analytic Representation of Binary Black Holes: The Brill-Lindquist Initial Value Solution

The model used in this article to describe compact objects are Dirac delta functions. In this section it will be shown that Dirac delta functions can indeed be used to describe black holes in interaction with each other.

An isolated black hole with mass $m$ is represented through the line element

$$ds^2 = -\left(\frac{1-\frac{Gm}{2rc^2}}{1+\frac{Gm}{2rc^2}}\right)^2 c^2dt^2 + \left(1+\frac{Gm}{2rc^2}\right)^4 \delta_{ij}dx^idx^j$$
$$= -\left(\frac{1-\frac{Gm}{2Rc^2}}{1+\frac{Gm}{2Rc^2}}\right)^2 c^2dt^2 + \left(1+\frac{Gm}{2Rc^2}\right)^4 \delta_{ij}dX^idX^j, \quad (20)$$

when isotropic coordinates $x^i$ with $r^2 = x^ix^i$ and $X^i$ with $R^2 = X^iX^i$ are employed. The two coordinate systems are related through the inversion map $R = \left(\frac{Gm}{2c^2}\right)^2/r$. Obviously, the line element shows isometry under this map. For binary black holes, initially at rest, the metric at that instant of time may have the form

$$ds^2 = -\left(\frac{1-\frac{\beta_1 G}{2r_1c^2}-\frac{\beta_2 G}{2r_2c^2}}{1+\frac{\alpha_1 G}{2r_1c^2}+\frac{\alpha_2 G}{2r_2c^2}}\right)^2 c^2dt^2 + \left(1+\frac{\alpha_1 G}{2r_1c^2}+\frac{\alpha_2 G}{2r_2c^2}\right)^4 d\mathbf{x}^2, \quad (21)$$

as shown by Brill and Lindquist in 1963, [34], for the space part of the given metric. The time part of the given metric has been derived later by Jaranowski and Schäfer [35]. The coefficients $\alpha_a$ and $\beta_a$, $a = 1, 2$ do depend on the masses and the relative coordinate distance of the two black holes $r_{12}^2 = (x_1^i - x_2^i)(x_1^i - x_2^i)$. Furthermore, $r_a^2 = (x^i - x_a^i)(x^i - x_a^i)$, where $x_a^i$ are the position vectors of the black holes. Notice, our variables are living in an euclidean space $x^i$ which is conformally related with the physical one. The geometrical interpretation of Brill-Lindquist black holes are two Einstein-Rosen bridges, both starting in the same physical space but ending in two different other ones. The energy of the Brill-Lindquist black hole configuration reads

$$E_{ADM} = -\frac{c^4}{2\pi G}\oint_{i_0} dS_i\partial_i\Psi = (\alpha_1+\alpha_2)c^2, \quad (22)$$

where $\Psi = 1 + \frac{\alpha_1 G}{2r_1c^2} + \frac{\alpha_2 G}{2r_2c^2}$, where $\Delta\Psi = 0$ for $x^i \neq x_1^i, x_2^i$ with $\Delta \equiv \nabla^2$. The inversion map of the three-metric of black hole, say 1, at its throat reads $r_1'^i = r_1^i\alpha_1^2G^2/4c^4r_1^2$, where $r_1'^i = x'^i - x_1^i$, $r_1^i = x^i - x_1^i$, $r_1 = |x^i - x_1^i|$. The three-metric line element $dl^2$ takes the forms

$$dl^2 = \Psi^4 d\mathbf{x}^2 = \left(1+\frac{\alpha_1 G}{2r_1c^2}+\frac{\alpha_2 G}{2r_2c^2}\right)^4 d\mathbf{x}^2$$
$$= \Psi'^4 d\mathbf{x}'^2 = \left(1+\frac{\alpha_1 G}{2r_1'c^2}\left(1+\frac{\alpha_2 G}{2r_2c^2}\right)\right)^4 d\mathbf{x}'^2, \quad (23)$$

with $\Psi' = 1 + \frac{\alpha_1 G}{2r_1'c^2} + \frac{\alpha_1\alpha_2 G^2}{4r_2r_1'c^4}$ and $r_2^i = \frac{\alpha_1^2G^2}{4c^4}\frac{r_1'^i}{r_1'^2} + r_{12}^i$, $r_{12}^i = r_1^i - r_2^i = x_2^i - x_1^i$. Hereof, by definition, the rest-mass of black hole 1 comes out to read

$$m_1 \equiv -\frac{c^2}{2\pi G} \oint_{i_0} dS'_i \partial'_i \Psi' = \alpha_1 + \frac{\alpha_1 \alpha_2 G}{2 r_{12} c^2}. \tag{24}$$

The calculations presented above treat the metric functions geometrically as pure vacuum solution without sources. No divergences occur. Infinities only mean infinite distances. In the following it will be shown how to reconstruct the Brill-Lindquist initial value solution with the aid of Dirac delta functions.

Let us have a look at the constraint equations with point-mass sources,

$$g^{1/2}\mathrm{R} - \frac{1}{g^{1/2}}\left(\pi^i_j \pi^j_i - \frac{1}{2}\pi^i_i \pi^j_j\right) = \frac{16\pi G}{c^3} \sum_a \left(m_a^2 c^2 + \gamma^{ij} p_{ai} p_{aj}\right)^{1/2} \delta_a \tag{25}$$

(identical with $2\sqrt{-g}G^{00} = \frac{16\pi G}{c^4}\sqrt{-g}T^{00}$ in standard notation),

$$-2\partial_j \pi^j_i + \pi^{kl}\partial_i g_{kl} = \frac{16\pi G}{c^3} \sum_a p_{ai}\delta_a, \tag{26}$$

($2\sqrt{-g}G^0_i = \frac{16\pi G}{c^4}\sqrt{-g}T^0_i$). $\delta_a$ is an abbreviation of $\delta(x^i - x^i_a)$ with $x^i_a$ the position vector of mass $a$ ($\int d^3x \delta_a = 1$). Using the ADM gauge in the form

$$g_{ij} = \left(1 + \frac{1}{8}\phi\right)^4 \delta_{ij} + h^{\mathrm{TT}}_{ij}, \quad [3\partial_j g_{ij} - \partial_i g_{jj} = 0],$$
$$\pi^{ii} = 0, \text{ or } \pi^{\mathrm{ij}} = \partial_{\mathrm{i}}\pi^{\mathrm{j}} + \partial_{\mathrm{j}}\pi^{\mathrm{i}} - \frac{2}{3}\delta_{\mathrm{ij}}\partial_{\mathrm{k}}\pi^{\mathrm{k}} + \pi^{\mathrm{ij}}_{\mathrm{TT}}, \tag{27}$$

the constraint equations simplify to the following equation, imposing $h^{\mathrm{TT}}_{ij} = \pi^{ij}_{\mathrm{TT}} = p_{ai} = 0$,

$$-\left(1 + \frac{1}{8}\phi\right)\Delta\phi = \frac{16\pi G}{c^2} \sum_a m_a \delta_a. \tag{28}$$

With the aid of the ansatz

$$\phi = \frac{4G}{c^2}\left(\frac{\alpha_1}{r_1} + \frac{\alpha_2}{r_2}\right), \tag{29}$$

after Hadamard partie finie regularization, the constraint equation yields,

$$m_a = \alpha_a + \frac{\alpha_a \alpha_b G}{2 r_{ab} c^2}, \quad a \neq b, \tag{30}$$

$$\alpha_a = m_a - \frac{m_a + m_b}{2} + \frac{c^2 r_{ab}}{G}\left(\sqrt{1 + \frac{m_a + m_b}{c^2 r_{ab}/G} + \left(\frac{m_a - m_b}{2c^2 r_{ab}/G}\right)^2} - 1\right). \quad (31)$$

The Hamiltonian clearly results in

$$H_{\rm BL} = (\alpha_1 + \alpha_2)\, c^2 = (m_1 + m_2)\, c^2 - G\,\frac{\alpha_1 \alpha_2}{r_{12}}. \quad (32)$$

The d-dimensional generalization of the above treatment runs as follows. The d-metric functions read,

$$g_{ij} = \Psi^{\frac{4}{d-2}} \delta_{ij}, \quad \Psi = 1 + \frac{1}{4}\frac{d-2}{d-1}\phi,$$
$$\phi = \frac{4G}{c^2}\frac{\Gamma(\frac{d-2}{2})}{\pi^{\frac{d-2}{2}}}\left(\frac{\alpha_1}{r_1^{d-2}} + \frac{\alpha_2}{r_2^{d-2}}\right). \quad (33)$$

The d-dimensional inverse Laplacian $\Delta^{-1}$ takes the form,

$$-\Delta^{-1}\delta_a = \frac{\Gamma((d-2)/2)}{4\pi^{d/2}} r_a^{2-d}, \quad (34)$$

where $\Gamma$ denotes the Euler gamma function. The ansatz for $\Psi$ thus reads

$$\Psi = 1 + \frac{G(d-2)\Gamma((d-2)/2)}{c^2(d-1)\pi^{(d-2)/2}}\left(\frac{\alpha_1}{r_1^{d-2}} + \frac{\alpha_2}{r_2^{d-2}}\right) \quad (35)$$

and the constraint equation takes the form

$$\left(1 + \frac{G(d-2)\Gamma((d-2)/2)}{c^2(d-1)\pi^{(d-2)/2}}\left(\frac{\alpha_1}{r_1^{d-2}} + \frac{\alpha_2}{r_2^{d-2}}\right)\right)\alpha_a \delta_a = m_a \delta_a. \quad (36)$$

Chosing $1 < d < 2$, a fully finite result comes in the form,

$$\left(1 + \frac{G(d-2)\Gamma((d-2)/2)}{c^2(d-1)\pi^{(d-2)/2}}\frac{\alpha_b}{r_{12}^{d-2}}\right)\alpha_a \delta_a = m_a \delta_a, \quad a \neq b. \quad (37)$$

Analytic continuation through $d = 3$ can now be performed without facing divergences.

## 3 Spin in Minkowski Space

Before entering into gravity in asymptotically flat spacetimes, a discussion of spin in Minkowski spacetime seems most convenient. Using canonical variables, the total angular momentum $\mathbf{J} = (J^i) = (J_i)$ of a particle with spin vector $\hat{\mathbf{S}}$ reads

$$\mathbf{J} = \hat{\mathbf{X}} \times \mathbf{P} + \hat{\mathbf{S}}, \tag{38}$$

where $\hat{\mathbf{X}}$ denotes the canonical position vector and $\mathbf{P}$ its canonical linear momentum. The Lorentz boost is given by

$$\mathbf{K} \equiv -t\mathbf{P} + \mathbf{G} = -t\mathbf{P} + \mathrm{H}\hat{\mathbf{X}} - \frac{1}{H+m}\hat{\mathbf{S}} \times \mathbf{P}, \tag{39}$$

with the center-of-energy vector $\mathbf{G}$. The free-particle Hamiltonian $H = \sqrt{m^2 + \mathbf{P}^2}$.

The center-of-energy position vector is given by

$$\bar{\mathbf{X}} = \hat{\mathbf{X}} - \frac{1}{(H+m)H}\,\hat{\mathbf{S}} \times \mathbf{P}, \tag{40}$$

thus, $\mathbf{G} = H\bar{\mathbf{X}}$. The center-of-spin vector, or Newton-Wigner position vector, is defined by $\hat{\mathbf{X}}$. The Poisson brackets of its components vanish $\{\hat{X}^i, \hat{X}^j\} = 0$. The center-of-energy position vector $\bar{\mathbf{X}} = \hat{\mathbf{X}} - \frac{1}{(H+m)H}\,\hat{\mathbf{S}} \times \mathbf{P}$ has non-vanishing Poisson brackets of its components and the center-of-inertia position vector $\mathbf{X} = \hat{\mathbf{X}} + \frac{1}{(H+m)m}\,\hat{\mathbf{S}} \times \mathbf{P}$ as well. Using the center-of-inertia position vector, in four-dimensional language, $S^{\mu\nu}P_\nu = 0$ holds, for the center-of-energy position vector $\bar{S}^{\mu\nu}n_\nu = 0$, $n_\mu = (-1, 0, 0, 0)$ is valid, and for the center-of-spin position vector $m\hat{S}^{\mu\nu}n_\nu + \hat{S}^{\mu\nu}P_\nu = 0$ happens. The reason for those different spin-supplementary conditions is rooted in the invariance of the total angular momentum against shift of coordinates.

For isolated systems, the Poincaré algebra is valid,

$$\begin{aligned}
&\{P_i, \mathrm{H}\} = \{J_i, H\} = 0, \quad \{J_i, P_j\} = \varepsilon_{ijk}\,P_k, \\
&\{J_i, J_j\} = \varepsilon_{ijk}\,J_k, \qquad\quad \{J_i, G_j\} = \varepsilon_{ijk}\,G_k, \quad \{G_i, H\} = P_i, \\
&\{G_i, P_j\} = \frac{1}{c^2}\,H\,\delta_{ij}, \qquad \{G_i, G_j\} = -\frac{1}{c^2}\,\varepsilon_{ijk}\,J_k.
\end{aligned} \tag{41}$$

$G_i$ is not a constant of motion, but $K_i$ is,

$$d\mathbf{K}/dt = \partial\mathbf{K}/\partial t + \{\mathbf{K}, H\} = -\mathbf{P} + \{\mathbf{G}, H\} = 0. \tag{42}$$

For many-particle systems with interaction, it generally holds,

$$\mathbf{P} = \sum_a \mathbf{p}_a, \quad \mathbf{J} = \sum_a (\mathbf{r}_a \times \mathbf{p}_a + \mathbf{s}_a), \tag{43}$$

$$M \equiv \sqrt{H^2 - \mathbf{P}^2}, \quad H = \sqrt{M^2 + \mathbf{P}^2}, \tag{44}$$

$$\hat{\mathbf{X}} = \frac{\mathbf{G}}{H} + \frac{1}{M(H+M)}(\mathbf{J} - \frac{\mathbf{G}}{H} \times \mathbf{P}) \times \mathbf{P}, \tag{45}$$

with

$$\{\hat{X}^i, \hat{X}^j\} = \{P^i, P^j\} = 0, \quad \{\hat{X}^i, P^j\} = \delta^{ij},$$
$$\{M, \hat{X}^j\} = \{M, P^j\} = \{M, H\} = 0. \tag{46}$$

For free particles with spin, one has, $H = \sum_a h_a$, with $h_a = \sqrt{m_a^2 + \mathbf{p}_a^2}$, and $\mathbf{G} = \sum_a (h_a \mathbf{r}_a - \frac{1}{h_a + m_a} \mathbf{s}_a \times \mathbf{p}_a)$.

## 4 Spin and Gravity: Asymptotic Flat Spacetimes

The treatment of spin in gravity is most conveniently achieved by the introduction of a tetrad field $e_a^\mu$ ($\mu = 0, 1, 2, 3; a = 0, 1, 2, 3$) having the properties $e_a^\mu e_{b\mu} = \eta_{ab}$ and $e_{a\mu} e_{b\nu} \eta^{ab} = g_{\mu\nu} = g_{\nu\mu}$. Local Lorentz transformations are defined by $e_a'^\mu = L^b{}_a e_b^\mu$ with $L^a{}_c \eta_{ab} L^b{}_d = \eta_{cd}$. The condition of homogeneous transformation of the spacetime derivative of a physical object $\phi$ under local Lorentz transformations introduces a linear connection $\omega_\mu^{ab}$,

$$D_\mu \phi \equiv \partial_\mu \phi + \frac{1}{2} \omega_\mu^{ab} G_{[ab]} \phi, \quad \partial_\mu \equiv \frac{\partial}{\partial x^\mu}, \tag{47}$$

with transformation property $\omega_\mu'^{ab} = L^a{}_c L^b{}_d \omega_\mu^{cd} + L^a{}_d \partial_\mu L^{bd}$. The object $G_{[ab]}$ is defined through infinitesimal local Lorentz transformations of $\phi$, $\delta\phi = -\delta\xi^{[ab]} G_{[ab]}\phi$, with infinitesimal group parameters $\delta\xi^{[ab]}$.

The curvature tensor $R_{\mu\nu}^{ab}$ is defined by

$$D_\mu D_\nu \phi - D_\nu D_\mu \phi = R_{\mu\nu}^{ab} G_{[ab]} \phi, \tag{48}$$

$$R_{\mu\nu}^{ab} = \partial_\mu \omega_\nu^{ab} - \partial_\nu \omega_\mu^{ab} + \omega_\nu^{ac} \omega_\mu^{bd} \eta_{cd} - \omega_\mu^{ac} \omega_\nu^{bd} \eta_{cd}. \tag{49}$$

The simplest Lagrangian density (apart from the herein trivial cosmological constant) for the gravitational field reads,

$$\mathscr{L}_G = \det(e_\gamma^c) e_a^\mu e_b^\nu R_{\mu\nu}^{ab}(\omega) + \partial_\mu \mathscr{C}^\mu, \tag{50}$$

where the exact divergence $\partial_\mu \mathscr{C}^\mu$ is needed to render the variational principle valid for variations with prescribed properties at the boundary of the four-dimensional integration area,

The vacuum field equations read

$$0 = \frac{\delta \mathscr{L}_G}{\delta e^\mu_a} = \det(e^c_\gamma)(2R^{ab}_{\mu\sigma} e^\sigma_b - e^a_\mu R^{cb}_{\rho\sigma} e^\rho_c e^\sigma_b), \tag{51}$$

$$0 = \frac{\delta \mathscr{L}_G}{\delta \omega^{ab}_\mu} \quad \Rightarrow \quad \omega^{ab}_\mu = \omega^{ab}_\mu(e, \partial_\nu e). \tag{52}$$

The latter equation reduces the gravitational field to the Einsteinian one with suppressed torsion. This property will be kept in the following when treating spinning sources of the gravitational field.

The matter action $W_M = \int d^4x[\mathscr{L}_M + \mathscr{L}_C]$ for a spinning classical object can be put into the form, e.g. [36], where the Lagrangian density of the dynamical part reads,

$$\mathscr{L}_M = \int d\tau \left[ \left( p_\mu - \frac{1}{2} S_{ab}\, \omega^{ab}_\mu \right) \frac{dz^\mu}{d\tau} + \frac{1}{2} S_{ab} \frac{d\theta^{ab}}{d\tau} \right] \delta_{(4)} \tag{53}$$

and where the constraints part is given by

$$\mathscr{L}_C = \int d\tau \left[ \lambda^a_1 p^b S_{ab} + \lambda_{2[i]} \Lambda^{[i]a} p_a - \frac{\lambda_3}{2}(p^2 + m^2) \right] \delta_{(4)}. \tag{54}$$

$\delta_{(4)}$ is a four-dimensional Dirac delta function, $\delta_{(4)} = \delta(x^\mu - z^\mu)$, ($\int d^4x \delta_{(4)} = 1$), $\tau$ is a proper time variable, $p_\mu$ and $z^\mu$ denote respectively the four-dimensional kinetic momentum form and position vector of the particle and $S_{ab}$ is its spin tensor. The angle variables $\theta^{ab}$ with $d\theta^{ab} = \Lambda^{a}_{C} d\Lambda^{Cb} = -d\theta^{ba}$ are anholonomic ones. Capital indices refer to body-fixed Lorentz-frame coordinates. $\lambda_1$, $\lambda_2$, and $\lambda_3$ are Lagrangian multipliers.

The equations of motion resulting hereof reads,

$$\frac{DS_{ab}}{D\tau} = 0, \tag{55}$$

$$\frac{Dp_\mu}{D\tau} = -\frac{1}{2} \mathrm{R}_{\mu\rho ab} u^\rho S^{ab}, \quad u^\mu \equiv \frac{dz^\mu}{d\tau} = \lambda_3 p^\mu. \tag{56}$$

They completely correspond to the divergence freeness or dynamics of the Tulczyjew stress-energy tensor for pole-dipole particles,

$$N\sqrt{g} T^{\mu\nu} = \int d\tau \left[ \lambda_3 p^\mu p^\nu \delta_{(4)} + \left( u^{(\mu} S^{\nu)\alpha} \delta_{(4)} \right)_{||\alpha} \right]. \tag{57}$$

The index symbol || denotes covariant four-dimensional derivative.

## 4.1 Hamiltonian for Self-Gravitating Spinning Compact Objects

Variation of the matter action with respect to the Lagrangian multipliers $\lambda_1$, $\lambda_2$, and $\lambda_3$ results in the relations, respectively,

$$nS_i \equiv n^\mu S_{\mu i} = \frac{p_k \gamma^{kj} S_{ji}}{np} = g_{ij} n S^j, \tag{58}$$

$$\Lambda^{[j](0)} = \Lambda^{[j](i)} \frac{p_{(i)}}{p^{(0)}}, \quad \Lambda^{[0]a} = -\frac{p^a}{m}, \tag{59}$$

$$np \equiv n^\mu p_\mu = -\sqrt{m^2 + \gamma^{ij} p_i p_j}, \quad \gamma^{ik} g_{kj} = \delta^i_j, \tag{60}$$

where $n^\mu = (1, -N^i)/N, \quad n_\mu = (-N, 0, 0, 0)$.

To fix the tetrad field, the so-called time gauge (Schwinger's coining, but introduced by Dirac earlier) proves extremely useful,

$$e^\mu_{(0)} = n^\mu, \quad \text{i.e. } e^0_{(0)} = \frac{1}{N}, \quad e^i_{(0)} = -\frac{N^i}{N}. \tag{61}$$

Then

$$g_{ij} = e^{(m)}_i e_{(m)j}. \tag{62}$$

The matter Lagrangian density is split into three parts, a kinetic part with time derivative of the matter and field variables, respectively denoted $\mathscr{L}_{MK}$ and $\mathscr{L}_{GK}$, and a constraint part, $\mathscr{L}_{MC}$,

$$\mathscr{L}_{MK}, \quad \mathscr{L}_{GK}, \quad \mathscr{L}_{MC} = -N\mathscr{H}^{\text{matter}} + N^i \mathscr{H}_i^{\text{matter}}. \tag{63}$$

The matter energy and momentum densities read,

$$\mathscr{H}^{\text{matter}} = -np\delta - K^{ij} \frac{p_i n S_j}{np} \delta - (nS^k \delta)_{;k}, \tag{64}$$

$$\mathscr{H}_i^{\text{matter}} = (p_i + K_{ij} n S^j)\delta + \left(\frac{1}{2}\gamma^{mk} S_{ik}\delta + \delta_i^{(k}\gamma^{l)m} \frac{p_k n S_l}{np}\delta\right)_{;m}. \tag{65}$$

The semicolon denotes covariant three-dimensional derivative; later on, the comma will denote partial derivative. The transformation to canonical matter variables, indicated by hats, is given by

$$z^i = \hat{z}^i - \frac{nS^i}{m - np}, \quad nS_i = -\frac{p_k \gamma^{kj} \hat{S}_{ji}}{m}, \tag{66}$$

$$S_{ij} = \hat{S}_{ij} - \frac{p_i nS_j}{m - np} + \frac{p_j nS_i}{m - np}, \tag{67}$$

$$\Lambda^{[i](j)} = \hat{\Lambda}^{[i](k)} \left( \delta_{kj} + \frac{p_{(k)} p^{(j)}}{m(m - np)} \right), \tag{68}$$

$$p_i = \hat{p}_i - K_{ij} nS^j - \hat{A}^{kl} e_{(j)k} e^{(j)}_{l,i} + \left( \frac{1}{2} S_{kj} + \frac{p_{(k} nS_{j)}}{np} \right) \Gamma^{kj}_{\;\;i}, \tag{69}$$

where

$$g_{ik} g_{jl} \hat{A}^{kl} = \frac{1}{2} \hat{S}_{ij} + \frac{mp_{(i} nS_{j)}}{np(m - np)} \tag{70}$$

and

$$S^{ab} S_{ab} = \hat{S}_{(i)(j)} \hat{S}_{(i)(j)} = 2\hat{S}_{(i)} \hat{S}_{(i)} = 2s^2 = \text{const}, \tag{71}$$

$$\hat{\Lambda}^{(i)}_{[k]} \hat{\Lambda}^{[k](j)} = \delta_{ij}, \tag{72}$$

$$d\hat{\theta}^{(i)(j)} \equiv \hat{\Lambda}^{(i)}_{[k]} d\hat{\Lambda}^{[k](j)} = -d\hat{\theta}^{(j)(i)}. \tag{73}$$

Putting $\mathscr{L}_{MK} + \mathscr{L}_{GK} = \hat{\mathscr{L}}_{MK} + \hat{\mathscr{L}}_{GK} + \text{(td)}$, one finds,

$$\hat{\mathscr{L}}_{MK} = \hat{p}_i \dot{\hat{z}}^i \delta + \frac{1}{2} \hat{S}_{(i)(j)} \dot{\hat{\theta}}^{(i)(j)} \delta, \tag{74}$$

$$\hat{\mathscr{L}}_{GK} = \hat{A}^{ij} e_{(k)i} e^{(k)}_{j,0} \delta. \tag{75}$$

Adding the Lagrangian of gravity, $\mathscr{L}_G$, results in a new Lagrangian for gravity, also see Deser and Isham [37],

$$\hat{\mathscr{L}}_{GK} + \mathscr{L}_G = [2\pi^{ij} + \hat{A}^{ij}\delta]e_{(k)i}e^{(k)}_{j,0} + \mathscr{L}_{GC} - \mathscr{E}_{i,i} \tag{76}$$

with

$$\mathscr{L}_{GC} = -N\mathscr{H}^{\text{field}} + N^i\mathscr{H}_i^{\text{field}} \tag{77}$$

and

$$\mathscr{H}^{\text{field}} = -\frac{1}{\sqrt{g}}\left[g\mathrm{R} + \frac{1}{2}\left(g_{ij}\pi^{ij}\right)^2 - g_{ij}g_{kl}\pi^{ik}\pi^{jl}\right], \tag{78}$$

$$\mathscr{H}_i^{\text{field}} = 2g_{ij}\pi^{jk}_{\ \ ;k}, \tag{79}$$

as well as,

$$\mathscr{E}_i = g_{ij,j} - g_{jj,i}. \tag{80}$$

The application of the crucial spatially symmetric time gauge for the tetrads introduced by Kibble, [13],

$$e_{(i)j} = e_{ij} = e_{ji}, \tag{81}$$

$$e_{ij}e_{jk} = g_{ik}, \quad e_{ij} = \sqrt{(g_{kl})} \quad \text{(matrix root!)}, \tag{82}$$

reduces the tetrads to the metric functions. The partial derivatives of the tetrads result in the expressions

$$e_{(k)i}e^{(k)}_{j,\mu} = B^{kl}_{ij}g_{kl,\mu} + \frac{1}{2}g_{ij,\mu}, \quad B^{kl}_{ij} = B^{(kl)}_{[ij]}, \tag{83}$$

where $(\cdot\cdot)$ and $[\cdot\cdot]$ denote symmetrization and antisymmetrization, respectively, and

$$2B^{kl}_{ij} = e_{mi}\frac{\partial e_{mj}}{\partial g_{kl}} - e_{mj}\frac{\partial e_{mi}}{\partial g_{kl}}. \tag{84}$$

From the new gravity action the canonical field momentum is easily read off to be

$$\pi^{ij}_{\text{can}} = \pi^{ij} + \frac{1}{2}\hat{A}^{(ij)}\delta + B^{ij}_{kl}\hat{A}^{[kl]}\delta. \tag{85}$$

The ADM spacetime coordinate conditions do take now the following forms,

$$3g_{ij,j} - g_{jj,i} = 0, \quad \pi^{ii}_{\text{can}} = 0, \tag{86}$$

$$g_{ij} = \Psi^4 \delta_{ij} + h^{\rm TT}_{ij}, \quad \pi^{ij}_{\rm can} = \tilde{\pi}^{ij}_{\rm can} + \pi^{ij\rm TT}_{\rm can}, \tag{87}$$

with the transverse-traceless objects $h^{\rm TT}_{ij}$ and $\pi^{ij\rm TT}_{\rm can}$,

$$h^{\rm TT}_{ii} = \pi^{ii\rm TT}_{\rm can} = h^{\rm TT}_{ij,j} = \pi^{ij\rm TT}_{{\rm can},j} = 0, \tag{88}$$

and longitudinal one $\tilde{\pi}^{ij}_{\rm can}$,

$$\tilde{\pi}^{ij}_{\rm can} = V^i_{{\rm can},j} + V^j_{{\rm can},i} - \frac{2}{3}\delta_{ij} V^k_{{\rm can},k}. \tag{89}$$

The constraint equations read

$$\mathcal{H}^{\rm field} + \mathcal{H}^{\rm matter} = 0, \quad \mathcal{H}^{\rm field}_i + \mathcal{H}^{\rm matter}_i = 0. \tag{90}$$

Finally, the total action in canonical form is given by, [36],

$$W = \int d^4x\, \pi^{ij\rm TT}_{\rm can} h^{\rm TT}_{ij,0} + \int dt \left[ \hat{p}_i \dot{\hat{z}}^i + \frac{1}{2} \hat{S}_{(i)(j)} \dot{\hat{\theta}}^{(i)(j)} - E \right], \tag{91}$$

with $E = \oint dS_i \mathcal{E}_i$, and the Hamiltonian reads,

$$E \equiv H_{\rm ADM} = -8 \int d^3x\, \Delta\Psi \left[ \hat{z}^i, \hat{p}_i, \hat{S}_{(i)(j)}, h^{\rm TT}_{ij}, \pi^{ij\rm TT}_{\rm can} \right], \tag{92}$$

with Poisson bracket commutation relations

$$\begin{aligned} \{\hat{z}^i, \hat{p}_j\} &= \delta_{ij}, \quad \{\hat{S}_{(i)}, \hat{S}_{(j)}\} = \epsilon_{ijk} \hat{S}_{(k)}, \\ \{h^{\rm TT}_{ij}(\mathbf{x}, t), \pi^{kl\rm TT}_{\rm can}(\mathbf{x}', t)\} &= \delta^{{\rm TT}kl}_{ij} \delta(\mathbf{x} - \mathbf{x}'). \end{aligned} \tag{93}$$

## 5 Post-Newtonian (PN) Hamiltonians

In this section, the spin will be counted of order $(1/c)^0$ and not $1/c$ as in the Introduction. For non-spinning compact objects, the binary dynamics is known up to the 3.5PN order,

$$\begin{aligned} H(t) &= m_1 c^2 + m_2 c^2 + H_N + H_{1PN} \\ &+ H_{2PN} + H_{3PN} + \cdots \\ &+ H_{2.5PN}(t) + H_{3.5PN}(t) + \cdots, \end{aligned} \tag{94}$$

where the $2.5PN$ and $3.5PN$ Hamiltonians are non-autonomous dissipative ones, [20]. Introducing the following quantities, $\hat{H} = (H - Mc^2)/\mu$, $\mu = m_1 m_2/M$,

$M = m_1 + m_2$, $\nu = \mu/M$ with $0 \le \nu \le 1/4$ (test particle case $\nu = 0$, equal mass case $\nu = 1/4$), $\mathbf{p} = \mathbf{p}_1/\mu$, $r = r_{12} = |\mathbf{x}_1 - \mathbf{x}_2|$, $p_r = (\mathbf{n} \cdot \mathbf{p})$, $\mathbf{q} = (\mathbf{x}_1 - \mathbf{x}_2)/GM$, and $\mathbf{n} = \mathbf{n}_{12} = \mathbf{q}/|\mathbf{q}|$, in the center-of-mass frame, $\mathbf{p}_1 + \mathbf{p}_2 = 0$, the following expressions hold,

$$\hat{H}_N = \frac{p^2}{2} - \frac{1}{q}, \tag{95}$$

$$c^2 \hat{H}_{1PN} = \frac{1}{8}(3\nu - 1)p^4 - \frac{1}{2}[(3+\nu)p^2 + \nu p_r^2]\frac{1}{q} + \frac{1}{2q^2}, \tag{96}$$

$$\begin{aligned} c^4 \hat{H}_{2PN} = & \frac{1}{16}(1 - 5\nu + 5\nu^2)p^6 \\ & + \frac{1}{8}[(5 - 20\nu - 3\nu^2)p^4 - 2\nu^2 p_r^2 p^2 - 3\nu^2 p_r^4]\frac{1}{q} \\ & + \frac{1}{2}[(5 + 8\nu)p^2 + 3\nu p_r^2]\frac{1}{q^2} - \frac{1}{4}(1 + 3\nu)\frac{1}{q^3}, \end{aligned} \tag{97}$$

$$\begin{aligned} c^6 \hat{H}_{3PN} = & \frac{1}{128}(-5 + 35\nu - 70\nu^2 + 35\nu^3)p^8 \\ & + \frac{1}{16}\Big[(-7 + 42\nu - 53\nu^2 - 5\nu^3)p^6 + (2 - 3\nu)\nu^2 p_r^2 p^4 \\ & + 3(1-\nu)\nu^2 p_r^4 p^2 - 5\nu^3 p_r^6\Big]\frac{1}{q} \\ & + \Big[\frac{1}{16}(-27 + 136\nu + 109\nu^2)p^4 + \frac{1}{16}(17 + 30\nu)\nu p_r^2 p^2 \\ & + \frac{1}{12}(5 + 43\nu)\nu p_r^4\Big]\frac{1}{q^2} \\ & + \Bigg[\left(-\frac{25}{8} + \left(\frac{1}{64}\pi^2 - \frac{335}{48}\right)\nu - \frac{23}{8}\nu^2\right)p^2 \\ & + \left(-\frac{85}{16} - \frac{3}{64}\pi^2 - \frac{7}{4}\nu\right)\nu p_r^2\Bigg]\frac{1}{q^3} \\ & + \left[\frac{1}{8} + \left(\frac{109}{12} - \frac{21}{32}\pi^2\right)\nu\right]\frac{1}{q^4}. \end{aligned} \tag{98}$$

To save space, from the dissipative Hamiltonians only the leading one is given, [16],

$$c^5 \hat{H}_{2.5PN}(\hat{t}) = \frac{2}{5}\left[p_i p_j - \frac{n^i n^j}{q}\right]\frac{d^3 \hat{Q}_{ij}(\hat{t})}{d\hat{t}^3}. \tag{99}$$

Here $\hat{Q}_{ij}(\hat{t}) = \nu(q'^i q'^j - \delta_{ij} q'^2/3)$ and its time derivatives ($\hat{t} = t/GM$) are allowed to be eliminated using the equations of motion. Only after the performance of the phase-space derivatives, the primed variables are allowed to be identified with the unprimed ones.

For convenience, the spin-gravity interaction Hamiltonians are given in the non-center-of-mass frame. To simplify notation, $\mathbf{S} \equiv \hat{\mathbf{S}}$ will be put. The leading order spin-orbit Hamiltonian reads, spin *not* counted in terms of $1/c$,

$$H_{\mathrm{SO}}^{1PN} = \frac{G}{c^2} \sum_a \sum_{b \neq a} \frac{1}{r_{ab}^2} (\mathbf{S}_a \times \mathbf{n}_{ab}) \cdot \left[ \frac{3m_b}{2m_a} \mathbf{p}_a - 2\mathbf{p}_b \right]. \tag{100}$$

The leading order spin(1)–spin(2) Hamiltonian takes the form,

$$H_{\mathrm{S}_1\mathrm{S}_2}^{1PN} = \frac{G}{c^2} \sum_a \sum_{b \neq a} \frac{1}{2r_{ab}^3} \left[ 3(\mathbf{S}_a \cdot \mathbf{n}_{ab})(\mathbf{S}_b \cdot \mathbf{n}_{ab}) - (\mathbf{S}_a \cdot \mathbf{S}_b) \right] \tag{101}$$

and the leading order spin(1)–spin(1) dynamics is given by, going beyond linear order in spin,

$$H_{\mathrm{S}_1\mathrm{S}_1}^{1PN} = \frac{G}{c^2} \frac{m_2}{2m_1 r_{12}^3} \left[ 3(\mathbf{S}_1 \cdot \mathbf{n}_{12})(\mathbf{S}_1 \cdot \mathbf{n}_{12}) - (\mathbf{S}_1 \cdot \mathbf{S}_1) \right]. \tag{102}$$

The next-to-leading order spin-orbit Hamiltonian reads,

$$\begin{aligned} H_{SO}^{2PN} = \frac{G}{c^4 r^2} \Bigg[ & - ((\mathbf{p}_1 \times \mathbf{S}_1) \cdot \mathbf{n}_{12}) \Bigg[ \frac{5m_2 \mathbf{p}_1^2}{8m_1^3} + \frac{3(\mathbf{p}_1 \cdot \mathbf{p}_2)}{4m_1^2} \\ & - \frac{3\mathbf{p}_2^2}{4m_1 m_2} + \frac{3(\mathbf{p}_1 \cdot \mathbf{n}_{12})(\mathbf{p}_2 \cdot \mathbf{n}_{12})}{4m_1^2} + \frac{3(\mathbf{p}_2 \cdot \mathbf{n}_{12})^2}{2m_1 m_2} \Bigg] \\ & + ((\mathbf{p}_2 \times \mathbf{S}_1) \cdot \mathbf{n}_{12}) \left[ \frac{(\mathbf{p}_1 \cdot \mathbf{p}_2)}{m_1 m_2} + \frac{3(\mathbf{p}_1 \cdot \mathbf{n}_{12})(\mathbf{p}_2 \cdot \mathbf{n}_{12})}{m_1 m_2} \right] \\ & + ((\mathbf{p}_1 \times \mathbf{S}_1) \cdot \mathbf{p}_2) \left[ \frac{2(\mathbf{p}_2 \cdot \mathbf{n}_{12})}{m_1 m_2} - \frac{3(\mathbf{p}_1 \cdot \mathbf{n}_{12})}{4m_1^2} \right] \Bigg] \\ & + \frac{G^2}{c^4 r^3} \Bigg[ - ((\mathbf{p}_1 \times \mathbf{S}_1) \cdot \mathbf{n}_{12}) \left[ \frac{11m_2}{2} + \frac{5m_2^2}{m_1} \right] \\ & + ((\mathbf{p}_2 \times \mathbf{S}_1) \cdot \mathbf{n}_{12}) \left[ 6m_1 + \frac{15m_2}{2} \right] \Bigg] + (1 \leftrightarrow 2) \end{aligned} \tag{103}$$

and the next-to-leading order spin(1)–spin(2) Hamiltonian is given by

$$
\begin{aligned}
H^{2PN}_{S_1S_2} = {} & (G/2m_1m_2c^4r^3)[3((\mathbf{p}_1 \times \mathbf{S}_1)\cdot \mathbf{n}_{12})((\mathbf{p}_2 \times \mathbf{S}_2)\cdot \mathbf{n}_{12})/2 \\
& + 6((\mathbf{p}_2 \times \mathbf{S}_1)\cdot \mathbf{n}_{12})((\mathbf{p}_1 \times \mathbf{S}_2)\cdot \mathbf{n}_{12}) \\
& - 15(\mathbf{S}_1 \cdot \mathbf{n}_{12})(\mathbf{S}_2 \cdot \mathbf{n}_{12})(\mathbf{p}_1 \cdot \mathbf{n}_{12})(\mathbf{p}_2 \cdot \mathbf{n}_{12}) \\
& - 3(\mathbf{S}_1 \cdot \mathbf{n}_{12})(\mathbf{S}_2 \cdot \mathbf{n}_{12})(\mathbf{p}_1 \cdot \mathbf{p}_2) + 3(\mathbf{S}_1 \cdot \mathbf{p}_2)(\mathbf{S}_2 \cdot \mathbf{n}_{12})(\mathbf{p}_1 \cdot \mathbf{n}_{12}) \\
& + 3(\mathbf{S}_2 \cdot \mathbf{p}_1)(\mathbf{S}_1 \cdot \mathbf{n}_{12})(\mathbf{p}_2 \cdot \mathbf{n}_{12}) + 3(\mathbf{S}_1 \cdot \mathbf{p}_1)(\mathbf{S}_2 \cdot \mathbf{n}_{12})(\mathbf{p}_2 \cdot \mathbf{n}_{12}) \\
& + 3(\mathbf{S}_2 \cdot \mathbf{p}_2)(\mathbf{S}_1 \cdot \mathbf{n}_{12})(\mathbf{p}_1 \cdot \mathbf{n}_{12}) - 3(\mathbf{S}_1 \cdot \mathbf{S}_2)(\mathbf{p}_1 \cdot \mathbf{n}_{12})(\mathbf{p}_2 \cdot \mathbf{n}_{12}) \\
& + (\mathbf{S}_1 \cdot \mathbf{p}_1)(\mathbf{S}_2 \cdot \mathbf{p}_2) - (\mathbf{S}_1 \cdot \mathbf{p}_2)(\mathbf{S}_2 \cdot \mathbf{p}_1)/2 + (\mathbf{S}_1 \cdot \mathbf{S}_2)(\mathbf{p}_1 \cdot \mathbf{p}_2)/2] \\
& + (3/2m_1^2r^3)[-((\mathbf{p}_1 \times \mathbf{S}_1)\cdot \mathbf{n}_{12})((\mathbf{p}_1 \times \mathbf{S}_2)\cdot \mathbf{n}_{12}) \\
& + (\mathbf{S}_1 \cdot \mathbf{S}_2)(\mathbf{p}_1 \cdot \mathbf{n}_{12})^2 - (\mathbf{S}_1 \cdot \mathbf{n}_{12})(\mathbf{S}_2 \cdot \mathbf{p}_1)(\mathbf{p}_1 \cdot \mathbf{n}_{12})] \\
& + (3/2m_2^2r^3)[-((\mathbf{p}_2 \times \mathbf{S}_2)\cdot \mathbf{n}_{12})((\mathbf{p}_2 \times \mathbf{S}_1)\cdot \mathbf{n}_{12}) \\
& + (\mathbf{S}_1 \cdot \mathbf{S}_2)(\mathbf{p}_2 \cdot \mathbf{n}_{12})^2 - (\mathbf{S}_2 \cdot \mathbf{n}_{12})(\mathbf{S}_1 \cdot \mathbf{p}_2)(\mathbf{p}_2 \cdot \mathbf{n}_{12})] \\
& + (6G^2(m_1+m_2)/c^4r^4)[(\mathbf{S}_1 \cdot \mathbf{S}_2) - 2(\mathbf{S}_1 \cdot \mathbf{n}_{12})(\mathbf{S}_2 \cdot \mathbf{n}_{12})]. \qquad (104)
\end{aligned}
$$

Finally, the next-to-leading order spin(1)–spin(1) dynamics reads,

$$
\begin{aligned}
H^{2PN}_{S_1S_1} = {} & \frac{G}{c^4r^3}\Bigg[\frac{m_2}{4m_1^3}(\mathbf{p}_1 \cdot \mathbf{S}_1)^2 + \frac{3m_2}{8m_1^3}(\mathbf{p}_1 \cdot n)^2\,\mathbf{S}_1^2 \\
& - \frac{3m_2}{8m_1^3}\mathbf{p}_1^2\,(\mathbf{S}_1 \cdot \mathbf{n})^2 - \frac{3m_2}{4m_1^3}(\mathbf{p}_1 \cdot \mathbf{n})\,(\mathbf{S}_1 \cdot \mathbf{n})\,(\mathbf{p}_1 \cdot \mathbf{S}_1) - \frac{3}{4m_1m_2}\mathbf{p}_2^2\mathbf{S}_1^2 \\
& + \frac{9}{4m_1m_2}\mathbf{p}_2^2\,(\mathbf{S}_1 \cdot \mathbf{n})^2 + \frac{3}{4m_1^2}(\mathbf{p}_1 \cdot \mathbf{p}_2)\,\mathbf{S}_1^2 - \frac{9}{4m_1^2}(\mathbf{p}_1 \cdot \mathbf{p}_2)\,(\mathbf{S}_1 \cdot \mathbf{n})^2 \\
& - \frac{3}{2m_1^2}(\mathbf{p}_1 \cdot \mathbf{n})\,(\mathbf{p}_2 \cdot \mathbf{S}_1)\,(\mathbf{S}_1 \cdot \mathbf{n}) \\
& + \frac{3}{m_1^2}(\mathbf{p}_2 \cdot \mathbf{n})\,(\mathbf{p}_1 \cdot \mathbf{S}_1)\,(\mathbf{S}_1 \cdot \mathbf{n}) + \frac{3}{4m_1^2}(\mathbf{p}_1 \cdot \mathbf{n})\,(\mathbf{p}_2 \cdot \mathbf{n})\,\mathbf{S}_1^2 \\
& - \frac{15}{4m_1^2}(\mathbf{p}_1 \cdot \mathbf{n})\,(\mathbf{p}_2 \cdot \mathbf{n})\,(\mathbf{S}_1 \cdot \mathbf{n})^2\Bigg] \\
& - \frac{G^2m_2}{2c^4r^4}\left[5\left(1+\frac{6m_2}{5m_1}\right)((\mathbf{S}_1 \cdot \mathbf{n})^2 - \mathbf{S}_1^2) + 4\left(1+\frac{2m_2}{m_1}\right)(\mathbf{S}_1 \cdot \mathbf{n})^2\right]. \qquad (105)
\end{aligned}
$$

Also this Hamiltonian goes beyond linear order in spin. For its derivation an extension of the Tulczyjew stress-energy tensor for pole-dipole particles was needed, [31]. As further examples, the next-to-leading order spin-orbit center-of-energy vector is given,

$$
\begin{aligned}
\mathbf{G}_{\mathrm{SO}}^{2PN} = &- \sum\nolimits_a \frac{\mathbf{p}_a^2}{8c^4 m_a^3} (\mathbf{p}_a \times \mathbf{S}_a) \\
&+ \sum\nolimits_a \sum\nolimits_{b \neq a} \frac{m_b G}{4c^4 m_a r_{ab}} \left[ ((\mathbf{p}_a \times \mathbf{S}_a) \cdot \mathbf{n}_{ab}) \frac{5\mathbf{x}_a + \mathbf{x}_b}{r_{ab}} - 5(\mathbf{p}_a \times \mathbf{S}_a) \right] \\
&+ \sum\nolimits_a \sum\nolimits_{b \neq a} \frac{G}{c^4 r_{ab}} \left[ \frac{3}{2} (\mathbf{p}_b \times \mathbf{S}_a) - \frac{1}{2} (\mathbf{n}_{ab} \times \mathbf{S}_a)(\mathbf{p}_b \cdot \mathbf{n}_{ab}) \right. \\
&\left. - ((\mathbf{p}_a \times \mathbf{S}_a) \cdot \mathbf{n}_{ab}) \frac{\mathbf{x}_a + \mathbf{x}_b}{r_{ab}} \right], \qquad (106)
\end{aligned}
$$

as well as the spin(1)–spin(2) one,

$$
\mathbf{G}_{\mathrm{SS}}^{2PN} = \frac{G}{2c^4} \sum_a \sum_{b \neq a} \left\{ [3(\mathbf{S}_a \cdot \mathbf{n}_{ab})(\mathbf{S}_b \cdot \mathbf{n}_{ab}) - (\mathbf{S}_a \cdot \mathbf{S}_b)] \frac{\mathbf{x}_a}{r_{ab}^3} + (\mathbf{S}_b \cdot \mathbf{n}_{ab}) \frac{\mathbf{S}_a}{r_{ab}^2} \right\}. \qquad (107)
$$

To summarize, for binary systems, and in part for many-body systems too, the following PN Hamiltonians are known fully explicitly,

$$
\begin{aligned}
H = &H_N + H_{1PN} + H_{2PN} + H_{2.5PN} + H_{3PN} + H_{3.5PN} \\
&+ H_{SO}^{1PN} + H_{SO}^{2PN} + H_{SO}^{3PN} + H_{SO}^{3.5PN} \\
&+ H_{S_1S_2}^{1PN} + H_{S_1S_2}^{2PN} + H_{S_1S_2}^{3PN} + H_{S_1S_2}^{3.5PN} \\
&+ H_{S_1S_1}^{1PN} + H_{S_2S_2}^{1PN} + H_{S_1S_1}^{2PN} + H_{S_2S_2}^{2PN} \\
&+ H_{p_1S_2^3} + H_{p_2S_1^3} + H_{p_1S_1S_2^2} + H_{p_2S_2S_1^2} \\
&+ H_{S_1S_2^3} + H_{S_1^3S_2} + H_{S_1^2S_2^2}. \qquad (108)
\end{aligned}
$$

Also the n-body Hamiltonian through linear order in Newton's gravitational constant $G$ is known $H = H_{1PM}$, as well as the test-spin Hamiltonian in the Kerr metric.

For selfgravitating objects, the Hamiltonians primarily come out in the form $H = H[p, q, h^{\mathrm{TT}}, \pi_{\mathrm{TT}}]$. The transition to a Routhian description of the type $H = H[p, q, h^{\mathrm{TT}}, \dot{h}^{\mathrm{TT}}]$ then allows the derivation of an autonomous Hamiltonian for the conservative dynamics in the form $H = H[p, q, h^{\mathrm{TT}}(x; p, q), \dot{h}^{\mathrm{TT}}(x; p, q)] = H(p, q)$ as well as a non-autonomous one given by $H(t) = H[p, q, h^{\mathrm{TT}}(x; p', q'), \dot{h}^{\mathrm{TT}}(x; p', q')] = H(p, q; p', q')$. The presented dynamical systems have found derivations with other methods too. Particularly the Effective Field Theory method by Goldberger and Rothstein, [38], has proven very powerful. For details the reader is referred to the literature.

**Acknowledgments** The author thanks Stanley Deser for useful discussions.

# References


1. Dirac, P.A.M.: The theory of gravitation in Hamiltonian form. Proc. R. Soc. Lond. A **246**, 333 (1958). doi:10.1098/rspa.1958.0142
2. Dirac, P.A.M.: Fixation of coordinates in the hamiltonian theory of gravitation. Phys. Rev. **114**, 924 (1959). doi:10.1103/PhysRev.114.924
3. Dirac, P.A.M.: Energy of the gravitational field. Phys. Rev. Lett. **2**, 368 (1959). doi:10.1103/PhysRevLett.2.368
4. Arnowitt, R., Deser, S.: Quantum theory of gravitation: general formulation and linearized theory. Phys. Rev. **113**, 745 (1959). doi:10.1103/PhysRev.113.745
5. Arnowitt, R., Deser, S., Misner, C.M.: Canonical variables in general relativity. Phys. Rev. **117**, 1959 (1960). doi:10.1103/PhysRev.117.1595
6. Arnowitt, R., Deser, S., Misner, C.M.: Consistency of the canonical reduction of general relativity. J. Math. Phys. **1**, 434 (1960). doi:10.1063/1.1703677
7. Arnowitt, R., Deser, S., Misner, C.W.: The dynamics of general relativity. In: Witten L. (ed.) Gravitation: An Introduction to Current Research, pp. 227–265. Wiley, New York (1962). doi:10.1007/s10714-008-0661-1
8. Schwinger, J.: Quantized gravitational field. Phys. Rev. **130**, 1253 (1963)
9. Regge, T., Teitelboim, C.: Role of surface integrals in the hamiltonian formulation of general relativity. Ann. Phys. (N.Y.) **88**, 286 (1974). doi:10.1016/0003-4916(74)90404-7
10. Wheeler, J.A.: Geometrodynamics and the issue of the final state. In: DeWitt, C., DeWitt, B.S. (eds.) Relativity, Groups, and Topology, pp. 315–520. Gordon and Breach, New York (1964)
11. DeWitt, B.S.: Quantum theory of gravity. I. The canonical theory. Phys. Rev. **160**, 1113 (1967). doi:10.1103/PhysRev.160.1113
12. Kimura, T.: Fixation of physical space-time coordinates and equation of motion of two-body problem. Prog. Theor. Phys. **26**, 157 (1961). doi:10.1143/PTP.26.157
13. Kibble, T.W.B.: Canonical variables for the interacting gravitational and Dirac fields. J. Math. Phys. **4**, 1433 (1963). doi:10.1063/1.1703923
14. Nelson, J.E., Teitelboim, C.: Hamiltonian formulation of the theory of interacting gravitational and electron fields. Ann. Phys. (N.Y.) **116**, 86 (1978). doi:10.1016/0003-4916(78)90005-2
15. Dirac, P.A.M.: Interacting gravitational and spinor fields. In: Recent Developments in General Relativity, pp. 191–207. Pergamon Press, Oxford (1962)
16. Schäfer, G.: The gravitational quadrupole radiation-reaction force and the canonical formalism of ADM. Ann. Phys. (N.Y.) **161**, 81 (1985). doi:10.1016/0003-4916(85)90337-9
17. Ohta, T., Okamura, H., Kimura, T., Hiida, K.: Coordinate condition and higher order gravitational potential in canonical formalism. Prog. Theor. Phys. **51**, 1598 (1974). doi:10.1143/PTP.51.1598
18. Damour, T., Schäfer, G.: Lagrangians for *n* point masses at the second post-Newtonian approximation of general relativity. Gen. Relativ. Gravit. **17**, 879 (1985). doi:10.1007/BF00773685
19. Damour, T., Jaranowski, P., Schäfer, G.: Dimensional regularization of the gravitational interaction of point masses. Phys. Lett. B **513**, 147 (2001). doi:10.1016/S0370-2693(01)00642-6
20. Jaranowski, P., Schäfer, G.: Radiative 3.5 post-Newtonian ADM Hamiltonian for many-body point-mass systems. Phys. Rev. D **55**, 4712 (1997). doi:10.1103/PhysRevD.55.4712
21. Schäfer, G.: The ADM Hamiltonian at the postlinear approximation. Gen. Relativ. Gravit. **18**, 255 (1986). doi:10.1007/BF00765886
22. Ledvinka, T., Schäfer, G., Bičák, J.: Relativistic closed-form Hamiltonian for many-body gravitating systems in the post-Minkowskian approximation. Phys. Rev. Lett. **100**, 251101 (2008). doi:10.1103/PhysRevLett.100.251101
23. Damour, T.: Coalescence of two spinning black holes: an effective one-body approach. Phys. Rev. D **64**, 124013 (2001). doi:10.1103/PhysRevD.64.124013
24. Damour, T., Jaranowski, P., Schäfer, G.: Hamiltonian of two spinning compact bodies with next-to-leading order gravitational spin-orbit coupling. Phys. Rev. D **77**, 064032 (2008). doi:10.1103/PhysRevD.77.064032

25. Hartung, J., Steinhoff, J.: Next-to-leading order spin-orbit and spin(a)-spin(b) Hamiltonians for n gravitating spinning compact objects. Phys. Rev. D **83**, 044008 (2011). doi:10.1103/PhysRevD.83.044008
26. Hartung, J., Steinhoff, J.: Next-to-next-to-leading order post-Newtonian spin-orbit Hamiltonian for self-gravitating binaries. Ann. Phys. (Berlin) **523**, 783 (2011). doi:10.1002/andp.201100094
27. Steinhoff, J., Wang, H.: Canonical formulation of gravitating spinning objects at 3.5 post-Newtonian order. Phys. Rev. D **81**, 024022 (2010). doi:10.1103/PhysRevD.81.024022
28. Steinhoff, J., Hergt, S., Schäfer, G.: The next-to-leading order gravitational spin(1)-spin(2) dynamics in Hamiltonian form. Phys. Rev. D **77**, 081501 (2008). doi:10.1103/PhysRevD.77.081501
29. Hartung, J., Steinhoff, J.: Next-to-next-to-leading order post-Newtonian spin(1)-spin(2) Hamiltonian for self-gravitating binaries. Ann. Phys. (Berlin) **523**, 919 (2011). doi:10.1002/andp.201100163
30. Wang, H., Steinhoff, J., Zeng, J., Schäfer, G.: Leading-order spin-orbit and spin(1)-spin(2) radiation-reaction Hamiltonians. Phys. Rev. D **84**, 124005 (2011). doi:10.1103/PhysRevD.84.124005
31. Steinhoff, J., Hergt, S., Schäfer, G.: Spin-squared Hamiltonian of next-to-leading order gravitational interaction. Phys. Rev. D **78**, 101503 (2008). doi:10.1103/PhysRevD.78.101503
32. Hergt, S., Schäfer, G.: Higher-order-in-spin interaction Hamiltonians for binary black holes from source terms of the Kerr geometry in approximate ADM coordinates. Phys. Rev. D **77**, 104001 (2008). doi:10.1103/PhysRevD.77.104001
33. Barausse, E., Racine, E., Buonanno, A.: Hamiltonian of a spinning test particle in curved spacetime. Phys. Rev. D **80**, 104025 (2009). doi:10.1103/PhysRevD.80.104025
34. Brill, D.R., Lindquist, R.W.: Interation energy in geometrostatics. Phys. Rev. **131**, 471 (1963). doi:10.1103/PhysRev.131.471
35. Jaranowski, P., Schäfer, G.: Lapse function for maximally sliced Brill-Lindquist initial data. Phys. Rev. D **65**, 127501 (2002). doi:10.1103/PhysRevD.65.127501
36. Steinhoff, J., Schäfer, G.: Canonical formulation of self-gravitating spinning-object systems. Europhys. Lett. **87**, 50004 (2009). doi:10.1209/0295-5075/87/50004
37. Deser, S., Isham, C.J.: Canonical vierbein form of general relativity. Phys. Rev. D **14**, 2505 (1976). doi:10.1103/PhysRevD.14.2505
38. Goldberger, W.D., Rothstein, I.Z.: An effective field theory of gravity for extended objects. Phys. Rev. D **73**, 104029 (2006). doi:10.1103/PhysRevD.73.104029

# Stability of Marginally Outer Trapped Surfaces and Geometric Inequalities

**Marc Mars**

**Abstract** Marginally outer trapped surfaces (MOTS) admit a notion of stability that in many respects generalizes a similar notion for minimal hypersurfaces. Stable MOTS play an interesting role in a number of geometric inequalities involving physical parameters such as area, mass, charge or, in the axially symmetric case, angular momentum. Some of those inequalities are global in nature while others are local, with interesting relationships between them. In this lecture the notion of stable MOTS will be reviewed and some of the geometric inequalities involving stable MOTS will be described.

## 1 Introduction

Geometric inequalities play a fundamental role in gravitation because they provide information on the relationship between physically relevant quantities in a robust way, independently of the details of the particular spacetime under consideration. It is often the case that not even field equations are necessary for the validity of such inequalities and that only energy conditions are required, which make their range of validity very broad and transverse to several theories of gravity. One of the most fundamental geometric inequalities in gravitation is the Positive Mass Theorem which, as is well-known establishes that the ADM mass is non-negative for any asymptotically flat spacetime satisfying the dominant energy condition (DEC), i.e. such that the Einstein tensor contracted with any pair of future directed causal vectors gives a non-positive quantity. Another very important geometric inequality, which so far has been proved only in special circumstances is the Penrose inequality [1].

M. Mars (✉)
Dpto. Física Fundamental, University of Salamanca, Pl. de la Merced s/n,
37008 Salamanca, Spain
e-mail: marc@usal.es

J. Bičák and T. Ledvinka (eds.), *General Relativity, Cosmology and Astrophysics*,
Fundamental Theories of Physics 177, DOI: 10.1007/978-3-319-06349-2_8,
© Springer International Publishing Switzerland 2014

In the asymptotically flat case, this inequality conjectures that the total ADM mass $M$ satisfies the inequality

$$|S_{\text{min}}| \leq 16\pi M^2.$$

where $|S_{\text{min}}|$ is the minimal area needed to enclose a given weakly future trapped surface $S$, i.e. a closed, spacelike, codimension-two surface with future directed, causal mean curvature vector. The notion of "enclosing" also needs a definition, see e.g. [2] for details. The Penrose inequality is a strengthening of the positive mass theorem when there are trapped surfaces present. In turn, the Penrose inequality can be strengthened when the spacetime is charged and the total charge $Q$ cannot be radiated away (e.g. in electrovacuum, or when all matter present in the spacetime is electrically neutral). In this case, the conjectured inequality becomes [3]

$$|S_{\text{min}}| \leq 8\pi \left( M^2 - \frac{Q^2}{2} + \sqrt{M^2 \left( M^2 - Q^2 \right)} \right).$$

It is clear that, in order for this inequality to even make sense, it is necessary that the total ADM mass satisfies the bound $M \geq |Q|$. This inequality was proved in [4] (see also [5] for a mathematically complete argument) and provides a direct strengthening of the positive mass theorem in charged spacetimes, irrespectively of whether a weakly future trapped surface is present in the spacetime or not.

Another global charge in asymptotically flat spacetimes is the ADM angular momentum $J$. In general, angular momentum can be radiated away by gravitational waves, so in general no strengthening of the Penrose inequality involving angular momentum should be expected (this is because the physical argument leading to the Penrose inequality involves, on the one hand, the weak cosmic censorship hypothesis and, on the other, the asymptotic values of mass, charge and angular momentum of the black hole that forms during the collapse). There is one interesting case, however, when angular momentum cannot be radiated away, namely when the spacetime is axially symmetric. Thus, in this case (and assuming again that electric charge is either absent or cannot be radiated away) the Penrose inequality can be strengthened to (see [6, 7] for a discussion)

$$|S_{\text{min}}| \leq 8\pi \left( M^2 - \frac{Q^2}{2} + \sqrt{M^4 - Q^2 M^2 - J^2} \right) := F_{Q,J}(M), \qquad (1)$$

where the last equality defines a function of $M$ for each choice of $Q$ and $J$. As before, in order for this inequality to make sense, it becomes necessary that $M^4 - Q^2 M^2 - J^2 \geq 0$ or, equivalently,

$$M^2 \geq \frac{1}{2}(Q^2 + \sqrt{Q^4 + 4J^2}).$$

This is a strengthening of the positive mass theorem that should be valid for any asymptotically flat, axially symmetric spacetime with conserved total electric charge. This result has been proved whenever the spacetime admits a maximal (i.e. with second fundamental form of vanishing trace), axially symmetric slice with two asymptotically flat ends, first in vacuum [8, 9] and then in electrovacuum [10, 11]. The assumption of having more than one asymptotically flat end is made because, in electrovacuum, the total electric charge and the total angular momentum can only be non-zero in the presence of non-trivial topology. In the case of vacuum with an arbitrary number of asymptotically flat ends, the inequality is not yet settled but a closely related inequality has been proved in [12], which possibly reduces to the previous one depending on the values of a certain functional defined on stationary and axially symmetric asymptotically flat spacetimes (see the review [13] for many more details).

All the geometric inequalities discussed so far are global in nature because they involve the total ADM mass, and the charge and angular momentum are also global quantities defined at infinity. Even the area term in the Penrose inequality is global quantity because of the need of taking the minimal area enclosure of the weakly future trapped surface. It is straightforward to check that $F_{Q,J}$ is an increasing function of $M$. Its minimum value is $4\pi\sqrt{Q^4+4J^2}$. It therefore follows that, whenever the minimal area enclosure $|S_{\text{min}}|$ satisfies $|S_{\text{min}}| \leq 4\pi\sqrt{Q^4+4J^2}$ then the Penrose inequality (1) is satisfied automatically. It follows that this inequality is non-trivial only if

$$|S_{\text{min}}| \geq 4\pi\sqrt{Q^4+4J^2}.$$

In particular, when the total charge is non-conserved or vanishes identically, this inequality becomes

$$|S_{\text{min}}| \geq 8\pi|J|$$

and when the total angular momentum is non-conserved, the inequality reads

$$|S_{\text{min}}| \geq 4\pi Q^2.$$

These inequalities are still of global nature, but now it makes sense to try and see whether a local version of them is still valid. Indeed, one may think of replacing the minimal area enclosure $|S_{\text{min}}|$ by the area $|S|$ of the weakly future trapped surface itself. Moreover, the notion of total charge enclosed by a closed, orientable surface makes sense, and in the axially symmetric case, the presence of a Killing vector allows one to define the Komar angular momentum of any closed, orientable surface. Moreover, in electrovacuum there is a modification [14] of the Komar definition of angular momentum involving the electromagnetic field which provides a conserved quantity in the sense that it gives an object depending only on the homology class of the surface under consideration.

There has been very interesting and remarkable progress in recent years towards the proof of local inequalities of this type on certain surfaces. They were first found in the case of stationary and axially symmetric black hole horizons admitting arbitrary matter outside the horizon but such that a neighbourhood of the horizon itself is vacuum or electrovacuum. The first result along those lines was for the degenerate case (i.e. when the surface gravity of the horizon vanishes) [15]. The non-degenerate case for vacuum horizons was dealt with in [16]. Finally, the inequality in the charged, rotating case and for electrovacuum horizons was solved in [17]. These results had important implications in the problem of non-existence of stationary and axially symmetric two-black hole configurations [18–21] (see also G. Neugebauer's and J. Hennig's contribution to this volume).

Remarkably, a purely local version of this inequality where only properties of suitable spacelike two-surfaces are used has also been obtained recently. The first case, proved by Dain and Reiris [22] involved stable minimal, axially symmetric surfaces embedded in maximal, axially symmetric hypersurfaces in a vacuum spacetime. In this setting, the universal inequality $|S| \geq 8\pi |J|$ was proved, where $J$ is the Komar angular momentum. This inequality was then extended [23] to arbitrary, axially symmetric, stable marginally outer trapped surfaces (defined below) embedded in a spacetime with arbitrary matter contents as long as the dominant energy condition is satisfied. The case with electric charge (and no angular momentum so that no need to restrict oneself to axially symmetric situations) was analyzed in [24] where the inequality $|S| \geq 4\pi Q^2$ was proved for suitable surfaces. The case involving both charge and angular momentum has been proved recently in [25].

The key underlying property of the local versions of the inequality is the notion of stability, both for minimal hypersurfaces and for marginally outer trapped surfaces. The aim of this lecture is to review the notion of stability for marginally outer trapped surfaces and discuss some of its consequences. Then, I will present in more detail the various inequalities and explain how does stability enter into the arguments. The final aim will be to relate the black hole-type inequalities to the purely local inequalities by summarizing recent results [26] on the stability properties of Killing horizons.

## 2 Basics on the Geometry of Spacelike Surfaces

Our framework will be a four-dimensional spacetime $(M, g)$, which we will take to be oriented and time-oriented. Scalar product with the spacetime metric will be denoted by $\langle\, ,\, \rangle$ and $S$ will refer to a closed (i.e. compact without boundary), spacelike, two-dimensional, orientable, connected, embedded surface in $(M, g)$ (simply *surface* from now on). The normal space to $S$ at any of its points is a Lorentzian vector space. The collection of all normal spaces is a vector bundle over $S$ that admits two global, smooth, nowhere zero cross-sections $\{\ell, k\}$ satisfying $\langle \ell, \ell \rangle = 0$, $\langle k, k \rangle = 0$ and $\langle \ell, k \rangle = -1$. We always take $\ell$ (and hence $k$) to be future directed. These sections are defined uniquely up to the usual boost freedom $\ell \to F\ell, k \to F^{-1}k$, where $F$ is a smooth, positive scalar function on $S$. The (positive definite) induced metric on $S$ will be denoted by $h$ and the corresponding covariant derivative by $D$. The null extrinsic curvatures are defined, as usual, by $\chi_{\ell AB} \equiv \langle e_A, \nabla_{e_B} \ell \rangle$, $\chi_{k\,AB} \equiv \langle e_A, \nabla_{e_B} k \rangle$, where $\{e_A\}$ is a basis of the tangent space of $S$. The null expansions are the traces of these tensors, i.e. $\theta_\ell = \mathrm{tr}_h(\chi_\ell), \theta_k = \mathrm{tr}_h(\chi_k)$. A relevant geometric object in the following is the mean curvature vector, defined in terms of the null expansions by $H = -\theta_\ell k - \theta_k \ell$. It is well-known (and straightforward) that this vector is independent of the choice of null basis $\{\ell, k\}$. Finally, the normal bundle admits a canonical connection with connection one-form given by $s_A \equiv -\langle k, \nabla_{e_A} \ell \rangle$ in the basis $\{\ell, k\}$.

The mean curvature is fundamental, among other things, because it contains full information on how the area of the surface changes to first order under general deformations. Denoting by $\delta_\xi |S|$ the first order variation of area along a deformation vector $\xi$, the following identity holds

$$\delta_\xi |S| = \int_S \langle H, \xi \rangle \eta_h,$$

where $\eta_h$ is the metric volume form of $(S, h)$. Hence, when $H$ is future causal then the area of $S$ does not increase for any future causal deformation $\xi$. This is generally taken as a clear signal of the presence of a strong gravitational field. A surface $S$ with this property is called weakly future trapped.

## 3 Marginally Outer Trapped Surfaces and Stability

As just mentioned, the future causal character for the mean curvature is sufficient to signal the presence of a strong gravitational field. However, if the surface admits a well-defined notion of exterior (for instance when $S$ is contained in a spacelike, asymptotically flat hypersurface and separates this hypersurface into an asymptotically flat exterior and a compact domain), then a non-increase of area along the exterior future null cone may also be taken as convincing indication that the gravitation field on the surface is strong. Such surfaces are defined by the property that $\theta_\ell \leq 0$, where $\ell$ is the future null normal pointing into the exterior domain.

The borderline case is given by surfaces satisfying the equality case $\theta_\ell = 0$. It turns out that studying this borderline case is interesting even when there is no clearly distinguished notion of exterior. This leads to the following standard definition:

**Definition 1** A marginally outer trapped surface (MOTS) is a surface which satisfies either $\theta_\ell = 0$ or $\theta_k = 0$ everywhere (after renaming we always take $\theta_\ell = 0$).

From the general expression for the first variation of area it follows that MOTS are stationary points for the area functional with respect to arbitrary variations tangent to $\ell$. Stationary points of any functional call immediately for analyzing their behaviour under second order variations. In this case, the result is a direct consequence of the Raychauduri equation and gives

$$\delta^2_{\psi\ell}|S| = -\int_S \psi^2 \left( G(\ell,\ell) + |\chi_\ell|^2 \right) \eta_S,$$

where $G$ is the Einstein tensor of the spacetime and $|\cdot|$ is the norm of tensors on $S$ with the metric $h$. In contrast to the case of minimal hypersurfaces in Riemannian ambient manifolds, the second order variation does not define a differential operator acting on $\psi$. Moreover, it is always non-positive definite, provided the spacetime satisfies the null energy condition NEC ($G(\ell', \ell') \geq 0$ for any null vector $\ell'$). Thus, the second variation of area along the direction $\ell$ for MOTS does not provide with any useful notion of stability. A suitable alternative is to use first variations of $\theta_\ell$ (which vanishes for a MOTS) along arbitrary normal directions not tangent to $\ell$. To be more precise, select a normal direction to $S$ nowhere tangent to $\ell$. This defines uniquely a vector $v$ satisfying $\langle \ell, v \rangle = 1$. Consequently, it can be decomposed uniquely as $v = -k + V\ell$ where $V$ is a scalar function on $S$. In other words, we parametrize the collection of directions orthogonal to $S$ and nowhere tangent to $\ell$ by a real function $V : S \to \mathbb{R}$ via the generator $v$ defined above.

Note that $V$ may a priori have any value. The sign of $V$ at any given point $p \in S$ is directly tied to the causal character of the normal direction selected at $p$. If $V > 0$ then this direction is spacelike, if $V = 0$ the direction is null and if $V < 0$ the direction is timelike. It is also convenient to define the vector $v^\star := k + V\ell$, which defines a second normal direction. This direction is linearly independent to the previous one except when $v$ is null, in which case they coincide. In any other case the causal character of $v^\star$ is always opposite (in the sense of spacelike vs. timelike) to the causal character of $v$. In particular, $v^\star$ is future causal if and only if $v$ is "achronal" (i.e. spacelike or null or, equivalently, $V \geq 0$).

Given a fixed direction defined by $v$ we can perform variations restricted to this direction. Any such variation is defined by a vector $\xi = \psi v$, where $\psi : S \to \mathbb{R}$. The first order variation of $\theta_\ell$ along $\psi v$ was first computed by Newman [27] (the calculation was performed assuming implicitly that $\psi \neq 0$ everywhere, but the result is generally valid, c.f. [28]). The resulting expression gives a differential operator $L_v$ acting $\psi$ via the definition $L_v(\psi) \equiv \delta_{\psi v}\theta_\ell$. Its explicit expression is

$$L_v(\psi) = -\Delta_h \psi + 2 s_A D^A \psi + \left( \frac{R(h)}{2} - G(\ell, v^\star) - V|\chi_\ell|^2 - s_A s^A + D_A s^A \right)\psi, \tag{2}$$

where $\Delta_h$ is the Laplacian of $(S, h)$ and $R(h)$ is the curvature scalar of the metric $h$.

As discussed in detail in [28] this operator is elliptic and, in general, not self-adjoint. However, any elliptic operator on a compact manifold (or on a bounded domain with Dirichlet boundary conditions) admits a principal eigenvalue $\lambda_v$. This is a real eigenvalue (i.e. such that there exists a real function $\phi_v$ satisfying $L_v(\phi_v) = \lambda_v \phi_v$) with the property that any other eigenvalue (which will be complex in general) satisfies $\mathrm{Re}(\lambda) > \lambda_v$. Moreover, as in the self-adjoint case, the principal eigenfunction $\phi_v$ does not change sign and hence can be taken to be positive everywhere. For self-adjoint operators, the principal eigenvalue admits a characterization in terms of the so-called Rayleigh-Ritz quotient which is very useful, firstly because it provides upper bounds for the principal eigenvalue and secondly, and even more importantly, because it gives lower bounds for certain integral functionals acting on arbitrary functions. This latter property is the key for translating sign conditions on the principal eigenvalues into useful analytic inequalities that, in turn, can be used to derive geometric properties of the surface. This is crucial, e.g., for studying stable minimal hypersurfaces in Riemannian manifolds. Such surfaces have the property that their own stability operator, which is now self-adjoint, has a non-negative principal eigenvalue.

The Rayleigh-Ritz characterization is no longer true for non self-adjoint operators. Given its importance, it is reasonable to ask whether there exists any analogue to this characterization valid for any elliptic operator. Donsker and Varadhan [29] found a number of characterizations of the principal eigenvalue of the min-max type, i.e. involving a minimization of a certain class of suprema. Such characterizations are, in general, difficult to work with. In [28] one of these characterizations was elaborated further and a Rayleigh-Ritz type characterization for the principal eigenvalue of any elliptic second order operator was found. In order to describe it, recall that any one-form in a compact Riemannian manifold without boundary can be decomposed as the sum of the differential of a function $f$ plus a divergence-free one-form. Such decomposition is usually called Helmholz decomposition in the physics literature. Recall that this decomposition is unique up to an additive constant in $f$ provided the manifold is connected.

In the case of the stability operator $L_v$, the characterization of $\lambda_v$ obtained in [28] reads as follows.

**Proposition 1** *Let $L_v$ be the stability operator* (2). *Decompose the normal connection one-form $s_A$ according to the Helmoltz decomposition as $s_A = D_A f + z_A$, where $z_A$ is divergence-free. Then the principal eigenvalue of $L_v$ is given by*

$$\lambda_v = \inf_{u>0} \frac{\int_S \left(|Du|^2 + \left(\frac{R(h)}{2} - G(\ell, v^\star) - V|\chi_\ell|^2\right) u^2 - |d\omega_u + z|^2 u^2\right) \eta_h}{\int_S u^2 \eta_h}, \quad (3)$$

*where $\omega_u$ is any solution of*

$$-\Delta_h \omega_u - \frac{2}{u} D_A \omega_u D^A u = \frac{2}{u} z_A D^A u. \quad (4)$$

It is straightforward to check that, given any positive function $u$ on $S$, the partial differential equation (4) always admits a solution, which is unique up to an additive constant.

An immediate consequence of this result is the so-called "symmetrized inequality along $v$" proved by Galloway and Schoen [30] using explicit estimates,

$$\int_S \left(|Du|^2 + \frac{1}{2} R(h) u^2\right) \eta_h \geq \int_S \left(\lambda_v + G(\ell, v^\star) + V|\chi_\ell|^2\right) u^2 \eta_h. \quad (5)$$

Indeed, by dropping the term $|d\omega_u + z|^2 u^2$ in (3) the right-hand side is never decreased, and hence so does infimum, which leads immediately to (5).

The stability operator allows us to define a notion of stability for MOTS [28], which generalizes a similar notion for minimal hypersurfaces.

**Definition 2** A MOTS $S$ is **stable along** $v$ if and only if the principal eigenvalue $\lambda_v$ of the stability operator satisfies $\lambda_v \geq 0$. $S$ is **strictly stable along** $v$ if and only if $\lambda_v > 0$.

One of the consequences of this definition is given by the following lemma, which links the stability of a MOTS with the existence of suitable variations which increase the value of the outer null expansion.

**Lemma 1** *A MOTS $S$ is stable along $v$ if there exists an outward variation $\psi v$ ($\psi \geq 0$, $\psi \not\equiv 0$) such that $\delta_{\psi v}\theta_\ell \geq 0$. $S$ is strictly stable if, in addition, $\delta_{\psi v}\theta_\ell > 0$.*

Although, by construction, there is a stability operator (and a notion of stability) for each direction $v$, the dependence of $L_v$ on $v$ is very simple, namely $L_v = L_{-k} - W\,V$, where $W$ is defined as $W := (G(\ell,\ell) + |\chi_\ell|^2)$. Note that under the NEC $W \geq 0$ and hence stability improves when $v$ is tilted away from $\ell$ (i.e. when $V$ is made larger at every point). In fact, when $W \not\equiv 0$, there always exists a direction sufficiently tilted away from $\ell$ for which the principal eigenvalue is positive, and hence the MOTS is strictly stable along this direction. However, it turns out that the stability of the MOTS along a direction $v$ gives, in general, useful information only when the direction $v$ is achronal. This is because the terms $G(\ell, v^\star)$ and $V|\chi_\ell|^2$ are non-negative (under the NEC) only when $V \geq 0$, i.e. when $v$ is spacelike or null at every point. This leads to the following definition, spelled out in [23] and closely related to the notion of future outer trapping horizon defined by Hayward [31].

**Definition 3** A MOTS $S$ is **spacetime stable** if it is stable along an achronal direction $v$.

It is now straightforward to check that, under the NEC, $S$ is spacetime stable if and only if it is stable along $-k$. Since the null direction $-k$ is privileged as a transverse direction for the MOTS, we will write $\lambda_{-k}$ simply as $\lambda_-$.

# 4 Area-Charge-Topology Inequalities for MOTS

As discussed in the Introduction, a class of local inequalities exists for suitable surfaces relating the area and the total charge enclosed by the surface. If the surface has non-trivial topology, its genus (recall that our surfaces are two-dimensional, connected and orientable, so that their topology is uniquely determined by their genus) also enters into the inequality. The basic assumption made in this context is that the energy-momentum contents of the spacetime splits into an electromagnetic field and the rest of matter in such a way that this rest satisfies the dominant energy condition. Allowing also a cosmological constant $\Lambda$, the Einstein tensor takes the form (assuming we are in General Relativity) $G_{\mu\nu} + \Lambda g_{\mu\nu} = T^{\mathrm{EM}}_{\mu\nu} + T^{\mathrm{mat}}_{\mu\nu}$, where $T^{\mathrm{EM}}_{\mu\nu} = 2(F_{\mu\alpha}F_{\nu}{}^{\alpha} - \frac{1}{4}g_{\mu\nu}F_{\alpha\beta}F^{\alpha\beta})$ is the energy-momentum tensor of the electromagnetic field $F_{\mu\nu}$ and $T^{\mathrm{mat}}$ is arbitrary except for the condition of satisfying DEC. The electric and magnetic charges on $S$ are defined as

$$Q_E = \frac{1}{4\pi}\int_S F(\ell, k)\eta_h, \qquad Q_M = \frac{1}{4\pi}\int_S F^{\star}(\ell, k)\eta_h,$$

where $F^{\star}$ is the Hodge dual of $F$. We now assume that $S$ is a MOTS. If we insert $u = 1$ in the symmetrized inequality (5) along $-k$ (i.e. with $V = 0$) it follows

$$4\pi(1-g) \geq (\lambda + \Lambda)\,|S| + \int_S (F(\ell, k))^2 + (F^{\star}(\ell, k))^2\eta_h. \tag{6}$$

Now, following [24], the Hölder inequality implies the inequality $\int_S f^2 \geq |S|^{-1} (\int_s f)^2$ valid for any $L^2$ function $f$. Thus, (6) implies

$$4\pi(1-g) \geq (\lambda + \Lambda)|S| + 16\pi^2|S|^{-1}(Q_E^2 + Q_M^2). \tag{7}$$

This inequality including the cosmological constant, arbitrary topology of $S$ and the principal eigenvalue has been obtained in [32], where a number of consequences have also been derived. Two immediate consequences are the following:

Assume that $S$ is spacetime stable. Since, under the assumptions above, the spacetime satisfies the NEC (irrespectively of the sign of $\Lambda$) it follows that $S$ is stable along the direction $-k$, i.e. $\lambda_- \geq 0$. If we assume further that $\Lambda \geq 0$, then it follows immediately that $S$ must be of spherical or toroidal topology. This recovers a well-known theorem on the topology of MOTS due to Hawking [33]. This theorem has been generalized to higher dimensions in [30] where it was shown that the Yamabe type of any stable MOTS must be non-negative. The case of vanishing Yamabe type (i.e. toroidal topology in the case of four dimensions) has been shown to be very rigid in [34] and to be excluded when the MOTS satisfies suitable barrier properties.

If $S$ is spacetime stable and there is a negative cosmological constant, then (7) implies an upper bound on the genus of $S$, namely $g \leq 1 - \frac{|S|\Lambda}{4\pi}$. This bound was first obtained in [35].

For the area-charge inequality, (7) immediately implies that, as long as $S$ is spacetime stable and $\Lambda \geq 0$, then $|S| \geq 4\pi(Q_E^2 + Q_M^2)$, as first proved in [24]. This inequality, in turn, is a generalization of a previous result by Gibbons which establishes this inequality in the case of minimal surfaces embedded a time-symmetric slice [36].

## 5 Axially Symmetric MOTS and Angular Momentum

We next discuss local geometric inequalities involving angular momentum. As mentioned in the Introduction, this case requires the surfaces to be axially symmetric in order to have a proper definition of angular momentum. In principle, one may think that this entails restricting oneself to axially symmetric spacetimes. In fact, fewer requirements are needed and the following definition, essentially put forward in [23], is sufficient for the purpose of writing down and proving the desired inequalities:

**Definition 4** A MOTS $S$ is axially symmetric if there exists a vector $\eta$ tangent to $S$ satisfying

1. $\mathscr{L}_\eta h = 0$.
2. $\mathscr{L}_\eta s = 0$, for some choice of basis $\{\ell, k\}$.
3. $\eta$ commutes with the stability operator $L_v$ for some choice of $v$.

It is clear that if the spacetime $(M, g)$ admits an axial Killing vector $\eta$ and $\eta$ is tangent to $S$, then the $S$ is also axially symmetric according to this definition.

The angular momentum $J$ of an axially symmetric MOTS is then defined by

$$J(S) := \frac{1}{8\pi}\int_S s(\eta)\eta_S.$$

It is straightforward to check that this definition is independent of the choice of basis $\{\ell, k\}$. Nevertheless, for some expressions below, it is necessary to restrict the basis to satisfy point 2 in definition 4. We will do so from now on.

The following theorem due to Jaramillo et al. [23] establishes a remarkable inequality involving the area and the angular momentum of an axially symmetric MOTS. The only requirement is that the MOTS is spacetime stable and that energy-momentum contents of the spacetime satisfies the dominant energy condition. It is therefore a very general and robust inequality which reveals a deep connection between the rotation and the shape of quasi-local black holes in four spacetime dimensions.

**Theorem 1** *Let $(S, h)$ be an axially symmetric, two-dimensional MOTS, stable with respect to an achronal direction $v$ in a spacetime satisfying DEC. Then*

$$|S| \geq 8\pi |J|. \tag{8}$$

*Moreover, equality can only happen if the following five conditions are simultaneously satisfied:*

*(i)* $S$ *is marginally stable,*

*(ii)* $h = |J|\left(1+\cos^2\theta\right)d\theta^2 + \frac{4|J|\sin^2\theta}{1+\cos^2\theta}d\varphi^2,$

*(iii)* $G(\ell, k) = 0$ *on* $S$,

*(iv)* $z = \dfrac{2J\sin^2\theta}{|J|\left(1+\cos^2\theta\right)^2}d\varphi,$

*(v)* *If* $v$ *is spacelike then* $G(\ell,\ell) = 0$ *and* $\chi_\ell = 0$.

The equality case corresponds to the geometry of the extreme Kerr horizon, i.e. the induced metric on any spacelike section of the degenerate horizon of the Kerr metric satisfying $M^2_{\mathrm{ADM}} = |J_{\mathrm{ADM}}|$. Previous to Theorem 1, the same inequality had been proved by Dain and Reiris [22] for minimal surfaces embedded in maximal slices of a vacuum spacetime

In the following I will describe the basic steps involved in the proof of this result. The argument has two parts. The first one consists in finding an inequality valid for arbitrary functions on $S$ and applying it to an appropriate function defined in terms of the geometry of $S$. The second step consists in showing that the resulting inequality can be related to the angular momentum, to conclude finally that (8) holds.

Step 1 was accomplished in [23] as a consequence of the spacetime stability of $S$ by performing suitable direct estimates. However, as mentioned before, stable MOTS satisfy general inequalities given by the Rayleigh-Ritz type characterization described in Proposition 1. This allows us to describe step one as follows.

First, using $u = 1$ in the general inequality (3) (notice that for $u = 1$, the solution to equation (4) is $\omega_u = \text{const}$) and the Gauss-Bonnet theorem shows that the genus of the surface must be positive, or else $z_A = 0$, which would imply $J = 0$ and hence a trivial area-angular momentum inequality. So, only the spherically symmetric case needs to be considered. In this setting axisymmetry implies that the divergence-free one form $z_A$ in the Helmholz decomposition of the connection one-form $s_A$ is proportional to the Killing vector $\eta$ (with its indexes lowered with the metric $h$). Consequently, the source term in equation (4) vanishes for any choice of axially symmetric $u$. The corresponding solution is again $\omega_u = \text{const}$. Since restricting the class of functions $u$ in (3) to being axially symmetric cannot decrease the infimum, it follows from stability that

$$\begin{aligned}\int_S \left(|Du|^2 + \frac{R(h)}{2}u^2\right)\eta_h &\geq \int_S \left(G(\ell, v^\star) + V|\chi_\ell|^2 + |z|^2\right)u^2\eta_h \\ &\geq \int_S |z|^2 u^2 \eta_h, \end{aligned} \tag{9}$$

where the second inequality follows because $\left(G(\ell, v^\star) + V|\chi_\ell|^2\right)u^2$ is non-negative under the assumptions of the theorem. Now, it is simple to see that the metric $h$ can be written in the form (c.f. [22])

$$h = \frac{1}{\cosh^2\tau}\left(e^{2c}e^{-\sigma(\tau)}d\tau^2 + e^{\sigma(\tau)}d\varphi^2\right), \quad -\infty < \tau < \infty, \ 0 < \phi \leq 2\pi, \tag{10}$$

where $c$ is a constant related to the total area by $|S| = 4\pi e^c$. Regularity on the axis of symmetry (i.e. where $\tau \to \pm\infty$) imposes the following asymptotic behaviour on $\sigma(\tau)$,

$$\lim_{\tau\to\pm\infty} \sigma = c, \qquad \lim_{\tau\to\pm\infty} \frac{d\sigma}{d\tau} = 0.$$

Now, a key insight of Dain and Reiris [22] was to use the function $u = e^{-\frac{\sigma}{2}}$ in the analytic inequality (9). This gives an inequality which involves only $\sigma$ and $z_A$. The second part of the proof consists in showing that this leads to an inequality involving only $J$ and the area $|S|$ in such a way that all details of the function $\sigma(\tau)$ and $z_A(\tau)$ disappear. To discuss this second part we use a simplified version of an argument due to [21], which, in turn, is a simplification of the original argument in [37] where the absolute minimum of a renormalized energy of a harmonic map was computed.

First of all we define a function $Y(\tau)$ (up to an arbitrary additive constant) by the equation $z(\eta)\eta_h = \frac{1}{2}\frac{dY}{d\tau} d\tau \wedge d\varphi$. From the definition of angular momentum we have

$$8J = \lim_{\tau\to\infty} (Y(\tau) - Y(-\tau)).$$

Now, inserting $u = e^{-\frac{\sigma}{2}}$ and the expression for $z_A$ in terms of $Y(\tau)$ in (9) and computing explicitly the curvature scalar for the metric (10) it is straightforward to see that the stability inequality becomes

$$0 \geq \int_S \left[ \frac{1}{X^2}\left( \left(\frac{dX}{d\tau}\right)^2 + \left(\frac{dY}{d\tau}\right)^2 \right) - 4 \right] d\tau d\phi, \tag{11}$$

where $X(\tau)$ has been defined as $X = |\eta|^2$ (i.e. $X = e^{\sigma}\cosh^{-2}(\tau)$). Now we can view the first term in the integrand as the energy-density of a path $\gamma(\tau) \equiv \{X(\tau), Y(\tau)\}$ in the hyperbolic space $(\mathbb{H}^2, g_{\mathbb{H}^2} = \frac{dX^2+dY^2}{X^2})$. Note that, since the variable $\tau$ takes values on the whole real line and $X(\tau) \to 0$ when $|\tau| \to \infty$ the total energy of the path diverges (this simply reflects the fact that while the function in the right-hand side of (11) is integrable on $S$, this property is obviously not true for the constant term in the expression). Rewriting the total integral as a definite integral from $\tau = -L$ to $\tau = L$ and sending $L$ to infinity, inequality (11) becomes, after performing the trivial angular integration,

$$0 \geq \lim_{L\to+\infty} \left( \int_{-L}^{L} g_{\mathbb{H}^2}(\dot{\gamma}, \dot{\gamma})\, d\tau - 8L \right),$$

where dot means derivative with respect to $\tau$. We now apply once again the Hölder inequality in the form $\int_{-L}^{L} g_{\mathbb{H}^2}(\dot{\gamma}, \dot{\gamma})\, d\tau \geq \frac{1}{2L}(\int_{-L}^{L} \sqrt{g_{\mathbb{H}^2}(\dot{\gamma}, \dot{\gamma})} d\tau)^2$ and use the obvious property that the total length of any curve is never smaller than the distance between its initial and final points to rewrite the stability inequality as

$$0 \geq \lim_{L\to+\infty} \left( \frac{1}{2L} \mathrm{dist}^2_{\mathbb{H}^2}[\gamma(L), \gamma(-L)] - 8L \right). \tag{12}$$

Now, the distance function between two arbitrary points $(X_1, Y_1)$ and $(X_2, Y_2)$ in the Poincaré upper-half plane is well-known to be [38]

$$\mathrm{dist}_{\mathbb{H}^2}[(X_1, Y_1), (X_2, Y_2)] = \operatorname{arccosh}\left[1 + \frac{(Y_2 - Y_1)^2 + (X_2 - X_1)^2}{2X_1X_2}\right].$$

The limit $L \to +\infty$ in (12) is straightforward to obtain after using $\sigma(\pm L) \to c$ and $Y(L) - Y(-L) \to 8J$ and gives

$$0 \geq 8 \ln\left(\frac{8\pi|J|}{|S|}\right) \iff |S| \geq 8\pi|J|,$$

which proves the inequality. The statements regarding the equality case are also straightforward in this framework and follow by imposing that all inequalities along the process become equalities. In particular, the curve $\gamma(\tau)$ must be a parametrized geodesic in $(\mathbb{H}^2, g_{\mathbb{H}^2})$. This fixes the metric $h$ and $z_A$ to be those in the extreme Kerr geometry. This proves points (ii) and (v) in the theorem. The rest follows directly from the non-negative terms discarded in the stability inequality.

## 6 Area-Angular Momentum Inequality for Black Holes

As mentioned in the Introduction, the first examples of area-angular momentum inequalities were obtained for black holes. The assumptions in this setting were that the spacetime is stationary and axially symmetric and that it contains a Killing horizon. Moreover, the spacetime was assumed to be vacuum in a neighbourhood of the horizon, but matter was allowed outside the black hole. The first case treated in the literature assumed the Killing horizon to be degenerate, i.e. with vanishing surface gravity. In this setting, Ansorg and Pfister [15] were able to show that the inequality $|S| = 8\pi|J|$ was universally valid. Here $|S|$ is the area of any axially symmetric spacelike section of the Killing horizon. The non-degenerate case was considered in [16] where the inequality $|S| > 8\pi|J|$ was proved under the additional condition that the black hole is *subextremal*. The definition of subextremal is due to [39] and requires the existence of a future directed null vector $k$ transverse to the Killing horizon for which the inequality $\delta_k\theta_\xi < 0$ holds, where $\xi$ is the Killing vector which generates the Killing horizon. It is clear from the definition of strict stability of MOTS that a Killing horizon is subextremal if and only if all of its spacelike sections are strictly stable along the transverse direction $k$. The condition of subextremality was used in [16] by working in an Eddington-Finkelstein advanced extension of the following metric in Boyer-Lindquist type coordinates

$$ds^2 = \hat{\mu}\left(\frac{dR^2}{R^2 - r_h^2} + d\theta^2\right) + \hat{u}\sin^2\theta\,(d\varphi - \omega dt)^2 - \frac{4}{\hat{u}}\left(R^2 - r_h^2\right)dt^2,$$

where the Killing horizon is located at $R = r_h$. Imposing the condition $\delta_m \theta_\xi < 0$, where $m$ is proportional to $\partial_R$ in Eddington-Finkelstein coordinates, and discarding a number of positive terms, the actual "subextremality" assumption made by the authors was

$$\int_{S_0} \partial_R(\hat{u}\hat{\mu})|_{R=r_h} \sin\theta d\theta > 0,$$

where $S_0$ is a section of the Killing horizon corresponding to a constant coordinate time $\hat{t}$ in the Eddington-Finkelstein advanced coordinate system. This equality is certainly implied by the geometric subextremality condition by Booth and Fairhurst, but it is not equivalent to it and, in principle, it is weaker. The purely local inequality in Theorem 1 requires stability, but not strict stability, so the connection between the two inequalities is not immediately obvious. The proofs are also quite different, so it became a problem of interest to try and relate the purely local and the black hole versions of the area-angular momentum inequality. A first step along this direction was made in [21] where the comparison was focused on the relationship between the proof in the minimal surface case in [22] and the behaviour of the Killing horizon on its bifurcation surface. However, no geometrically clear reason of why both inequalities work in seemingly different regimes was given. In [26] a detailed study of the stability properties of Killing horizons was performed. As a by product, a clear relationship between the two types of area-angular momentum inequality was obtained. The next section is devoted to reviewing these results briefly.

## 7 MOTS and Killing Horizons

Recall that, in a spacetime $(M, g)$ admitting a Killing vector $\xi$, a Killing horizon $\mathscr{H}$ is a null hypersurface where the Killing vector $\xi$ is null, tangent and nowhere zero. The integral lines of $\xi$ are null geodesics on $\mathscr{H}$ and the surface gravity is the scalar function on $\mathscr{H}$ defined by $\kappa \colon \nabla_\xi \xi \stackrel{\mathscr{H}}{=} \kappa \xi$. It is well-known that $\kappa$ is constant if $(M, g)$ satisfies the DEC. In the following we will assume that $\mathscr{H}$ has topology $S \times \mathbb{R}$ with $S$ closed and that the $\mathbb{R}$ factor is tangent to the integral lines of $\xi$.

The Killing equations imply immediately that all spacelike sections $S_0$ in $\mathscr{H}$ are MOTS and, in fact, with vanishing second fundamental form along $\xi$, i.e. $\chi_\xi(S_0) = 0$. The Raychaudhuri equation then implies $G(\xi, \xi) \stackrel{\mathscr{H}}{=} 0$. These two properties say that the function $W$ introduced before vanishes identically in this case and, hence, that the stability operator of $S_0$ is independent of the transverse direction $v$. Both the stability operator and its principal eigenvalue are therefore properties of $S_0$ alone. We will denote them simply by $L_{S_0}$ and $\lambda_{S_0}$ respectively. A natural question is then whether the stability is a property of the horizon itself or whether it depends on the choice of section $S_0$. To address this issue it is convenient to obtain first an explicit form of the stability operator. This was done in [26] for general totally geodesic null hypersurfaces. Here we restrict ourselves to Killing horizons for definiteness.

**Theorem 2** *Let $S_0$ be a spacelike section of a Killing horizon $\mathscr{H}$ of topology $S \times \mathbb{R}$. Let $k$ be a null vector field on $S_0$, orthogonal to $S_0$ and satisfying $\langle \xi, k \rangle = -1$. The stability operator $L_{S_0}$ takes the form*

$$L_{S_0}(\psi) = -D_A\left(uD^A\left(\frac{\psi}{u}\right)\right) + 2z^A D_A\psi - \kappa\theta_k\psi,$$

*where the positive scalar function $u$ and the one-form $z_A$ are defined via the Helmholz decomposition of the normal connection one-form $s_A$ as $s = z + \frac{du}{2u}$.*

Notice that $u$ is defined up to an arbitrary multiplicative constant, which obviously has no effect in the expression for $L_{S_0}$. Combining this theorem with the behaviour of $s_A$ and $\theta_k$ under a change of section the following theorem follows [26].

**Theorem 3** *If $\mathscr{H}$ has constant surface gravity, then $\lambda_{S_0}$ is independent of $S_0$. Moreover there exist Killing horizons $\mathscr{H}$ with non-constant $\kappa$ for which $\lambda_{S_0}$ depends on $S_0$.*

Regarding the analysis of the area-angular momentum inequality for Killing horizons, it turns out that the stability of $S_0$ has implications on certain integral of the transverse null expansion $\theta_k$ on $S_0$. More precisely [26].

**Proposition 2** *Let $\mathscr{H}$ be a Killing horizon with topology $S \times \mathbb{R}$ and let $S_0$ any spacelike section of $\mathscr{H}$. Assume that $z_A$ and $du$ in the decomposition $s = z + \frac{du}{2u}$ are orthogonal (this occurs automatically if $\mathscr{H}$ is axially symmetric and $S_0$ respects the axial symmetry and has spherical topology). Then the following holds*

- *If $S_0$ is stable then $\int_{S_0} \kappa\theta_k u \leq 0$.*
- *If $S_0$ is strictly stable then $\int_{S_0} \kappa\theta_k u < 0$.*
- *If $S_0$ is stable and $\kappa$ is constant and non-zero, then $\int_{S_0} \kappa\theta_k u = 0 \Longleftrightarrow \theta_k = 0$.*

This Proposition is the key property that allows one to find the link between the proof of the area-angular momentum inequality in the black hole case [16] and the purely local inequality in [23]. This relationship is given in the following result [26].

**Theorem 4** *Assume that the spacetime $(M, g)$ satisfies the DEC and admits a Killing vector $\xi$ with a Killing horizon $\mathscr{H}$. Assume that $\mathscr{H} \simeq \mathbb{S}^2 \times \mathbb{R}$ is axially symmetric with $\kappa \neq 0$ constant. Write the metric of spacelike sections of $\mathscr{H}$ in the form*

$$h = e^{2c}e^{-\sigma}d\theta^2 + e^{\sigma}\sin^2\theta d\phi^2. \tag{13}$$

*Then there exists a section $S_1$ for which $s = z - \frac{1}{2}d\sigma$. Moreover, if $\int_{S_1} \kappa\theta_k e^{-\sigma} \leq 0$, then $|S| \geq 8\pi|J|$.*

Notice that $S_1$ is *not* assumed to be stable in this theorem. The inequality involving $\theta_k$ is sufficient for proving the area-angular momentum inequality. Proposition 2 shows that stability of the Killing horizon (or equivalently stability of any of its spacelike sections) is sufficient to imply the validity of the integral inequality assumed in the theorem. It turns out that the surface $S_0$ used by [16] is *precisely* the surface $S_1$ in this theorem. Moreover, their inequality $\int_{S_0} \partial_R(\hat{u}\hat{\mu})|_{R=r_h} \sin\theta d\theta > 0$ is precisely the geometric inequality $\int_S \kappa\theta_k e^{-\sigma} < 0$, and the link between the two approaches becomes clear. A similar clarification for the degenerate case has been recently obtained in [40].

# 8 Further Results and Open Problems

As mentioned in the Introduction, when the spacetime has an electromagnetic field, the area-energy momentum inequality can be strengthened to include the electric charge. This was first done for stationary and axially symmetric degenerate black holes in [15]. The main assumption was that the spacetime is electrovacuum in a neighbourhood of the horizon and it was shown that the equality $|S| = 4\pi\sqrt{Q^4 + 4J^2}$ always holds. In the same setting, but allowing non-degenerate subextremal black holes, the strict inequality $|S| > 4\pi\sqrt{Q^4 + 4J^2}$ was proved in [17]. The local version of this inequality (i.e. for stable, axially symmetric MOTS) has been proved in [25].

Everything we have said so far involves four-dimensional spacetimes. Recent work by Hollands [41] establishes the following generalization to arbitrary dimension.

**Theorem 5** *Let $(M^{n+1}, g)$ be a vacuum spacetime with a non-negative cosmological constant $\Lambda$, $n \geq 3$ and which admits a stable Killing horizon $\mathscr{H}$ of topology $S \times \mathbb{R}$ with $S$ closed. Assume further that $(M^{n+1}, g)$ admits an additional isometry group $U(1)^{n-2}$ leaving $\mathscr{H}$ invariant and consider a section $S_0$ of $\mathscr{H}$ respecting the $U(1)^{n-2}$ symmetries. Let $\eta_\pm$ be the Killing vectors which vanish, respectively, at the "north" and "south" poles of $S_0$. Then*

$$|S_0| \geq 8\pi |J(\eta_+)J(\eta_-)|^{1/2},$$

*where* $J(\eta) := \frac{1}{8\pi}\int_{S_0} s(\eta)\eta_{S_0}$
(for the precise definition of the vectors $\eta_\pm$, see in [41]).

Before concluding this lecture, let me present a brief list of open problems. The first one refers to the higher dimensional case. The statement above requires that the spacetime contains a Killing horizon and that the spacetime is vacuum, possibly with a positive cosmological constant. It would be of interest to relax this and admit a general spacetime satisfying the DEC. It is also of interest to prove the statement directly at the local level, i.e. for stable MOTS. Another interesting problem in this context is whether the symmetry assumptions can be relaxed to a small number of linearly independent Killing vectors.

In relation to the structure of the proof of the area-angular momentum inequality, it would be of interest to find a deeper reason of why the choice of $u = e^{-\frac{\sigma}{2}}$, where $\sigma$

is defined as a metric coefficient, is the appropriate choice and what is the underlying reason for the role played by the hyperbolic space (and other highly symmetric spaces in the charged case) in the proof of the inequality.

Another interesting problem is to understand why the surface $\tilde{t} = \text{const}$ in Eddington-Finkesltein coordinates for black holes is precisely the surface $S_1$ in the area-angular momentum inequality for Killing horizons, i.e. why for these coordinates, the normal connection one-form is linked to the induced metric via the condition spelled out in Theorem 4.

## References


1. Penrose, R.: Naked singularities. Ann. N.Y. Acad. Sci. **224**, 125 (1973). doi:10.1111/j.1749-6632.1973.tb41447.x
2. Mars, M.: Present status of the penrose inequality. Class. Quantum Grav. **26**, 193001 (2009). doi:10.1088/0264-9381/26/19/193001
3. Gibbons, G.W.: The isoperimetric and bogomolny inequalities for black holes in global riemannian geometry. In: Willmore, T.J., Hitchin, N.J. (eds.) Ellis Horwood Series in Mathematics and Its Applications, pp. 194–202. Ellis Horwood, Chichester (1984)
4. Gibbons, G.W., Hawking, S.W., Horowitz, G.T., Perry, M.J.: Positive mass theorems for black holes. Commun. Math. Phys. **88**, 295 (1983). doi:10.1007/BF01213209
5. Herzlich, M.: The positive mass theorem for black holes revisited. J. Geom. Phys. **26**, 97 (1998). doi:10.1016/S0393-0440(97)00040-5
6. Dain, S., Lousto, C.O., Takahashi, R.: New conformally flat initial data for spinning black holes. Phys. Rev. D **66**, 104038 (2002). doi:10.1103/PhysRevD.65.104038
7. Szabados L.B.: Quasi-local energy-momentum and angular momentum in GR: a review article. Living Rev. Relativity 7, lrr-2004-4 (2004). http://www.livingreviews.org/lrr-2004-4
8. Dain, S.: The inequality between mass and angular momentum for axially symmetric black holes. Int. J. Mod. Phys. D **17**, 519 (2008). doi:10.1142/S021827180801219X
9. Dain, S.: Proof of the angular momentum-mass inequality for axisym-metric black holes. J. Differ. Geom. **79**, 33 (2008)
10. Chruściel P.T., Lopes Costa J.: Mass, angular-momentum, and charge inequalities for axisymmetric initial data. Class. Quant. Grav. 26, 235013 (2009). doi:10.1088/0264-9381/26/23/235013
11. Costa, J.L.: Proof of a Dain inequality with charge. J. Phys. A: Math. Theor. **43**, 285202 (2010). doi:10.1088/1751-8113/43/28/285202
12. Chruściel P.T., Li Y., Weinstein G.: Mass and angular-momentum inequalities for axi-symmetric initial data sets: II. Angular-momentum. Ann. Phys. (N.Y.) **323**, 2591 (2008). doi:10.1016/j.aop.2007.12.011
13. Dain, S.: Geometric inequalities for axially symmetric black holes. Class. Quantum Grav. **29**, 073001 (2012). doi:10.1088/0264-9381/29/7/073001
14. Simon, W.: Gravitational field strength and generalized komar integral. Gen. Relativ. Gravit. **17**, 439 (1985). doi:10.1007/BF00761903
15. Ansorg, M., Pfister, H.: A universal constraint between charge and rotation rate for degenerate black holes surrounded by matter. Class. Quantum Grav. **25**, 035009 (2008). doi:10.1088/0264-9381/25/3/035009
16. Hennig, J., Ansorg, M., Cederbaum, C.: A universal inequality between the angular momentum and the horizon area for axisymmetric and stationary back holes with surrounding matter. Class. Quantum Grav. **25**, 162002 (2008). doi:10.1088/0264-9381/25/16/162002

17. Hennig, J., Cederbaum, C., Ansorg, M.: A universal inequality for axisymmetric and stationary black holes with surrounding matter in the einstein-maxwell theory. Commun. Math. Phys. **293**, 449 (2010). doi:10.1007/s00220-009-0889-y
18. Neugebauer, G., Hennig, J.: Non-existence of stationary two-black-hole configurations. Gen. Relativ. Gravit. **41**, 2113 (2009). doi:10.1007/s10714-009-0840-8
19. Hennig, J., Neugebauer, G.: Non-existence of stationary two-black-hole configurations: the degenerate case. Gen. Relativ. Gravit. **43**, 3139 (2011). doi:10.1007/s10714-011-1228-0
20. Hennig, J., Neugebauer, G.: Stationary two-black-hole configurations: a non-existence proof. J. Geom. Phys. **62**, 613 (2012). doi:10.1016/j.geomphys.2011.05.008
21. Chruściel, P.T., Eckstein, M., Nguyen, L., Szybka, S.: Existence of singularities in two-kerr black holes. Class. Quantum Grav. **28**, 245017 (2011). doi:10.1088/0264-9381/28/24/245017
22. Dain, S., Reiris, M.: Area—Angular momentum inequality for axisymmetric black holes. Phys. Rev. Lett. **107**, 051101 (2011). doi:10.1103/PhysRevLett.107.051101
23. Jaramillo, J.L., Reiris, M., Dain, S.: Black hole area-angular momentum inequality in non-vacuum spacetimes. Phys. Rev. D **84**, 121503 (2011). doi:10.1103/PhysRevD.84.121503
24. Dain, S., Jaramillo, J.L., Reiris, M.: Area-charge inequality for black holes. Class. Quantum Grav. **29**, 035013 (2012)
25. Gabach-Clement M.E., Jaramillo J.L., Reiris M.: Proof of the area-angular momentum-charge inequality for axisymmetric black holes, ArXiv e-prints 1207.6761 [gr-qc] (2012)
26. Mars, M.: Stability of mots in totally geodesic null horizons. Class. Quantum Grav. **29**, 145019 (2012). doi:10.1088/0264-9381/29/14/145019
27. Newman, R.P.A.C.: Topology and stability of marginal 2-surfaces. Class. Quantum Grav. **4**, 277 (1987). doi:10.1088/0264-9381/4/2/011
28. Andersson, L., Mars, M., Simon, W.: Stability of marginally outer trapped surfaces and existence of marginally outer trapped tubes. Adv. Theor. Math. Phys. **12**, 853 (2008)
29. Donsker, M.D., Varadhan, S.R.S.: On the principal eigenvalue of second order elliptic differential operators. Commun. Pure Appl. Math. **29**, 591 (1976). doi:10.1002/cpa.3160290606
30. Galloway, G.J., Schoen, R.: A generalization of hawking's black hole topology theorem to higher dimensions. Commun. Math. Phys. **266**, 571 (2006). doi:10.1007/s00220-006-0019-z
31. Hayward, S.: General laws of black-hole dynamics. Phys. Rev. D **49**, 6467 (1994). doi:10.1103/PhysRevD.49.6467
32. Simon, W.: Bounds on area and charge for marginally trapped surfaces with a cosmological constant. Class. Quantum Grav. **29**, 062001 (2012). doi:10.1088/0264-9381/29/6/062001
33. Hawking, S.W.: The event horizon. In: DeWitt, C., DeWitt, B.S. (eds.) Black Holes, pp. 1–56. Gordon and Breach, New York (1973)
34. Galloway, G.J.: Rigidity of marginally trapped surfaces and the topology of black holes. Comm. Anal. Geom. **16**, 217 (2008)
35. Woolgar, E.: Bounded area theorems for higher-genus black holes. Class. Quantum Grav. **16**, 3005 (1999). doi:10.1088/0264-9381/16/9/316
36. Gibbons, G.W.: Some comments on gravitational entropy and the inverse mean curvature flow. Class. Quantum Grav. **16**, 1677 (1999). doi:10.1088/0264-9381/16/6/302
37. Aceña, A., Dain, S., Gabach Clément, M.E.: Horizon area-angular momentum inequality for a class of axially symmetric black holes. Class. Quantum Grav. **28**, 105014 (2011). doi:10.1088/0264-9381/28/10/105014
38. Anderson, J.W.: Hyperbolic Geometry. Springer Undergraduate Mathematics Series. Springer, London (2007)
39. Booth, I., Fairhurst, S.: Extremality conditions for isolated and dynamical horizons. Phys. Rev. D **77**, 084005 (2008). doi:10.1103/PhysRevD.77.084005
40. Jaramillo, J.L.: A note on degeneracy, marginal stability and extremality of black hole horizons. Class. Quantum Grav. **29**, 177001 (2012). doi:10.1088/0264-9381/29/17/177001
41. Hollands, S.: Horizon area-angular momentum inequality in higher dimensional spacetimes. Class. Quantum Grav. **29**, 065006 (2012). doi:10.1088/0264-9381/29/6/065006

# Stationary Black-Hole Binaries: A Non-existence Proof

**Gernot Neugebauer and Jörg Hennig**

**Abstract** We resume former discussions of the question, whether the spin-spin repulsion and the gravitational attraction of two aligned black holes can balance each other. Based on the solution of a boundary problem for disconnected (Killing) horizons and the resulting violation of characteristic black hole properties, we present a non-existence proof for the equilibrium configuration in question. From a mathematical point of view, this result is a further example for the efficiency of the inverse ("scattering") method in non-linear theories.

## 1 Introduction

The examination of time-independent two-body systems dates back to the early days of General Relativity. In a 1922 paper, Bach and Weyl [1] discussed the superposition of two exterior Schwarzschild solutions in Weyl coordinates as a characteristic example for an equilibrium configuration consisting of two "sphere-like" bodies at rest. Bach noted that this *static* solution develops a singularity on the portion of the symmetry axis between the two bodies, which violates the elementary flatness on this interval. In a supplement to Bach's contribution, Weyl focused on the interpretation of this type of singularity and used stress components of the energy-momentum tensor to define a *non-gravitational* repulsion between the bodies which compensates the gravitational attraction. Weyl's result is based on some artificial assumptions but

G. Neugebauer
Theoretisch-Physikalisches Institut, Friedrich-Schiller-Universität Jena, Max-Wien-Platz 1, 07749 Jena, Germany
e-mail: neugebauer@tpi.uni-jena.de

J. Hennig (✉)
Department of Mathematics and Statistics, University of Otago, P.O. Box 56, Dunedin 9054, New Zealand
e-mail: jhennig@maths.otago.ac.nz

J. Bičák and T. Ledvinka (eds.), *General Relativity, Cosmology and Astrophysics*, Fundamental Theories of Physics 177, DOI: 10.1007/978-3-319-06349-2_9,

© Springer International Publishing Switzerland 2014

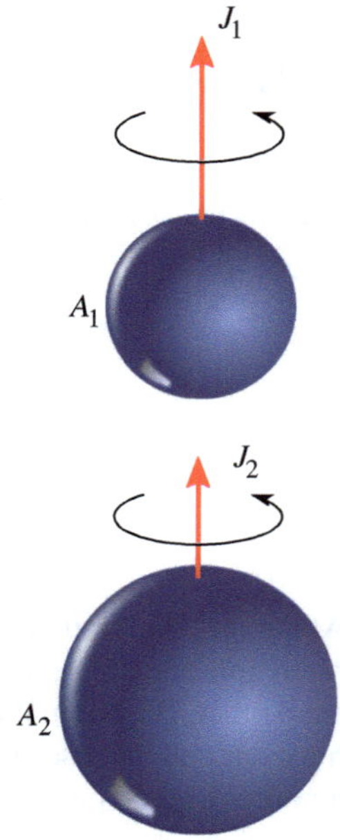


**Fig. 1** Illustration of two aligned rotating black holes with horizon areas $A_1$, $A_2$ and angular momenta $J_1$, $J_2$

implies an interesting question: Are there repulsive effects of *gravitational* origin which could counterbalance the omnipresent mass attraction?

Post-Newtonian approximations tell us that the interaction of the angular momenta of rotating bodies ("spin-spin interaction") could indeed generate repulsive effects. This is a good motivation to study, in a rigorous way, *stationary* two-body problems.

In this contribution we shall summarize the results that we obtained for a stationary two-black-hole system consisting of two aligned rotating black holes with parallel (or anti-parallel) spins, see Fig. 1. The representation is based on three recent papers which contain the *details* of the analysis [2–4]. The *idea* of our non-existence proof is to construct the exterior gravitational fields of two disconnected Killing horizons, see Fig. 2, via a boundary problem for the *nonlinear* Ernst equation, which is essentially equivalent to the vacuum Einstein equations. Fortunately, this equation belongs to a class of completely integrable differential equations, which can be mapped to *linear* structures ("Linear Problems"). This fact is the source of powerful solution generating methods such as Bäcklund transformations. It can be shown that a single Bäcklund transformation [5, 6] applied to Minkowski space creates a Kerr-NUT spacetime which includes the spacetime of the rotating black hole. Since iterative Bäcklund transformations act as a "nonlinear superposition principle", the double-Kerr-NUT solution [5, 7] was considered to be a good candidate for the solution of the two-horizon problem and extensively discussed in the literature [7–17]. However, there was no argument ensuring that this particular solution be the *only* candidate. We have removed this objection and shown that the solution procedure for the *boundary problem* necessarily leads to a subclass of the double-Kerr-NUT solution.

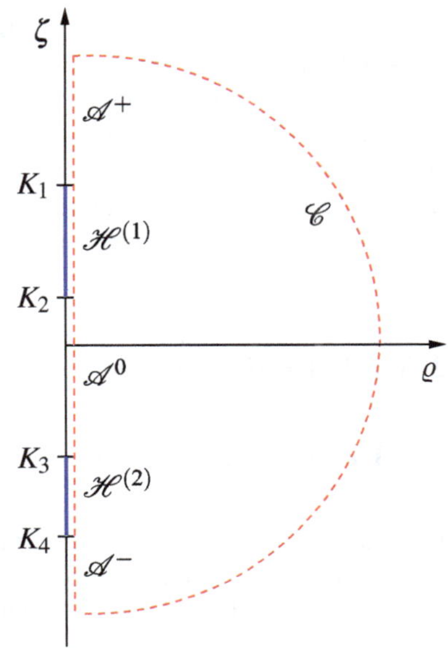


**Fig. 2** A two-black-hole equilibrium configuration in Weyl-Lewis-Papapetrou coordinates (Adapted from [2])

Thus we could make use of results derived for the double-Kerr-NUT solution. The result is in line with a theorem of Varzugin [18, 19], which says that the $2N$-soliton solution by Belinskiĭ and Zakharov [20, 21] contains all possible solutions (if any existed) corresponding to an equilibrium configuration of black holes. The subclass is characterized by a set of restrictions for the parameters of the general double-Kerr-NUT solution. These restrictions, first derived and discussed by Tomimatsu and Kihara [11, 16], ensure the regularity of the double-Kerr-NUT solution on the axis of symmetry and on the horizons. An elegant reformulation of the Tomimatsu-Kihara regularity conditions by Manko et al. [15] made it possible to express black hole quantities such as mass, angular momentum and surface area in terms of independent parameters. We have made use of these results. Another condition to be satisfied is the positivity of the total mass. Combining the restrictions with symmetry arguments, Hoenselaers and Dietz [8, 10] and Krenzer [13] could show that the double-Kerr-NUT solution cannot describe a configuration consisting of two *identical* black holes. Manko and Ruiz [14] generalized this result by showing that the regularity conditions imply that at least one of the two horizons has a negative Komar mass. They argued, without giving any explanation, that this peculiarity casts out the double-Kerr-NUT solution. Remarkably, Ansorg and Petroff [22], who described an equilibrium configuration with a positive total mass and a component which has a negative Komar mass, came to an opposite interpretation. However, considerations like these stimulated us to examine further black hole inequalities. Fundamental *local* "state variables" of a rotating black hole are its area $A$ and its angular momentum $J$. Indeed, for a single black hole these quantities are restricted by the inequality $8\pi|J| \leq A$. Based on results of Ansorg and Pfister [23], who examined extremal black holes,

Hennig et al. [24], who, following Booth and Fairhurst [25], studied sub-extremal black holes defined by existence of trapped surfaces (surfaces with a negative expansion of outgoing null rays) in every sufficiently small interior neighbourhood of the event horizon, and Chruściel et al. [26], we can assume that each of the two black holes has to satisfy the angular momentum-area inequality *individually*. Surprisingly, Dain and Reiris [27] were able to extend its range of application to non-stationary black holes, see also the overview article by Dain [28] and references therein.

# 2 Mathematical Tools

## 2.1 Metric and Horizons

The exterior vacuum gravitational field of axially symmetric and stationary gravitational sources can be described in cylindrical Weyl-Lewis-Papapetrou coordinates $(\rho, \zeta, \varphi, t)$,[1] in which the line element takes the form

$$ds^2 = e^{-2U}\left[e^{2k}(d\rho^2 + d\zeta^2) + \rho^2 d\varphi^2\right] - e^{2U}(dt + a\, d\varphi)^2, \qquad (1)$$

where the "Newtonian" gravitational potential $U$, the gravitomagnetic potential $a$ and the "superpotential" $k$ are functions of $\rho$ and $\zeta$ alone. At large distances $r = \left|\sqrt{\rho^2 + \zeta^2}\right| \to \infty$ from isolated sources located around the origin of the coordinate system, $r = 0$, the spacetime has to be Minkowskian,

$$r \to \infty : ds^2 = d\rho^2 + d\zeta^2 + \rho^2 d\varphi^2 - dt^2. \qquad (2)$$

Metric (1) admits an Abelian group of motions $G_2$ with the generators (Killing vectors)

$$\xi^i = \delta^i_t, \quad \text{(stationarity)}, \qquad (3)$$

$$\eta^i = \delta^i_\varphi, \quad \text{(axisymmetry)}, \qquad (4)$$

where the Kronecker symbols $\delta^i_t$, $\delta^i_\varphi$ indicate that $\xi^i$ has only a time $t$-component whereas $\eta^i$ points in the azimuthal $\varphi$-direction. $\eta^i$ has closed compact trajectories about the axis of symmetry and is therefore spacelike off the axis (and the horizons). $\xi^i$ is timelike sufficiently far from the black holes but can become spacelike inside ergoregions. Obviously,

$$e^{2U} = -\xi^i \xi_i, \qquad a = -e^{-2U}\eta_i \xi^i \qquad (5)$$

[1] In the following, we also use the complex coordinates $z = \rho + i\zeta$ and $\bar{z} = \rho - i\zeta$. $t$ is the time coordinate.

is a coordinate-free representation of the two relativistic gravitational potentials $U$ and $a$.

In stationary (and axisymmetric) spacetimes, the *event horizon* as the central black hole property is a *local* concept. Consider the Killing vector $\xi'$,

$$\xi' = \xi + \Omega\eta \tag{6}$$

with the norm

$$e^{2V} = -(\xi', \xi') = e^{2U}\left[(1 + \Omega a)^2 - \rho^2\Omega^2 e^{-4U}\right], \tag{7}$$

where $\Omega$ is a real constant. A connected component of the set of points with $e^{2V} = 0$, which is a null hypersurface, $(de^{2V}, de^{2V}) = 0$, is called a *Killing horizon* $\mathscr{H}(\xi')$,

$$\mathscr{H}(\xi') : e^{2V} = -(\xi', \xi') = 0, \quad (de^{2V}, de^{2V}) = 0. \tag{8}$$

Since the Lie derivative $\mathscr{L}_{\xi'}$ of $e^{2V}$ vanishes, we have $(\xi', de^{2V}) = 0$. Being null vectors on $\mathscr{H}(\xi')$, $\xi'$ and $de^{2V}$ are proportional to each other,

$$\mathscr{H}(\xi') : de^{2V} = -2\kappa\xi'. \tag{9}$$

Using the (vacuum) field equations one can show that the *surface gravity* $\kappa$ is a constant on $\mathscr{H}(\xi')$. $\Omega$ is the *angular velocity* of the horizon. In black hole thermodynamics, $\kappa$ and $\Omega$ are conjugate to the extensive quantities $A$ (area) and $J$ (angular momentum), respectively.

In the $\rho$-$\zeta$ plane ($t$ = constant, $\varphi$ = constant) of the Weyl-Lewis-Papapetrou coordinate system (1), the horizons cover a finite portion on the $\zeta$-axis ($\rho = 0$), see Fig. 2, or shrink to a point [29]. It turns out that extended horizons ("sub-extremal horizons") and point-like horizons ("degenerate horizons") require different considerations. Note that a Killing horizon is always a two-surface in the time slice $t$ = constant. The degeneracy to a line or a point is a peculiarity of the special coordinate system.

In this paper, we explain the non-existence proof for extended (sub-extremal) horizons and end up with a brief comment on degenerate horizons.

The dashed line in Fig. 2 sketches the boundaries of the vacuum region: $\mathscr{A}^+$, $\mathscr{A}^0$, $\mathscr{A}^-$ are the vacuum parts of the $\zeta$-axis (axis of symmetry), $\mathscr{H}^{(1)}$ and $\mathscr{H}^{(2)}$ denote the two Killing horizons, which are located in the intervals $[K_2, K_1]$ and $[K_4, K_3]$ on the $\zeta$-axis, and $\mathscr{C}$ stands for spatial infinity. The gravitational fields $a$, $k$, $U$ have to satisfy the following boundary conditions

$$\mathscr{A}^{\pm}, \mathscr{A}^0 : \ a = 0, \quad k = 0, \tag{10}$$

$$\mathscr{H}^{(i)} : \ 1 + \Omega_i a = 0, \quad i = 1, 2, \tag{11}$$

$$\mathscr{C} : \ U \to 0, \quad a \to 0, \quad k \to 0, \tag{12}$$

where $\Omega_1$ and $\Omega_2$ are the angular velocities of the two horizons. Equations (10) characterize the axis of symmetry (rotation axis). The first relation originates from the second equation in (5), since the compact trajectories of $\eta$ with the standard periodicity $2\pi$ become infinitesimal circles with the consequence $\eta \to 0$. The second relation is a necessary condition for elementary flatness (Lorentzian geometry in the vicinity of the rotation axis). Equation (11) is a reformulation of Eqs. (8) ($e^{2V} = 0$) and (7) since the horizons are located on the $\zeta$-axis ($\rho = 0$), see Fig. 2. Finally, Eq. (12) ensures the asymptotic flatness of the metric (1), see (2).

## 2.2 Field Equations and Linear Problem

The stationary and axisymmetric vacuum Einstein equations for the metric potentials $U$ and $a$ are equivalent to the Ernst equation

$$(\Re f)\Big(f_{,\rho\rho} + f_{,\zeta\zeta} + \frac{1}{\rho} f_{,\rho}\Big) = f_{,\rho}^2 + f_{,\zeta}^2 \tag{13}$$

for the complex function

$$f(\rho, \zeta) = e^{2U(\rho,\zeta)} + ib(\rho, \zeta), \tag{14}$$

where the twist potential $b$ is defined by

$$a_{,\rho} = \rho e^{-4U} b_{,\zeta}, \quad a_{,\zeta} = -\rho e^{-4U} b_{,\rho}. \tag{15}$$

The potential $k$ can be calculated from

$$k_{,\rho} = \rho\Big[U_{,\rho}^2 - U_{,\zeta}^2 + \frac{1}{4} e^{-4U}(b_{,\rho}^2 - b_{,\zeta}^2)\Big], \tag{16}$$

$$k_{,\zeta} = 2\rho\Big[U_{,\rho}U_{,\zeta} + \frac{1}{4} e^{-4U} b_{,\rho} b_{,\zeta}\Big]. \tag{17}$$

As a consequence of the Ernst equation (13), the integrability conditions $a_{,\rho\zeta} = a_{,\zeta\rho}$ and $k_{,\rho\zeta} = k_{,\zeta\rho}$ are satisfied such that the metric potentials $a$ and $k$ may be calculated via line integration from the Ernst potential $f$. Since $e^{2U} = \Re f$, all metric coefficients in (1) can uniquely be determined from $f$. Thus the boundary problem for the vacuum Einstein equations reduces to a boundary problem for the Ernst equation. However, we have to cope with non-local boundary conditions (10)–(12), (15)–(17) for the Ernst potential $f$. Fortunately, these boundary conditions are well-adapted to the "inverse method", which we applied to solve the boundary value problem.

The Ernst equation is the integrability condition $\boldsymbol{\Phi}_{,z\bar{z}} = \boldsymbol{\Phi}_{,\bar{z}z}$ of the *Linear Problem* (LP) [30, 31]

$$\boldsymbol{\Phi}_{,z} = \left[\begin{pmatrix} N & 0 \\ 0 & M \end{pmatrix} + \lambda \begin{pmatrix} 0 & N \\ M & 0 \end{pmatrix}\right]\boldsymbol{\Phi}, \tag{18}$$

$$\boldsymbol{\Phi}_{,\bar{z}} = \left[\begin{pmatrix} \bar{M} & 0 \\ 0 & \bar{N} \end{pmatrix} + \frac{1}{\lambda} \begin{pmatrix} 0 & \bar{M} \\ \bar{N} & 0 \end{pmatrix}\right]\boldsymbol{\Phi}, \tag{19}$$

where the *pseudopotential* $\boldsymbol{\Phi}(z, \bar{z}, \lambda)$ is a $2 \times 2$ matrix depending on the spectral parameter

$$\lambda = \sqrt{\frac{K - \mathrm{i}\bar{z}}{K + \mathrm{i}z}} \tag{20}$$

as well as on the complex coordinates

$$z = \rho + \mathrm{i}\zeta, \quad \bar{z} = \rho - \mathrm{i}\zeta, \tag{21}$$

whereas $M$, $N$ and the complex conjugate quantities $\bar{M}$, $\bar{N}$ are functions of $z$, $\bar{z}$ (or $\rho$, $\zeta$) alone and do not depend on the constant complex parameter $K$. Since the integrability condition must hold identical in $\lambda$ (or $K$) it yields a system of first order differential equations for $N$ and $M$ which is equivalent to the Ernst equation. The first order system has the "first integrals"

$$M = \frac{f_{,z}}{f + \bar{f}}, \qquad N = \frac{\bar{f}_{,z}}{f + \bar{f}}. \tag{22}$$

Vice versa, any solution $f$ of the Ernst equation admits the unique determination of the pseudopotential $\boldsymbol{\Phi}$ up to constants of integration. Thus the Ernst equation (13) and the LP (18), (19) are equivalent to each other.

Multiplying (18) by $\mathrm{d}z$ and adding (19) multiplied by $\mathrm{d}\bar{z}$ one obtains the reformulation $\mathrm{d}\boldsymbol{\Phi} = (\cdots)\boldsymbol{\Phi}$ of the LP in the form of a system of (overdetermined) total differential equations.

Without loss of generality we choose the standard representation

$$\boldsymbol{\Phi} = \begin{pmatrix} \psi(\rho, \zeta, \lambda) & \psi(\rho, \zeta, -\lambda) \\ \chi(\rho, \zeta, \lambda) & -\chi(\rho, \zeta, -\lambda) \end{pmatrix} \tag{23}$$

where

$$\bar{\psi}\left(\rho, \zeta, \frac{1}{\bar{\lambda}}\right) = \chi(\rho, \zeta, \lambda) \tag{24}$$

due to the special structure of the coefficient matrices of the LP. For $K \to \infty$ and $\lambda \to -1$, the functions $\psi$, $\chi$ can be normalized by

$$\psi(\rho, \zeta, -1) = \chi(\rho, \zeta, -1) = 1. \tag{25}$$

As a consequence of the LP, the Ernst potential and the gravitomagnetic potential can be read off from the pseudopotential $\boldsymbol{\Phi}$ at $K \to \infty$ and $\lambda \to +1$,

$$f(\rho, \zeta) = \chi(\rho, \zeta, 1) = \boldsymbol{\Phi}_{21}(\rho, \zeta, 1), \tag{26}$$

$$a(\rho, \zeta) = \mathrm{i}e^{-2U} \left( K^2 \frac{\partial}{\partial K} [\chi(-\lambda) - \psi(-\lambda)] \right) \Bigg|_{\substack{\lambda = 1 \\ K \to \infty}} - C, \tag{27}$$

where $C$ is an arbitrary constant. The idea of the inverse (scattering) method is to discuss $\boldsymbol{\Phi}$, for fixed but arbitrary values of $\rho$, $\zeta$ $(z, \bar{z})$ as a holomorphic function of $\lambda$ or $K$. In the latter case, $\boldsymbol{\Phi}$ "lives" on the two sheets of the Riemann surface associated with (20). As this mapping depends on the parameters $\rho$, $\zeta$, the position of the branch points $K_{\mathrm{B}} = \mathrm{i}\bar{z}$, $\bar{K}_{\mathrm{B}} = -\mathrm{i}z$ and the branch cut between them changes with the coordinates.

## 3 Non-existence Proof

### *3.1 Integration of the Linear Problem*

In order to solve the boundary problem (13), (10)–(12), we shall integrate ("solve") the LP along the dashed line in Fig. 2 which marks, in the $\rho$-$\zeta$ plane, the boundary of the vacuum domain outside the horizons. Starting from and returning to any axis point, say $\rho = 0$, $\zeta \in \mathscr{A}^+$, we shall make use of the boundary conditions and finally arrive at a representation of the Ernst potential on the axis of symmetry. It turns out that this representation is sufficient to express all black hole quantities (such as areas $A_1$, $A_2$ and angular momenta $J_1$, $J_2$ of the black holes) in terms of three independent real parameters (plus two additional scaling parameters) and to establish the equations of state of the black hole thermodynamics of the equilibrium configuration under discussion. Furthermore, the axis values $f(\zeta) = f(\rho = 0, \zeta)$ fix the solution $f(\rho, \zeta)$ of the Ernst equation uniquely [32].

Since $\lambda(K)$ as defined in (20) degenerates on the $\zeta$-axis, $\lambda = \pm 1$, the LP $\mathrm{d}\boldsymbol{\Phi} = (\cdots)\,\boldsymbol{\Phi}$ can easily be integrated. For $\lambda = +1$ one obtains

$$\mathscr{A}^{\pm}, \mathscr{A}^0, \mathscr{H}^{(i)} : \boldsymbol{\Phi} = \begin{pmatrix} \bar{f} & 1 \\ f & -1 \end{pmatrix} \mathbf{L}, \quad \mathbf{L} = \begin{pmatrix} A(K) & B(K) \\ C(K) & D(K) \end{pmatrix}. \tag{28}$$

The representation for $\lambda = -1$ follows from (23) by exchanging column elements. Remarkably, the $\boldsymbol{\Phi}$-matrix separates. The first factor depends on the path of integration whereas $\mathbf{L}$ representing the "integration constants" is a function of the spectral parameter $K$ alone. There is a difference between the case of two extended horizons and that of one or two degenerate horizons. In the first case one can parametrize the dashed curve by the coordinate $\zeta$ everywhere on the $\zeta$-axis, i.e.

$f{=}f(\rho=0,\zeta){=}f(\zeta)$ on $\mathscr{A}^{\pm}, \mathscr{A}^{0}, \mathscr{H}^{(i)}$. This is clearly impossible if the dashed curve runs around a point-like horizon. A path like this can be described by an infinitesimal semicircle which brings (local) polar coordinates $(R,\theta)$ into play [33]. Then we have $f = f(R\to 0,\theta) = f(\theta), \theta\in[0,\pi]$ in (28).

To exploit characteristic properties of the horizons such as (8) and (9), it is helpful to introduce corotating frames of reference defined by

$$\rho' = \rho, \quad \zeta' = \zeta, \quad \varphi' = \varphi - \Omega t, \quad t' = t, \tag{29}$$

where $\Omega = \Omega_1, \Omega_2$. This coordinate transformation induces transformations of the gravitational potentials in (1) such that

$$e^{2U'} \equiv e^{2V} = e^{2U}[(1+\Omega a)^2 - \Omega^2\rho^2 e^{-4U}], \tag{30}$$

$$(1-\Omega a')e^{2U'} = (1+\Omega a)e^{2U}, \tag{31}$$

where a prime denotes "corotating" quantities. To determine the corotating Ernst potential $f'$ one has to apply (15) to $a'$ and $e^{2U'}$. Finally, using the Eq. (22) for $N', M'$, one obtains the corotating pseudopotential [31]

$$\boldsymbol{\Phi}' = \mathbf{T}_\Omega \boldsymbol{\Phi}, \tag{32}$$

where

$$\mathbf{T}_\Omega = \begin{pmatrix} 1+\Omega a - \Omega\rho e^{-2U} & 0 \\ 0 & 1+\Omega a + \Omega\rho e^{-2U} \end{pmatrix} + i(K+iz)\Omega e^{-2U}\begin{pmatrix} -1 & -\lambda \\ \lambda & 1 \end{pmatrix}. \tag{33}$$

The validity of the Ernst equations of the non-rotating and corotating system at the points of intersection $\mathscr{A}/\mathscr{H}$ (axis of symmetry/extended horizon) and $\mathscr{A}/\mathscr{C}$ (axis/circle at infinity), see Fig. 2, implies that $\boldsymbol{\Phi}$ and $\boldsymbol{\Phi}'$ must be continuous there as well. By way of example let us consider the continuity of $\boldsymbol{\Phi}$ in (28) and $\boldsymbol{\Phi}'$ in (32) at the point $\rho=0$, $\zeta = K_1$. It immediately leads to a connection of the horizon and axis values of the "integration constants" $\mathbf{L}^{(1)}$ and $\mathbf{L}^{+}$, cf. (28),

$$\mathbf{L}^{(1)} = \left(\mathbf{1} - \frac{\mathbf{F}_1}{2i\Omega^{(1)}(K-K_1)}\right)\mathbf{L}^{+}, \tag{34}$$

where $\mathbf{F}_1$ is is a special case of $\mathbf{F}_i$ as needed later,

$$\mathbf{F}_i := \begin{pmatrix} -f_i & 1 \\ -f_i^2 & f_i \end{pmatrix}, \quad f_i = f(K_i), \quad i = 1,\dots,4. \tag{35}$$

Note that $\bar{f}_1 = -f_1$ since $e^{2V}$ and $e^{2U}$ are continuous at the points of intersection and $e^{2V} = 0$ from the side of the horizon. If one continues the interlinking procedure along the closed contour, one returns to the starting point with the result

$$\mathbf{L}^+ \begin{pmatrix} 0 & 1 \\ 1 & 0 \end{pmatrix} (\mathbf{L}^+)^{-1} = \mathbf{R}^+, \tag{36}$$

where

$$\mathbf{R}^+ := \prod_{i=1}^{4} \left( \mathbf{1} - (-1)^i \frac{\mathbf{F}_i}{2\mathrm{i}\Omega^{(i)}(K - K_i)} \right) \begin{pmatrix} 0 & 1 \\ 1 & 0 \end{pmatrix} \tag{37}$$

with $\Omega^{(1)} = \Omega^{(2)} = \Omega_1$, $\Omega^{(3)} = \Omega^{(4)} = \Omega_2$.

*Point-like* (degenerate) horizons can be involved without any difficulty by setting $K_1 = K_2$ or/and $K_3 = K_4$ in these equations [33].

We shall show that (36) with (37) as the result of the integration of the LP along the closed (dashed) contour in Fig. 2 yields the Ernst potential on the axis.

## 3.2 Ernst Potential on the Axis

At the branch points $K = K_B = \mathrm{i}\bar{z}$, $K = \bar{K}_B = -\mathrm{i}z$ of the Riemann $K$-surface, where $\lambda = 0$, $\lambda = \infty$, respectively, one finds from (23) $\boldsymbol{\Phi}_{11} = \boldsymbol{\Phi}_{12}$, $\boldsymbol{\Phi}_{21} = -\boldsymbol{\Phi}_{22}$,

$$K = K_B : \boldsymbol{\Phi} \begin{pmatrix} 0 & 1 \\ 1 & 0 \end{pmatrix} \boldsymbol{\Phi}^{-1} = \begin{pmatrix} 1 & 0 \\ 0 & -1 \end{pmatrix}. \tag{38}$$

On the $\zeta$-axis one has *confluent* branch points, $K_B = \bar{K}_B = \zeta$. For this choice one obtains from (38), (28) and (36) in terms of the Ernst potential $f^+$ on $\mathscr{A}^+$

$$\mathbf{R}^+(\zeta) = \begin{pmatrix} \bar{f}^+(\zeta) & 1 \\ f^+(\zeta) & -1 \end{pmatrix}^{-1} \begin{pmatrix} 1 & 0 \\ 0 & -1 \end{pmatrix} \begin{pmatrix} \bar{f}^+(\zeta) & 1 \\ f^+(\zeta) & -1 \end{pmatrix} \tag{39}$$

and so

$$[\mathbf{R}^+(\zeta) - \mathbf{1}] \begin{pmatrix} 1 \\ f^+(\zeta) \end{pmatrix} = \mathbf{0} \tag{40}$$

for $\mathbf{R}^+(\zeta) := \mathbf{R}^+(K = \zeta)$, cf. (37). Note that (40) is equivalent to (39): a second column resulting from (39) and (40) are complex conjugate.

We shall now discuss properties of the Ernst potential on the axis which can be derived from the eigenvalue Eq. (40). First of all, let us point out that similar equations can be derived for all intervals $\mathscr{A}$, $\mathscr{H}$ by the interlinking procedure as explained in (34). At the first glance, $f^+(\zeta)$ seems to be a quotient of two polynomials of fourth degree in $\zeta$. However, attention must be paid to the fact that the characteristic determinant has to vanish,

$$|\mathbf{R}^+(\zeta) - \mathbf{1}| = \mathbf{0}. \tag{41}$$

This condition tells us that the numerator and the denominator of $f^+(\zeta)$ must have two common zeros such that *the axis potential is a quotient of two (normalized) polynomials of second degree in* $\zeta$,

$$f^+(\zeta) = \frac{n_2(\zeta)}{d_2(\zeta)} = \frac{\zeta^2 + q\zeta + r}{\zeta^2 + s\zeta + t}, \tag{42}$$

where $q, r, s, t$ are complex constants which can be expressed in terms of $f_i$, $K_i$ and $\Omega^{(i)}$, $(i = 1, \ldots, 4)$. For *extended* (sub-extremal) horizons the following reparametrization is useful:

Defining

$$\alpha_i := \frac{\bar{d}_2(K_i)}{d_2(K_i)}, \quad \alpha_i \bar{\alpha}_i = 1, \quad \beta_i := \frac{\bar{n}_2(K_i)}{n_2(K_i)}, \quad \beta_i \bar{\beta}_i = 1 \tag{43}$$

and using

$$\overline{f^+(K_i)} = -f^+(K_i), \qquad i = 1, \ldots, 4, \tag{44}$$

one obtains

$$\beta_i = -\alpha_i. \tag{45}$$

The Eqs. (43) form a linear algebraic system for the parameters $q, r, s, t$. Eliminating them in (42) one arrives at a determinant representation for the *Ernst potential on the axis* $\mathscr{A}^+$, $f^+(\zeta)$, which can be written in the form

$$f^+(\zeta) = \frac{\begin{vmatrix} s_{12} - 1 & s_{14} - 1 \\ s_{23} - 1 & s_{34} - 1 \end{vmatrix}}{\begin{vmatrix} s_{12} + 1 & s_{14} + 1 \\ s_{23} + 1 & s_{34} + 1 \end{vmatrix}}, \qquad s_{ij} := \frac{\alpha_i(\zeta - K_i) - \alpha_j(\zeta - K_j)}{K_{ij}}, \tag{46}$$

where

$$K_{ij} := K_i - K_j, \qquad i, j = 1, \ldots, 4. \tag{47}$$

The continuation of $f^+(\zeta)$ to all space is unique [32] and leads to the representation

$$f(\rho,\zeta)=\frac{\begin{vmatrix} R_{12}-1 & R_{14}-1\\ R_{23}-1 & R_{34}-1\end{vmatrix}}{\begin{vmatrix} R_{12}+1 & R_{14}+1\\ R_{23}+1 & R_{34}+1\end{vmatrix}},\qquad R_{ij}:=\frac{\alpha_i r_i-\alpha_j r_j}{K_{ij}},\tag{48}$$

where

$$r_i:=\sqrt{(\zeta-K_i)^2+\rho^2}\geq 0,\qquad i,j=1,\dots,4.\tag{49}$$

$f(\rho,\zeta)$ is the Ernst potential of the *double-Kerr-NUT solution* which was originally generated by a two-fold Bäcklund transformation of Minkowski space [5, 7] in the form of a quotient of two $5\times 5$ determinants. According to a rule of Yamazaki [17], this type of determinants can be reduced to $2\times 2$ determinants as used in (48). Making use of (15) [or (27)], (16), (17) and $e^{2U}=\Re f$, one finds determinant representations for all metric coefficients, i.e. for $a$, $k$, $e^{2U}$ in (1), see [4].

As a condition identical in $K$, Eq. (41) yields four constraints among the parameters $\Omega_1$, $\Omega_2$; $K_1-K_2$, $K_2-K_3$, $K_3-K_4$; $f_1$, …, $f_4$. It can be shown [4] that this system of algebraic equations guarantees that $a=0$ on $\mathscr{A}^{\pm}$, $\mathscr{A}^0$ and $\mathscr{C}$. As a consequence, it eliminates NUT parameters from the Ernst potential. In consideration of (40) and introducing dimensionless coordinates $\tilde{\rho}$, $\tilde{\zeta}$ via

$$\tilde{\rho}=\frac{\rho}{K_{23}},\qquad \tilde{\zeta}=\frac{\zeta-K_1}{K_{23}}\tag{50}$$

one realizes that $f^+(\zeta)$ from (40) and therefore $f(\rho,\zeta)$ depend on four real parameters.

### 3.3 Weyl-Bach Force Between the Black Holes

So far we have examined the Ernst equation as a classical field equation. It is questionable whether the parameter conditions (41) alone could rule out the Ernst potential under discussion and lead to a non-existence proof. Consider the static Ernst equation which is an axisymmetric Laplace equation for the "Newtonian" gravitational potential $U$ in Weyl coordinates. The superposition of two solutions with aligned rod-shaped sources (as "classical" precursors of horizons) is regular outside the sources. It is the gravitational interaction ("force") between the rods that forbids equilibrium. Bach, who examined this example in the already mentioned Bach and Weyl paper [1], noted that the metric function $k$ cannot vanish on the portion of the axis between the two sources and that this fact violates the regularity of the solution. Weyl's remarks (published as a supplement to Bach's paper) focused on the interpretation of this type of singularity. He used fictitious stresses described by an energy-momentum tensor to define a force of attraction between the sources. This "Weyl-Bach force" turned out to be proportional to a constant value of $k$ on the portion of the axis between the two sources (in our notation $k^0$) provided that $k^+ = 0$, $k^- = 0$, which is a possible gauge. Note that $k$ as defined in (16), (17) has constant values on *all* intervals of the $\zeta$-axis. One of them, considered to be the arbitrary integration constant, can be chosen so that, say, $k^+ = 0$. Integration of (16), (17) along $\mathscr{C}$ in Fig. 2 results in $k^- = 0$.

Equipped with that physical as well as geometrical interpretation of the superpotential $k^0$ ("attractive force" that violates the Lorentzian geometry in the vicinity of the rotation axis) we shall examine the boundary condition

$$k^0 = 0 \tag{51}$$

for the *gauge*

$$k^+ = k^- = 0. \tag{52}$$

Our discussion is based on the original parametrization of the double Kerr-NUT solution, see (43), (45), (46), and Kramer's representation of $e^{2k}$ [12]. (In principle, one could determine the axis values of this gravitational potential from the axis values of the Ernst potential by integrating the Eqs. (16), (17) along the axis, i.e. by operations on the dashed contour in Fig. 2.) It turns out that the boundary conditions for non-overlapping extended horizons (51), (52) are satisfied by only one parameter condition:

$$\alpha_1\alpha_2 + \alpha_3\alpha_4 = 0. \tag{53}$$

Two of the four conditions (41) can be used to eliminate $\Omega_1$ and $\Omega_2$. The two remaining equations turn out to be equivalent to the equations

$$\frac{(1-\alpha_4)^2}{\alpha_4} w^2 = \frac{(1-\alpha_3)^2}{\alpha_3}, \quad w := \sqrt{\frac{K_{14}K_{24}}{K_{13}K_{23}}} \in [1, \infty), \tag{54}$$

$$\frac{(1+\alpha_2)^2}{\alpha_2} w'^2 = \frac{(1+\alpha_1)^2}{\alpha_1}, \quad w' := \sqrt{\frac{K_{23}K_{24}}{K_{13}K_{14}}} \in (0, 1]. \tag{55}$$

The three restrictions (53)–(55) are nothing else but a reformulation of the original Tomimatsu-Kihara conditions [11, 16]. This reformulation is due to essential examinations of the double-Kerr-NUT solution by Manko, Ruiz and Sanabria-Gómez [14, 15]. As was particularly shown in [14], the restrictions (53)–(55) can be solved to express the parameters $\alpha_1, \ldots, \alpha_4$ in terms of the three real parameters $w$, $w'$ and $\phi$:

$$\alpha_1 = \frac{w'\alpha^2 + \mathrm{i}\varepsilon\alpha}{w' - \mathrm{i}\varepsilon\alpha}, \quad \alpha_2 = \frac{\alpha^2 + \mathrm{i}w'\varepsilon\alpha}{1 - \mathrm{i}w'\varepsilon\alpha}, \tag{56}$$

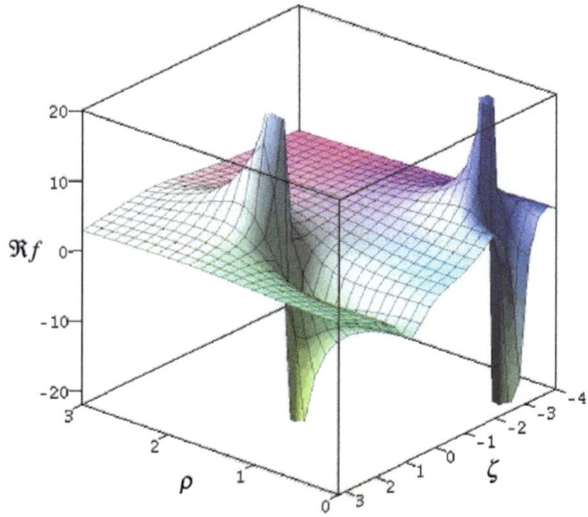


**Fig. 3** The singular Ernst potential for a particular example configuration. Parameters: $\phi = -0.1$, $w = 1.3$, $w' = 0.5$, $K_1 = 2$, $K_{23} = 2$, $\varepsilon = 1$

$$\alpha_3 = \frac{w\alpha^2 - \alpha}{w - \alpha}, \quad \alpha_4 = \frac{\alpha^2 - w\alpha}{1 - w\alpha}, \tag{57}$$

where $\varepsilon = \pm 1$ and

$$\alpha_3\alpha_4 = -\alpha_1\alpha_2 \equiv \alpha^2 \qquad (\alpha = \mathrm{e}^{\mathrm{i}\phi},\ \phi \in [0, 2\pi)). \tag{58}$$

Now we have arrived at the final form of the solution of the boundary problem (10)–(12) for the Ernst equation (13). Eliminating the $\alpha_i$ in favour of $w$, $w'$, $\phi$ and introducing dimensionless coordinates (50), $f$ becomes a function of two coordinates and three real parameters,[2]

$$f = f(\tilde{\rho}, \tilde{\zeta}; w, w', \phi). \tag{59}$$

Note that the relative horizon "lengths" $K_{12}/K_{23}$, $K_{34}/K_{23}$ can be expressed in terms of $w$, $w'$ as well. At this point we cannot guarantee that this Ernst potential is well-behaved. Computer experiments show that the regularity of the Ernst potential on the axis of symmetry must be "paid of" in the form of singular rings outside the horizons. Figure 3 conveys an impression of the structure of this type of singularity. Irregularities on the horizons could also rule out the solution. Since area $A$ and angular momentum $J$ are characteristic parameters for each axisymmetric black hole,

[2] From this point of view, the Ernst potential of the Kerr solution depends on one real parameter and two dimensionless coordinates.

it was obvious to examine restrictions of these parameters. Based on the literature as commented on in the introduction, we could take for granted that the inequality $8\pi|J| < A$ has to hold on both horizons of a regular spacetime with two sub-extremal black holes.

## *3.4 Angular Momentum-Area Inequality and Non-existence Proof*

In order to examine the inequalities

$$\mathscr{H}^{(i)} : 8\pi|J_i| < A_i, \quad i = 1, 2, \tag{60}$$

we calculate the quantities

$$p_i := \frac{8\pi J_i}{A_i}, \qquad i = 1, 2, \tag{61}$$

in terms of the parameters $w$, $w'$ and $\phi \in [0, 2\pi)$. Note that this computation can be performed with the aid of the axis values of the Ernst potential alone, see [2]. The result is

$$p_1 = \varepsilon \frac{1 + \Phi w'}{w'(\Phi + w')}, \quad p_2 = \varepsilon \frac{w(w - \Phi)}{1 - w\Phi} \tag{62}$$

where

$$\Phi := \cos\phi + \varepsilon \sin\phi, \quad \varepsilon = \pm 1. \tag{63}$$

From this we have

$$p_1^2 - 1 = (1 - w'^2) \frac{w'^2 + 2\Phi w' + 1}{w'^2(\Phi + w')^2} < 0 \tag{64}$$

and

$$p_2^2 - 1 = (w^2 - 1) \frac{w^2 - 2\Phi w + 1}{(w\Phi - 1)^2} < 0. \tag{65}$$

For the allowed parameter ranges $w \in [1, \infty)$, $w' \in (0, 1]$, which follow from the definitions (54), (55) of $w$ and $w'$, these inequalities can only hold if

$$w'^2 + 2\Phi w' + 1 < 0 \quad \text{and} \quad w^2 - 2\Phi w + 1 < 0. \tag{66}$$

This, however, implies $\Phi w' < 0$ and $\Phi w > 0$ in contradiction to $w' > 0$ and $w > 0$. Thus *we have proved the non-existence of stationary and axisymmetric*

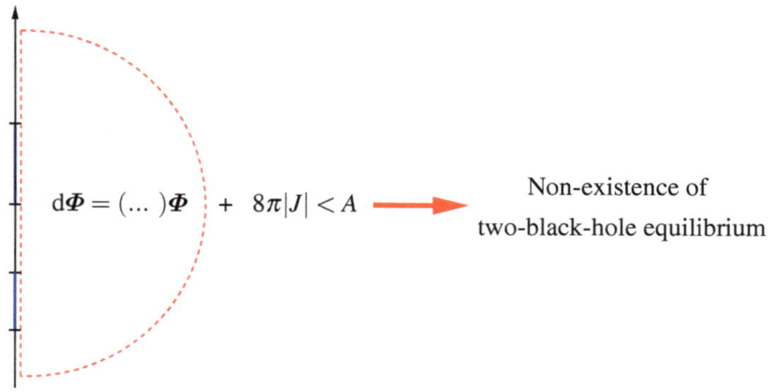


**Fig. 4** Summary of the non-existence proof for two sub-extremal black holes

*configurations consisting of two aligned sub-extremal black holes* and we conclude that the spin-spin repulsion cannot compensate for the gravitational attraction.

## 4 Summary

As a characteristic example for the ongoing discussion about existence or non-existence of stationary equilibrium configurations within the theory of General Relativity, we have studied the question whether two aligned, sub-extremal black holes can be in equilibrium. The result of our above analysis, whose details can be found in [2–4], is negative: *there are no two-black-hole equilibrium configurations!* The idea of the non-existence proof is illustrated in Fig. 4 and can be summarized as follows.

Equilibrium configurations with two aligned rotating black holes, if any existed, can be described by a boundary value problem for two separate (Killing-) horizons. Remarkably, this problem can be solved by integrating the Linear Problem $d\boldsymbol{\Phi} = (\cdots)\boldsymbol{\Phi}$ along the dashed contour as sketched in Fig. 2 or Fig. 4. Thus we arrive necessarily at particular Kerr-NUT solutions which have two horizons [at least according to the definitions (8) and (9)] and show the correct regular behaviour at infinity and on the symmetry axis, whereas regularity off the axis is not guaranteed. On the contrary, we find that all candidate solutions indeed do suffer from irregularities. One of them is the violation of the angular momentum-area inequality $8\pi|J| < A$, which must hold for any regular sub-extremal black hole. We could show that there is no choice of parameters for which angular momentum and area of the two horizons jointly satisfy the inequality. Hence, there exists no regular solution of the vacuum field equations for stationary two-black-hole configurations. For brief comments on the extension of the non-existence proof to *degenerate* black holes, see the following supplement and Fig. 5a.

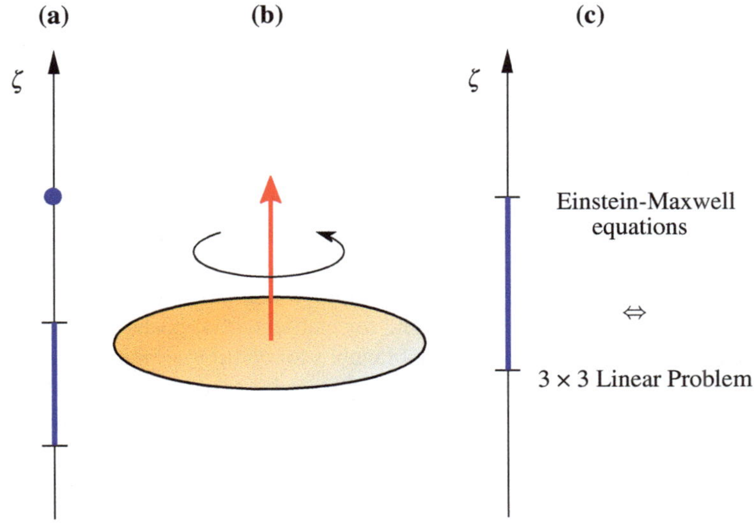


**Fig. 5** Illustration of further BVPs that can be solved with the inverse scattering method: **a** two-horizon problem with one degenerate black hole, **b** rigidly rotating disk of dust, **c** constructive uniqueness proof for the Kerr-Newman black hole

# 5 Supplement

## *5.1 Degenerate Horizons*

The analysis as presented above only applies to configurations with two *sub-extremal ("extended") horizons*. But we have already indicated several times that it is possible to extend the proof to configurations containing one or even two *degenerate black holes* (with "point-like" horizons, where $K_1 = K_2$ or/and $K_3 = K_4$), see Fig. 5a. In the first part of this supplement we give an outline of this generalisation. For details we refer to [3].

As already mentioned, the degeneracy of a Killing horizon to a point is merely a peculiarity of the Weyl-Lewis-Papapetrou coordinates. In order to resolve the internal structure of the horizon, we have discussed the LP in polar coordinates centred at the point-like horizon. Integrating the LP, we found that all possible equilibrium configurations with degenerate black holes can be obtained as particular limits of the double-Kerr-NUT family of solutions. It turned out that there are two families of (two-parametric) solutions describing configurations with one degenerate and one extended horizon and three (one-parametric) solution families for configurations with two degenerate horizons. In order to exclude these families as acceptable equilibrium configurations, we showed that they suffer from unphysical singularities.

In the case of one degenerate and one sub-extremal black hole, the angular momentum-area inequality becomes an *equality* for the degenerate horizon (which turned out to be satisfied identically and, therefore, did not provide any new information). The inequality for the sub-extremal black hole restricts the parameters, but does not yet exclude the possibility of a regular equilibrium configuration. Hence an additional ingredient was required for the desired non-existence proof, which was the positivity of the total mass (ADM mass) of the spacetime. As first shown by Schoen and Yau [34, 35], the mass of a regular, asymptotically flat spacetime satisfying the dominant energy condition (which is certainly satisfied for a black hole vacuum spacetime) is positive. However, for the configuration in question, we found that the entire parameter range in which the sub-extremal black hole satisfies the angular momentum-area inequality has a *negative* mass. Thus we could conclude that the solutions with one degenerate and one sub-extremal horizon are singular.

Finally, we studied the three solution branches for configurations with two degenerate horizons. One branch could be excluded since it has a negative ADM mass for all possible parameter values. The other two branches have negative masses for most parameter values, but there are small parameter regions with positive mass. Fortunately, these solution branches are relatively simple and, by studying the solutions of a certain quartic equation, it was possible to demonstrate explicitly that singularities (singular rings around the $\zeta$-axis) are present even in the parameter range with positive ADM mass.

Hence we could extend the non-existence proof to *all* forms of horizons.

## 5.2 Further Applications of the Inverse Method

In the following we briefly comment on some other applications of the inverse (scattering) method to rotating objects. The integration of the Linear Problem of the Ernst equation $\mathrm{d}\boldsymbol{\Phi} = (\ldots)\boldsymbol{\Phi}$ along a suitable closed contour was first practised to determine the gravitational field of a rigidly rotating disk of dust [36–38], see Fig. 5b. Among other things, the solution (expressed in terms of theta functions) describes a parametric collapse of the disk with a final phase transition to an extreme black hole.

Obviously, the integration method under discussion can be used to construct the Kerr solution as the unique solution of the one-horizon boundary problem [39]. This corresponds to the methods of electrostatics and is an alternative to "complex tricks" and other formal derivations of the Kerr metric in the textbooks.

Extending the vacuum examinations to Einstein-Maxwell fields, Meinel [40] has recently constructed the Kerr-Newman solution by integrating a $3 \times 3$ electrovacuum LP, see Fig. 5c and thus proved the uniqueness of the solution.

**Acknowledgments** We would like to thank Reinhard Meinel, Marcus Ansorg and Andreas Kleinwächter for many valuable discussions and Ben Whale for commenting on the manuscript.

## References


1. Bach, R., Weyl, H.: Neue Lösungen der Einsteinschen Gravitationsgleichungen. Mathemat. Z. **13**, 134 (1922). [Republication in English: Gen. Relativ. Gravit. **44**, 817, (2012)]
2. Neugebauer, G., Hennig, J.: Non-existence of stationary two-black-hole configurations. Gen. Relativ. Gravit. **41**, 2113 (2009). doi:10.1007/s10714-009-0840-8
3. Hennig, J., Neugebauer, G.: Non-existence of stationary two-black-hole configurations: the degenerate case. Gen. Relativ. Gravit. **43**, 3139 (2011). doi:10.1007/s10714-011-1228-0
4. Neugebauer, G., Hennig, J.: Stationary two-black-hole configurations: a non-existence proof. J. Geom. Phys. **62**, 613 (2012). doi:10.1016/j.geomphys.2011.05.008
5. Neugebauer, G.: A general integral of the axially symmetric stationary Einstein equations. J. Phys. A **13**, L19 (1980). doi:10.1088/0305-4470/13/2/003
6. Neugebauer, G.: Gravitostatics and rotating bodies. In: Hall, G.S., Pulham J.R. (eds.) Proceedings of 46th Scottish Universities Summer School in Physics (Aberdeen), Copublished by SUSSP Publications, Edinburgh, and Institute of Physics Publishing, London (1996)
7. Kramer, D., Neugebauer, G.: The superposition of two Kerr solutions. Phys. Lett. A **75**, 259 (1980). doi:10.1016/0375-9601(80)90556-3
8. Dietz, W., Hoenselaers, C.: Two mass solution of Einstein's vacuum equations: the double Kerr solution. Ann. Phys. **165**, 319 (1985). doi:10.1016/0003-4916(85)90301-X
9. Hoenselaers, C.: Remarks on the double-Kerr-solution. Prog. Theor. Phys. **72**, 761 (1984). doi:10.1143/PTP.72.761
10. Hoenselaers, C., Dietz, W.: Talk given at the GR10 meeting, Padova (1983)
11. Kihara, M., Tomimatsu, A.: Some properties of the symmetry axis in a superposition of two Kerr solutions. Prog. Theor. Phys. **67**, 349 (1982). doi:10.1143/PTP.67.349
12. Kramer, D.: Two Kerr-NUT constituents in equilibrium. Gen. Relativ. Gravit. **18**, 497 (1980). doi:10.1007/BF00770465
13. Krenzer, G.: Schwarze Löcher als Randwertprobleme der axialsymmetrisch-stationären Einstein-Gleichungen, PhD Thesis, University of Jena (2000)
14. Manko, V.S., Ruiz, E.: Exact solution of the double-Kerr equilibrium problem. Class. Quantum Grav. **18**, L11 (2001). doi:10.1088/0264-9381/18/2/102
15. Manko, V.S., Ruiz, E., Sanabria-Gómez, J.D.: Extended multi-soliton solutions of the Einstein field equations: II. Two comments on the existence of equilibrium states. Class. Quantum Grav. **17**, 3881 (2000). doi:10.1088/0264-9381/18/2/102
16. Tomimatsu, A., Kihara, M.: Conditions for regularity on the symmetry axis in a superposition of two Kerr-NUT solutions. Prog. Theor. Phys. **67**, 1406 (1982). doi:10.1143/PTP.67.1406
17. Yamazaki, M.: Stationary line of *N* Kerr masses kept apart by gravitational spin-spin interaction. Phys. Rev. Lett. **50**, 1027 (1983). doi:10.1103/PhysRevLett.50.1027
18. Varzugin, G.: Equilibrium configuration of black holes and the inverse scattering method. Theoret. Math. Phys. **111**, 667 (1997). doi:10.1007/BF02634055
19. Varzugin, G.: The interaction force between rotating black holes in equilibrium. Theoret. Math. Phys. **116**, 1024 (1998). doi:10.1007/BF02557144
20. Belinskiĭ, V.A., Zakharov, V.E.: Integration of the Einstein equations by means of the inverse scattering problem technique and construction of exact soliton solutions. Pis'ma Zh. Eksp. Teor. Fiz. (in Russian) **75**, 1955 (1978). [English translation: Sov. Phys. JETP **48**, 985 (1978)]
21. Belinskiĭ, V.A., Zakharov, V.E.: Stationary gravitational solitons with axial symmetry. Pis'ma Zh. Eksp. Teor. Fiz. (in Russian) **77**, 3 (1979). [English translation: Sov. Phys. JETP **50**, 1 (1979)]
22. Ansorg, M., Petroff, D.: Negative Komar mass of single objects in regular, asymptotically flat spacetimes. Class. Quantum Grav. **23**, L81 (2006). doi:10.1088/0264-9381/23/24/L01
23. Ansorg, M., Pfister, H.: A universal constraint between charge and rotation rate for degenerate black holes surrounded by matter. Class. Quantum Grav. **25**, 035009 (2008). doi:10.1088/0264-9381/25/3/035009

24. Hennig, J., Ansorg, M., Cederbaum, C.: A universal inequality between the angular momentum and horizon area for axisymmetric and stationary black holes with surrounding matter. Class. Quantum Grav. **25**, 162002 (2008). doi:10.1088/0264-9381/25/16/162002
25. Booth, I., Fairhurst, S.: Extremality conditions for isolated and dynamical horizons. Phys. Rev. D **77**, 084005 (2008). doi:10.1103/PhysRevD.77.084005
26. Chruściel, P.T., Eckstein, M., Nguyen, L., Szybka, S.J.: Existence of singularities in two-Kerr black holes. Class. Quantum Grav. **28**, 245017 (2011). doi:10.1088/0264-9381/28/24/245017
27. Dain, S., Reiris, M.: Area-angular momentum inequality for axisymmetric black holes. Phys. Rev. Lett. **107**, 051101 (2011). doi:10.1103/PhysRevLett.107.051101
28. Dain, S.: Geometric inequalities for axially symmetric black holes. Class. Quantum Grav. **29**, 073001 (2012). doi:10.1088/0264-9381/29/7/073001
29. Carter, B.: Black hole equilibrium states, In: deWitt, C., deWitt, B. (eds.) Black Holes (Les Houches). Gordon and Breach, London (1973)
30. Neugebauer, G.: Recursive calculation of axially symmetric stationary Einstein fields. J. Phys. A **13**, 1737 (1980). doi:10.1088/0305-4470/13/5/031
31. Neugebauer, G., Meinel, R.: Progress in relativistic gravitational theory using the inverse scattering method. J. Math. Phys. **44**, 3407 (2003). doi:10.1063/1.1590419
32. Hauser, I., Ernst, F.J.: Proof of a Geroch conjecture. J. Math. Phys. **22**, 1051 (1981). doi:10.1063/1.525012
33. Meinel, R., Ansorg, M., Kleinwächter, A., Neugebauer, G., Petroff, D.: Relativistic Figures of Equilibrium. Cambridge University Press, Cambridge (2008)
34. Schoen, R.M., Yau, S.-T.: On the proof of the positive mass conjecture in general relativity. Commun. Math. Phys. **65**, 45 (1979). doi:10.1007/BF01940959
35. Schoen, R., Yau, S.T.: Proof of the positive mass theorem. II. Commun. Math. Phys. **79**, 231 (1981). doi:10.1007/BF01942062
36. Neugebauer, G., Meinel, R.: The Einsteinian gravitational field of a rigidly rotating disk of dust. Astrophys. J. **414**, L97 (1993). doi:10.1086/187005
37. Neugebauer, G., Meinel, R.: General relativistic gravitational field of a rigidly rotating disk of dust: axis potential, disk metric and surface mass density. Phys. Rev. Lett. **73**, 2166 (1994). doi:10.1103/PhysRevLett.73.2166
38. Neugebauer, G., Meinel, R.: General relativistic gravitational field of a rigidly rotating disk of dust: solution in terms of ultraelliptic functions. Phys. Rev. Lett. **75**, 3046 (1995). doi:10.1103/PhysRevLett.75.3046
39. Neugebauer, G.: Rotating bodies as a boundary value problems. Ann. Phys. (Leipzig) **9**, 342 (2000)
40. Meinel, R.: Constructive proof of the Kerr-Newman black hole uniqueness including the extreme case. Class. Quantum Grav. **29**, 035004 (2012). doi:10.1088/0264-9381/29/3/035004

# Dynamic and Thermodynamic Stability of Black Holes and Black Branes

**Robert M. Wald**

**Abstract** I describe recent work with Stefan Hollands that establishes a new criterion for the dynamical stability of black holes in $D \geq 4$ spacetime dimensions in general relativity with respect to axisymmetric perturbations: Dynamical stability is equivalent to the positivity of the canonical energy, $\mathscr{E}$, on a subspace of linearized solutions that have vanishing linearized ADM mass, momentum, and angular momentum at infinity and satisfy certain gauge conditions at the horizon. We further show that $\mathscr{E}$ is related to the second order variations of mass, angular momentum, and horizon area by $\mathscr{E} = \delta^2 M - \sum_i \Omega_i \delta^2 J_i - (\kappa/8\pi)\delta^2 A$, thereby establishing a close connection between dynamical stability and thermodynamic stability. Thermodynamic instability of a family of black holes need not imply dynamical instability because the perturbations towards other members of the family will not, in general, have vanishing linearized ADM mass and/or angular momentum. However, we prove that all black branes corresponding to thermodynamically unstable black holes are dynamically unstable, as conjectured by Gubser and Mitra. We also prove that positivity of $\mathscr{E}$ is equivalent to the satisfaction of a "local Penrose inequality," thus showing that satisfaction of this local Penrose inequality is necessary and sufficient for dynamical stability.

## 1 Introduction

It is of considerable interest to determine the linear stability of black holes in ($D$-dimensional) general relativity. It is also of interest to determine the linear stability of the corresponding black branes in $(D + p)$-dimensions, i.e., spacetimes with metric of the form

R. M. Wald (✉)
Department of Physics, Enrico Fermi Institute, The University of Chicago, Chicago, IL 60637, USA
e-mail: rmwa@uchicago.edu

J. Bičák and T. Ledvinka (eds.), *General Relativity, Cosmology and Astrophysics*, Fundamental Theories of Physics 177, DOI: 10.1007/978-3-319-06349-2_10,

© Springer International Publishing Switzerland 2014

$$d\tilde{s}^2_{D+p} = ds^2_D + \sum_{i=1}^{p} dz_i^2, \tag{1}$$

where $ds^2_D$ is a black hole metric. In this paper, I will describe some recent general results, obtained in collaboration with Stefan Hollands, on the stability of black holes and black branes. In our work, we restrict consideration to vacuum general relativity without a cosmological constant, but our methods are applicable to general theories of gravity derived from a diffeomorphism covariant Lagrangian. A full account of our results can be found in [1].

One can analyze the stability of a black hole or black brane by writing out the linearized Einstein equation of the black hole or black brane background spacetime. One can establish linear stability by finding a positive definite conserved norm for perturbations. Linear instability can be established by finding a solution with (gauge independent) unbounded growth in time. However, even in the very simplest cases—such as the Schwarzschild black hole [2, 3] and the Schwarzschild black string [4]—it is quite nontrivial to carry out the decoupling of equations and the fixing of gauge needed to determine stability or instability directly from the equations of motion. Furthermore, since this analysis depends on the details of the equations of motion, it must be done on a case-by-case basis. Thus, it would be useful to have a much simpler criterion for stability that can be applied to any black hole or black brane.

## 2 Dynamic and Thermodynamic Stability of Black Holes and Black Branes

In ordinary thermodynamics, one has a very simple and general criterion for thermodynamic instability of a homogeneous system in thermal equilibrium. Consider such a system, whose entropy, $S$, is a function of energy, $E$, and other extensive state parameters $X_i$, so that $S = S(E, X_i)$. The condition for thermodynamic instability is that the Hessian matrix

$$\mathbf{H}_S = \begin{pmatrix} \frac{\partial^2 S}{\partial E^2} & \frac{\partial^2 S}{\partial X_i \partial E} \\ \frac{\partial^2 S}{\partial E \partial X_i} & \frac{\partial^2 S}{\partial X_i \partial X_j} \end{pmatrix}. \tag{2}$$

admit a positive eigenvalue.[1] This criterion arises from the fact that if the Hessian had a positive eigenvalue, then one could increase total entropy by exchanging $E$ and/or $X_i$ between different parts of the system. To see this more explicitly, let $\xi_0 = (E_0, X_{i0})$ denote the parameter values of a particular thermodynamic state, let $\mathbf{v}$ be an arbitrary vector in the thermodynamic state space, and consider the one-

[1] Note that for the case where $E$ is the only state parameter, this criterion is equivalent to the system having a negative heat capacity.

parameter family $\xi(\lambda) = \xi_0 + \lambda \mathbf{v}$ of thermodynamic states. It is obvious that for this family, we have $d^2E/d\lambda^2 = d^2X_i/d\lambda^2 = 0$, whereas $d^2S/d\lambda^2|_0 = \text{Hess}_S|_{\xi_0}(\mathbf{v}, \mathbf{v})$. Suppose now that, for our homogeneous system, we change the state parameters by $\lambda\mathbf{v}$ in one part of the system and compensate for this by changing the state parameters by $-\lambda\mathbf{v}$ in a different part of the system (of the same "size"). To first order in $\lambda$, there will be no change in the total entropy. To second order in $\lambda$, the change in total entropy will be proportional to $\text{Hess}_S|_{\xi_0}(\mathbf{v}, \mathbf{v})$. Thus, if $\mathbf{H}_S$ admits a positive eigenvalue, one can choose $\mathbf{v}$ so as to find a state of higher entropy at fixed $(E, X_i)$ arbitrarily close to the original thermal equilibrium state.

An equivalent statement of the criterion for thermodynamic instability of the state $\xi_0 = (E_0, X_{i0})$ is (assuming $T > 0$) that one can find a one parameter family $\xi(\lambda)$ of thermal equilibrium states with $\xi(0) = \xi_0$ such that

$$\delta^2 E - T\delta^2 S - \sum_i Y_i \delta^2 X_i < 0, \tag{3}$$

where $Y_i \equiv (\partial E/\partial X_i)_S$ and $\delta^2$ denotes $d^2/d\lambda^2$ evaluated at $\lambda = 0$. To see this, we note that for a one parameter family of the form $\xi_0 + \lambda\mathbf{v}$, the left side is just $-T\text{Hess}_S|_{\xi_0}(\mathbf{v}, \mathbf{v})$. However, by the first law of thermodynamics (i.e., the definitions of $Y_i$ and $T^{-1} \equiv (\partial S/\partial E)_{X_i}$), the left side does not depend on the second order change in the state, so Eq. (3) holds for the family $\xi(\lambda)$ if and only if it holds for the family $\xi_0 + \lambda\mathbf{v}$ with $\mathbf{v} = d\xi/d\lambda|_0$.

Black holes are thermodynamic systems, with

$$\begin{aligned} E &\leftrightarrow M, \\ S &\leftrightarrow \frac{A}{4}, \\ X_i &\leftrightarrow J_i, Q_i. \end{aligned} \tag{4}$$

Thus, in the vacuum case ($Q_i = 0$) a black hole would be said to be thermodynamically unstable if the Hessian matrix

$$\mathbf{H}_A = \begin{pmatrix} \frac{\partial^2 A}{\partial M^2} & \frac{\partial^2 A}{\partial J_i \partial M} \\ \frac{\partial^2 A}{\partial M \partial J_i} & \frac{\partial^2 A}{\partial J_i \partial J_j} \end{pmatrix} \tag{5}$$

admits a positive eigenvalue. This is equivalent to finding a perturbation for which

$$\delta^2 M - \frac{\kappa}{8\pi}\delta^2 A - \sum_i \Omega_i \delta^2 J_i < 0. \tag{6}$$

One might expect that this condition for thermodynamic instability might imply dynamical instability. However, this is clearly false: The Schwarzschild black hole has negative heat capacity ($A = 16\pi M^2$, so $\partial^2 A/\partial M^2 = 32\pi > 0$) but is well known to be dynamically stable. A black hole is not "homogeneous" in a manner that would allow one to borrow energy and/or angular momentum from one part of it and give it to another part in such a way as to increase the total entropy (area) at fixed total energy and angular momentum in the manner described above for thermodynamic systems.

However, a black brane is potentially homogeneous in this sense, and the Schwarzschild black string is known to be unstable [4]. The Gubser-Mitra conjecture [5, 6] states that the above thermodynamic criterion for instability is a valid criterion dynamical instability for black branes. As described further below, our work provides a proof of the Gubser-Mitra conjecture, which follows as a consequence of a more fundamental stability criterion that we shall establish.

Another simple possible stability criterion that is applicable to black holes is the "local Penrose inequality," discussed in [7]. We reformulate this criterion as follows: Suppose one has a family of stationary, axisymmetric black holes parametrized by $M$ and angular momenta $J_1, \ldots, J_N$. Consider a one-parameter family $g_{ab}(\lambda)$ of axisymmetric spacetimes, with $g_{ab}(0)$ being a member of this family with surface gravity $\kappa > 0$. Consider initial data on a hypersurface $\Sigma$ passing through the bifurcation surface $B$. By the linearized Raychaudhuri equation, to first order in $\lambda$, the event horizon coincides with the apparent horizon on $\Sigma$. They need not coincide to second order in $\lambda$, but since $B$ is an extremal surface in the background spacetime, their areas must agree to second order. Let $\mathscr{A}(\lambda)$ denotes the area of the apparent horizon of $g_{ab}(\lambda)$, and let $\bar{A}(\lambda)$ denote the area of the event horizon of the stationary black hole in the family with the same mass and angular momentum as $g_{ab}(\lambda)$. Suppose that to second order, we have

$$\delta^2 \mathscr{A} > \delta^2 \bar{A}.$$

Since (i) the area of the event horizon can only increase with time (by cosmic censorship), (ii) the final mass of the black hole cannot be larger than the initial total mass (by positivity of Bondi flux), (iii) its final angular momenta must equal the initial angular momenta (by axisymmetry), and (iv) $\bar{A}(M, J_1, \ldots, J_N)$ is an increasing function of $M$ at fixed $J_i$ (by the first law of black hole mechanics with $\kappa > 0$), it follows that there would be a contradiction if the perturbed black hole solution were to settle down to a stationary black hole in the family. This implies that satisfaction of this inequality implies instablity—although it does not imply stability if $\delta^2 \mathscr{A} \leq \delta^2 \bar{A}$ always holds. As discussed further below, our more fundamental stability criterion implies that satisfaction of $\delta^2 \mathscr{A} \leq \delta^2 \bar{A}$ is necessary and sufficient for the dynamical stability of black holes with respect to axisymmetric perturbations.

Our results are based upon identities arising from the Lagrangian formulation of general relativity. Although we restrict consideration here to vacuum general relativity, these formulas can be generalized to allow for the presence of matter fields and, indeed, they can be generalized to an arbitrary diffeomorphism covariant theory of gravity [8], provided only that the field equations are derived from a Lagrangian. The key identities we use are obtained as follows:

The Lagrangian $D$-form for vacuum general relativity in $D$ dimensions is

$$L_{a_1\cdots a_D} = \frac{1}{16\pi} R\, \varepsilon_{a_1\cdots a_D}. \tag{7}$$

Its first variation yields

$$\delta L = E \cdot \delta g + d\theta, \tag{8}$$

where $E = 0$ is the vacuum Einstein field equation and the $(D-1)$-form $\theta(g, \delta g)$ is the "boundary term" that is usually discarded when the variation of $L$ is performed under an integral sign. Explicitly, we have

$$\theta_{a_1\cdots a_{D-1}} = \frac{1}{16\pi} g^{ac} g^{bd} (\nabla_d \delta g_{bc} - \nabla_c \delta g_{bd}) \varepsilon_{aa_1\cdots a_{D-1}}. \tag{9}$$

The symplectic current $(D-1)$-form is defined by

$$\omega(g; \delta_1 g, \delta_2 g) = \delta_1 \theta(g; \delta_2 g) - \delta_2 \theta(g; \delta_1 g). \tag{10}$$

The symplectic form, $W_\Sigma(g; \delta_1 g, \delta_2 g)$, is obtained by integrating $\omega$ over a Cauchy surface $\Sigma$

$$W_\Sigma(g; \delta_1 g, \delta_2 g) \equiv \int_\Sigma \omega(g; \delta_1 g, \delta_2 g). \tag{11}$$

It can be shown to be given by [9]

$$W_\Sigma(g; \delta_1 g, \delta_2 g) = -\frac{1}{32\pi} \int_\Sigma (\delta_1 h_{ab} \delta_2 p^{ab} - \delta_2 h_{ab} \delta_1 p^{ab}), \tag{12}$$

where

$$p^{ab} \equiv h^{1/2} (K^{ab} - h^{ab} K), \tag{13}$$

where $K_{ab}$ is the extrinsic curvature of $\Sigma$.

The Lagrangian $L$, Eq. (7), is diffeomorphism covariant. An arbitrary vector field $X^a$ is the generator of an infinitesimal diffeomorphism, and associated to $X^a$ is a conserved Noether current $(D-1)$-form, defined by

$$\mathscr{J}_X \equiv \theta(g, \mathscr{L}_X g) - X \cdot L, \tag{14}$$

where $X \cdot L$ denotes the $(D-1)$-form $X^a L_{aa_1\cdots a_{D-1}}$. It can be shown quite generally [10] that $\mathscr{J}_X$ always can be written in the form

$$\mathscr{J}_X = X \cdot C + dQ_X, \tag{15}$$

where $C = 0$ are the constraint equations [11] of the theory, and where the $(D-2)$-form $Q_X$ is called the *Noether charge*.

We now take the first variation of $\mathscr{J}_X$, using Eqs. (14) and (15) as well as Eqs. (8) and (10). We thereby obtain the following fundamental variational identity:

$$\omega(g; \delta g, \mathscr{L}_X g) = X \cdot [E(g) \cdot \delta g] + X \cdot \delta C + d\,[\delta Q_X(g) - X \cdot \theta(g; \delta g)]\,. \quad (16)$$

It should be emphasized that Eq. (16) holds for an arbitrary metric $g_{ab}$ (not necessarily a solution to the field equations), an arbitrary metric perturbation $\delta g_{ab}$ (not necessarily a solution to the linearized field equations) and an arbitrary vector field $X^a$.

By definition, a Hamiltonian, $h_X$, for the "time evolution" generated by $X^a$ is a function on phase space whose first variation satisfies

$$\delta h_X = W_\Sigma(g; \delta g, \mathscr{L}_X g) \quad (17)$$

if and only if $g_{ab}$ satisfies the equations of motion $E = 0$. By Eq. (16), if a Hamiltonian $h_X$ conjugate to $X^a$ exists, its first variation must satisfy

$$\delta h_X = \int_\Sigma (X \cdot \delta C + d\,[\delta Q_X(g) - X \cdot \theta(g; \delta g)])\,. \quad (18)$$

For asymptotically flat spacetimes, this motivates the definition of the ADM conserved quantity, $H_X$, associated with an asymptotic symmetry $X^a$, as the quantitiy defined for solutions whose first variation is given by

$$\delta H_X = \int_\infty [\delta Q_X(g) - X \cdot \theta(g; \delta g)]. \quad (19)$$

Now consider a stationary black hole solution ($E = 0$) with surface gravity $\kappa > 0$, so the event horizon is of "bifurcate type," with bifurcation surface $B$. Let $\Sigma$ be a Cauchy surface for the exterior region, so that it extends from spatial infinity to $B$. We choose $X$ to be the horizon Killing field

$$K^a = t^a + \sum \Omega_i \phi_i^a. \quad (20)$$

Finally, let $\gamma = \delta g$ satisfy the linearized constraint equations $\delta C = 0$. Integration of the fundamental identity (16) over $\Sigma$—using $\mathscr{L}_X g = 0$, $E = 0$, and $\delta C = 0$—then yields the first law of black hole mechanics [8]

$$0 = \delta M - \sum_i \Omega_i \delta J_i - \frac{\kappa}{8\pi}\delta A. \quad (21)$$

To proceed further, we impose two gauge conditions at $B$ on our perturbation $\gamma = \delta g$. The first condition,

$$\delta\vartheta|_B = 0, \quad (22)$$

ensures that the location of the horizon does not change to first order. (Here $\vartheta|_B$ denotes the expansion of the outgoing null geodesics from $B$.) The second condition is

$$\delta\varepsilon|_B = \frac{\delta A}{A}\varepsilon, \tag{23}$$

where $\varepsilon|_B$ denotes the surface area element on $B$. The imposition of these conditions does not involve any loss of generality, i.e., they can be imposed for arbitrary perturbations [1].

We define the *canonical energy* of a perturbation $\gamma$ by

$$\mathscr{E} \equiv W_\Sigma\left(g; \gamma, \mathscr{L}_t\gamma\right). \tag{24}$$

The second variation of our fundamental identity (16) then yields (for axisymmetric perturbations)

$$\mathscr{E} = \delta^2 M - \sum_i \Omega_i \delta^2 J_i - \frac{\kappa}{8\pi}\delta^2 A. \tag{25}$$

Thus positivity of $\mathscr{E}$ for all perturbations $\gamma$ is equivalent to thermodynamic stability (see Eq. (6)).

Our results on dynamical stability follow from various properties of $\mathscr{E}$. To establish these properties, it is useful to view $\mathscr{E}$ as a quadratic form on perturbations:

$$\mathscr{E}(\gamma_1, \gamma_2) = W_\Sigma(g; \gamma_1, \mathscr{L}_t\gamma_2). \tag{26}$$

In [1], we proved that $\mathscr{E}$ satisfies the following properties:

- $\mathscr{E}$ is conserved, i.e., it takes the same value if evaluated on another Cauchy surface $\Sigma'$ extending from spatial infinity to $B$.
- $\mathscr{E}$ is symmetric, $\mathscr{E}(\gamma_1, \gamma_2) = \mathscr{E}(\gamma_2, \gamma_1)$.
- When restricted to perturbations for which $\delta A = 0$ and $\delta P_i = 0$ (where $P_i$ is the ADM linear momentum), $\mathscr{E}$ is gauge invariant.
- When restricted to the subspace, $\mathscr{V}$, of perturbations for which $\delta M = \delta J_i = \delta P_i = 0$ (and hence, by the first law of black hole mechanics $\delta A = 0$), we have $\mathscr{E}(\gamma', \gamma) = 0$ for all $\gamma' \in \mathscr{V}$ if and only if $\gamma$ is a perturbation towards another stationary and axisymmetric black hole.

Thus, if we restrict to perturbations in the subspace, $\mathscr{V}'$, of perturbations in $\mathscr{V}$ modulo perturbations towards other stationary black holes, then $\mathscr{E}$ is a non-degenerate quadratic form. Consequently, on $\mathscr{V}'$, either (a) $\mathscr{E}$ is positive definite or (b) there is a $\psi \in \mathscr{V}'$ such that $\mathscr{E}(\psi) < 0$. If (a) holds, then $\mathscr{E}$ provides a positive definite conserved norm on perturbations. *Thus, if (a) holds, we have stability.*

To analyze case (b), we must consider the flux of $\mathscr{E}$ through null infinity, $\mathscr{I}^+$, and through the black hole horizon, $\mathscr{H}$. Let $\delta N_{ab}$ denote the perturbed Bondi news tensor at null infinity and let $\delta\sigma_{ab}$ denote the perturbed shear on the horizon. If the perturbed black hole were to "settle down" to another stationary black hole at late times, then $\delta N_{ab} \to 0$ and $\delta\sigma_{ab} \to 0$ at late times. In [1], we showed that—for axisymmetric perturbations—the change in canonical energy is then given by

$$\Delta\mathscr{E} = -\frac{1}{16\pi}\int_{\mathscr{I}} \delta\tilde{N}_{cd}\delta\tilde{N}^{cd} - \frac{1}{4\pi}\int_{\mathscr{H}} (K^a\nabla_a u)\,\delta\sigma_{cd}\delta\sigma^{cd} \leq 0. \qquad (27)$$

Thus, $\mathscr{E}$ can only decrease. Therefore if one has a perturbation $\psi \in \mathscr{V}$ such that $\mathscr{E}(\psi) < 0$, then $\psi$ cannot "settle down" to a stationary solution at late times because $\mathscr{E} = 0$ for stationary perturbations with $\delta M = \delta J_i = \delta P_i = 0$. *Thus, in case (b) we have instability.*

The above results show that the necessary and sufficient condition for stability of a black hole (or black brane) with respect to axisymmetric perturbations is positivity of $\mathscr{E}$ on a Hilbert space, $\mathscr{V}$, of perturbations with vanishing perturbed mass, angular momentum, and linear momentum, $\delta M = \delta J_i = \delta P_i = 0$. This is our fundamental criterion for the dynamical stability of black holes and black branes. In view of Eqs. (6) and (25), it follows that *dynamical stability is equivalent to thermodynamic stability on the subspace of perturbations that satisfy* $\delta M = \delta J_i = \delta P_i = 0$.

The restriction that $\delta M = \delta J_i = \delta P_i = 0$ can be removed for the case of black branes as a consequence of the following theorem [1]:

**Theorem 1** *Suppose a family of black holes parametrized by* $(M, J_i)$ *is thermodynamically unstable at* $(M_0, J_{0i})$*, i.e., there exists a perturbation within the black hole family for which* $\mathscr{E} < 0$*. Then, for any black brane corresponding to* $(M_0, J_{0i})$ *one can find a sufficiently long wavelength perturbation for which* $\tilde{\mathscr{E}} < 0$ *and* $\delta\tilde{M} = \delta\tilde{J}_i = \delta\tilde{P}_i = \delta\tilde{A} = \delta\tilde{T}_i = 0$ *(where* $\tilde{T}_i$ *denotes the momenta conjugate to the translational symmetries of the brane).*

This theorem is proven by starting with the initial data for the perturbation to another black hole with $\mathscr{E} < 0$, multiplying it by $\exp(ikz)$—where "$z$" denotes a brane coordinate as in Eq. (1)—and then re-adjusting the initial data so that it satisfies the constraints. The new data will automatically satisfy $\delta\tilde{M} = \delta\tilde{J}_A = \delta\tilde{P}_i = \delta\tilde{A} = \delta\tilde{T}_i = 0$ because of the $\exp(ikz)$ factor. For sufficiently small $k$, it can be shown to satisfy $\tilde{\mathscr{E}} < 0$.

The above theorem, together with our fundamental criterion for dynamical stability, proves the Gubser-Mitra conjecture. To illustrate the nature of this result, consider the one-parameter family of Schwarzschild black holes, parametrized by mass $M$. It is easily seen that the "change of mass" perturbation has $\mathscr{E} < 0$. However, this tells one nothing about the stability of Schwarzschild black holes because, obviously, for these perturbations we have $\delta M \neq 0$, so they do not "count" for testing stability of the Schwarzschild black hole. However, the fact that $\mathscr{E} < 0$ for this "change of mass" perturbation proves the instability of Schwarzschild black branes to sufficiently long wavelength perturbations.

The equivalence of the satisfaction of the local Penrose inequality to our fundamental stability criterion for black holes can be seen as follows. As above, let $\bar{g}_{ab}(M, J_i)$ be a family of stationary, axisymmetric, and asymptotically flat black hole metrics on $M$. Let $g_{ab}(\lambda)$ be a one-parameter family of axisymmetric metrics such that $g_{ab}(0) = \bar{g}_{ab}(M_0, J_{0i})$. Let $M(\lambda)$, $J_i(\lambda)$ denote the mass and angular momenta of $g_{ab}(\lambda)$ and let $\mathscr{A}(\lambda)$ denote the area of its apparent horizon. Let $\bar{g}_{ab}(\lambda) = \bar{g}_{ab}(M(\lambda), J_i(\lambda))$ denote the one-parameter family of stationary black holes with the same mass and angular momenta as $g_{ab}(\lambda)$. We have the following result:

**Theorem 2** *There exists a one-parameter family $g_{ab}(\lambda)$ for which*

$$\mathscr{A}(\lambda) > \bar{\mathscr{A}}(\lambda) \tag{28}$$

*to second order in $\lambda$ if and only if there exists a perturbation $\gamma'_{ab}$ of $\bar{g}_{ab}(M_0, (J_{0i})$ with $\delta M = \delta J_i = \delta P_i = 0$ such that $\mathscr{E}(\gamma') < 0$.*

*Proof* The first law of black hole mechanics implies $\mathscr{A}(\lambda) = \bar{\mathscr{A}}(\lambda)$ to first order in $\lambda$, so what counts are the second order variations. Since the families have the same mass and angular momenta, we have

$$\begin{aligned} \frac{\kappa}{8\pi}\left[\frac{d^2\mathscr{A}}{d\lambda^2}(0) - \frac{d^2\bar{\mathscr{A}}}{d\lambda^2}(0)\right] &= \mathscr{E}(\bar{\gamma}, \bar{\gamma}) - \mathscr{E}(\gamma, \gamma) \\ &= -\mathscr{E}(\gamma', \gamma') + 2\mathscr{E}(\gamma', \bar{\gamma}) \\ &= -\mathscr{E}(\gamma', \gamma'), \end{aligned}$$

where $\gamma' = \bar{\gamma} - \gamma$. □

In summary, the remarkable relationship between the laws of black hole physics and the laws of thermodynamics has been shown to extend to dynamical stability.

This research was sponsored in part by NSF grants PHY-0854807 and PHY-1202718 to the University of Chicago.

## References

1. Hollands, S, Wald, R.M.: Stability of black holes and black branes. Commun. Math. Phys. **313**, 257–290 (2012). doi:10.1007/s00220-012-1638-1
2. Regge, T, Wheeler, J.A.: Stability of a Schwarzschild singularity. Phys. Rev. **108**, 1063 (1957). doi:10.1103/PhysRev.108.1063
3. Zerilli, F.J.: Effective potential for even-parity Regge-Wheeler gravitational perturbation equations. Phys. Rev. Lett. **24**, 737 (1970). doi:10.1103/PhysRevLett.24.737
4. Gregory, R., Laflamme, R.: Black strings and $p$-branes are unstable. Phys. Rev. Lett. **70**, 2837 (1993). doi:10.1103/PhysRevLett.70.2837
5. Gubser, S.S., Mitra, I.: The Evolution of unstable black holes in anti-de Sitter space. J. High Energy Phys. **2001**(08), 018 (2001). doi:10.1088/1126-6708/2001/08/018
6. Gubser, S.S., Mitra, I.: Instability of charged black holes in anti-de Sitter space. ArXiv e-prints arxiv:hep-th/0009126 (2000)

7. Figueras, P., Murata, K., Reall, H.S.: Black hole instabilities and local Penrose inequalities. Class. Quant. Grav. **28**, 225030 (2011). doi:10.1088/0264-9381/28/22/225030
8. Iyer, V., Wald, R.M.: Some properties of the Noether charge and a proposal for dynamical black hole entropy. Phys. Rev. D **50**, 846 (1994). doi:10.1103/PhysRevD.50.846
9. Burnett, G.A., Wald, R.M.: A conserved current for perturbations of Einstein-Maxwell space-times. Proc. R. Soc. Lond. A **430**, 57 (1990). doi:10.1098/rspa.1990.0080
10. Iyer, V., Wald, R.M.: Comparison of the Noether charge and Euclidean methods for computing the entropy of stationary black holes. Phys. Rev. D **52**, 4430 (1995). doi:10.1103/PhysRevD.52.4430
11. Seifert, M.D., Wald, R.M.: General variational principle for spherically symmetric perturbations in diffeomorphism covariant theories. Phys. Rev. D **75**, 084029 (2007). doi:10.1103/PhysRevD.75.084029

# Instability of Anti-de Sitter Spacetime

**Piotr Bizoń and Andrzej Rostworowski**

**Abstract** In this talk we summarize our recent numerical and perturbative calculations which indicate that AdS spacetime is unstable. Namely, we study spherically symmetric Einstein-massless-scalar field equations with negative cosmological constant and show that this system is unstable against black hole formation for a large class of initial data arbitrarily close to the AdS solution. We conjecture that this instability is triggered by a resonant mode mixing which gives rise to diffusion of energy from low to high frequencies.

## 1 Introduction

Anti-de Sitter (AdS) spacetime is the unique maximally symmetric solution of the vacuum Einstein equations $G_{\alpha\beta} + \Lambda g_{\alpha\beta} = 0$ with negative cosmological constant $\Lambda$. In $d+1$ dimensions the AdS metric in global dimensionless coordinates ($t \in \mathbb{R}, x \in [0, \pi/2), \omega \in S^{d-1}$) reads

$$ds^2 = \frac{\ell^2}{\cos^2 x}\left(-dt^2 + dx^2 + \sin^2 x \, d\omega^2_{S^{d-1}}\right),$$

where $\ell^2 = -d(d-1)/2\Lambda$ sets the length scale. Conformal infinity $x = \pi/2$ is the timelike cylinder $\mathscr{I} = \mathbb{R} \times S^{d-1}$ with the boundary metric $ds^2_{\mathscr{I}} = -dt^2 + \sin^2 x \, d\omega^2_{S^{d-1}}$.

Asymptotically AdS spacetimes (that is spacetimes which share the conformal boundary with AdS but may be very different in the bulk, in particular may contain

P. Bizoń · A. Rostworowski (✉)
Institute of Physics, Jagiellonian University, Kraków, Poland
e-mail: arostwor@th.if.uj.edu.pl

P. Bizoń
Max Planck Institute for Gravitational Physics (Albert Einstein Institute), Golm, Germany

J. Bičák and T. Ledvinka (eds.), *General Relativity, Cosmology and Astrophysics*, Fundamental Theories of Physics 177, DOI: 10.1007/978-3-319-06349-2_11,

© Springer International Publishing Switzerland 2014

horizons) have come to play a central role in theoretical physics, prominently due to the AdS/CFT correspondence which conjectures a duality between gravity in the AdS bulk and a quantum field theory on the conformal boundary at infinity. By the positive energy theorem, AdS spacetime is a ground state among asymptotically AdS spacetimes, much as Minkowski spacetime is a ground state among asymptotically flat spacetimes. However, the evolutions of small perturbations of these ground states are different. In the case of Minkowski, small perturbations disperse to infinity and the spacetime is asymptotically stable [1]. In contrast, asymptotic stability of AdS is precluded because the conformal boundary acts like a mirror at which perturbations propagating outwards bounce off and return to the bulk that results in complex nonlinear wave interactions in an effectively bounded domain. Understanding of these interactions is the key to the problem of stability of AdS spacetime.

## 2 Instability of Anti-de Sitter Spacetime

In this talk we summarize our recent numerical and perturbative calculations [2, 3] which indicate that AdS spacetime is unstable. To make the problem tractable we assume spherical symmetry. Since by Birkhoff's theorem spherically symmetric vacuum solutions are static, we need to add matter to generate dynamics. A simple matter model is the minimally coupled massless scalar field

$$G_{\alpha\beta} + \Lambda g_{\alpha\beta} = 8\pi G \left( \partial_\alpha \phi \, \partial_\beta \phi - \frac{1}{2} g_{\alpha\beta} (\partial\phi)^2 \right), \qquad g^{\alpha\beta} \nabla_\alpha \nabla_\beta \phi = 0\,. \tag{1}$$

Let us recall that in the asymptotically flat case ($\Lambda = 0$) this model has led to important insights, such as the proof of the weak cosmic censorship by Christodoulou [4, 5] and the discovery of critical phenomena at the threshold for black hole formation by Choptuik [6].

We use the following parametrization of asymptotically AdS spacetimes

$$ds^2 = \frac{\ell^2}{\cos^2 x} \left( -Ae^{-2\delta} dt^2 + A^{-1} dx^2 + \sin^2 x \, d\omega^2_{S^{d-1}} \right),$$

where $A$ and $\delta$ are functions of $(t, x)$. As our results are qualitatively the same for all $d \geq 3$, for concreteness we set $d = 3$ herefater. Under the above assumptions the Einstein-massless scalar field equations (1) reduce to the quasilinear hyperbolic-elliptic system consisting of the scalar wave equation

$$\partial_t \left( A^{-1} e^{\delta} \partial_t \phi \right) = \frac{1}{\tan^2 x} \partial_x \left( \tan^2 x \, A \, e^{-\delta} \partial_x \phi \right) \tag{2}$$

and two constraint equations (we set $4\pi G = 1$)

$$\partial_x A = \frac{1+2\sin^2 x}{\sin x \cos x}(1-A) - \sin x \cos x\, A\, \rho, \qquad \partial_x \delta = -\sin x \cos x\, \rho\,, \tag{3}$$

where

$$\rho = A^{-2} e^{2\delta} (\partial_t \phi)^2 + (\partial_x \phi)^2$$

is the scalar field energy density. This system has a one-parameter family of static solutions ($\phi = 0, \delta = 0, A = 1 - M\cos^3 x/\sin x$) which are Schwarzschild-AdS black holes for $M > 0$ and the pure AdS for $M = 0$.

We restrict our attention to smooth solutions with finite mass

$$M := \frac{1}{2}\int_0^{\pi/2} A\rho \tan^2 x\, dx.$$

It follows that near $x = \pi/2$ we must have (using $y = \pi/2 - x$)

$$\phi(t,x) = f_\infty(t)\, y^3 + \mathcal{O}\left(y^5\right), \;\; \delta(t,x) = \delta_\infty(t) + \mathcal{O}\left(y^6\right), \;\; A(t,x) = 1 - 2My^3 + \mathcal{O}\left(y^6\right).$$

The local well-posedness of the above initial-boundary value problem was proved in [7].

The dynamics of solutions starting from small initial data

$$(\phi, \dot{\phi})_{|t=0} = (\varepsilon f(x), \varepsilon g(x))$$

can be approximated using weakly nonlinear perturbation analysis. To this end we expand the solution in the perturbation series

$$\phi = \varepsilon\phi_1 + \varepsilon^3\phi_3 + \cdots, \qquad \delta = \varepsilon^2\delta_2 + \varepsilon^4\delta_4 + \cdots, \qquad 1 - A = \varepsilon^2 A_2 + \varepsilon^4 A_4 + \cdots,$$

where $(\phi_1, \dot{\phi}_1)_{|t=0} = (f(x), g(x))$ and $(\phi_j, \dot{\phi}_j)_{|t=0} = (0,0)$ for $j > 1$. Inserting this expansion into the field equations (2) and (3) and collecting terms of the same order in $\varepsilon$, we obtain a hierarchy of linear equations which can be solved order-by-order. At the first order we get the linear wave equation

$$\ddot{\phi}_1 + L\phi_1 = 0,$$

where

$$L = -\frac{1}{\tan^2 x}\, \partial_x \left(\tan^2 x\, \partial_x\right)$$

is an essentially self-adjoint operator on $L^2([0,\pi/2], \tan^2 x\, dx)$. The eigenvalues of $L$ are $\omega_j^2 = (3+2j)^2$ ($j = 0, 1, \ldots$) which implies that AdS is linearly stable.

The corresponding orthonormal eigenfunctions are

$$e_j(x) = d_j \cos^3 x \, P_j^{(\frac{1}{2},\frac{3}{2})}(\cos 2x)\,,$$

where $d_j$ is a normalization factor. Thus, at the linear level the solution is

$$\phi_1(t,x) = \sum_{j=0}^{\infty} a_j \cos(\omega_j t + \beta_j)\, e_j(x),$$

where amplitudes $a_j$ and phases $\beta_j$ are determined by the initial data. Using this solution at the second order we get perturbations of the metric functions $A_2$ and $\delta_2$ (so called backreaction) and at the third order we obtain an inhomogeneous linear wave equation of the form $\ddot{\phi}_3 + L\phi_3 = S(\phi_1, A_2, \delta_2)$. A calculation shows that in general $\phi_3$ contains secular terms that grow linearly in time. They are due to four-wave resonances present in the Fourier decomposition of the source $S$. We interpret this breakdown of the perturbation analysis as indicating the onset of instability at time of order $\mathcal{O}(\varepsilon^{-2})$. We believe that the secular terms appearing in $\phi_3$ are progenitors of the higher-order resonant mode mixing which shifts the energy spectrum to higher frequencies. This heuristics is corroborated by numerical simulations which show that, indeed, generic perturbations start to grow rapidly after a time that scales as $\varepsilon^{-2}$. This growth eventually leads to the formation of a horizon.

To demonstrate the transfer of energy to higher frequencies we define the Fourier coefficients

$$\Phi_j := (A^{1/2}\, \partial_x \phi, \partial_x e_j) \quad \text{and} \quad \Pi_j := (A^{-1/2} e^{\delta}\, \partial_t \phi, e_j)$$

and express the mass as the Parseval sum

$$M = \sum_{j=0}^{\infty} E_j(t),$$

where

$$E_j := \Pi_j^2 + \omega_j^{-2} \Phi_j^2$$

is the $j$-mode energy. The evolution of the energy spectrum, that is the distribution of mass among the modes, is depicted in Fig. 1 for Gaussian initial data. Initially, the energy is concentrated in low modes; the exponential cutoff of the spectrum expresses the smoothness of initial data. During the evolution the range of excited modes increases and the spectrum becomes broader. Just before horizon formation the spectrum exhibits the power-law scaling $E_j \sim j^{-\alpha}$ with exponent $\alpha \approx 1.2$. This value seems to be universal, i.e., the same for all initial data (but it changes with dimension $d$). Note that the formation of a black hole provides a cutoff for

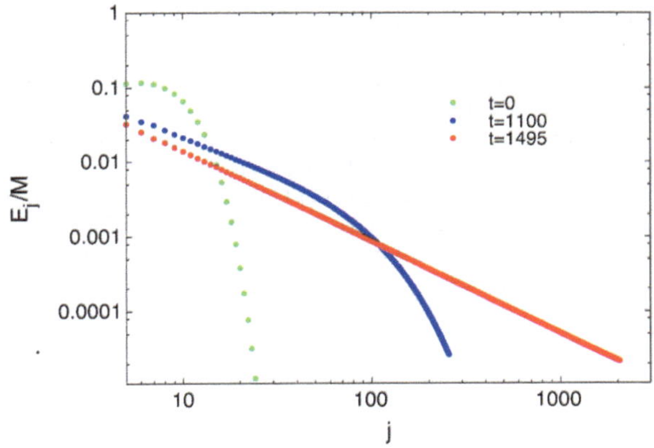


**Fig. 1** Log-log plot of the energy spectrum at three moments of time: initial, intermediate, and just before collapse. The fit of the power law $E_j \sim j^{-\alpha}$ at time $t = 1495$ gives the slope $\alpha \approx 1.2$

the turbulent energy cascade (in amusing analogy to viscosity for the turbulent cascade in fluids). Clearly, the formation of the power-law spectrum reflects the loss of smoothness of the solution during collapse; it would be very interesting compute $\alpha$ analytically.

To summarize, our numerical simulations and formal nonlinear perturbation analysis lead us to conjecture that anti-de Sitter space is unstable against the formation of a black hole under arbitrarily small generic perturbations. We wish to stress the genericity condition in the above conjecture: we do not claim that *all* perturbed solutions end up as black holes. On the contrary, in [2] we gave evidence for the existence of non-generic solutions that remain non-singular for very long (possibly infinite) time. In particular, preliminary calculations based on the Poincaré-Lindstedt method indicate the existence of time periodic solutions. A similar conjecture (existence of geons) was put forward by Dias et al. [8] for the vacuum Einstein equations.

The results described above have opened up new and unexpected research paths lying at the interface of classical general relativity and turbulence theory. Exploration of these paths will hopefully lead to better understanding of the dynamics of asymptotically AdS spacetimes, which in turn may have interesting implications in gauge/gravity dualities.

## References

1. Christodoulou, D., Klainerman, S.: The Global Nonlinear Stability of the Minkowski Space, Princeton Mathematical Series, vol. 41. Princeton University Press, Princeton (1993)
2. Bizoń, P., Rostworowski, A.: Weakly turbulent instability of anti-de Sitter spacetime. Phys. Rev. Lett. **107**, 031102 (2011). doi:10.1103/PhysRevLett.107.031102
3. Jałmużna, J., Rostworowski, A., Bizoń, P.: AdS collapse of a scalar field in higher dimensions. Phys. Rev. D **84**, 085021 (2011). doi:10.1103/PhysRevD.84.085021

4. Christodoulou, D.: The problem of a self-gravitating scalar field. Commun. Math. Phys. **105**, 337 (1986). doi:10.1007/BF01205930
5. Christodoulou, D.: A mathematical theory of gravitational collapse. Commun. Math. Phys. **109**, 613 (1987). doi:10.1007/BF01208960
6. Choptuik, M.W.: Universality and scaling in gravitational collapse of a massless scalar field. Phys. Rev. Lett. **70**, 9 (1993). doi:10.1103/PhysRevLett.70.9
7. Holzegel, G., Smulevici, J.: Self-gravitating Klein-Gordon fields in asymptotically anti-de-Sitter spacetimes. Ann. Henri Poincaré **13**, 991 (2012). doi:10.1007/s00023-011-0146-8
8. Dias, Ó.J.C., Horowitz, G.T., Santos, J.E.: Gravitational turbulent instability of anti-de Sitter space. Class. Quantum Gravity **29**(19), 194002 (2012). doi:10.1088/0264-9381/29/19/194002

# Higher-Dimensional Black Holes

**Harvey S. Reall**

**Abstract** This article reviews black hole solutions of higher-dimensional General Relativity. The focus is on stationary vacuum solutions and recent work on instabilities of such solutions.

## 1 Introduction

General Relativity (GR) in $D > 4$ spacetime dimensions has been actively investigated for more than a decade. There are several reasons for this interest in higher dimensions, and higher-dimensional black holes in particular.

1. Statistical calculation of black hole entropy using string theory. This was first done for certain $D = 5$ black holes [1]. Each entropy calculation is a check on the theory, irrespective of the dimension. Hence the study of higher-dimensional black holes is a worthwhile contribution to developing a theory of quantum gravity.
2. The gauge/gravity correspondence [2] relates the properties of black holes in $D$ dimensions to strongly coupled, finite temperature, quantum field theory in $D-1$ dimensions. This provides a way of calculating certain field theory quantities which are very hard to determine by more traditional methods.
3. The possibility of producing tiny higher-dimensional black holes at colliders in certain "brane-world" scenarios [3].
4. Higher-dimensional black hole spacetimes might have useful mathematical properties. For example, analytically continued versions of black hole solutions have been used to obtain explicit metrics on compact Sasaki-Einstein spaces [4].

---

H. S. Reall (✉)
Department of Applied Mathematics and Theoretical Physics, University of Cambridge, Wilberforce Road, Cambridge CB3 0WA, UK
e-mail: hsr1000@cam.ac.uk

J. Bičák and T. Ledvinka (eds.), *General Relativity, Cosmology and Astrophysics*, Fundamental Theories of Physics 177, DOI: 10.1007/978-3-319-06349-2_12,

© Springer International Publishing Switzerland 2014

5. An explicit higher-dimensional solution might provide a clean example of some important effect in GR. A nice example of this is the frame-dragging effect exhibited by the "black Saturn" solution (see below).
6. Progress in quantum field theory has been made by considering $D$ different from 4, and fields different from those of the Standard Model. In the same spirit, perhaps we will arrive at a better understanding of GR by allowing the parameter $D$ to take values other than 4 [5].

This article is a brief, selective, review of higher-dimensional black hole solutions. The scope is limited to solutions of the vacuum Einstein equation without cosmological constant. I shall discuss two types of black holes. (i) Black hole solutions of Kaluza-Klein theory that are static in the higher-dimensional sense. This includes black string solutions. (ii) Asymptotically flat black hole solutions. In each case, there has been recent progress in demonstrating the existence of instabilities of certain solutions and so special attention is given to this topic.

We close this introduction by presenting the simplest higher-dimensional black hole, the $D$-dimensional Schwarzschild solution:

$$ds^2 = -f dt^2 + \frac{dr^2}{f} + r^2 d\Omega_{D-2}^2, \qquad f = 1 - \left(\frac{r_+}{r}\right)^{D-3} \tag{1}$$

where $d\Omega_{D-2}^2$ is the line-element on a unit round $S^{D-2}$ and the event horizon is at $r = r_+$.

## 2 Black Holes in Kaluza-Klein Theory

Consider vacuum GR with a compact Kaluza-Klein circle. We are interested in black hole solutions of this theory which are asymptotically flat in a Kaluza-Klein sense, which means that, at large distance in the non-compact directions, the metric approaches that of Minkowski space with a compact circle of circumference $L$:

$$ds^2 \approx -dt^2 + dr^2 + r^2 d\Omega_{D-2}^2 + dz^2, \qquad z \sim z + L \tag{2}$$

I shall describe the known *static* black hole solutions of this theory. For a more detailed recent review see [6].

The simplest such black hole solution is the product of the $(D-1)$-dimensional Schwarzschild solution with a circle of circumference $L$. This gives a *black string* solution for which cross-sections of the event horizon have topology $S^1 \times S^{D-3}$. In the decompactified limit $L \to \infty$ it gives a black string of infinite length.

Another solution of this theory describes a black hole of topology $S^{D-2}$ localized on the Kaluza-Klein (KK) circle. Such solutions are not known explicitly. Solutions describing black holes with radius much smaller than $L$ have been constructed perturbatively [7, 8]. Larger black hole solutions have been constructed numerically for

$D = 5, 6$ [9, 10]. They cannot become arbitrarily large: there is an upper bound on their mass determined by $L$. In fact, as one moves along this family of solutions, starting from a small black hole, the mass increases to a maximum and then decreases [10, 11].

The higher-dimensional Schwarzschild solution is stable against linearized gravitational perturbations [12]. However, black strings suffer from the Gregory-Laflamme (GL) instability [13]. If $r_+/L$ is less than a certain $D$-dependent critical value then there exist linearized gravitational perturbations which grow exponentially with time. These perturbations break the translational symmetry around the KK circle. If $r_+/L$ exceeds the critical value then the string is believed to be stable. So "thin" strings are unstable and "fat" strings are stable. Taking the limit $L \to \infty$ shows that uncompactified black strings are unstable.

Since its discovery, the nonlinear evolution of this instability has been the subject of considerable debate. It was first proposed that the black string would "pinch off" to form a localized KK black hole or holes [13]. This cannot happen classically owing to the result that black holes cannot bifurcate. Instead, a singularity would have to form at the horizon at the moment of the pinching off, but perhaps this is resolved by quantum effects.

Recent numerical simulations (with $D = 5$) support this picture [14]. These start from a perturbation which is sinusoidal along the flat direction. When evolved, the perturbation becomes more inhomogeneous, with the configuration reaching a transient state resembling a line of localized black holes connected by thin threads of black string. However, these threads are in turn unstable and suffer the Gregory-Laflamme instability, but on a shorter timescale, leading to smaller black holes connected by even thinner strings. The process appears to continue on smaller and smaller scales, in a self-similar manner, but in a finite total time as measured by an observer far from the string. The curvature of (parts of) the horizon becomes large in this process, so it seems that a naked singularity does indeed form. This is strong evidence against the validity of the cosmic censorship hypothesis in higher dimensions.

The black strings described above are invariant under translations around the KK circle. For this reason they are called *uniform* black strings. Some time ago, it was conjectured that there should exist a 1-parameter family of static *nonuniform* black strings which lack this translational symmetry, and bifurcate from the uniform black string family at the critical value of $r_+/L$ discussed above [15]. Such solutions were subsequently constructed perturbatively, for infinitesimal non-uniformity [16]. Fully nonlinear solutions have been constructed numerically [17–20]. Just like the localized KK black holes, there is an upper bound on their mass determined by $L$.

Kol [21] conjectured that the families of localized KK black hole solutions and non-uniform black strings should merge at a common limiting solution. The idea is that, moving along the family of localized black holes, they increase in size until they fill the KK circle, and then transition to a black string of high non-uniformity. Numerical evidence supporting this suggestion was obtained in [9–11].

The perturbative construction of non-uniform black strings reveals that, for $D \leq 13$, infinitesimally non-uniform black strings have lower horizon area than a uniform black string of the same mass, whereas for $D > 13$, they have greater hori-

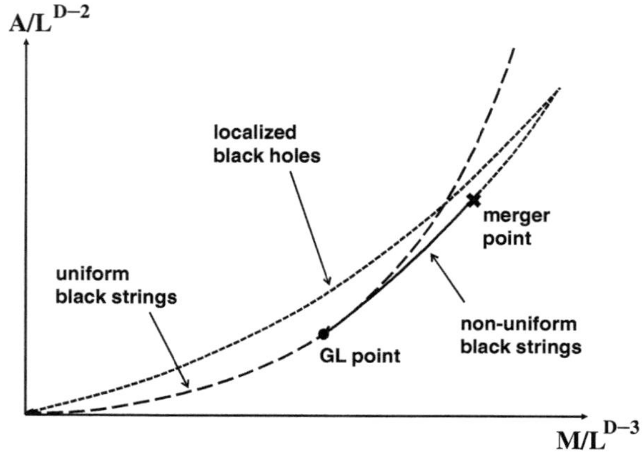


**Fig. 1** Schematic plot of horizon area against mass for Kaluza-Klein black holes/strings with $D = 5, 6$ based on results of Kudoh and Wiseman [10], Headrick et al. [11]. The *dashed curve* is the uniform black string branch, the *solid curve* the non-uniform string branch and the *dotted curve* the localized black hole branch. It seems likely that this will be the qualitative behaviour for all $D \leq 11$ (Plot reproduced from [20])

zon area [22]. This suggests that infinitesimally non-uniform black strings should be classically unstable for $D \leq 13$ but stable for $D > 13$. It also suggests that a non-uniform black string could be an endpoint of the GL instability for $D > 13$.

Very recently, Figueras et al. [20] has performed the first study of the stability of non-uniform black strings with finite non-uniformity. The results confirm the perturbative results for infinitesimal non-uniformity. It was found that the instability persists to large non-uniformity for $D \leq 11$. Strong evidence that non-uniform strings with $D > 13$ are all stable was presented. Solutions with $D = 12, 13$ were constructed for the first time in [20]. It was found that there is a maximum mass as one moves along the family of non-uniform black strings. Solutions before the maximum are unstable whereas solutions after the maximum appear to be stable, and can have greater horizon area than a uniform string of the same mass.

Properties of these different solutions can be summarized in a plot of horizon area against mass, for fixed $L$. This is useful in understanding possible time evolution of instabilities of the various solutions. Horizon area increases, and energy decreases (via emission of gravitational waves) so the final state of an instability must lie "up and left" of the initial state on such a diagram.

Figure 1 shows the qualitative form of a plot of horizon area against mass for the cases $D = 5, 6$ (see [6] for a more quantitative plot). Recall that there is a maximum mass solution along the localized black hole branch of solutions. By the first law, an extremum of the mass must also be an extremum of horizon area, and this results in the cusp shown in the figure.

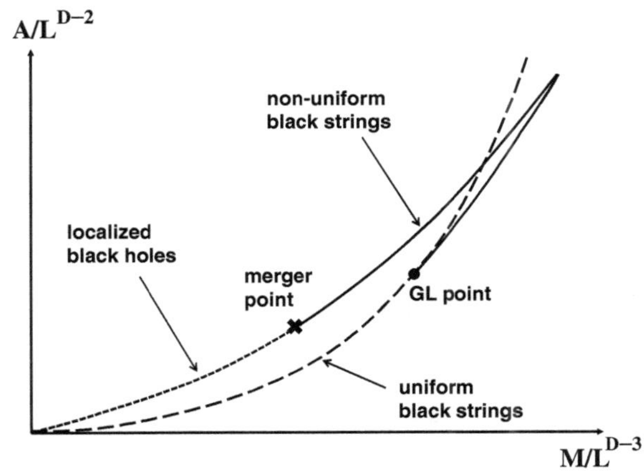


**Fig. 2** Schematic plot of horizon area against mass for Kaluza-Klein black holes/strings with $D = 12, 13$ based on results of Figueras et al. [20]. The localized black hole curve is conjectural (Plot reproduced from [20])

The stability of localized black holes has not been studied. This could be done using the method of Figueras et al. [20]. This method shows that an instability appears as one moves through a maximum of mass along a branch of solutions. Since small localized black holes are expected to be stable, it seems likely that, moving along the branch of solutions, they will be stable until the maximum is reached and unstable thereafter. Hence, in Fig. 1, unstable solutions lie on the part of the uniform string curve extending from the origin to the GL point, along the non-uniform string curve from the GL point to the merger point, and then along the localized black hole curve from the merger point to the cusp.

For $D = 12, 13$, the change in behaviour of the non-uniform string branch implies that the diagram must change to that of Fig. 2 where the cusp now appears along the non-uniform black string branch. (Note that localized black hole solutions have not been constructed for $D > 6$ so this part of the is conjectural.) In this case, unstable solutions lie on the part of the uniform string curve extending from the origin to the GL point and along the non-uniform string curve from the GL point to the cusp.

The results of Figueras et al. [20] suggest that there is no maximum mass non-uniform black string for $D = 11$. Hence it seems likely that Fig. 1 gives the behaviour for all $D \leq 11$.

For $D > 13$, the results for non-uniform black strings, and the simplest guess for the behaviour of localized black holes, results in the phase diagram shown in Fig. 3. Unstable solutions lie on the part of the uniform string curve extending from the origin to the GL point.

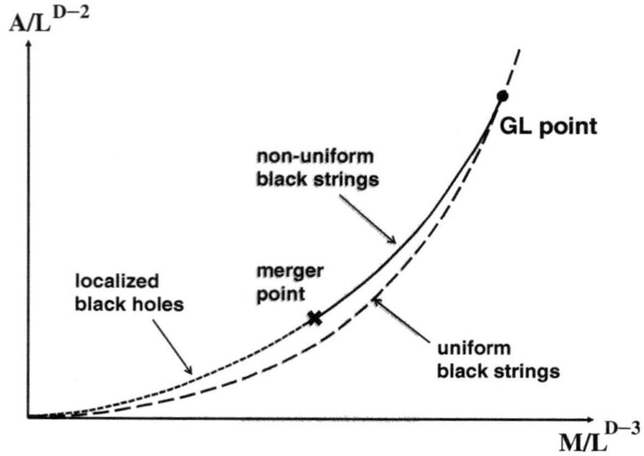


**Fig. 3** Schematic plot of horizon area against mass for Kaluza-Klein black holes/strings with $D > 13$ based on results of Figueras et al. [20]. The localized black hole curve is conjectural (Plot reproduced from [20])

# 3 Asymptotically Flat Black Holes

## *3.1 Introduction*

For a comprehensive 2008 review of higher-dimensional black holes, see [5]. For more recent reviews, see [23, 24].

Recall that angular momentum is defined in terms of an antisymmetric matrix $J_{ij}$ where $i, j$ run over the $D-1$ spatial dimensions (e.g. for a particle, $J_{ij} = x_i p_j - x_j p_i$). For $D = 4$, $J_{ij}$ is equivalent to a vector $J_i$ and so, by choosing $z$-axis aligned with this vector, one can write $J_{ij}$ in terms of a single component $J$. For $D > 4$ dimensions, the best one can achieve by a choice of axes is a block-diagonal form for $J_{ij}$ where each block is a $2 \times 2$ antisymmetric matrix specified by a component $J_I$, where $I = 1, \ldots, N = [(D-1)/2]$.

There are two families of explicit black hole solutions of the vacuum Einstein equation in $D > 4$ spacetime dimensions: Myers-Perry black holes [25] and black rings [26, 27].

## *3.2 Myers-Perry Black Holes*

The Myers-Perry solution is the generalization of the Kerr solution to $D$ spacetime dimensions. See [28] for a more detailed review. Myers-Perry black holes share many properties with the Kerr solution. Cross-sections of the event horizon have spherical topology $S^{D-2}$. The solution is uniquely parameterized by its mass $M$ and its angular

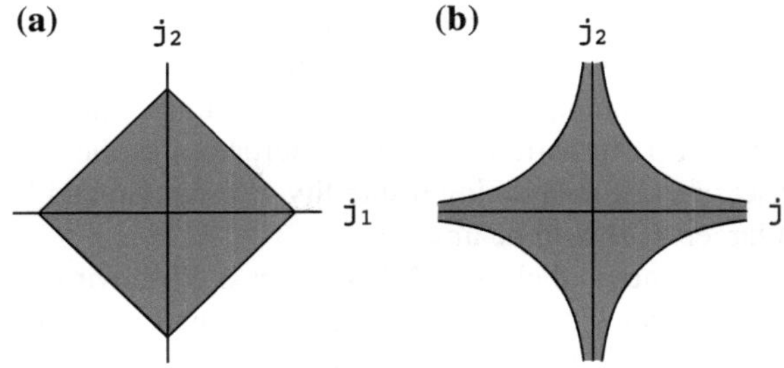


**Fig. 4** Parameter space of Myers-Perry black holes for **a** $D = 5$ and **b** $D = 6$. The axes are dimensionless angular momentum $j_I \sim J_I M^{-(D-2)/(D-3)}$. Non-extreme black holes correspond to the *shaded region*. The boundary of this region corresponds to extreme black holes, except for the vertices of the square, which describe singular solutions (Plot reproduced from [5])

momenta $J_I$. The general Myers-Perry solution has $N = [(D-1)/2]$ commuting rotational Killing vector fields $\partial/\partial\phi_I$. The Myers-Perry solution with all $J_I = 0$ reduces to the $D$-dimensional Schwarzschild solution.

Recall that, for given $M$, the Kerr solution has an upper bound on its angular momentum $|J| \leq M^2$ and saturating this bound gives the extreme Kerr solution with a regular, but degenerate horizon. For $D = 5$, there is a similar upper bound on the angular momenta of the MP solution: for given $M$, regular black holes have $J_1, J_2$ confined to a square region centred on the origin in the $(J_1, J_2)$ plane, see Fig. 4. Saturating this bound gives a black hole with a degenerate horizon except when one of the angular momenta vanishes (the vertices of the square), in which case the spacetime is singular with no horizon.

For $D > 5$, there is a qualitative difference between MP and Kerr. It is possible for the angular momenta to be arbitrarily large for fixed $M$. A "singly spinning" black hole, i.e., one with $J_2 = J_3 = \cdots = J_N = 0$, has no upper bound on $J_1$. More generally, it is possible for some of the angular momenta to be very large if others are small. See Fig. 4 for the $D = 6$ case. Emparan and Myers [29] studied the geometry of singly spinning MP black holes in the "ultraspinning" limit of very large rotation. It was found that the black hole becomes flattened into the plane of rotation, so that it resembles a rotating pancake. The geometry near the intersection of the axis of rotation with the horizon approaches that of a black *membrane*: the product of a $(D-2)$-dimensional Schwarzschild solution with two flat directions. Black membranes suffer from the Gregory-Laflamme instability. Hence it was conjectured in [29] that rapidly rotating $D > 5$ MP black holes are classically unstable.

Confirmation of this conjecture required a study of linearized gravitational perturbations of MP solutions. For $D = 4$, the study of gravitational perturbations of a Kerr black hole is simplified by the remarkable "decoupling" phenomenon discovered by Teukolsky [30], which reduces the problem to a PDE for a single scalar quantity. Unfortunately, decoupling does not occur for $D > 4$ [31] and so one has to solve a large set of coupled PDEs instead. This was done in [32], which studied a class of linearized perturbations of a singly spinning MP solution, restricting to perturbations that preserve the symmetries of the MP solution, i.e., stationarity and the rotational symmetries. For fixed $M$, it was found (numerically) that there exists a critical value of $J_1$ for which a singly spinning $D > 5$ MP solution admits a non-trivial stationary linearized gravitational perturbation. This was interpreted as the "threshold mode" indicating the onset of the ultraspinning instability of Emparan and Myers [29], i.e., black holes with larger $J_1$ should be unstable.

There is a more symmetrical class of MP solutions: those with odd $D$ and $J_1 = J_2 = \cdots = J_N \equiv J$. Such solutions are "cohomogeneity-1": they depend non-trivially only on the radial coordinate. This implies that the equations governing gravitational perturbations are ODEs rather than PDEs [33]. However, in this case, there is an upper (extremality) bound on $J$ for given $M$, i.e., there is no reason to expect an ultraspinning instability. The $D = 5$ case was studied in [34] and no evidence of any instability was found. However, [35] showed that, for $D = 9$, with $J$ close to the upper bound, there are linearized gravitational perturbations which grow exponentially with time, i.e., an instability. This was extended to $D = 7$ in [36], which also considered a class of MP solutions interpolating between singly spinning and cohomogeneity-1 and determined the threshold of instability in this case.

Although decoupling of perturbations does not occur for higher-dimensional black holes, it does occur for the *near-horizon geometry* of an extreme vacuum black hole [31]. Durkee and Reall [37] used this to argue that the instability of near-extreme cohomogeneity-1 MP solutions can be predicted analytically, thereby extending the result to any odd $D > 5$. Tanahashi and Murata [38] used the same approach to show that MP solutions with even $D$ and $J_1 = J_2 = \cdots = J_N$ (which are cohomogeneity-2) also are unstable near extremality.

The perturbations discussed so far are invariant under the Killing vector field $\Omega_I \partial/\partial\phi_I$ where $\Omega_I$ are the angular velocities of the horizon.[1] This makes the resulting equations easier to solve. In the singly spinning case, this means that the perturbations are axisymmetric. *Non*axisymmetric perturbations of singly spinning MP have been studied using full-blown numerical relativity [39, 40].

Shibata and Yoshino [39] studied the $D = 5$ case and found that, for $J_1$ near to the upper bound, an initially small non-axisymmetric perturbation grows in amplitude. It was not possible to evolve the system long enough to determine the end-point of this instability. The corresponding problem for $D = 6, 7, 8$ was studied in [40]. An instability was found for large enough dimensionless angular momentum $j_1 \equiv J_1 M^{-(D-2)/(D-3)}$. This instability appears at a lower value of $j_1$ than the axisymmetric instability discussed above, i.e., the nonaxisymmetric instability is the first one to appear as the angular momentum is increased. In this case, it was possible to follow the long time evolution of the instability. It was found that the perturbed black hole emits gravitional waves, which carry away angular momentum (and energy), and the black hole finally settles down to a (presumably stable) Myers-Perry black hole with a lower value of $j_1$. It is not clear whether this also hap-

[1] The exception is the analysis in [34] of the cohomogeneity-1 $D = 5$ case, i.e., $J_1 = J_2$. The results are consistent with stability of this solution.

pens for $D = 5$ or whether the evolution of the instability is qualitatively different in this case.

Finally we should note that there is an instability that afflicts *extreme* black holes, including extreme Myers-Perry. Consider a massless scalar field in the background of a Kerr black hole. In the non-extreme case, it has been proved that, for any initial data (decaying at infinity), the scalar field and all its derivatives decay on, and outside, the horizon [41]. However, the extreme case is qualitatively different. For axisymmetric initial data, the scalar field decays on, and outside, the event horizon [42]. However, a transverse derivative of the scalar field at the horizon generically does not decay (hence the energy-momentum tensor does not decay), and higher transverse derivatives at the horizon blow up, i.e., they become large at late time [43]. Therefore the scalar field is stable in the background of an arbitrarily non-extreme black hole, but unstable in the background of an exactly extreme black hole. Note that the instability involves power-law, rather than exponential, growth in time.

The scalar field here should be regarded as a toy model for linearized gravitational perturbations, which suggests that an extreme Kerr black hole should suffer a gravitational instability. Lucietti and Reall [44] showed that this is indeed the case. This reference also showed that similar non-decay and blow-up results hold for a massless scalar field in *any* extreme black hole spacetime. Hence extreme black holes generically are unstable, as conjectured in [45].

### 3.3 Black Rings

A black ring is an asymptotically flat black hole for which horizon cross-sections have topology $S^1 \times S^{D-3}$. There is a heuristic argument for the existence of such objects in vacuum gravity. Take a finite segment of (uniform) black string and imagine forming it into a loop. The loop would collapse under its own gravity and tension. However, if it rotates then Newtonian arguments suggest that the resulting centrifugal repulsion can balance the gravitational and tension forces, resulting in a stationary black ring.

This heuristic argument for the existence of black rings is confirmed by explicit solutions, which are known only for the special case $D = 5$ [26, 27]. These solutions are the first examples of asymptotically flat black holes with horizon cross-sections of non-spherical topology. They form a 3-parameter family and have 2 commuting rotational symmetries. Unlike the Myers-Perry solutions, black rings are *not* uniquely labelled by $M$ and $J_I$: it is possible for there to be two black rings with the same values for these quantities. Furthermore, a black ring can have the same value for $M$, $J_I$ as a Myers-Perry black hole. See Fig. 5. Hence the existence of black rings shows that black hole uniqueness cannot be straightforwardly extended to higher dimensions.

An important difference between black rings and $D = 5$ MP solutions is that, for given $M$, there is a *lower* bound on the angular momentum $J_1$, and (for $J_2 = 0$) no upper bound. Black rings do not admit a regular static limit, as expected from the heuristic argument for their existence.

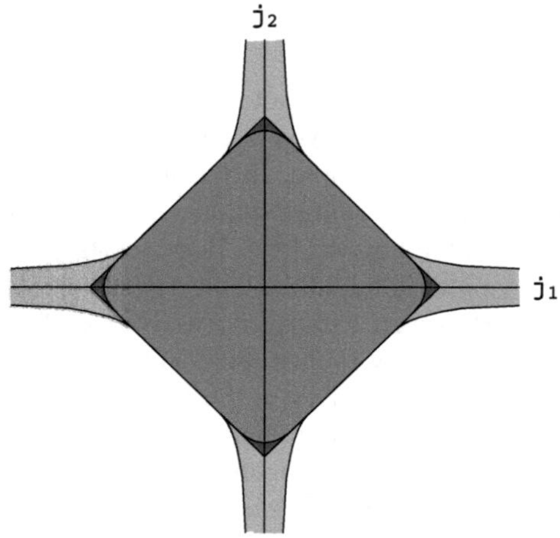


**Fig. 5** Phase space of $D = 5$ Myers-Perry black holes and black rings. See Fig. 4 for notation. For each point of the *light grey* regions there exists a "thin" black ring. For each point of the *mid-grey* region there exists a MP black hole. For each point of the *dark grey* region there exists a MP black hole, a fat black ring and a thin black ring (Plot reproduced from [5])

It is convenient to divide black rings into two subclasses according to the sign of the heat capacity at constant angular momenta $c_J$. Black rings with $c_J < 0$ are called "thin"; and those with $c_J > 0$ are called "fat". The terminology arises from the geometry of the horizon. Thin rings look more like hula-hoops and fat rings more like bagels. Rings of each type are uniquely labelled by $M$, $J_I$.

Heuristic arguments indicate that fat black rings probably are unstable. Arcioni and Lozano-Tellechea [46] used the Poincaré turning point method to argue that fat black rings with small $c_J$ should have one more "unstable mode" than thin rings. Elvang et al. [47] considered certain singular deformations of the black ring solution to determine an effective potential for radial deformations of the ring. It was found that thin rings sit at a local minimum of the potential but fat rings correspond to a local maximum, suggesting instability.

These results were confirmed by the analysis of Figueras et al. [48], which introduced a new method for studying black hole stability. If a black hole is stable then any small perturbation of it must eventually settle down to a black hole belonging to the same family, with a small change in its parameters. If one restricts to rotationally symmetric perturbations, so that angular momentum is conserved, then, by using increase of horizon area and decrease of (Bondi) energy, one can deduce that the initial data describing the perturbed black hole must satisfy a certain inequality relating its mass, angular momenta and horizon area. If one can find an initial perturbation that violates this inequality then the black hole cannot be stable. The nice thing about this method is that it requires only the construction of initial data, rather than determining the full time-evolution of the perturbation. Using this approach, it was shown that fat black rings suffer from a rotationally symmetric instability.

Rings with large $J_1$ (for given $M$) are very thin. As $J_1 \to \infty$, the geometry near a section of the ring approaches that of a boosted uniform black string. Since the latter suffers from the Gregory-Laflamme instability, it seems very likely that black rings with large $J_1$ will be classically unstable. This instability breaks rotational symmetry so it cannot be studied using the approach of Figueras et al. [48]. Demonstrating the existence of this instability will require a study of linearized gravitational perturbations of black rings (or full-blown numerical GR). This has not yet been attempted.

In summary, fat black rings are known to be unstable and very thin black rings are believed to be unstable. But it is not known whether all black rings are unstable or whether some thin, but not too thin, rings are stable.

So far we have been discussing $D = 5$ black rings. Explicit black ring solutions are not known for $D > 5$. However, approximate solutions describing very thin black rings have been constructed using the perturbative "blackfold" approach to be described below [49]. Recently, [50] reported a breakthrough in determining $D > 5$ dimensional black ring solutions by numerical solution of the Einstein equation. Results were presented for black rings with $D = 6, 7$ with a single non-vanishing angular momentum $J_1$. Their properties appear similar to those of $D = 5$ black rings, and agree with the predictions of the blackfold approach when the radius becomes large.

## *3.4 Black Saturn and Generalizations*

Given the existence of black rings and Myers-Perry black holes, it is natural to ask whether one can "superpose" these solutions to construct a "Black Saturn" describing a MP black hole with a concentric black ring. Of course the Einstein equation is nonlinear so this is highly non-trivial. Nevertheless, solution generating techniques have been used to construct such a solution with $D = 5$ [51].

Black Saturn is the first example of an explicit stationary, asymptotically flat, vacuum, regular, multi-black hole solution. It provides an interesting demonstration of the frame-dragging effect: if one sets the (Komar) angular momentum of the MP black hole to zero then its angular velocity is non-zero because the horizon generators are dragged around by the rotation of the black ring.

Solution generating techniques have also been used to construct solutions with multiple concentric black rings. For example the di-ring of Iguchi and Mishima [52], Evslin and Krishnan [53] describes a pair of concentric black rings lying in the same plane. Elvang and Rodriguez [54] gave a solution describing a pair of concentric black rings lying in orthogonal planes.

## *3.5 Classification of Asymptotically Flat Black Holes*

For $D = 4$, the black hole uniqueness theorem provides a complete classification of stationary vacuum black holes. For $D > 4$, the known solutions show that the situation is much more complicated and the uniqueness theorem does not generalize in a simple way. However, it is useful to explore whether some aspects of the $D = 4$ theorem can be generalized to $D > 4$.

Uniqueness of static vacuum black holes turns out to generalize straightforwardly to $D > 4$: the only asymptotically flat static vacuum black holes solution is the Schwarzschild solution, for any $D \geq 4$ [55].

For non-static black holes, the first logical step in the $D = 4$ uniqueness theorem is Hawking's topology theorem [56], stating that horizon cross-sections must have $S^2$ topology. This result has been generalized to $D > 4$ dimensions, with the result that horizon cross-sections must admit a metric of positive scalar curvature [57]. This is a topological restriction on the horizon. For $D = 4$ it reduces to Hawking's result. For $D = 5$ it implies that the horizon cross-section must be either $S^3$ (or a quotient), $S^1 \times S^2$, or a connected sum of these. It is striking that $S^3$ and $S^1 \times S^2$ are precisely the topologies realized by the known Myers-Perry and black ring solutions. However, the possibility of taking quotients and connected sums implies that there are infinitely many topologies consistent with this theorem. For $D > 6$, there is no simple description of the possible topologies.

The next step in the $D = 4$ uniqueness proof for non-static black holes is Hawking's rigidity theorem [56], the statement that a stationary, rotating, analytic, black hole solution must be axisymmetric. This has been generalised to $D > 4$: a stationary, rotating, analytic, black hole solution must admit a rotational symmetry (a $U(1)$ isometry acting as a rotation at infinity) [58, 59]. However, this theorem guarantees only *one* rotational symmetry whereas the known explicit solutions have $N > 1$ commuting rotational symmetries, i.e., more symmetry than guaranteed by the theorem. This suggests that there may exist other $D > 4$ black hole solutions which have less symmetry than the known solutions. We will discuss this further in the next section.

Progress with classifying black holes can be made if one restricts attention to the case of multiple rotational symmetries. The classification of $D = 5$ stationary rotating vacuum black holes with two rotational symmetries was studied in [60] (extending [61]). It was shown that two such solutions are isometric if, and only if, they have the same mass, angular momenta, and *rod structure*. The latter (introduced in [62, 63]) encodes the angular velocity of the horizon and the nature of "axes of rotation" in the spacetime. The assumed symmetries imply that the horizon topology must be $S^3$, $S^1 \times S^2$ or a lens space (a quotient of $S^3$). It is not known whether (asymptotically flat) solutions of the latter kind exist.

### *3.6 Perturbative Solutions*

Given the difficulty in finding explicit solutions of the Einstein equation, perturbative techniques have been used to obtain some insight into what other solutions might exist. Of particular interest are the possible topologies of higher-dimensional black holes, and the question of whether there exist higher-dimensional black holes with just one rotational symmetry. Two techniques have been used to investigate these questions.

The "blackfold" technique is a method for constructing black hole solutions whose horizons exhibit a large hierarchy of length scales. An example is a black ring for which the $S^1$ radius is much greater than the $S^{D-3}$ radius [49]. Such large-radius black ring solutions are constructed perturbatively as an expansion in the ratio of these radii. More generally, the blackfold method has been used to construct approximate higher-dimensional solutions with topologies of the form $S^{p_1} \times S^{p_2} \times \cdots \times S^{p_K} \times s^q$ where $p_1, \ldots p_K$ are odd, $q \geq 2$, $S^{p_1}, \ldots S^{p_k}$ are large radius spheres and $s^q$ is a small radius sphere [64].

The blackfold approach has also been used to construct solutions with just one rotational symmetry. Emparan et al. [65] presented a perturbative solution describing a $D = 5$ "helical" black ring. This solutions can be visualised by imagining a spring formed into a loop.

Another perturbative approach is to study linearized perturbations of an explicit solution. Non-uniform black strings provide a nice example of this. A uniform black string admits a time-independent linearized perturbation corresponding to the "threshold mode" of the Gregory-Laflamme instability. This time-independent perturbation exists precisely at the point where the non-uniform black string family bifurcates from the uniform string family. It corresponds to a non-uniform string solution with infinitesimal non-uniformity. Thus by studying linearized perturbations of the uniform string one can infer the existence of the non-uniform string family.

This method has been applied to perturbations of Myers-Perry black holes. Consider a singly spinning MP black hole with $D > 5$. As explained above, [32] showed that there is a critical $J_1$ (for given $M$) at which such a solution admits a non-trivial time-independent linearized perturbation, corresponding to the threshold mode of the ultraspinning instability. Earlier, [29] had suggested that such a perturbation should be interpreted as evidence for a new family of black holes that bifurcates from the MP family. This new family would possess the same symmetries as the MP solution but with a slightly deformed horizon, corresponding to a depression at the poles of the sphere. It was suggested that moving along this new branch of solutions, the depression would increase, corresponding to an increasingly "pinched" sphere. Eventually, the sphere is expected to "pinch off" completely, and merge with the black ring family [49].

The same strategy has been applied to perturbations of cohomogeneity-1 MP black holes (odd $D$ with $J_1 = J_2 = \cdots = J_N$). Recall that [35, 36] demonstrated an instability of such solutions near extremality. Again there is a time-independent threshold mode for this instability and so this was interpreted as evidence for a new branch of solutions bifurcating from the MP family. However, in this case, the threshold mode does not preserve the symmetries of the background geometry. In general it preserves only the rotational symmetry whose existence is guaranteed by the rigidity theorem. Hence the new family of solutions should have just one rotational symmetry. Note that these black holes would have spherical topology. Furthermore, in this case, there is not just one threshold mode but a multi-parameter set of them. If each of these extends to the new branch of solutions then these new solutions would have many more parameters than the MP family. For example, in $D = 9$ they would have 70 parameters [35], many more than the 5 parameters of the MP solution.

**Acknowledgments** I would like to thank Pau Figueras and Keiju Murata for permission to reproduce figures from [20] and Roberto Emparan for permission to reproduce figures from [5]. I am supported by a Royal Society University Research Fellowship and by European Research Council grant no. ERC-2011-StG 279363-HiDGR.

## References

1. Strominger, A., Vafa, C.: Microscopic origin of the bekenstein-hawking entropy. Phys. Lett. **B379**, 99 (1996). doi:10.1016/0370-2693(96)00345-0
2. Maldacena, J.M.: The large n limit of superconformal field theories and supergravity. Adv. Theor. Math. Phys. **2**, 231 (1998)
3. Kanti, P.: Black holes at the LHC. In: Papantonopoulos, E. (eds.) Physics of Black Holes: A Guided Tour, Lecture notes in physics, vol. 769, pp. 387–423. Springer, Berlin (2009). doi:10.1007/978-3-540-88460-6_10
4. Cvetic, M., Lü, H., Page, D.N., Pope, C.N.: New Einstein-Sasaki and Einstein spaces from Kerr-de Sitter. J. High Energy Phys. **2009**(07), 082 (2009). doi:10.1088/1126-6708/2009/07/082
5. Emparan, R., Reall, H.S.: Black holes in higher dimensions. Living Rev. Relativ. **11**(6), lrr-2008-6 (2008). http://www.livingreviews.org/lrr-2008-6
6. Horowitz, G.T., Wiseman, T.: General black holes in Kaluza-Klein theory. In: Horowitz, G.T. (ed.) Black Holes in Higher Dimensions, pp. 69–98. Cambridge University Press, Cambridge (2012)
7. Harmark, T.: Small black holes on cylinders. Phys. Rev. D **69**, 104015 (2004). doi:10.1103/PhysRevD.69.104015
8. Gorbonos, D., Kol, B.: A dialogue of multipoles: matched asymptotic expansion for caged black holes. J. High Energy Phys. **2004**(06), 053 (2004). doi:10.1088/1126-6708/2004/06/053
9. Sorkin, E., Kol, B., Piran, T.: Caged black holes: black holes in compactified spacetimes. ii. 5d numerical implementation. Phys. Rev. D **69**, 064032 (2004). doi:10.1103/PhysRevD.69.064032
10. Kudoh, H., Wiseman, T.: Connecting black holes and black strings. Phys. Rev. Lett. **94**, 161102 (2005). doi:10.1103/PhysRevLett.94.161102
11. Headrick, M., Kitchen, S., Wiseman, T.: A new approach to static numerical relativity, and its application to kaluza-klein black holes. Class. Quantum Grav. **27**, 035002 (2010). doi:10.1088/0264-9381/27/3/035002
12. Ishibashi, A., Kodama, H.: Stability of higher dimensional Schwarzschild black holes. Prog. Theor. Phys. **110**, 901 (2003). doi:10.1143/PTP.110.901
13. Gregory, R., Laflamme, R.: Black strings and p-branes are unstable. Phys. Rev. Lett. **70**, 2837 (1993). doi:10.1103/PhysRevLett.70.2837
14. Lehner, L., Pretorius, F.: Black strings, low viscosity fluids, and violation of cosmic censorship. Phys. Rev. Lett. **105**, 101102 (2010). doi:10.1103/PhysRevLett.105.101102
15. Horowitz, G.T., Maeda, K.: Fate of the black string instability. Phys. Rev. Lett. **87**, 131301 (2001). doi:10.1103/PhysRevLett.87.131301
16. Gubser, S.S.: On non-uniform black branes. Class. Quantum Grav. **19**, 4825 (2002). doi:10.1088/0264-9381/19/19/303
17. Wiseman, T.: Static axisymmetric vacuum solutions and nonuniform black strings. Class. Quantum Grav. **20**, 1137 (2003). doi:10.1088/0264-9381/20/6/308
18. Kleihaus, B., Kunz, J., Radu, E.: New nonuniform black string solutions. J. High Energy Phys. **2006**(06), 016 (2006). doi:10.1088/1126-6708/2006/06/016

19. Sorkin, E.: Non-uniform black strings in various dimensions. Phys. Rev. D **74**, 104027 (2006). doi:10.1103/PhysRevD.74.104027
20. Figueras, P., Murata, K., Reall, H.S.: Stable non-uniform black strings below the critical dimension. J. High Energy Phys. **2012**(11), 071 (2012). doi:10.1007/JHEP11(2012)071
21. Kol, B.: Topology change in general relativity, and the black hole black string transition. J. High Energy Phys. **2005**(10), 049 (2005). doi:10.1088/1126-6708/2005/10/049
22. Sorkin, E.: Critical dimension in the black-string phase transition. Phys. Rev. Lett. **93**, 031601 (2004). doi:10.1103/PhysRevLett.93.031601
23. Horowitz, G.T. (ed.): Black Holes in Higher Dimensions. Cambridge University Press, Cambridge (2012)
24. Maeda, K., Shiromizu, T., Tanaka T. (eds.): Higher Dimesional Black Holes, Progress of Theoretical Physics Supplement, vol. 189, Oxford University Press, Oxford (2011). http://ptps.oxfordjournals.org/content/189.toc
25. Myers, R.C., Perry, M.J.: Black holes in higher dimensional space-times. Ann. Phys. (N.Y.) **172**, 304 (1986). doi:10.1016/0003-4916(86)90186-7
26. Emparan, R., Reall, H.S.: A rotating black ring solution in five dimensions. Phys. Rev. Lett. **88**, 101101 (2002). doi:10.1103/PhysRevLett.88.101101
27. Pomeransky, A.A., Sen'kov, R.A.: Black ring with two angular momenta. ArXiv e-prints arXiv:0612005 [hep-th] (2006)
28. Myers, R.C.: Myers-Perry black holes. In: Horowitz, G.T. (ed.) Black Holes in Higher Dimensions, pp. 101–133. Cambridge University Press, Cambridge (2012)
29. Emparan, R., Myers, R.C.: Instability of ultra-spinning black holes. J. High Energy Phys. **2003**(09), 025 (2003). doi:10.1088/1126-6708/2003/09/025
30. Teukolsky, S.A.: Rotating black holes: separable wave equations for gravitational and electromagnetic perturbations. Phys. Rev. Lett. **29**, 1114 (1972). doi:10.1103/PhysRevLett.29.1114
31. Durkee, M., Reall, H.S.: Perturbations of higher-dimensional spacetimes. Class. Quantum Grav. **28**, 035011 (2011). doi:10.1088/0264-9381/28/3/035011
32. Dias, O.J.C., Figueras, P., Monteiro, R., Santos, J.E., Emparan, R.: Instability and new phases of higher-dimensional rotating black holes. Phys. Rev. D **80**, 111701(R) (2009). doi:10.1103/PhysRevD.80.111701
33. Kunduri, H.K., Lucietti, J., Reall, H.S.: Gravitational perturbations of higher dimensional rotating black holes: tensor perturbations. Phys. Rev. D **74**, 084021 (2006). doi:10.1103/PhysRevD.74.084021
34. Murata, K., Soda, J.: Stability of five-dimensional Myers-Perry black holes with equal angular momenta. Prog. Theor. Phys. **120**, 561 (2008). doi:10.1143/PTP.120.561
35. Dias, Ó.J.C., Figueras, P., Monteiro, R., Reall, H.S., Santos, J.E.: An instability of higher-dimensional rotating black holes. J. High Energy Phys. **2010**(05), 076 (2010). doi:10.1007/JHEP05(2010)076
36. Dias, Ó.J.C., Monteiro, R., Santos, J.E.: Ultraspinning instability: the missing link. J. High Energy Phys. **2011**(08), 139 (2011). doi:10.1007/JHEP08(2011)139
37. Durkee, M., Reall, H.S.: Perturbations of near-horizon geometries and instabilities of Myers-Perry black holes. Phys. Rev. D **83**, 104044 (2011). doi:10.1103/PhysRevD.83.104044
38. Tanahashi, N., Murata, K.: Instability in near-horizon geometries of even-dimensional Myers-Perry black holes. Class. Quantum Grav. **29**, 235002 (2012). doi:10.1088/0264-9381/29/23/235002
39. Shibata, M., Yoshino, H.: Nonaxisymmetric instability of rapidly rotating black hole in five dimensions. Phys. Rev. D **81**, 021501 (2010). doi:10.1103/PhysRevD.81.021501
40. Shibata, M., Yoshino, H.: Bar-mode instability of rapidly spinning black hole in higher dimensions: numerical simulation in general relativity. Phys. Rev. D **81**, 104035 (2010). doi:10.1103/PhysRevD.81.104035
41. Dafermos, M., Rodnianski, I.: The black hole stability problem for linear scalar perturbations. In: Damour, T., Jantzen, R.T., Ruffini R. (eds.) The Twelfth Marcel Grossmann Meeting On Recent Developments in Theoretical and Experimental General Relativity, Astrophysics and Relativistic Field Theories, pp. 132–189. World Scientific, Singapore (2012). doi:10.1142/9789814374552_0008

42. Aretakis, S.: Decay of axisymmetric solutions of the wave equation on extreme Kerr backgrounds. ArXiv e-prints arxiv:1110.2006 [gr-qc] (2011)
43. Aretakis, S.: Horizon instability of extremal black holes. ArXiv e-prints arxiv:1206.6598 [gr-qc] (2012)
44. Lucietti, J., Reall, H.S.: Gravitational instability of an extreme Kerr black hole. Phys. Rev. D **86**, 104030 (2012). doi:10.1103/PhysRevD.86.104030
45. Marolf, D.: The dangers of extremes. Gen. Relativ. Gravit. **42**, 2337 (2010). doi:10.1007/s10714-010-1027-z
46. Arcioni, G., Lozano-Tellechea, E.: Stability and critical phenomena of black holes and black rings. Phys. Rev. D **72**, 104021 (2005). doi:10.1103/PhysRevD.72.104021
47. Elvang, H., Emparan, R., Virmani, A.: Dynamics and stability of black rings. J. High Energy Phys. **2006**(12), 074 (2006). doi:10.1088/1126-6708/2006/12/074
48. Figueras, P., Murata, K., Reall, H.S.: Black hole instabilities and local penrose inequalities. Class. Quantum Grav. **28**, 225030 (2011). doi:10.1088/0264-9381/28/22/225030
49. Emparan, R., Harmark, T., Niarchos, V., Obers, N.A., Rodriguez, M.J.: The phase structure of higher-dimensional black rings and black holes. J. High Energy Phys. **2007**(10), 110 (2007). doi:10.1088/1126-6708/2007/10/110
50. Kleihaus, B., Kunz, J., Radu, E.: Black rings in six dimensions. Phys. Lett. B **718**, 1073 (2013). doi:10.1016/j.physletb.2012.11.015
51. Elvang, H., Figueras, P.: Black saturn. J. High Energy Phys. **2007**(05), 050 (2007). doi:10.1088/1126-6708/2007/05/050
52. Iguchi, H., Mishima, T.: Black di-ring and infinite nonuniqueness. Phys. Rev. D **75**(064018), 2008 (2007). doi:10.1103/PhysRevD.75.064018 [Erratum: ibid. 78, 069903]
53. Evslin, J., Krishnan, C.: The black di-ring: an inverse scattering construction. Class. Quant. Grav. **26**, 125018 (2009). doi:10.1088/0264-9381/26/12/125018
54. Elvang, H., Rodriguez, M.J.: Bicycling black rings. J. High Energy Phys. **2008**(04), 045 (2008). doi:10.1088/1126-6708/2008/04/045
55. Gibbons, G.W., Ida, D., Shiromizu, T.: Uniqueness and nonuniqueness of static black holes in higher dimensions. Phys. Rev. Lett. **89**, 041101 (2002). doi:10.1103/PhysRevLett.89.041101
56. Hawking, S.W., Ellis, G.F.R.: The Large Scale Structure of Space-Time. Cambridge Monographs on Mathematical Physics. Cambridge University Press, Cambridge (1973)
57. Galloway, G.J., Schoen, R.: A generalization of hawking's black hole topology theorem to higher dimensions. Commun. Math. Phys. **266**, 571 (2006). doi:10.1007/s00220-006-0019-z
58. Hollands, S., Ishibashi, A., Wald, R.M.: A higher dimensional stationary rotating black hole must be axisymmetric. Commun. Math. Phys. **271**, 699 (2007). doi:10.1007/s00220-007-0216-4
59. Moncrief, V., Isenberg, J.: Symmetries of higher dimensional black holes. Class. Quantum Grav. **25**, 195015 (2008). doi:10.1088/0264-9381/25/19/195015
60. Hollands, S., Yazadjiev, S.: Uniqueness theorem for 5-dimensional black holes with two axial killing fields. Commun. Math. Phys. **283**, 749 (2008). doi:10.1007/s00220-008-0516-3
61. Morisawa, Y., Ida, D.: A boundary value problem for the five-dimensional stationary rotating black holes. Phys. Rev. D **69**, 124005 (2004). doi:10.1103/PhysRevD.69.124005
62. Emparan, R., Reall, H.S.: Generalized Weyl solutions. Phys. Rev. D **65**, 084025 (2002). doi:10.1103/PhysRevD.65.084025
63. Harmark, T.: Stationary and axisymmetric solutions of higher-dimensional general relativity. Phys. Rev. D **70**, 124002 (2004). doi:10.1103/PhysRevD.70.124002
64. Emparan, R., Harmark, T., Niarchos, V., Obers, N.A.: World-volume effective theory for higher-dimensional black holes. Phys. Rev. Lett. **102**, 191301 (2009). doi:10.1103/PhysRevLett.102.191301
65. Emparan, R., Harmark, T., Niarchos, V., Obers, N.A.: New horizons for black holes and branes. J. High Energy Phys. **2010**(04), 046 (2010). doi:10.1007/JHEP04(2010)046

# Black Holes, Hidden Symmetry and Complete Integrability: Brief Review

**Valeri P. Frolov**

**Abstract** This chapter contains a brief review of the remarkable properties of higher dimensional rotating black holes with the spherical topology of the horizon. We demonstrate that these properties are connected with and generated by a special geometrical object, the Principal Conformal Killing-Yano tensor (PCKYT). The most general solution, describing such black holes, Kerr-NUT-ADS metric, admits this structure. Moreover a solution of the Einstein Equations with (or without) a cosmological constant which possesses PCKYT is the Kerr-NUT-ADS metric. This object (PCKYT) is responsible for such remarkable properties of higher dimensional rotating black holes as: (i) complete integrability of geodesic equations and (ii) complete separation of variables of the important field equations.

## 1 Introduction

Main motivations for study higher dimensional black holes are nicely summarized in the chapter by Harvey Reall. String theory, brane world models and ADS/CFT correspondence naturally involve higher dimensional gravity, and black holes play role of natural probes of extra dimensions. Possibility of creation of microscopic black holes in high energy colliders was (and still is) a subject which attracts a lot of intension. Besides these 'rather technical' reasons there exists another one of more general nature. The driving force of our scientific knowledge is quite often our curiosity. We keep asking ourselves questions: "What happens if ...". From time to time by answering such questions one discovers interesting and non-trivial results.

V. P. Frolov (✉)
Theoretical Physics Institute, University of Alberta, Edmonton, AB T6G 2E1, Canada
e-mail: vfrolov@ualberta.ca

J. Bičák and T. Ledvinka (eds.), *General Relativity, Cosmology and Astrophysics*, Fundamental Theories of Physics 177, DOI: 10.1007/978-3-319-06349-2_13,

© Springer International Publishing Switzerland 2014

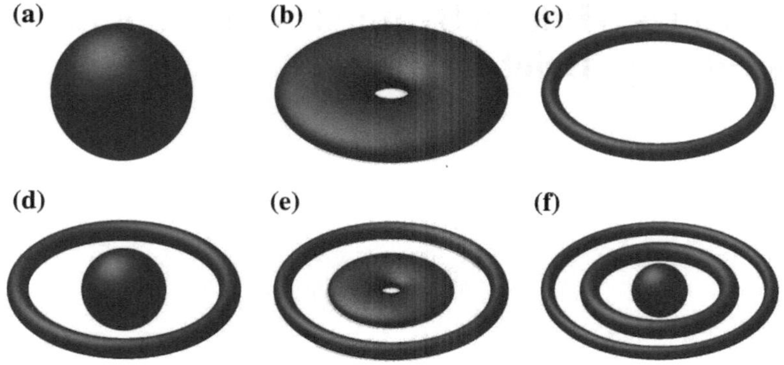


**Fig. 1** Stationary 5D black holes

By answering the question "Do black hole exist in higher dimensional gravity and what are their properties?" one understands better which of the properties of four-dimensional black holes are generic (valid in any number of dimensions) and which are specific in only our four-dimensional world. One of the most surprising discoveries was that stationary black holes in higher dimensions can have large variety of the horizon topology. The first of such 5-dimensional black objects with toroidal topology of the horizon $S^1 \times S^2$, called black ring, was found by Emparan and Reall [1]. By now there are known many other stationary exact solutions describing vacuum black objects in 5D, the horizon of which is schematically shown at Fig. 1 (For general discussion see the chapter by Reall in this volume and a nice review [2]).

Let us emphasize, that according to the general definition all such black objects, strictly speaking, are *black holes*, since the region beyond their event horizon is causally separated from the future null infinity. However, in order to distinguish black objects with different horizon topology they received special nicknames (black ring, black Saturn, etc), while the name of a black hole is reserved for the black object with spherical topology of the horizon. In this talk we shall follow this tradition. This means that from now on any time when we speak about a black hole we have in mind the object presented at the Fig. 1a. To be more concrete we focus our attention on isolated higher dimensional rotating black holes with spherical topology of the horizon, obeying equations

$$R_{\mu\nu} = \Lambda g_{\mu\nu}. \tag{1}$$

The corresponding spacetime is asymptotically either flat ($\Lambda = 0$) or (anti) DeSitter one.

## 2 Four Dimensional Kerr-NUT-ADS Metric and Its Higher Dimensional Generalization

### *2.1 Four Dimensional Kerr-NUT-ADS Spacetime*

The simplest example of a vacuum static black hole in 4D is the Schwarzschild metric

$$ds^2 = -f dt^2 + dr^2/f + r^2 d\Omega_2^2\,, \quad f = 1 - r_0/r. \tag{2}$$

It contains only one parameter, the mass $M$ of the black hole, which determines the gravitational radius (size of the horizon) $r_0 = 2M$. Here and later we denote by $d\Omega_m^2$ the metric on a unit $m$-dimensional round sphere.

A vacuum *stationary* black hole is characterized by one more parameter, $J$. Such a black hole is rotating and $J$ is the value of its angular momentum. In fact the angular momentum (measured at infinity) is described by $3 \times 3$ antisymmetric matrix $J_{ij}$. By rigid 3-dimensional rotations this matrix can be put in a standard form

$$\mathbf{J} = \begin{pmatrix} 0 & J & 0 \\ -J & 0 & 0 \\ 0 & 0 & 0 \end{pmatrix}. \tag{3}$$

Kerr metric (with $J \le M^2$) is the most general vacuum stationary solution describing a regular black hole in an asymptotically flat spacetime.

In 1963 Newman, Tamburino and Unti [3] found another generalization of the Schwarzschild solution which besides mass $M$ contained another parameter $N$. This parameter $N$, called NUT-parameter, describes "gravitomagnetic monopole" [4]. The corresponding NUT solution contains an analogue of the Dirac strings for a magnetic monopole, but in the gravitational case it cannot be excluded and does affect spacetime. The most general stationary black hole solution of the Einstein equation (1) in an asymptotically (anti)deSitter spacetime contains all these parameters $(M, J, N, \Lambda)$. It was obtained in [5, 6] and it is called Kerr-NUT-ADS metric.

### *2.2 Higher Dimensional Solutions*

Let us discuss now higher dimensional black hole solutions of the Einstein equation (1). We denote by $D$ the total number of (space and time) dimensions. Since some of relations have slightly different form in odd and even dimensions, we write

$$D = 2n + \varepsilon. \tag{4}$$

It is quite easy to generalize the Schwarzschild solution to any number of spacetime dimensions. It is sufficient to substitute in (2) $d\Omega^2_{D-2}$ by the line element on $(D-2)-$ dimensional sphere $S^{D-2}$, and put $f = 1 - (r_0/r)^{D-3}$. This metric is known as Tangherlini solution [7]. The cosmological constant can also be easily included by using $f$ in the form

$$f = 1 - (r_0/r)^{D-3} - \frac{\Lambda}{D-1} r^2. \tag{5}$$

Before discussing higher dimensional rotating black holes let us make the following remark. In the asymptotically flat spacetime the total angular momentum of the objects, as measured at infinity, it described by an antisymmetric tensor $J_{ij}$, where $i$ and $j$ are spatial indices. By suitable rigid rotations of the spatial coordinates, this $(D-1) \times (D-1)$-matrix can be transformed into the following canonical form:

$$\mathbf{J} = \begin{pmatrix} 0 & J_1 & 0 & 0 & \dots \\ -J_1 & 0 & 0 & 0 & \dots \\ 0 & 0 & 0 & J_2 & \dots \\ 0 & 0 & -J_2 & 0 & \dots \\ \dots & \dots & \dots & \dots & \dots \end{pmatrix}. \tag{6}$$

It is easy to see that the total number of independent $2 \times 2$ blocks is equal to $n - 1 + \varepsilon$. This means that there exist $n - 1 + \varepsilon$ independent components of the angular momentum $J_i$, associated with $n - 1 + \varepsilon$ asymptotic independent two-dimensional spatial planes of rotation. The most general solution for a vacuum stationary rotating black hole was found in 1986 [8]. It contains $n + \varepsilon$ independent constants: mass$M$ and components of the angular momentum $J_i$.

It took 20 years to discover the most general higher dimensional black hole solution, which besides mass and angular momentum contains also the cosmological constant and generalization of the NUT parameters [9]. This solution is called a 'general Kerr-NUT-ADS metric'. It is a natural generalization of the 4D Kerr-NUT-ADS metric. It contains $D - \varepsilon$ arbitrary parameters, which include mass, $(n - 1 + \varepsilon)$ rotation parameters, cosmological constant and $(n - 1 - \varepsilon)$ NUT parameters.

## 3 Principal Conformal Killing-Yano Tensor

### *3.1 Definition*

The remarkable fact is: Properties of higher dimensional rotating black holes and their four dimensional 'cousins' are very similar. There exist a very deep geometrical reason for this similarity. All these metrics admit a special geometric object, the Principal Conformal Killing-Yano tensor (PCKYT), which is a generator of a complete set of explicit and hidden symmetries, that uniquely specifies the solution up to some constants. This solution coincides with Kerr-NUT-ADS metric.

In fact this object is special case of a closed conformal Killing-Yano tensor. We shall give a general definition of Killing-Yano tensors later. We shall explain also in more detail what are the hidden symmetries and how are they generated. At the moment, we focus on some remarkable properties of PCKYT.

Let us first give a definition of the PCKYT. Consider a rank 2 antisymmetric tensor (2-form) **h** which obeys the equation

$$\nabla_c h_{ab} = g_{ca}\xi_b - g_{cb}\xi_a. \tag{7}$$

By contracting this equation one finds

$$\xi_a = \frac{1}{D-1}\nabla^b h_{ba}. \tag{8}$$

If one antisymmetrizes the indices $a, b, c$ in (7), the right-hand side of this equation vanishes. This means that **h** is closed form, and (at least locally) can be presented in the form

$$\boldsymbol{h} = d\boldsymbol{b}, \tag{9}$$

where **b** is a potential one-form.

Another important relation can be obtained if one takes the covariant derivative of (8) and symmetrises it. Using an expression for commutator of covariant derivatives, and the Einstein equation (1) one proves that

$$\xi_{(a;b)} = 0. \tag{10}$$

In other words a spacetime admitting the 2-form **h** which is a solution of (7) always has a Killing vector. We call it a *primary Killing vector.*

We call **h** obeying (7) the Principal Conformal Killing-Yano tensor (PCKYT) if it is non-degenerate. More concretely this means the following:

- The matrix rank of $(D \times D)$-antisymmetric matrix $h_{ab}$ is the largest possible, that is equal to $2n$.
- Consider eigenvalue problem for a matrix $H^a_{\ b} = h^{ac}h_{cb}$

$$H^a_{\ b}e^b_{(i)} = x_i e^a_{(i)}. \tag{11}$$

It is easy to see that $h^a_{\ b}e^b_{(i)}$ is again an eigenvector with the same eigenvalue $x_{(i)}$. We assume that $\boldsymbol{H}$ has largest possible number, $n$, of different eigenvalues, and hence $n$ linearly independent eigen 2-planes.

### 3.2 Remarkable Properties of PCKYT

The most general higher dimensional black hole metric, Kerr-NUT-ADS solution, admits the Principal Conformal Killing-Yano tensor. This result was first proved in [10] for Myers-Perry metrics, and later in [11] it was proved for the general Kerr-NUT-ADS spacetimes.

Moreover, a solution of the Einstein equation (1) in any number of dimensions ($D \geq 4$), which admits the Principal Conformal Killing-Yano tensor, is the Kerr-NUT-ADS spacetime. This result under special assumptions was first proved in [12]. A general proof was given in [13, 14].

In other words, the existence of PCKYT is a characteristic property of higher dimensional rotating black holes, with the spherical topology of horizon. Namely the existence of this object explains many remarkable properties of these spaces which sometimes are called 'miraculous'. Before discussing this subject let us briefly remind some properties of completely integrable dynamical systems.

## 4 Complete Integrability

### 4.1 Liouville Theorem

Particle and light motion in a curved spacetime is described by geodesic equations. These equations are of the second order. Let $x^a(\tau)$ be a trajectory. By introducing a momentum $p_a = g_{ab}\dot{x}^b$ as an independent variable, it is possible to rewrite the geodesic equations in the first order form. These equations has the Hamiltonian form. This means that the general theory of dynamical systems can be applied for this problem. This approach is well known and its tools are very useful. Let us demonstrate this for the special problem: motion of a particle in a spacetime of a higher dimensional rotating black hole.

Consider a dynamical system. It is described by an even dimensional phase space $M^{2m}$ with a closed non-degenerate symplectic 2-form $\boldsymbol{\Omega}$ and a Hamiltonian $H$, which is a scalar function on the phase space. If $z^A$ are coordinates, then the dynamical equation is

$$\dot{z}^A = \Omega^{AB} H_{,B}. \tag{12}$$

Poisson bracket for two functions $F$ and $G$ on the phase space is defined as

$$\{F, G\} = -\Omega^{AB} F_{,A} G_{,B}. \tag{13}$$

These two functions are called to be in involution if their Poisson bracket vanishes. A scalar function $F(z^A)$ on the phase space is a first integral of motion if its Poisson bracket with the Hamiltonian vanishes $\{F, H\} = 0$.

Liouville proved the following theorem: If a system with a Hamiltonian $H$ in the $2m$ dimensional phase space has $m$ independent first integrals in involution, $F_1 = H, F_2, \ldots, F_m$, then the system can be integrated by quadratures. Such a system is called *completely integrable*.

### 4.2 Relativistic Particle as Dynamical System

Motion of a particle in a curved spacetime is special case of a dynamical system. If $D$ is the dimension of the spacetime and its coordinates are $x^a$, a particle trajectory is a line $x^a(\tau)$. The canonical coordinates in the corresponding phase space are $(x^a, p_a = g_{ab}\dot{x}^b)$. The canonical symplectic form and the Hamiltonian are

$$\boldsymbol{\Omega} = dx^a \wedge dp_a, \quad H = \frac{1}{2} g^{ab} p_a p_b. \tag{14}$$

It is easy to show that the Hamilton equations of motion in the phase space are equivalent to geodesic equations in the spacetime.

Let us assume that the Hamilton equations have an integral of motion of the form $\mathscr{K} = K^{a_1 \ldots a_q}(x) p_{a_1} \ldots p_{a_q}$, then

$$K_{(a_1 \ldots a_q ; a_{q+1})} = 0. \tag{15}$$

Such a tensor in the spacetime is called a *Killing tensor*. The Killing tensor of the rank 1, $\xi_a$, is a Killing vector. The metric $g_{ab}$ is a trivial Killing tensor.

It is well known that Killing vectors generate symmetry transformation on the spacetime manifold with metric $g_{ab}$. Usually this symmetry is called an *explicit symmetry*. The Killing tensors of higher rank more directly connected with the properties of the phase space, and their meaning is not so transparent in the spacetime itself. For this reason, they are often called *hidden symmetries*.

Poisson brackets in the phase space after reduction to the spacetime determine the *Schouten-Nijenhuis brackets*. When two integrals of motion are in the involution and each of them is a monomial in momentum, the Schouten-Nijenhuis bracket for the corresponding Killing tensors vanish. In such a case we simply say that the Killing tensors commute.

The Liouville theorem being applied to relativistic particles implies that the geodesic equations are completely integrable (can be solved in quadratures) if there exist $D$ independent commuting Killing tensors. Some of them can be Killing vectors. $D$ dimensional Kerr-NUT-ADS admits $n+\varepsilon$ Killing vectors. For complete integrability there must exist additionally $n$ Killing tensors. One of them is trivial, $g_{ab}$.

### 4.3 Page's Proposal

Let $u^a = dx^a/d\tau$ be a unit tangent vector to a particle trajectory, then $P^a_b = \delta^a_b + u^a u_b$ is a projector to a plane, orthogonal to $u^a$. Using the definition of the PCKYT (Eq. (7)) it is easy to show that $F_{ab} = P^c_a h_{cd} P^d_b$ is parallel transported along the geodesic. Hence any scalar invariant constructed from this tensor is a constant of motion. To obtain such scalars it is sufficient to take a trace of powers of the matrix $F_a{}^b$. In [15] it was shown that these integrals of motion are independent and commuting, and their total number is sufficient for the complete integrability of the geodesic equations in the Kerr-NUT-ADS spacetime. However these set of integral of motion is reducible. In other words most of these integrals are of high rank,while there exist a complete set of lower rank integrals of motion. In fact, in a general case in the presence of the PCKYT the irreducible set of integrals of motion is determined by $(n + \varepsilon)$ Killing vectors and $n$ Killing tensors of the second rank. To demonstrate this we need to consider Killing and Killing-Yano tensors in more detail.

## 5 PCKYT and Killing-Yano Tower

### 5.1 Killing-Yano Tensors

Let us first introduce two objects. The *Killing-Yano (KY) tensor* $k_{a_1 \ldots a_q}$ is an antisymmetric $q$−form on the spacetime, which obeys the equation

$$\nabla_a k_{a_1 \ldots a_q} = \nabla_{[a} k_{a_1 \ldots a_q]}. \tag{16}$$

On the other hand, the *closed conformal Killing-Yano (CCKY) tensor* $h_{a_1 \ldots a_q}$ is an antisymmetric $q$—form the covariant derivative of which is determined by its divergence

$$\nabla_a h_{a_1 \ldots a_q} = q g_{a[a_1} \xi_{a_2 \ldots a_q]}, \tag{17}$$

$$\xi_{a_2 \ldots a_q} = \frac{1}{D - r + 1} \nabla_b h^b{}_{a_2 \ldots a_q}. \tag{18}$$

KY and CCKY tensors are related to each other through the Hodge duality: the Hodge dual of a KY form is a CCKY tensor, and vice versa. It is easy to check that if $k_{a_1 \ldots a_q}$ is a Killing-Yano tensor, then

$$K_{ab} = k_{a a_2 \ldots a_q} k_b{}^{a_2 \ldots a_q} \tag{19}$$

is a Killing tensor. We shall use the following schematic notation for this operation $\mathbf{K} = \mathbf{k} \cdot \mathbf{k}$.

CCKY tensors possess the following remarkable property: *An external product of two CCKY tensors is again a CCKY tensor.* This property was at first proved in the tensorial form in [16]. Slightly later a simple direct proof of this result was obtained in [17] using the formalism of differential forms.

## 5.2 Killing-Yano Tower

Let us return to our main 'hero'—Principal Conformal Killing-Yano tensor. This is a special case of CCKY tensor. Its additional properties are that it has tensor rank 2 and it is non-degenerate. Namely these properties make it so useful. If the spacetime admits a PCKYT **h** it also has a whole set of other objects, which we call *Killing-Yano tower*. First of all one can define a set of external powers of **h**

$$\boldsymbol{h}^{\wedge(j)} = \underbrace{\boldsymbol{h} \wedge \ldots \wedge \boldsymbol{h}}_{j \text{ times}} . \tag{20}$$

All of these objects are CCKY tensors of different matrix rank, starting from 2 for $j = 1$ till $2(n-1)$ for $j = n-1$. Taking Hodge dual tensors for each of $\boldsymbol{h}^{\wedge(j)}$ with $j = 1, \ldots, n-1$ one obtains a set $n-1$ Killing-Yano tensors $\boldsymbol{k}^{(j)}$, and their squares

$$\mathbf{K}^{(j)} = \mathbf{k}^{(j)} \cdot \mathbf{k}^{(j)} \tag{21}$$

determine $n-1$ Killing tensors of the rank 2.

For $j > n$ $\boldsymbol{h}^{\wedge(j)} = 0$. The case $j = n$ is special. In the even dimensional case for $j = n$ one obtains an absolutely antisymmetric object. Its Hodge dual is a scalar. For the odd dimensional case and $j = n$, the Hodge dual object is a Killing vector. One can also show that action of the Killing tensors $\mathbf{K}^{(j)}$ on the primary Killing vector gives a new independent (secondary) Killing vectors.

As a result of this construction starting with the PCKYT **h** one obtains $(n-1)$ second rank Killing tensors and $(n+\varepsilon)$ Killing vectors. This, together with metric gives $D = 2n + \varepsilon$ integrals of motion. Additional check shows that they are independent and in involution. This proves that the geodesic motion in the spacetime with PCKYT (and hence in the most general Kerr-NUT-ADS metric) is completely integrable.

Complete integrability of physically interesting finite dimensional dynamical systems is quite rare property. Examples of such systems include: motion in Euclidean space under central potential; motion in the two Newtonian fixed centers; geodesics on an ellipsoid; Jacobi motion of a rigid body about a fixed point (Euler, Lagrange and Kowalevski); Neumann model (for more details and examples see e.g. the monograph [18]). Characteristic property of a completely integrable dynamical system is that its trajectories are regular and can be used to construct regular foliations of the phase space. The opposite case are chaotic dynamical systems. It should be emphasized that the study of geodesic motion in higher dimensional black holes provides one with new wide class completely integrable dynamical systems, which might have interesting applications both in the theoretical and mathematical physics.

# 6 Separation of Variables

## 6.1 General Remarks

The complete integrability of geodesic equations in the higher dimensional black holes is closely related to the property of the *complete separation of variables* in some field equations in the same spacetime.

Consider a Hamiltonian $H(\mathbf{p}, \mathbf{q})$, where $\mathbf{p} = (p_1, \ldots, p_m)$ and $\mathbf{q} = (q^1, \ldots, q^m)$. The Hamilton-Jacobi equation for this Hamiltonian is a first order partial differential equation for function $S(\mathbf{q})$ of the form

$$H(S_{,\mathbf{q}}, \mathbf{q}) = 0. \tag{22}$$

Here $S_{,\mathbf{q}} = (S_{,q^1}, \ldots, S_{,q^m})$. Suppose $q^1$ and $S_{,q^1}$ enter this equation only in a special combination $\Phi_1(S_{,q^1}, q^1)$. Then the variable $q^1$ can be separated and a solution $S$ can be written in the form

$$S = S_1(q^1, C_1) + S'(q^2, \ldots, q^m), \quad \Phi_1(S_{,q^1}, q^1) = C_1, \tag{23}$$

and the new function $S'(q^2, \ldots, q^m)$ obeys a reduced Hamilton-Jacobi

$$H_1(S'_{,q^2}, \ldots, q^2, \ldots) = 0. \tag{24}$$

*Complete separation of variables* implies that the solution $S(\mathbf{q})$ can be written in the form

$$S = S_1(q^1, C_1) + S_2(q^2, C_1, C_2) + \ldots + S_m(q^m, C_1, \ldots, C_m). \tag{25}$$

The constants $C_i$ generate first integrals on the phase space. When these integrals are independent and in involution, the system is integrable in the Liouville sense (see e.g. [19]).

## 6.2 Complete Separation of Variables in Kerr-NUT-ADS Spacetime

Complete integrability of geodesic equations in the four-dimensional Kerr metric was discovered by Carter [20] who succeeded to separate variables in the corresponding Hamilton-Jacobi equation. Similar approach does work also in five dimensional case for the Myers-Perry metric, written in a similar Boyer-Lindquist coordinates [21]. However numerous attempts to separate variable in the higher dimensional Myers-Perry metric were not successful, except some special cases when additional restrictions were imposed on the rotating parameters.

Nevertheless, the general Kerr-NUT-ADS spacetime with arbitrary number of dimensions allows complete separation of variables for main field equations. The reason is the following. The very property of the separation of variables implies that there exist such a special coordinate system, in which this property is valid. For example, the Boyer-Lindquist in the Myers-Perry spacetime with $D > 5$ are not good for this purpose. The separation is possible only if one makes additional restrictions of the parameters of the solution. The existence coordinates in which the separability takes place in the Kerr-NUT-ADS in a general case is connected with presence of the Principal Conformal Killing-Yano tensor.

Let us describe these special coordinates. We already mention that eigen-values $x_i$, determined by Eq. (11), are independent, and they can be used as $n$ coordinates (at least in some spacetime domain). Moreover the spacetime with PCKYT has $(n+\varepsilon)$ independent **commuting** Killing vectors $\xi_{(j)}$, $j = 0, \dots, n-1+\varepsilon$. Consider integral lines of these vector fields

$$\frac{dy^a}{d\psi_j} = \xi^a_{(j)}. \tag{26}$$

It can be shown that the set of $(D = 2n + \varepsilon)-$quantities $(x_i, \psi_j)$ can be used as coordinates. Namely in these coordinates the complete separation of variables takes place. This was first demonstrated for Klein-Gordon and Hamilton-Jacobi equations in the higher dimensional Kerr-NUT-ADS in [22]. The massive Dirac equation has a similar property [23]. In fact these three different types of equations are closely related to on another. The Hamilton-Jacobi equation for $S$ can be obtained as the eikonal equation in the lowest order of WKB approximation, by substituting $\varphi \sim \exp(iS)$ into the Klein-Gordon equation

$$(\Box - m^2)\varphi = 0. \tag{27}$$

On the other hand, the massive Dirac equation is just 'a square root' of the Klein-Gordon equation.

## 6.3 *Complete Integrability and Separation of Variables in Weakly Charged Black Holes*

Till now we discussed black hole solutions of the Einstein equation (1). It is interesting that their nice properties still remain valid for a wider class of the black holes which are slightly charged. We assume that the cosmological constant vanishes, so that a Killing vector $\xi^a$ field obeys the equation

$$\Box\xi_a = 0. \tag{28}$$

This equation coincides with the Maxwell equation for the electromagnetic field potential $A_a$ in the Lorentz gauge $A^a{}_{;a} = 0$. Hence, one can consider $\xi^a$ as a test electromagnetic field on a given Kerr-NUT background and include the interaction of charged particles with it. For the primary Killing vector $\xi_{(0)}$ such a system describes a weakly charged black hole. The secondary fields can be used to describe weakly magnetized ones.

It is interesting that equations of motion of charged particles in weakly charged higher dimensional black holes are completely integrable and the corresponding Hamilton-Jacobi and Klein-Gordon equations are completely separable [24]. Moreover, the complete separability also takes place for charged Dirac equations in a weakly charged black hole [25].

In these notes I focused on the remarkable properties of higher dimensional rotating black holes with the spherical topology of the horizon. It was demonstrated that many of their properties are quite similar to the properties of their 4 dimensional Kerr-NUT-ADS 'cousins'. The reason of this is the existence of the Principal Conformal Killing-Yano tensor, which determines quite 'rigid' structure of the solutions and serves as a 'seed' generating their explicit and hidden symmetries. Many recent interesting generalization of this approach and its application is discussed in the review article of Marco Cariglia included in this volume. It also contains plenty of important references. I need also to mention three general reviews on this subject [26–28] which contain a lot of additional information and references.

**Acknowledgments** The author thanks David Kubiznak for useful remarks. The author thanks the Natural Sciences and Engineering Research Council of Canada and the Killam Trust for the financial support. He also appreciate the financial support from CNRS and the hospitality of the University of Tours, where this chapter was written.

## References

1. Emparan, R., Reall, S.: A rotating black ring solution in five dimensions. Phys. Rev. Lett. **88**, 101101 (2002). doi:10.1103/PhysRevLett.88.101101
2. Emparan, R., Reall, S.: Black holes in higher dimensions. Living Rev. Relativ. **11**(6), lrr-2008-6 (2008). http://www.livingreviews.org/lrr-2008-6
3. Newman, E., Tamburino, L., Unti, T.: Empty-space generalization of the Schwarzschild metric. J. Math. Phys. **4**, 915 (1963). doi:10.1063/1.1704018
4. Bonnor, W.B.: Physical interpretation of vacuum solutions of Einstein's equations. Part I. Time-independent solutions. Gen. Relativ. Gravit. **24**, 551 (1992). doi:10.1007/BF00760137
5. Carter, B.: Hamilton-Jacobi and Schrödinger separable solutions of Einstein's equations. Commun. Math. Phys. **10**, 280 (1968)
6. Carter, B., A new family of Einstein spaces, Physics Letters A, **26**, 399 (1968). doi:10.1016/0375-9601(68)90240-5
7. Tangherlini, F.R.: The flatness problem and the pulsating universe. Nuovo Cimento B **108**(1253), 1995 (1993). doi:10.1007/BF02741278. Erratum: ibid. 110, 1505

8. Myers, R.C., Perry, M.J.: Black holes in higher dimensional space-times. Ann. Phys. (N.Y.) **172**, 304 (1986). doi:10.1016/0003-4916(86)90186-7
9. Chen, W., Lü, H., Pope, C.N.: General Kerr-NUT-Ads metrics in all dimensions. Class. Quantum Grav. **23**, 5323 (2006). doi:10.1088/0264-9381/23/17/013
10. Frolov, V.P., Kubiznák, D.: Hidden symmetries of higher-dimensional rotating black holes. Phys. Rev. Lett. **98**, 011101 (2007). doi:10.1103/PhysRevLett.98.011101
11. Kubizňák, D., Frolov, V.P.: The hidden symmetry of higher dimensional kerr-nut-ads spacetimes. Class. Quantum Grav. **24**, F1 (2007). doi:10.1088/0264-9381/24/3/F01
12. Houri, T., Oota, T., Yasui, Y.: Closed conformal Killing-Yano tensor and Kerr-NUT-de Sitter space-time uniqueness. Phys. Lett. B **656**, 214 (2007). doi:10.1016/j.physletb.2007.09.034
13. Krtouš, P., Frolov, V.P., Kubizňák, D.: Hidden symmetries of higher-dimensional black holes and uniqueness of the Kerr-NUT-(A)ds spacetime. Phys. Rev. D **78**, 064022 (2008). doi:10.1103/PhysRevD.78.064022
14. Houri, T., Oota, T., Yasui, Y.: Closed conformal Killing-Yano tensor and the uniqueness of generalized Kerr-Nut-de Sitter spacetime. Class. Quantum Grav. **26**, 045015 (2009). doi:10.1088/0264-9381/26/4/045015
15. Page, D.N., Kubizňák, D., Vasudevan, M., Krtouš, P.: Complete integrability of geodesic motion in general higher-dimensional rotating black-hole spacetimes. Phys. Rev. Lett. **98**, 061102 (2007). doi:10.1103/PhysRevLett.98.061102
16. Krtouš, P., Kubizňák, D., Page, D.N., Frolov, V.P.: Killing-Yano tensors, rank-2 Killing tensors, and conserved quantities in higher dimensions. J. High Energy Phys. **2007**(02), 004 (2007). doi:10.1088/1126-6708/2007/02/004
17. Frolov, V.P.: Hidden symmetries of higher-dimensional black hole spacetimes. Prog. Theor. Phys. Suppl. **172**, 210 (2008). doi:10.1143/PTPS.172.210
18. Babelon, O., Bernard, D., Talon, M.: Introduction to Classical Integrable Systems. Cambridge Monographs on Mathematical Physics. Cambridge University Press, Cambridge (2003)
19. Arnold, V.I.: Mathematical Methods of Classical Mechanics, Graduate Texts in Methematics, vol. 60. Springer, New York (1989)
20. Carter, B.: Global structure of the Kerr family of gravitational fields. Phys. Rev. **174**, 1559 (1968). doi:10.1103/PhysRev.174.1559
21. Frolov, V.P., Stojkovic, D.: Quantum radiation from a 5-dimensional rotating black hole. Phys. Rev. D **67**, 084004 (2003). doi:10.1103/PhysRevD.67.084004 (2 in 1, see FS2: Phys. Rev. D 68, 064011)
22. Frolov, V.P., Krtouš, P., Kubizňák, D.: Separability of Hamilton-Jacobi and Klein-Gordon equations in general Kerr-NUT-Ads spacetimes. J. High Energy Phys. **2007**(02), 005 (2007). doi:10.1088/1126-6708/2007/02/005
23. Oota, T., Yasui, Y.: Separability of Dirac equation in higher dimensional Kerr-NUT-de Sitter spacetime. Phys. Lett. B **659**, 688 (2008). doi:10.1016/j.physletb.2007.11.057
24. Frolov, P., Krtouš, P.: Charged particle in higher dimensional weakly charged rotating black hole spacetime. Phys. Rev. D **83**, 024016 (2011). doi:10.1103/PhysRevD.83.024016
25. Cariglia, M., Frolov, V.P., Krtouš, P., Kubizňák, D.: Electron in higher-dimensional weakly charged rotating black hole spacetimes, ArXiv e-prints. arXiv:1211.4631 [gr-qc] (2012)
26. Frolov, V.P., Kubizňák, D.: Higher-dimensional black holes: hidden symmetries and separation of variables. Class. Quantum Grav. **25**, 154005 (2008). doi:10.1088/0264-9381/25/15/154005
27. Cariglia, M., Krtouš, P., Kubizňák, D.: Hidden symmetries and integrability in higher dimensional rotating black hole spacetimes. Fortschr. Phys. **60**, 947 (2012). doi:10.1002/prop.201200005
28. Yasui, Y., Houri, T.: Chapter 5. Hidden symmetry and exact solutions in Einstein gravity. Prog. Theor. Phys. Suppl. **189**, 126 (2011). doi:10.1143/PTPS.189.126

# Part III
# Cosmology and Quantum Gravity

# Cosmological Models and Stability

**Lars Andersson**

*I would already have concluded my researches about world harmony, had not Tycho's astronomy so shackled me that I nearly went out of my mind.*

Johannes Kepler
Letter to Herwart, quoted in [1, p. 127]

**Abstract** Principles in the form of heuristic guidelines or generally accepted dogma play an important role in the development of physical theories. In particular, philosophical considerations and principles figure prominently in the work of Albert Einstein. As mentioned in the talk by Jiří Bičák at this conference, Einstein formulated the equivalence principle, an essential step on the road to general relativity, during his time in Prague 1911–1912. In this talk, I would like to discuss some aspects of cosmological models. As cosmology is an area of physics where "principles" such as the "cosmological principle" or the "Copernican principle" play a prominent role in motivating the class of models which form part of the current standard model, I will start by comparing the role of the equivalence principle to that of the principles used in cosmology. I will then briefly describe the standard model of cosmology to give a perspective on some mathematical problems and conjectures on cosmological models, which are discussed in the later part of this paper.

## 1 Introduction

As stated by Einstein in his paper from 1912 [2], submitted just before his departure from Prague, the equivalence principle is "*eine Natürliche Extrapolation einer der allgemeinsten Erfahrungssätze der Physik*",[1] and can consequently be claimed to

[1] "A natural extrapolation of one of the most general empirical propositions of physics"

L. Andersson (✉)
Albert Einstein Institute, Am Mühlenberg 1, 14476 Potsdam, Germany
e-mail: laan@aei.mpg.de

J. Bičák and T. Ledvinka (eds.), *General Relativity, Cosmology and Astrophysics*, Fundamental Theories of Physics 177, DOI: 10.1007/978-3-319-06349-2_14,

© Springer International Publishing Switzerland 2014

be exactly valid on all scales. Since the equivalence principle is compatible with Einstein's relativity principle of 1905 only in the limit of constant gravitational potential, accepting the principle of equivalence meant that a new foundation for the theory of gravitation must be sought. The challenge of doing so, which Einstein in his 1912 paper poses to his colleagues: *"Ich möchte alle Fachgenossen bitten, sich an diesem wichtigen Problem zu versuchen!"*, is one that he himself devoted the coming years to, finally arriving at the 1915 theory of general relativity.

General Relativity describes the universe as a 4-manifold $\mathbf{M}$ with a metric $\mathbf{g}_{\alpha\beta}$ of Lorentzian signature. The Einstein equations,

$$\mathbf{R}_{\alpha\beta} - \frac{1}{2}\mathbf{R}\mathbf{g}_{\alpha\beta} + \Lambda \mathbf{g}_{\alpha\beta} = 8\pi G \mathbf{T}_{\alpha\beta}\,, \tag{1}$$

originally given in [3], relate the geometry of spacetime $(\mathbf{M}, \mathbf{g}_{\alpha\beta})$ to matter fields with energy-momentum tensor $\mathbf{T}_{\alpha\beta}$. By the correspondence principle, the stress energy tensor $\mathbf{T}_{\alpha\beta}$ should correspond to the stress energy tensor of a special relativistic matter model, and in particular be divergence free. For "ordinary matter" one expects $\mathbf{T}_{\alpha\beta}$ to satisfy energy conditions such as the dominant energy conditon. Here I have included the "cosmological constant term" $\Lambda \mathbf{g}_{\alpha\beta}$ in (1), which was not present in the equations given in [3]. The left hand side of (1), where $\mathbf{R}_{\alpha\beta}$ is the Ricci tensor, $\mathbf{R}$ is the Ricci scalar and $\Lambda$ is a constant, is the most general covariant tensor expression of vanishing divergence, depending on $\mathbf{g}_{\alpha\beta}$ and its derivatives up to second order, and linear in second derivatives. Further, its left hand side is the most general second order Euler-Lagrange equation, derived by varying a covariant Lagrange density defined in $\mathbf{g}_{\alpha\beta}$ and its first two derivatives, see [4, 5] and references therein. The covariance of the equations of general relativity under spacetime diffeomorphisms, makes the theory compatible with the strong version of the equivalence principle.

Since it can be claimed to be exactly valid, the equivalence principle is subject to empirical tests and there is a long history of experiments testing various versions of the (weak or strong) equivalence principles, see e.g. [6], see also [7] in this volume. Until the present, the equivalence principle has survived all experimental tests, and an experiment clearly demonstrating a deviation from the predictions based on the equivalence principle would necessitate a revision of the foundations of modern physics.

The arguments of the physicist and philosopher Ernst Mach played an important role in the development of Einstein's ideas leading up to general relativity, including the formulation of the equivalence principle. The fact that in general relativity, matter influences the motion of test particles via its effect on spacetime curvature means that in contrast to Newtonian gravity, the "action at a distance" which was criticized by Mach is not present in general relativity, which hence agrees with the guiding idea which Einstein referred to as "Mach's principle", i.e. loosely speaking the idea that the distribution of matter in the universe determines local frames of inertia, see [8], see also [9]. The role of Mach's principle in the context of cosmology is discussed in [10]. This played a central role in Einstein's development of general relativity, and also in his discussion of general relativistic cosmology, but it appears difficult to formulate

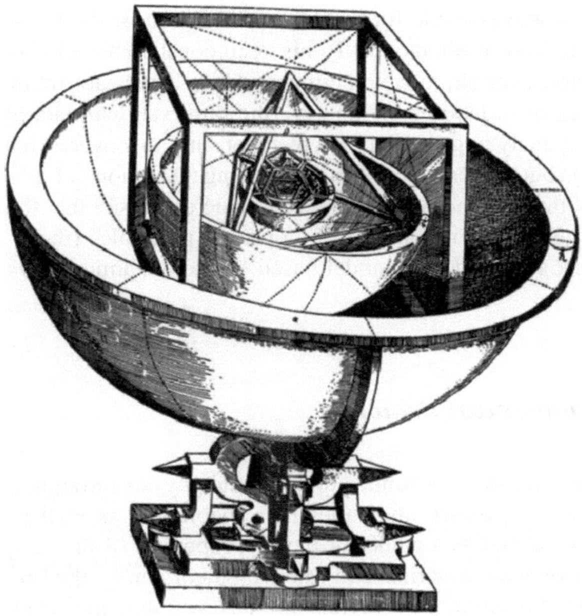

**Fig. 1** Kepler's model of the solar system based on platonic solids, from Mysterium Cosmographicum (1596)

experimentally testable consequences, cf. [11], although Mach's principle has of course been brought up in connection with "Newton's bucket" and frame dragging. The book [12] gives an excellent overview of issues related to Mach's principle. However, the principles which are most relevant for the present discussion are the hierarchy of "cosmological principles", for example the cosmological principle of Einstein and the perfect cosmological principle of Bondi, Gold and Hoyle. See [13, Sect. 2.1] for an overview of the cosmological principles. These principles play a role which is fundamentally different from that of the equivalence principle, in the sense that they do not make predictions which are expected to be exactly true at all scales. At best, they can be viewed as simplifying assumptions that enable one to construct testable physical models.

The work of Kepler, who is perhaps more intimately connected with Prague than Einstein, provides an interesting illustration of the relationship between theoretical principle and observation. In the time of Kepler, the world-model of Copernicus had placed the sun at the center of the universe and described the planets as moving on circular orbits around it. Not long before his move to Prague in 1600, Kepler believed himself to have completed the Copernican world-model based on the mathematical perfection of circles, by adding to it an element of equal perfection and beauty, namely the geometry of the Platonic solids, which according to Kepler's expectations would determine the sizes of the planetary orbits (Fig. 1).

Fortunately, it was possible for Kepler to use Tycho Brahe's observational data to test the predictions of his model. To his deep consternation Kepler realized that the planets do not, after all, move on circular orbits. The beautiful principles which had inspired Kepler to laboriously analyze the observational data of Tycho had to be discarded. In analyzing the data, Kepler not only discovered his three laws of planetary motion but also came close to introducing the notion of force which became fully clear only through the work of Newton. One could say that through the work of Kepler and later Newton, one set of "a priori" principles (those of Copernicus and Kepler) were replaced by a model based on the dynamical laws of Newtonian gravity.

### *1.1 The Cosmological Principle*

Although Newtonian ideas continued to dominate physics throughout the 19th century, there were well known anomalies of a theoretical as well as observational nature, and these served as a guide for the developments of the early 20th century. The conflict between the covariance of Maxwell theory under the Lorentz group and the more restricted invariance properties of the Newtonian laws led to the introduction of special relativity. Similarly, as discussed above, the incompatibility of special relativity and gravitation led to the development of general relativity. The explanation of the anomalous precession of the perihelion of Mercury [14][2] by general relativity was, together with its new prediction for the deflection of light by the sun, confirmed by subsequent observations [17], were among the factors which led to its rapid acceptance.

Among the main paradoxes of Newtonian physics and world view in applications to cosmology were Olbers' paradox and the incompatibility of Newtonian gravity with infinitely extended homogenous matter distributions, which had prevented the construction of a cosmological model consistent with Newtonian ideas. This latter fact, which had been elucidated by von Seeliger and others, see [18] for discussion and references, played an important role in Einstein's reasoning about cosmological models in his 1917 paper [19], in particular in motivating the introduction of the cosmological constant in that paper.

As has already been mentioned, the philosophy of Mach, albeit firmly based in Newtonian physics, was an important source of inspiration for Einstein. However, incorporating Machian ideas in a general relativistic cosmology presented serious difficulties. After some early attempts had been discarded, Einstein in [19] adopted a spatially homogenous model of the universe as a means of making a general relativistic cosmology compatible with Machian ideas. Introducing a "cosmological constant" term $\Lambda \mathbf{g}_{\alpha\beta}$ in the field equation of general relativity, which Einstein first

---

[2] From the perspective of the current situation in physics, it is amusing to recall that attempts had been made in the 19th century to explain the observed precession of Mercury both by dark matter [15] (the planet Vulcan hypothesis) as well as modifications of gravity [16].

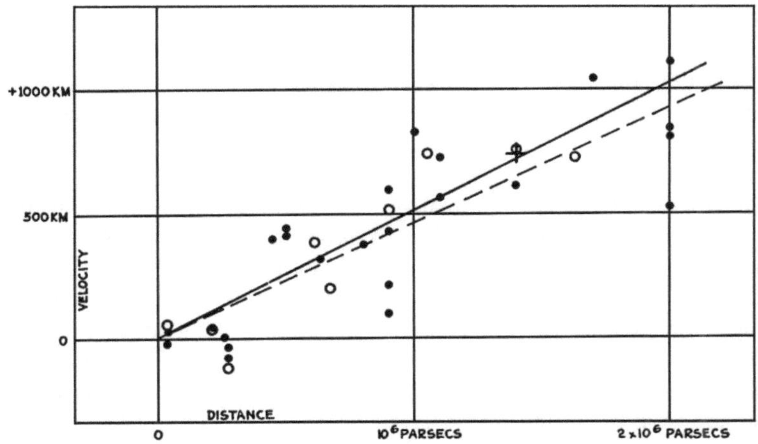


**Fig. 2** Hubble's original 1929 graph [21]

motivated through a discussion of homogenous matter distributions in Newtonian gravity, and assuming that there is a family of observers who see the same matter density everywhere, led to a static universe filled with a homogenous and isotropic matter distribution. The spacetime of the Einstein model is a Lorentzian cylinder. The line element takes, up to a rescaling, the form

$$ds^2 = \mathbf{g}_{\alpha\beta} dx^\alpha dx^\beta = -dt^2 + g_{S^3} .$$

This give a solution to (1) with positive $\Lambda$, and with matter consisting of a pressureless fluid with everywhere constant energy density.

Shortly after Einstein's initial work on a static general relativistic cosmology, Friedmann [20] proposed a model of an expanding universe

$$ds^2 = -dt^2 + a^2(t) g_\kappa \tag{2}$$

where $a(t)$ is a scale factor, $g_\kappa$ for $\kappa = +1, 0, -1$ is the sperical, flat or hyperbolic metric. Line elements of the form (2) are also called Robertson-Walker line elements, see below. During the 1920s, Lemaître and Hubble showed, based on observational work of Slipher, Humason and others, that redshift increases with distance leading to the Hubble law, see Fig. 2, which fits with the expanding Friedmann models. In the context of the expanding Friedmann models, Olbers' paradox can be resolved. Expanding Friedmann models containing ordinary matter have $a \searrow 0$ at some time in the past, where spacetime curvature and matter densities diverge. These models led, via the work of Lemaître, Gamow, Hoyle and others, to the hot big bang model which is the basis for the cosmological models in use today.

Milne criticized the big bang models on the basis that they introduced an extraneous "cosmic time" and also that they lacked explanatory power (e.g. the sign of the spatial curvature is a priori undetermined). Instead, he proposed an extension of what he termed "Einstein's cosmological principle", to the effect that "The universe must appear the same to all observers" [22]. Milne added to this the postulate that observations are interpreted by each observer according to the principles of special relativity and argued that this "extended relativity principle" led to an essentially unique cosmological model.

The derivation of the general form of the line element compatible with the isotropy of the universe, and also with Einstein's cosmological principle in the sense discussed by Milne was given by Robertson [23] and Walker [24] around the same time, and found to be of the same form as that used by Friedmann and Lemaître in their cosmological models. As pointed out by Robertson [25], the general relativistic line element compatible with Milne's cosmology is a special case of (2), namely the empty $\kappa = -1$ universe, which is locally isometric to Minkowski space. This is therefore known as the Milne model.

It was a similar dissatisfaction with the lack of predictivity of general relativistic cosmology that led Bondi and Gold [26] and Hoyle [27] to introduce the "perfect cosmological principle", which is essentially a version of the postulate of Milne, but viewed from the perspective of general relativity. By allowing for creation of matter, they showed that it is possible to construct an expanding cosmological model satisfying this principle. However, the perfect cosmological principle tightly constrains the possible models of the universe and the resulting steady state model is considered to be incompatible with observations. The book of Kragh [28] contains an interesting discussion of the conflict between the steady state model and the now-standard "big bang" cosmology.

From the current perspective, it may be said that the introduction of what Milne called Einstein's cosmological principle led to a class of general relativistic cosmological models. By introducing a collection of perfect fluids, a much simplified version of the problem of cosmological modelling reduces to the problem of fitting a relatively small number of parameters to observational data, which could be said to put cosmology on a similar footing as high energy particle physics. Indeed, as mentioned by Peebles [29, Chap. 1], it was Weinberg [30] who introduced the notion, borrowed from high energy particle physics, of a "standard model" into cosmology.

At present, with the tremendous influx of data from observations of many different types and at many different wavelengths, including observations of the cosmic microwave background and galaxy surveys, it is often stated that we are entering an era of precision cosmology. However, the widening range of observational methods makes the process from observations to parameter estimation increasingly complex. In particular, the prominent role of simplifying assumptions or principles in the formulation of cosmological models and the model dependence in the analysis of astronomical data, makes it important to keep in mind the difference between a model which fits data to a high degree of precision and a model which accurately describes the actual universe [31].

## 2 Cosmological Models

For a Friedmann model, with line element of the form (2), the stress energy tensor has the form

$$\mathbf{T}_{\alpha\beta} = \rho \mathbf{u}_\alpha \mathbf{u}_\beta + p(\mathbf{g}_{\alpha\beta} + \mathbf{u}_\alpha \mathbf{u}_\beta)\,,$$

which is compatible with perfect fluid matter. Here $u^\alpha$ is the unit timelike normal to the $t$ level sets, which in the special case of the Friedmann model coincides with the normalized 4-velocity of the fluid particles, $\rho$ is the energy density of the matter and $p$ is the pressure. We consider matter and radiation as described by a collection of fluids, indexed by $i$, with linear equations of state,

$$p_i = \omega_i \rho_i\,.$$

The Hubble constant (i.e. up to a constant factor the mean curvature of the $t$ level sets) is

$$H = \dot{a}/a$$

In the special case of a Friedmann model, the contribution of the curvature of the $t$ level sets in the Einstein equations can be described in terms of a fluid with equation of state $p = -\rho/3$, while the effect of the cosmological constant can be described by a fluid satisfying $p = -\rho$. Thus if we consider a simple model containing a fluid with pressure zero (dust), and with a cosmological constant $\Lambda$, this can be described by introducing the dimensionless density parameters

$$\Omega_m = \frac{8\pi}{3H^2}\rho_m, \qquad \text{“Matter”: } \omega = 0,$$

$$\Omega_\kappa = -\frac{\kappa}{a^2H^2}, \qquad \text{“Curvature”: } \omega = -1/3,$$

$$\Omega_\Lambda = \frac{8\pi}{3H^2}\rho_\Lambda, \qquad \text{“Vacuum”: } \omega = -1.$$

The model can be parametrized by the present values

$$\Omega_{m0}, \Omega_{\kappa 0}, \Omega_{\Lambda 0},$$

of the density parameters. The conservation of matter and equation of state implies that the fluid densities $\rho_i$ depend only on the scale factor

$$\rho_i \propto a^{-3(1+\omega_i)}. \tag{3}$$

The Hamiltonian constraint (i.e. the projection of the Einstein equations (1) on $u^\alpha$) takes the form

$$\Omega_m + \Omega_\kappa + \Omega_\Lambda = 1, \tag{4}$$

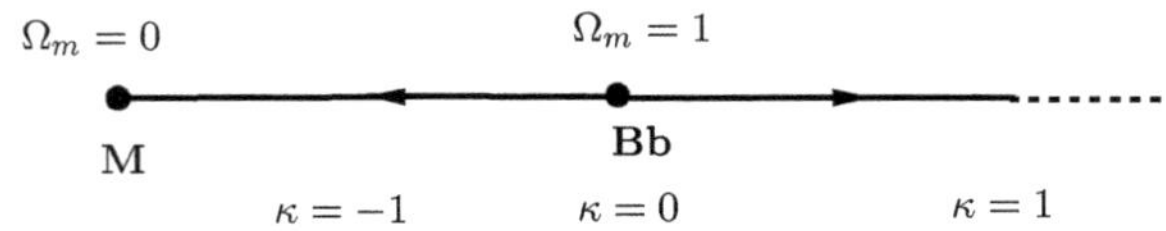


**Fig. 3** The dynamics of Friedmann dust models for $\Lambda = 0$

which, using (3), can be written as

$$\frac{H^2}{H_0^2} = \Omega_{0m}\left(\frac{a_0}{a}\right)^3 + \Omega_{0\Lambda} + \Omega_{0\kappa}\left(\frac{a_0}{a}\right)^2. \tag{5}$$

Here $H_0, a_0$ are the present value of the Hubble constant and of the scale factor respectively. Due to the uncertainty in the value of $H_0$, it is usually given in terms of a dimensionless parameter $h$ as

$$H_0 = 100\,h\mathrm{km\ s^{-1}Mpc^{-1}}.$$

Equation (5) can be integrated to relate observable quantities, e.g. redshift and luminosity distance, for given values of the parameters $H_0, \Omega_{m0}, \Omega_{\kappa 0}, \Omega_{\Lambda 0}$.

It is convenient to study the global behavior of Friedmann models in terms of the dimensionless density parameters. This analysis is explained in [32, Chap. 2], see also [33, 34]. Due to the Hamiltonian constraint (4), we have $\Omega_\kappa = 1 - \Omega_m - \Omega_\Lambda$.

The fixed points of the dynamical system in the $(\Omega_m, \Omega_\Lambda)$ plane are the Einstein-de Sitter big-bang model **Bb** $= (1, 0)$ and the spatially flat de Sitter model **dS** $= (0, 1)$, as well as the empty $\kappa = -1$ Milne model **M** $= (0, 0)$. One finds that **Bb** is a source and **dS** is a sink, while **M** is a saddle point. The static Einstein universe has $H = 0$, so the dimensionless parameters $\Omega_m$ and $\Omega_\Lambda$ are ill-defined, but this point may be represented in an extended phase space as $\mathscr{E} = (\infty, \infty)$. This point is unstable, but is connected to the source **Bb** by an exceptional trajectory, which separates the models which recollapse from those which expand forever.

Restricting to $\Lambda = 0$, the only fixed points are **Bb** and **M**, with **Bb** a source and **M** a sink, see Fig. 3. The unstable Einstein-de Sitter universe **Bb** has slow volume growth $a \sim t^{2/3}$, while the stable Milne universe **M** has volume growth $a \sim t$. In fact, this growth rate is maximal among $\Lambda = 0$ models. This indicates that rapid volume growth goes together with stability.

Now we can give an extremely simplified description of the current situation in cosmology by saying that the laws of general relativity together with the cosmological principle and observations leads to the "standard model" with the cosmological parameters

$$\Omega_{\kappa 0} \sim 0, \quad \Omega_{m0} \sim 0.3, \quad \Omega_{\Lambda 0} \sim 0.7, \quad h \sim 0.7.$$

The standard model is a big bang model. There is an initial singularity, $a \searrow 0$ as $t \searrow 0$ and the universe expands indefinitely to the future, $a \nearrow \infty$ as $t \nearrow \infty$. The

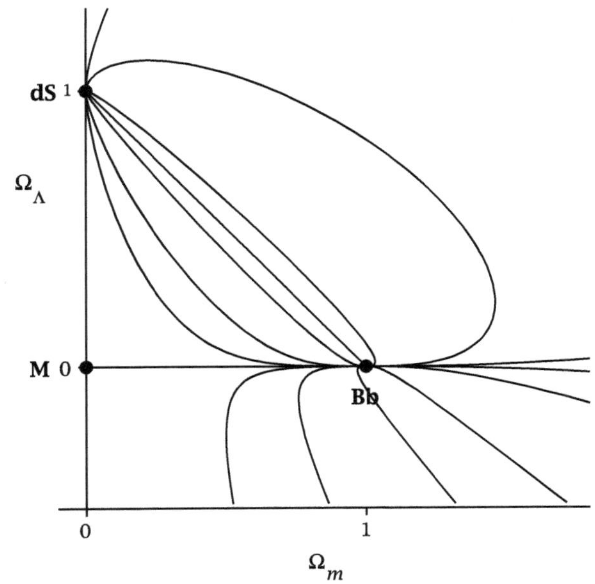


**Fig. 4** This figure shows some orbits for Friedmann cosmologies with dust and dark energy ($\Lambda$) in the ($\Omega_m$, $\Omega_\Lambda$) plane. The Einstein-de Sitter point **Bb** $= (1, 0)$ is a source, the Milne point **M** $= (0, 0)$ is a saddle node, and the de Sitter point **dS** $= (0, 1)$ is a sink. See [33] for background

model predicts a hot big bang, which leads to the prediction of cosmic background radiation [35, 36]. The observation of a highly homogenous cosmic background radiation with a spectrum close to that of a black body is a major success of the big bang models of cosmology.

Most of the energy density in the standard model consists at present of as yet unknown "dark matter" (accounting for approximately 85 % of the matter density) and "dark energy" in the form of the cosmological constant. Dark matter, which for a long time has been broadly accepted in astronomy and cosmology, cf. [37], is distinguished from dark energy by the fact that its existence is motivated by studies of the dynamics of galaxy clusters and galactic rotation curves, which are independent of the Friedmann model which forms the basis of the standard model in cosmology. On the other hand, the cosmological constant was deemed unacceptable on philosophical grounds and entered the standard model fairly recently, shortly before the year 2000; the effects of dark energy being seen only indirectly via cosmological models and eg. studies of structure formation in the universe (Fig. 4).

The acceptance of $\Lambda$ came about only after the observation of the dimming of type Ia supernovae. The observations are interpreted as saying that the rate of expansion is accelerated, i.e. $\ddot{a} > 0$, which is incompatible with a Friedmann model filled with ordinary matter and $\Lambda = 0$. Figure 5 shows the supernova data compared to the standard model and Einstein-de Sitter. The horizontal axis is the Milne model ($\Omega_m = \Omega_\Lambda = 0$).

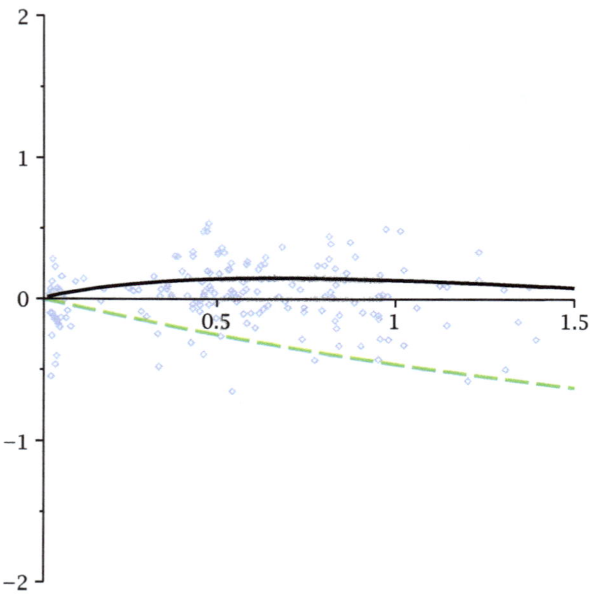


**Fig. 5** Magnitude residual for SNe Ia Gold data [38] (dots) relative to the Milne model, plotted against redshift $z$. The *black, solid curve* is the standard model, while the *green, dashed curve* is the Einstein-de Sitter model. The *horizontal* axis is the Milne model

## 2.1 Cosmological Problems

One of the important arguments against introducing the cosmological constant (apart from the difficulty of explaining the value $\Lambda$ which appears motivated by cosmology from the point of view of particle physics) has been the *coincidence problem*, which might also be termed the "why now" problem. Figure 6 shows the time evolution of the dark energy density $\Omega_\Lambda$. We see that it is only close to the present epoch that $\Omega_\Lambda$ becomes significant, and in the later universe it will dominate the dynamics. Due to the different scaling behavior of the matter and $\Lambda$ densities in view of (3), the fact that these are both of order unity at the present epoch is a coincidence that could be argued to be contrary to the idea that we are not "special observers". In contrast, in the Einstein-de Sitter model the matter density is time independent.

A related problem is the *flatness problem*. Roughly speaking, this is the question why $\Omega_\kappa \sim 0$ at present. In case $\Lambda = 0$ this can be seen to be problematic simply from Fig. 6. Since **Bb** is unstable, fine tuning of the initial conditions is required in order to have $\Omega_\kappa \sim 0$ at present. Lake [33] argues, using the presence of a conserved quantity for the dynamics in the $(\Omega_m, \Omega_\Lambda)$-plane, that fine tuning is not needed to have $\Omega_\kappa \sim 0$ throughout the history of the universe.

The universe is not exactly homogenous or isotropic; this holds at best in an approximate sense on sufficiently large scales. This raises the problem of whether

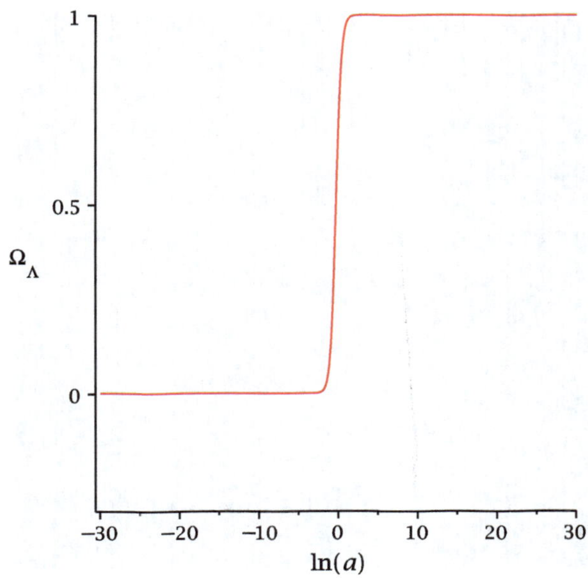


**Fig. 6** The time evolution of the dark energy density, see [39] for discussion

it is possible to determine from observations, which are necessarily restricted to our past light cone, to what extent, and at what scales, the assumption of homogeneity and isotropy is valid. A problem here is that local isotropy (i.e. isotropy around the world line of one observer) does not imply global homogeneity (Fig. 7).

The Ehlers-Gehren-Sachs theorem gives conditions under which it is possible to conclude from exact isotropy of the cosmic microwave background that the universe is exactly isotropic. However, this result can fail in several ways. For example, there are homogenous but non-isotropic models where the CMB is exactly isotropic at one instant in time. Extensions of the EGS theorem to situations where only approximate isotropy of the CMB holds are problematic, see [40–42] and references therein. This raises the problem of determining to what degree observations of the actual universe can be modelled and analyzed in the framework of Friedmann models (and perturbations thereof). One aspect of this problem is the question whether there is a scale at which (statistical) homogeneity and isotropy can be said to hold. Current estimates place this scale at approximately $150\,h^{-1}$ Mpc, see e.g. [43], see also [44]. However, recent observations indicate the existence of inhomogenous structures of a dimension which may be in conflict with isotropy at this scale, see [45]. It is conceivable that observations which extend to ever higher redshifts continue to yield evidence of structures in the universe of a size comparable to the homogeneity scale. Some aspects of inhomogeneity in cosmology were recently surveyed in a focus issue of CQG, see [46] and references therein.

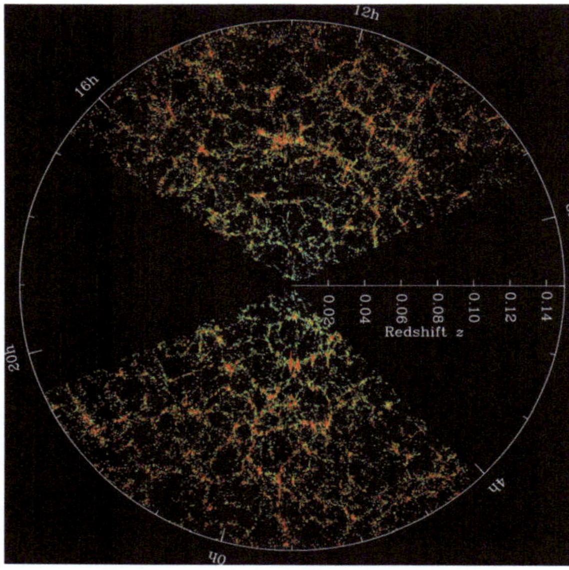


**Fig. 7** Sloan digital sky survey galaxy map, from www.sdss.org

The question of how the potential effects of large scale inhomogeneities on observations should be analyzed raises several important issues. Ellis has formulated the "fitting problem", see [47] and references therein, which asks about the effect of analyzing observations from an inhomogenous universe via a Friedmann model which is in some sense the "best fit" to the actual universe. The effect on observations of the fact that the model universe used to analyze data is only an approximation of the actual universe is sometimes referred to as "backreaction". An important question here is whether perturbation theory can be applied to take into account the deviation of the model from the actual universe. Kolb et al. have argued [48] that this analysis should take into account the peculiar velocities due to the different expansion rate in the model and the actual universe. Another effect of inhomogeneities which also sometimes is referred to as backreaction, is the *dynamical* effect of the inhomogeneities on the expansion of the universe. A possible approach is to use averaging [49] or coarse-graining [50] to derive a set of effective equations modelling the universe. In order to carry out such a scheme, one must introduce closure relations which allow one to extract an autonomous system. It is here worth mentioning the ideas on multi scale averaging, see e.g. [51, 52]. In particular, Wiltshire [53] argues that one should consider modifying the Copernican principle to take into account the idea that we reside in a gravitationally bound structure in a universe which has both bound systems and voids.

It is apparent that the matter distribution in the universe is "lumpy" due to the matter concentrations in stars, galaxies and other structures, and inhomogenous due to the presence of large scale voids and bound structures, and the effect of these must be taken into account when analyzing observations, see Clarkson et al. [54] for discussion. The optical properties of the universe are, in the Friedmann models which form the basis for the standard model of cosmology, calculated using the properties of a fluid which is used to approximate the actual matter distribution. Thus it is necessary to analyze whether the optical properties of a lumpy matter distribution differ in a significant way from the optical properties of a fluid. Light from distant stars passes through the gravitational wells of bound objects as well as voids on the way to the observer, and the effect of this process must be analyzed and compared to light passing through the fluid in a Friedmann model. This problem has been studied by among others Clifton et al. [55], see also [56]. In this context, we also mention the so-called swiss cheese models, in which one attempts to analyze the optical effect of voids and structure in the universe by introducing under-densities in a background Friedmann model, see e.g. [57] and references therein. The swiss cheese models generally suffer from the limitation that the over-all expansion of the model is determined by the chosen background Friedmann geometry.

In this situation one may contemplate introducing weaker cosmological principles, incorporating ideas of statistical homogeneity, or weakening the Copernican principle by restricting to matter bound observers as suggested by Wiltshire.

As we have seen, the standard cosmological model is not located at a fixed point for the dynamical system governing the evolution of the dimensionless parameters $\Omega_m$, $\Omega_\Lambda$, rather it is close to the spatially flat orbit connecting the source **Bb** to the sink **dS**. Further, in that orbit, $\Omega_m/\Omega_\Lambda$ takes on all positive real values. Thus, we as observers are not in an asymptotic regime, but rather, as mentioned above, at a special moment where $\Omega_m$ and $\Omega_\Lambda$ are both of order unity. Thus, from this point of view, we are neither in the "early universe" or the "late universe" and we cannot argue that our current universe is singled out as the asymptotic state of the evolution of the universe.

This makes the situation in cosmology rather different from the situation in many branches of physics where asymptotically stable objects are those which one expects to find in nature. As an example, the Kerr black hole solution is expected to be the unique stationary, asymptotically flat black hole spacetime. In order to establish the astrophysical significance of this solution, it is essential to prove that it is stable. This leads to the black hole stability problem, one of the central open problems in general relativity. The problem of determining from observations whether or not for example the supermassive black holes expected to be found at the center of most galaxies are Kerr black holes or not is being actively studied.

As was just mentioned, from the point of view of the current standard model in cosmology, questions about the asymptotics of cosmological models do not appear to be the right ones to ask. Nevertheless, such questions give rise to interesting mathematical problems which we shall discuss in the rest of this paper. The questions about the asymptotic behavior of cosmological models include the structure of the big-bang singularity and questions about the behavior in the expanding direction. In particular we can ask: What does an observer in the late universe see?

# 3 Asymptotics of Cosmological Models

In this section we will describe a scenario for the asymptotic future behavior of cosmological models with vanishing cosmological constant. Recall that the Milne model with line element

$$ds^2 = -dt^2 + t^2 g_{\mathbb{H}^3}$$

where $g_{\mathbb{H}^3}$ is the hyperbolic 3-metric with sectional curvature $-1$, is isometric to the flat interior of the lightcone in Minkowski space. The Milne universe may be viewed as the future of $O$, the origin in Minkowski space. This point represents the big bang singularity in the Milne universe and is in the past of all spacetime points (i.e. all observers). The cosmological time at a point $P$ is the proper time elapsed from the origin to $P$. The level surfaces of cosmological time are simply the hyperboloids. We next consider a flat, but non-isotropic model, which may be viewed as a deformation of Milne. Let $I$ be a spacelike interval in Minkowski space and consider the future of $I$. The resulting spacetime can be constructed by cutting the Milne spacetime by a timelike hyperplane through $O$ and gluing in a spacetime of the form $\mathbb{R}^{2+1} \times I$ with line element

$$-dt^2 + t^2 g_{\mathbb{H}^2} + dz^2 .$$

The deformed Milne spacetime has a big-bang singularity given by the interval $I$, and defining the cosmological time at $P$ as the maximal proper time of any past inextendible geodesic starting at $P$ the level sets of cosmological time are as in Fig. 8; it is flat and empty, but not homogenous and isotropic. Measuring the volume of co-moving regions in the deformed Milne universe we see that in the deformed regions, the volume of the cosmic time levels grows asymptotically as $t^{2/3}$, i.e. the growth rate of the Einstein-de Sitter universe, while in the undeformed regions, the growth rate is asymptotically as $t$. The behavior is similar for the level sets of the Hubble (mean curvature) time. On the other hand, asymptotically as $t \nearrow \infty$, the volume fraction in the undeformed region tends to 1, while in the asymptotic past (near the big bang) these regions have a negligible volume fraction.

More general flat spacetimes may be constructed as the future of sets (e.g. fractals) in Minkowski space, and quotients of these by the action of discrete groups of isometries. Flat, or more generally, constant curvature spacetimes are examples of $G$-structures and such spacetimes admitting compact Cauchy surfaces have been completely analyzed, starting with the work of Mess [58], see also [59], who analyzed the class of constant curvature $2+1$ dimensional spacetimes admitting a compact Cauchy surface. For example, one may show that the space of flat $2+1$ dimensional spacetimes with Cauchy surface of genus $g > 1$ is isomorphic to $\partial\mathscr{M} \times \mathscr{M}$, where $\mathscr{M}$ is Teichmuller space of surfaces of genus $g$ and $\partial\mathscr{M}$ is the Thurston boundary. The particular case of constant curvature spacetimes with compact Cauchy surface has been analyzed in [60]. In particular, it was shown there that such flat spacetimes can be globally foliated by Cauchy surfaces of constant mean curvature (i.e. constant Hubble time).

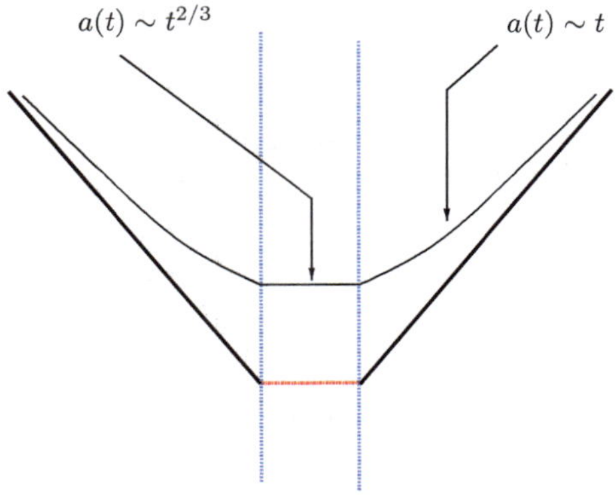


**Fig. 8** A flat cosmological spacetime not isometric to Milne. A level set of cosmological time $t$ is shown. The *vertical lines* indicate the flat wedge which has been glued in

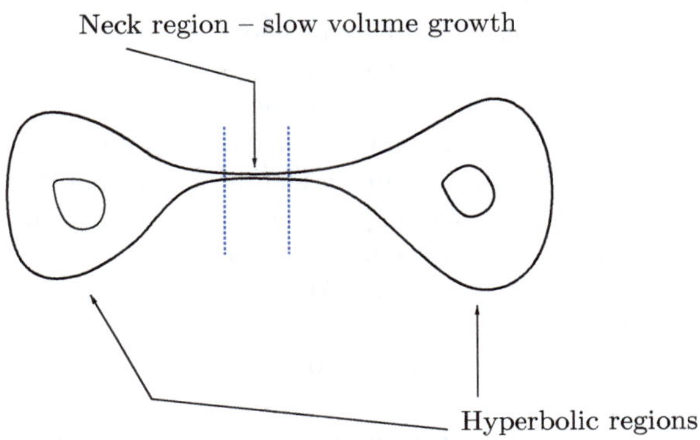


**Fig. 9** Qualitative shape of Hubble level set

The level sets of Hubble time can be related to the level sets of the cosmological time by an application of a maximum principle, and one may show that the volume growth of these level sets is comparable to that of the level sets of the cosmological time. This leads to a generalization of the statements made above for the simple deformed Milne universe, see [61].

In view of the above mentioned work, these generalized Milne spacetimes may have a very complex (e.g. fractal) big bang type initial singularity. In some cases their future asymptotics can be analyzed, see [62]. One finds that the level sets of Hubble time decompose into "neck regions" with slow volume growth, and "hyperbolic regions" with fast volume growth. The scale free geometry of these level sets may may be depicted as in Fig. 9.

In particular, one finds that in the asymptotically expanding direction, the volume fraction of asymptotically, hyperbolic (thick) regions dominate while the neck regions

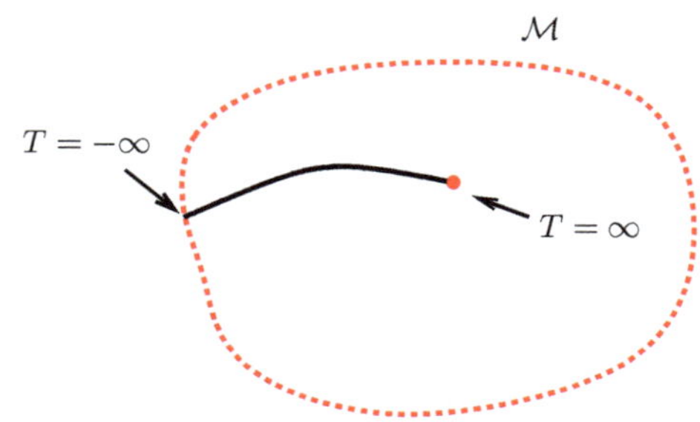


**Fig. 10** The Einstein flow in the $2 + 1$ dimensional case

(thin) become insignificant. Therefore, a "typical" (volume averaged) observer at late time lives in a thick region.

It is interesting to compare the relation between the thin and thick regions to the overdense and void regions in an inhomogenous universe containing matter, in particular in view of the fact that the thin regions have volume growth approximating that of Einstein-de Sitter universe which has critical matter density.

We now consider the generalization of the above picture to the case of general, inhomogenous universes. We start by noting that the Lorentzian Einstein equations define a flow on the space of (scale free) geometries. By analogy with the Ricci flow of Riemannian geometries, this may be termed the Einstein flow.

For simplicity, we consider spacetimes $(\mathbf{M}, \mathbf{g}_{ab})$ of dimension $D = d + 1$ which are vacuum, i.e. with

$$\mathbf{R}_{\alpha\beta} = 0\,.$$

Suppose $\mathbf{M}$ admits a foliation by Cauchy surfaces of constant mean curvature $H$. Introduce the dimensionless logarithmic constant mean curvature (Hubble) time $T = -\ln(H/H_0)$, and consider the evolution of the scale free geometry $[g] = H^2 g$. The Lorentzian Einstein equations define a flow $T \mapsto [g](T)$, on the space of scale free geometries. In particular, in the $2 + 1$ dimensional case, the Einstein equations correspond to a time dependent Hamiltonian system on Teichmüller space [63], and each universe corresponds to a curve connecting a point on the boundary of Teichmuller space to an interior point, see Fig. 10.

One arrives at the following heuristic scenario [64, 65]. Consider spacetimes with Cauchy surface $M$. The non-collapsing case corresponds to the case where $M$ has negative Yamabe type. For $T \nearrow \infty$, $(M, [g])$ decomposes into hyperbolic pieces and Seyfert fibered pieces, and this decomposition corresponds to a (weak) geometrization, cf. [65]. The Einstein flow in CMC time results in a thick/thin decomposition of $M$, where the thick (hyperbolic) pieces have full volume growth. As a consequence we have that in the far future, the hyperbolic pieces represent most of the volume of $M$, cf. Fig. 11. Proving statements along the lines described above appears to be very difficult, and one must therefore start by considering sub-problems.

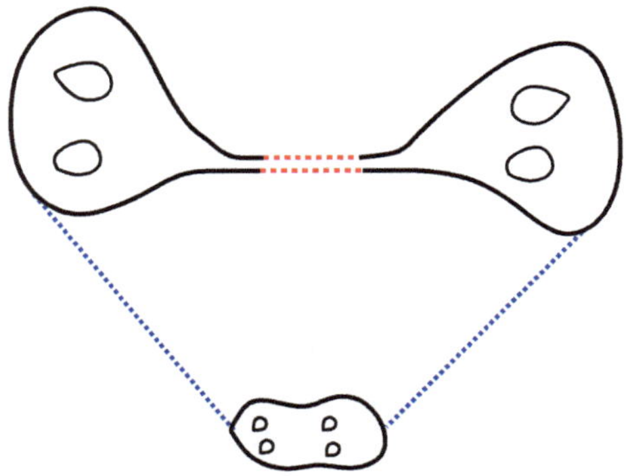

**Fig. 11** The collapse of necks in the Einstein flow

# 4 Results on Nonlinear Stability

To give some perspective on the nonlinear stability problems introduced above, we discuss some results on other stability problems in general relativity. These are organized according to the asymptotic model spacetime. The black hole stability problem, cf. [66] for discussion and references, is not mentioned here. In the following, we mention only the cases with conformally flat background spacetimes.

## *4.1 Minkowski*

First we consider the nonlinear stability of Minkowski space, i.e. $\mathbb{R}^4$ with line element

$$ds^2 = -dt^2 + dx^2 + dy^2 + dz^2.$$

The conformal type of Minkowski space is that of the Minkowski diamond, see Fig. 12. In this causal diagram, each interior point represents a 2-sphere.

Nonlinear stability holds, in the sense that for Cauchy data near Minkowski data, the maximal development is geodesically complete and asymptotically Minkowskian. A key fact is that radiation carries energy through the conformal boundary $\mathscr{I}$. Due to the fact that the nonlinearity in the Einstein equations is quadratic, it is necessary to exploit a cancellation in the equations in order to prove stability.

The first result in this direction is due to Friedrich [67], who proved that for data close to the data induced on a hyperboloid in Minkowski space, one has nonlinear stability to the future, and with suitable asymptotic regularity for the data, the maximal development has a regular $\mathscr{I}^+$ to the future of the initial slice. The full nonlinear stability result was proved by Christodoulou and Klainerman [68]. This work was extended to include the full peeling at $\mathscr{I}$ by Klainerman and Nicolo [69].

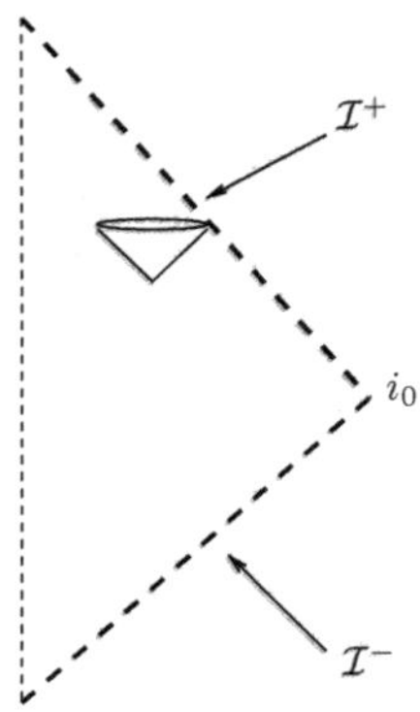


**Fig. 12** Conformal diagram of Minkowski space

A simpler proof of nonlinear stability, using wave coordinates (spacetime harmonic coordinates) gauge was given by Lindblad and Rodnianski [70]. Using both of these methods, the proof of nonlinear stability can be readily adapted to the Einstein-matter system, provided that the matter fields do not destroy the conformal properties of the Einstein equations. Examples include a massless scalar field, which was included in the work of Lindblad and Rodnianski, and a Maxwell field, see [71].

## *4.2 de Sitter*

Next we consider cosmological models with positive $\Lambda$. The canonical example is de Sitter space with line element

$$ds^2 = -dt^2 + \cosh^2(t) g_{S^3} .$$

This is conformal to a finite cylinder with spacelike conformal boundary, and hence one has future horizons and "locality" at $\mathscr{I}^+$ (Fig. 13). Due to this fact, topology does not matter for the future dynamics (but cf. [72]). Due to the locality at $\mathscr{I}^+$, we have that a suitable notion for smallness in the stability argument can be defined locally in space. We mention some results in this setting. Friedrich proved global nonlinear stability of de Sitter space for the Einstein-Yang-Mills system with positive cosmological constant [73]. Ringström proved a "local in space" small data global existence results for the Einstein-$\Lambda$-scalar field system [74, 75]. The case of fluid matter was considered in this situation by Rodnianski and Speck [76] for the irrotational case, see Speck [77] for the Einstein-Euler system. Finally, the Einstein-$\Lambda$-Vlasov system has been studied by Ringström [78] (Fig. 13).

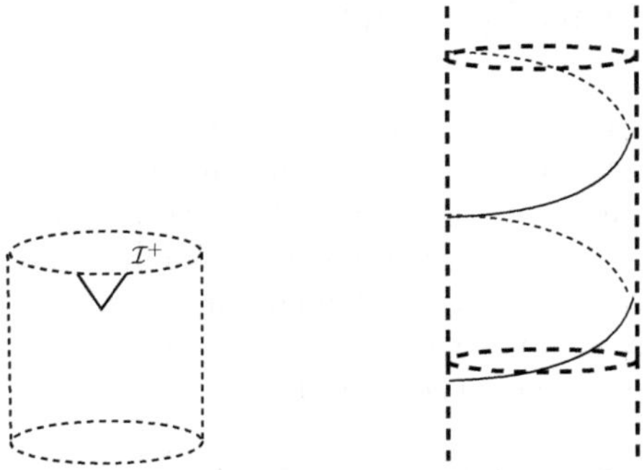


**Fig. 13** Conformal diagrams of de Sitter and Milne spacetimes

## *4.3 Milne*

Finally we consider the stability problem for a cosmological models with $\Lambda = 0$. Here, the only general results are for the vacuum case. By passing to a quotient of the Milne spacetime, we may consider a flat spacetime which has a Cauchy surface isometric to a compact hyperbolic 3-manifold. The line element is

$$ds^2 = -dt^2 + t^2 g_{\mathbb{H}^3}$$

($\kappa = -1$ empty Friedmann) and the spacetime is conformal to an infinite cylinder

$$-d\tau^2 + g_{\mathbb{H}^3}$$

In this case topology does matter, in the sense that an observer is able to see the whole past of his spacetime. Since there is no future conformal boundary, it is not possible to localize the future evolution problem.

Future stability for Milne with compact Cauchy surface as described above was proven by the author in collaboration with Moncrief for spacetime dimension $d+1$, $d \geq 3$, cf. [79, 80], see also [81, 82]. For the $2+1$ dimensional case, see [63]. Concerning the stability problem for the Einstein-matter systems in this setting, much less is known than in the case with positive $\Lambda$. Some sub-problems have been considered for the Einstein-Vlasov system in Bianchi symmetry (spacetimes with a 3-dimensional Lie group acting by isometries on Cauchy surfaces), see [83–85]. Finally, we mention the work concerning test fluids on Friedmann backgrounds by Speck [86].

The case of vacuum spacetimes with $U(1)$ symmetry leads after a Kaluza-Klein reduction to $2+1$ dimesional gravity with wave maps matter. The nonlinear stability of the flat cones over surfaces of genus $g > 1$ in this setting has been studied by studied by Choquet-Bruhat and Moncrief, see [87, 88].

# 5 Generalized Kasner Spacetimes

In Sect. 4.3 we discussed a stability theorem for the future of a Cauchy surface in a class of spacetimes. The background spacetime in that case is a Lorentz cone over a compact Einstein space with negative scalar curvature, i.e. a generalized Milne space. In particular these are warped products of the line with an Einstein space. In this section we shall discuss a class of double warped product spacetimes, with two scale factors. These spacetimes which were considered in [89] may be viewed as generalized Kasner spacetimes. They have the form

$$\mathbf{M} \cong \mathbb{R} \times M \times N \,,$$

with $(M, g)$, $(N, h)$, compact negative Einstein spaces of dimensions $m$, $n$, respectively. The dimension of $\mathbf{M}$ is $D = d + 1 = m + n + 1$. We assume $\mathrm{Ric}_g = -(m + n - 1)g$, $\mathrm{Ric}_h = -(m + n - 1)h$. and consider a line element on $\mathbf{M}$ of the form

$$ds^2 = -dt^2 + a^2(t)g + b^2(t)h \,.$$

Let $p = -\dot{a}/a, \quad q = -\dot{b}/b$, and introduce the scale invariant variables

$$P = p/H, \quad Q = q/H, \quad A = \frac{1}{aH}, \quad B = \frac{1}{bH} \,.$$

The Einstein equations imply an autonomous system for $(P, Q, A, B)$ with 2 constraints. A dynamical systems analysis shows that the generic orbit has generalized Kasner behavior, i.e. $a \sim t^p, b \sim t^q$ at singularity, and is asymptotically Friedmann (in fact asymptotic to a Lorentz cone spacetime) in the expanding direction

$$a, b = t + O(t^{1-\lambda^*}), \quad \lambda^* > 0 \,.$$

Friedmann is a stable node only in spacetime dimension $D \geq 11$ (Fig. 14).

## 5.1 From $\alpha$ to $\omega$

Belinskiĭ et al. [90] argued that a generic cosmological singularities in $3 + 1$ dimensions in spacetimes with ordinary matter is oscillatory. The picture developed by Belinskiĭ et al. is often referred to as the BKL proposal. BKL type behavior has been proved rigorously so far only for the Bianchi VIII and IX models, see [91], where also strong cosmic censorship for this class of models was shown. On the other hand, Belinskiĭ and Khalatnikov [92] pointed out that cosmological singularities in spacetimes containing stiff fluid or scalar field can be non-oscillatory, or *quiescent*. The heuristic analysis of Belinskiĭ and Khalatnikov was extended to the higher dimensional case by Demaret et al. [93] who showed that quiescent behavior at singularity

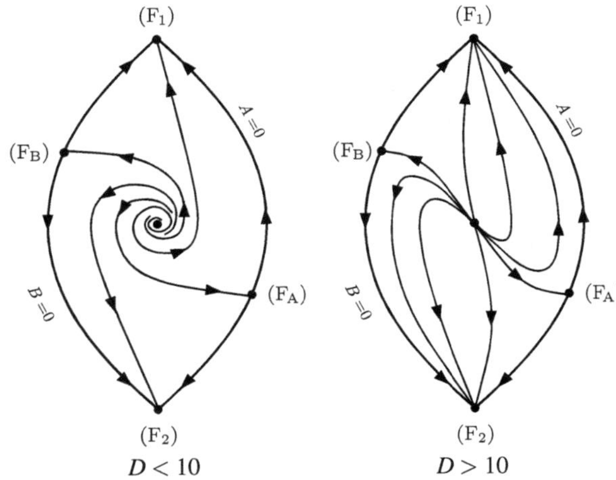


**Fig. 14** The dynamics of the generalized Kasner models [89]. The *arrows* point in the past direction. There are five fixed points, one of which is the Friedmann point in the interior of the phase space. The Friedmann point is a past unstable node for $D > 10$ and a unstable spiral point for $D < 10$. The past stable fixed points $F_{1,2}$ satisfy condition (6) for $D > 10$. This implies quiescent behavior at the singularity for inhomogenous deformations of the generalized Kasner models in $D > 10$

in $D = d + 1$ dimensions holds if the condition

$$1 + p_1 - p_d - p_{d-1} > 0 \tag{6}$$

holds, where $p_a$ are the generalized Kasner exponents at the singularity. This heuristic analysis shows that (6) holds in vacuum *only* if $D \geq 11$, and hence one expects that generic vacuum, $D < 11$ spacetimes have oscillatory singularity, while generic vacuum, $D \geq 11$ spacetime have quiescent singularity. It was shown in [89, Sect. 4] that (6) holds for generalized Kasner spacetimes if $D \geq 11$, in agreement with the result of Demarat et al. [93].

As a step towards making this heuristic scenario rigorous, the author showed with Rendall [94] that generic $D = 4$ spacetime with scalar field has quiescent singularity. In that paper we constructed a full parameter family of Einstein-scalar field and Einstein-stiff fluid spacetimes with quiescent singularity using Fuchsian analysis. This work was extended to the case of $D \geq 11$ vacuum spacetimes by Damour et al. [95], again using a Fuchsian analysis.

One may use the techniques discussed above to prove that a type of global non-linear stability holds for a class of generalized Kasner spacetimes. It was shown in [96] that for generalized Kasner spacetimes as above, with $D \geq 11$, satisfying the additional condition that the moduli space of negative Einstein metrics on $M, N$ is integrable (which is expected to hold in general), there is a full-parameter family of $C^\omega$ Cauchy data on $M \times N$, such that the maximal Cauchy development $(\mathbf{M}, \mathbf{g})$ has a global CMC time function, and has quiescent, crushing singularity. Further

$(\mathbf{M}, \mathbf{g})$ is future causally complete and is asymptotically Friedmann to the future, with $g(T) \to \gamma_\infty^M + \gamma_\infty^N$, as $T \to \infty$, where $\gamma_\infty^M$ and $\gamma_\infty^N$ are negative Einstein metrics on $M$, $N$, respectively. This applies to a large variety of factors $M$, $N$, and can easily be generalized to multiple factors.

## 6 Concluding Remarks

In this paper we have given brief overview of some of the ideas underlying the general relativistic cosmological models which form the core of the standard model of cosmology, and pointed out the need for an improved analysis, both from the physical and mathematical point of view, of the effect of deviations from homogeneity and isotropy in the dynamics of cosmological models, and consequently in the analysis of cosmological data. Motivated by this, we have discussed some results on nonlinear stability for cosmological models. We end by listing some open problems.

The exponential expansion caused by the presence of the cosmological constant in the case $\Lambda > 0$ and also in the presence of certain self-gravitating scalar field models for inflation makes the large data future behavior of these models tractable and here there are several results which do not require any symmetry assumptions, see Sect. 4.2.

For the case $\Lambda = 0$ and ordinary matter, the situation is more delicate. The global behavior of cosmological models is well understood in highly symmetric cases, including the $3+1$ dimensional Friedmann, Bianchi, Gowdy (spatial $T^2$ symmetric, with symmetry action generated by hypersurface orthogonal Killing fields) and so-called surface symmetric cases, see [97] and references therein. For the Bianchi case, see the remarks in Sect. 5.1 and [98, 99], and for the Gowdy case see [100] and references therein. However, for large data, the asymptotic behavior of the general $T^2$, $U(1)$ (circle symmetric) and the full $3+1$ case are mostly open. Similarly, future stability is open in the $3+1$ dimensional case for Einstein-matter models without symmetry assumptions in the case $\Lambda = 0$. As an example, one would like to prove nonlinear stability of Milne for Einstein-Vlasov. This is work in progress by the author with Fajman.

Our understanding of the behavior of cosmological models in the direction of the initial singularity is also limited. The BKL proposal provides a heuristic scenario which has been verified only in the Bianchi case, where also strong cosmic censorship has been shown to hold, see above. However, in spite of some recent progress [101–103], even the question whether the singularity in generic Bianchi models is local, is open. See [99, 104] for references and discussion. For Gowdy models with $T^3$ Cauchy surface, Ringström has proved that strong cosmic censorship holds, see [100] for an overview, while for Gowdy with Cauchy surfaces diffeomorphic to $S^3$ or $S^2 \times S^1$, and the general $T^2$ symmetric case (dropping the condition on hypersurface orthogonality) the situation is much more complicated and cosmic censorship is open. In particular, in the $T^2$ symmetric case, one has the new phenomenon of dynamical spikes, see [105, 106].

The work by the author and Rendall, and by Damour et al. on quiescent singularities, see Sect. 5.1 opens up the problem of proving quiescent behavior at the singularity as well as global nonlinear stability for an open set of Cauchy data (in a suitable topology). This is work in progress by the author and Ringström. Work on this type of stability problem for the Friedmann case was mentioned in a recent talk by Speck [107]. For the case $D < 11$ one may consider suitable Einstein-scalar field models and for $D \geq 11$ one may formulate the global nonlinear stability problem for the generalized Kasner backgrounds as discussed in Sect. 5. Here it should be pointed out that the global stability result mentioned there relies on Fuchsian methods and therefore suffers from the same weakness as the work by the author and Rendall, and Damour et al. on quiescent singularities. It would be interesting to prove a true nonlinear stability result, stating that for an open set of Cauchy data close to the generalized Kasner background data, the maximal development is geodesically complete to the future, asymptotically Friedmann, and with crushing singularity with geometry close to, in a suitable sense, the singularity in the generalized Kasner spacetime.

For the near future, I expect that numerical studies of cosmological models in GR, with less symmetry than the 2 Killing field models including LTB, $T^2$ and spherically symmetric models studied in detail so far, will play an important role in exploring the future behavior of cosmological models. One can expect that such investigations will have an impact on both physical cosmology and the mathematical analysis of cosmological models.

**Acknowledgments** I would like to thank Jiří Bičák and Bernd Schmidt for their comments on an early version of the paper.

## References

1. Caspar, M.: Kepler. Dover Publications, New York (1993)
2. Einstein, A.: Relativität und Gravitation. Erwiderung auf eine Bemerkung von M. Abraham. Ann. Phys. **343**, 1059 (1912). doi:10.1002/andp.19123431014
3. Einstein, A.: Näherungsweise Integration der Feldgleichungen der Gravitation, Sitzungsberichte der Königlich Preußischen Akademie der Wissenschaften, (Berlin), 844–847 (1915)
4. Lovelock, D.: The uniqueness of the Einstein field equations in a four-dimensional space. Arch. Ration Mech. Anal. **33**, 54 (1969). doi:10.1007/BF00248156
5. Lovelock, D.: Divergence-free tensorial concomitants. Aequationes Math. **4**, 127 (1970)
6. Haugan, M.P., Lämmerzahl, C.: Principles of equivalence: their role in gravitation physics and experiments that test them. In: Lämmerzahl, C., Everitt, C.W.F., Hehl, F.W. (eds.) Gyros, Clocks, Interferometers: Testing Relativistic Gravity in Space. Lecture Notes in Physics, vol. 562, p. 195. Springer, Berlin (2001)
7. Bičák, J.: Einstein in Prague: relativity then and now. In: Bičák, J., Ledvinka, T. (eds.) General Relativity, Cosmology and Astrophysics: Perspectives 100 Years After Einstein in Prague, pp. 271–297. Springer, Berlin (2014)
8. Einstein, A.: Prinzipielles zur allgemeinen Relativitätstheorie. Ann. Phys. **360**, 241 (1918). doi:10.1002/andp.19183600402
9. Barbour, J.B.: The part played by Mach's Principle in the genesis of relativistic cosmology. p. 47, Cambridge University Press, Cambridge (1990)

10. Bičák, J., Katz, J., Lynden-Bell, D.: Cosmological perturbation theory, instantaneous gauges, and local inertial frames. Phys. Rev. D **76**(6), 063501 (2007). doi:10.1103/PhysRevD.76.063501
11. Will, C.: Testing Machian effects in laboratory and space experiments. In: Barbour, J.B., Pfister, H. (eds.) Mach's Principle: From Newton's Bucket to Quantum Gravity, p. 365. (1995)
12. Barbour, J.B., Pfister, H. (eds.): Mach's Principle: From Newton's Bucket to Quantum Gravity. Birkhäuser, Basel (1995)
13. Lahav, O., Suto, Y.: Measuring our universe from galaxy redshift surveys. Living Rev. Relativ. **7**, 8 (2004)
14. Le Verrier, U.J.: Theorie du mouvement de mercure. Ann. l'Obs. Paris **5**, 1 (1859)
15. Lescarbault, M., Le Verrier, U.J.: Passage d'une planete sur le disque du soleil. Ann. l'Obs. Paris **5**, 394 (1860)
16. Hall, A.: A suggestion in the theory of mercury. Astron. J. **14**, 49 (1894). doi:10.1086/102055
17. Dyson, F.W., Eddington, A.S., Davidson, C.: A determination of the deflection of light by the sun's gravitational field, from observations made at the total eclipse of may 29, 1919. R. Soc. Lond. Philos. Trans. Ser. A **220**, 291 (1920). doi:10.1098/rsta.1920.0009
18. Norton, J.D.: The Cosmological Woes of Newtonian Gravitation Theory, p. 271. Birkhäuser, Boston (1999)
19. Einstein, A.: Kosmologische Betrachtungen zur allgemeinen Relativitätstheorie, Sitzungsberichte der Königlich Preußischen Akademie der Wissenschaften (Berlin). 142–152 (1917)
20. Friedman, A.: Über die Krümmung des Raumes. Z. Phys. **10**, 377 (1922). doi:10.1007/BF01332580
21. Hubble, E.: A relation between distance and radial velocity among extra-galactic nebulae. Proc. Nat. Acad. Sci. **15**, 168 (1929). doi:10.1073/pnas.15.3.168
22. Milne, E.A.: World-structure and the expansion of the universe. Mit 6 Abbildungen. Z. Angew. Phys. **6**, 1 (1933)
23. Robertson, H.P.: Relativistic cosmology. Rev. Mod. Phys. **5**, 62 (1933). doi:10.1103/RevModPhys.5.62
24. Walker, A.G.: On Riemanntan spaces with spherical symmetry about a line, and the conditions for isotropy in Genj relativity. Q. J. Math. **6**, 81 (1935). doi:10.1093/qmath/os-6.1.81
25. Robertson, H.P.: On E. A. Milne's theory of world structure. Z. Angew. Phys. **7**, 153 (1933)
26. Bondi, H., Gold, T.: The steady-state theory of the expanding universe. Mon. Not. R. Astron. Soc. **108**, 252 (1948)
27. Hoyle, F.: A new model for the expanding universe. Mon. Not. R. Astron. Soc. **108**, 372 (1948)
28. Kragh, H.: Cosmology and Controversy: The Historical Development of Two Theories of the Universe. Princeton University Press, Princeton (1996)
29. Peebles, P.J.E.: Principles of Physical Cosmology. Princeton University Press, Princeton (1993)
30. Weinberg, S.: Gravitation and Cosmology: Principles and Applications of the General Theory of Relativity. Wiley, New York (1972)
31. Peebles, P.J.E.: From Precision Cosmology to Accurate Cosmology. ArXiv e-prints astro-ph/0208037 (2002)
32. Wainwright, J., Ellis, G.F.R.: Dynamical Systems in Cosmology. Cambridge University Press, Cambridge (2005)
33. Lake, K.: The flatness problem and $\lambda$. Phys. Rev. Lett. **94**(20), 201102 (2005). doi:10.1103/PhysRevLett.94.201102
34. Helbig, P.: Is there a flatness problem in classical cosmology? Mon. Not. R. Astron. Soc. **421**, 561 (2012). doi:10.1111/j.1365-2966.2011.20334.x
35. Alpher, R.A., Bethe, H., Gamow, G.: The origin of chemical elements. Phys. Rev. **73**, 803 (1948). doi:10.1103/PhysRev.73.803
36. Alpher, R.A., Herman, R.: Evolution of the universe. Nature **162**, 774 (1948). doi:10.1038/162774b0

37. van den Bergh, S.: The early history of dark matter. Publ. Astron. Soc. Pac. **111**, 657 (1999). doi:10.1086/316369
38. Riess, A.G., Strolger, L.G., Casertano, S., et al.: New Hubble Space Telescope discoveries of type Ia supernovae at $z \geq 1$: narrowing constraints on the early behavior of dark energy. Astrophys. J. **659**, 98 (2007). doi:10.1086/510378
39. Carroll, S.M.: The cosmological constant. Living Rev. Relativ. **4**, 1 (2001). doi:10.12942/lrr-2001-1
40. Clarkson, C.: Establishing homogeneity of the universe in the shadow of dark energy. C. R. Phys. **13**, 682 (2012). doi:10.1016/j.crhy.2012.04.005
41. Maartens, R.: Is the universe homogeneous? R. Soc. Lond. Philos. Trans. Ser. A **369**, 5115 (2011). doi:10.1098/rsta.2011.0289
42. Räsänen, S.: Relation between the isotropy of the cmb and the geometry of the universe. Phys. Rev. D **79**(12), 123522 (2009). doi:10.1103/PhysRevD.79.123522
43. Marinoni, C., Bel, J., Buzzi, A.: The scale of cosmic isotropy. J. Cosmol. Astropart. Phys. **10**, 036 (2012). doi:10.1088/1475-7516/2012/10/036
44. Sylos, F.: Labini. Inhomogen. Universe Class. Quantum Gravity **28**(16), 164003 (2011). doi:10.1088/0264-9381/28/16/164003
45. Clowes, R.G., Harris, K.A., Raghunathan, S., et al.: A structure in the early universe at z$\sim$1.3 that exceeds the homogeneity scale of the R-W concordance cosmology. ArXiv e-prints arXiv:1211.6256 (2012)
46. Andersson, L., Coley, A.: Inhomogeneous cosmological models and averaging in cosmology: overview. Class. Quantum Gravity **28**(16), 160301 (2011). doi:10.1088/0264-9381/28/16/160301
47. Ellis, G.F.R.: Inhomogeneity effects in cosmology. Class. Quantum Gravity **28**(16), 164001 (2011). doi:10.1088/0264-9381/28/16/164001
48. Kolb, E.W., Marra, V., Matarrese, S.: Cosmological background solutions and cosmological backreactions. Gen. Rel. Grav. **42**, 1399 (2010). doi:10.1007/s10714-009-0913-8
49. Buchert, T.: Toward physical cosmology: focus on inhomogeneous geometry and its non-perturbative effects. Class. Quantum Gravity **28**(16), 164007 (2011). doi:10.1088/0264-9381/28/16/164007
50. Korzyński, M.: Covariant coarse graining of inhomogeneous dust flow in general relativity. Class. Quantum Gravity **27**(10), 105015 (2010). doi:10.1088/0264-9381/27/10/105015
51. Räsänen, S.: Evaluating backreaction with the peak model of structure formation. J. Cosmol. Astropart. Phys. **4**, 026 (2008). doi:10.1088/1475-7516/2008/04/026
52. Wiegand, A., Buchert, T.: Multiscale cosmology and structure-emerging dark energy: a plausibility analysis. Phys. Rev. D **82**(2), 023523 (2010). doi:10.1103/PhysRevD.82.023523
53. Wiltshire, D.L.: Exact solution to the averaging problem in cosmology. Phys. Rev. Lett. **99**(25), 251101 (2007). doi:10.1103/PhysRevLett.99.251101
54. Clarkson, C., Ellis, G.F.R., Faltenbacher, A., et al.: (mis)interpreting supernovae observations in a lumpy universe. Mon. Not. R. Astron. Soc. **426**, 1121 (2012). doi:10.1111/j.1365-2966.2012.21750.x
55. Clifton, T., Rosquist, K., Tavakol, R.: An exact quantification of backreaction in relativistic cosmology. Phys. Rev. D **86**(4), 043506 (2012). doi:10.1103/PhysRevD.86.043506
56. Bentivegna, E., Korzyński, M.: Evolution of a periodic eight-black-hole lattice in numerical relativity. Class. Quantum Gravity **29**(16), 165007 (2012). doi:10.1088/0264-9381/29/16/165007
57. Marra, V., Kolb, E.W., Matarrese, S.: Light-cone averages in a swiss-cheese universe. Phys. Rev. D **77**(2), 023003 (2008). doi:10.1103/PhysRevD.77.023003
58. Mess, G.: Lorentz spacetimes of constant curvature. Technical Report IHES/M/90/28. Institute des Hautes Etudes Scientifiques (1990)
59. Andersson, L., Barbot, T., Benedetti, R., et al.: Notes on: "Lorentz spacetimes of constant curvature" [geom. dedicata 126 (2007), 3–45] by G. Mess. Geom. Dedicata **126**, 47 (2007). doi:10.1007/s10711-007-9164-6

60. Andersson, L., Barbot, T., Beguin, F., Zeghib, A.: Cosmological time versus cmc time. Asian J. Math. **16**, 37 (2012)
61. Andersson, L.: Constant mean curvature foliations of flat space-times. Comm. Anal. Geom. **10**(5), 1125 (2002)
62. Andersson, L.: Constant mean curvature foliations of simplicial flat spacetimes. Comm. Anal. Geom. 13(5), 963 (2005). http://projecteuclid.org/getRecord?id=euclid.cag/1144438303
63. Andersson, L., Moncrief, V., Tromba, A.J.: On the global evolution problem in $2+1$ gravity. J. Geom. Phys. **23**(3–4), 191 (1997)
64. Fischer, A.E., Moncrief, V.: Hamiltonian reduction of Einstein's equations and the geometrization of three-manifolds. In: International Conference on Differential Equations, vol. 1, 2, pp. 279–282. World Scientific Publishing, River Edge (2000) (Berlin, 1999)
65. Anderson, M.T.: On long-time evolution in general relativity and geometrization of 3-manifolds. Comm. Math. Phys. **222**(3), 533 (2001). doi:10.1007/s002200100527
66. Andersson, L., Blue, P.: Hidden symmetries and decay for the wave equation on the Kerr spacetime. ArXiv e-prints arXiv:0908.2265 (2009)
67. Friedrich, H.: On the existence of $n$-geodesically complete or future complete solutions of Einstein's field equations with smooth asymptotic structure. Comm. Math. Phys. **107**(4), 587 (1986)
68. Christodoulou, D., Klainerman, S.: The Global Nonlinear Stability of the Minkowski Space. Princeton University Press, Princeton (1993)
69. Klainerman, S., Nicolò, F.: Peeling properties of asymptotically flat solutions to the Einstein vacuum equations. Class. Quantum Gravity **20**, 3215 (2003). doi:10.1088/0264-9381/20/14/319
70. Lindblad, H., Rodnianski, I.: The global stability of Minkowski space-time in harmonic gauge. Ann. Math. **171**(2–3), 1401 (2010). doi:10.4007/annals.2010.171.1401
71. Bieri, L., Zipser, N.: AMS/IP Studies in Advanced Mathematics. Extensions of the stability theorem of the Minkowski space in general relativity, vol. 45. American Mathematical Society, Providence (2009)
72. Andersson, L., Galloway, G.J.: Ds/cft and spacetime topology. Adv. Theor. Math. Phys. **6**(2), 307 (2002)
73. Friedrich, H.: On the global existence and the asymptotic behavior of solutions to the Einstein-Maxwell-Yang-Mills equations. J. Differ. Geom. **34**(2), 275 (1991)
74. Ringström, H.: Future stability of the Einstein-non-linear scalar field system. Invent. Math. **173**, 123 (2008). doi:10.1007/s00222-008-0117-y
75. Ringström, H.: Power law inflation. Commun. Math. Phys. **290**, 155 (2009). doi:10.1007/s00220-009-0812-6
76. Rodnianski, I., Speck, J.: The stability of the irrotational Euler-Einstein system with a positive cosmological constant. ArXiv e-prints arXiv:0911.5501 (2009)
77. Speck, J.: The nonlinear future-stability of the flrw family of solutions to the Euler-Einstein system with a positive cosmological constant. ArXiv e-prints arXiv:1102.1501 [math.AP] (2011)
78. Ringström, H.: On the topology and future stability of models of the universe. Oxford University Press, Oxford, New York (2013)
79. Andersson, L., Moncrief, V.: Future complete vacuum spacetimes. In: The Einstein equations and the large scale behavior of gravitational fields, pp. 299–330. Birkhäuser, Basel (2004)
80. Andersson, L., Moncrief, V.: Einstein spaces as attractors for the Einstein flow. J. Differ. Geom. **89**(1), 1 (2011). http://projecteuclid.org/getRecord?id=euclid.jdg/1324476750
81. Reiris, M.: Aspects of the long time evolution in general relativity and geometrizations of three-manifolds (ProQuest LLC, Ann Arbor, MI). Ph. D. Thesis. State University of New York at Stony Brook http://gateway.proquest.com/openurl?urlver=Z39.88-2004&rftvalfmt=info:ofi/fmt:kev:mtx:dissertation&resdat=xri:pqdiss&rftdat=xri:pqdiss:3206447 (2005)
82. Reiris, M.: The ground state and the long-time evolution in the cmc Einstein flow. Ann. Henri Poincaré **10**(8), 1559 (2010). doi:10.1007/s00023-010-0027-6

83. Rendall, A.D., Tod, K.P.: Dynamics of spatially homogeneous solutions of the Einstein-Vlasov equations which are locally rotationally symmetric. Class. Quantum Gravity **16**, 1705 (1999). doi:10.1088/0264-9381/16/6/305
84. Heinzle, J.M., Uggla, C.: Dynamics of the spatially homogeneous Bianchi type I Einstein-Vlasov equations. Class. Quantum Gravity **23**, 3463 (2006). doi:10.1088/0264-9381/23/10/016
85. Nungesser, E.: Late-time behaviour of the Einstein-Vlasov system with Bianchi I symmetry. J. Phys. Conf. Ser. **314**(1), 012097 (2011). doi:10.1088/1742-6596/314/1/012097
86. Speck, J.: The stabilizing effect of spacetime expansion on relativistic fluids with sharp results for the radiation equation of state. ArXiv e-prints arXiv:1201.1963 (2012)
87. Choquet-Bruhat, Y., Moncrief, V.: Future global in time Einsteinian spacetimes with u(1) isometry group. Ann. Henri Poincaré **2**(6), 1007 (2001)
88. Choquet-Bruhat, Y.: Future complete U(1) symmetric Einsteinian spacetimes, the unpolarized case. In: The Einstein equations and the large scale behavior of gravitational fields, pp. 251–298. Birkhäuser, Basel (2004)
89. Andersson, L., Heinzle, J.M.: Eternal acceleration from M-theory. Adv. Theor. Math. Phys. **11**(3), 371 (2007). http://projecteuclid.org/getRecord?id=euclid.atmp/1185303966
90. Belinskiǐ, V.A., Khalatnikov, I.M., Lifshitz, E.M.: Oscillatory approach to a singular point in the relativistic cosmology. Adv. Phys. **19**, 525 (1970). doi:10.1080/00018737000101171
91. Ringström, H.: Curvature blow up in bianchi viii and ix vacuum spacetimes. Class. Quantum Gravity **17**, 713 (2000). doi:10.1088/0264-9381/17/4/301
92. Belinskiǐ, V.A., Khalatnikov, I.M.: Effect of scalar and vector fields on the nature of the cosmological singularity. Soviet J. Exp. Theor. Phys. **36**, 591 (1973)
93. Demaret, J., Henneaux, M., Spindel, P.: Nonoscillatory behaviour in vacuum Kaluza-Klein cosmologies. Phys. Lett. B **164**(1–3), 27 (1985)
94. Andersson, L., Rendall, A.D.: Quiescent cosmological singularities. Comm. Math. Phys. **218**(3), 479 (2001)
95. Damour, T., Henneaux, M., Rendall, A.D., Weaver, M.: Kasner-like behaviour for subcritical einstein-matter systems. NASA STI/Recon Technical Report N **2**, 87809 (2002)
96. Andersson, L.: Stability of doubly warped product spacetimes. In: Sidoravicius, V. (ed.) New Trends Math. Phys., pp. 23–32. Springer, Dordrecht (2009)
97. Andersson, L.: The global existence problem in general relativity. In: The Einstein equations and the large scale behavior of gravitational fields, pp. 71–120. Birkhäuser, Basel (2004)
98. Ringström, H.: The future asymptotics of Bianchi VIII vacuum solutions. Class. Quantum Gravity **18**, 3791 (2001). doi:10.1088/0264-9381/18/18/302
99. Heinzle, J.M., Ringström, H.: Future asymptotics of vacuum Bianchi type $\mathrm{VI}_0$ solutions. Class. Quantum Gravity **26**, 145001 (2009). doi:10.1088/0264-9381/26/14/145001
100. Ringström, H.: Cosmic censorship for Gowdy spacetimes. Living Rev. Relativ. **13**, 2 (2010)
101. Béguin, F.: Aperiodic oscillatory asymptotic behavior for some Bianchi spacetimes. Class. Quantum Gravity **27**(18), 185005 (2010). doi:10.1088/0264-9381/27/18/185005
102. Liebscher, S., Härterich, J., Webster, K., Georgi, M.: Ancient dynamics in Bianchi models: approach to periodic cycles. Commun. Math. Phys. **305**, 59 (2011). doi:10.1007/s00220-011-1248-3
103. Reiterer, M., Trubowitz, E.: The BKL conjectures for spatially homogeneous spacetimes. ArXiv e-prints arXiv:1005.4908 (2010)
104. Heinzle, J.M., Uggla, C.: Mixmaster: fact and belief. Class. Quantum Gravity **26**(7), 075016 (2009). doi:10.1088/0264-9381/26/7/075016
105. Andersson, L., van Elst, H., Lim, W.C., Uggla, C.: Asymptotic silence of generic cosmological singularities. Phys. Rev. Lett. **94**(5), 051101 (2005). doi:10.1103/PhysRevLett.94.051101
106. Heinzle, J.M., Uggla, C., Lim, W.C.: Spike oscillations. Phys. Rev. D **86**(10), 104049 (2012). doi:10.1103/PhysRevD.86.104049
107. Speck, J.: On big bang spacetimes. In Dafermos, M. et al (eds.) Mathematical Aspects of General Relativity, Oberwolfach Reports, vol. 37, p. 2308. (2012). doi:10.4171/OWR/2012/37

# Inflation and Birth of Cosmological Perturbations

**Misao Sasaki**

**Abstract** We review recent developments in the theory of inflation and cosmological perturbations produced from inflation. After a brief introduction of the standard, single-field slow-roll inflation, and the curvature and tensor perturbations produced from it, we discuss possible sources of nonlinear, non-Gaussian perturbations in other models of inflation. Then we describe the so-called $\delta N$ formalism, which is a powerful tool for evaluating nonlinear curvature perturbations on super Hubble scales.

## 1 Introduction

One of the most successful applications of the theory of general relativity is cosmology. Over the past half century the big-bang theory of the universe, that the universe was born in an extremely hot and dense state, expanded explosively and cooled down to the present state, was observationally tested from various aspects and it is now firmly established. According to the big-bang theory, our universe is about 14 Giga years old, and the universe was radiation-dominated in the beginning. It became matter-dominated when the universe was about 100,000 years old, which happens to be about the same time when the photons decoupled from baryons, and started to travel freely until today, which are observed as the cosmic microwave background (CMB) radiation. The epoch when the CMB photons were scattered last before they reach us forms a 3-dimensional hypersurface, and it is called the last scattering surface (LSS).

Despite its tremendous success, there are still a couple of very basic problems that the big-bang theory cannot explain. One of them is the horizon problem or perhaps better to be called the causality problem, and the other the flatness problem or the entropy problem.

M. Sasaki (✉)
Yukawa Institute for Theoretical Physics, Kyoto University, Kyoto 606-8502, Japan
e-mail: misao@yukawa.kyoto-u.ac.jp

J. Bičák and T. Ledvinka (eds.), *General Relativity, Cosmology and Astrophysics*, Fundamental Theories of Physics 177, DOI: 10.1007/978-3-319-06349-2_15,

© Springer International Publishing Switzerland 2014

## 1.1 Horizon Problem

Let us first consider the horizon problem. The big-bang theory assumes an homogeneous and isotropic universe on large scales. So the metric is assumed to be in the form

$$ds^2 = -dt^2 + a^2(t)d\sigma_{(3)}^2, \tag{1}$$

where $d\sigma_{(3)}^2$ is the 3-metric of a constant curvature space with $K$ being the curvature, ${}^{(3)}R^{ij}{}_{km} = K(\delta^i_k\delta^j_m - \delta^i_m\delta^j_k)$. A coordinate system that spans $d\sigma_{(3)}^2$ is said to be comoving because an observer staying at a fixed point on the 3-space is comoving with the expansion of the universe. In this spacetime, the time-time component of the Einstein equations, the Friedmann equation, is

$$H^2 = \frac{\rho}{3M_{pl}^2} - \frac{K}{a^2}; \quad H \equiv \frac{\dot{a}}{a}, \tag{2}$$

where $M_{pl}^2 = (8\pi G)^{-1}$ in the units $\hbar = c = 1$, and the trace of the space-space components of the Einstein equations gives

$$\frac{\ddot{a}}{a} = -\frac{\rho + 3P}{3M_{pl}^2}, \tag{3}$$

where $\rho$ is the energy density and $P$ is the pressure in the universe. This latter equation shows that the expansion of the universe is always decelerating as long as $\rho + 3P > 0$, which holds for both radiation $P = \rho/3$ and matter $P = 0$. For simplicity, if we assume a simple equation of state $P/\rho = w =$ constant and $K = 0$ (which should be a good approximation in the early universe when $w = 1/3$ since $\rho \propto a^{-3(1+w)} = a^{-4}$), one finds

$$a \propto t^n; \quad n = \frac{2}{3(1+w)} < 1 \quad \text{for } w > -\frac{1}{3}. \tag{4}$$

This result may be regarded as a consequence of the attractive nature of the gravitational force.

Now we introduce the conformal time $d\eta = dt/a(t)$, and rewrite the metric as

$$ds^2 = a^2(\eta)d\hat{s}^2; \quad d\hat{s}^2 = -d\eta^2 + d\sigma_{(3)}^2. \tag{5}$$

Since the conformal transformation of the metric does not change the causal structure, the static metric $d\hat{s}^2$ perfectly describes the causal structure of the universe. If the range of $\eta$ were infinite to the past, there would be no horizon problem. The problem is that the conformal time is finite in the past if $w > -1/3$ or $\rho + 3P > 0$, because

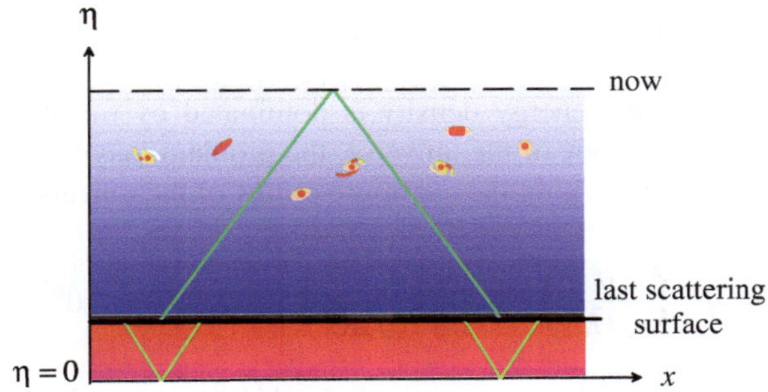


**Fig. 1** Horizon problem. The conformal time of the last scattering surface $\eta_{LSS}$ from $\eta = 0$ is about 1/30 of that of today $\eta_0$

$$\eta = \int_0^t \frac{dt'}{a(t')} \propto \int_0^t \frac{dt'}{t'^n}; \quad n = \frac{2}{3(1+w)}. \tag{6}$$

This implies that the size of lightcone emanating from a point at the beginning of the universe when $\eta = 0$ will cover only a finite fraction of spacetime. Since the comoving distance traveled by light is equal to the corresponding conformal time interval, the comoving radius of the causally connected region on the LSS is equal to its conformal time $\eta_{LSS}$. From the fact that the LSS is located at redshift $z \sim 10^3$ and the universe is approximately matter-dominated since then, one finds that this region will cover only a tiny fraction (about $10^{-3}$ sr) of the sky. This is the horizon problem (see Fig. 1).

The solution is clear: The horizon problem disappears if the conformal time is either infinite in the past or the beginning of the universe $\eta = 0$ is extended sufficiently back in time to cover the whole visible universe. Since the comoving radius of the visible universe on the LSS is $\eta_0 - \eta_{LSS}$, where $\eta_0$ is the conformal time today, the problem is solved if $\eta_{LSS} > \eta_0 - \eta_{LSS}$. In Einstein gravity, this means that the equation of state must be $w < -1/3$ or the expansion of the universe must be accelerating ($\ddot{a} > 0$) for a sufficient lapse of time in the very early universe.

Here we should note that solving the horizon problem does *not* mean explaining the homogeneity and isotropy of the universe. As it is clear from the above argument, we had to assume the homogeneity and isotropy of the universe to pose the horizon problem. This point is very often misunderstood in the literature.

## 1.2 Flatness Problem

Again we assume a spatially homogeneous and isotropic universe, Eq. (1). The Friedmann equation (2) tells us that the curvature term $K/a^2$ is completely negligible in the early universe when $\rho \propto a^{-4}$. Conversely, if the curvature term was of the same order of magnitude as the density at an epoch in the early universe, the universe must have either collapsed (if $K > 0$) or become completely empty (if $K < 0$) by now.

Alternatively, since the energy density is dominated by radiation in the early universe and so is the entropy of the universe, the problem may be rephrased as the existence of huge entropy within the curvature radius of the universe,

$$S = T^3 \left(\frac{a}{\sqrt{|K|}}\right)^3 \approx T_0^3 \left(\frac{a_0}{\sqrt{|K|}}\right)^3 > T_0^3 H_0^3 \approx 10^{87}, \tag{7}$$

where $T_0 \approx 2.7\,\mathrm{K}$ is the CMB temperature today [1] and $H_0 \approx 72\,\mathrm{km/s/Mpc}$ is the Hubble constant [2]. Hence the flatness problem may be called the entropy problem.

It is then apparent that the solution to the flatness problem needs huge entropy production at a sufficiently early stage of the universe.

## 1.3 Inflation as a Solution to Horizon and Flatness Problems

A simple and perhaps the best solution to the horizon and flatness problems is given by the inflationary universe [3, 4]. Let us assume that the universe was dominated by a spatially homogeneous scalar field. For a minimally coupled canonical scalar field $\phi$, we have

$$\rho = \frac{1}{2}\dot{\phi}^2 + V(\phi), \quad P = \frac{1}{2}\dot{\phi}^2 - V(\phi), \tag{8}$$

so $\rho + 3P = 2\big(\dot{\phi}^2 - V(\phi)\big)$. Hence if $\dot{\phi}^2 < V(\phi)$, we may have accelerated expansion. In particular, if the energy density is dominated by the potential energy, $\dot{\phi}^2 \ll V(\phi)$, the motion of the scalar field can be ignored within a few expansion times $\sim H^{-1}$, and the universe expands almost exponentially,

$$H^2 \approx \frac{\rho}{3M_{pl}^2} \approx \text{constant}. \tag{9}$$

The curvature term $K/a^2$ becomes completely negligible.

Thus if the universe is dominated by the potential energy, or the *vacuum energy*, and the potential energy is converted to radiation after a sufficient lapse of time of such a stage, a huge entropy is produced and the horizon and flatness problems are solved simultaneously.

## 2 Slow-Roll Inflation and Vacuum Fluctuations

There have been a number of proposals for inflationary models. Among others, a simplest class of models, and which explains the observational data almost perfectly, is the slow-roll inflation [5–7]. The field equation for $\phi$ and the Friedmann equation are

$$\ddot{\phi} + 3H\dot{\phi} + V'(\phi) = 0, \quad H^2 = \frac{1}{3M_{pl}^2}\left[\frac{1}{2}\dot{\phi}^2 + V(\phi)\right], \tag{10}$$

where we have justifiably neglected the curvature term.

The standard slow-roll condition consists of two assumptions. One is that $\ddot{\phi}$ is negligible compared to $3H\dot{\phi}$ in the field equation, that is, the equation of motion is friction-dominated. The other is that the kinetic term $\dot{\phi}^2/2$ is negligible compared to the potential term $V$ in the energy density. Under this condition we have

$$\dot{\phi} = -\frac{V'(\phi)}{3H}; \quad H^2 = \frac{V}{3M_{pl}^2}. \tag{11}$$

Then the potential energy dominance implies

$$\varepsilon \equiv -\frac{\dot{H}}{H^2} = \frac{\frac{3}{2}\dot{\phi}^2}{\frac{1}{2}\dot{\phi}^2 + V} \approx \frac{3\dot{\phi}^2}{2V} \approx \frac{M_{pl}^2}{2}\frac{V'^2}{V^2} \equiv \varepsilon_V \ll 1, \tag{12}$$

that is, the universe is expanding almost exponentially, and the friction-dominated equation of motion $|\ddot{\phi}/(3H\dot{\phi})| \ll 1$ implies

$$\frac{V''}{3H^2} \approx M_{pl}^2 \frac{V''}{V} \equiv \eta_V \ll 1. \tag{13}$$

The single-field slow-roll inflation satisfies these conditions.

The important property of slow-roll inflation is that Eq. (11) is completely integrable since $H$ is a function of $\phi$. In particular, there is one-to-one correspondence between $\phi$ and $t$. So instead of the cosmic time $t$ we may measure the time in terms of the value of the scalar field.

Here we introduce a quantity which plays a very important role in the dynamics of slow-roll inflation, namely the number of $e$-folds counted *backward* in time, say from the end of inflation to an epoch during inflation,

$$\frac{a(t_{end})}{a(t)} = \exp[N(t \to t_{end})] \;\to\; N = N(\phi) = \int_{t(\phi)}^{t_{end}} H dt. \tag{14}$$

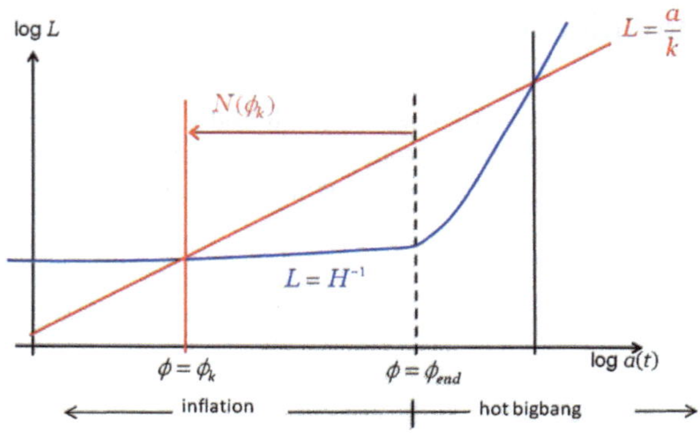


**Fig. 2** The Hubble radius $L = H^{-1}$ and the length scale $L = a/k$ of a comoving wavenumber $k$ in the inflationary cosmology, and the definition of the number of $e$-folds $N(\phi)$

Its important property is that by definition it does not depend on how and when the inflation began. As shown in Fig. 2, $N$ is uniquely determined in terms of the value of the scalar field (up to a constant which depends on the choice of an epoch from which $N$ is computed), and one can associate $N$ with the time at which a given comoving wavenumber $k$ crossed the Hubble radius, $k = aH$, at which the value of the scalar field was $\phi_k$; $N = N(\phi_k)$. As we shall see below, this turns out to be an essential quantity for the evaluation of the curvature perturbation from inflation.

## 2.1 Curvature Perturbation

Let us now consider the curvature perturbation produced from inflation. It arises from the quantum vacuum fluctuations of the inflaton field $\phi$. Since a rigorous derivation would take too much space, here we give an intuitive, rather hand-waving derivation. We caution that it could well lead to an incorrect result if used blindly.

The vacuum fluctuations of the inflaton field with a comoving wavenumber $k$ is given simply by its positive frequency function, $\varphi_k$. Because of the condition $V''/H^2 \ll 1$, on scales $k/a \gg H$, the inflaton field fluctuation behaves like a minimally coupled massless scalar. Hence we have

$$|\langle\delta\phi|\mathbf{k}\rangle|^2 = |\varphi_k|^2, \quad \varphi_k \sim \frac{1}{a^{3/2}\sqrt{2\omega_k}} e^{-i\omega_k t}; \quad \omega_k = \frac{k}{a} \gg H. \tag{15}$$

As the universe expands the physical wavenumber decreases exponentially and becomes smaller than the Hubble parameter, $k/a < H$, or the physical wavelength exceed the Hubble radius. Then the oscillations of $\varphi_k$ are frozen. This could be regarded as "classicalization" of the quantum fluctuations. Note that this is merely an interpretation. In a more rigorous sense, freezing of the mode function is a process toward infinite squeezing of the vacuum state.

Setting $a = k/H$ in Eq. (15) gives

$$\varphi_k \sim \frac{H}{\sqrt{2k^3}}; \quad \frac{k}{a} \ll H. \tag{16}$$

Therefore the mean square amplitude in unit logarithmic interval of $k$ is

$$\langle \delta\phi^2 \rangle_k \equiv \frac{4\pi k^3}{(2\pi)^3} |\varphi_k|^2 \approx \left( \frac{H}{2\pi} \right)^2_{k/a=H}. \tag{17}$$

Inclusion of the non-trivial evolution of the background spacetime and the coupling of the scalar field fluctuation with the metric fluctuation do not change the above estimate if we interpret $\delta\phi$ in the above as those evaluated on the flat slicing, that is, on hypersurfaces on which the spatial scalar curvature remains unperturbed.

It is known that the curvature perturbation on the comoving hypersurface $\mathscr{R}_c$ is conserved if the perturbation is adiabatic [8]. The comoving hypersurface is defined as a surface of uniform $\phi$. Then the gauge transformation from the flat slicing to the comoving slicing gives the relation between $\mathscr{R}_c$ and $\delta\phi$,

$$\mathscr{R}_c = -\frac{H}{\dot{\phi}} \delta\phi. \tag{18}$$

Since this is conserved for $k/a < H$, the spectrum of the comoving curvature perturbation in unit logarithmic interval of $k$ is given by

$$\mathscr{P}_{\mathscr{R}}(k) \equiv \frac{4\pi k^3}{(2\pi)^3} |\mathscr{R}_k|^2 \approx \left( \frac{H^2}{2\pi\dot{\phi}} \right)^2_{k/a=H}. \tag{19}$$

A rigorous, first-principle derivation of the above result was first done in [9, 10].

The important relation of the above result with the number of $e$-folds was first pointed out in [11]: If we rewrite Eq. (14) as

$$N = \int_t^{t_{end}} H dt = \int_\phi^{\phi_{end}} \frac{H}{\dot{\phi}} d\phi, \tag{20}$$

we find

$$\delta N(\phi_k) = \left[ \frac{\partial N}{\partial \phi} \delta\phi \right]_{k/a=H} = \left[ -\frac{H}{\dot{\phi}} \delta\phi \right]_{k/a=H} = \mathscr{R}_c, \tag{21}$$

provided that we identify $\delta\phi$ with the scalar field fluctuation evaluated on the flat hypersurface. This is called the $\delta N$ formula.

The $\delta N$ formula implies that we only need the knowledge of the background evolution to obtain the power spectrum of the comoving curvature perturbation, once we know the amplitude of the quantum fluctuations of the scalar field at the horizon crossing (i.e. when $k/a = H$). It is quite generally given by $H/(2\pi)$ in slow-roll inflation. With careful geometrical considerations, the $\delta N$ formula can be extended to general multi-field inflation [12],

$$\mathscr{P}_{\mathscr{R}}(k) = \left(\frac{H}{2\pi}\right)^2 ||\nabla N||^2; \quad ||\nabla N||^2 \equiv G^{ab}(\phi)\frac{\partial N}{\partial \phi^a}\frac{\partial N}{\partial \phi^b}, \tag{22}$$

where $G^{ab}$ is the field space metric and it is assumed that the vacuum expectation values are given by

$$\langle \delta\phi^a \delta\phi^b \rangle = G^{ab}\left(\frac{H}{2\pi}\right)^2. \tag{23}$$

The nonlinear generalization of the $\delta N$ formalism will be discussed in Sect. 4.

## *2.2 Tensor Perturbation*

There are not only vacuum fluctuations of the inflaton field but also those of the transverse-traceless part of the metric, $\partial^i h_{ij}^{TT} = \delta^{ij} h_{ij}^{TT} = 0$, that is, the tensor perturbation or gravitational wave degrees of freedom. If we construct the second-order action for $h_{ij}^{TT}$, we find

$$S \sim \frac{M_{pl}^2}{8}\int d^4x\sqrt{-g}(\dot{h}_{ij}^{TT})^2 + \cdots. \tag{24}$$

To quantize $h_{ij}^{TT}$ it is convenient to normalize the kinetic term to the canonical form. This gives

$$S \sim \frac{1}{2}\int d^4x\sqrt{-g}(\dot{\phi}_{ij})^2 + \cdots; \quad \phi_{ij} \equiv \frac{M_{pl}}{2}h_{ij}^{TT}. \tag{25}$$

If one writes down the field equation for $\phi_{ij}$, one finds its mode function $\phi_k$ obeys exactly the same equation as the one for a minimally coupled massless scalar field,

$$\ddot{\phi}_k + 3H\dot{\phi}_k + \frac{k^2}{a^2}\phi_k = 0. \tag{26}$$

Since there are two independent degrees of freedom in $\phi_{ij}$, the power spectrum of the tensor perturbation $h_{ij}^{TT}$ is obtained as

$$\mathscr{P}_T(k) = \frac{4}{M_{pl}} \times 2 \times \frac{4\pi k^3}{(2\pi)^3}|\phi_k|^2 = \frac{8H^2}{(2\pi)^2 M_{pl}^2}. \tag{27}$$

Taking the ratio of the tensor spectrum to the curvature perturbation spectrum, we find [12]

$$r \equiv \frac{\mathscr{P}_T}{\mathscr{P}_{\mathscr{R}}} \leq 8|n_T| = -2\frac{\dot{H}}{H^2}, \tag{28}$$

where $n_T$ is the tensor spectral index, $n_T = d \ln \mathscr{P}_T(k)/d \ln k$, and the equality holds for the case of single-field slow-roll inflation. This is a consistency relation in general slow-roll inflation. As a prototype example, if we consider chaotic inflation [7], we expect to have $r \sim 0.1$.

The important point to be kept in mind is that the existence of the vacuum fluctuations of the tensor part of the metric is a proof of the existence of quantum gravity. These fluctuations exist in any theory of gravity that respects general covariance, apart from possible inessential modifications of the spectrum. Thus a clear detection of the tensor spectrum will be a confirmation of not only the inflationary universe but also of quantum gravity.

## 3 Origin of Non-Gaussianity

The standard, single-field slow-roll inflation predicts that the curvature perturbation is a Gaussian random field and it has an almost scale-invariant spectrum. This seems to fit the current observational data quite well [13], it is quite possible that the actual model turns out to be non-standard. Maybe it is multi-field, maybe non-slow-roll and/or non-canonical. In such a case, the curvature perturbation may become non-Gaussian. Search for possible non-Gaussian signatures in the primordial curvature perturbation has become one of the important directions in observations in recent years [14, 15].

Here we consider possible origins of non-Gaussianity in the curvature perturbation. Essentially one can classify the origins into three categories: (1) Self-interactions of the inflaton field and/or non-trivial vacua, (2) multi-field dynamics, and (3) non-linearity in gravity.

The non-Gaussianities of the first category are generated on subhorizon scales during inflation, hence they are of quantum field theoretical origin. Those of the second category are usually generated on superhorizon scales either during or after inflation, and they are due to nonlinear coupling of the scalar field to gravity. Since they are generated on superhorizon scales, they are of classical origin. Finally those of the third category are due to nonlinear dynamics in general relativity. Hence they are generated after the scale of interest re-enters the Hubble horizon. Since the last category is not really primordial in nature, let us focus on the first two categories.

### 3.1 Non-Gaussianity From Self-Interaction/Non-Trivial Vacuum

It is known that conventional self-interactions by the potential are ineffective [16]. This can be seen by considering chaotic inflation, for example. In the simplest case of a quadratic potential, $V = m^2\phi^2/2$, the inflaton is actually a free field apart from the interaction through gravitation perturbations. But the gravitational interaction is Planck-suppressed, i.e., it is always suppressed by a factor $O(M_{pl}^{-2})$. In the case of a quartic potential, $V = \lambda\phi^4$, it is known that $\lambda$ should be extremely small $\lambda \sim 10^{-15}$ in order for it to be consistent with observation.

Thus some kind of unconventional self-interaction is necessary. A popular example is the case of a scalar field with a non-canonical kinetic term such as DBI inflation [17]. In this case the kinetic term takes the form,

$$K \sim f^{-1}(\phi)\sqrt{1 - f(\phi)\dot{\phi}^2} \equiv f^{-1}\gamma^{-1}. \tag{29}$$

If we expand this perturbatively,

$$K = K_0 + \delta_1 K + \delta_2 K + \delta_3 K + \cdots , \tag{30}$$

we will find

$$\delta_2 K \propto \gamma^3, \quad \delta_3 K \propto \gamma^{3+2}, \tag{31}$$

since $\delta\gamma = \gamma^3 \delta X$ where $X \equiv f\dot{\phi}^2/2$. If we regard the third order part as the interaction, the above implies that the scalar field fluctuation will be expressed qualitatively as

$$\delta\phi \sim \delta\phi_0 + \gamma^2 \delta\phi_0^2 + \cdots , \tag{32}$$

where $\delta\phi_0$ is the free, Gaussian fluctuation. Thus the non-Gaussianity in $\delta\phi$ may become large if $\gamma$, which mimics the Lorentz factor, is large [18].

A non-trivial vacuum state is another source of non-Gaussianity. If the universe were a pure de Sitter spacetime, gravitational interaction would be totally negligible in vacuum, except for the effect due to graviton (tensor mode) loops. This may be regarded as due to the maximally symmetric nature of the de Sitter space, $SO(4, 1)$, which has the same number of degrees of symmetry as the Poincare (Minkowski) symmetry. In slow-roll inflation, the de Sitter symmetry is slightly broken. Nevertheless the effect induced by this symmetry breaking is small because it is suppressed by the slow-roll parameter $\varepsilon = -\dot{H}/H^2$.

However, if the vacuum state does not respect the de Sitter symmetry, there can be a large non-Gaussianity. Such a deviation from the quasi-de Sitter vacuum, usually called the Bunch-Davies vacuum, may occur in various situations, studied e.g. in [19, 20].

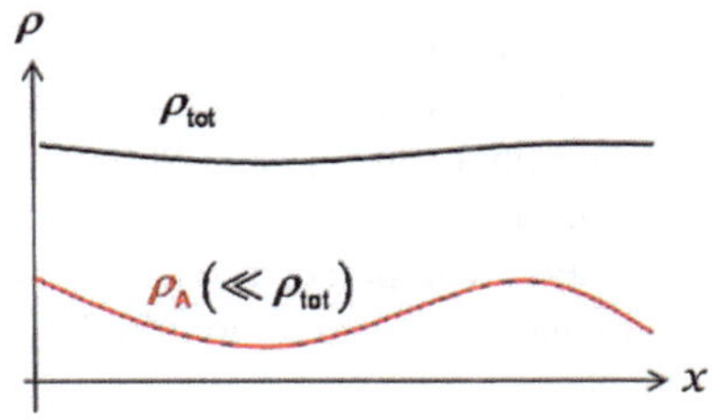


**Fig. 3** An illustration of the energy density configuration in the multi-field case. The density of the $A$-matter/field $\rho_A$ may vary nonlinearly without significantly affecting the total energy density

## *3.2 Non-Gaussianity From Multi-Field Dynamics*

Non-Gaussianity may appear if the energy momentum tensor depends nonlinearly on the scalar field even if the fluctuation of the scalar field itself is Gaussian. This effect is generally important when the fluctuations are on superhorizon scales, i.e., the characteristic wavelength is larger than the Hubble radius. It is small in single-field slow-roll models because the linear approximation is valid to high accuracy [21], generically suppressed by the slow-roll parameter $\eta_V$ defined in Eq. (13).

For multi-field models, however, the contribution to the energy momentum tensor from some of the fields can be highly nonlinear as depicted in Fig. 3. The important property of non-Gaussianity in this case is that it is always of the spatially local type. Namely, to second order in nonlinearity, the curvature perturbation will take the form [22],

$$\mathscr{R}_c(x) = \mathscr{R}_{c,0}(x) + \frac{3}{5} f_{NL}^{\rm local} \mathscr{R}_{c,0}^2(x), \tag{33}$$

where $\mathscr{R}_{c,0}$ is the Gaussian random field and $f_{NL}^{\rm local}$ is a constant representing the amplitude of non-Gaussianity. The factor 3/5 in front of $f_{NL}^{\rm local}$ is due to a historical reason. The reason why it is of local type is simply causality: No information can propagate over a length scale greater than the Hubble horizon scale.

Observationally, this type of non-Gaussianity can be tested by using the so-called squeezed type templates where one of the wavenumbers, say $k_1$, in the bispectrum $B(\mathbf{k}_1, \mathbf{k}_2, \mathbf{k}_3)$ is much smaller than the other two, $k_1 \ll k_2 \approx k_3$ [14], and there are a few observational indications that $f_{NL}^{\rm local}$ is actually non-vanishing. For example, the WMAP 7 year data analysis gave a one-sigma bound $11 < f_{NL}^{\rm local} < 53$ (68 % CL) [13].

# 4 $\delta N$ Formalism

As mentioned in Sect. 2, the $\delta N$ formalism is a powerful tool to evaluate the comoving curvature perturbation on superhorizon scales. It then turned out that it can be easily extended to the evaluation of nonlinear, non-Gaussian curvature perturbations [23, 24]. Let us recapitulate its definition and properties:

(1) $\delta N$ is the perturbation in the number of $e$-folds counted *backward* in time from a fixed final time, say $t = t_f$, to some initial time $t = t_i$.
(2) The final time $t_f$ should be chosen such that the evolution of the universe has become unique by that time, i.e., the universe has reached the adiabatic limit. Then the hypersurface $t = t_f$ should be identified with a comoving (or uniform density) slice, and the initial hypersurface $t = t_i$ should be identified with a flat slice.
(3) $\delta N$ is equal to the conserved (nonlinear) comoving curvature perturbation on superhorizon scales at $t > t_f$.
(4) By definition, it is nonlocal in time. However, because of its purely geometrical definition, it is valid independent of which theory of gravity one considers, provided that the adiabatic limit is reached by $t = t_f$.

There are various kinds of sources that generate $\delta N$. They may be classified into three types, as depicted in Fig. 4. The left one describes a perturbation along the evolutionary trajectory of the universe. This case is the same as that of single-field slow-roll inflation, in which the comoving curvature perturbation is conserved all the way until it re-enters the horizon. The middle one is the case when a small difference in the initial data develops into a substantial difference in $\delta N$. Typically this is realized when there is some instability orthogonal to the trajectory, like the case when the scalar field moves along a ridge. This type of sources of $\delta N$ usually induces a feature in the spectrum and/or bispectrum of the curvature perturbation. The right one represents the case when the perturbation orthogonal to the trajectory does not contribute to the curvature perturbation until or after the end of inflation, but $\delta N$ is generated due to a sudden transition that brings the universe into an adiabatic stage. Typical examples are curvaton models [25–27] and multi-brid inflation models [28, 29].

Here, for the sake of completeness, let us present the precise definition of the nonlinear $\delta N$ formula. See Fig. 5. It is based on the leading order approximation in the spatial gradient expansion or the separate universe approach [23], where spatial derivatives are assumed to be negligible in comparison with time derivatives. At leading order of the spatial gradient expansion, if we express the spatial volume element as $\sqrt{^{(3)}\gamma} = a^3(t)\exp[3\mathscr{R}(t,x)]$ where $a(t)$ is the scale factor of a fiducial homogeneous and isotropic universe, we easily find that the perturbation in the number of $e$-folds along a comoving trajectory between two hypersurfaces $t = t_1$ and $t = t_2$ is given by

$$\delta N(t_2, t_1; x^i) = \mathscr{R}(t_2, x^i) - \mathscr{R}(t_1, x^i), \tag{34}$$

where $x^i$ are the comoving coordinates. Here we note that this is purely a geometrical relation. It has nothing to do with any equations of motion.

First we fix the final hypersurface $t = t_2$. It should be taken at the stage when the evolution of the universe has become unique. That is, there exists no isocurvature perturbation any longer that could develop into an adiabatic perturbation at later epochs. Thus the comoving curvature perturbation is conserved at $t > t_2$. In the

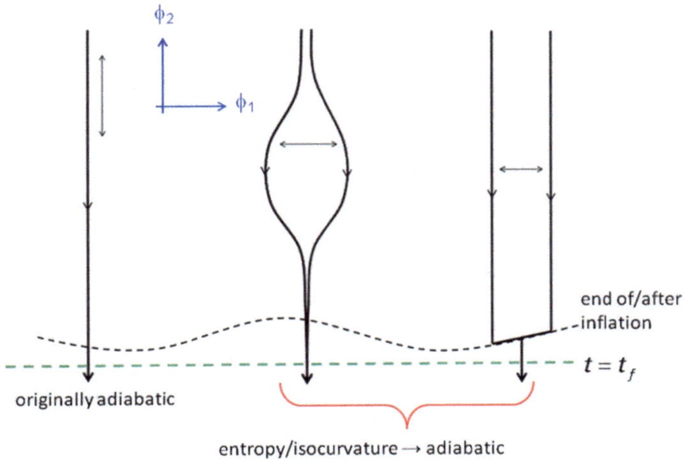


**Fig. 4** Three different types of $\delta N$. The field space $(\phi_1, \phi_2)$ in the figure represents the degrees of freedom in the initial condition of the universe. The adiabatic limit is defined to be the stage by which all the trajectories converge to a unique one

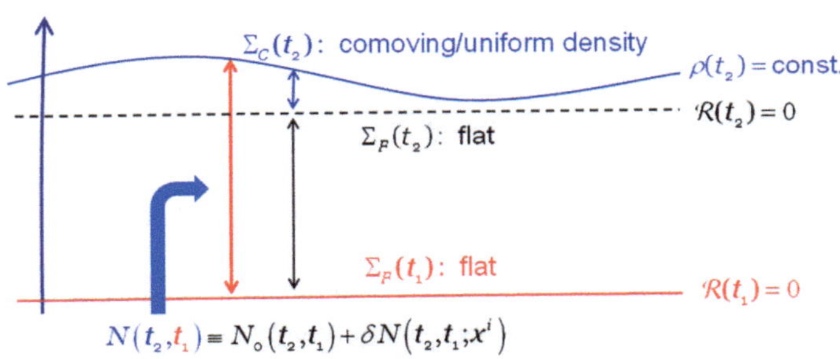


**Fig. 5** Definition of nonlinear $\delta N$. It is defined as the perturbation in the number of $e$-folds from an initial flat slice to a final comoving slice

context of the concordance ΛCDM model of the universe, this corresponds to the final radiation-dominated stage of the universe.

Next we choose the initial slice $t = t_1$. It should be chosen to be flat. Here 'flat' means that the perturbation in the spatial volume element vanishes. Namely, the flat slice is defined as a hypersurface on which $\mathscr{R} = 0$. We note that despite its name, the scalar curvature vanishes only in the linear theory limit: It is non-vanishing in general in the nonlinear case.

Applying the above choice of the initial and final hypersurfaces to Eq. (34), it is trivial to see that we have

$$\delta N(t_2, t_1; x^i) = \mathscr{R}_c(t_2, x^i). \tag{35}$$

Now by assumption $\mathscr{R}_c$ is conserved at $t > t_2$. So it is the quantity we want to evaluate. This completes the derivation of the nonlinear $\delta N$ formula.

As mentioned above, since Eq. (34) is a pure geometrical relation, so is the nonlinear $\delta N$ formula (35). This is the reason why it can be applied to any theory of gravity as long as it is a geometrical (i.e. general covariant) theory.

Of course, the above definition tells us nothing about how to evaluate it in practice. In this respect, we have a very fortunate situation in the case of inflationary cosmology. It is the fact that the evaluation of the quantum fluctuations of the inflaton field, whether it is single- or multi-component, can be most easily done in a gauge in which the time slicing is chosen to be flat [12]. Thus we can choose the initial slice to be an epoch when the scale of our interest has just exited the horizon during inflation. Let the fluctuations of a multi-component scalar field on the flat slice at $t = t_1$ be $\delta\phi^a$. Then assuming that the values of the scalar field determine the evolution of the universe completely, which is the case for slow-roll inflation, the nonlinear $\delta N$ can be simply evaluated as

$$\delta N = N(\phi^a + \delta\phi^a) - N(\phi^a), \tag{36}$$

where $N(\phi^a)$ is the $e$-folding number of the fiducial background. In particular, to second order in $\delta\phi^a$, we obtain

$$\mathscr{R}_c = \delta N = \frac{\partial N}{\partial\phi^a}\delta\phi^a + \frac{1}{2}\frac{\partial^2 N}{\partial\phi^a\partial\phi^b}\delta\phi^a\delta\phi^b + \cdots\,. \tag{37}$$

Comparing this with Eq. (33), we see that the curvature perturbation takes a bit more complicated form that the simplest form. Nevertheless if we consider the bispectrum, i.e., the Fourier component of the three-point function $\langle\mathscr{R}_c(x_1)\mathscr{R}_c(x_2)\mathscr{R}_c(x_3)\rangle$, we find there is a quantity that exactly corresponds to $f_{NL}^{\text{local}}$ defined in Eq. (33). Namely [24],

$$\frac{3}{5}f_{NL}^{\text{local}} = \frac{G^{ab}G^{cd}N_aN_{bc}N_d}{2\left(||\nabla N||^2\right)^2}; \quad N_a \equiv \frac{\partial N}{\partial\phi^a}, \quad N_{ab} \equiv \frac{\partial^2 N}{\partial\phi^a\partial\phi^b}. \tag{38}$$

Before concluding this section, we mention the fact that the $\delta N$ formalism does not require the scalar field fluctuations to be Gaussian. In fact, except for the last equation in the above, Eq. (38) which assumes the Gaussianity of $\delta\phi^a$, the general $\delta N$ formula (36) or its second order version (37) can be used for non-Gaussian $\delta\phi^a$ [30]. Such a case may happen, for example, in multi-field DBI inflation.

# 5 Summary

It has been about 30 years since the inflationary universe was first proposed, and there is increasing observational evidence that inflation did take place in the very early universe. Among others, the measured CMB temperature anisotropy is fully consistent with the predictions of inflation that the primordial curvature perturbation spectrum is almost scale-invariant and it is statistically Gaussian.

Inflation also predicts a scale-invariant tensor spectrum, and if the energy scale of inflation is high enough as in the case of chaotic inflation, the tensor-scalar ratio $r$ can be as large as 0.1. If this is the case, the tensor perturbation will be detected in the near future, and it will confirm not only the inflationary universe but also quantum gravity.

Even if the tensor perturbation will not be detected, there may be other interesting signatures of inflation. Non-Gaussianity from inflation is attracting attention as one of those signatures that can distinguish or constrain models of inflation significantly.

We discussed that the origins of primordial non-Gaussianities may be classified into three categories, according to different length scales on which different mechanisms are effective:

(1) Quantum theoretical origin on subhorizon scales during inflation.
(2) Classical nonlinear scalar field dynamics on superhorizon scales during or after inflation.
(3) Nonlinear gravitational dynamics after the horizon re-entry.

In particular we argued that non-Gaussianities in the second case are always of spatially local type. We then mentioned that there are three different kinds of situations in which such local non-Gaussianities can be generated, and described in some detail a very efficient method to compute them, namely, the $\delta N$ formalism.

Apparently identifying properties of primordial non-Gaussianities in the observational data is extremely important for understanding the physics of the early universe. Here we mentioned only the bispectrum or the 3-point function. But if it is detected, higher order $n$-point functions may become important as a model discriminator. Other types of non-Gaussianity discriminators may also become necessary.

What is important is that we are now beginning to test observationally the physics of the very early universe, the physics at an energy scale closer to the Planck scale, at a scale that can never be attained in high energy accelerator experiments.

Cosmology has become not only a precision science, but now it constitutes a truly indispensable part of fundamental physics. General relativity is the backbone of cosmology. I wonder what Einstein would say if he were here in this very exciting era—100 years after he visited Prague.

**Acknowledgments** I am very grateful to the organizers of the conference, "Relativity and Gravitation, 100 years after Einstein in Prague", particularly to Jiří Bičák, who kindly invited me to this meeting, and who accorded me a warm hospitality. I am also grateful to Laila Alabidi for careful reading of the manuscript and very useful comments. This work was supported in part by

JSPS Grant-in-Aid for Scientific Research (A) No. 21244033, and by Monbukagaku-sho Grant-in-Aid for the Global COE programs, "The Next Generation of Physics, Spun from Universality and Emergence" at Kyoto University.

## References

1. Mather, J.C., Cheng, E.S., Cottingham, D.A., et al.: Measurement of the cosmic microwave background spectrum by the COBE FIRAS instrument. Astrophys. J. **420**, 439 (1994). doi:10.1086/173574
2. Freedman, W.L., Madore, B.F.: The Hubble constant. Annu. Rev. Astron. Astrophys. **48**, 673 (2010). doi:10.1146/annurev-astro-082708-101829
3. Sato, K.: First-order phase transition of a vacuum and the expansion of the universe. Mon. Not. R. Astron. Soc. **195**, 467 (1981)
4. Guth, A.H.: Inflationary universe: a possible solution to the horizon and flatness problems. Phys. Rev. D **23**, 347 (1981). doi:10.1103/PhysRevD.23.347
5. Linde, A.D.: A new inflationary universe scenario: a possible solution of the horizon, flatness, homogeneity, isotropy and primordial monopole problems. Phys. Lett. B **108**, 389 (1982). doi:10.1016/0370-2693(82)91219-9
6. Albrecht, A., Steinhardt, P.J.: Cosmology for grand unified theories with radiatively induced symmetry breaking. Phys. Rev. Lett. **48**, 1220 (1982). doi:10.1103/PhysRevLett.48.1220
7. Linde, A.D.: Chaotic inflation. Phys. Lett. B **129**, 177 (1983). doi:10.1016/0370-2693(83)90837-7
8. Kodama, H., Sasaki, M.: Cosmological perturbation theory. Prog. Theor. Phys. Suppl. **78**, 1 (1984). doi:10.1143/PTPS.78.1
9. Mukhanov, V.F.: Gravitational instability of the universe filled with a scalar field. JETP Lett. **41**, 493 (1985)
10. Sasaki, M.: Large scale quantum fluctuations in the inflationary universe. Prog. Theor. Phys. **76**, 1036 (1986). doi:10.1143/PTP.76.1036
11. Starobinskii, A.A.: Multicomponent de Sitter (inflationary) stages and the generation of perturbations. JETP Lett. **42**, 152 (1985)
12. Sasaki, M., Stewart, E.D.: A general analytic formula for the spectral index of the density perturbations produced during inflation. Prog. Theor. Phys. **95**, 71 (1996). doi:10.1143/PTP.95.71
13. Komatsu, E., Smith, K.M., Dunkley, J., et al.: Seven-year Wilkinson microwave anisotropy probe (WMAP) observations: cosmological interpretation. Astrophys. J. Suppl. Ser. **192**, 18 (2011). doi:10.1088/0067-0049/192/2/18
14. Komatsu, E., Afshordi, N., Bartolo, N et al.: Non-Gaussianity as a probe of the physics of the primordial universe and the astrophysics of the low redshift universe. In: Proceedings of Astro 2010: The Astronomy and Astrophysics Decadal Survey—Science White Papers, National Academy of Sciences (2009). http://sites.nationalacademies.org/bpa/bpa_050603
15. Sasaki, M., Wands, D.: Focus section editorial: Non-linear and non-Gaussian cosmological perturbations, Class. Quantum Gravity **27**(12) 120301 (2010). doi:10.1088/0264-9381/27/12/120301
16. Maldacena, J.M.: Non-Gaussian features of primordial fluctuations in single field inflationary models. J. High Energy Phys. **2003**(05), 013 (2003). doi:10.1088/1126-6708/2003/05/013
17. Alishahiha, M., Silverstein, E., Tong, D.: DBI in the sky: non-Gaussianity from inflation with a speed limit. Phys. Rev. D **70**, 123505 (2004). doi:10.1103/PhysRevD.70.123505
18. Mizuno, S., Arroja, F., Koyama, K., Tanaka, T.: Lorentz boost and non-Gaussianity in multifield DBI inflation. Phys. Rev. D **80**, 023530 (2009). doi:10.1103/PhysRevD.80.023530
19. Chen, X., Easther, R., Lim, E.A.: Generation and characterization of large non-Gaussianities in single field inflation. J. Cosmol. Astropart. Phys. **2008**(04), 010 (2008). doi:10.1088/1475-7516/2008/04/010

20. Flauger, R., McAllister, L., Pajer, E., Westphal, A., Xu, G.: Oscillations in the CMB from axion monodromy inflation. J. Cosmol. Astropart. Phys. **2010**(06), 009 (2010). doi:10.1088/1475-7516/2010/06/009
21. Salopek, D.S., Bond, J.R.: Nonlinear evolution of long-wavelength metric fluctuations in inflationary models. Phys. Rev. D **42**, 3936 (1990). doi:10.1103/PhysRevD.42.3936
22. Komatsu, E., Spergel, D.N.: Acoustic signatures in the primary microwave background bispectrum. Phys. Rev. D **63**, 063002 (2001). doi:10.1103/PhysRevD.63.063002
23. Lyth, D.H., Malik, K.A., Sasaki, M.: A general proof of the conservation of the curvature perturbation. J. Cosmol. Astropart. Phys. **2005**(05), 004 (2005). doi:10.1088/1475-7516/2005/05/004
24. Lyth, D.H., Rodrǵuez, Y.: Inflationary prediction for primordial non-Gaussianity. Phys. Rev. Lett. **95**, 121302 (2005). doi:10.1103/PhysRevLett.95.121302
25. Lyth, D.H., Wands, D.: Generating the curvature perturbation without an inflaton. Phys. Lett. B **524**, 5 (2002). doi:10.1016/S0370-2693(01)01366-1
26. Moroi, T., Takahashi, T.: Effects of cosmological moduli fields on cosmic microwave background. Phys. Lett. B **522**(215), 2002 (2001). doi:10.1016/S0370-2693(01)01295-3.Erratum: ibid.539,303
27. Sasaki, M., Valiviita, J., Wands, D.: Non-Gaussianity of the primordial perturbation in the curvaton model. Phys. Rev. D **74**, 103003 (2006). doi:10.1103/PhysRevD.74.103003
28. Sasaki, M.: Multi-brid inflation and non-Gaussianity. Prog. Theor. Phys. **120**, 159 (2008). doi:10.1143/PTP.120.159
29. Naruko, A., Sasaki, M.: Large non-Gaussianity from multi-brid inflation. Prog. Theor. Phys. **121**, 193 (2009). doi:10.1143/PTP.121.193
30. Byrnes, C.T., Koyama, K., Sasaki, M., Wands, D.: Diagrammatic approach to non-Gaussianity from inflation. J. Cosmol. Astropart. Phys. **2007**(11), 027 (2007). doi:10.1088/1475-7516/2007/11/027

# Loop Quantum Gravity and the Planck Regime of Cosmology

**Abhay Ashtekar**

**Abstract** The very early universe provides the best arena we currently have to test quantum gravity theories. The success of the inflationary paradigm in accounting for the observed inhomogeneities in the cosmic microwave background already illustrates this point to a certain extent because the paradigm is based on quantum field theory on the curved cosmological space-times. However, this analysis excludes the Planck era because the background space-time satisfies Einstein's equations all the way back to the big bang singularity. Using techniques from loop quantum gravity, the paradigm has now been extended to a self-consistent theory from the Planck regime to the onset of inflation, covering some 11 orders of magnitude in curvature. In addition, for a narrow window of initial conditions, there are departures from the standard paradigm, with novel effects, such as a modification of the consistency relation involving the scalar and tensor power spectra and a new source for non-Gaussianities. The genesis of the large scale structure of the universe can be traced back to quantum gravity fluctuations in the Planck regime. This report provides a bird's eye view of these developments for the general relativity community.

## 1 Introduction

In this conference, Professor Bičák and others described the ideas that Einstein developed in Prague during 1911–12. From then until 1915 he worked largely by himself on the grand problem of extending the reach of special relativity to encompass gravity. Finally, in November 1915, he provided us with the finished theory. For almost a century, the relativity community has been engaged in understanding the astonishingly rich physics the theory contains, testing it ever more accurately, and

A. Ashtekar (✉)
Institute for Gravitation and the Cosmos and Physics Department, Penn State, University Park, PA16802, USA
e-mail: Ashtekar@gravity.psu.edu

J. Bičák and T. Ledvinka (eds.), *General Relativity, Cosmology and Astrophysics*, Fundamental Theories of Physics 177, DOI: 10.1007/978-3-319-06349-2_16,
© Springer International Publishing Switzerland 2014

applying it to greater and greater domains of astrophysics and cosmology. The theory has so many marvelous features. Amazingly, the field equations turned out to provide an elliptic-hyperbolic system with a well-posed initial value problem. After many decades, we realized that the total mass of an isolated system is a well defined geometric invariant and, furthermore, positive if the local energy density of matter is positive. The theory naturally admits cosmological solutions in which the universe is expanding, just as the observations tell us. It admits black hole solutions that serve as engines for the most energetic phenomena seen in the universe. None of these fascinating features that we now regard as fundamental consequences were part of Einstein's motivation during his quest which he described as "*one of the most exciting and exacting times of my life*" [1]. He essentially handed to us the finished product on a platter. We have been engaged in uncovering the numerous hidden treasures it contains by working out the philosophical, mathematical, physical, astronomical and cosmological consequences of the new paradigm.

But we know that the theory is incomplete. Indeed, it exhibits its own fundamental limitations through singularities where space-time ends and general relativistic physics comes to a halt. We also understand that this occurs because general relativity ignores quantum physics. Perhaps the most outstanding example is the prediction of the big bang. If we go back in time, *much* before we reach the singularity, matter densities exceed the nuclear density, $\sim 10^{14}$–$10^{15}$ gms/cc, where we definitely know that quantum properties of matter dominate. Since gravity couples to matter, the conceptual paradigm of general relativity becomes inadequate. If we go further back in time, general relativity presents us with an epoch in which densities reach $\sim 10^{94}$ gms/cc. This is the Planck scale and now physics of general relativity becomes inadequate not only conceptually but also in practice. In this regime we expect gross departures from Einstein's theory. Just as it is totally inadequate to use Newtonian mechanics to explore physics near the horizon of a solar mass black hole, it is incorrect to trust general relativity once the matter density and space-time curvature enter the Planck regime. Thus, *big bang is a prediction of general relativity in a domain in which it is simply invalid.* Normally physicists do not advertise such predictions of theories. But unfortunately they often seem to make an exception for the big bang. One hears statements like 'the cosmic microwave background (CMB) is a fingerprint of the big bang'. But in the standard scenario, CMB refers to a time some 380,000 years after the putative big bang. Existence or even the detailed features of CMB have no bearing on whether the big bang with *infinite* matter density and curvature ever occurred. Indeed, as we will see, loop quantum cosmology (LQC) has no big bang singularity and yet reproduces these features. What about inflation? In the standard scenario, it is supposed to have commenced 'only' $10^7$ Planck seconds after the big bang. Does its success not imply that there was a big bang? It does *not* because the matter density and curvature at the onset of inflation are only $10^{-11}$–$10^{-12}$ times the Planck scale. Indeed, this is why one can use Einstein's equations and quantum field theory (QFT) on Friedmann, Lemaître, Robertson, Walker (FLRW) solutions in the analysis of inflation. Inflationary physics by itself cannot say what really happened in the Planck regime and, again, as we will see, is compatible with the LQC prediction that there was no big bang singularity.

Thus, to know what really happened in the Planck regime and go beyond the singularities predicted by general relativity, we need a viable quantum theory of gravity. Since the search for this theory has been ongoing for decades, justifiably, there is sometimes a sentiment of pessimism in the general relativity circles. In my view, this is largely because one judges progress using the criterion of general relativity. In a masterful stroke, Einstein gave us the final theory and we have been happily engaged in investigating its content. It seems disappointing that this has not happened with quantum gravity. But progress of physical theories has more often mimicked the development of quantum theory rather than general relativity. More than a century has passed since Planck's discovery that launched the quantum. Yet, the theory is incomplete. We do not have a satisfactory grasp of the foundational issues, often called the 'measurement problem', nor do we have a single example of an interacting QFT in 4 dimensions. A far cry from what Einstein offered us in 1915! Yet, no one would deny that quantum theory has been extremely successful; indeed, much more so than general relativity.

Thus, while it is tempting to wait for another masterful stroke like Einstein's to deliver us a finished quantum gravity theory, it is more appropriate to draw lessons from quantum theory. There, progress occurred by focussing not on the 'final, finished' theory, but on concrete physical problems where quantum effects were important. It would be more fruitful to follow this path in quantum gravity. Indeed, even though we are far from a complete theory, advances can occur by focusing on specific physical problems and challenges.

Over the last several years, research in loop quantum gravity (LQG) has been driven by this general philosophy. In addition to seeking a completion of the general program based on connection variables, spin networks and spin foams, more and more effort is now focused on specific physical problems where quantum gravity effects are expected to be important. The idea behind this research is to *first truncate general relativity (with matter) to sectors tailored to specific physical problems, and then pass to quantum theory using the background independent methods based on the specific quantum geometry that underlies LQG.* This strategy of focusing on specific problems of quantum gravity also distinguishes LQG from string theory in terms of their main trust in the last few years. In string theory, the focus has shifted to *using* the well-understood parts of gravity to explore other areas of physics—use of the AdS/CFT hypothesis to understand the strong coupling regime of QCD, to gain new insights into hydrodynamics and tackle the strong coupling problems in mathematical physics to better understand condensed matter systems such as high temperature super-conductivity. The LQG community, on the other hand, has continued to tackle the long standing problems of quantum gravity per se—absence of a space-time in the background, the problem of time, fate of cosmological singularities in the quantum theory, quantum geometry of horizons, and derivation of the graviton propagator in a background independent setting.

The goal of my talk was to report the advances in the cosmology of the very early universe that have resulted from a continued application of the truncation strategy in LQG. Of course, both the talk and this report can only provide a bird's eye view of these developments. The results I reported are based largely on joint work with

Corichi, Pawlowski and Singh [2–6] on the singularity resolution in cosmology; with David Sloan [7, 8] on effective LQC dynamics tailored to inflation, with Kaminski and Lewandowski [9] on QFT on quantum space-times; and especially with Ivan Agullo and William Nelson on extension of the cosmological perturbation theory to the Planck regime and its application to inflation [10–12]. Therefore, there is a large overlap with the material covered in these original references. Finally, by now there are well over a 1000 papers on LQC which include several investigations of inflationary dynamics. What I can cover constitutes only a very small fraction of what is known. For reviews on results until about a year ago, see, e.g. [13, 14].

## 2 Setting the Stage

Perhaps the most significant reason behind the rapid and spectacular success of quantum theory, especially in its early stage, is the fact that there was already a significant accumulation of relevant experimental data, and further experiments to weed out ideas could be performed on an ongoing basis. Unfortunately this is *not* the case for quantum gravity simply because theory has raced far ahead of technology. Indeed, even in the classical regime, we still lack detailed tests of general relativity in the strong field regime!

Currently, the early universe offers by far the best arena to test various ideas on quantum gravity. Most scenarios assume that the early universe is well described by a FLRW solution to Einstein's equations with suitable matter, *together with* first order perturbations. The background is treated classically, as in general relativity, and the perturbations are described by *quantum fields*. Thus, the main theoretical ingredient in the analysis are: cosmological perturbation theory and QFT on FLRW space-times. It is fair to say that among the current scenarios, the inflationary paradigm has emerged as the leading candidate. In addition to the common assumption described above, this scenario posits:

- Sometime in its early history, the universe underwent a phase of rapid expansion. This was driven by the slow roll of a scalar field in a suitable potential causing the Hubble parameter to be nearly constant.
- Fourier modes of the quantum fields representing perturbations were initially in a specific state, called the Bunch-Davies (BD) vacuum, for a certain set of co-moving wave numbers $(k_o, 2000k_o)$ where the physical wave length of the mode $k_o$ equals the radius $R_{\rm LS}$ of the observable universe at the surface of last scattering.[1]

[1] Strictly speaking, the BD vacuum refers to deSitter space; it is the unique 'regular' state which is invariant under the full deSitter isometry group. During slow roll, the background FLRW geometry is only approximately deSitter whence there is some ambiguity in what one means by the BD vacuum. One typically assumes that all the relevant modes are in the BD state (tailored to) a few e-foldings before the mode $k_o$ leaves the Hubble horizon. Throughout this report, by BD vacuum I mean this state.

- Soon after any mode exits the Hubble radius, its quantum fluctuation can be regarded as a classical perturbation and evolved via linearized Einstein's equations.

One then evolves the perturbations from the onset of the slow roll till the end of inflation using QFT on FLRW space-times and calculates the power spectrum (see, e.g., [15–19]). When combined with standard techniques from astrophysics to further evolve the results to the surface of last scattering, one finds that they are in excellent agreement with the inhomogeneities seen in the CMB. Supercomputer simulations have shown that these inhomogeneities serve as seeds for the large scale structure in the universe. Thus, in a precise sense, the origin of the qualitative features of the observed large scale structure can be traced back to the fluctuations in the quantum vacuum at the onset of inflation. This is both intriguing and very impressive.

Over the years, the inflationary paradigm has witnessed criticisms from the relativity community, most eloquently expressed by Roger Penrose (see, e.g., [20]). However, these criticisms refer to the motivations that were originally used by the proponents, rather than to the methodology underlying its success in accounting for the CMB inhomogeneities. There are plenty of examples in fundamental physics where the original motivations turned out not to be justifiable but the idea was highly successful. I share the view that, while the basic assumptions, listed above, have not been justified from first principles, the success of the inflationary paradigm with CMB measurements is impressive because one 'gets much more out than what one puts in'.

In spite of this success, however, the inflationary scenario is conceptually incomplete in several respects. (For a cosmology perspective on these limitations see e.g. [21, 22].) In particular, as Borde, Guth and Vilenkin [23] showed, inflationary space-times inherit the big-bang singularity in spite of the fact that the inflation violates the standard energy conditions used in the original singularity theorems [24]. As we discussed in Sect. 1, this occurs because one continues to use general relativity even in the Planck regime in which it is simply not applicable. One expects new physics to play a dominant role in this regime, thereby resolving the singularity and significantly changing the very early history of the universe. One is therefore led to ask: Will inflation arise naturally in the resulting deeper theory? Or, more modestly, can one at least obtain a consistent quantum gravity extension of this scenario?

The open-ended nature of the inflationary paradigm has three facets. First, there are issues whose origin lies in particle physics. Where does the inflaton come from? How does potential arise? Is there a single inflaton or many? If many, what are the interactions between them? Since the required mass of the inflaton is very high, above $10^{12}$ Gev, the fact that we have not seen it at CERN does not mean it cannot exist. But in the inflationary scenario this is the only matter field in the early universe and particles of the standard model are supposed to be created during 'reheating' at the end of inflation when the inflaton is expected to roll back and forth around its minimum. However, how this happens is not at all well-understood. What are the admissible interactions between the inflaton and the standard model particles which causes this decay? Does the decay produce the correct abundance of the standard model particles? These questions with origin in particle physics are wide open.

The second issue is the quantum to classical transition referred to in the last assumption of standard inflation. In practice one calculates the expectation values of perturbations and the two point function at the end of inflation and assumes one can replace the actual quantum state of perturbations with a Gaussian statistical distribution of classical perturbations with the mean and variance given by the quantum expectation value and the 2-point function. As a calculational devise this strategy works very well. However, what happens physically? While this issue has drawn attention, we do not yet have a clear consensus on the actual, detailed physics that is being approximated in the last assumption.

The third set of issues have their origin in quantum gravity. In the standard inflationary scenario, one specifies initial conditions at the onset of inflation and then evolves the quantum perturbations. As a practical strategy, something like this is unavoidable within general relativity. Ideally one would like to specify the initial conditions at 'the beginning', but one simply cannot do this because the big bang is singular. Furthermore, since the curvature at the onset of inflation is some $10^{-11}$–$10^{-12}$ times the Planck scale, by starting calculations there, one bypasses the issue of the correct Planck scale physics. But this is just an astute stopgap measure. Given any candidate quantum gravity theory, one can and *has to* ask whether one can do better. Can one meaningfully specify initial conditions in the Planck regime? In a viable quantum gravity theory, this should be possible because there would be no singularity and the Planck scale physics would be well-controlled. If so, in the systematic evolution from there, does a slow roll phase compatible with the 7 year WMAP data [25] arise *generically* or is an enormous fine tuning needed? One could argue that it is acceptable to use fine tuning because, after all, the initial state is very spatial. If so, can one provide physical principles that select this special state? In the standard inflationary scenario, if we evolve the modes of interest back in time, they become trans-Planckian. Is there a QFT on *quantum* cosmological space-times needed to adequately handle physics at that stage? Can one *arrive at* the BD vacuum (at the onset of the WMAP slow roll) starting from the initial conditions at the Planck scale?

In this report, I will not address the first two sets of issues. Rather, the focus will be on the incompleteness related to the third set, i.e., on quantum gravity. Systematic advances within LQC over the past six years have provided a viable extension of the inflationary scenario all the way to the Planck regime. This extension enables us to answer in detail most of the specific questions posed above. To arrive at a coherent extension, LQC had to develop a conceptual framework, mathematical tools and high precision numerical simulations because the issues are so diverse: The meaning of time in the Planck regime; the nature of quantum geometry in the cosmological context; QFT on *quantum* cosmological space-times; renormalization and regularization of composite operators needed to compute stress energy and back reaction; and, relation between theory and the WMAP data.

A consistent theoretical framework to deal with cosmological perturbations on quantum FLRW space-times now exists [11]. Starting with 'natural' initial conditions in the Planck regime, one can evolve the quantum perturbations on quantum FLRW backgrounds and study in detail the pre-inflationary dynamics [10, 12]. Detailed numerical simulations have shown that the *predictions are in agreement with the power spectrum and the spectral index reported in the 7 year WMAP data.* However, there is also a small window in the parameter space where the initial state at the onset of inflation differs sufficiently from the BD vacuum assumed in standard inflation to give rise to new effects. These are the prototype observable signatures of pre-inflationary dynamics. In this sense, *LQC offers the possibility of extending the reach of cosmological observations to the deep Planck regime of the early universe.*

# 3 Why Pre-inflationary Dynamics Matters

It is often claimed that pre-inflationary dynamics will not change the observable predictions of the standard inflationary scenario. Indeed, this belief is invoked to justify why one starts the analysis just before the onset of the slow roll. The belief stems from the following argument, sketched in the left panel of Fig. 1. If one evolves the modes that are seen in the CMB *back* in time starting from the onset of slow roll, their physical wave lengths $\lambda_{\rm phy}$ continue to remain within the Hubble radius $1/H_{\rm GR}$ all the way to the big bang. Therefore, one argues, they would not experience curvature and their dynamics would be trivial all the way from the big bang to the onset of inflation; because they are not 'exited', all these modes would be in the BD vacuum at the onset of inflation. However, this argument is flawed on two accounts. First, if one examines the equation governing the evolution of these modes, one finds that what matters is the curvature radius $R_{\rm curv} = \sqrt{6/\mathfrak{R}}$ determined by the Ricci scalar $\mathfrak{R}$, and not the Hubble radius. The two scales are equivalent only during slow roll on which much of the intuition in inflation is based. However, in general they are quite different from one another. Thus we should compare $\lambda_{\rm phy}$ with $R_{\rm curv}$ in the pre-inflationary epoch. The second and more important point is that the pre-inflationary evolution should not be computed using general relativity, as is done in the argument given above. One has to use an appropriate quantum gravity theory since the two evolutions are expected to be *very* different in the Planck epoch. Then modes that are seen in the CMB could well have $\lambda_{\rm phy} \gtrsim R_{\rm curv}$ in the pre-inflationary phase. If this happens, these modes *would be* excited and the quantum state at the onset of the slow roll could be quite different from the BD vacuum. Indeed, the difference could well be so large that the amplitude of the power spectrum and the spectral index are incompatible with WMAP observations. In this case, that particular quantum gravity scenario would be ruled out. On the other hand, the differences could be more subtle: the new power spectrum for scalar modes could be compatible with observations but there may be departures from the standard predictions that involve tensor modes or higher order correlation functions of scalar modes, changing the standard conclusions on non-Gaussianities [26–30]. In this case, the quantum gravity theory would have interesting predictions for future observational missions. Thus,

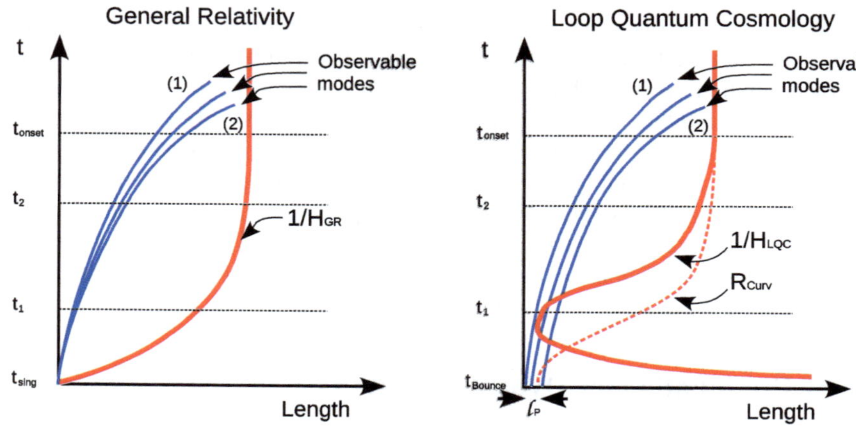


**Fig. 1** Schematic time evolution of the Hubble radius (*red solid line* on the *right* in each panel) and of wave lengths of three modes seen in the CMB (three *solid blue lines* in each panel). Credits: W. Nelson. *Left Panel: General relativity.* The modes of interest have wave lengths less than the Hubble radius $1/H_{\rm GR}$ all the way from the big bang ($t_{\rm sing}$) until after the onset of slow roll. *Right Panel: LQC.* The Hubble radius diverges at the big bounce ($t_{\rm Boun}$), decreases rapidly to reach its minimum in the deep Planck era and then increases monotonically. Because of this, modes seen in the CMB can have wave lengths larger than the Hubble radius $1/H_{\rm LQC}$ in the very early universe. Detailed analysis shows that what really matters is the curvature radius $R_{\rm curv}$ shown schematically by the *dashed red line* rather than the Hubble radius $1/H_{\rm LQC}$. But again the modes can exit the curvature radius in the Planck regime and, if they do, they are excited during the pre-inflationary evolution. They will not be in the BD vacuum at the onset of slow roll inflation

pre-inflationary dynamics can provide an avenue to confront quantum gravity theories with observations.

These are not just abstract possibilities. The right panel of Fig. 1 shows schematically the situation in LQC. (For the precise behavior obtained from numerical simulations, see [12, Fig. 1].) The wave lengths of some of the observable modes *can* exit the curvature radius during pre-inflationary dynamics, whence there are departures from the standard predictions (which turn out to be of the second type in the discussion above).

So far we have focused only on why a common argument suggesting that pre-inflationary dynamics cannot have observational consequences is fallacious. At a deeper level, pre-inflationary dynamics matters because of a much more general reason: It is important to know if inflationary paradigm is part of a conceptually coherent framework encompassing the quantum gravity regime. Can one trust the standard scenario in spite of the fact that the modes it focuses on become trans-Planckian in the pre-inflationary epoch? Does one have to artificially fine-tune initial conditions in the Planck regime to arrive at the BD vacuum? Do initial conditions for the background in the Planck regime naturally give rise to solutions that encounter the desired inflationary phase some time in the future evolution? To investigate any one of these issues, one needs a reliable theory for pre-inflationary dynamics and also good control on its predictions.

# 4 The LQG Strategy

LQG offers an attractive framework to investigate pre-inflationary dynamics because its underlying quantum geometry becomes important at the Planck scale and leads to the resolution of singularities in a variety of cosmological models. In particular the following cosmologies have been investigated in detail: the k = 0 and k = 1 FLRW models are discussed in [2–5, 31–35]; a non-zero cosmological constant is included in [6, 36, 37]; anisotropic are discussed via Bianchi I, II and IX models in [1, 38–40]; and the inhomogeneous Gowdy models, which have been analyzed in detail in classical general relativity, have been studied in [41–46]. In all cases, the big bang singularity is resolved and replaced by quantum bounces. It is therefore natural to use LQC as the point of departure for extending the cosmological perturbation theory.

In the standard perturbation theory, one begins with linearized solutions of Einstein's equations on a FLRW background. Unfortunately, we cannot mimic this procedure because in LQG we do not yet have the analog of full Einstein's equations that one would have to perturb. But one can adopt the *truncation strategy* discussed in Sect. 1. Thus, one starts with a truncation $\Gamma_{\rm Trun}$ of the phase space $\Gamma$ of general relativity, tailored to the linear perturbations off FLRW backgrounds. Furthermore since we are interested in the issue of whether the inflationary framework admits a quantum gravity extension, the matter source will be just a scalar field $\phi$ with the simplest, i.e. quadratic, potential $V(\phi) = (1/2)m^2\phi^2$. Thus, $\Gamma_{\rm Trun}$ is given by $\Gamma_{\rm Trun} = \Gamma_o \times \Gamma_1$ where $\Gamma_o$ is the 4-dimensional FLRW phase space, with the scale factor $a$ and the homogeneous inflaton $\phi$ as configuration variables, and $\Gamma_1$ is the phase space of gauge invariant first order perturbations consisting of a scalar mode and two tensor modes. Since the background fields are homogeneous, it is simplest to assume that the perturbations are purely inhomogeneous. Thus, regarded as a sub-manifold of the full phase space $\Gamma$, $\Gamma_{\rm Trun}$ is the normal bundle over $\Gamma_o$.

As usual, for perturbations one can freely pass between real space and momentum space using Fourier transforms of fields in co-moving coordinates. For pre-inflationary dynamics, we work with the Mukhanov–Sasaki variables, denoted by $\mathscr{Q}_{\mathbf{k}}$, because they are well-defined all the way from the bounce to the onset of slow roll.[2] We denote the two tensor modes collectively by $\mathscr{T}_{\mathbf{k}}$. This structure is the same as that used in standard inflation [47].

New features appear in the next step: In the passage to quantum theory, we work with the *combined system*, i.e., with all of $\Gamma_{\rm Trun}$. Therefore, we are naturally led a theory in which not only the perturbations but even the background geometry is quantum. Rather than having quantum fields $\hat{\mathscr{Q}}$ and $\hat{\mathscr{T}}$ propagating on a classical FLRW space-time, they now propagate on a *quantum* FLRW geometry.

[2] The curvature perturbations $\mathscr{R}_{\mathbf{k}}$ fail to be well-defined at the 'turning point' where $\dot{\phi} = 0$, which occurs during pre-inflationary dynamics. However, they are much more convenient for relating the spectrum of perturbations at the end of inflation with the CMB temperature fluctuations. Therefore, we first calculate the power spectrum $\mathscr{P}_{\mathscr{Q}}$ for Mukhanov–Sasaki variable $\mathscr{Q}_{\mathbf{k}}$ and then convert it to $\mathscr{P}_{\mathscr{R}}$, reported in Fig. 3.

Thus, the strategy to truncate the classical phase space and then pass to quantum theory using LQG techniques leads to a novel quantum theory. The total Hilbert space is a tensor product, $\mathscr{H} = \mathscr{H}_o \otimes \mathscr{H}_1$, where $\mathscr{H}_o$ is the space of wave functions $\Psi_o$ describing a quantum FLRW geometry and $\mathscr{H}_1$ is the space of quantum states $\psi$ of perturbations. The first task is to construct the Hilbert space $\mathscr{H}_o$ of physical states $\Psi_o(a, \phi)$, by imposing the Hamiltonian constraint on the homogeneous sector $\Gamma_o$. The second task is to study quantum dynamics of fields $\hat{\mathscr{Q}}$ and $\hat{\mathscr{T}}$ on the quantum geometry encapsulated in $\Psi_o(a, \phi)$. In particular we have to introduce the Hilbert space $\mathscr{H}_1$ of wave functions $\psi(\mathscr{Q}_{\mathbf{k}}, \mathscr{T}_{\mathbf{k}})$ of perturbations and develop techniques to calculate the 2-point functions on $\mathscr{H}_1$ that are needed to obtain the scalar and the tensor power spectra. The final task is to check the self-consistency of the truncation strategy with which we began. Already in the classical theory, the truncated phase space $\Gamma_{\mathrm{Trun}}$ is useful only so long as the back reaction can be neglected. Therefore, in the quantum theory, we have to check that the $\mathscr{H}$ admits solutions $\Psi_o \otimes \psi$ in which the energy density of perturbations is negligible compared to that in the background all the way from the LQC bounce to the onset of slow roll. On the analytical side, this requires the introduction of suitable regularization and renormalization techniques for quantum fields $\hat{\mathscr{Q}}$ and $\hat{\mathscr{T}}$ propagating on the quantum background $\Psi_o$. On the numerical side, one has to devise accurate numerical methods to calculate the energy density in perturbations with sufficient precision during the evolution all the way from the bounce to the onset of inflation, as the background energy density falls by some 11 orders of magnitude.

These tasks have been carried out in [10–12] using earlier results obtained in [4–9]. The next two sections provide a flavor of this analysis.

## 5 Analytical Aspects

**Background Quantum Geometry** In the classical theory, dynamics on $\Gamma_o$ is generated by the single, homogeneous, Hamiltonian constraint, $\mathbb{C}_o = 0$. Each dynamical trajectory on $\Gamma_o$ represents a classical FLRW space-time. In quantum theory, physical states are represented by wave functions $\Psi_o(a, \phi)$ satisfying the quantum constraint $\hat{\mathbb{C}}_o \Psi_o = 0$. Each of these solutions represents a quantum FLRW geometry.
We are interested in those solutions $\Psi_o$ which remain sharply peaked on classical FLRW solutions at late times. In the sector of the theory that turns out to be physically most interesting [12], these states remain sharply peaked all the way up to the bounce but in the Planck regime they follow certain effective trajectories which include quantum corrections [14]. In particular, rather than converging on the big bang singularity, as classical FLRW solutions do, they exhibit a bounce when the density reaches $\rho_{\max} \approx 0.41\rho_{\mathrm{Pl}}$ (see Fig. 2). It turns out that each (physically distinct) effective solution is completely characterized by the value $\phi_{\mathrm{B}}$ that the inflaton assumes at the bounce. *This value turns out to be the key free parameter of the theory.* Finally, we need full quantum evolution from the bounce only until the density and curvature

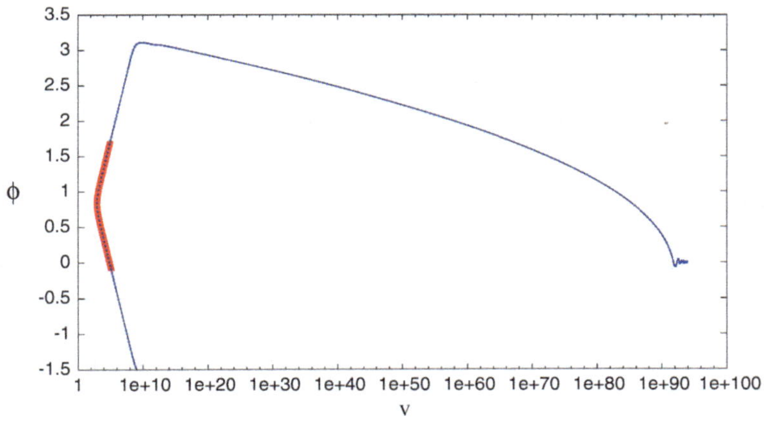


**Fig. 2** An effective LQC trajectory in presence of an inflation with a quadratic potential $(1/2)m^2\phi^2$, where the value $m = 6.1 \times 10^{-6} m_{Pl}$ of the mass is calculated from the 7 year WMAP data (source). Here $V \sim a^3$ is the volume of a fixed fiducial region. The long (*blue*) sloping line at the *top* depicts slow roll inflation. As $V$ decreases (*right* to *left*), we go back in time and the inflaton $\phi$ first climbs up the potential, then turns around and starts going descending. In classical general relativity, volume would continue to decrease until it becomes zero, signalling the big bang singularity. In LQC, the trajectory bounces at $\phi \sim 0.95$ and volume never reaches zero; the entire evolution is non-singular

fall by a factor of, say, $10^{-6}$–$10^{-7}$. After that, the background can be taken to follow the general relativity trajectory to a truly excellent approximation.[3] (For details, see [4, 5]).

**Dynamics of Perturbations** There is an important subtlety which is often overlooked in the quantum gravity literature: Dynamics of perturbations is *not* generated by a constraint, or, indeed by *any* Hamiltonian. On the truncated phase space $\Gamma_{\rm Trun}$, the dynamical trajectories are tangential to a vector field $X^\alpha$ of the form $X^\alpha = \Omega_o^{\alpha\beta}\partial_\beta \mathbb{C}_o + \Omega_1^{\alpha\beta}\partial_\beta \mathbb{C}_2'$ where $\Omega_o$ and $\Omega_1$ are the symplectic structures on $\Gamma_o$ and $\Gamma_1$, and $\mathbb{C}_2'$ is the part of the second order Hamiltonian constraint function in which only terms that are quadratic in the first order perturbations are kept (ignoring terms which are linear in the second order perturbations). $X^\alpha$ fails to be Hamiltonian on $\Gamma_{\rm Trun}$ because $\mathbb{C}_2'$ depends not only on perturbations but also background quantities. However, given a dynamical trajectory $\gamma_o(t)$ on $\Gamma_o$ and a perturbation at a point thereon, $X^\alpha$ provides a canonical lift of $\gamma_o(t)$ to the total space $\Gamma_{\rm Trun}$, describing the evolution of that perturbation along $\gamma_o(t)$.

Therefore, in the quantum theory, dynamics of the combined system *cannot be obtained by simply imposing a quantum constraint* on the wave functions $\Psi_o \otimes \psi$

[3] During this phase, the scalar field is monotonic in time in the effective trajectory. Therefore we can use the scalar field as an 'internal' or 'relational' time variable with respect to which the background scale factor (and curvature) as well as perturbations evolve. This interpretation is not essential but very helpful in practice because of the form of the Hamiltonian constraint $\hat{\mathbb{C}}_o\, \Psi_o = 0$ (for details, see e.g. [14]).

of the combined system. One has to follow a procedure similar to what is done in the classical theory. Thus, one first obtains a background quantum geometry $\Psi_o$ by solving $\hat{\mathbb{C}}_o\, \Psi_o(a, \phi) = 0$, specifies the quantum state $\psi(\mathcal{Q}_{\mathbf{k}}, \mathcal{T}_{\mathbf{k}})$ of the perturbation at, say, the bounce time, and evolves it using the operator $\hat{\mathbb{C}}_2'$. The resulting state $\Psi(a, \mathcal{Q}_{\mathbf{k}}, \mathcal{T}_{\mathbf{k}}, \phi)$ describes the evolution of the quantum perturbations $\psi$ on the quantum geometry $\Psi_o$ in the Schrödiger picture. (For details, see [11]).

**Trans-Planckian Issues** Quantum perturbations $\hat{\mathcal{Q}}$, $\hat{\mathcal{T}}$ propagate on quantum geometries $\Psi_o$ which are all regular, free of singularities. Thus, the *the framework is tailored to cover the Planck regime.* What is the status of the 'trans-Planckian problems' which are associated with modes of trans-Planckian frequencies in heuristic discussions? To probe this issue one has to first note that the quantum Riemannian geometry underlying LQG is quite subtle [48–50]: in particular, while there is a minimum non-zero eigenvalue of the area operators, the area gap, there is no volume gap, even though their eigenvalues are also discrete [51, 52].[4] As a consequence, there is no fundamental obstacle preventing the existence of trans-Planckian modes of perturbations in our truncated theory. Indeed, in the homogeneous LQC models that have been analyzed in detail, the momentum $p_{(\phi)}$ of the scalar field $\phi$ is generally *huge* in Planck units. This poses no problem and, in particular, on the physical Hilbert space the total energy density is still guaranteed to be bounded by $\rho_{\max}$ (see, e.g. [14]). Similarly, perturbations $\hat{\mathcal{Q}}$, $\hat{\mathcal{T}}$ of our truncated theory are permitted to acquire trans-Planckian momenta. The real danger is rather that, in presence of such modes, the *energy density* in perturbations may fail to be negligible compared to that in the quantum background geometry. This issue is extremely non-trivial, especially in the Planck regime. If the energy density does become comparable to that in the background, then we would not be able to neglect the back-reaction and our truncation would fail to be self-consistent.[5] *This is the trans-Planckian problem we face in our theory of quantum perturbations on inflationary quantum geometries.* To address it we need regularization and renormalization methods to compute energy density for quantum fields on quantum FLRW geometries. (For details, see [11, 12]).

**An Unforeseen Simplification** As we just noted, the underlying FLRW quantum geometry provides the necessary control on calculations in the deep Planck regime. However, it confronts us with a new challenge of developing the mathematical theory of quantum fields on quantum geometries. At first this problem seems formidable. But fortunately there is a key simplification within the test field approximation we are using in the truncated theory [9, 11]: Mathematically the evolution of $\hat{\mathcal{Q}}$, $\hat{\mathcal{T}}$ on any one of our quantum geometries $\Psi_o$ is completely equivalent to that of these

---

[4] Properties of the eigenvalues of length operators [53–55] have not been analyzed in comparable detail. But since their definitions involve volume operators, it is expected that there would be no 'length gap'.

[5] Of course, this would not imply that the inflationary scenario does not admit an extension to the Planck regime. But to obtain it one would then have to await the completion of a *full* quantum gravity theory.

fields propagating on a dressed, effective metric $\tilde{g}_{ab}$ constructed from $\Psi_o$.[6] Note that $\tilde{g}_{ab}$ contains quantum corrections and does not satisfy Einstein's equation. Indeed, it does not even satisfy the effective equations of LQC because, whereas the effective trajectories follow the 'peak of $\Psi_o$', $\tilde{g}_{ab}$ also knows about certain fluctuations encoded in $\Psi_o$.[7] Nonetheless, since $\tilde{g}_{ab}$ is a smooth metric with FLRW symmetries, it is now possible to use the rich machinery of QFT on cosmological space-times to analyze the dynamics of $\hat{\mathcal{Q}}$, $\hat{\mathcal{T}}$ in detail. In addition, one can now make use of the powerful technique of adiabatic regularization that has been developed over some three decades [56–61]. In particular, by restricting ourselves to states $\psi$ of perturbations which are of fourth adiabatic order, one can compute the expectation values of energy density. This provides a clear avenue to face the true trans-Planckian problem, i.e., to systematically test if the truncation approximation is valid.

This remarkable simplification occurs because the dynamics of test quantum fields is not sensitive to all the details of the probability amplitude for various FLRW metrics encapsulated in $\Psi_o$; it experiences only to a few moments of this distribution. The phenomenon is analogous to the propagation of light in a medium where all the complicated interactions of the Maxwell field with the atoms in the medium can be captured just in a few parameters such as the refractive index. (For details, see [9, 11, 12]).

**Initial Conditions** In the Schrödinger picture, the above simplification enables us to evolve the quantum state $\psi$ of perturbations. But we still have to specify the initial conditions. Since the big bang of general relativity is replaced by the big bounce in LQC, it is natural to specify them at the bounce. Now, in the truncation approximation, perturbation is treated as a test field. Therefore, it is appropriate to assume that the initial state has the form $\Psi_o \otimes \psi$ at the bounce. Furthermore this tensor product form will be preserved under dynamics so long as the back reaction due to the perturbation remains negligible.

Let us begin with $\Psi_o$. In the effective theory, phase space variables are subject to certain constraints at the bounce. We assume that $\Psi_o$ is sharply peaked at a point on this constraint surface (with small fluctuations in each of the two 'conjugate' variables). At the bounce, the allowed range of $\phi$ is finite but large, $|\phi_{\rm B}| \in (0, 7.47 \times 10^5)$ in Planck units, and a detailed analysis of effective solutions has shown that unless $|\phi_{\rm B}| < 0.93$, the effective trajectory necessarily encounters a slow roll phase compatible with WMAP sometime in the future [8]. Thus, the peak of initial $\Psi_o$ is almost unconstrained. However, the requirement that $\Psi_o$ be peaked is very strong and makes the initial state of background geometry very special.

For perturbations, we assume the following three conditions on $\psi$ at the bounce: (i) Symmetry: $\psi$ should be invariant under the FLRW isometry group, i.e., under spatial

---

[6] For scalar modes, the classical equation of motion involves also 'an external potential' $\mathfrak{A}$. This has also to be replaced by a dressed effective potential $\tilde{\mathfrak{A}}$. For details, see [12].

[7] While this difference is conceptually important, because the states $\Psi_o$ of interest are so sharply peaked, in practice the deviations from effective trajectories are small even in the Planck regime. Of course the deviations from classical solutions are enormous in the Planck regime because $\tilde{g}_{ab}$ is non-singular.

translations and rotations. This condition is natural because these are the symmetries of the background $\Psi_o$ and hence also of $\tilde{g}_{ab}$ it determines; (ii) Regularity: $\psi$ should be of forth adiabatic order so that the Hamiltonian operator has a well-defined action on it; and, (iii) The initial renormalized energy density $\langle\psi|\,\hat{\rho}\,|\psi\rangle_{\rm ren}$ in the perturbation should be negligible compared to the energy density $\rho_{\rm max}$ in the background. We have an explicit example, $|\psi\rangle = |0_{\rm obv}\rangle$, of such a state called the 'obvious vacuum of forth adiabatic order' which has several attractive properties [12]. Furthermore we also know that, given a state satisfying these properties, there are 'infinitely many' such states in its neighborhood. Thus, the existence of the desired states is assured. However, in view of the large freedom that remains, it is desirable to develop clear-cut physical criteria to cut down this freedom significantly. This is an open issue, currently under investigation. (For details, see [10–12]).

Let us summarize the analytical framework. The initial condition for the quantum state $\Psi_o \otimes \psi$ of the combined system is specified at the bounce in such a manner that a slow roll inflation compatible with the 7 year WMAP data is guaranteed in the background geometry. Thanks to an unforeseen simplification, we can use techniques from QFT on cosmological space-times to evolve the perturbations $\hat{\mathcal{Q}}$ and $\hat{\mathcal{T}}$ on the quantum background geometry $\Psi_o$. Finally, the initial conditions guarantee that the truncation approximation does hold at the bounce: $\psi$ can be regarded as a perturbation whose back reaction on $\Psi_o$ is negligible initially. Furthermore, states are sufficiently regular to enable us to calculate the energy density in the background and in the perturbation at all times. Therefore, one can carry out the entire evolution numerically, calculate the power spectra and spectral indices and check if the truncation approximation continues to hold under evolution all the way from the bounce to the onset of the slow roll.

As discussed in Sect.3, a priori there are several possible outcomes. Pre-inflationary dynamics could have such a strong effect that the power spectra and the spectral indices that result from these calculations are incompatible with the WMAP observations. In this case, the LQC extension would be ruled out by observations. It is also possible that the Planck scale dynamics is such that the back reaction ceases to be negligible very soon after the bounce making the truncation strategy inconsistent. In this case, our truncation strategy would fail to be self-consistent. Finally, even if these possibilities do not occur, we may find that, for observable modes, the state at the onset of inflation is sufficiently different from the BD vacuum that there are departures from the standard inflationary predictions for future observations. One needs explicit numerical simulations to find out which of these various a priori possibilities are realized.

## 6 Numerical Aspects, Observations and Self-Consistency

In this section, numerical values of all physical quantities will be given in natural Planck units $c = \hbar = G = 1$ (as opposed to the reduced Planck units used in the cosmology literature where one sets $8\pi G = 1$). We will use both the conformal time $\tilde{\eta}$ and the proper (or cosmic) time $\tilde{t}$ determined by the dressed effective metric $\tilde{g}_{ab}$ via $d\tilde{s}^2 := \tilde{g}_{ab}dx^a dx^b = a^2(-d\tilde{\eta}^2 + d\mathbf{x}^2) = -d\tilde{t}^2 + a^2 d\mathbf{x}^2$ (where, as usual, $x^a$ are the co-moving coordinates). This is because the cosmology literature generally uses conformal time but comparison with general relativity can be made more transparent in cosmic time by setting it equal to zero at the big bang in general relativity and at the big bounce in LQC.

## 6.1 WMAP Phenomenology

The 7 year WMAP data [25] uses a reference mode $k_\star \approx 8.58 k_o$ where, as before, $k_o$ is the co-moving wave number of the mode whose physical wave length equals the radius of the observable universe at the surface of last scattering. The WMAP analysis provides us with the amplitude $\mathscr{P}_{\mathscr{R}}(k_\star)$ of the power spectrum and the spectral index $n_s(k_\star)$ which encodes the small deviation from scale invariance, both for the scalar perturbations. The values are given by

$$\mathscr{P}_{\mathscr{R}}(k_\star) = (2.430 \pm 0.091) \times 10^{-9} \quad \text{and} \quad n_s(k_\star) = 0.968 \pm 0.012\,. \qquad (1)$$

For the quadratic potential considered here, these observational data provide the following values of the Hubble parameter $H$ and the slow roll parameter $\varepsilon = -\dot{H}/H^2$:

$$H(\tilde{\eta}(k_\star)) = 7.83 \times 10^{-6} \quad \text{and} \quad \varepsilon(\tilde{\eta}(k_*)) = 8 \times 10^{-3}\,, \qquad (2)$$

where $\tilde{\eta}(k_\star)$ is the conformal time in our dressed effective metric $\tilde{g}_{ab}$ at which the mode $k_\star$ exited the Hubble radius and the 'dot' refers to the derivative w.r.t. $\tilde{t}$. Since the physical wave length of the mode $k_o$ is 8.58 times larger, it must have left the Hubble radius ~2 e-foldings before $\tilde{\eta}(k_\star)$. Onset of slow roll inflation is taken to commence a little before the $k_o$ exits its Hubble horizon. The value of the Hubble parameter at this time is so low that the total energy density is less than $10^{-11}\rho_{\rm Pl}$. Therefore throughout the inflationary era general relativity is an excellent approximation to LQC. Equations of general relativity (or, LQC) determine the mass $m$ of the inflaton as well as values of the inflaton $\phi$ at $\tilde{\eta}(k_\star)$:

$$m = 1.21 \times 10^{-6} \quad \text{and} \quad \phi(\tilde{t}(k_\star)) = \pm 3.15\,. \qquad (3)$$

Because of the observational error bars, these quantities are uncertain by about 2 %. In the numerical simulations we use the value of $m$ given in (3). (For details, see [8]).

## 6.2 Evolution of the Background

So far numerical evolutions of the background wave function $\Psi_o$ are feasible only for kinetic dominated bounces, i.e., bounces for which $\phi_{\rm B}$ is small. This is because the required time over which one has to integrate to arrive in the general relativity regime increases rapidly with $\phi_{\rm B}$. Fortunately, as we will see below, this is the most interesting portion of the allowed values of $\phi_{\rm B}$. These simulations show that $\Psi_o$ remains sharply peaked on an effective trajectory. Since there is no obvious reason why this should not continue for higher $\phi_B$ values, it is instructive to examine all effective trajectories without restricting ourselves to kinetic energy dominated bounces. The trajectory would be compatible with the 7 year WMAP data *only if at the point at which* $H$ *takes the value* $7.83 \times 10^{-6}$, *within the margin given by observational errors,* $\varepsilon = 8 \times 10^{-3}$, *and* $\phi = 3.15$. A surprising result is that this is in fact the case under a *very* mild condition: $\phi_{\rm B} \geq 0.93$ [8]. Note that this result is much stronger than the qualitative 'attractor behavior' of inflationary trajectories because it is quantitative and tuned to the details of the WMAP observations. (For details, see [8]).

To make contact with the WMAP observations, we need to find $k_\star$ and the time $\tilde{\eta}(k_\star)$ at which the mode with co-moving wave number $k_\star$ exits the Hubble horizon during inflation. For this, it is simplest to fix the scale factor at the bounce and we will choose the convention $a_{\rm B} = 1$. (Note that this is very different from $a_{\rm today} = 1$ often used in cosmology.) Then, along each dynamical trajectory one locates the point at which the Hubble parameter takes the value $H = 7.83 \times 10^{-6}$ (and makes sure that at this time $\varepsilon$ and $\phi$ are given by (2) and (3) within observational errors). One calls the conformal time at which this occurs $\tilde{\eta}(k_\star)$ and numerically calculates the scale factor $a(\tilde{\eta}(k_\star))$ at this time. Then, the value of the co-moving momentum $k_\star$ of this mode is determined by the fact that this mode exits the Hubble radius at time $\tilde{\eta}(k_\star)$. Thus, one asks that the *physical* wave number of this mode should equal the Hubble parameter: $k/a(\tilde{\eta}(k_\star)) = H(\tilde{\eta}(k_\star))$. Table 1 shows the values of $k_\star$, the physical wave length of the mode at the bounce time, the *proper* time $\tilde{t}(k_\star)$ at which the mode exits the Hubble horizon, and the number of e-foldings between the bounce and time $\tilde{t}(k_\star)$ for a range of values of $\phi_{\rm B}$ which turns out to be physically most interesting. (For details, see [12]).

## 6.3 Evolution of Perturbations

Preliminary numerical simulations were first carried out using four different states $\psi$ at the bounce, satisfying the initial conditions discussed in Sect. 3. They showed that the results are essentially insensitive to the choice. Then detailed and much higher precision simulations were carried out using $|\psi\rangle = |0_{\rm obv}\rangle$, the 'obvious vacuum of fourth adiabatic order', at the bounce because, as mentioned before, this state has a number of attractive properties. These simulations revealed an unforeseen behavior:

**Table 1** This table from [12] shows the value of the reference co-moving momentum $k_\star$ used in the WMAP data, the corresponding physical wavelength $\lambda_\star(\tilde{t}_B)$ at the bounce, the time $\tilde{t}(k_\star)$ at which the mode $k_\star$ exits the Hubble radius during inflation, and $\ln[a(\tilde{t}(k_\star))/a(\tilde{t}_B)]$, the number of e-folds of expansion between the bounce and $\tilde{t}(k_\star)$

| $\phi(t_B)$ | $k_*$ | $\ln k_*$ | $\lambda_*(t_B)$ | $t_{k_*}$ | $\ln[a(t_{k_*})/a(t_B)]$ |
|---|---|---|---|---|---|
| 0.934 | 0.0016 | −6.4 | 4008 | $1.8 \times 10^5$ | 5.2 |
| 1 | 0.024 | −3.7 | 261 | $5.2 \times 10^5$ | 8.0 |
| 1.05 | 0.17 | −1.8 | 37.1 | $7.6 \times 10^5$ | 10 |
| 1.1 | 1.2 | 0.2 | 5.1 | $1.0 \times 10^6$ | 12 |
| 1.15 | 9.17 | 2.83 | 0.63 | $1.25 \times 10^6$ | 13.9 |
| 1.2 | 70.7 | 4.2 | 0.09 | $1.48 \times 10^6$ | 16 |
| 1.3 | $4.58 \times 10^3$ | 8.43 | $1.36 \times 10^{-3}$ | $1.97 \times 10^6$ | 20.2 |
| 1.5 | $2.7 \times 10^7$ | 17.1 | $2.3 \times 10^{-7}$ | $2.9 \times 10^6$ | 28.9 |

We focus on the range for $\phi_B$ that is relevant to explore whether pre-inflationary dynamics can lead to deviations from the BD vacuum at the onset of the slow roll

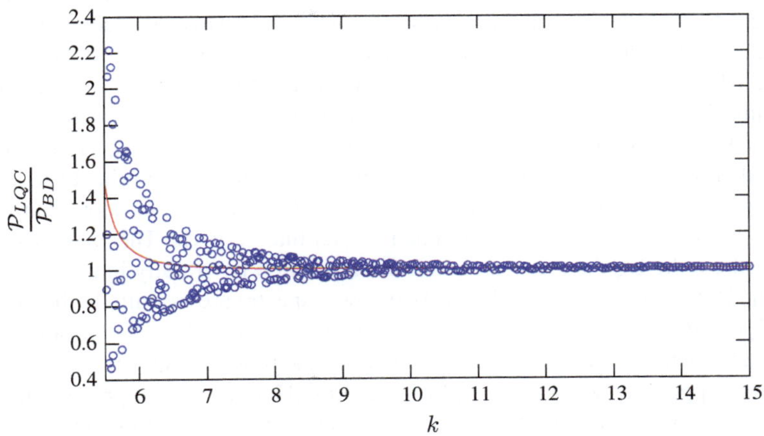


**Fig. 3** Ratio of the LQC power spectrum for curvature perturbations in the scalar modes to that predicted by standard inflation (source [12]). For small $k$, the ratio oscillates very rapidly. The (*red*) *solid curve* shows averages over (co-moving) bins with width $0.5\,\ell_{\rm Pl}{}^{-1}$

the power spectra for scalar and tensor perturbations are largely insensitive to the value of $\phi_B$. However, recall that there is finite window $(k_o,\ 2000k_o)$ of co-moving modes that can be seen in the CMB. Because of the pre-inflationary dynamics, the value of $k_\star$—and hence of $k_o$—does depend on $\phi_B$ and rapidly increases with $\phi_B$. (See Table 1.) Therefore, the window of observable modes *is sensitive to the value of* $\phi_B$ *and moves steadily to the right as* $\phi_B$ *increases.*

Figure 3 shows the plot of the ratio $\mathscr{P}_{\mathscr{R}}^{\rm LQC}/\mathscr{P}_{\mathscr{R}}^{\rm BD}$ of the LQC power spectrum to the standard inflationary one for curvature perturbations $\mathscr{R}$ of the scalar modes. The (blue) circles are the data points. The LQC power spectrum has very rapid oscillations (whose amplitudes decay quickly with $k$) which descend to the ratio that is plotted. Since observations have only a finite resolution, to compare with data it is simplest to average over small bins. We used bins which, at the bounce, correspond to a band-width in physical wave numbers of $0.5t_{\mathrm{Pl}}^{-1}$. The result is the solid (red) line. We see that the two power spectra agree for $k \gtrsim 6.5$ but LQC predicts an enhancement for $k \lesssim 6.5$. We will now comment on these features.

Let us first note that the LQC power spectrum in this plot uses the value $\phi_{\mathrm{B}} = 1.15$. As Table 1 shows, the corresponding $k_\star$ is 9.17. At this value, the two power spectra are identical, whence the amplitude and the spectral index obtained from the LQC evolution at $k = k_\star$ agrees with the values (1) observed by WMAP. However, as we remarked, for $k \lesssim 6.5$, the LQC prediction departs from that of standard inflation. These low $k$ values correspond to $\ell \lesssim 22$ in the angular decomposition used by WMAP for which the error bars are quite large. Therefore, although the LQG power spectrum differs from the standard one in this range, both are admissible as far as the current observations are concerned.

What is the physics behind the enhancement of the LQC power spectrum for $k \lesssim 6.5$? And where does this specific scale come from? This enhancement is due to pre-inflationary dynamics. At the bounce, the scalar curvature has a universal value in LQC which sets a scale $k_{\mathrm{LQC}} \approx 3.21$. Modes with $k \gg k_{\mathrm{LQC}}$ experience negligible curvature during their pre-inflationary evolution while those with $k$ comparable to $k_{\mathrm{LQC}}$ or less do experience curvature and therefore get excited. These are general physical arguments and one needs numerical simulations to determine exactly what 'much greater than' and 'comparable to' means. The simulations show that modes with $k \gtrsim 2k_{\mathrm{LQC}}$ already satisfy the 'much greater than' criteria. They are not excited and for them the LQC state $\psi$ at the onset of inflation is virtually indistinguishable from the BD vacuum. That is why the two power spectra are essentially the same for $k \gtrsim 2k_{\mathrm{LQC}}$. But for modes with $k \lesssim 2k_{\mathrm{LQC}}$ the LQC state $\psi$ has excitations over the BD vacuum whence there is an enhancement of the power spectrum.

What happens if we change $\phi_{\mathrm{B}}$? As we remarked above, the prediction of the LQC power spectrum is pretty insensitive to the value of $\phi_{\mathrm{B}}$ but the window in the $k$ space spanned by modes which are observable in the CMB changes, moving to the right as $\phi_{\mathrm{B}}$ increases. Now, as Table 1 shows, if $\phi_{\mathrm{B}} > 1.2$, we have $k_o > 6.5$, whence none of the observable modes would be excited during the pre-inflationary evolution. In this case, at the onset of the slow roll, the LQC sate $\psi$ would be indistinguishable from the BD vacuum, whence all LQC predictions would agree with those of standard inflation. Thus, there is a narrow window, $0.93 \leq \phi_{\mathrm{B}} \leq 1.2$ for which the background $\Psi_o$ admits the desired slow roll phase and yet LQC predictions for future observations can differ from the standard ones. One example is given by a consistency relation $r = -8n_t$ in standard inflation, where $r = 2\mathscr{P}_{\mathscr{T}}^{\mathrm{BD}}/\mathscr{P}_{\mathscr{R}}$ is the tensor to scalar ratio and $n_t$ is the spectral index for tensor modes. This relation is significant because it does not depend on the form of inflationary potential. It turns out that $r$ does not change in LQC but $n_t$ does, whence this standard consistency relation is modified. Future observations would be able to test for such departures. There is also a systematic study of the effect that excitations over the BD vacuum can have on non-Gaussianities [26–30]. Furthermore, it has been recently pointed out that these non-Gaussianities could be seen in the galaxy correlation functions and also in certain distortions in the CMB [62–64]. Thus, there are concrete directions in which cosmological observations could soon start probing effects that originate at the Planck scale. (For further details, see [12]).

### *6.4 Self Consistency*

Finally, let us discuss the issue of self-consistency of the truncation scheme, i.e., the issue of whether the test field approximation continues to hold under evolution. This issue is quite intricate and had remained unexplored because of two different issues. The first issue is conceptual: it was not clear how to compute the renormalized energy density for the quantum fields $\hat{\mathscr{Q}}$, $\hat{\mathscr{T}}$ in a manner that is meaningful in the Planck regime. As discussed in Sect. 5, we were able to construct this framework by 'lifting' the adiabatic renormalization theory on classical cosmological space-times to that on quantum geometries $\Psi_o$. The second set of difficulties comes from numerics: one requires very high accuracy and numerical precision. This is because (i) the rapid oscillations of integrand of $\langle\psi|\,\hat{\rho}\,|\psi\rangle_{\rm ren}$ in the $k$ space make it difficult to evaluate the exact value of the renormalized energy density; and, (ii) the background energy density itself decreases from Planck scale to $10^{-11}$ times that scale. Indeed, so far we have only managed to find an upper bound on the energy density in the perturbations, shown in Fig. 4. But this suffices to show that, for $\phi_B > 1.22$, our initial conditions at the bounce do give rise to a self-consistent solution $\Psi_o \otimes \psi$ throughout the evolution from the big bounce to the onset of slow roll. These solutions provide a viable extension of the standard inflationary scenario all the way to the Planck scale. The issue of whether one can push the value of $\phi_{\rm B}$ to include the interesting domain $\phi_{\rm B} < 1.2$ is still under investigation. (There are several aspects to this problem, including a better handling of the infrared regime, briefly discussed in [12].)

## 7 Summary and Discussion

I began in Sect. 1 by making some suggestions: (i) Progress in quantum gravity should be gauged by the degree to which an approach succeeds in overcoming limitations of general relativity; (ii) The development of quantum theory, rather than general relativity, offers a better example to emulate in this endeavor; and, (iii) As in quantum theory, it may be more fruitful to resolve concrete physical problems at the interface of gravity and quantum theory rather than focusing all efforts on obtaining a complete quantum gravity theory in one stroke. In Sects. 2 and 3 we saw that the very early universe offers an obvious arena for this task for both conceptual and practical reasons. Conceptually, the big bang is a prediction of general relativity in a regime in which the theory is not applicable, whence it is important to find out what really happened in the Planck regime. In practical terms, currently the early universe offers the

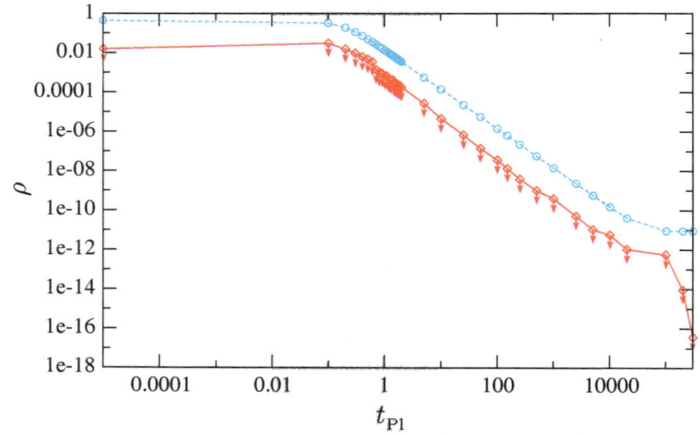


**Fig. 4** For $\phi_B = 1.23 m_{Pl}$, energy density in the background (*upper curve*) and an *upper bound* on the energy density in perturbations (*lower curve*) are plotted against time from the bounce to the onset of slow roll using Planck units (source [10]). The test field approximation holds across a change of 11 orders of magnitude in both quantities

best hope to confront quantum gravity theories with observations. In particular, we saw that the inflationary paradigm has been highly successful in accounting for the inhomogeneities in the CMB—and hence accounting for the large scale structure of the universe—but it has several limitations. In Sects. 4–6, I summarized how the limitations related to the Planck scale physics are being addressed in LQG. Specifically, by using the truncation strategy of LQG, over the last 6 years it has been possible to extend the inflationary paradigm all the way to the deep Planck regime. (For other treatments of pre-inflationary dynamics within LQG, see e.g. [65, 66].)

The first finding is that the big bang singularity is resolved in LQC and replaced by the big bounce. Since quantum physics—including quantum geometry—is regular at the big bounce, it is natural to specify initial conditions for the quantum state $\Psi_o$ that encodes the background, homogeneous quantum geometry, as well as for $\psi$ that describes the quantum state of perturbations. Physically, the initial conditions amount to assuming that the state $\Psi_o \otimes \psi$ at the bounce should satisfy 'quantum homogeneity'. More precisely, at the bounce one focuses just on that region which expands to become the observable universe and demands that it be homogeneous except for the inevitable quantum fluctuations that one cannot get rid of even in principle. Now, because of the pre-inflationary and inflationary expansion, the region of interest has a radius smaller than $\sim 10\ell_{Pl}$ at the bounce. But as has been emphasized in the relativity literature, this creates a huge fine tuning problem. For, to account for the impressive fact that inhomogeneities in the CMB are really tiny—just one part in $10^5$—the required homogeneity at the bounce has to be truly extraordinary. The standard inflationary paradigm is not really applicable at the Planck scale and, even if one were to ignore this fact, it does not have a natural mechanism to achieve this degree of homogeneity. In LQC, on the other hand, the big bang singularity is resolved precisely because there is an in-built repulsive force with its origin in the specific quantum geometry that underlies LQG. While this force is negligible when curvature is less than, say, $10^{-6}$ in Planck units, it rises spectacularly in the Planck regime, overcomes the huge classical gravitational attraction and prevents the big bang singularity. In more general models referred to in Sect. 4, one finds a pattern: every time a curvature scalar enters the Planck regime, this repulsive force becomes dominant and dilutes that curvature scalar, preventing a singularity (see e.g. [14]). This opens the possibility that the 'dilution effect' of the repulsive force may be sufficient to create the required degree of homogeneity on the scale of about $10\ell_{\rm Pl}$, thereby accounting for the assumed 'quantum homogeneity'. If this idea could to be developed in detail, dynamics of the pre-bounce universe will leave no observable effects, providing a clear-cut case for specifying initial conditions at the bounce. Of course, the pre-bounce dynamics will still lead to inhomogeneities at larger scales on the bounce surface but they would have Fourier modes whose physical wave length is much larger than the radius of the observable universe. Therefore, they would not be in the observable range; in the truncated theory considered here, they would be absorbed in the quantum geometry of the homogeneous background. This 'dilution mechanism' and other issues related to initial conditions are likely to be a center of activity in the coming years.

As we saw in Sects. 5 and 6, we now have a conceptual framework and numerical tools to evolve these initial conditions all the way from the bounce to the onset of slow roll. The result depends on where one is in the parameter space that is labeled by the value $\phi_{\rm B}$ of the inflaton at the bounce. For a very large portion of the parameter space we obtain the following three features: (i) Some time in its future evolution, the background geometry encounters a slow roll phase that is compatible with the 7 year WMAP observations; (ii) At the onset of this slow roll, the state $\psi$ of perturbations is essentially indistinguishable from the BD vacuum used in standard inflation; and (iii) the back reaction due to perturbations remains negligible throughout pre-inflationary dynamics in which the background curvature falls by some 11 orders of magnitude, justifying the underlying 'truncation approximation'. Thus, for this portion of the parameter space, we have a self-consistent extension of the standard inflationary paradigm.

There is, however, a small window in the parameter space for which the feature (i) is realized but the initial state at the onset of inflation contains an appreciable number of BD excitations. This number is within the current observational limits. But the presence of these excitations signals new effects such as a departure from the inflationary 'consistency relation' involving both scalar and tensor modes and a new source of non-Gaussianities. These could be seen in future observational missions [62–64]. The physical origin of these effects can be traced back to a new energy scale $k_{\rm LQC}$ defined by the universal value of the scalar curvature at the bounce. Excitations with $k \lesssim 2k_{\rm LQC}$ are created in the Planck regime near the bounce. It turns out that if the number $N$ of e-foldings in the scale factor $a$ between the bounce and $\tilde{\eta} = \tilde{\eta}_{k^\star}$ is less than 15, then the modes which are excited would be seen in the CMB. This occurs only in the small window of parameter space referred to above. Since the window is very small, the 'a priori probability' that one of these values of $\phi_{\rm B}$ is realized in

Nature would seem to be tiny. However, one can turn this argument around. Should these effects be seen, the parameter space would be narrowed down so much that very detailed calculations would become feasible. In either case, it is rather exciting that the analysis relates initial conditions and Planck scale dynamics with observations, thereby expanding the reach of cosmology to the earliest moment in the deep Planck regime.

Even when a self-consistent solution $\Psi_o \otimes \psi$ to the truncated theory exists, how would it fit in full LQG? Recall the situation in classical general relativity. In cosmology as well as black hole physics, one routinely expects first order perturbations whose back reaction is negligible to provide excellent approximations to the phenomenological predictions of the exact theory. I see no obvious reason why the situation would be different in quantum gravity. As a simple example to illustrate the general viewpoint, consider the Dirac solution of the hydrogen atom. Since one assumes spherical symmetry prior to quantization, this truncation excludes photons from the beginning. Therefore, at a conceptual level, the Dirac description is *very* incomplete. Yet, as far as experiments are concerned, it provides excellent approximations to answers provided by full QED until one achieves the accuracy needed to detect the Lamb shift. I expect the situation to be similar for our truncated theory: Conceptually it is surely quite incomplete vis a vis full LQG, but the full theory will provide only small corrections to the observable effects.

To conclude, let me emphasize that there was no a priori reason to anticipate either of the two main conclusions—the extension of standard inflation to the Planck regime for much of the parameter space and deviations from some of its predictions in a narrow window. Indeed, it would *not* have been surprising if the pre-inflationary dynamics of LQC was such that the predicted power spectra were observationally ruled out for the 'natural' initial conditions we used at the bounce, or, if the self-consistency of truncation had failed quite generally because of the Planck scale dynamics. Indeed, this could well occur in generic bouncing scenarios, e.g. in situations in which the expansion between the bounce and the surface of large scattering is not sufficiently large for the modes observed in the CMB to have wave lengths smaller than the curvature radius throughout this evolution.

**Acknowledgments** This report is based on joint work with Ivan Agullo, Alejandro Corichi, Wojciech Kaminski, Jerzy Lewandowski, William Nelson, Tomasz Pawlowski, Parampreet Singh and David Sloan over the past six years. I am most grateful for this collaboration. I am also indebted to a very large number of colleagues especially in the LQG community for discussions, comments, questions and criticisms. This work was supported by the NSF grant PHY-1205388 and the Eberly research funds of Penn state.

## References

1. Einstein A.: Letter to Arnold Sommerfeld, dated November 28th, 1915, reporting his discovery of general relativity (1915)
2. Ashtekar, A., Pawlowski, T., Singh, P.: Quantum nature of the big bang. Phys. Rev. Lett. **96**, 141301 (2006). doi:10.1103/PhysRevLett.96.141301

3. Ashtekar, A., Pawlowski, T., Singh, P.: Quantum nature of the big bang: An analytical and numerical investigation I. Phys. Rev. D **73**, 124038 (2006). doi:10.1103/PhysRevD.73.124038
4. Ashtekar, A., Pawlowski, T., Singh, P.: Quantum nature of the big bang: improved dynamics. Phys. Rev. D **74**, 084003 (2006)
5. Ashtekar, A., Corichi, A., Singh, P.: Robustness of predictions of loop quantum cosmology. Phys. Rev. D **77**, 024046 (2008). doi:10.1103/PhysRevD.77.024046
6. Pawlowski, T., Ashtekar, A.: Positive cosmological constant in loop quantum cosmology. Phys. Rev. D **85**, 064001 (2012). doi:10.1103/PhysRevD.85.064001
7. Ashtekar, A., Sloan, D.: Loop quantum cosmology and slow roll inflation. Phys. Lett. B **694**, 108 (2010)
8. Ashtekar, A., Sloan, D.: Probability of inflation in loop quantum cosmology. Gen. Relativ. Gravit. **43**, 3619 (2011). doi:10.1007/s10714-011-1246-y
9. Ashtekar, A., Kaminski, W., Lewandowski, J.: Quantum field theory on a cosmological, quantum space-time. Phys. Rev. D **79**, 064030 (2009). doi:10.1103/PhysRevD.79.064030
10. Agullo, I., Ashtekar, A., Nelson, W.: A quantum gravity extension of the inflationary scenario. Phys. Rev. Lett. **109**, 251301 (2012). doi:10.1103/PhysRevLett.109.251301
11. Agullo, I., Ashtekar, A., Nelson, W.: An extension of the quantum theory of cosmological perturbations to the Planck era. ArXiv e-prints arXiv:1211.1354 [gr-qc] (2012)
12. Agullo, I., Ashtekar, A., Nelson, W.: The pre-inflationary dynamics of loop quantum cosmology: confronting quantum gravity with observations. Class. Quant. Grav. **30**, 085014 (2013). doi:10.1088/0264-9381/30/8/085014
13. Bojowald, M.: Loop quantum cosmology. Living Rev. Relativ. **8**, 11. http://www.livingreviews.org/lrr-2005-11 (2005)
14. Ashtekar, A., Singh, P.: Loop quantum cosmology: a status report. Class. Quantum Grav. **28**, 213001 (2011). doi:10.1088/0264-9381/28/21/213001
15. Liddle, A.R., Lyth, D.H.: Cosmological Inflation and Large-Scale Structure. Cambridge University Press, Cambridge (2000)
16. Dodelson, S.: Modern Cosmology. Academic Press, Amsterdam (2003)
17. Mukhanov, V.: Physical Foundations of Cosmology. Cambridge University Press, Cambridge (2005)
18. Weinberg, S.: Cosmology. Oxford University Press, Oxford (2008)
19. Gorbunov, D.S., Rubokov, V.A.: Introduction to the Theory of the Early Universe: Cosmological Perturbations and Inflationary Theory. World Scientific, Singapore (2011)
20. Penrose, R.: Faith, Fashion and Fantasy in the New Physics of the Universe (2003). http://lectures.princeton.edu/2006/roger-penrose/, http://hulk03.princeton.edu:8080/WebMedia/lectures/
21. Brandenberger, R.H.: Introduction to early universe cosmology. ArXiv e-prints arXiv:1103:2271 [astro-ph.CO], (PoS(ICFI 2010)001) (2011)
22. Brandenberger, R.H., Martin, J.: Trans-Planckian issues for inflationary cosmology, arXiv:1211.6753
23. Borde, A., Guth, A., Vilenkin, A.: Inflationary space-times are not past-complete. Phys. Rev. Lett. **90**, 151301 (2003). doi:10.1103/PhysRevLett.90.151301
24. Hawking, S.W., Ellis, G.F.R.: Large Scale Structure of Space-Time. Cambridge University Press, Cambridge (1973)
25. Komatsu, E., et al.: Seven-year Wilkinson microwave anisotropy probe (WMAP) observations: cosmological interpretation. Astrophys. J. Suppl. Ser. **192**, 18 (2011). doi:10.1088/0067-0049/192/2/18
26. Holman, R., Tolley, A.: Enhanced non-Gaussianity from excited states. J. Cosmol. Astropart. Phys. **2008**(05), 001 (2008). doi:10.1088/1475-7516/2008/05/001
27. Agullo, I., Parker, L.: Non-gaussianities and the stimulated creation of quanta in the inflationary universe. Phys. Rev. D **83**, 063526 (2011a). doi:10.1103/PhysRevD.83.063526
28. Agullo, I., Parker, L.: Stimulated creation of quanta during inflation and the observable universe Gen. Relativ. Gravit. **43**, 2541 (2011b). doi:10.1007/s10714-011-1220-8

29. Ganc, J.: Calculating the local-type fNL for slow-roll inflation with a non-vacuum initial state. Phys. Rev. D **84**, 063514 (2011). doi:10.1103/PhysRevD.84.063514
30. Agullo, I., Navarro-Salas, J., Parker, L.: Enhanced local-type inflationary trispectrum from a non-vacuum initial state. J. Cosmol. Astropart. Phys. **2012**(05), 019 (2012). doi:10.1088/1475-7516/2012/05/019
31. Bojowald, M.: Absence of singularity in loop quantum cosmology. Phys. Rev. Lett. **86**, 5227 (2001). doi:10.1103/PhysRevLett.86.5227
32. Ashtekar, A., Bojowald, M., Lewandowski, J.: Mathematical structure of loop quantum cosmology. Adv. Theor. Math. Phys. **7**, 233 (2003)
33. Singh, P.: Are loop quantum cosmologies never singular? Class. Quantum Grav. **26**, 125005 (2009). doi:10.1088/0264-9381/26/12/125005
34. Ashtekar, A., Pawlowski, T., Singh, P., Vandersloot, K.: Loop quantum cosmology of k=1 FRW models. Phys. Rev. D **75**, 0240035 (2006)
35. Szulc, L., Kaminski, W., Lewandowski, J.: Closed FRW model in loop quantum cosmology. Class. Quantum Gravity **24**, 2621 (2007)
36. Bentivegna, E., Pawlowski, T.: Anti-deSitter universe dynamics in LQC. Phys. Rev. D **77** 124025 (2008). doi:10.1103/PhysRevD.77.124025
37. Kaminski, W., Pawlowski, T.: The LQC evolution operator of FRW universe with positive cosmological constant. Phys. Rev. D **81**, 024014 (2010)
38. Ashtekar, A., Wilson-Ewing, E.: Loop quantum cosmology of Bianchi type I models, Phys. Rev. D **79**, 083535 (2009). doi:10.1103/PhysRevD.79.083535
39. Martin-Benito, M., Mena Marugan, G.A., Pawlowski, T.: Loop quantization of vacuum Bianchi I cosmology. Phys. Rev. D **78**, 064008 (2008). doi:10.1103/PhysRevD.78.064008
40. Wilson-Ewing, E.: Loop quantum cosmology of Bianchi type IX models. Phys. Rev. D **82**, 043508 (2010). doi:10.1103/PhysRevD.82.043508
41. Martin-Benito, M., Garay, L.J., Mena, G.A.: Hybrid quantum Gowdy cosmology: combining loop and Fock quantizations. Phys. Rev. D **78**, 083516 (2008). doi:10.1103/PhysRevD.78.083516
42. Garay, L.J., Martn-Benito, M., Mena Marugan, G.A.: Inhomogeneous loop quantum cosmology: Hybrid quantization of the Gowdy model. Phys. Rev. D **82** 044048 (2010). doi:10.1103/PhysRevD.82.044048
43. Brizuela, D., Mena Marugan, G.A., Pawlowski, T.: Big bounce and inhomogeneities. Class. Quantum Grav. **27**, 052001 (2010). doi:10.1088/0264-9381/27/5/052001
44. Martin-Benito, M., Mena Marugan, G.A., Wilson-Ewing, E.: Hybrid quantization: from Bianchi I to the Gowdy model. Phys. Rev. D **82**, 084012 (2010). doi:10.1103/PhysRevD.82.084012
45. Martn-Benito, M., Martin-de Blas, D., Mena Marugan, G.A.: Matter in inhomogeneous loop quantum cosmology: the Gowdy $T^3$ model. ArXiv e-prints arXiv:1012.2324 [gr-qc] (2010)
46. Brizuela, D., Mena Marugan, G.A., Pawlowski, T.: Effective dynamics of the hybrid quantization of the Gowdy T3 universe. ArXiv e-prints arXiv:1106.3793 [gr-qc] (2011)
47. Langlois, D.: Hamiltonian formalism and gauge invariance for linear perturbations in inflation. Class. Quant. Grav. **11**, 389 (1994). doi:10.1088/0264-9381/11/2/011
48. Ashtekar, A., Lewandowski, J.: Background independent quantum gravity: a status report. Class. Quantum Grav. **21**, R53 (2004). doi:10.1088/0264-9381/21/15/R01
49. Rovelli, C.: Quantum Gravity. Cambridge University Press, Cambridge (2004)
50. Thiemann, T.: Introduction to Modern Canonical Quantum General Relativity. Cambridge University Press, Cambridge (2007)
51. Rovelli, C., Smolin, L.: Discreteness of area and volume in quantum gravity. Nucl. Phys. B **442**(593), 1995 (1995). doi:10.1016/0550-3213(95)00150-Q. Erratum: ibid. 456, 753
52. Ashtekar, A., Lewandowski, J.: Quantum theory of geometry I: area operators. Class. Quantum Grav. **14**, A55 (1997). doi:10.1088/0264-9381/14/1A/006
53. Thiemann, T.: A length operator for canonical quantum gravity. J. Math. Phys. **39**, 3372 (1998)
54. Bianchi, E.: The length operator in loop quantum gravity. Nucl. Phys. B **807**, 591 (2009). doi:10.1016/j.nuclphysb.2008.08.013

55. Ma, Y., Soo, C., Yang, J.: New length operator for loop quantum gravity. Phys. Rev. D **81**, 124026 (2010). doi:10.1103/PhysRevD.81.124026
56. Parker, L.: The creation of particles in an expanding universe. Ph.D. thesis, Harvard University, Harvard (1966)
57. Parker, L., Fulling, S.A.: Adiabatic regularization of the energy-momentum tensor of a quantized field in homogeneous spaces. Phys. Rev. D **9**, 341 (1974). doi:10.1103/PhysRevD.9.341
58. Birrell, N.D.: The application of adiabatic regularization to calculations of cosmological interest. Proc. R. Soc. Lond. Ser. A **361**, 513 (1978)
59. Anderson, P.R., Parker, L.: Adiabatic regularization in closed Rebertson–Walker universes. Phys Rev. D **36**, 2963 (1987). doi:10.1103/PhysRevD.36.2963
60. Fulling, S.: Aspects of Quantum Field Theory in Curved Space-Times. Cambridge University Press, Cambridge (1989)
61. Parker, L., Toms, D.: Quantum Field Theory in Curved Space-Time. Cambridge University Press, Cambridge (2009)
62. Ganc, J., Komatsu, E.: Scale dependent bias of galaxies and $\mu$-type disruption of the cosmic microwave background spectrum from a single field inflation with a modified initial state. Phys. Rev. D **86**, 023518 (2012). doi:10.1103/PhysRevD.86.023518
63. Agullo, I., Shandera, S.: Large non-Gaussian halo bias from single field inflation. J. Cosmol. Astropart. Phys. **2012**(09), 007 (2012). doi:10.1088/1475-7516/2012/09/007
64. Schmidt, F., Hui, L.: CMB power asymmetry from Gaussian modulation. ArXiv e-prints arXiv:1210.2965 [astro-ph.CO] (2012)
65. Linsefors, L., Cailleteau, T., Barrau, A., Grain, J.: Primordial tensor power spectrum in holonomy corrected Omega-LQC. ArXiv e-prints arXiv:1212.2852 [gr-qc] (2012)
66. Fernandez-Mendez, M., Mena Marugan, G.A., Olmedo, J.: Hybrid quantization of an inflationary universe. Phys. Rev. D **86**, 024003 (2012). doi:10.1103/PhysRevD.86.024003

# The Inflationary Origin of the Seeds of Cosmic Structure: Quantum Theory and the Need for Novel Physics

**Daniel Sudarsky**

**Abstract** The Inflationary account for the emerging of the seeds of cosmic structure from quantum fluctuations is a central part of our current views of cosmology. It is, on the one hand, extremely successful at the phenomenological level, and yet, it retains an aspect that is generally regarded as controversial: The exact mechanism by which quantum fluctuations transmute into actual inhomogeneities. We will review the considerations that lead us to conclude that the fully satisfactory resolution of the issue requires novel physics, and we will discuss an option we have been considering in this regard.

## 1 Introduction

This conference commemorates the time spent by Einstein in Prague, which was instrumental in his development of General Relativity, a subject which has been the focus of the majority of the other presentations. I will be touching on the other great question that preoccupied Einstein at the time: Quantum Theory. We note that, the subject of this manuscript; inflation, represents the only generally accepted example of an instance in which General Relativity, Quantum Theory and observations come together. It is, therefore, quite remarkable that it is precisely here where we must confront the conceptual difficulties of quantum theory itself. In fact, the ideas I will be exploring are strongly motivated by the arguments that Penrose [1] and Diósi [2] have been advancing regarding the collapse of the wave function as a dynamical process to be incorporated in a modified Schrödinger's equation, and the role that gravity might play in this.

D. Sudarsky (✉)
Institute for Nuclear Sciences, National Autonomous University of Mexico, Apdo. Postal 70-543, 04510 Mexico, D.F., Mexico
e-mail: sudarsky@nucleares.unam.mx

J. Bičák and T. Ledvinka (eds.), *General Relativity, Cosmology and Astrophysics*, Fundamental Theories of Physics 177, DOI: 10.1007/978-3-319-06349-2_17, 
© Springer International Publishing Switzerland 2014

In the present manuscript, and in contrast to other works [3, 4] where I have focused on the difficulties or shortcomings of the postures advocated in standard treatments of the emergence of the seeds of cosmic structure, I will be focusing on aspects that would be encountered independently of the conceptual approach one takes, as long as one attempts to provide a specific characterization of the various stages in the cosmological evolution. Moreover, I will show the connection, sometimes not easily recognized, between the actual treatment we have been using, and other approaches to deal with the characterization of space-time when the matter content is taken to be described in terms of quantum fields.

## 2 The General Setting

The first simplification we will be using is justified by the fact that, despite the many important and sometimes spectacular advances, at this point in time we still do not have a fully workable and completely satisfactory theory of quantum gravity. For instance, we do not know how to construct a quantum state representing Minkowski space-time. In fact, it is well known that any canonical approach to quantum gravity inevitably leads to a timeless theory where the recovery of fully covariant space-time notions becomes, by itself, a nontrivial task. Those approaches usually require selecting a physical observable to play the role of a clock, and while this can be achieved quite satisfactorily in certain cases, the resolution of the problem in full generality is not available.

Therefore, although we will adhere to the view that the fundamental description of everything, including space-time, ought to be always quantum mechanical, we will be using a classical description of the space-time metric.

In general, we will view the so called *classical regimes* in connection with some physical variable as those where such quantities can be described to a sufficient accuracy by their classical counterparts representing the corresponding quantum expectation values. The paradigmatic example here is provided by the coherent states of a harmonic oscillator which correspond to minimal wave-packets with expectation values of position and momentum following the classical equations of motion. In the specific case of the space-time, we will accept that, at the fundamental level, it would have a quantum description in terms of some unspecified variables (they might be those of loop quantum gravity (LQG), the "causal set" approach, or the "dynamical triangulations" approach, etc) but we will be characterizing them, according to this view, using effectively classical terms. We might consider such description in analogy with the hydrodynamical characterization of a fluid: as representing a good enough description at certain scales, but having a radically different description at a more fundamental level. Einstein's equations would correspond to the Navier-Stokes equations, the space-time metric to the fluid velocity and density fields, and the atomic and molecular characterization of the matter making the fluid would correspond to the fundamental degrees of freedom of quantum gravity.

According to this view, Einstein's equations would be of limited validity and there would be conditions where they will not hold. In fact, in such situations one can expect a more general failure of the characterization of the situation in terms of a space-time metric, just as, in the case of a fluid, one might expect not only the violation of Navier-Stokes equations but also of the hydrodynamic characterization of matter, when something like a phase transition from liquid to gas is taking place. There is one example coming from LQG where one faces precisely this kind of breakdown: It corresponds to the situation where a collapsing star forms a black hole that eventually evaporates completely [5]. In that work, it is argued that, although the classical singularity would be resolved in terms of the LQG degrees of freedom, the region of space-time corresponding to the singularity is characterized, in terms of the fundamental variables, by a situation that has no metric counterpart.

On the other hand, the situations I want to consider are those that require a full quantum treatment of the matter fields. For this, we will rely on the standard quantum field theory in curved space-time treatments, such as described for instance in [6]. The setting is therefore that of semiclassical gravity where Einstein's equations read:

$$G_{\mu\nu} = 8\pi G \langle \hat{T}_{\mu\nu} \rangle \tag{1}$$

and matter is described in terms of states of a quantum field, which, in our case, will be the scalar field of inflationary cosmology.

The validity of such semiclassical treatment is, as we have indicated, of limited scope, and would require, among other things, the scalar curvature of space-time to be small compared to $l_p^{-2}$ ($l_p$ is the Planck length). The inflationary regime is thought to be associated to scales that are way below the Planck mass and thus this part of the requirement should be easily satisfied in the case we want to consider.

## 3 The Stern-Gerlach Analogy

The general issue we want to consider is the emergence of the seeds of quantum structure from the quantum fluctuations of the inflaton vacuum. The aspect that will be guiding our inquest is the required change in symmetry between the conditions characterized by a classical and unperturbed Robertson Walker space-time, where the inflaton field's zero mode is slowly rolling down the potential, and where the other modes are in the the Bunch-Davies vacuum, to a stage characterized by a slightly perturbed RW space-time containing small anisotropies and inhomogeneities, and a quantum state of the inflaton where the expectation value of the energy momentum tensor has the corresponding anisotropies and inhomogeneities.

The question is how to characterize the evolution in time (because, after all, *emergence* is a word that has very clear time connotations[1]), from a situation corresponding to a homogeneous and isotropic (H&I) background and a quantum aspect characterized by a H&I state, to a stage lacking such symmetries. This issue has

[1] Something emerges when it is not present at a certain time but it is present at a later time.

been the central focus of the discussion in previous works: Is there a measurement involved? Can we account for it using just decoherence? Should we rely on the many worlds interpretation (MWI) or must we call upon a novel gravity-induced collapse of the wave function following the ideas of Penrose and Diósi? I have extensively argued that among those the only option is the last one, however, here I will focus on aspects that should be dealt with, even if one is intent on sticking with some of the alternative views mentioned above.

In order to clarify the issue, I shall consider a much simpler problem, and point to the parallels with the inflationary problem, as well as to those aspects where the analogy breaks down.

Consider a standard Stern-Gerlach experiment: The setting involves an electron that has been prepared moving along the $x$ axis from the $x < 0$ region, towards a magnet placed at the origin of coordinates. The inhomogeneous magnetic field points along the $y$ axis so that the electron will be diverted towards the $+y$ or $-y$ directions depending on whether the spin state of the electron is $|+\rangle$ or $|-\rangle$ (we are taking the basis to be that of eigen-states of the spin along the y axis). If the spin was prepared initially in the direction $+x$ (eigenstate of the x component of the spin) we know that there is a 50 % probability that the electron will be diverted towards the $+y$ direction.

Let us imagine for a moment that we do not fully understand the theory, that there are aspects of the electromagnetic interaction that still elude us (the allegory here alludes to the quantum theory of gravity, of course). Now, let us consider the theoretical analysis of the said experiment: If we do not invoke any sort of reduction, or collapse of the wave function, the result of the unitary evolution will be a state that corresponds to a superposition of the electron going up and the electron going down. However, suppose we want to investigate in depth what happens when the electron is deflected: Say, we want to understand exactly the details of the momentum transferred from the magnetic field to the electron. We could, for instance, find that the momentum transfer has components predominantly in the $Y$ direction (depending on the deflection, with a sign that depends on the alternative) but is accompanied by momentum components in the $X$ direction (of a specific and correlated magnitude), and use this to study the change in the kinetic energy. We can even inquire about the rate of transfer as the electron moves along the $X$ direction (by considering an appropriate wave packet characterization of the electron and, say, following the expectation value of the center of mass). How can we do that if we maintain that the electron is, even after the passing through the magnet, in the superposition of moving up and moving down? In that case, if we try to compute the momentum transfer from the EM field to the electron we will find that it is zero.

Suppose we want to further inquire about the back reaction of the electron on the EM field. It would seem very difficult to do so without incorporating the collapse. Note that, if we are so inclined, we could even adopt the many worlds interpretation (MWI) but still concern ourselves with one of the realizations of the electron's path, that which corresponds to "our branch of the many worlds". Now, suppose we wanted to do this before we had a fully workable quantum theory of the electromagnetic field, but instead we had very refined experimental data about the back reaction acting on the magnet as a result of the electron scattering. Could we not hope to investigate some of the properties of the quantum electromagnetic field using a combination of the data, some rough classical characterization of the EM field, taken to be only valid to a certain degree (evidently not in the full description of the back reaction, but, perhaps, as it applies to the "after" and "before" state of the EM field)? Could we hope to do that if we had never even been able to consider the back reaction, as a result of our failure to acknowledge that the full superposition (in which the expectation value of the momentum transfer was zero) was not the appropriate description? It seems we could start considering something like the dispersion of the momentum transfer but, that would be very difficult due to our lacking of a workable quantum theory of Maxwell's field.

The situation we face regarding the problem of the emergence of seeds of structure in inflationary cosmology is, in a sense, analogous to the one above: the symmetry of homogeneity and isotropy in the cosmological case has in the example above a simple counterpart: the symmetry $y \to -y$.

The most important aspect where the analogy breaks down is the fact that in contrast with the Stern-Gerlach example above, in cosmology we can not call upon external observers, and, what is even worse, the emergence of structure, which is what we want to explain, is a prerequisite for the subsequent emergence of anything one can consider as an "observer".

The issue of symmetry is brought in because, as we all know, when dealing with complicated problems, symmetry arguments are often one of the few paths available to arrive to clear and definite conclusions, and thus they provide the only hope to make progress.

## 4 A Word About Collapse Theories

The idea of modifying quantum theory by adding to it a mechanism for explicit dynamical reduction has a long history. The existing work in this direction includes: GRW [7], Pearle [8], Diósi [2], Penrose [1], Bassi [9] (where its worthwhile noting some recent advances towards making collapse theories compatible with special relativity [10, 11]), and recently Weinberg [12].

As an example, let us consider the modification of the Schrödinger equation that underlies the Continuous Spontaneous Localization (CSL) theory, developed in [13]:

$$d|\psi\rangle = -\{[i\hat{H} - \frac{\lambda^2}{2}(\hat{A} - \langle\psi|\hat{A}|\psi\rangle)^2]dt + \lambda(\hat{A} - \langle\psi|\hat{A}|\psi\rangle)dW_t\}|\psi\rangle, \quad (2)$$

where $W_t$ is a Wiener process ($\overline{W_t^2} = t$).

Its merit is that it includes the unitary Schrödinger evolution $U$ and the nondeterministic, non-unitary reduction process $R$ (for measuring $\hat{A}$) in a unified fashion. The proposal for particles assumes that the fundamental localization takes place in position or configuration space, thus $\hat{A} = \hat{\mathbf{X}}$. The value of the parameter $\lambda$ is taken to be small enough so that particle physics is not strongly affected, but large enough so that it leads to a rapid localization of macroscopic objects.

In the reminder of this manuscript, we will not follow any of those proposals, but take from them only some essential aspects. What we want to do, at this point, is to present a generic formalism capable of treating the cosmological problem that motivates this line of research, within the context of semiclassical gravity with a full quantum treatment for the inflaton field (including its zero mode), incorporating a collapse process. On the other hand, we should mention the work [14], where precisely a version of CSL has been adopted to this problem, as well as an ongoing research project by our group in which a different implementation of those ideas is studied [15].

## 5 The Self Consistent Semiclassical Configurations

In this section, we will describe the precise formalism that we consider as appropriate to characterize, at the desired level, the situation we will be studying. As we said, the setting is that of semiclassical gravity and quantum field theory in curved space-time. Such setting is often considered as a context in which the space time is fixed and given, and where the quantum fields can, at most, produce some small modifications which are referred to as the back reaction of space-time to the effects of the quantum fields. That point of view will not be sufficient for our purposes, as we want to be able to, in principle, treat the case where the only matter content is represented by the inflation field, and where the space-time is fundamentally tied to its properties: Inflation is supposed to be the result of the non vanishing value of the inflation potential. In such situation, the field is generically treated at a classical level, and only its perturbations are quantized. We want to be able to explore the setting in which the classical quantum partition in the description is, in principle, not tied to a perturbative treatment.[2]

We have formalized these ideas in [16] based on the notion of the "Self Consistent Semiclassical Configurations" (SSC) provided by the following,

**Definition 1** The set $g_{\mu\nu}(x), \hat{\varphi}(x), \hat{\pi}(x), \hat{\mathscr{H}}, |\xi\rangle \in \hat{\mathscr{H}}$ represents a SSC if and only if $\hat{\varphi}(x)$ and $\hat{\pi}(x)$ correspond to quantum field operators over the Hilbert space and $\hat{\mathscr{H}}$ is constructed acceding to the standard QFT over the curved space-time with metric $g_{\mu\nu}(x)$ (as described in, say [6]), and the state $|\xi\rangle$ in $\hat{\mathscr{H}}$ is such that:

$$G_{\mu\nu}[g(x)] = 8\pi G\langle\xi|\hat{T}_{\mu\nu}[g(x), \hat{\varphi}(x), \hat{\pi}(x)]|\xi\rangle \tag{3}$$

for all the points $x$ in the space-time manifold.

[2] The point is that a perturbative treatment should be considered as an approximation to something else, and it is very useful when one can establish explicitly what is that which the perturbative treatment is approximately trying to describe.

It is, in a sense, the GR version of Schrödinger-Newton equation [17] where one considers the Schrödinger equation for the wave function $\psi$ of a particle subject to the gravitational interaction described in terms of a Newtonian potential $\Phi_N$:

$$i\frac{\partial\psi}{\partial t} = -\frac{1}{2M}\nabla^2\psi + M\Phi_N\psi, \tag{4}$$

where the wave function of the particle is taken as a gravitating mass distribution, therefore

$$\nabla^2\Phi_N = 4\pi GM|\psi|^2. \tag{5}$$

The non linearity implied by these equations is known to lead to interesting and suggestive behavior [18].

## 6 Collapse

The point, however, is that this setting will not, by itself, be enough to describe the situations involving a collapse of the wave function. As we have argued, in order to be able to describe the evolution in time from an early inflationary era characterized by an H&I situation to a later regime characterized by a situation that is not H&I, we will be relying on the collapse of the wave function, represented here in the simplest fashion: an instantaneous jump in the state of the quantum field.

*The collapse process* reflecting some remanent signature from a fundamental quantum gravity regime (as suggested by Penrose and Doisi's ideas) will be described here as an instantaneous jump: i.e., besides the standard smooth unitary evolution of a quantum field characterized by the Schrödinger's dynamics (the so called $U$ process, which in the QFT theory setting we are considering here, is usually incorporated using the "Heisenberg picture"), there are, sometimes, spontaneous jumps in the quantum state:

$$\ldots|0\rangle_{k_1}\otimes|0\rangle_{k_2}\otimes|0\rangle_{k_3}\otimes\ldots\ldots\to\ldots\ldots|\Xi\rangle_{k_1}\otimes|0\rangle_{k_2}\otimes|0\rangle_{k_3}\otimes\ldots\ldots \tag{6}$$

We might view the collapse as triggered by an aspect of the dynamics which is not susceptible of description in standard Hamiltonian terms, but which is nonetheless taken as an interaction treated here in the interaction picture. That is, we take the Hamiltonian part of the evolution and absorb it in the quantum field operators as in the standard Heisenberg picture, but we leave the reminder, viewed as an interaction, to be treated using the interaction picture.

## 7 Relation to Other Approaches

This setting might seem very novel and unusual, however, the fact is that it can be seen to lie, unsuspectedly, underneath some more conventional approaches, such as the stochastic gravity formalism [19].

In order to see this, let us consider one of such jumps or collapses: $|\psi(t)\rangle = \theta(t_0 - t)|0\rangle + \theta(t - t_0)|\xi\rangle$, and its gravitational effects.

Now Einstein's semiclassical equations read:

$$G_{\mu\nu} = 8\pi G\langle\psi(t)|\hat{T}_{\mu\nu}|\psi(t)\rangle \tag{7}$$

which we can write as:

$$G_{\mu\nu} = 8\pi G\langle 0|\hat{T}_{\mu\nu}|0\rangle + 8\pi G\xi_{\mu\nu}, \tag{8}$$

where

$$\xi_{\mu\nu} \equiv \theta(t - t_0)(\langle\xi|\hat{T}_{\mu\nu}|\xi\rangle - \langle 0|\hat{T}_{\mu\nu}|0\rangle) \tag{9}$$

might be seen as corresponding to an individual stochastic step. Stochastic gravity might correspond to a continuous version of dynamical collapses (like CSL).

Note: the equation can not be valid on the jump, but might well be so before and after. We take the view, motivated in part by the black hole singularity example in LQG, that during the jump the degrees of freedom of the quantum space-time are excited. In the fluid analogy, this might be thought as corresponding to some chemical reaction or phase transition occurring in the fluid. It is clear that during such processes, which generally involve energy flux between the atomic or molecular degrees of freedom to the macroscopic degrees of freedom characterized in terms of the fluid variables, the Navier-Stokes equations can not be valid. If, however, the phase transition takes place rapidly, one can assume such equation to be valid before and after the chemical reaction or phase transition.

Next, consider the inflationary problem at hand, and assume one adopts one of the more popular postures regarding the emergence of classicality, or more precisely, the generation of primordial inhomogeneities and anisotropies. These include, for instance (i) the notion that after a given mode exits the horizon (its physical wave-length as seen in a co-moving frame, becomes larger than the Hubble radius) the fluctuation corresponding to that mode becomes classical, or (ii) that, due to some decoherence effect, we can at a certain point adopt the Many Worlds Interpretation of quantum theory, and consider the state of the quantum field as characterizing not our universe, but an ensemble of universes of which ours is just a typical element. Now, let us say that in case (i) we want to produce a description (even an approximate one) of our universe concentrating, for simplicity, on a single mode, but we want a description that is valid before the mode exits the horizon and afterwards. In that case, the approach I will present, seems to be the best one can do, as long as we do not have a workable theory of quantum gravity which allows us to characterize space-time in a full quantum language. In case (ii) we might also be interested in putting together the characterization of our space-time before the decoherence is taken to be effective, and the one describing the particular branch of the many worlds, or the particular element of the enabled universes in which we happen to find ourselves. Again, in that situation the analysis I will present would offer perhaps the furthest one can go in achieving the said goal, given the present stage of the development of candidate theories of quantum gravity.

## 8 Application to Inflation

As we discussed in the previous sections, space-time will be treated as classical. In the case of interest, working in a specific gauge, and ignoring the tensor perturbations the metric is taken as:

$$ds^2 = a^2(\eta)\left[-(1+2\psi)d\eta^2 + (1-2\psi)\delta_{ij}dx^i dx^j\right], \psi(\eta, \mathbf{x}) \ll 1 \tag{10}$$

with $a(\eta)$ the scale factor and $\psi(\eta, x)$ representing (to the extent that it is nonzero) a possible slight departure from homogeneity and isotropy of the space-time, the so called Newtonian potential. We will use the notation $\mathcal{H} \equiv a^{-1}da/d\eta$ (not to be confused with the hatted quantity that stands for a Hilbert space).

Also, as explained before, the scalar field, which we take here to be described by the simple action $S = 1/2\int d^4x(\nabla_\mu\phi\nabla^\mu\phi - m^2\phi^2)$, including the zero mode (which in standard discussions of inflation is usually treated at a classical level) is treated here using quantum field theory on curved space times, so we write:

$$\hat{\phi}(x) = \sum_\alpha \left(\hat{a}_\alpha u_\alpha(x) + \hat{a}^\dagger_\alpha u^*_\alpha(x)\right), \tag{11}$$

with the functions $u_\alpha(x)$ a complete set of normal modes, orthonormal w.r.t. the symplectic product:

$$((\phi_1, \pi_1), (\phi_2, \pi_2))_{\text{Sympl}} \equiv -i\int_\Sigma \left[\phi_1\pi_2^* - \pi_1\phi_2^*\right] d^3x. \tag{12}$$

For simplicity, we set the problem in a co-moving coordinate box of size $L$.

Finally, one constructs the state such that Einstein semiclassical equations hold. This is nontrivial, but ts is a well defined problem. In what follows, the discussion will omit some complications that are required for the rigorous analysis, and discussed in detail in [16], but which are not central to the issue at hand. They have to do with the hermiticity of the operators that play a central role in the collapse, an issue that will be overlooked here to avoid nonessential complications in the presentation.

## 8.1 The Homogeneous and Isotropic Case: SSC I

We assume an almost de Sitter slow-roll expansion characterized by the parameters $H_0^{(I)}$ and $\epsilon^{(I)}$ (using standard inflationary notation [20]).

The quantum field theory construction requires a complete set of modes, which we take to be of the form $u_{\mathbf{k}}^{(\mathrm{I})}(x) = v_{\mathbf{k}}^{(\mathrm{I})}(\eta)e^{i\mathbf{k}\cdot\mathbf{x}}/L^{3/2}$.

The field equation of motion then leads to:

$$\ddot{v}_{\mathbf{k}}^{(\mathrm{I})} + 2\mathscr{H}^{(\mathrm{I})}\dot{v}_{\mathbf{k}}^{(\mathrm{I})} + \left(k^2 + a^{2(\mathrm{I})}m^2\right)v_{\mathbf{k}}^{(\mathrm{I})} = 0, \tag{13}$$

for modes, which must be normalized according to

$$v_{\mathbf{k}}^{(\mathrm{I})}\dot{v}_{\mathbf{k}}^{(\mathrm{I})*} - \dot{v}_{\mathbf{k}}^{(\mathrm{I})}v_{\mathbf{k}}^{(\mathrm{I})*} = i\hbar a^{-2(\mathrm{I})}. \tag{14}$$

For the modes with $k \neq 0$, the most general solution to the evolution equation is a linear combination of the functions: $\eta^{3/2}H_\nu^{(1)}(-k\eta)$ and $\eta^{3/2}H_\nu^{(2)}(-k\eta)$, (the Hankel functions of first and second kind), with $(\nu^{(\mathrm{I})})^2 = (9/4) - (m/H_0^{(\mathrm{I})})^2$. However, these functions are not well behaved at the origin and thus the zero mode is not included. For $k = 0$ the general solution to the equation is a linear combination of the functions $\eta^{(3-2\nu)/2}$ and $\eta^{(3+2\nu)/2}$. The choice can be made arbitrarily provided it has a positive symplectic norm. We take:

$$v_0^{(\mathrm{I})}(\eta) = \sqrt{\frac{\hbar}{H_0^{(\mathrm{I})}}}\left[1 - \frac{i}{6}\left(-H_0^{(\mathrm{I})}\eta\right)^3\right]\left(-H_0^{(\mathrm{I})}\eta\right)^{m^2/3H_0^{2(\mathrm{I})}}. \tag{15}$$

For the $k \neq 0$ modes, we make the Bunch-Davies choice: i.e., we use modes that, in the asymptotic past, behave as purely "positive frequency solutions". This fixes $\hat{\mathscr{H}}^{(\mathrm{I})}$ as the Fock space of the SSC-I construction.

To complete the SSC construction we still need to find a state $|\xi^{(\mathrm{I})}\rangle \in \hat{\mathscr{H}}^{(\mathrm{I})}$ such that its expectation value for the energy-momentum tensor leads to the desired nearly de Sitter expansion. Consider a state in which all the modes with $k \neq 0$ are in their vacuum state, while the zero mode is excited in a coherent state:

$$|\xi^{(\mathrm{I})}\rangle = ce^{\xi_0^{(\mathrm{I})}\hat{a}_0^{(\mathrm{I})\dagger}}|0^{(\mathrm{I})}\rangle, \tag{16}$$

Using Einstein's equations for the metric with a vanishing Newtonian potential, and the fact that for a coherent state we have

$$\langle\xi^{(\mathrm{I})}| : (\phi^{(\mathrm{I})})^2 : |\xi^{(\mathrm{I})}\rangle = (\langle\xi^{(\mathrm{I})}|(\phi^{(\mathrm{I})})|\xi^{(\mathrm{I})}\rangle)^2, \tag{17}$$

one finds that the expectation value of the field should satisfy:

$$\langle \xi^{(\mathrm{I})} | \langle \phi^{(\mathrm{I})} \rangle | \xi^{(\mathrm{I})} \rangle \propto \eta^{\sqrt{\epsilon^{(\mathrm{I})} m^2 / 3 (H_0^{(\mathrm{I})})^2}}. \tag{18}$$

On the other hand, taking the parameter $\xi_0^{(\mathrm{I})}$ as real we find:

$$\langle \xi^{(\mathrm{I})} | \hat{\phi}^{(\mathrm{I})}(x) | \xi^{(\mathrm{I})} \rangle = \frac{2\xi_0^{(\mathrm{I})}}{L^{3/2}} \sqrt{\frac{\hbar}{H_0^{(\mathrm{I})}}} \left( -H_0^{(\mathrm{I})} \eta \right)^{m^2 / 3 H_0^{2(\mathrm{I})}}. \tag{19}$$

That is, we will have compatibility if we set:

$$\epsilon^{(\mathrm{I})} = \frac{m^2}{3 H_0^{2(\mathrm{I})}}, \quad H_0^{(\mathrm{I})} = 16\pi G \hbar \epsilon^{(\mathrm{I})} \frac{(\xi_0^{(\mathrm{I})})^2}{L^3}. \tag{20}$$

This completes the explicit SSC -I construction representing an H&I state and space-time metric, corresponding to the early stages of inflation.

Next, we want to consider a situation where the universe is no longer H&I but has been excited in the $\mathbf{k}_0$ mode: We will denote this new SSC by SSC-II.

## 8.2 A Simple Inhomogeneous and Anisotropic Case: SSC II

It will be characterized by the parameters $H_0^{(\mathrm{II})}$ and $\epsilon^{(\mathrm{II})}$ (which might, in principle, differ slightly from those corresponding to the SSC-I discussed in the previous section), and a Newtonian potential described by an (in principle) arbitrary function $\psi(\eta, \mathbf{x}) = \varepsilon P(\eta) Cos(\mathbf{k}_0 . \mathbf{x})$, where $P(\eta)$ is an (in principle) arbitrary function, and $\varepsilon$ is a small (expansion) parameter (please do not confuse with $\epsilon$).

*The strategy:* We first construct the "generic" Hilbert space assuming that $P(\eta)$ is given. Then, make an "educated" guess for the form of the quantum state, and by requiring that our construction be a SSC we will find what the function $P(\eta)$ ought to be.

The first step is to find the complete set of modes, which we write as:

$$u_{\mathbf{k}}^{(\mathrm{II})}(x) = \frac{1}{L^{3/2}} [v_{\mathbf{k}}^{(\mathrm{II})0}(\eta)\, e^{i\mathbf{k}\cdot\mathbf{x}} + \varepsilon (\delta v_{\mathbf{k}}^{(\mathrm{II})-}(\eta)\, e^{i(\mathbf{k}-\mathbf{k}_0)\cdot\mathbf{x}} + \delta v_{\mathbf{k}}^{(\mathrm{II})+}(\eta)\, e^{i(\mathbf{k}+\mathbf{k}_0)\cdot\mathbf{x}})] \tag{21}$$

to the zeroth order in $\varepsilon$, the evolution equation is given by

$$\ddot{v}_{\mathbf{k}}^{(\mathrm{II})0} + 2\mathscr{H}^{(\mathrm{II})} \dot{v}_{\mathbf{k}}^{(\mathrm{II})0} + \left( k^2 + a^{2(\mathrm{II})} m^2 \right) v_{\mathbf{k}}^{(\mathrm{II})0} = 0, \tag{22}$$

with normalization condition

$$v_{\mathbf{k}}^{(\mathrm{II})0} \dot{v}_{\mathbf{k}}^{(\mathrm{II})0*} - \dot{v}_{\mathbf{k}}^{(\mathrm{II})0} v_{\mathbf{k}}^{(\mathrm{II})0*} = i\hbar a^{-2(\mathrm{I})}, \tag{23}$$

which is identical to the construction we have already done. Thus, we take the $v_{\mathbf{k}}^{(\mathrm{II})0}(\eta)$ as before.

At first order in $\varepsilon$ the corresponding evolution equation takes the form

$$\delta\ddot{v}_{\mathbf{k}}^{(\mathrm{II})\pm} + 2\mathscr{H}^{(\mathrm{II})}\delta\dot{v}_{\mathbf{k}}^{(\mathrm{II})\pm} + \left[(\mathbf{k}\pm\mathbf{k}_0)^2 + a^{2(\mathrm{II})}m^2\right]\delta v_{\mathbf{k}}^{(\mathrm{II})\pm} = F_{\mathbf{k}}(\eta) \tag{24}$$

where

$$F_{\mathbf{k}}(\eta) \equiv 4\dot{v}_{\mathbf{k}}^{(\mathrm{II})0}\dot{P} - 2\left(2k^2 + a^{2(\mathrm{II})}m^2\right)v_{\mathbf{k}}^{(\mathrm{II})0}P. \tag{25}$$

The normalization condition (needed only at one time) is:

$$\dot{v}_{\mathbf{k}+\mathbf{k}_0}^{(\mathrm{II})0*}\delta v_{\mathbf{k}}^{(\mathrm{II})+} - v_{\mathbf{k}+\mathbf{k}_0}^{(\mathrm{II})0*}\delta\dot{v}_{\mathbf{k}}^{(\mathrm{II})+} - \dot{v}_{\mathbf{k}}^{(\mathrm{II})0}\delta v_{\mathbf{k}+\mathbf{k}_0}^{(\mathrm{II})-*} + v_{\mathbf{k}}^{(\mathrm{II})0}\delta\dot{v}_{\mathbf{k}+\mathbf{k}_0}^{(\mathrm{II})-*} \tag{26}$$

$$= 4\left(v_{\mathbf{k}}^{(\mathrm{II})0}\dot{v}_{\mathbf{k}+\mathbf{k}_0}^{(\mathrm{II})0*} - \dot{v}_{\mathbf{k}}^{(\mathrm{II})0}v_{\mathbf{k}+\mathbf{k}_0}^{(\mathrm{II})0*}\right)P. \tag{27}$$

If we had $P(\eta)$ and the initial conditions for the $\delta v_{\mathbf{k}}$, the equation above would define a unique solution. As we said, we will assume that $P(\eta)$ is given and take the initial conditions to be

$$\delta\dot{v}_{\mathbf{k}}^{(\mathrm{II})\pm}(\eta_c) = 0, \quad \delta v_{\mathbf{k}}^{(\mathrm{II})\pm}(\eta_c) = 4v_{\mathbf{k}}^{(\mathrm{II})0}(\eta_c)P(\eta_c). \tag{28}$$

This finishes the generic (i.e. for arbitrary $P$) construction of $\hat{\mathscr{H}}^{(\mathrm{II})}$.

Next, we need to find the state $|\zeta^{(\mathrm{II})}\rangle \in \hat{\mathscr{H}}^{(\mathrm{II})}$ that completes the SSC-II construction. The symmetries of the space-time led us to consider the "ansatz":

$$|\zeta^{(\mathrm{II})}\rangle = \dots|\zeta_{-2\mathbf{k}_0}^{(\mathrm{II})}\rangle \otimes |\zeta_{-\mathbf{k}_0}^{(\mathrm{II})}\rangle \otimes |\zeta_0^{(\mathrm{II})}\rangle \otimes |\zeta_{\mathbf{k}_0}^{(\mathrm{II})}\rangle \otimes |\zeta_{2\mathbf{k}_0}^{(\mathrm{II})}\rangle \dots. \tag{29}$$

The vector in Fock space is characterized by the specific modes that are excited (all other modes are assumed to be in the vacuum of the corresponding oscillator) and the parameters $\zeta_{\mathbf{k}}^{(\mathrm{II})}$ indicate the coherent state for the mode $\mathbf{k}$.

The expectation value of the field operator in such a state is given by

$$\phi_{\zeta}^{(\mathrm{II})}(x) = \phi_{\zeta,0}^{(\mathrm{II})}(\eta) + \left(\delta\phi_{\zeta,\mathbf{k}_0}^{(\mathrm{II})}(\eta)e^{i\mathbf{k}_0\cdot\mathbf{x}}\right) + \left(\delta\phi_{\zeta,2\mathbf{k}_0}^{(\mathrm{II})}(\eta)e^{i2\mathbf{k}_0\cdot\mathbf{x}}\right) + \cdots. \tag{30}$$

We note that the coefficients $\delta\phi_{\zeta,n\mathbf{k}_0}^{(\mathrm{II})}(\eta)$ have a contribution from the modes $n\mathbf{k}$, $(n-1)\mathbf{k}$ and $(n+1)\mathbf{k}$. We set $\delta\phi_{\zeta,n\mathbf{k}_0}^{(\mathrm{II})}(\eta) = 0$ for all $n \geq 2$, simply by imposing the required relations between the parameters $\zeta_{\pm\mathbf{k}_0}^{(\mathrm{II})}$, $\zeta_{\pm2\mathbf{k}_0}^{(\mathrm{II})}$, $\zeta_{\pm3\mathbf{k}_0}^{(\mathrm{II})}$, etc. It is easy to see that $|\zeta_{\pm n\mathbf{k}_0}^{(\mathrm{II})}| \sim \varepsilon^n|\zeta_0^{(\mathrm{II})}|$.

The conditions above ensure that there are no terms in $e^{\pm in\mathbf{k}_0\cdot\mathbf{x}}$ (with $n \geq 2$) appearing in the expectation value of the energy-momentum tensor. That is necessary for the compatibility of our state ansatz with the semiclassical Einstein's equations. We studied these in detail up to to the first order in $\varepsilon$.

The zero order equations are identical to those we found in constructing the SSC-I. They fix the construction of SSC to the lowest order, i.e. they determine the relation between $a^{(\mathrm{II})}$ and $\zeta_0^{(\mathrm{II})}$.

Considering the next order one obtains after a lengthy calculation the key result, that enables us to carry out the construction in a complete manner: That the equations can be combined into a simple dynamical equation for the Newtonian potential, which is independent of the first order quantities and where, at the level of precision we are working at, the equation above becomes simply:

$$\ddot{P} + \epsilon^{(\mathrm{II})}\mathscr{H}^{(\mathrm{II})}\dot{P} + \left[k_0^2 - \epsilon^{(\mathrm{II})}\mathscr{H}^{2(\mathrm{II})}\right]P = 0. \tag{31}$$

The general solution

$$P(\eta) = C_1\,\eta^{\frac{1}{2}[1+\epsilon^{(\mathrm{II})}]}J_\alpha(-k\eta) + C_2\,\eta^{\frac{1}{2}[1+\epsilon^{(\mathrm{II})}]}Y_\alpha(-k\eta), \tag{32}$$

where $J_\alpha(-k\eta)$ and $Y_\alpha(-k\eta)$ are the Bessel functions of the first and second kind, $\alpha = [1 + 3\epsilon^{(\mathrm{II})}]/2$.

Einstein's equations lead, as is well known, to constraints which, at this order provide relations involving the initial values that would determine the specific solution $P(\eta)$:

$$\varepsilon\begin{pmatrix} P \\ \dot{P} \end{pmatrix} = \frac{\sqrt{4\pi G\epsilon^{(\mathrm{II})}}\mathscr{H}^{(\mathrm{II})}}{k_0^2 - \mathscr{H}^{2(\mathrm{II})}\epsilon^{(\mathrm{II})}} \times \begin{pmatrix} (3\mathscr{H}^{(\mathrm{II})} - am\sqrt{3/\epsilon^{(\mathrm{II})}}) & 1 \\ (am\sqrt{3/\epsilon^{(\mathrm{II})}}\mathscr{H}^{(\mathrm{II})} - k_0^2 + (\epsilon^{(\mathrm{II})} - 3)\mathscr{H}^{2(\mathrm{II})}) & -\mathscr{H}^{(\mathrm{II})} \end{pmatrix} \cdot \begin{pmatrix} \delta\phi^{(\mathrm{II})}_{\zeta,\mathbf{k}_0} \\ \delta\dot{\phi}^{(\mathrm{II})}_{\zeta,\mathbf{k}_0} \end{pmatrix}. \tag{33}$$

Thus, given $\delta\phi^{(\mathrm{II})}_{\zeta,\mathbf{k}_0}(\eta_c)$ and $\delta\dot{\phi}^{(\mathrm{II})}_{\zeta,\mathbf{k}_0}(\eta_c)$, we have a completely determined space-time metric. In particular, we have a completely determined function $P(\eta)$ and thus, as discussed around Eq. (26), a completely specified set of mode functions for the expansion of the field operator. Furthermore, those determine the state parameters $\zeta_{\mathbf{k}_0}$ (and thus the rest as well). Thus, we have a complete SSC-II (to this order in $\varepsilon$).

## 8.3 The Collapse: Joining SSC-I and SSC-II

Next, we want to consider a space-time that includes a collapse. That is, a space-time that results from the *matching* of the two constructions. We will consider here that the collapse corresponds to a hypersurface that is matched to the hypersurfaces $\eta = \eta_c$

of SSC-I and SSC-II. Note that this gives such hypersurface $\Sigma_c$ a preferred status in the resulting space-time, and is not something to be thought as related to a gauge freedom: To the past of that hypersurface $\Sigma_c$ the space-time is H&I, and to the future it is not. We will assume here the induced metric is continuous on $\Sigma_c$. This requires $P(\eta_c) = 0$ and thus

$$(3\mathscr{H}^{(\mathrm{II})} - am\sqrt{3/\epsilon^{(\mathrm{II})}})\delta\phi^{(\mathrm{II})}_{\zeta,\mathbf{k}_0}(\eta_c) + \delta\dot{\phi}^{(\mathrm{II})}_{\zeta,\mathbf{k}_0}(\eta_c) = 0, \tag{34}$$

and therefore

$$\varepsilon\dot{P} = -\sqrt{4\pi G\epsilon^{(\mathrm{II})}}\mathscr{H}^{(\mathrm{II})}\delta\phi^{(\mathrm{II})}_{\zeta,\mathbf{k}_0}(\eta_c), \tag{35}$$

which indicates a discontinuity in the extrinsic curvature of the hypersurface $\Sigma_c$.

Assume that the collapse is characterized by a loose analogy with "an imprecise measurement" (of the operators $\hat{\phi}^{(\mathrm{I})}_{\mathbf{k}_0}(\eta)$) in standard QT: Before the collapse, the operator had zero expectation value but an uncertainty $\Delta\hat{\phi}^{(\mathrm{I})}_{\mathbf{k}_0}(\eta_c)$, and thus we assume that after the collapse, the new expectation value will fall in that range. We thus consider what would be the energy momentum tensor computed using a state that results from such measurement (on the SSC-I side of $\Sigma_c$) and demand that the state on the SSC-II side be such that the energy momentum tensor on the SSC-II side of $\Sigma_c$ be exactly that. The final result is then:

$$\delta\phi^{(\mathrm{II})}_{\zeta_t,\mathbf{k}_0}(\eta_c) = x_{\mathbf{k}_0}\sqrt{\langle 0^{(\mathrm{I})}_{\mathbf{k}_0}|\left[\Delta\hat{\phi}^{(\mathrm{I})}_{\mathbf{k}_0}(\eta_c)\right]^2|0^{(\mathrm{I})}_{\mathbf{k}_0}\rangle} \approx x_{\mathbf{k}_0}a(\eta_c)^{-1}\sqrt{\frac{\hbar}{2k}}$$

with $x_{\mathbf{k}_0}$ a random number taken from a distribution characterized by a Gaussian function centered at zero with unit-spread. A choice of the random number $x_{\mathbf{k}_0}$ then determines the SSC-II.

Thus, we have a well defined framework where one could, in principle, carry out all the analysis of the collapse approach to the inflationary origin of the seeds of cosmic structure.

## 9 Phenomenological Studies

In a realistic situation, we need to consider a collapse of not just one, but of all the modes. Thus, in contrast with the previous analysis, we have adopted for this purpose a simplified treatment that makes the realistic problem manageable. We note that considering simultaneously all modes is required if we want to compare the theory with observations.

In this simplified treatment, one avoids the complications of the previous treatment by ignoring the multiplicity of Hilbert spaces and considering the quantum states that result from collapses to be elements of the Hilbert space based on the initial homogeneous and isotropic space-time (as is done in standard treatments). We thus split the treatment into that of a classical homogeneous ("background") part, and an in-homogeneous part ("fluctuation"), i.e. $g = g_0 + \delta g$, $\phi = \phi_0 + \delta\phi$.

The background is taken again to be Friedmann-Robertson universe (with vanishing Newtonian potential), and the homogeneous scalar field $\phi_0(\eta)$. In the previous, more precise treatment this would have corresponded to the zero mode of the quantum field.

The main difference, with respect to the ordinary approach, will be in the spatially dependent perturbations. Here, our approach indicates we should quantize the scalar field but not the metric perturbation.

We will set $a = 1$ at the "present cosmological time", and assume that the inflationary regime ends at a value of $\eta = \eta_0$, negative and very small in absolute terms. Again, in our case the semiclassical Einstein's equations, at lowest order lead to

$$\nabla^2\Psi = 4\pi G\dot{\phi}_0\langle\delta\dot{\phi}\rangle = s\langle\delta\dot{\phi}\rangle, \tag{36}$$

where $s \equiv 4\pi G\dot{\phi}_0$.

Consider the quantum theory of the field $\delta\phi$. In this practical treatment it is convenient to work with the rescaled field variable $y = a\delta\phi$ and its conjugate momentum $\pi = \delta\dot{\phi}/a$. We decompose the field and momentum operators as:

$$y(\eta, \mathbf{x}) = \frac{1}{L^3}\sum_{\mathbf{k}} e^{i\mathbf{k}\cdot\mathbf{x}}\hat{y}_k(\eta), \qquad \pi_y(\eta, \mathbf{x}) = \frac{1}{L^3}\sum_{\mathbf{k}} e^{i\mathbf{k}\cdot\mathbf{x}}\hat{\pi}_k(\eta), \tag{37}$$

where $\hat{y}_k(\eta) \equiv y_k(\eta)\hat{a}_k + \bar{y}_k(\eta)\hat{a}^{+}_{-k}$ and $\hat{\pi}_k(\eta) \equiv g_k(\eta)\hat{a}_k + \bar{g}_k(\eta)\hat{a}^{\dagger}_{-k}$. The usual choice of modes $y_k(\eta) = \frac{1}{\sqrt{2k}}\left(1 - \frac{i}{\eta k}\right)\exp(-ik\eta)$, $g_k(\eta) = -i\sqrt{\frac{k}{2}}\exp(-ik\eta)$, which leads to what is known as the Bunch-Davies vacuum: the state defined by $\hat{a}_k|0\rangle = 0$ . At this point it is worthwhile to remind the reader that this state is translationally and rotationally invariant, as can be easily checked by applying the corresponding rotation and displacement operators to it. Note also that $\langle 0|\hat{y}_k(\eta)|0\rangle = 0$ and $\langle 0|\hat{\pi}_k(\eta)|0\rangle = 0$ . The collapse will modify the state and thus expectation values of the operators $\hat{y}_k(\eta)$ and $\hat{\pi}_k(\eta)$.

Next, we specify the rules according to which the collapse happens, and thus the state $|\Theta\rangle$ after the collapse. We assume that after the collapse, the expectation values of the field and momentum operators in each mode will be related to the uncertainties of the pre-collapse state (these quantities for the vacuum are *not* zero).

In the vacuum state, $\hat{y}_k$ and $\hat{\pi}_k$ are characterized by Gaussian wave functions centered at 0 with spread $\Delta y_k$ and $\Delta\pi_{y_k}$, respectively.

We will want to consider various possibilities for the detailed form of this collapse. Thus, for their generic form, associated with the ideas above, we assume that at time $\eta^c_k$ the part of the state corresponding to the mode $\mathbf{k}$ undergoes a sudden jump so that, immediately afterwards, the state describing the system is such that

$$\langle \hat{y}_k(\eta_k^c) \rangle_\Theta = x_{k,1}\sqrt{\Delta \hat{y}_k}, \qquad \langle \hat{\pi}_k(\eta_k^c) \rangle_\Theta = x_{k,2}\sqrt{\Delta \hat{\pi}_k^y}, \tag{38}$$

where $x_{k,1}$, $x_{k,2}$ are (single specific values) selected randomly from within a Gaussian distribution centered at zero with spread one.

Finally, using the evolution equations for the expectation values (i.e. using Ehrenfest's Theorem), we obtain $\langle \hat{y}_k(\eta) \rangle$ and $\langle \hat{\pi}_k(\eta) \rangle$ for the state that resulted from the collapse for all later times.

## 9.1 Analysis of the Phenomenology

The semi-classical version of the perturbed Einstein's equation that, in our case, leads to $\nabla^2 \Psi = 4\pi G \dot{\phi}_0 \langle \delta\dot{\phi} \rangle$ indicates that the Fourier components at the conformal time $\eta$ are given by:

$$\Psi_k(\eta) = -(s/ak^2)\langle \hat{\pi}_k(\eta) \rangle. \tag{39}$$

Prior to the collapse, the state is the BD vacuum, and it is easy to see that $\langle 0|\hat{\pi}_k(\eta)|0\rangle = 0$, so in that situation we would have $\Psi_k(\eta) = 0$. However, after the collapse has occurred, we have instead: $\Psi_k(\eta) = -(s/ak^2)\langle \Theta|\hat{\pi}_k(\eta)|\Theta\rangle \neq 0$. From those quantities, we can reconstruct the Newtonian potential (for times after the collapse):

$$\Psi(\eta, \mathbf{x}) = \frac{1}{L^3}\sum_{\mathbf{k}} e^{i\mathbf{k}\cdot\mathbf{x}}\Psi_k(\eta) = \sum_{\mathbf{k}} \frac{sU(k)}{k^2}\sqrt{\frac{\hbar k}{L^3}}\frac{1}{2a}F(\mathbf{k})e^{i\mathbf{k}\cdot\mathbf{x}}, \tag{40}$$

where $F(\mathbf{k})$ contains, besides the random quantities $x_{k,i}$, $i = 1, 2$, the information about the time at which the collapse of the wave function for the mode $\mathbf{k}$ occurs.

We now focus our attention on the "Newtonian potential" on the surface of last scattering: $\Psi(\eta_D, \mathbf{x}_D)$, where $\eta_D$ is the conformal time at decoupling and $\mathbf{x}_D$ are co-moving coordinates of points on the last scattering surface corresponding to us as observers. The quantity is identified with the temperature fluctuations on the surface of last scattering. Thus:

$$\alpha_{lm} = \int \Psi(\eta_D, \mathbf{x}_D) Y_{lm}^* d^2\Omega. \tag{41}$$

The factor $U(k)$ is called the transfer function and represents known physics like the acoustic oscillations of the plasma. Now, putting all this together we find,

$$\alpha_{lm} = s\sqrt{\frac{\hbar}{L^3}}\frac{1}{2a}\sum_{\mathbf{k}} \frac{U(k)\sqrt{k}}{k^2} F(\mathbf{k}) 4\pi i^l j_l(|\mathbf{k}|R_D) Y_{lm}(\hat{k}), \tag{42}$$

where $j_l(x)$ is the spherical Bessel function of the first kind, $R_D \equiv ||\mathbf{x_D}||$, and $\hat{k}$ indicates the direction of the vector $\mathbf{k}$. Note that in the usual approaches it is impossible to produce an explicit expression for this quantity, other than zero.

Thus $\alpha_{lm}$ is the sum of complex contributions from all the modes, i.e., the equivalent to a two dimensional random walk, whose total displacement corresponds to the observational quantity. We then evaluate the most likely value of such quantity, and then pass to the continuum obtaining:

$$|\alpha_{lm}|^2_{M.L.} = \frac{s^2\hbar}{2\pi a^2} \int \frac{U(k)^2 C(k)}{k^4} j_l^2(|\mathbf{k}| R_D)\, k^3 dk. \tag{43}$$

The function $C(k)$ encodes information contained in $F(k)$. For each model of collapse it has a slightly different functional form.

It turns out that in order to get a reasonable spectrum, we have one single simple option: $z_k$ must be almost independent of $k$. That is: $\eta_k^c = z/k$.

This result shows that the details of the collapse have observational consequences! In fact, we have

$$C(k) = 1 + \frac{2}{z_k^2} \sin^2 \Delta_k + \frac{1}{z_k} \sin(2\Delta_k), \tag{44}$$

where $\Delta_k = k\eta - z_k$, $z_k = \eta_k^c k$ with $\eta$ representing the conformal time of observation, and $\eta_k^c$ the conformal time of collapse of the mode $k$.

If $z_k$ is independent of $k$, this will not modify the form of the spectrum because these functions become constants. We can consider simple departures from the pattern $\eta_k^c = z/k$, say, assuming $\eta_k^c = A/k + B$. These can now be compared with observations! We have carried out a preliminary exploration [21] considering the departures from the HZ spectrum, and a more detailed analysis [22] incorporating the well understood late time physics (acoustic oscillations, etc.) and comparing directly with the observational data. Those represent the first limits on a collapse model coming from the CMB observations. This analysis can now be used to constrain a more specific version of the collapse theories, particularly those schemes which indicate specific ranges for the collapse times and the specific operators involved in the collapse process.

## 10 Conclusions

We have argued elsewhere extensively about the need to deal with the fact that the standard accounts for the generation of the primordial seeds of cosmic structure from quantum fluctuations during inflation are not completely satisfactory. In the present work, I have focused on the formal implementation of such ideas, and on the fact that, even if one wanted to ignore the conceptual issues that we have pointed out in previous works, but at the same time, one wanted to consider the possibility of a maximal characterization of our universe, one would be making a similar kind of description as that used in the collapse approach we favor.

We have presented a formalism that allows the incorporation of a collapse process within a semiclassical treatment of gravity interacting with quantum fields. Finally, I have made a brief overview of the phenomenological analysis that relied on a simplified treatment, and which made it manageable to consider the realistic situation involving collapses in all modes of the quantum field.

There is, in fact, an interesting possibility of connection of the ideas presented here to some appearing in the context of the singularity resolution LQG, where one expects a failure of the approach to lead to even an approximate characterization, in terms of classical geometry. In this case, we have found a specific kind of breakdown of the space-time description at the collapsing hypersurface: A discontinuity in the extrinsic curvature of such hypersurface. The amount of such discontinuity is related to the details of the collapse. The investigation of this issue in the context of candidate theories for quantum space-time would be very interesting.

We believe that one of the best ways to inquire about the interface of quantum and gravitation is by pushing our attempts to describe space-time in the context where quantum effects become important. The inflationary situation offers us a unique opportunity. In order for us to be able to take full advantage of such window into the unknown, we need to start by recognizing the shortcomings in our current treatments. We trust that the program here outlined represents the first steps in that direction.

**Acknowledgments** This work is supported in part by the CONACYT grant No 101712. and by UNAM-PAPIIT IN107412-3 grant. The author thanks the organizers for the kind hospitality during the conference in Prague.

## References

1. Penrose, R.: On gravity's role in quantum state reduction. Gen. Relativ. Gravit. **28**, 581 (1996). doi:10.1007/BF02105068
2. Diósi, L.: A universal master equation for the gravitational violation of quantum mechanics. Phys. Lett. A **120**, 337 (1987). doi:10.1016/0375-9601(87)90681-5
3. Perez, A., Sahlmann, H., Sudarsky, D.: On the quantum origin of the seeds of cosmic structure. Class. Quantum Grav. **23**, 2317 (2006). doi:10.1088/0264-9381/23/7/008
4. Sudarsky, D.: Shortcomings in the understanding of why cosmological perturbations look classical. Int. J. Mod. Phys. D **20**, 509 (2011). doi:10.1142/S0218271811018937
5. Ashtekar, A., Bojowald, M.: Quantum geometry and the Schwarzschild singularity. Class. Quantum Grav. **23**, 391 (2006). doi:10.1088/0264-9381/23/2/008
6. Wald, R.M.: Quantum Field Theory in Curved Spacetime and Black Hole Thermodynamics. Chicago Lectures in Physics. University of Chicago Press, Chicago (1994)
7. Ghirardi, G.C., Rimini, A., Weber, T.: Unified dynamics for microscopic and macroscopic systems. Phys. Rev. D **34**, 470 (1986). doi:10.1103/PhysRevD.34.470
8. Pearle, P.: Reduction of the state vector by a nonlinear Schrödinger equation. Phys. Rev. D. **13**, 857 (1976). doi:10.1103/PhysRevD.13.857
9. Bassi, A., Ghirardi, G.: Dynamical reduction models. Phys. Rep. **379**, 257 (2003). doi:10.1016/S0370-1573(03)00103-0

10. Tumulka, R.: On spontaneous wave function collapse and quantum field theory. Proc. R. Soc. London, Ser. A **462**, 1897 (2006). doi:10.1098/rspa.2005.1636
11. Bedingham, D.J.: Relativistic state reduction dynamics. Found. Phys. **41**, 686 (2011). doi:10.1007/s10701-010-9510-7
12. Weinberg, S.: Collapse of the state vector. Phys. Rev. A **85**, 062116 (2012). doi:10.1103/PhysRevA.85.062116
13. Pearle, P.: Combining stochastic dynamical state-vector reduction with spontaneous localization. Phys. Rev. A **39**, 2277 (1989). doi:10.1103/PhysRevA.39.2277
14. Martin, J., Vennin, V., Peter, P.: Cosmological inflation and the quantum measurement problem. Phys. Rev. D **86**, 103524 (2012). doi:10.1103/PhysRevD.86.103524
15. Cañate, P., Pearle, P., Sudarsky, D.: CSL wave function collapse model as a mechanism for the emergence of cosmological asymmetries in inflation. ArXiv e-prints arXiv:1211.3463 [gr-qc] (2012)
16. Diez-Tejedor, A., Sudarsky, D.: Towards a formal description of the collapse approach to the inflationary origin of the seeds of cosmic structure. J. Cosmol. Astropart. Phys. **2012**(07), 045 (2012). doi:10.1088/1475-7516/2012/07/045
17. Diósi, L.: Gravitation and quantum-mechanical localization of macro-objects. Phys. Lett. A **105**, 199 (1984). doi:10.1016/0375-9601(84)90397-9
18. van Meter, J.R.: Schrödinger-Newton 'collapse' of the wavefunction. Class. Quantum Grav. **28**, 215013 (2011). doi:10.1088/0264-9381/28/21/215013
19. Hu, B.L., Verdaguer, E.: Stochastic gravity: theory and applications. Living Rev. Relativ. **11**(3), lrr-2008-3 (2008). http://www.livingreviews.org/lrr-2004-3
20. Mukhanov, V.: Physical Foundations of Cosmology. Cambridge University Press, Cambridge (2005)
21. de Unánue, A., Sudarsky, D.: Phenomenological analysis of quantum collapse as source of the seeds of cosmic structure. Phys. Rev. D **16**, 043510 (2008). doi:10.1103/PhysRevD.78.043510
22. Landau, S.J., Scóccola, C.G., Sudarsky, D.: Cosmological constraints on nonstandard inflationary quantum collapse models. Phys. Rev. D **85**, 123001 (2012). doi:10.1103/PhysRevD.85.123001

# Quantum Gravity: The View From Particle Physics

**Hermann Nicolai**

**Abstract** This lecture reviews aspects of and prospects for progress towards a theory of quantum gravity from a particle physics perspective, also paying attention to recent findings of the LHC experiments at CERN.

## 1 Introduction

First of all I would like to thank Jiří Bičák for inviting me to this prestigious conference in commemoration of Einstein's stay in Prague a 100 years ago. Although it was only a short stay, as Einstein left Prague again after little more than 1 year, it was here that he made major progress towards the final version of General Relativity, and surely the beauty of this city must have played an important inspirational part in this endeavor.

In view of the more general nature of this conference, I have decided not to give a technical talk on my current work, but rather to present some thoughts on the state of quantum gravity from the point of view of a particle physicist, but with an audience of general relativists in mind. Taking such a point of view is quite appropriate, as LHC is about to end its first phase of experiments, with the solid evidence for a scalar boson that has all the requisite properties of a Higgs boson as the main outcome so far. This boson was the final missing link in the Standard Model of Particle Physics (or SM, for short), and therefore its discovery represents the final step in a story that has been unfolding for almost 50 years. Equally important, as the CERN experiments continue to confirm the Standard Model to ever higher precision, with (so far) no indications of 'new physics', it is also a good time to ask whether these results can

H. Nicolai (✉)
Max Planck Institute for Gravitational Physics (Albert Einstein Institute),
Am Mühlenberg 1, 14476 Golm, Germany
e-mail: nicolai@aei.mpg.de

J. Bičák and T. Ledvinka (eds.), *General Relativity, Cosmology and Astrophysics*, Fundamental Theories of Physics 177, DOI: 10.1007/978-3-319-06349-2_18,
© Springer International Publishing Switzerland 2014

possibly offer any insights into quantum gravity. So my main message will be that we should not ignore the hints from particle physics in our search for quantum gravity!

I do not think I need to tell you *why* a theory of quantum gravity is needed, as some of the key arguments were already reviewed in other talks at this conference. There is now ample evidence that both General Relativity (GR) and Quantum Field Theory (QFT) are incomplete theories, and both are expected to break down at sufficiently small distances. The generic occurrence of space-time singularities in GR is an unavoidable feature of the theory, indicating that classical concepts of space and time must be abandoned at distances of the order of the Planck scale. Likewise, there are indications of a breakdown of conventional QFT in this regime. Accordingly, and in line with the title of this lecture, I would therefore like to concentrate on *the lessons from particle physics* pointing beyond QFT and conventional concepts of space and time.

In its current incarnation, QFT mainly relies on perturbation theory. The ultraviolet (UV) divergences that inevitably appear in higher order Feynman diagrams require a carefully crafted procedure for their removal, if one is to arrive at testable predictions. This renormalization prescription in essence amounts to an order by order tuning of a finite number of parameters by infinite factors. Although mathematically on very shaky grounds, this procedure has produced results in stunning agreement with experimental findings, with a precision unmatched by any other scheme in the physical sciences. The most famous example is, of course, the QED prediction of the anomalous magnetic moment of the electron, but the agreement between very recent precision measurements at LHC and the theoretical predictions of the Standard Model is now equally impressive. Yet, in spite of this extraordinary success there is good reason to believe that neither the SM in its present form nor any of its quantum field theoretic extensions (such as the supersymmetric versions of the SM) are likely to exist in a strict mathematical sense. The ineluctable conclusion therefore seems to be that the UV completion of the SM requires something beyond QFT as we know it.

The difficulties in both GR and conventional QFT have a common origin. In both frameworks space-time is assumed to be a *continuum*, that is, a differentiable manifold. As a consequence, there should exist no obstacle of principle in going to arbitrarily small distances if either of these theories were universally valid. Nevertheless, the very nature of quantum mechanics suggests that its principles should ultimately also apply to space-time itself, whence one would expect the emergence of a grainy structure at the Planck scale. Indeed, and in spite of their disagreements, almost all approaches to quantum gravity[1] are united in their expectation that something dramatic must happen to space-time at Planck scale distances, where the continuum should thus give way to some kind of *discretuum*.

A second, and related, source of difficulties is the assumption that elementary particles are to be treated as *point-like* excitations. And indeed, there is not a shred of a hint so far that would point to an extended structure of the fundamental constituents

---

[1] With the possible exception of the Asymptotic Safety program [1].

of matter (quarks, leptons and gauge bosons), so this assumption seems well supported by experimental facts. Nevertheless, it is at the root of the ultraviolet infinities in QFT. Moreover, it is very hard to do away with, because the point-likeness of particles and their interactions seems to be required by both relativistic invariance and locality/causality—building a (quantum) theory of relativistic extended objects is not an easy task! In classical GR, the very notion of a point-particle is problematic as well, because any exactly point-like mass would have to be a mini black hole surrounded by a tiny horizon, and thus the putative point particle at the center would move on a space-like rather than a time-like trajectory. Again, one is led to the conclusion that these concepts must be replaced by more suitable ones in order to resolve the inconsistencies of GR and QFT.

Current approaches to quantum gravity can be roughly put into one of the two following categories (for a general overviews see e.g. [2–4][2]).

- According to the first hypothesis quantum gravity in essence is nothing but the *non-perturbative* quantization of Einstein Gravity (in metric/connection/loop or discrete formalism). Thus GR, suitably treated and eventually complemented by the Standard Model of Particle Physics or one of its possible extensions, should correctly describe the physical degrees of freedom also at the very smallest distances. The first attempt of quantizing gravity relied on canonical quantization, with the spatial metric components and their conjugate momenta as the canonical variables, and the Wheeler-DeWitt equation governing the dynamics [3]. Superimposing Schrödinger-type wave mechanics on classical GR, this scheme was still rather close to classical concepts of space and time. By contrast, modern versions of this approach look quite different, even though their starting point is still the standard Einstein-Hilbert action in four dimensions: for instance, the discrete structure that emerges from the loop quantum gravity program relies on holonomies and fluxes as the basic variables, leading to a discretuum made of spin networks or spin foams [5, 6].
- According to the opposite hypothesis (most prominently represented by string theory [7–9]) GR is merely an effective (low energy) theory arising at large distances from a more fundamental Planck scale theory whose basic degrees of freedom and whose dynamics are very different from either GR or conventional QFT, and as yet unknown. In this view, classical geometry and space-time itself, as well as all matter degrees of freedom are assumed to be 'emergent', in analogy with the emergence of classical macroscopic physics from the completely different quantum world of atoms and molecules. Likewise, concepts such as general covariance and even background independence might only emerge in the large distance limit and not necessarily be features of the underlying theory. Consequently, attempts to unravel the quantum structure of space and time by directly quantizing Einstein's theory would seem as futile as trying to derive microscopic physics by

[2] As there is a vast literature on this subject, I here take the liberty of citing only a few representative introductory texts, where more references can be found.

**Fig. 1** The steady progress of quantum gravity?

applying canonical quantization procedures to, say, the Navier-Stokes equation. The fundamental reality might then be something like the abstract space of all conformal field theories, only a small subset of which would admit a geometrical interpretation. The occasional 'condensation' of a classical space-time out of this pre-geometrical framework would then appear as a rare event.

Pursuing different and independent ideas is certainly a good strategy as long as we do not know the final answer, but it is a bit worrisome (at least to me) that the proponents of the different approaches not only base their approaches on very different assumptions, but continue to speak languages that are foreign to one another. Surely, when zeroing in on the 'correct' theory there should be *a convergence of ideas and concepts*: when Schrödinger proposed wave mechanics and Heisenberg formulated matrix mechanics, these were initially regarded as very different, but it did not take long before it became clear that they were just equivalent descriptions of the *same* theory. Unfortunately, at this time there is no such convergence in existing approaches to quantum gravity—a sign that we are probably still very far from the correct answer! So let us hope that our noble search will not end like the historic event in the Breughel painting shown in Fig. 1.

## 2 The Divergence Problem

From the point of view of perturbative QFT the basic difference between gravity and matter interactions is the non-renormalizability of perturbatively treated GR. For instance, at two loops the Einstein-Hilbert action must be supplemented by the following counterterm cubic in the Weyl tensor [10, 11] [3]

$$\Gamma_{div}^{(2)} = \frac{1}{\varepsilon}\frac{209}{2880}\frac{1}{(16\pi^2)^2}\int dV\, C_{\mu\nu\rho\sigma}C^{\rho\sigma\lambda\tau}C_{\lambda\tau}{}^{\mu\nu}\,, \tag{1}$$

if the calculations are to produce *finite* predictions for graviton scattering at this order (the parameter $\varepsilon$ here is the deviation from four dimensions in dimensional regularization, and must be taken to zero at the end of the calculation). At higher orders there will arise similar infinities that likewise must be cancelled by counterterms of higher and higher order in the Riemann tensor. Because one thus has to introduce an unlimited number of counterterms in order to make predictions at arbitrary loop orders and therefore has to fix an infinite number of coupling constants, the theory looses all predictive power.

From the non-renormalizability of perturbatively quantized gravity, one can draw quite different conclusions, in particular reflecting the two opposite points of view cited above. According to the string/supergravity 'philosophy', a consistent quantization of gravity necessarily requires a modification of Einstein's theory at short distances, in order to cancel the infinities. This entails the necessity of (possibly supersymmetric) matter and in particular fermions, thus furnishing a possible *raison d'être* for the existence of matter in the world. It was originally thought that the UV finiteness requirements might single out the unique maximally supersymmetric field theory—*maximal $N = 8$ supergravity*—as the prime candidate for a unified theory of quantum gravity, but that theory was eventually abandoned in favor of superstring theory as it became clear that maximal supersymmetry by itself may not suffice to rule out all possible counterterms. Superstring theory gets rid of the divergences in a different way, by resolving the point-like interactions of QFT into extended vertices, relying not only on supersymmetry, but also on a specifically 'stringy' symmetry, *modular invariance*. Nevertheless, very recent developments [12] have rekindled the debate whether $N = 8$ supergravity could, after all, be a purely field theoretic extension of Einstein's theory that is UV finite to all orders.

On the other side, one can argue that the UV divergences of perturbative quantum gravity are merely an artifact of the perturbative treatment, and will disappear upon a proper *non-perturbative quantization* of Einstein's theory. In this view, perturbative quantization is tantamount to 'steamrollering Einstein's beautiful theory into flatness and linearity' (R. Penrose): by giving up the core features of GR, namely general covariance and background independence, one cannot expect to get any sensible

---

[3] There is no need here to distinguish between the Riemann tensor and the Weyl tensor, as all terms containing the Ricci scalar or the Ricci tensor can be absorbed into (possibly divergent) redefinitions of the metric.

complete answer. This is the point of view adopted by most of the 'non-string' approaches, see e.g. [5, 6]. The concrete technical implementation of this proposal invokes unusual properties which are very different from familiar QFT concepts; for instance, the finiteness properties of canonical loop quantum gravity hinge on the non-separability of the kinematical Hilbert space.[4] These features are at the origin of the difficulties that this approach encounters in recovering a proper semi-classical limit, and make it difficult to link up with established QFT results. Also for this reason there is so far no clue from non-perturbative quantization techniques as to what the detailed mechanism is that could dispose of the divergence (1).

There is a third (and more conservative) possibility that has lately received considerable attention, namely *asymptotic safety* [1, 14]. This is the proposal that the non-renormalizability of quantum gravity can be resolved by a kind of *non-perturbative renormalizability*, in the sense that there might exist a non-trivial fixed point to which the theory flows in the UV. Such a behavior would be similar to QCD, which flows to an asymptotically free theory in the UV, but the UV fixed point action for Einstein's theory would not be free, but rather characterized by higher order contributions in the Riemann tensor. In this case there would be no such thing as a 'smallest distance', and space-time would remain a continuum below the Planck scale. If it works, asymptotic safety is probably the only way to tame the divergences of perturbatively quantized gravity *without* resorting to the cancellation mechanisms invoked by supergravity and superstring theory. The hypothetical non-perturbative renormalizability of gravity would also have to come to the rescue to resolve the inconsistencies of standard QFT.

However, independently of which point of view one prefers, it should be clear that *no approach to quantum gravity can claim complete success that does not explain in full and convincing detail the ultimate fate of the divergences of perturbative quantum gravity.*

## 3 The Role of Matter

A main point of disagreement between the different approaches concerns the role of matter degrees of freedom. At least up to now, in modern loop and spin foam quantum gravity or other discrete approaches 'matter does not matter', in the sense that matter degrees of freedom are usually treated as more of an accessory that can be added at will once the quantization of pure gravity has been achieved. By contrast, to a supergravity/string practitioner the matter content of the world must play a key role in the search for quantum gravity. String theory goes even further, positing that the graviton is but one excitation (although a very distinguished one) among an infinite tower of quantized vibrational modes that should also include all the constituents of

---

[4] The non-separability of the kinematical Hilbert space is also a crucial ingredient in proposals to resolve space-time singularities in loop quantum cosmology [13].

matter, and that all these degrees of freedom are required for the consistency of the theory.

Perhaps it is fitting at this point to recall what Einstein himself remarked on the different character of the two sides of his field equations: the left hand side is pure geometry and beautifully unique, thus made of marble, whereas the right hand side has no share in this beauty:

$$\underbrace{R_{\mu\nu} - \frac{1}{2} g_{\mu\nu} R}_{\text{Marble}} = \underbrace{\kappa T_{\mu\nu}}_{\text{Timber?}} \tag{2}$$

Indeed, the question that occupied Einstein until the end of his life was this: can we understand the right hand side geometrically, thereby removing its arbitrariness? Put differently, is there a way of massaging the right hand side and moving it to the left hand side, in such a way that everything can be understood as coming from some sort of generalized geometry?

Over the last 90 years there has been some remarkable progress in this direction (see e.g. the reprint volume [15]), but we still do not know whether these ideas really pan out. Already in 1921, T. Kaluza noticed that electromagnetism (Maxwell's theory) can be understood as originating from a five-dimensional theory of pure gravity; later O. Klein extended this proposal to non-abelian gauge interactions. The idea of higher dimensions and of finding a geometrical explanation for the existence of matter continues to hold fascination to this day, most recently with the idea of *large* extra dimensions (whereas the original Kaluza-Klein proposal assumed the extra dimensions to be of Planck size in extension). Supersymmetry and supergravity may likewise be viewed as variants of the Kaluza-Klein program: they generalize ordinary geometry by including *fermionic* dimensions. This leads to the replacement of ordinary space-time by a superspace consisting of bosonic (even) and fermionic (odd) coordinates, thus incorporating *fermionic matter* into the geometry [16]. Accordingly, the possible discovery of supersymmetric particles at LHC could be interpreted as evidence of new dimensions of space and time.

There is not so much discussion of such ideas in the 'non-string' context, where neither unification nor extra dimensions feature prominently (the Ashtekar variables exist only in three and four space-time dimensions), and the focus is on Einstein gravity in four dimensions. Although loop and spin foam quantum gravity are thus very much tuned to four dimensions, there have nevertheless been attempts to extend the framework to higher dimensions, specifically by replacing the groups $SU(2)$ (for loop quantum gravity) or $SO(4)$ or $SO(3,1)$ (for spin foam models) by bigger groups, with higher dimensional analogues of the Ashtekar variables, but it is not clear whether one can arrive in this way at a unification properly incorporating the SM degrees of freedom.

# 4 The Hierarchy Problem

A problem that is not so much in the focus of the GR community, but much discussed in the particle physics community concerns the question of scales and hierarchies. The gravitational force is much weaker than the other forces (as one can see immediately by comparing the gravitational attraction between the electron and the nucleus in an atom with the electric Coulomb force, which differ by a factor $10^{-40}$). The so-called hierarchy problem, then, is the question whether this huge difference in scales can be 'naturally' understood and explained.[5] In particle physics the problem is reflected in the mass hierarchies of elementary particles. Already by itself, the observed particle spectrum covers a large range of mass values: light neutrinos have masses of less than 1 eV, the lightest quarks have masses of a few MeV while the top quark, which is the heaviest quark discovered so far, has a mass of around 173 GeV, so even quark masses differ by factors on the order of $10^5$, presenting a 'little hierarchy problem'. But all these mass values are still tiny in comparison with the Planck scale, which is at $10^{19}$ GeV! This, then, is the distance that theory has to bridge: at the lower end it is the electroweak scale that is now being explored at LHC, while at the higher end it is Planck scale quantum gravity.

A much advertized, but very QFT specific, indication of the problem is the occurrence of quadratic divergences in radiative corrections to the scalar (Higgs) boson mass, which require an enormous fine-tuning to keep the observed value so small in comparison with the Planck mass, the 'natural' value. The absence of quadratic divergences in supersymmetric theories, where divergences are at most logarithmic, is widely considered as a strong argument for low energy supersymmetry, and the prediction that each SM particle should be accompanied by a supersymmetric partner.

The smallness of the gravitational coupling in comparison with the other couplings in nature is the main obstacle towards the verification or falsification of any proposed model of quantum gravity. Unless there is a dramatic evolution of the strength of the gravitational coupling over experimentally reachable energy scales there is no hope of 'seeing' quantum gravity effects in the laboratory. So one needs to find ways and means to reason indirectly in order to identify low energy hints of Planck scale physics. One possibility might be to look for signatures of quantum gravity in the detailed structure of CMB fluctuations. The other possibility (which is more in line with this talk) is to try to read the signs and hints from the observed structure of the low energy world. In the final consequence, this would require a more or less unique prediction for low energy physics and the observed matter content of the world.[6] A more exotic possibility, advocated by proponents of large extra dimension scenarios, could be an (enormous) increase in the gravitational coupling strength in

[5] Of course, the biggest and most puzzling hierarchy problem concerns the smallness of the observed cosmological constant.

[6] This option is not very popular with aficionados of the multiverse or the anthropic principle but, interestingly, the hope for a *unique* path from quantum gravity to the SM is also prominently visible in the very first papers on the heterotic superstring [17, 18].

the TeV range that would make quantum gravity and quantum string effects directly accessible to experiment ('TeV scale quantum gravity').

At any rate, *it remains a key challenge for any proposed theory of quantum gravity to offer quantifiable criteria for its confirmation or falsification.* And the emphasis here is on 'quantitative', not on qualitative features that might be shared by very different approaches and thus may not suffice to discriminate between them (for instance, I would suspect this to be the case for specific properties of the CMB fluctuations, which may not contain enough information for us to 'read off' quantum gravity). So the challenge is to come up with criteria that allow to *unambiguously discriminate a given proposal against alternative ones*!

# 5 From the Standard Model to the Planck Scale

By now, the SM of particle physics is an extremely well tested theory. It is based on the (Yang-Mills) gauge principle with Yang-Mills group $G_{SM} = SU(3)_c \times SU(2)_w \times U(1)_Y$. Forces are mediated by spin-one gauge bosons. Matter is made up of spin-$\frac{1}{2}$ fermions: at this time, we know of 48 fundamental fermions which are grouped into three generations (families) of 16 quarks and leptons each (including right-chiral neutrinos). There is no evidence so far from LHC of any new fundamental fermions.

However, after decades of theoretical research, we still do not know what distinguishes the SM gauge group from other possible choices. Apart from anomaly cancellations (see below), the same ignorance prevails with regard to the observed matter content of the SM. Why does Nature repeat itself with three generations of quarks and leptons (another fact confirmed by CERN experiments, as well as cosmological observations)? What causes symmetry breaking and what is the origin of mass? And, returning to the question of hierarchies, what keeps the electroweak scale stable with regard to the Planck scale? (More on this below...) And, finally, why do we live in four space-time dimensions? To tackle these questions, numerous proposals have been put forward for physics beyond the Standard Model (or 'BSM physics', for short): Grand Unification (or GUTs, for short), technicolor, low energy supersymmetry, large extra dimensions, TeV scale gravity, excited gauge bosons, and so on.

The main recent progress is the discovery of a scalar boson by LHC and the strong evidence that this particle has all the requisite properties of the Higgs boson, especially with the most recent data indicating that it has indeed spin zero and even parity (a remaining uncertainty concerns the coupling to the SM fermions, which must be proportional to their masses). In addition the symmetry breaking mechanism giving mass to gauge bosons ('Brout-Englert-Higgs mechanism') has now been confirmed. However, much to the dismay of many of my colleagues, no signs of 'new physics' have shown up so far at LHC. It is therefore not excluded that there may be nothing more than the SM, augmented by right-chiral neutrinos, right up to the Planck scale, a scenario that is usually referred to as the 'Grand Desert'.

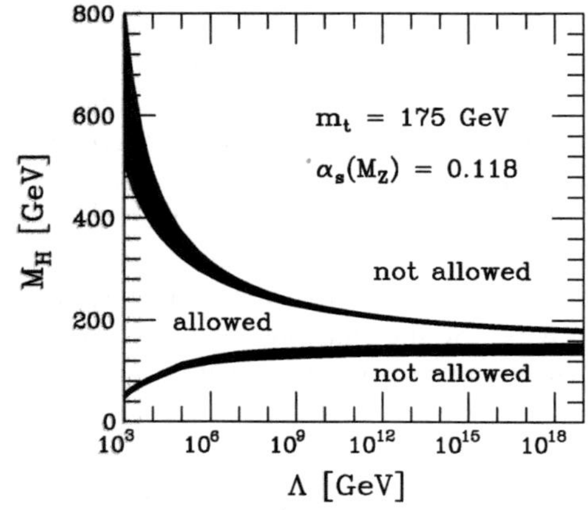


**Fig. 2** Can the SM survive up to the Planck scale? The upper envelope enforces avoidance of Landau pole for the scalar self-coupling, while the lower envelope ensures avoidance of vacuum instability [19] (with an assumed top quark mass of 175 GeV this plot is not quite up to date, but this does not affect our main conclusions)

Since the Higgs boson is partly responsible[7] for the generation of mass, and mass measures the strength of gravitational coupling, one can reasonably ask whether these data contain indications of Planck scale physics reaching down to the electroweak scale. It is here that the question of stability of the electroweak scale comes into play. Actually, the stability of the Standard Model is under menace from two sides. Following the renormalization group evolution of the scalar self-coupling up to the Planck scale, one danger is the Landau pole where the scalar self-coupling diverges (as happens for IR free theories with scalar fields) and the theory breaks down. The other danger is the potential instability caused by the negative contribution to the effective potential from the top quark (the effective potential includes perturbatively computable quantum corrections to the classical potential). In Homer's tale, Ulysses has to maneuver his ship between two formidable obstacles, *Skylla* and *Charybdis*: on the one side he must steer it away from the rock against which it will crash, and on the other side must avoid the sea monster that will swallow the ship. The plot of Fig. 2 illustrates the situation. The vertical direction is the Higgs mass, which grows proportionally with the Higgs self-coupling. Through the RG evolution the scalar self-coupling will eventually hit the Landau pole, at which point the theory crashes. Exactly where this happens depends very delicately on the Higgs mass. The upper curve (the rock) in Fig. 2 [19] relates the location of the Landau pole directly to the Higgs mass. The lower curve (the sea monster) is the constraint from the negative contribution to the effective Higgs potential. If this contribution is too negative, it will make the potential unbounded from below. For the SM, the negative contribution is mainly due to the top quark, and it is a danger precisely because the top quark mass is so large. As a result, if you want to salvage the SM up to the Planck scale, there remains only a very narrow strip for the SM parameters (masses and couplings). The recent results and data from LHC indicate that Nature might indeed avail itself of this possibility: with a Higgs mass of about 125 GeV, the Landau pole can safely

[7] Only partly, as for instance the larger part of the proton mass is due to non-perturbative QCD effects!

hide behind the Planck scale, but this value is so low that the SM hovers on the brink of instability! See also [20, 21] for interesting alternative interpretations of this value from the points of view of asymptotic safety, and non-commutative geometry, respectively.

To be sure, the potential instability of the effective potential is the worse of the two dangers. Namely, the occurrence of a Landau pole can always be interpreted as signalling the onset of 'new physics' where new degrees of freedom open up and thereby cure the problem. A well known example of this phenomenon is the old Fermi theory of weak interactions, where the non-renormalizable four-fermion vertex is dissolved at sufficiently high energies by new degrees of freedom ($W$ and $Z$ bosons) into a renormalizable and unitary theory. Another example would be the (still conjectural) appearance of supersymmetry in the TeV range, which would remove the Landau pole and also ensure full stability, as the effective potential in a globally supersymmetric theory is always bounded from below (this is no longer true for local supersymmetry).

If we find out whether or not there are genuine new degrees of freedom in the TeV range of energies, we may also get closer to answering the old question of the ultimate divisibility of matter, namely the question whether the known particles possess further substructures, sub-substructures, and so on, as we probe smaller and smaller distances. Translated into the UV, the question can be rephrased as the question whether there are any 'screens' ($\equiv$ scales of 'new physics') between the electroweak scale and the Planck scale. The more of such screens there were between the electroweak scale and the Planck scale, the less one would be be able to 'see' of Planck scale physics. On the other hand, the fewer there are, the harder becomes the challenge of explaining low energy physics from Planck scale physics.

LHC is now testing a large number of 'BSM' proposals, and actually eliminating many of them.[8] Figure 3 shows the latest exclusion plot from May 2013 on the search for various signatures of supersymmetry. Figure 4 shows a similar plot from May 2013 for 'exotica' such as large extra dimensions, mini black holes, excited $W$ and $Z$ bosons, quark substructure, and so on, with some exclusions already reaching up to 10 TeV. As you can see, even to refute only a representative subset of the proposals on the market and to keep up with the flood of theoretical ideas is a painstaking effort for the experimentalists, requiring teams of thousands of people and thousands of computers!

There are theoretical indications that LHC may not reveal much new beyond the SM Higgs boson, and thus *no* screens between the electroweak scale and the Planck scale. It is a remarkable fact that the SM Lagrangian is *classically conformally invariant except for a single term*, the explicit mass term in the Higgs potential. But in a classically conformal theory mass terms can in principle be generated by the conformal anomaly, the quantum mechanical breaking of conformal invariance, which could also trigger spontaneous symmetry breaking [23]. Maybe no explicit mass terms are needed in the SM Lagrangian, and the mechanism stabilizing the

---

8 The two plots shown below have been downloaded from the CERN website [22] where also a summary of many further results can be found.

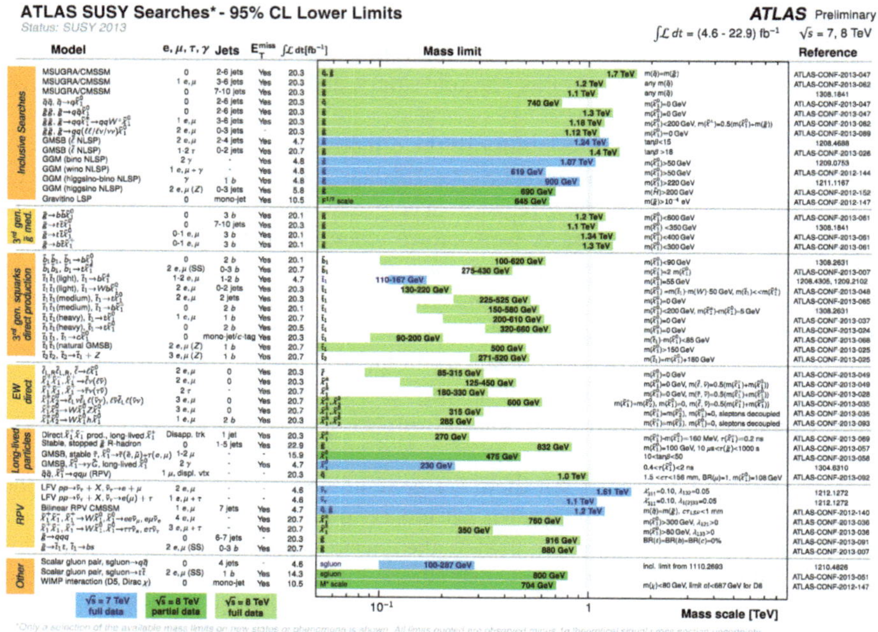


**Fig. 3** Low energy supersymmetry? [22]

electroweak scale is *conformal symmetry* rather than low energy supersymmetry [24, 25]? A further hint in this direction comes from the flows of the SM couplings under the renormalization group: it almost looks like these couplings could 'keep each other under control' so as to prevent both Landau poles and instabilities right up to the Planck scale! This is because bosons and fermions contribute with opposite signs to the corresponding $\beta$-functions. The scalar self-coupling would normally blow up under the flow, but is kept under control by the top quark contribution which delays the appearance of the Landau pole until after the Planck scale. The same mechanism is at work for the top quark (Yukawa) coupling which is asymptotically not free either: it, too, would blow up, but is kept under control by the strong coupling $\alpha_s$, again shifting the Landau pole beyond the Planck scale. Finally $\alpha_s$ itself is kept under control in the UV by asymptotic freedom.

In summary, it could just be that *the mass patterns and the couplings in the Standard Model precisely conspire to make the theory survive to the Planck scale.* In this case there would be no 'new physics' beyond the electroweak scale and the theory would have to be embedded directly *as is* into a Planck scale theory of quantum gravity. In my opinion, this may actually be our best chance to gain direct access to the Planck scale, both theoretically and experimentally!

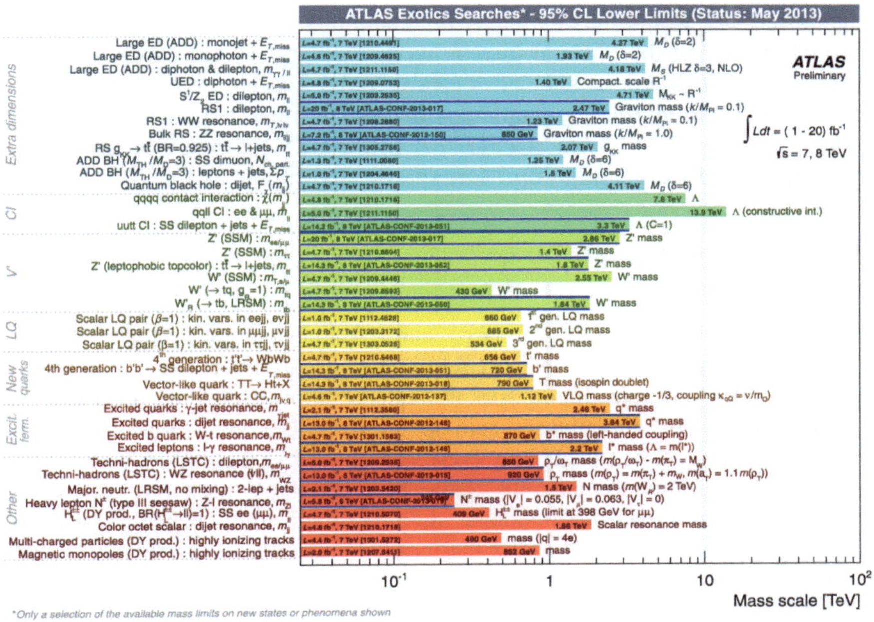


**Fig. 4** Low energy exotics? [22]

## 6 Anomalies

There is another remarkable property of the SM which may be interpreted as a hint of how Planck scale physics could affect low energy physics, and this is the complete cancellation of gauge anomalies (see [26] for an introduction and many references to the original work). Anomalies occur generically when a classical Lagrangian is invariant under a symmetry, but that symmetry cannot be preserved by the regularization that quantization requires. When the regulator is removed there is a finite remnant, and this is referred to as the anomaly, an $\mathcal{O}(\hbar)$ violation of a classical conservation laws. The classic example of such a symmetry is chiral invariance that explains the (near-)masslessness of fermions, but cannot be regulated, leading to the famous axial anomaly in QED that accounts for the decay of the $\pi^0$ meson.

When anomalous currents are coupled to gauge fields, the anomaly can deal a fatal blow to the theory. Recall that the coupling of a gauge field $A_\mu$ to charged matter generally takes the Noether form $\propto A_\mu J^\mu$, where $J^\mu$ is the classically conserved matter current. In the presence of an anomaly the variation of this term would give

$$\int \delta A_\mu J^\mu = \int \partial_\mu \omega J^\mu = - \int \omega \partial_\mu J^\mu \propto \mathcal{O}(\hbar) \neq 0 . \tag{3}$$

Gauge invariance would thus no longer hold, and this violation would destroy the renormalizability of the SM and thereby its predictivity. To verify that all gauge anomalies and gravitational anomalies cancel in the Standard Model requires the computation of various triangle diagrams with chiral fermions circulating in the loop, and involves traces of the form $\mathrm{Tr}\, T^a\{T^b, T^c\}$, where $T^a$ belong to the Lie algebra of the Standard Model gauge group. More specifically, the calculation reduces to the evaluation of

$$\sum \pm \mathrm{Tr}\, YYY = \sum \pm \mathrm{Tr}\, ttY = \sum \pm \mathrm{Tr}\, Y = 0, \tag{4}$$

where $Y$ is the electroweak hypercharge, and $t$ denotes any generator of $SU(2)_w$ or $SU(3)_c$; the sum runs over all SM fermions, with '+' for positive and '−' for negative chirality fermions. If you work through the whole list of such diagrams you will find that they all 'miraculously' sum up to zero [26]. From (4) it is obvious that the cancellation would be trivial if the SM were a vector-like theory with no preferred handedness or chirality. Remarkably, Nature prefers to break parity invariance, and to do so subtly in a way that maintains the renormalizability, hence consistency. In fact, the anomaly cancellations fix the fermion content almost uniquely to what it is, separately for each generation. Therefore, despite its 'messy' appearance *the Standard Model is surprisingly unique, and also surprisingly economical for what it does!*

There are two crucial features that must be emphasized here. The first is that *a proper anomaly does not and must not depend on how the theory is regulated.* Secondly, anomalies are often regarded as a perturbative phenomenon, but this is not strictly true. The famous Adler-Bardeen theorem asserts that the anomaly is entirely due to the one-loop contribution, and that there are thus no further contributions beyond one loop. In other words, the one-loop result is *exact to all orders*, hence non-perturbative!

Anomalies should also be expected to play a role in quantum gravity, and in determining whether a specific proposal is ultimately consistent or not. For instance, the classical constraint algebra of General Relativity in the Hamiltonian formulation has the schematic form

$$\{D, D\} \sim D\,, \quad \{D, H\} \sim H\,, \quad \{H, H\} \sim D, \tag{5}$$

where $D$ and $H$, respectively denote the diffeomorphism constraints and the Hamiltonian constraint, and this algebra is expected to be modified by quantum corrections. This expectation is borne out by the simplest example, matter-coupled quantum gravity in two space-time dimensions. Here the most general form of the space-time diffeomorphism algebra including anomalies is known to take the form

$$[T_{\pm\pm}(x), T_{\pm\pm}(y)] = \delta'(x, y)\Big(T_{\pm\pm}(x) + T_{\pm\pm}(y)\Big) + \hbar c \delta'''(x, y)\,, \tag{6}$$

where $x, y \in \mathbb{R}$, $T_{\pm\pm} := H \pm D$ and $c$ is the central charge. As is well known, virtually all of string theory hinges on the non-zero value of the central charge $c$! Unfortunately in higher dimensions, there exists neither an analogous uniqueness result, nor even a classification of what the anomalies may be. The main difficulty here is that higher-dimensional diffeomorphism algebras are 'soft', which means that Lie algebra structure 'constants' are not really constant, but field dependent.

# 7 Outlook

So where do we stand? At this time there is a growing array of proposals for quantum gravity, based on a variety of different and even mutually contradictory assumptions and hypotheses. The following is a selection of current approaches (to which you may add your own favorite):

- Supergravity, Superstrings and M-Theory
- AdS/CFT and Holography
- Path integrals: Euclidean, Lorentzian, matrix models, ...
- Canonical Quantization (metric formalism)
- Loop Quantum Gravity (with either connections or holonomies)
- Discrete Quantum Gravity: Regge calculus, (causal) dynamical triangulations
- Discrete Quantum Gravity: spin foams, group field theory
- Non-commutative geometry and space-time
- Asymptotic Safety and RG Fixed Points
- Emergent (quantum) gravity from thermodynamics
- Causal Sets
- Cellular Automata ('computing quantum space-time')

Among these string theory remains the leading contender, not least because it naturally incorporates (and even requires) matter degrees of freedom. Nevertheless, we still do not have a single hint from experiment and observation (for instance, in the form of supersymmetric partners to the known elementary particles) that it is indeed the right theory. Perhaps it is thus not so surprising that 'non-string approaches' have been gaining in popularity over the past few years.

Having grown out of particle physics and being modeled on its basic concepts, string theory has no problem of principle in connecting to low energy physics; being a perturbative approach, it also has no difficulties in reproducing the correct semi-classical limit and the Einstein field equations. But after more than two decades of effort, string theory is still struggling to reproduce the Standard Model *as is*, that is, without the heavy extra baggage that comes with (for instance) the supersymmetric extensions of the SM referred to as 'MSSM', 'CMSSM' or 'NMSSM', and so on. Moreover, it has considerable difficulties in incorporating a *positive* cosmological constant—in fact, like supergravity, superstring theory has an overwhelming preference for negative $\Lambda$! String theory, as originally formulated, is a background dependent and perturbative theory. However, there have been important advances and recent developments, especially in connection with the AdS/CFT correspondence and gauge/gravity or weak/strong dualities, that transcend perturbation theory and have provided important insights into the non-perturbative functioning of the theory (see e.g. [27] for a recent update). Nevertheless, in its present form string theory does not offer a convincing scenario for the resolution of (cosmological) space-time singularities, and so far cannot tell us what really 'happens' to space-time at the Planck scale.

I have already mentioned the impressive recent advances in perturbative QFT techniques [12], yielding evidence that $N = 8$ supergravity may be finite to all orders, contrary to expectations held for more than 30 years. If this theory could be shown to be a purely quantum field theoretic extension of Einstein's theory *without* UV singularities, this would partially undermine one of string theory's chief arguments why QFT must be abandoned. Of course, this would not relieve us of the task of working towards a *non-perturbative* understanding of physics at the very shortest distances, as the putative finiteness by itself would not tell us why and how the space-time continuum is dissolved at the Planck scale. And even if the theory turned out to be UV finite, many would doubt whether $N = 8$ supergravity has anything to do with 'real world physics'. Yet, there is a curious coincidence here: when supersymmetry is completely broken, eight spin-$\frac{1}{2}$ fermions are converted into Goldstinos in order render the eight gravitinos massive, leaving us with 48 spin-$\frac{1}{2}$ fermions, exactly the right number! Most likely a mirage, but who knows?[9]

In contrast to string theory the non-perturbative approaches put the main emphasis on GR concepts from the very beginning, to wit, (spatial) background independence and diffeomorphism invariance. Following this avenue has led to intriguing new ideas and proposals as to what a quantum space-time might actually 'look like'. Nevertheless, it is hard to see how such ideas could ever be put to a real test (other than internal consistency checks). A main criticism from the point of view taken here is that these approaches have not incorporated essential insights from particle physics up to now, such as the restrictions from anomaly cancellations. Furthermore, the ambiguities related to quantization and the incorporation of matter couplings have not been resolved in a satisfactory fashion in my opinion, and the recovery of the proper semi-classical limit remains an outstanding challenge.

To conclude let me restate my main worry. In one form or another the existing approaches to quantum gravity suffer from a very large number of ambiguities, so far preventing any kind of prediction with which the theory will stand or fall. Even at the risk of sounding polemical, I would put this ambiguity at $10^{500}$ (or even more)—in any case a number too large to cut down for any conceivable kind of experimental or observational advance.

- **Superstring theory** predicts the existence of myriads of 'consistent' vacua, all of which are supposed to be realized somewhere in the multiverse (or 'megaverse')—leading to the conclusion that essentially *anything goes* when it comes to answering

[9] On this point, see also [28].

the questions raised at the beginning of Sect. 5 (most notably, it is claimed that the multiverse also 'solves' the cosmological constant problem).

- **Loop quantum gravity** and related approaches are compatible with many 'consistent' Hamiltonians (or spin foam models), and with an essentially arbitrary menu of matter fields. Even disregarding technical issues such as quantization ambiguities, it looks again like almost *anything goes*. Idem for models of lattice and discrete quantum gravity.
- **Asymptotic Safety** is an assumption that, according to its proponents, works almost *generically*—that is, independently of the specific 'initial' conditions for the RG flows, of the matter content and even the number of space-time dimensions (if that number is not extremely large), leaving us with numerous 'consistent' RG flows.

In my view the real question is this: if there are all these 'consistent' (according to your definition) ansätze, does Nature simply pick the 'right' answer at random from a huge variety of possibilities, or are there criteria to narrow down the number of choices? Being exposed to many talks from the different 'quantum gravity camps' I am invariably struck by the success stories I keep hearing, and the implicit or explicit claims that 'we are almost there'. I, for one, would much prefer to hear once in a while that something does *not* work, and to see some indications of *inconsistencies* that might enable us to discriminate between a rapidly growing number of diverging ideas on quantum gravity [29, 30]. If, however, the plethora of theory ambiguities were to stay with us I would conclude that our search for an ultimate explanation, and with it the search for quantum gravity, may come to an ignominious end (like in Breughel's painting). I cannot imagine that this is what Einstein had in mind during his stay in Prague, nor in the later years of his life when he was striving to figure out "the old one's tricks" (or, in the original German, "dem Alten auf die Schliche kommen").

So let me repeat my main message: the incompleteness of the Standard Model is one of the strongest arguments in favor of quantizing gravity and searching for new concepts replacing classical notions of space and time. The observed features of SM may contain important hints of its possible UV completion and Planck scale physics, and these hints should be given due consideration in the search for a consistent theory of quantum gravity.

**Acknowledgments** I would like to thank Jianwei Mei for his help in turning my talk into a (hopefully) readable text and Krzysztof Meissner for many enjoyable and illuminating discussions on the state of the art.

## References

1. Weinberg, S.: Living with Infinities, ArXiv e-prints 0903.0568 [hep-th] (2009)
2. Hawking, S.W., Israel, W. (eds.): General Relativity: An Einstein Centenary Survey. Cambridge University Press, Cambridge (1979)

3. Kiefer, C.: Quantum gravity, International Series of Monographs on Physics. Clarendon Press, Oxford (2004)
4. Ashtekar, A. (ed.): 100 Years of Relativity. Space-Time Structure: Einstein and Beyond. World Scientific, Singapore (2005)
5. Rovelli, C.: Quantum Gravity, Cambridge University Press, Cambridge (2004)
6. Thiemann, T.: Modern Canonical Quantum General Relativity. Cambridge Monographs on Mathematical Physics. Cambridge University Press, Cambridge (2007)
7. Green, M.B., Schwarz, J.H., Witten, E.: Superstring Theory, vol. 1 & 2. Cambridge Monographs on Mathematical Physics. Cambridge University Press, Cambridge (1987)
8. Polchinski, J.: String Theory, vol. 1 & 2. Cambridge Monographs on Mathematical Physics. Cambridge University Press, Cambridge (1998)
9. Blumenhagen, R., Lüst, D., Theisen, S.: Basic Concepts of String Theory. Theoretical and Mathematical Physics. Springer, Berlin (2012)
10. Goroff, M.H., Sagnotti, A.: The ultraviolet behavior of Einstein gravity. Nucl. Phys. B **266**, 709 (1986). doi:10.1016/0550-3213(86)90193-8
11. van de Ven, A.: Two-loop quantum gravity. Nucl. Phys. B **378**, 309 (1992). doi:10.1016/0550-3213(92)90011-Y
12. Bern, Z., Carrasco, J.J., Dixon, L.J., Johansson, H., Roiban, R.: Amplitudes and ultraviolet behavior of $N = 8$ supergravity. Fortschr. Phys. **59**, 561 (2011). doi:10.1002/prop.201100037
13. Bojowald, M.: Absence of a singularity in loop quantum cosmology. Phys. Rev. Lett. **86**, 5227 (2001). doi:10.1103/PhysRevLett.86.5227
14. Reuter, M., Saueressig, F.: Quantum Einstein gravity. New J. Phys. **14**, 055022 (2012). doi:10.1088/1367-2630/14/5/055022
15. Applequist, T., Chodos, A., Freund, P.G.O. (eds.): Modern Kaluza-Klein Theories, Frontiers in Physics. Addison-Wesley, Reading (1987)
16. Wess, J., Bagger, J.: Supersymmetry and Supergravity. Princeton University Press, Princeton (1983)
17. Gross, D.J., Harvey, J.A., Martinec, E.J., Rohm, R.: Heterotic string. Phys. Rev. Lett. **54**, 502 (1985). doi:10.1103/PhysRevLett.54.502
18. Candelas, P., Horowitz, G.T., Strominger, A., Witten, E.: Vacuum configurations for superstrings. Nucl. Phys. B **B258**, 46 (1985). doi:10.1016/0550-3213(85)90602-9
19. Hambye, T., Riesselmann, K.: Matching conditions and Higgs boson mass upper bounds reexamined. Phys. Rev. D **55**, 7255 (1997). doi:10.1103/PhysRevD.55.7255 See also Proceedings of the 'ECFA/DESY Study on Physics and Detectors for the Linear Collider', DESY 97-123E, ed. R. Settles
20. Shaposhnikov, M., Wetterich, C.: Asymptotic safety of gravity and the Higgs boson mass. Phys. Lett. B **683**, 196 (2010). doi:10.1016/j.physletb.2009.12.022
21. Chamseddine, A.H., Connes, A.: Resilience of the spectral standard model. J. High Energy Phys. **9**, 104 (2012). doi:10.1007/JHEP09(2012)104
22. ATLAS Physics Summary Plots. (ATLAS Experiment 2013 CERN). URL https://twiki.cern.ch/twiki/bin/view/AtlasPublic/CombinedSummaryPlots
23. Coleman, S., Weinberg, E.: Radiative corrections as the origin of spontaneous symmetry breaking. Phys. Rev. D **7**, 1888 (1973). doi:10.1103/PhysRevD.7.1888
24. Bardeen, W.A.: On naturalness in the standard model. In: FERMILAB-CONF-95-391-T, C95–08-27.3 (1995)
25. Meissner, K.A., Nicolai, H.: Conformal symmetry and the standard model. Phys. Lett. B **648**, 312 (2007). doi:10.1016/j.physletb.2007.03.023
26. Bertlmann, R.A.: Anomalies in Quantum Field Theory, International Series of Monographs on Physics. Clarendon Press, Oxford (1996)
27. Blau, M., Theisen, S.: String theory as a theory of quantum gravity: a status report. Gen. Relativ. Gravit. **41**, 743 (2009). doi:10.1007/s10714-008-0752-z
28. Nicolai, H., Warner, N.P.: The SU(3)×U(1) invariant breaking of gauged $N = 8$ supergravity. Nucl. Phys. B **259**, 412 (1985). doi:10.1016/0550-3213(85)90643-1

29. Nicolai, H., Peeters, K., Zamaklar, M.: Loop quantum gravity: an outside view. Class. Quantum Gravity **22**, R193 (2005). doi:10.1088/0264-9381/22/19/R01
30. Alexandrov, S., Roche, P.: Critical overview of loops and foams. Phys. Rep. **506**, 41 (2011). doi:10.1016/j.physrep.2011.05.002

# Part IV
# Numerical Relativity and Relativistic Astrophysics

# Three Little Pieces for Computer and Relativity

**Luciano Rezzolla**

**Abstract** Numerical relativity has made big strides over the last decade. A number of problems that have plagued the field for years have now been mostly solved. This progress has transformed numerical relativity into a powerful tool to explore fundamental problems in physics and astrophysics, and I present here three representative examples. These "three little pieces" reflect a personal choice and describe work that I am particularly familiar with. However, many more examples could be made.

## 1 Introduction

Numerical relativity has hardly seen better times before. Over the last few years, in fact, a truly remarkable development has shaken the field. Starting from the first simulations showing that black-hole binaries could be evolved for a few orbits [1–3], or that black-hole formation could be followed stably using simple gauges and without excision [4], new results, some awaited for decades, have been obtained steadily. As a direct consequence of this "Renaissance", it is now possible to simulate binary black holes [5] and binary neutron stars [6] accurately for dozens of orbits, from the weak-field inspiral, down to the final black-hole ringdown (see also [7, 8] for recent reviews).

There are several reasons behind this rapid progress. These include the use of more advanced and accurate numerical techniques [9, 10], the availability of larger computational facilities, but also the development of formulations of the Einstein equations and gauges that are particularly well-suited for numerical evolutions [11–21]. The phase transition that has taken during this year has radically changed numerical rel-

L. Rezzolla (✉)
Max-Planck-Institut für Gravitationsphysik, Albert-Einstein-Institut, Potsdam, Germany
e-mail: rezzolla@th.physik.uni-frankfurt.de

L. Rezzolla
Institut für Theoretische Physik, Frankfurt, Germany

J. Bičák and T. Ledvinka (eds.), *General Relativity, Cosmology and Astrophysics*, Fundamental Theories of Physics 177, DOI: 10.1007/978-3-319-06349-2_19,

© Springer International Publishing Switzerland 2014

ativity, freeing it from the corner of idealised investigations. Most importantly, it has transformed numerical relativity into a research area where long-standing problems can found a quantitative and accurate solution, and into a tool by means of which it is possible to *explore* fundamental aspects of physics and astrophysics.

Numerous examples could be given to testify this transformation, although I will report here only those that I am particular familiar with. More specifically, in what follows I will discuss: (i) how numerical simulations of magnetised neutron stars provide convincing evidence that this process leads to the conditions that are expected behind the phenomenology of short gamma ray burst; (ii) how numerical simulations of the head-on collision of selfgravitating fluids boosted at relativistic speeds can be used to understand the conditions leading to the formation of a black hole and provide a dynamical version of the hoop conjecture; (iii) how the study of the local properties of apparent horizons can be used to explain bizarre behaviours in binary black-hole simulations and can be effectively correlated with a portion of the spacetime infinitely far away: $\mathscr{I}^+$. This selection is by no means comprehensive, but rather a very personal one, and I apologise in advance for not discussing all the excellent work that cannot find space in this contribution.

## 2 First Piece: From Neutrons Star to Gamma-Ray Bursts

The numerical investigation of the inspiral and merger of binary neutron stars in full general relativity has seen enormous progress made in recent years. Crucial improvements in the formulation of the equations and numerical methods, along with increased computational resources, have extended the scope of early simulations. These developments have made it possible to compute the full evolution, from large binary-separations up to black-hole formation, without and with magnetic fields [6, 22–26], and with idealised or realistic equations-of-state [27, 28]. This tremendous advancement is also providing information about the entire gravitational waveform, from the early inspiral up to the ringing of the black hole (see, e.g., [29–31]). Advanced interferometric detectors starting from 2014 are expected to observe these sources at a rate of $\sim$40–400 events per year [32].

These simulations also probe whether the end-product of mergers can serve as the "central engine" of short gamma-ray bursts (SGRBs) [33–35]. The prevalent scenario invoked to explain SGRBs involves the coalescence of a binary system of compact objects, e.g., a black hole and a neutron star or two neutron stars [36–39]. After the coalescence, the merged object is expected to collapse to a black hole surrounded by an accretion torus. An essential ingredient in this scenario is the formation of a central engine, which is required to launch a relativistic outflow with an energy of $\sim 10^{48}$–$10^{50}$ erg on a timescale of $\sim$0.1–1 s [38, 39].

The qualitative scenario described above is generally supported by the association of SGRBs with old stellar populations, distinct from the young massive star associations for long GRBs [40, 41]. It is also supported to a good extent by fully general-relativistic simulations, which show that the formation of a torus of mass

$M_{\rm tor} \lesssim 0.4\,M_\odot$ around a black hole with spin $J/M^2 \simeq 0.7-0.8$, is inevitable [27]. In addition, recent simulations have also provided the first evidence that the merger of a binary of modestly magnetised neutron stars naturally forms many of the conditions needed to produce a jet of ultrastrong magnetic field, with properties that are broadly consistent with SGRB observations. This *missing link* between the astrophysical phenomenology of GRBs and the theoretical expectations is a genuine example of the new potential of numerical relativity and I will discuss it in detail below.[1]

## 2.1 The Numerical Setup

It is not useful to discuss here in detail the numerical setup and the technical details of the numerical codes used in these calculations. These details can be found in Refs. [26, 43], while a description of the physical initial data was presented in Ref. [42]. It is sufficient to recall here that the evolution of the spacetime is obtained using a three-dimensional finite-differencing code providing the solution of a conformal traceless formulation of the Einstein equations [44] (i.e., the `CCATIE` code). The equations of general-relativistic magnetohydrodynamics (GRMHD) in the ideal-MHD limit are instead solved using a code code [43, 45, 46] which adopts a flux-conservative formulation of the equations as presented in [47] and high-resolution shock-capturing schemes (i.e., the `Whisky` code). In order to guarantee the divergence-free character of the MHD equations the flux-CD approach described in [48] was employed, although with the difference that the vector potential is used as evolution variable rather than the magnetic field. Both the Einstein and the GRMHD equations are solved using the vertex-centred adaptive mesh-refinement (AMR) approach provided by the `Carpet` driver [49]. In essence, the highest-resolution refinement level is centred around the peak in the rest-mass density of each star and in moving the "boxes" following the position of this maximum as the stars orbit. The boxes are evolved as a single refinement level when they overlap. The calculations were carried out using six levels of mesh refinement with the finest level having a resolution of $\Delta = 0.1500\,M_\odot \simeq 221\,\mathrm{m}$.

From a physical point of view, the initial data is given by a configuration that could represent the properties of a neutron star-binary a few orbits before their coalescence. More specifically, we simulate two equal-mass neutron stars, each with a gravitational mass of $1.5\,M_\odot$ (i.e., sufficiently large to produce a black hole soon after the merger), an equatorial radius of 13.6 km, and on a circular orbit with initial separation of $\simeq$45 km between the centres (all lengthscales are coordinate scales). Confined in each star is a poloidal magnetic field with a maximum strength of $10^{12}$ G. At this separation, the binary loses energy and angular momentum via emission of gravitational waves, thus rapidly proceeding on tighter orbits as it evolves.

---

[1] Much of what follows is taken from the discussion presented in Ref. [42].

## 2.2 The Basic Dynamics

After about 8 ms (~3 orbits) the two neutron stars merge forming a hypermassive neutron star (HMNS), namely, a rapidly and differentially-rotating neutron star, whose mass, $3.0\,M_\odot$, is above the maximum mass, $2.1\,M_\odot$, allowed with uniform rotation by an ideal-fluid equation of state (EOS),[2] $p = (\Gamma - 1)\rho\varepsilon$, where $\rho$ is the baryonic density, $\epsilon$ the specific internal energy, and $\Gamma = 2$ with an adiabatic index of 2. Being metastable, a HMNS can exist as long as it is able to resist against collapse via a suitable redistribution of angular momentum (e.g., deforming into a "bar" shape [6, 22]), or through the increased pressure-support coming from the large temperature-increase produced by the merger. However, because the HMNS is also losing angular momentum through gravitational waves, its lifetime is limited to a few ms, after which it collapses to a black hole with mass $M = 2.91\,M_\odot$ and spin $J/M^2 = 0.81$, surrounded by a hot and dense torus with mass $M_{\rm tor} = 0.063\,M_\odot$ [26].

These stages of the evolution can be seen in Fig. 1, which shows snapshots of the density colour-coded between $10^9$ and $10^{10}\,{\rm gr/cm^3}$, and of the magnetic field lines (green on the equatorial plane and white outside the torus). Soon after the black hole formation the torus reaches a quasi-stationary regime, during which the density has maximum values of $\sim 10^{11}\,{\rm g/cm^3}$, while the accretion rate settles to $\dot{M} \sim 0.2\,M_\odot/{\rm s}$. Using the measured values of the torus mass and of the accretion rate, and assuming the latter will not change significantly, such a regime could last for $t_{\rm accr} \simeq M_{\rm tor}/\dot{M} \simeq 0.3$ s, after which the torus is fully accreted; furthermore, if the two neutron stars have unequal masses, tidal tails are produced which provide additional late-time accretion [27]. This accretion timescale is close to the typical observed SGRB durations [38, 50]. It is also long enough for the neutrinos produced in the torus to escape and annihilate in its neighbourhood; estimates of the associated energy deposition rate range from $\sim 10^{48}$ erg/s [51] to $\sim 10^{50}$ erg/s [52], thus leading to a total energy deposition between a few $10^{47}$ erg and a few $10^{49}$ erg over a fraction of a second. This energy would be sufficient to launch a relativistic fireball, but because radiative losses are not yet accounted for the large reservoir of thermal energy in the torus cannot be extracted in these simulations.

The gravitational wave signal of the whole process is shown in the top part of the left panel in Fig. 2, while the bottom part exhibits the evolution of the MHD luminosity, $L_{\rm MHD}$, as computed from the integrated Poynting flux (solid line) and of the corresponding energy, $E_{\rm MHD}$, (dashed line). Clearly, the MHD emission starts only at the time of merger and increases exponentially after black-hole formation, when the gravitational wave signal essentially shuts off. Assuming that the quasi-stationary MHD luminosity is $\sim 4 \times 10^{48}$ erg/s, the total MHD energy released during the lifetime of the torus is $\sim 1.2 \times 10^{48}$ erg, which, if spread over an opening half-angle of $\sim 30°$ (see discussion below), suggests a lower limit to the isotropic equivalent energy in the outflow of $\sim 9 \times 10^{48}$ erg. While this is at the low end of the observed distribution of gamma-ray energies for SGRBs, larger MHD luminosities

---

[2] The use of a simplified EOS does not influence particularly the results besides determining the precise time when the HMNS collapses to a black hole.

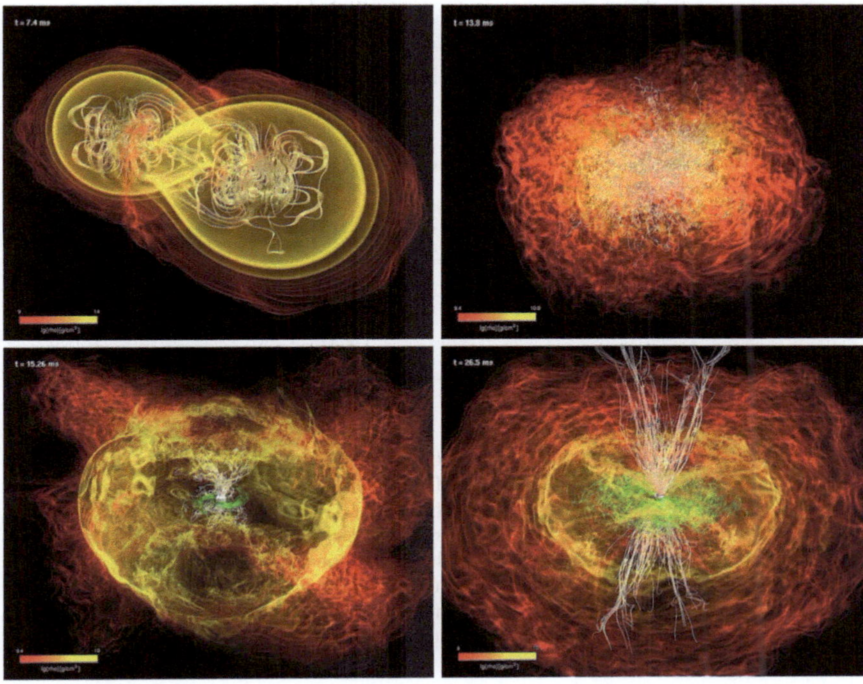

**Fig. 1** Snapshots at representative times of the evolution of the binary and of the formation of a large-scale ordered magnetic field. Shown with a colour-code map is the density, over which the magnetic-field lines are superposed. The panels in the *upper row* refer to the binary during the merger ($t = 7.4$ ms) and *before* the collapse to black hole ($t = 13.8$ ms), while those in the *lower row* to the evolution *after* the formation of the black hole ($t = 15.26$ ms, $t = 26.5$ ms). *Green lines* sample the magnetic field in the torus and on the equatorial plane, while *white lines* show the magnetic field outside the torus and near the black hole spin axis. The inner/outer part of the torus has a size of $\sim$90/170 km, while the horizon has a diameter of $\simeq$9 km

are expected either through the additional growth of the magnetic field via the winding of the field lines in the differentially-rotating disk (the simulation covers only one tenth of $t_{\rm accr}$), or when magnetic reconnection (which cannot take place within an ideal-MHD approach), is also accounted for (which may also increase the gamma-ray efficiency, e.g., [53]).

The last two panels of Fig. 1 offer views of the accreting torus after the black-hole formation. Although the *matter* dynamics is quasi-stationary, the last two panels clearly show that the *magnetic-field* is not and instead evolves significantly. It is only when the system is followed well after the formation of a black hole, that MHD instabilities develop and generate the central, low-density, poloidal-field funnel. This regime, which was not accessible to previous simulations [23, 24, 54], is essential for the jet formation [55, 56]. Because the strongly magnetised matter in the torus is highly conductive, it shears the magnetic-field lines via differential rotation.

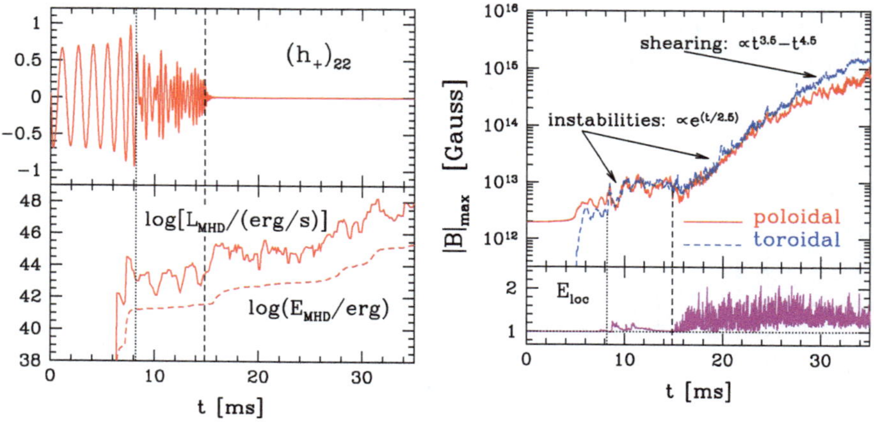


**Fig. 2** *Left panel* gravitational wave signal shown through the $\ell = 2, m = 2$ mode of the +polarization, $(h_+)_{22}$, (*top part*) and of the MHD luminosity, $L_{\rm MHD}$, (*bottom part*) as computed from the integrated Poynting flux and shown with a *solid line*. The corresponding energy, $E_{\rm MHD}$, is shown with a *dashed line*. The *dotted* and *dashed vertical lines* show the times of merger (as deduced from the first peak in the evolution of the gravitational wave amplitude) and black-hole formation, respectively. *Right panel* Evolution of the maximum of the magnetic field in its poloidal (*red solid line*) and toroidal (*blue dashed line*) components. The *bottom panel* shows the maximum local fluid energy indicating that an unbound outflow (i.e., $E_{\rm loc} > 1$) develops and is sustained after black-hole formation

A measurement of the angular velocity in the torus indicates that it is essentially Keplerian and thus unstable to the magneto-rotational instability [57], which develops $\simeq$5 ms after black-hole formation and amplifies exponentially both the poloidal and the toroidal magnetic fields; the e-folding time of the instability is $\simeq$2.5 ms and in good agreement with the one expected in the outer parts of the torus [57]. Because of this exponential growth, the final value of the magnetic field is largely insensitive to the initial strength and thus a robust feature of the dynamics (see also [58] for a similar behaviour recently computed in a HMNS).

A quantitative view of the magnetic-field growth is shown in the right panel of Fig. 2, which shows the evolution of the maximum values in the poloidal and toroidal components. Note that the latter is negligibly small before the merger, reaches equipartition with the poloidal field as a result of a Kelvin-Helmholtz instability triggered by the shearing of the stellar surfaces at merger [54, 59], and finally grows to $\simeq 10^{15}$ G by the end of the simulation. At later times ($t \gtrsim 22$ ms), when the instability is suppressed, the further growth of the field is due to the shearing of the field lines and it increases only as a power-law with exponent 3.5 (4.5) for the poloidal (toroidal) component. Although the magnetic-field growth essentially stalls after $t \simeq 35$ ms, further slower growths are possible [60], yielding correspondingly larger Poynting fluxes. Indeed, when the ratio between the magnetic flux across the horizon and the mass accretion rate becomes sufficiently large, a Blandford-Znajek mechanism [61] may be ignited [62]; such conditions are not met over the timescale

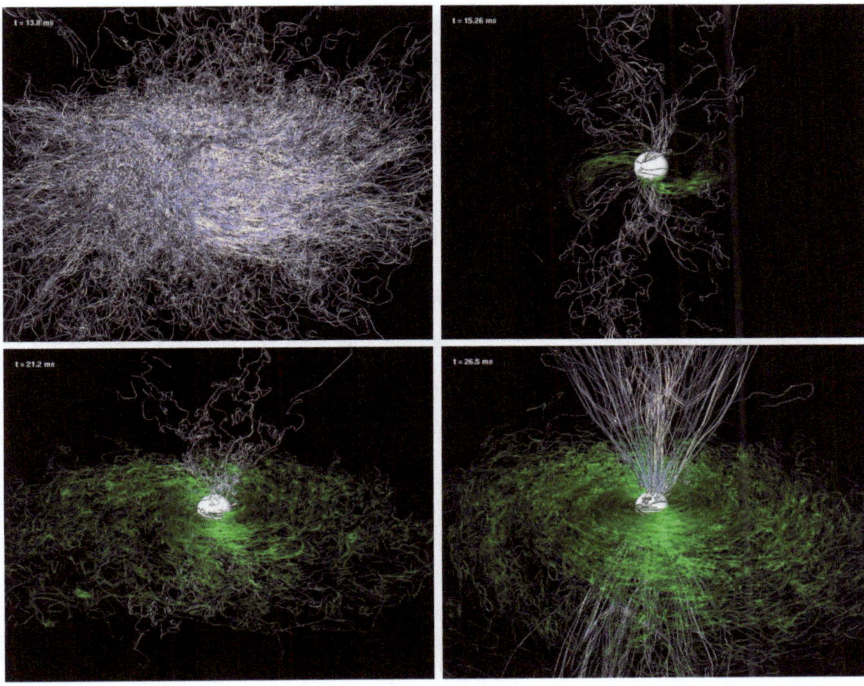


**Fig. 3** Magnetic-field structure in the HMNS (*first panel*) and after the collapse to black hole (*last three panels*). *Green* refers to magnetic-field lines inside the torus and on the equatorial plane, while *white* refers to magnetic-field lines outside the torus and near the axis. The highly turbulent, predominantly poloidal magnetic-field structure in the HMNS ($t = 13.8$ ms) changes systematically as the black hole is produced ($t = 15.26$ ms), leading to the formation of a predominantly toroidal magnetic field in the torus ($t = 21.2$ ms). All panels have the same linear scale, with the horizon's diameter being of $\simeq$9 km

of the simulations, but could develop over longer timescales. Also shown in the right panel of Fig. 2 is the maximum local fluid energy, highlighting that an *unbound outflow* (i.e., $E_{\rm loc} > 1$) develops after black-hole formation along the outer walls of the torus and persists for the whole duration of the simulation.

Finally, Fig. 3 provides a summary of the magnetic-field dynamics. It shows the magnetic field in the HMNS formed after the merger and its structure and dynamics after the collapse to black hole. In particular, in the last three panels it shows the magnetic-field structure inside the torus and on the equatorial plane (green), and outside the torus and near the axis (white). It is apparent that the highly turbulent magnetic field in the HMNS ($t = 13.8$ ms) changes systematically as the black hole is produced ($t = 15.26$ ms), leading to the formation of a toroidal magnetic field in the torus ($t = 21.2$ ms).[3] As the MRI sets in, the magnetic field is not only amplified,

[3] Turbulence in relativistic flows is an extremely challenging problem that is also essentially unexplored. Also in this case, the first relativistic simulations have been performed only recently [63, 64].

but also organises itself into a dual structure, which is mostly toroidal in the accretion torus with $B_{\rm tor}{\sim}2 \times 10^{15}$ G, but predominantly poloidal and jet-like along the black-hole spin axis, with $B_{\rm pol} \sim 8 \times 10^{14}$ G ($t = 26.5$ ms).[4] Note that the generation of an ordered large-scale field is far from trivial and a nonlinear dynamo may explain why the MRI brings a magnetic field self-organization, as it has been also suggested in case of MRI-mediated growth of the magnetic field in the conditions met in the collapse of massive stellar cores [60, 66]. However, the jet-like structure produced in the simulation is not yet the highly collimated ultrarelativistic outflow expected in SGRBs (see also below).

The magnetic hollow jet-like structure has an opening half-angle of $\sim 30°$, which sets an upper limit for the opening half-angle of any potential outflow, either produced by neutrino energy deposition [55] or by electromagnetic processes [56]. In these simulations most of the outflow develops along the edges of the jet-like structure, via a turbulent layer of electromagnetic driven matter, which shields the central funnel from excessive baryonic pollution. It is reasonable to expect that such a layer is crucial to set the opening angle of any ultrarelativistic jet, to shape both the radial and transverse structure of the jet, as well as to determine its stability properties. The Lorentz factors of the outflow measured in these simulations are not very high ($\gamma \lesssim 4$), but can potentially be amplified by several orders of magnitude in the inner baryon-poor regions through special-relativistic effects [67], the variability of the flow [68], or when resistive-MHD effects are taken into account [69]. Such accelerations will be produced as a more realistic and general-relativistic treatment of the radiative losses will become computationally affordable.

### 2.3 Comparison with Observations

Below I briefly discuss how the results presented above broadly match the properties of the central engine as deduced from the observations.

*Duration*: The observed duration of the prompt gamma-ray emission GRBs is energy dependent and is usually determined through $T_x$, the time over which $x$ % of the total counts are observed, between the $(100-x)/2$ and $(100+x)/2$ percentiles. The most common intervals used are $T_{90}$ (or $T_{50}$), initially defined [50] between 20 keV and 2 MeV. The GRB duration distribution is bimodal [50], where the durations of SGRBs (approximately 25 % of GRBs) are well-fit by a fairly wide log-normal distribution centred around $T_{90} \approx 0.8$ s with a FWHM of 1.4 dex [38]. The typical redshifts of the SGRBs observed with *Swift* are in the range $z \sim 0.3 - 1$, suggesting a central value of the intrinsic duration distribution of $\approx \langle 1 + z \rangle^{-1} 0.8$ s $\sim 0.5$ s, and a comparably wide distribution around this value. This is in close agreement with our accretion time of $\sim 0.3$ s.

---

[4] A similar magnetic-field configuration has been recently reproduced also when simulating the merger of a magnetised neutron star onto a black hole [65].

*Energy*: The isotropic equivalent energy output in the prompt gamma-ray emission of SGRBs, $E_{\gamma,\mathrm{iso}}$, spans a wide range, from $(2.7 \pm 1) \times 10^{48}$ erg (in the observed energy range 15–350 keV) for GRB 050509B at a redshift of $z = 0.225$ [70], up to $(1.08 \pm 0.06) \times 10^{53}$ erg (in the observed energy range 10 keV–30 GeV) for GRB 090510 at $z = 0.903$ [71]. However, the most typical values are in the range $E_{\gamma,\mathrm{iso}} \sim 10^{49} - 10^{51}$ erg [38]. In this model, the highly relativistic outflow may be powered either by neutrino-anti neutrino annihilation, or by the Blandford-Znajek mechanism. For the former one might expect a total energy release between a few $10^{47}$ erg and $\sim 10^{49}$ erg [72, 73], into a bipolar relativistic jet of opening half-angle $\theta_{\mathrm{jet}} \sim 8 - 30°$, corresponding to a fraction $f_b \sim 0.01 - 0.13$ of the total solid angle, and isotropic equivalent energies, $E_{\nu\bar{\nu},\mathrm{iso}}$, between a few $10^{48}$ erg and $\sim 10^{51}$ erg. For the latter mechanism, instead, and if the magnetisation near the event horizon becomes sufficiently high, the jet power for these values for the black-hole mass and spin is [74]

$$L_{\mathrm{BZ}} \sim 3.0 \times 10^{50} \left(\frac{f_{\mathrm{rel}}}{0.1}\right) \left(\frac{B}{2 \times 10^{15}\,\mathrm{G}}\right)^2 \mathrm{erg/s}, \tag{1}$$

where $f_{\mathrm{rel}}$ is the fraction of the total Blandford-Znajek power that is channelled into the resulting relativistic jet (and $f_{\mathrm{rel}} \sim 0.1$ might be expected for ejecta with asymptotic Lorentz factors above 100). This relativistic outflow is launched over a timescale of $\sim 0.2$ s and corresponds to

$$E_{\mathrm{BZ,iso}} \sim 1.2 \times 10^{51} \left(\frac{f_{\mathrm{rel}}}{0.1}\right) \left(\frac{f_b}{0.05}\right)^{-1} \left(\frac{B}{2 \times 10^{15}\,\mathrm{G}}\right)^2 \mathrm{erg}\,. \tag{2}$$

Comparing the X-ray afterglow luminosity (after 10 or 11 h) and $E_{\gamma,\mathrm{iso}}$ suggests that the efficiency of the prompt gamma-ray emission in SGRBs is typically high [70], and similar to that of long GRBs [75], with $E_{\gamma,\mathrm{iso}} \sim (0.1 - 0.9) E_{\mathrm{iso}}$, radiating between $\sim 10$ and $\sim 90$ % of the initial energy of the ultrarelativistic outflow. Therefore, this model is able to accommodate the observed $E_{\gamma,\mathrm{iso}}$ values.

*Lorentz factor*: The Fermi Gamma-Ray Space Telescope has detected GeV emission from SGRBs [76], suggesting typical lower limits of $\gamma_{\mathrm{min}} \sim 10^2 - 10^3$. In particular, $\gamma_{\mathrm{min}} \approx 1200$ was obtained for GRB 090510 [71]. However, a more realistic model [77] results in $\gamma_{\mathrm{min}}$ values lower by a factor of $\sim 3$. Therefore, the central engine should be capable of producing outflow Lorentz factors of at least a few hundred. The fact that our simulation produces a strongly magnetised mildly relativistic outflow at angles near $\sim 30°$ from the black-hole spin axis would help shield the inner region near the spin axis from excessive baryon loading, and thus assist in achieving high asymptotic Lorentz factors at large distance from the source, after the outflow in this region is triggered by neutrinos and/or the Blandford-Znajek mechanism.

*Jet angular structure*: This is poorly constrained by observations (even more so than for long GRBs). The only compelling case for a jet break in the afterglow light-curve is for GRB 090510 [78], which occurred very early on (after $\sim$1,400 s), and would thus imply an extremely narrow jet ($\theta_{\mathrm{jet}} \sim 0.2 - 0.4°$) and modest true energy output in gamma-rays ($\sim 10^{48}$ erg). If this is indeed a jet break, it might correspond to a line of sight near a very narrow and bright core of a jet, which also has significantly wider wings. Observers with lines of sight along these wings would then see a much dimmer and more typical SGRB [79, 80]; without such wings, however, the observations would suggest a very large intrinsic and beaming-corrected event rate per unit volume. In most cases there are only lower limits on a possible jet break time [38], resulting in typical limits of $f_b \gtrsim 10^{-2}$ or $\theta_{\rm jet} \gtrsim 8°$. This is consistent with the expectation of $\theta_{\rm jet} \sim 8 - 30°$ for the ultrarelativistic ejecta capable of producing a SGRB (which would also imply a reasonable SGRBs intrinsic event rate per unit volume).

### *2.4 Summary*

The calculations reported above demonstrate that a binary merger of two neutron stars inevitably leads to the formation of a relativistic jet-like and ultrastrong magnetic field, which could serve as a central engine for SGRBs. Because the magnetic-field growth is exponential, the picture emerging from these simulations is rather general and applies equally even to mildly magnetised neutron stars. Overall, this first "little piece" of numerical relativity removes a significant uncertainty as to whether such binary mergers can indeed produce the central engines of SGRBs. While the electromagnetic energy release is already broadly compatible with the observations, the simulations discussed above lack a proper treatment of the energy losses via photons and neutrinos or resistive dissipation, which can provide a fundamental contribution to the energy-input necessary to launch the fireball and cool the torus [51, 52]. This additional energy input, whose self-consistent inclusion in general relativity remains extremely challenging, may help to launch an ultrarelativistic outflow very early after the black hole forms and complete the picture of the central engine of a SGRB.

## 3 Second Piece: A Dynamical Hoop Conjecture

The second "little piece" of numerical relativity that I will discuss aims at addressing the issue of necessary conditions for the formation of a black hole, which still represents one of the most intriguing and fascinating predictions of classical general relativity. There is abundant astronomical evidence that black holes exist, and a number of considerations supporting the idea that under suitable conditions gravitational collapse is inevitable [81]. In addition, there is overwhelming numerical evidence that black-hole formation does take place in a variety of environments [42]. Yet, a rigorous definition of the sufficient conditions for black-hole formation is still lacking. Hence, it is not possible to predict whether the collision of two compact objects, either stars or elementary particles, will lead to the formation of a black hole.

The *hoop conjecture* proposed by Thorne in the '70s, provides some reasonable and intuitive guidelines [82]. I recall that the conjecture states that a black hole is formed if an amount of "mass-energy" $E$ can be compressed to fit within a hoop with radius equal or smaller than the corresponding Schwarzschild radius, i.e., if $R_{\rm hoop} \leq R_{\rm s} = 2GE/c^4$, where $G$ is gravitational constant and $c$ the speed of light. Even though it can be made precise under particular circumstances [83], the hoop conjecture is not meant to be a precise mathematical statement and, in fact, it is difficult to predict if the above-mentioned collision will compress matter sufficiently to fit within the limiting hoop. Loosely speaking, what is difficult is to determine which part of the "kinetic energy" of the system can be accounted to fit within the hoop. Since at the collision the conversion of kinetic energy into internal energy is a highly nonlinear process, any quantitative prediction becomes rapidly inaccurate as the speeds involved approach that of light.

As stated above, the hoop conjecture is purely classical. A quantum-mechanical equivalent is not difficult to formulate, although not very stringent, as it simply implies that a black hole will be formed at Planck-energy scales. The predicting power does not improve significantly when considering the conditions of black-hole formation in higher-dimensional theories of gravity (see, e.g., [84–86]). In these frameworks, the energy required for black-hole formation might be significantly smaller [84], thus providing the possibility of producing them in the Large Hadron Collider (LHC) [87], but no firm conclusion has been reached yet.

Clearly, although numerical simulations represent a realistic route to shed some light on this issue (see, e.g., [88–90]), even the simplest scenario of the collision of two compact objects at ultrarelativistic speeds is far from being simple and it is actually very challenging. A first step was taken by Eardley and Giddings [91], who have studied the formation of a black hole from the head-on collision of two plane-fronted gravitational waves with nonzero impact parameter (previous work of D'Eath and Payne [92–94] using different methods had considered a zero impact parameter). In all of these analyses each incoming particle is modelled as a point particle accompanied by a plane-fronted gravitational shock wave corresponding to the Lorentz-contracted longitudinal gravitational field of the particle. At the instant of collision the two shock waves pass through one another and interact through a nonlinear focusing and shearing. As a result of their investigation, a lower bound was set on the cross-section for black-hole production, i.e., $\sigma > 32.5(GE/2c^4)^2$, where $E$ is the centre-of-mass (lab) energy. More recently, and in a framework which is closer to the one considered here, this problem has been investigated by Choptuik and Pretorius [95], who studied the collision of two classical spherical solitons, with a total energy of the system in the lab frame $E = 2\gamma_{\rm b} m_0 c^2$, where $m_0$ is the "rest-mass", $\gamma_{\rm b} \equiv 1/\sqrt{1 - v_{\rm b}^2/c^2}$ and $v_{\rm b}$ the boost velocity. They were then able to show that for collisions with sufficiently high boost, i.e., $\gamma_{\rm b} \gtrsim 2.9$, a black hole can be formed.

In what follows I discuss what has been recently reported on the first calculations of black-hole production from the collision of two compact, selfgravitating, fluid objects boosted at ultrarelativistic speeds[5] (A similar investigation by East and Pretorius [97] has also appeared at about the same time).

I start by pointing out that there are several important differences with the previous investigations in [91–95]. Differently from [91–94], in fact, I will consider colliding objects that are not in vacuum and are not treated as point particles. Rather, they are relativistic stars, which obviously extended and selfgravitating objects, thus with a behaviour that is intrinsically different. Also, differently from [95], these objects are not described as scalar fields, but as fluids and thus represent a more realistic description of baryonic matter, such as the one employed when simulating relativistic heavy-ion collisions [98]. These intrinsic differences also make the comparison with the works of [91–94] very hard if possible at all. On the other hand, many analogies exist with the collision of bosons stars considered in [95], and that, as I will discuss below, can be interpreted within the more general description of black-hole production from ultrarelativistic collisions.

Overall, the most important and distinguishing feature in the collision of two selfgravitating stars is that a black hole can be produced even from zero initial velocities if the initial masses are large enough; this behaviour is clearly absent in all previous results, where instead a critical initial boost is necessary [91–95]. In addition, for each value of the effective Lorentz factor, $\langle\gamma\rangle$, a critical initial mass exists, $M_{\rm c}$, above which a black hole is formed and below which matter, at least in part, selfgravitates. More importantly, both $M_{\rm c}$ follows a simple scaling with $\langle\gamma\rangle$, thus allowing to extrapolate the results to the masses and energies of modern particle accelerators and to deduce that black-hole production is unlikely at LHC scales.

## 3.1 The Numerical Setup

The numerical setup employed in the simulations is the same presented in [99], and it uses an axisymmetric code to solve in two spatial dimensions, $(x, z)$, the set of the Einstein and of the relativistic-hydrodynamic equations [100]. The axisymmetry of the spacetime is imposed exploiting the "cartoon" technique, while the hydrodynamics equations are written explicitly in cylindrical coordinates. All the simulations use an ideal-fluid EOS with $\Gamma = 2$. The initial configurations consist of spherical stars, constructed as in [99, 101] after specifying the central density, $\rho_c$, where the latter also serves as parameter to determine the critical model. The stars have an initial separation $D$ and are boosted along the $z$-direction via a Lorentz transformation with boost $v_{\rm b}/c$. To limit the initial violation in the constraints, $D$ is chosen to be sufficiently large, i.e., $D = 240\,M_\odot$, and an optimal superposition of the two isolated-star solutions that will be presented in a longer paper. The grid has uniform spacing $\Delta = 0.08(0.06)\,M_\odot$ with extents $x/M_\odot \in [0, 80]$ and $z/M_\odot \in [0, 150(200)]$, where the round brackets refer to the more demanding high-boost cases. Reflection boundary conditions are applied on the $z = 0$ plane, while outgoing conditions are used elsewhere.

5 Much of what follows is taken from the discussion presented in Ref. [96].

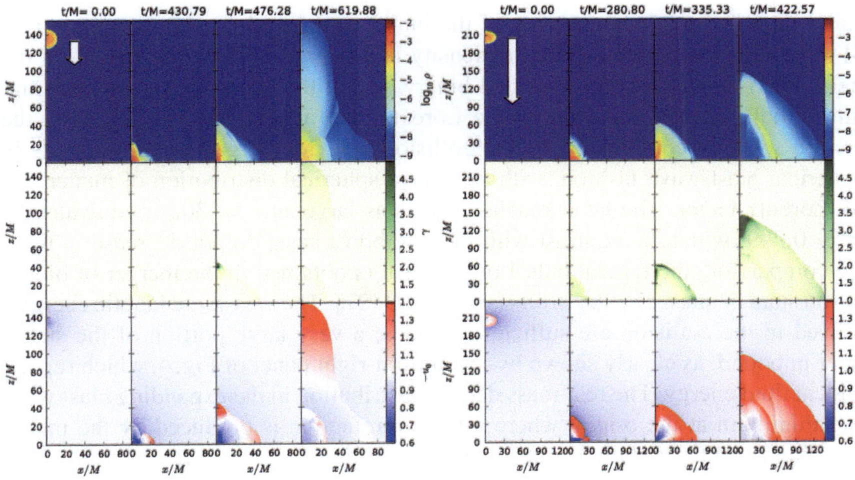

**Fig. 4** Representative snapshots of the rest-mass density, $\rho$ in units where $c = 1 = M_{\odot}$ (*top row*), of the Lorentz factor, $\gamma$ (*middle row*), and of the local fluid energy, $-u_0$ (*bottom row*), for subcritical models with an initial small boost $v_b/c = 0.3$ (*left panel*) or a large one $v_b/c = 0.8$ (*right panel*). Note that the post-collision flow is essentially jet-like for the low-boost case (*left panel*), while essentially spherical for the high-boost case (*right panel*); in this latter case, most of the matter is unbound

## 3.2 The Basic Dynamics

The dynamics of the process is rather simple. As the two stars approach each other, the initial boost velocity increases as a result of the gravitational attraction, leading to a strong shock as the surfaces of the stars collide. In the case of *supercritical* initial data, i.e., of stars with masses above a critical value, $M_c$, a black hole is promptly produced and most of the matter is accreted. Conversely, in the case of *subcritical* initial data, i.e., of stars with masses below $M_c$, the product of the collision is a hot and extended object with large-amplitude oscillations. Part of the stellar matter is unbound and leaves the numerical grid as the product of the collision reaches an equilibrium.

Figure 4 shows snapshots at representative times of the rest-mass density, $\rho$ (top row), of the Lorentz factor, $\gamma \equiv (1-v^i v_i/c^2)^{-1/2}$ (middle row), and of the local fluid energy, $-u_0$ (bottom row), for two subcritical models. The left panel, in particular, refers to a binary boosted at $v_b/c = 0.3$. Note that the stars are strongly compressed by the collision, with the rest-mass density increasing exponentially. The merged object expands in a jet-like fashion along the $z$-direction, with the bulk of the matter being accelerated up to $\gamma \sim 16$, or equivalently, $v/c \sim 0.998$, but then settling on much slower flows with $\gamma \lesssim 2.1$. Furthermore, the front of the jet has $-u_0 > 1$ indicating that part of the shocked matter has sufficient energy to have become gravitationally unbound. As a result, the rest-mass density at the center of the merged object is smaller than the maximum density of the initial configuration, although the origin still represents the region where the density is the largest. The right panel, on the other hand, refers to a highly-boosted binary, i.e., with $v_b/c = 0.8$, with each star being initially highly distorted by the Lorentz contraction. Also in this case, the stars are strongly compressed by the collision, but the merged object expands in a spherical blast-wave fashion, with an almost spherical distribution of matter and bulk Lorentz factor. The latter reaches values as large as $\gamma \sim 30$, or equivalently, $v/c \sim 0.999$, which, in contrast with the low-boost case, do not decrease in time. As a comparison, the typical bulk Lorentz factors obtained in the merger of binary neutron stars in quasi circular orbits is $\gamma \sim 1.03$ [27]. The very large kinetic energies involved in the collision are sufficient to make a very large portion of the stellar matter unbound, as clearly shown by the bottom-right panel of Fig. 4, which reports the local fluid energy. The rest-mass density distribution in the expanding blast wave has a minimum at the origin, where a large rarefaction is produced by the matter expanding as an ultrarelativistic thick shell.

The marked transition from a jet-like outflow, not too dissimilar from the simple Bjorken flow used to model the very early states of relativistic ion-collisions [102], to a shell-like structure, not too dissimilar from "transverse expansion" modelled in the subsequent stages of relativistic ion-collisions (see [103] and references therein), signals that it is not unreasonable to extrapolate some of the results presented here also to the collision of ultrarelativistic elementary particles.

The transition from the two qualitatively-different regimes discussed above is further confirmed by the evolution of the rest-mass normalized to the initial value $M_0$. The simulations in fact reveal that the unbound fraction is just a few percent of the total rest-mass in the case of a low-boost collision, with most of the matter being confined in the selfgravitating "star". This is to be contrasted with what happens for a high-boost collision, where the unbound fraction is $\sim$100 % of the total rest-mass. This behaviour provides a strong indication that, at least for subcritical collisions, the role played by gravitational forces is a minor one as the kinetic energy is increased. This is what happens in the collision of two particles at ultrarelativistic speeds, where all of the matter is obviously unbound.

### 3.3 *Critical Behaviour and Scaling*

A remarkable property of the head-on collision of compact stars is the existence of type-I critical behaviour, which was first pointed out in [104] and subsequently extended in [100]. In essence, in these works it was found that when considering stars with initial zero velocity at infinity, it is possible to fine-tune the initial central density $\rho_c$ (and hence the mass) near a critical value, $\rho_c^\star$, so that stars with $\rho_c > \rho_c^\star$ would collapse *eventually* to a black hole, while the models with $\rho_c < \rho_c^\star$ would *eventually* lead to a stable stellar configuration. As a result, the head-on collision of two neutron stars near the critical threshold can be seen as a transition in the space of configurations from an initial stable solution over to a critical metastable

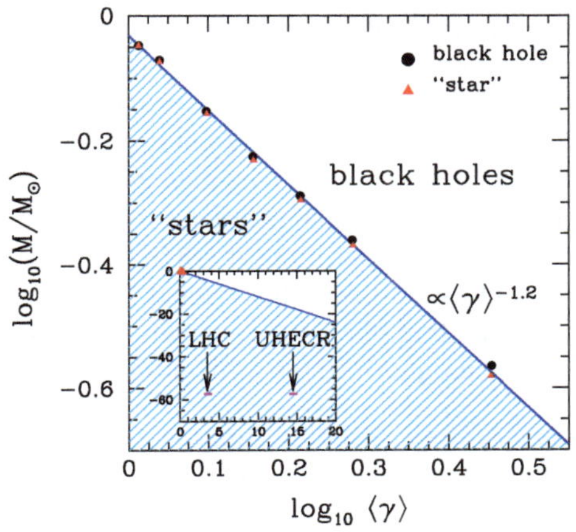


**Fig. 5** Critical line as a function of the average Lorentz factor, with *circles* indicating *black holes* and *triangles* selfgravitating objects. The inset shows the regimes explored at LHC and measured in UHECR

one which can either migrate to a stable solution or collapse to a black hole [101]. As the critical limit is approached, the survival time of the metastable object, $\tau_{\rm eq}$, increases as $\tau_{\rm eq} = -\lambda \ln|\rho_c - \rho_c^\star|$, with $\lambda \sim 10$ [100, 104].

Although the free-fall velocities considered in [100, 104] were very small, the critical behaviour continues to hold also when the stars are boosted to ultrarelativistic velocities. Interestingly, the threshold $\rho_{\rm c}^\star$ becomes now a function of the initial effective boost. Determining $\rho_{\rm c}^\star$ becomes especially challenging as the Lorentz factor is increased and the dynamics of the matter becomes extremely violent, with very strong shocks and rarefaction waves. However, it was possible to determine the threshold for all the range of initial boosts considered, i.e., $v_{\rm b}/c \in [0,\ 0.95]$, $\gamma_b \in [1,\ 3.2]$, and even to a reasonable accuracy, e.g., $\rho_c^\star = (3.288023 \pm 0.000003) \times 10^{14}\ {\rm g/cm}^3$, for the initial boost of $v_{\rm b}/c = 0.3$.

The existence of critical behaviour near which the details of the initial conditions become irrelevant and which is the *same* at different boosts, i.e., $\lambda$ does not depend on $\gamma$ nor on $\rho_c$ (Refs. [104, 105] have shown there is "universality" when varying $\gamma$ and fixing $\rho_c$), gives us a wonderful tool to explore the conditions of black-hole formation also far away from the masses and Lorentz factors considered in this paper. This is illustrated in Fig. 5, which reports the gravitational mass of the isolated spherical star as a function of the effective initial Lorentz factor

$$\langle\gamma\rangle \equiv \frac{\int dV\, T_{\mu\nu} n^\mu n^\nu}{\left(\int dV\, T_{\mu\nu} n^\mu n^\nu\right)_0}, \qquad (3)$$

where $T_{\mu\nu}$ is the stress-energy tensor, $n^\mu$ is the unit normal to the spatial hyperspace with proper volume element $dV$, and the index 0 refers to quantities measured in the initial unboosted frame. I should stress that the definition of the effective Lorentz factor (3) is necessary because the stars are extended and thus the Lorentz factor will be different in different parts of the star. Expression (3), on the other hand, can be taken as ratio of the energies measured in the boosted and unboosted frames, and hence a generalisation of the Lorentz factor for a point particle (Indeed $\langle\gamma\rangle \to 1$ for $v_b \to 1$). Of course, other parametrizations are possible, still leading to scaling laws, but with slightly different exponents. Filled circles indicate initial data leading to a black hole, while triangles indicate initial data leading to a "star", whereby I mean an object which is at least in part selfgravitating (orange errorbars provide an approximate upper limit of $\sim$8 % to the error in the measurements). Also indicated as a blue solid line is the critical line separating the two regions of black hole and star formation (the latter is shown as a shaded region). Clearly, the numerical results provide a tight fit of the critical line with a power law

$$\frac{M_c}{M_\odot} = K \frac{1}{\langle\gamma\rangle^n} \approx 0.92 \frac{1}{\langle\gamma\rangle^{1.03}} . \tag{4}$$

Expression (4) offers itself to a number of considerations. First, it essentially expresses the conservation of energy. Second, in the limit of zero initial velocities, $\langle\gamma\rangle \to 1$, one obtains that $M_c \simeq 0.92\, M_\odot$, so that the corresponding total mass, $2M_c$, is only $\sim$ 12 % larger than the maximum mass of the relative spherical-star sequence, i.e., $M_{\rm max} = 1.637\, M_\odot$. Third, in the opposite limit of $\langle\gamma\rangle \to \infty$, expression (4) predicts that the critical mass will go zero. This is indeed what one would expect: as the kinetic energy diverges, no room is left for selfgravitating matter, which will all be ejected but for an infinitesimal amount which will go into building the zero-mass critical black hole. Fourthly, (4) is also in agreement with the results in [95, 97], whereby one can recognize the black-hole formation as the crossing of the critical line when moving to larger Lorentz factors while keeping the rest-mass constant.

Finally, using (4) it is possible to probe whether the kinetic energies achieved by modern particle accelerators, such as the LHC, are sufficient to produce micro black holes from the collision of two ultrarelativistic particles. Using the results reported in Ref. [106], the expected energies achieved by LHC in the next couple of years will be 4 – 7 TeV, so that a proton, whose mass is $\sim$ 938 MeV $\sim 8.41 \times 10^{-58}\, M_\odot$, can be accelerated up to $\gamma \sim 7.5 \times 10^3$. I have therefore reported the range of masses and Lorentz factors accessible to LHC in the inset in Fig. 5, where it appears as a small magenta box. Note that the calculations reported here do not intend to be a realistic approximation of the dynamics of ultrarelativistic particle collisions. However, these calculations and the presence of a critical behaviour can be used to deduce that the ranges reachable at the LHC are well below the critical line and thus in the region where a partially-confined collided object is expected.

Of course, this line of arguments wildly extrapolates our results by almost 60 orders of magnitude in mass (11 in Lorentz factor) and neglects quantum effects and extra-dimension effects that might be important at Planck-energy scales. Bearing in mind these caveats, our calculations suggest that the production of micro black holes at LHC will be unlikely. An additional confirmation that our estimates are not unreasonable comes from considering the corresponding energy and Lorentz factors for the observed ultra-high energy cosmic rays (UHECR), that are observed with energies up to $\sim 10^{20}$ eV (i.e., $\gamma \sim 10^{11}$) and for which there is no evidence of black-hole formation when interacting with the atmosphere [107]. Also in this case, the relevant range of masses and Lorentz factors is shown in the inset and falls in the region where no black holes should be produced.

As a final remark I note that the scaling relation (4) can be expressed equivalently in terms of the original stellar compactness, $M/R$ as

$$\left(\frac{M}{R}\right)_{\rm c} = K' \frac{1}{\langle\gamma\rangle^{n'}} \approx 0.08 \frac{1}{\langle\gamma\rangle^{1.13}} . \tag{5}$$

Since $M_{\rm lab} \equiv \langle\gamma\rangle M$ is the mass in the lab frame, and since $R$ is the largest dimension in that frame being the transverse one to the motion, the ratio

$$\left(\frac{M_{\rm lab}}{R}\right)_{\rm c} = K' \frac{1}{\langle\gamma\rangle^{n'-1}} \sim K' \frac{1}{\langle\gamma\rangle^{0.13}} , \tag{6}$$

provides the condition for the amount of energy that, when confined in a hoop of radius $R$, would lead to a black hole. Hence, expression (6) extends the spirit of the hoop conjecture to the case in which a kinetic energy is present. Note that the limiting value $\langle\gamma\rangle = 1$ does not corresponds to a static configuration (as in the hoop conjecture) but to a binary that is at rest at infinity. This explains why in this limit $(M_{\rm lab}/R)_{\rm c} = (M/R)_{\rm c} \simeq 0.08$, which is considerably smaller than the value $1/2$ predicted by the hoop conjecture.

### 3.4 Summary

The calculations reported above demonstrate that it is possible to find a criterion for the conditions leading to black-hole formation in the collision of two selfgravitating fluids moving at ultrarelativistic velocities. The Lorentz factors reached in these simulations are considerably larger than those encountered in merging neutron-star binaries, especially if the inspiral is along quasi-circular orbits. The properties of the flow after the collision change with Lorentz factor, with most of the matter being ejected in a spherical blast wave for large boosts. Interestingly, the collided object exhibits a critical behaviour of type I, which is found to persist also as the initial boost is increased. This allows one to derive a simple scaling law and extrapolate these results to the energies of elementary particles at LHC and conclude that black-hole production is unlikely in that case.

## 4 Third Piece: Horizons as Probes of Black-Hole Dynamics

The third and last "little piece" of numerical relativity that I will discuss is instead about calculations in vacuum spacetimes and focuses on the merger of two black holes. This process, which represents one of the most important source of gravitational waves, is generally accompanied by the recoil of the final black hole as a result of anisotropic gravitational wave emission. While this scenario has been investigated for decades [108] and first estimates have been made using approximated and semi-analytical methods such as a particle approximation [109], post-Newtonian methods [110] and the close-limit approximation (CLA) [111], it is only thanks to the recent progress in numerical relativity that accurate values for the recoil velocity have been computed [112–120].

Besides being a genuine nonlinear effect of general relativity, the generation of a large recoil velocity during the merger of two black holes has a direct impact in astrophysics. Depending on its size and its variation with the mass ratio and spin, in fact, it can play an important role in the growth of supermassive black holes via mergers of galaxies and on the number of galaxies containing black holes. Numerical-relativity simulations of black holes inspiralling on quasi-circular orbits have already revealed many of the most important features of this process showing, for instance, that asymmetries in the mass can lead to recoil velocities $v_{\rm k} \lesssim 175\,{\rm km/s}$ [112, 113], while asymmetries in the spins can lead respectively to $v_{\rm k} \lesssim 450\,{\rm km/s}$ or $v_{\rm k} \lesssim 4000\,{\rm km/s}$ if the spins are aligned [115, 116, 118] or perpendicular to the orbital angular momentum [114, 121, 122] (see [123] for a review and [124] for the most recent results).

At the same time, however, there are a number of aspects of the nonlinear processes leading to the recoil that are far from being clarified even though interesting work has been recently carried out to investigate such aspects [125–127]. One of these features, and possibly the most puzzling one, is the generic presence of an "*anti-kick*", namely, of one (or more) decelerations experienced by the recoiling black hole. Such anti-kicks take place after a single apparent horizon has been found and have been reported in essentially all of the mergers simulated so far (see Fig. 8 of Ref. [118] for some examples).

What follows discusses a phenomenological framework which provides a novel description of the stages during which the anti-kick is generated, and that can be used to formulate a simple and qualitative interpretation of the physics underlying this process. I will focus on the head-on collision of two nonspinning black holes with different mass. Although this is the simplest scenario for a black-hole merger, it contains all the important aspects that can be encountered in more generic conditions.[6]

## 4.1 The Basic Picture

I will start by presenting a qualitative interpretation of the antikick by considering the simple head-on collision of two Schwarzschild black holes with unequal masses. This qualitative picture will be made quantitative and gauge-invariant by studying the logical equivalent of this process in the evolution of a Robinson-Trautman spacetime,

[6] Much of what follows is taken from the discussion presented in Refs. [128–130].

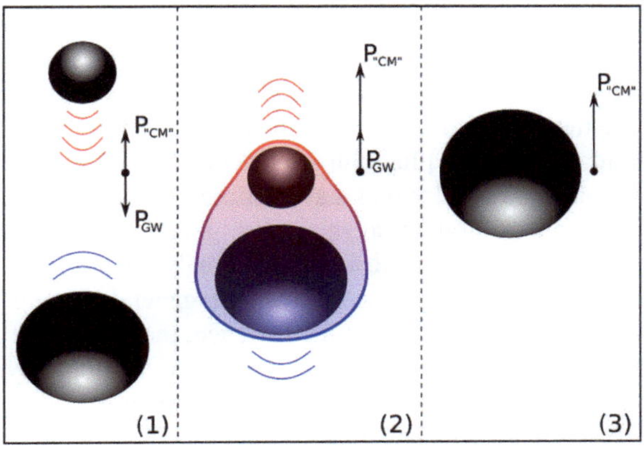


**Fig. 6** Cartoon of the generation of the anti-kick in the head-on collision of two unequal-mass Schwarzschild black holes. Initially the smaller *black hole* moves faster and linear momentum is radiated mostly downwards, thus leading to an upwards recoil of the system [stage (*1*)]. At the merger the curvature is higher in the upper hemisphere of the distorted black hole (cf., *red-blue* shading) and linear momentum is radiated mostly upwards leading to the anti-kick [stage (*2*)]. The black hole decelerates till a uniform curvature is restored on the horizon [stage (*3*)]

with measurements of the recoil made at future null infinity. The insight gained with this spacetime will be valuable to explain the anti-kick under generic conditions and to contribute to the understanding of nonlinear black-hole physics.

Figure 6 illustrates the dynamics of the head-on collision using a schematic cartoon where I am considering a coordinate system centred in the total centre of mass of the system and where the smaller black hole is initially on the positive $z$-axis, while the larger one is on the negative axis. As the two black holes free-fall towards each other, the smaller one will move faster and will be more efficient in "forward-beaming" its gravitational wave emission [110]. As a result, the linear momentum will be radiated mostly downwards, thus leading to an upwards recoil of the black hole binary [cf., stage (1) in Fig. 6]. At the merger, the black-hole velocities will be the largest and so will also be the anisotropic gravitational wave emission and the corresponding recoil of the system. However, when a single apparent horizon is formed comprising the two black holes, the curvature distribution on this 2-surface will be highly *anisotropic*, being higher in the upper hemisphere (cf., red-blue shading in stage (2) of Fig. 6). Because the newly formed black hole will want to radiate all of its deviations away from the final Schwarzschild configuration, it will do so more effectively there where the curvature is larger, thus with a stronger emission of gravitational waves from the northern hemisphere. As a result, after the merger the linear momentum will be emitted mostly upwards and this sudden change in sign will lead to the anti-kick. The anisotropic gravitational wave emission will decay exponentially as the curvature gradients are erased and the quiescent black hole reaches its final and decelerated recoil velocity [cf., stage (3)].[7]

Although this picture refers to a head-on collision, it is supported by the findings in the CLA (where the direction of the ringdown kick is approximately opposite to that of the accumulated inspiral plus plunge kick) [127] and it can be generalized to a situation in which the black holes have different masses, different spins and are merging through an inspiral. Also in a more generic case, the newly-formed apparent horizon will have a complicated but globally anisotropic distribution of the curvature, determining the direction (which is in general varying in time) along which the gravitational waves will be emitted. Hence, the geometric properties in a dynamical horizon (of a black or white hole) determine its global dynamics. I next use the Robinson-Trautman spacetime to validate this picture.

## 4.2 A Useful Playground

The Robinson-Trautman spacetime represents a class of vacuum solutions admitting a congruence of null geodesics which are twist and shear-free [133], with a future stationary horizon and a dynamical past (outer trapping) horizon [134–138] (past apparent horizon hereafter). A Robinson-Trautman spacetime is thus regarded as an isolated nonspherical white hole emitting gravitational waves, where the evolution of the apparent horizon curvature-anisotropies and the total spacetime momentum dynamics can be related unambiguously. The metric is given by [139]

$$ds^2 = -\left(K - \frac{2M_\infty}{r} - \frac{2r\partial_u Q}{Q}\right)du^2 - 2du dr + \frac{r^2}{Q^2}d\Omega^2\,, \tag{7}$$

where $Q = Q(u, \Omega)$, $u$ is the standard null coordinate, $r$ is the affine parameter of the outgoing null geodesics, and $\Omega = \{\theta, \phi\}$ are the angular coordinates on the unit sphere $S^2$. Here $M_\infty$ is a constant and is related to the asymptotic mass, while the function $K(u, \Omega)$ is the Gaussian curvature of the surface corresponding to $r = 1$ and $u =$ constant, $K(u, \Omega) \equiv Q^2(1 + \nabla^2_\Omega \ln Q)$, where $\nabla^2_\Omega$ is the Laplacian on $S^2$. The Einstein equations then lead to the evolution equation

$$\partial_u Q(u, \Omega) = -Q^3 \nabla^2_\Omega K(u, \Omega)/(12M_\infty). \tag{8}$$

Any regular initial data $Q = Q(0, \Omega)$ will smoothly evolve according to (8) until it achieves a stationary configuration corresponding to a Schwarzschild black hole at rest or moving with a constant speed [140]. Equation (8) implies the existence of

[7] I should remark that other explanations have also been suggested. One of them makes use of the Landau-Lifshitz pseudotensor and explains the recoil in terms of the cancellation of large and opposite fluxes of momentum, part of which are "swallowed" by the black hole [131]. Another one is even more essential and explains the antikick is in terms of the spectral features of the signal at large distances, quite independently of the presence of a black-hole horizon [132]. All of these views serve the scope of providing an intuitive description and are in my view equally valid and useful.

the constant of motion $\mathscr{A} \equiv \int_{S^2} d\Omega/Q^2$, which clearly represents the area of the surface $u =$ const., $r =$ const. and can be used to normalise $Q$ so that $\mathscr{A} = 4\pi$. All the physically relevant information is contained in the function $Q(u, \Omega)$, and this includes the gravitational radiation, which can be extracted by relating $Q(u, \Omega)$ to the radiative part of the Riemann tensor [141, 142].

The past apparent horizon radius $R(u, \Omega)$ is given by the vanishing expansion of the future ingoing null geodesics, satisfying [134, 135]

$$Q^2 \nabla^2_{\Omega} \ln R = K - 2M_{\infty}/R. \tag{9}$$

The mass and momentum of the black hole are computed at future null infinity using the Bondi four-momentum [139]

$$P^{\alpha}(u) \equiv \frac{M_{\infty}}{4\pi} \int_{S^2} \frac{\eta^{\alpha}}{Q^3} d\Omega\,, \tag{10}$$

with $(\eta^{\alpha}) = (1, \sin\theta\cos\phi, \sin\theta\sin\phi, \cos\theta)$. Given smooth initial data, the spacetime will evolve to a stationary non-radiative solution which, in axisymmetry, has the form $Q(\infty, \theta) = (1 \mp vx)/\sqrt{1 - v^2}$, with $x \equiv \cos\theta$ [139]. The Bondi four-momentum associated to $Q(\infty, \theta)$ has components

$$(P(\infty))^{\alpha} = \left(M_{\infty}/\sqrt{1 - v^2}\right)(1, 0, 0, \pm v)\,, \tag{11}$$

so that the parameter $v$ in $Q(\infty, \theta)$ can be interpreted as the velocity of the Schwarzschild black hole in the $z$-direction.

One of the difficulties with Robinson-Trautman spacetimes is the definition of physically meaningful initial data. Although this is meant more as a proof-of-principle than a realistic configuration, it is possible to adopt the prescription suggested in Ref. [142]

$$Q(0, \theta) = Q_0 \left[\frac{1}{\sqrt{1 - wx}} + \frac{q}{\sqrt{1 + wx}}\right]^{-2}, \tag{12}$$

and which was interpreted to represent the final stages (i.e., after a common apparent horizon is formed) of a head-on collision of two boosted black holes with opposite velocities $w$ and mass ratio $q$ [142]. In practice, to reproduce the situation shown in Fig. 6, it is sufficient to choose $w < 0$ and take $q \in [0, 1]$; a more general class of initial data and the corresponding phenomenology can be found in [129, 143]. Note that $Q_0$ is chosen so that to $\mathscr{A} = 4\pi$ and that in general the deformed black hole will not be initially at rest. As a result, given the initial velocity $v_0 \equiv P^3(0)/P^0(0)$, a boost is performed transformation $\overline{P}^{\alpha} = \Lambda^{\alpha}{}_{\beta}(v_0)P^{\beta}$ so that $\overline{P}^3(0) = 0$ by construction. The numerical solution of Eq. (8) with initial data (12) is performed using a Galerkin decomposition as discussed in detail in [139].

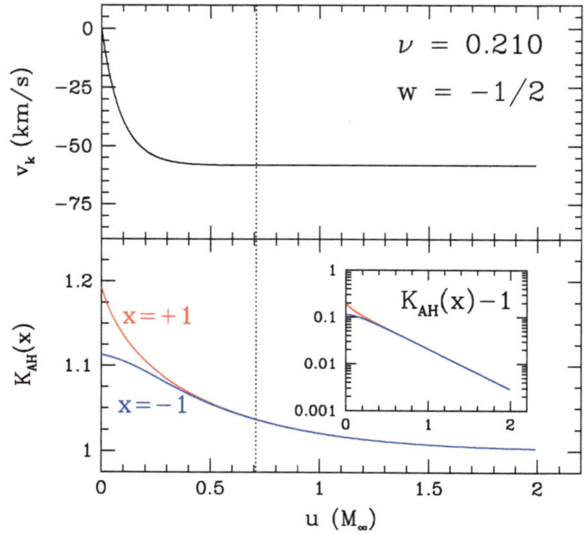


**Fig. 7** Typical evolution of a Robinson-Trautman spacetime. Shown in the *lower panel* is the evolution of the curvature $K_{\rm AH}$ at the north ($x = 1$) and south pole ($x = -1$). Shown in the *upper panel* is the evolution of the recoil, which stops decreasing when the curvature difference is erased by the emitted radiation (*dotted line*). Note that the curvature decays exponentially to that of a Schwarzschild black hole (inset)

Figure 7 reports the typical evolution of a Robinson-Trautman spacetime with the lower panel showing the evolution of the curvature of the past apparent horizon $K_{\rm AH} \equiv 2M_\infty/R^3(x)$ at the north ($x = 1$) and south pole ($x = -1$), and with the upper panel showing the evolution of the recoil velocity. Note that the two local curvatures are different initially, with the one in the upper hemisphere being larger than the one in the lower hemisphere (cf., Fig. 6). However, as the gravitational radiation is emitted, this difference is erased. When this happens, the deceleration stops and the black hole attains its asymptotic recoil velocity. The inset reports the curvature difference relative to the asymptotic Schwarzschild one, $K_{\rm AH} - 1$, whose exponentially decaying behaviour is the one expected in a ringing black hole (see also Fig. 7 of Ref. [129]).

As mentioned before, the one shown in Fig. 7 is a typical evolution of a Robinson-Trautman spacetime and is not specific of the initial data (12). By varying the values of $w$, in fact, it is possible to increase or decreases the final recoil and a sign change in $w$ simply inverts the curvature at the poles so that, for instance, initial data with $w > 0$ would yield a black hole accelerating in the positive $z$-direction. Interestingly, it is even possible to fine-tune the parameter $w$ so that the recoil produced for a Robinson-Trautman spacetime mimics the anti-kick produced by the quasi-circular inspiral of nonspinning binaries. This is shown in Fig. 8, which reports the recoil as a function of the symmetric mass ratio $\nu \equiv q/(1+q)^2$, and where the dashed line refers to the anti-kick for the inspiral of nonspinning binaries in the CLA [127] (the parameters chosen, i.e., $w = -0.425$ and $r_{12} = 2\,M$, are those minimising the differences). Considering that the two curves are related only logically and that the CLA one contains all the information about inspiralling black holes, including the orbital rotation, the match is surprisingly good.

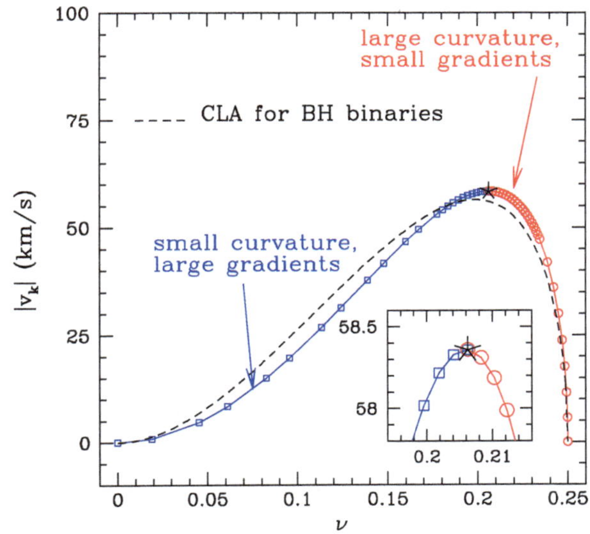


**Fig. 8** Recoil velocity shown as a function of the symmetric mass ratio $\nu$ when $w = -0.425$, with the *dashed line* refers to the anti-kick from the inspiral of nonspinning binaries in the CLA [127]. Note that the *curve* can be thought as composed of two different branches

It is also suggestive to think that the curve in Fig. 8 is actually composed of two different branches, one of which is characterized by large curvature gradients across the apparent horizon but small values of the curvature (this is the low-$\nu$ branch and is indicated with squares), while the other is characterized by small curvature gradients and large values of the curvature (this is the high-$\nu$ branch and is indicated with circles). The same recoil velocity can then be produced by two different values of $\nu$, for which the effects of large curvature gradients and small local curvatures are the same as those produced by small curvature gradients but large local curvatures.

To go from this intuition to a mathematically well-defined measure one can compute the mass multipoles of the intrinsic curvature of the initial data using the formalism developed in [144] for dynamical horizons. Namely, it is possible to compute the mass moments as (the mass-current are obviously zero)

$$M_n \equiv \oint \frac{P_n(\tilde{x})}{Q^2(\theta)R(\theta)} d\Omega\,, \tag{13}$$

where $P_n(\tilde{x})$ is the Legendre polynomial in terms of the coordinate $\tilde{x}(\theta)$ which obeys $\partial_\theta \tilde{x} = -\sin\theta R(\theta)^2/(R^2_{\rm AH}\, Q(\theta)^2)$, with $R_{\rm AH} \equiv \sqrt{\mathscr{A}_{\rm AH}/(4\pi)}$ and $\tilde{x}(0) = 1$. Using these multipoles it is possible to construct an effective-curvature parameter $K_{\rm eff}$ that represents a measure of the global curvature properties of the initial data and

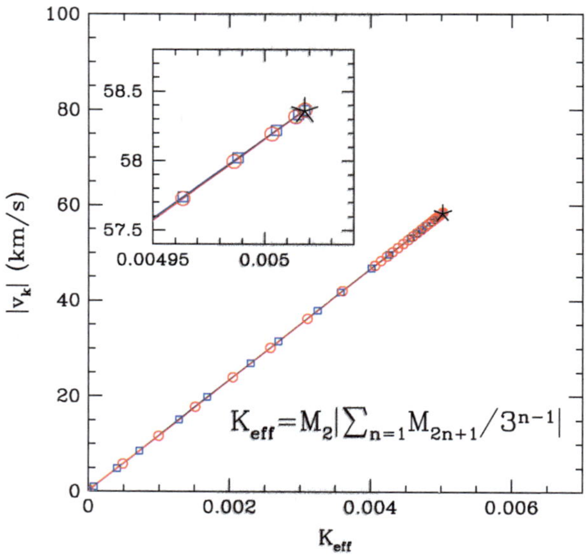


**Fig. 9** Recoil velocity shown as a function of the effective curvature. In contrast with Fig. 8, which uses the same symbols employed here, the relation between the curvature and the recoil is now injective

from which the recoil depends in an injective way. Because this effective-curvature parameter has to contain the contribution from the even and odd multipoles, the expression

$$K_{\text{eff}} = M_2 \Big| \sum_{n=1} M_{2n+1}/3^{n-1} \Big| , \tag{14}$$

was found to reproduce exactly what expected (note $M_1 = 0$ to machine precision).

This is shown in Fig. 9, which reports the recoil velocity as a function of $K_{\text{eff}}$. As predicted, and in contrast with Fig. 8, the relation between the curvature and the recoil is now injective, with the maximum recoil velocity being given by the maximum value of $K_{\text{eff}}$ (see inset), and with the two branches coinciding. The expression (14) suggested above for $K_{\text{eff}}$ is not unique and indeed a more generic one will have to include also the mass-current multipoles to account for the spin contributions (see discussion below). However, lacking a rigorous mathematical guidance, the phenomenological $K_{\text{eff}}$ is a reasonable, intuitive approximation.

## 4.3 A More General View

Despite the valuable insight, the treatment summarised above and presented in Ref. [128] had obvious limitations. First, the Ansatz (14) for $K_{\text{eff}}$, i.e., $K_{\text{eff}} = f_{\text{even}}(M_{2\ell}) \times f_{\text{odd}}(M_{2\ell+1})$ is not straightforwardly generalizable to the non-axisymmetric case. Second, the functions $f_{\text{even}}$ and $f_{\text{odd}}$ can be written in the

simplest
possible form, i.e., as a linear expansion in $M_\ell$'s, i.e., $K_{\rm eff} = (a_2 M_2 + a_4 M_4 + \cdots) \times (a_3 M_3 + a_5 M_5 + \cdots)$, where the phenomenological coefficients $a_\ell$'s depend on the details of the employed initial data. Finally, the white-hole horizon analysis in Robinson-Trautman spacetimes needs to be extended to the genuine black-hole horizon case.

While the focus in what discussed above (and presented in Ref. [128]) was on expressing the difference between the *final* kick velocity $v_\infty$ and the instantaneous kick velocity $v_{\rm k}(u)$ at an (initial) given time $u$, in terms of the geometry of the common apparent horizon at that time $u$, it is possible to derive a more generic view based on geometric quantities that are evaluated at a given time during the evolution. More specifically, it is possible to consider the variation of the Bondi linear momentum vector in time $(dP_i^{\rm B}/du)(u)$ as the relevant geometric quantity to monitor at null infinity $\mathscr{I}^+$. This quantity can then be correlated with a counterpart on the black-hole horizon $\mathscr{H}^+$, e.g., a vector $\tilde{K}^i_{\rm eff}(v)$ (function of an advanced time $v$), which represents an extension of the effective curvature introduced in the previous section.[8]

In the case of a Robinson-Trautman spacetime, the causal relation between the white-hole horizon $\mathscr{H}^-$ and null infinity $\mathscr{I}^+$ made possible to establish an explicit functional relation between $dv_{\rm k}/du$ and $K'_{\rm eff}(u)$. In the case of a generic black-hole horizon, such a direct causal relation between the inner horizon and $\mathscr{I}^+$ is lost. However, since the corresponding causal pasts of $\mathscr{I}^+$ and $\mathscr{H}^-$ coincide in part, non-trivial *correlations* are still possible and expected. These correlations can be measured by comparing geometric quantities $h_{\rm inn}(v)$ at $\mathscr{H}^+$ and $h_{\rm out}(u)$ at $\mathscr{I}^+$, both considered here as two timeseries.[9] In particular, it is reasonable to take $\tilde{K}^i_{\rm eff}(v)$ as $h_{\rm inn}(v)$ and $(dP_i^{\rm B}/du)(u)$ as $h_{\rm out}(u)$.

This approach resembles therefore the methodology adopted in *scattering* experiments. Gravitational dynamics in a given spacetime region affects the geometry of appropriately-chosen *outer* and *inner* hypersurfaces of the black-hole spacetime. These hypersurfaces are then understood as *test screens* on which suitable *geometric quantities* must be constructed. The correlations between the two encode geometric information about the dynamics in the bulk, providing information useful for an *inverse-scattering* approach to the near-horizon dynamics. In asymptotically flat black-hole spacetimes, null infinity $\mathscr{I}^+$ and the (event) black-hole horizon $\mathscr{H}^+$ provide natural choices for the outer and inner screens. This is summarised in the Carter-Penrose diagram in Fig. 10, which illustrates the cross-correlation approach to near-horizon gravitational dynamics. The event horizon $\mathscr{H}^+$ and null infinity $\mathscr{I}^+$ provide spacetime screens on which geometric quantities, accounting respectively for horizon deformations and wave emission, are measured. Their cross-correlation encodes information about the bulk spacetime dynamics.

---

[8] Another appealing approach that has a similar goal of correlating strong-fields effects with (the visualization of) spacetime curvature has been proposed recently by the group in Caltech [145, 146].

[9] Note that the meaningful definition of timeseries cross-correlations requires the introduction of a (gauge-dependent) relation between advanced and retarded time coordinates $v$ and $u$. In an initial value problem this is naturally provided by the $3+1$ spacetime slicing by time $t$.

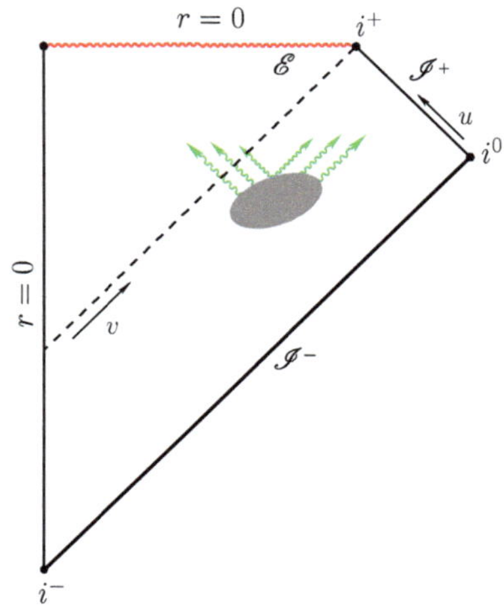


**Fig. 10** Carter-Penrose diagram illustrating the *scattering* approach to near-horizon gravitational dynamics in a generic spherically symmetric collapse. The event horizon $\mathscr{H}^+$ and null infinity $\mathscr{I}^+$ provide spacetime canonical screens on which *geometric quantities*, respectively accounting for horizon deformations and wave emission, are defined. Their cross-correlation encodes nontrivially information about the bulk spacetime dynamics

The picture offered by Fig. 10 can be easily adapted to the 3+1 approach commonly adopted in numerical relativity. Since neither the black-hole event horizon nor null infinity are in general available during the evolution,[10] it is possible to adopt as inner and outer screens a dynamical horizon $\mathscr{H}^+$ (future outer trapping horizon [147–149]) and a timelike tube $\mathscr{B}$ at large spatial distances, respectively. In this case, the time function $t$ associated with the $3+1$ spacetime slicing provides a (gauge) mapping between the retarded and advanced times $u$ and $v$, so that cross-correlations between geometric quantities at $\mathscr{H}^+$ and $\mathscr{B}$ can be calculated as standard timeseries $h_{\rm inn}(t)$ and $h_{\rm out}(t)$. This is summarised in the Carter-Penrose diagram in Fig. 11, which is the same as in Fig. 10, but where the $3+1$ slicing sets an in-built common time $t$ for cross-correlations between the dynamical horizon $\mathscr{H}^+$ (i.e., the inner screen) and a large-distance timelike hypersurface $\mathscr{B}$ (i.e., the outer screen).

Within this conceptual framework it is then possible to define a phenomenological curvature vector $\tilde{K}_i^{\rm eff}(t)$ in terms of the mass multipoles of the Ricci scalar curvature ${}^2R$ at $\mathscr{H}^+$ and show that this is closely correlated with a geometric quantities $(dP_i^{\mathscr{B}}/dt)(t)$, representing the variation of the Bondi linear momentum time on $\mathscr{I}^+$. How to do this in practice for a black-hole spacetime requires much more space that I can take in this contribution and therefore refer the interested reader to Refs. [129, 130], where this is discussed in great detail.

---

[10] The latter would properly require either characteristic or a hyperboloidal evolution approach.

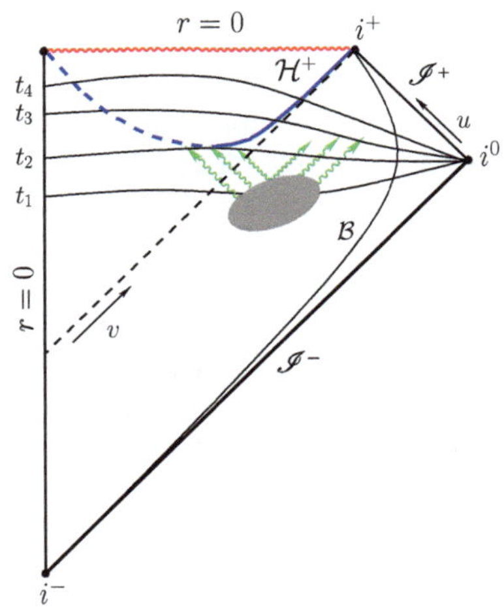


**Fig. 11** Carter-Penrose diagram for the *scattering* picture in a Cauchy initial value approach. The dynamical horizon $\mathscr{H}^+$ and a large-distance timelike hypersurface $\mathscr{B}$ provide *inner* and *outer* screens. Note that the dynamical horizon is split in two portions: *outer* and *inner* (*solid* and *dashed blue lines*, respectively) and that the 3+1 slicing sets a common time $t$ for cross-correlations

## *4.4 Summary*

The discussion reported above demonstrates that qualitative aspects of the post-merger recoil dynamics at infinity can be understood in terms of the evolution of the geometry of the common horizon of the resulting black hole. Moreover, suitably-built quantities defined on inner and outer worldtubes (represented either by dynamical horizons or by timelike boundaries) can act as test screens responding to the spacetime geometry in the bulk, thus opening the way to a cross-correlation approach to probe the dynamics of spacetime. This picture was shown to hold both for a simple Robinson-Trautman spacetime, but also for more generic binary black-hole spacetimes. In this latter case, this is possible through the construction of a phenomenological vector $\tilde{K}_i^{\mathrm{eff}}(t)$ from the Ricci curvature scalar ${}^2R$ on the dynamical horizon sections, which then captures the global properties of the flux of Bondi linear momentum $(dP_i^{\mathrm{B}}/dt)(t)$ at infinity, namely the acceleration of the BH.

A geometric framework looking at the horizon's properties offers a number of connections with the literature developing around the use of horizons to study the dynamics of black holes, as well as with the interpretations of such dynamics in terms of a viscous-hydrodynamics analogy. Much of the machinery developed using dynamical trapping horizons as inner screens can be extended also when a common horizon is not formed (as in the calculations reported in Ref. [150]). While in such cases the identification of an appropriate hypersurface for the inner screen can be more difficult, once this is found its geometrical properties can be used along the lines of the cross-correlation approach discussed here for dynamical horizons.

## 5 Conclusions

The "three little piece" for numerical computer and relativity presented in the sections above ought to provide a reasonable idea of the "Renaissance" that numerical relativity is now experiencing. More importantly, they should be able to convey the enormous potential that numerical-relativity simulations have in revealing aspects of the theory that cannot be handled analytically, or in exploring nonlinear regimes that cannot be investigated through perturbative approaches. As remarked repeatedly, the examples brought represent only a personal (and biased) selection of the intense work carried out recently and surely are not exhaustive in terms of the physical scenarios that can be explored. Much more can be said about this and surely it will not have to wait for the bicentenary of Einstein's stay in Prague.

**Acknowledgments** The work discussed here has been carried out in collaboration with M. A. Aloy, L. Baiotti, B. Giacomazzo, J. Granot, J. L. Jaramillo, C. Kouveliotou, R. P. Macedo K. Takami, whom I am indebted with. My thanks go also to the numerical-relativity group of the AEI for providing such a stimulating and productive environment. Support comes through the DFG grant SFB/Trans-regio 7 and "CompStar", a Research Networking Programme of the ESF. The calculations have been performed on the clusters at the AEI.

## References

1. Pretorius, F.: Evolution of binary black hole spacetimes. Phys. Rev. Lett. **95**, 121101 (2005). doi:10.1103/PhysRevLett.95.121101
2. Campanelli, M., Lousto, C.O., Marronetti, P., Zlochower, Y.: Accurate evolutions of orbiting black-hole binaries without excision. Phys. Rev. Lett. **96**, 111101 (2006). doi:10.1103/PhysRevLett.96.111101
3. Baker, J.G., Centrella, J., Choi, D.I., Koppitz, M., van Meter, J.: Gravitational wave extraction from an inspiraling configuration of merging black holes. Phys. Rev. Lett. **96**, 111102 (2006). doi:10.1103/PhysRevLett.96.111102
4. Baiotti, L., Rezzolla, L.: Challenging the paradigm of singularity excision in gravitational collapse. Phys. Rev. Lett. **97**, 141101 (2006). doi:10.1103/PhysRevLett.97.141101
5. Chu, T., Pfeiffer, H.P., Scheel, M.A.: High accuracy simulations of black hole binaries: spins anti-aligned with the orbital angular momentum. Phys. Rev. D **80**(12), 124051 (2009). doi:10.1103/PhysRevD.80.124051
6. Baiotti, L., Giacomazzo, B., Rezzolla, L.: Accurate evolutions of inspiralling neutron-star binaries: prompt and delayed collapse to a black hole. Phys. Rev. D **78**(8), 084033 (2008). doi:10.1103/PhysRevD.78.084033
7. Centrella, J., Baker, J.G., Kelly, B.J., van Meter, J.R.: Black-hole binaries, gravitational waves, and numerical relativity. Rev. Mod. Phys. **82**, 3069 (2010). doi:10.1103/RevModPhys.82.3069
8. Pfeiffer, H.P.: Numerical simulations of compact object binaries. Class. Quant. Grav. **29**(12), 124004 (2012). doi:10.1088/0264-9381/29/12/124004
9. Radice, D., Rezzolla, L.: Discontinuous Galerkin methods for general-relativistic hydrodynamics: formulation and application to spherically symmetric spacetimes. Phys. Rev. D **84**(2), 024010 (2011). doi:10.1103/PhysRevD.84.024010
10. Radice, D., Rezzolla, L.: THC: a new high-order finite-difference high-resolution shock-capturing code for special-relativistic hydrodynamics. Astron. Astrophys. **547**, A26 (2012). doi:10.1051/0004-6361/201219735

11. Nakamura, T., Oohara, K., Kojima, Y.: General relativistic collapse to black holes and gravitational waves from black holes. Prog. Theor. Phys. Suppl. **90**, 1 (1987)
12. Bona, C., Massó, J., Seidel, E., Stela, J.: New formalism for numerical relativity. Phys. Rev. Lett. **75**, 600 (1995). doi:10.1103/PhysRevLett.75.600
13. Shibata, M., Nakamura, T.: Evolution of three-dimensional gravitational waves: harmonic slicing case. Phys. Rev. D **52**, 5428 (1995). doi:10.1103/PhysRevD.52.5428
14. Baumgarte, T.W., Shapiro, S.L.: On the numerical integration of Einstein's field equations. Phys. Rev. D **59**, 024007 (1999). doi:10.1103/PhysRevD.59.024007
15. Alcubierre, M., Brügmann, B., Diener, P., et al.: Gauge conditions for long-term numerical black hole evolutions without excision. Phys. Rev. D **67**, 084023 (2003). doi:10.1103/PhysRevD.67.084023
16. Pretorius, F.: Numerical relativity using a generalized harmonic decomposition. Class. Quant. Grav. **22**, 425 (2005). doi:10.1088/0264-9381/22/2/014
17. Alic, D., Bona, C., Bona-Casas, C.: Towards a gauge-polyvalent numerical relativity code. Phys. Rev. D **79**(4), 044026 (2009). doi:10.1103/PhysRevD.79.044026
18. Bernuzzi, S., Hilditch, D.: Constraint violation in free evolution schemes: comparing BSSNOK with a conformal decomposition of Z4. Phys. Rev. D **81**, 084003 (2010). doi:10.1103/PhysRevD.81.084003
19. Müller, D., Grigsby, J., Brügmann, B.: Dynamical shift condition for unequal mass black hole binaries. Phys. Rev. D **82**, 064004 (2010). doi:10.1103/PhysRevD.82.064004
20. Schnetter, E.: Time step size limitation introduced by the BSSN Gamma driver. Class. Quant. Grav. **27**, 167001 (2010). doi:10.1088/0264-9381/27/16/167001
21. Alic, D., Bona-Casas, C., Bona, C., Rezzolla, L., Palenzuela, C.: Conformal and covariant formulation of the Z4 system with constraint-violation damping. Phys. Rev. D **85**, 064040 (2012)
22. Shibata, M., Taniguchi, K.: Merger of binary neutron stars to a black hole: disk mass, short gamma-ray bursts, and quasinormal mode ringing. Phys. Rev. D **73**, 064027 (2006). doi:10.1103/PhysRevD.73.064027
23. Anderson, M., Hirschmann, E.W., Lehner, L., et al.: Magnetized neutron star mergers and gravitational wave signals. Phys. Rev. Lett. **100**, 191101 (2008)
24. Liu, Y.T., Shapiro, S.L., Etienne, Z.B., Taniguchi, K.: General relativistic simulations of magnetized binary neutron star mergers. Phys. Rev. D **78**, 024012 (2008). doi:10.1103/PhysRevD.78.024012
25. Giacomazzo, B., Rezzolla, L., Baiotti, L.: Can magnetic fields be detected during the inspiral of binary neutron stars? Mon. Not. R. Astron. Soc. **399**, L164 (2009). doi:10.1111/j.1745-3933.2009.00745.x
26. Giacomazzo, B., Rezzolla, L., Stergioulas, N.: Collapse of differentially rotating neutron stars and cosmic censorship. Phys. Rev. D **84**(2), 024022 (2011). doi:10.1103/PhysRevD.84.024022
27. Rezzolla, L., Baiotti, L., Giacomazzo, B., Link, D., Font, J.A.: Accurate evolutions of unequal-mass neutron-star binaries: properties of the torus and short GRB engines. Class. Quant. Grav. **27**(11), 114105 (2010). doi:10.1088/0264-9381/27/11/114105
28. Kiuchi, K., Sekiguchi, Y., Shibata, M., Taniguchi, K.: Exploring binary-neutron-star-merger scenario of short-gamma-ray bursts by gravitational-wave observation. Phys. Rev. Lett. **104**(14), 141101 (2010). doi:10.1103/PhysRevLett.104.141101
29. Baiotti, L., Damour, T., Giacomazzo, B., Nagar, A., Rezzolla, L.: Analytic modelling of tidal effects in the relativistic inspiral of binary neutron stars. Phys. Rev. Lett. **105**, 261101 (2010). doi:10.1103/PhysRevLett.105.261101
30. Bernuzzi, S., Nagar, A., Thierfelder, M., Brügmann, B.: Tidal effects in binary neutron star coalescence. Phys. Rev. D **86**(4), 044030 (2012). doi:10.1103/PhysRevD.86.044030
31. Hotokezaka, K., Kyutoku, K., Shibata, M.: Exploring tidal effects of coalescing binary neutron stars in numerical relativity. Phys. Rev. D **87**(4), 044001 (2013). doi:10.1103/PhysRevD.87.044001

32. Abadie, J., Abbott, B.P., Abbott, R., et al.: Topical review: predictions for the rates of compact binary coalescences observable by ground-based gravitational-wave detectors. Class. Quant. Grav. **27**(17), 173001 (2010). doi:10.1088/0264-9381/27/17/173001
33. Paczynski, B.: Gamma-ray bursters at cosmological distances. Astrophys. J. Lett. **308**, L43 (1986). doi:10.1086/184740
34. Eichler, D., Livio, M., Piran, T., Schramm, D.N.: Nucleosynthesis, neutrino bursts and gamma-rays from coalescing neutron stars. Nature **340**, 126 (1989). doi:10.1038/340126a0
35. Narayan, R., Paczynski, B., Piran, T.: Gamma-ray bursts as the death throes of massive binary stars. Astrophys. J. **395**, L83 (1992)
36. Ruffert, M., Janka, H.T.: Gamma-ray bursts from accreting black holes in neutron star mergers. Astron. Astrophys. **344**, 573 (1999)
37. Rosswog, S., Ramirez-Ruiz, E., Davies, M.B.: High-resolution calculations of merging neutron stars - III. Gamma-ray bursts. Mon. Not. R. Astron. Soc. **345**, 1077 (2003). doi:10.1046/j.1365-2966.2003.07032.x
38. Nakar, E.: Short-hard gamma-ray bursts. Phys. Rep. **442**, 166 (2007). doi:10.1016/j.physrep.2007.02.005
39. Lee, W.H., Ramirez-Ruiz, E.: The progenitors of short gamma-ray bursts. New J. Phys. **9**, 17 (2007). doi:10.1088/1367-2630/9/1/017
40. Fox, D.B., Frail, D.A., Price, P.A., et al.: The afterglow of GRB050709 and the nature of the short-hard gamma-ray bursts. Nature **437**, 845 (2005). doi:10.1038/nature04189
41. Prochaska, J.X., Bloom, J.S., Chen, H.W., et al.: The galaxy hosts and large-scale environments of short-hard gamma-ray bursts. Astrophys. J. **642**, 989 (2006). doi:10.1086/501160
42. Rezzolla, L., Giacomazzo, B., Baiotti, L., et al.: The missing link: merging neutron stars naturally produce jet-like structures and can power short gamma-ray bursts. Astrophys. J. **732**(11), L6 (2011). doi:10.1088/2041-8205/732/1/L6
43. Giacomazzo, B., Rezzolla, L.: WhiskyMHD: a new numerical code for general relativistic magnetohydrodynamics. Class. Quant. Grav. **24**, S235 (2007). doi:10.1088/0264-9381/24/12/S16
44. Pollney, D., Reisswig, C., Rezzolla, L., et al.: Recoil velocities from equal-mass binary black-hole mergers: a systematic investigation of spin-orbit aligned configurations. Phys. Rev. D **76**, 124002 (2007)
45. Baiotti, L., Hawke, I., Montero, P., Rezzolla, L.: A new three-dimensional general-relativistic hydrodynamics code. In: Capuzzo-Dolcetta R. (ed.) Computational Astrophysics in Italy: Methods and Tools, vol. 1, p. 210. MSAIt, Trieste (2003)
46. Baiotti, L., Hawke, I., Montero, P.J., et al.: Three-dimensional relativistic simulations of rotating neutron star collapse to a Kerr black hole. Phys. Rev. D **71**, 024035 (2005)
47. Antón, L., Zanotti, O., Miralles, J.A., et al.: Numerical 3+1 general relativistic magnetohydrodynamics: a local characteristic approach. Astrophys. J. **637**, 296 (2006). doi:10.1086/498238
48. Toth, G.: The div $B = 0$ constraint in shock-capturing magnetohydrodynamics codes. J. Comput. Phys. **161**, 605 (2000). doi:10.1006/jcph.2000.6519
49. Schnetter, E., Hawley, S.H., Hawke, I.: Evolutions in 3D numerical relativity using fixed mesh refinement. Class. Quant. Grav. **21**(6), 1465 (2004). doi:10.1088/0264-9381/21/6/014
50. Kouveliotou, C., Meegan, C.A., Fishman, G.J., et al.: Identification of two classes of gamma-ray bursts. Astrophys. J. **413**, L101 (1993). doi:10.1086/186969
51. Dessart, L., Ott, C.D., Burrows, A., Rosswog, S., Livne, E.: Neutrino signatures and the neutrino-driven wind in binary neutron star mergers. Astrophys. J. **690**, 1681 (2009). doi:10.1088/0004-637X/690/2/1681
52. Setiawan, S., Ruffert, M., Janka, H.: Non-stationary hyperaccretion of stellar-mass black holes in three dimensions: torus evolution and neutrino emission. Mon. Not. R. Astron. Soc. **352**, 753 (2004). doi:10.1111/j.1365-2966.2004.07974.x
53. McKinney, J.C., Uzdensky, D.A.: A reconnection switch to trigger gamma-ray burst jet dissipation. Mon. Not. R. Astron. Soc. **419**, 573 (2012). doi:10.1111/j.1365-2966.2011.19721.x

54. Price, R.H., Rosswog, S.: Producing ultrastrong magnetic fields in neutron star mergers. Science **312**, 719 (2006). doi:10.1126/science.1125201
55. Aloy, M.A., Janka, H., Müller, E.: Relativistic outflows from remnants of compact object mergers and their viability for short gamma-ray bursts. Astron. Astrophys. **436**, 273 (2005). doi:10.1051/0004-6361:20041865
56. Komissarov, S.S., Vlahakis, N., Königl, A., Barkov, M.V.: Magnetic acceleration of ultrarelativistic jets in gamma-ray burst sources. Mon. Not. R. Astron. Soc. **394**, 1182 (2009). doi:10.1111/j.1365-2966.2009.14410.x
57. Balbus, S.A., Hawley, J.F.: Instability, turbulence, and enhanced transport in accretion disks. Rev. Mod. Phys. **70**, 1 (1998). doi:10.1103/RevModPhys.70.1
58. Siegel, D.M., Ciolfi, R., Harte, A.I., Rezzolla, L.: On the magnetorotational instability in relativistic hypermassive neutron stars. ArXiv e-prints [arXiv:1302.4368[gr-qc]] (2013)
59. Giacomazzo, B., Rezzolla, L., Baiotti, L.: Accurate evolutions of inspiralling and magnetized neutron stars: equal-mass binaries. Phys. Rev. D **83**(4), 044014 (2011). doi:10.1103/PhysRevD.83.044014
60. Obergaulinger, M., Cerdá-Durán, P., Müller, E., Aloy, M.A.: Semi-global simulations of the magneto-rotational instability in core collapse supernovae. Astron. Astrophys. **498**, 241 (2009). doi:10.1051/0004-6361/200811323
61. Blandford, R.D., Znajek, R.L.: Electromagnetic extraction of energy from Kerr black holes. Mon. Not. R. Astron. Soc. **179**, 433 (1977)
62. Komissarov, S.S., Barkov, M.V.: Activation of the Blandford-Znajek mechanism in collapsing stars. Mon. Not. R. Astron. Soc. **397**, 1153 (2009). doi:10.1111/j.1365-2966.2009.14831.x
63. Radice, D., Rezzolla, L.: Universality and intermittency in relativistic turbulent flows of a hot plasma. Astrophys. J. **766**, L10 (2013). doi:10.1088/2041-8205/766/1/L10
64. Zrake, J., MacFadyen, A.I.: Spectral and intermittency properties of relativistic turbulence. Astrophys. J. **763**, L12 (2013). doi:10.1088/2041-8205/763/1/L12
65. Etienne, Z.B., Paschalidis, V., Shapiro, S.L.: General-relativistic simulations of black-hole-neutron-star mergers: effects of tilted magnetic fields. Phys. Rev. D **86**(8), 084026 (2012). doi:10.1103/PhysRevD.86.084026
66. Lesur, G., Ogilvie, G.I.: Localized magnetorotational instability and its role in the accretion disc dynamo. Mon. Not. R. Astron. Soc. **391**, 1437 (2008). doi:10.1111/j.1365-2966.2008.13993.x
67. Aloy, M.A., Rezzolla, L.: A powerful hydrodynamic booster for relativistic jets. Astrophys. J. **640**, L115 (2006). doi:10.1086/503608
68. Granot, J., Komissarov, S.S., Spitkovsky, A.: Impulsive acceleration of strongly magnetized relativistic flows. Mon. Not. R. Astron. Soc. **411**, 1323 (2011). doi:10.1111/j.1365-2966.2010.17770.x
69. Dionysopoulou, K., Alic, D., Palenzuela, C., Rezzolla, L., Giacomazzo, B.: General-relativistic resistive magnetohydrodynamics in three dimensions: formulation and tests. ArXiv e-prints 1208.3487 [gr-qc] (2012)
70. Bloom, J.S., Prochaska, J.X., Pooley, D., et al.: Closing in on a short-hard burst progenitor: constraints from early-time optical imaging and spectroscopy of a possible host galaxy of GRB 050509b. Astrophys. J. **638**, 354 (2006). doi:10.1086/498107
71. Ackermann, M., Asano, K., Atwood, W.B., et al.: Fermi observations of GRB 090510: a short-hard gamma-ray burst with an additional, hard power-law component from 10 keV TO GeV energies. Astrophys. J. **716**, 1178 (2010). doi:10.1088/0004-637X/716/2/1178
72. Oechslin, R., Janka, H.T.: Torus formation in neutron star mergers and well-localized short gamma-ray burst. Mon. Not. R. Astron. Soc. **368**, 1489 (2006)
73. Birkl, R., Aloy, M.A., Janka, H.T., Müller, E.: Neutrino pair annihilation near accreting, stellar-mass black holes. Astron. Astrophys. **463**, 51 (2007). doi:10.1051/0004-6361:20066293
74. Lee, H.K., Wijers, R.A.M.J., Brown, G.E.: The Blandford-Znajek process as a central engine for a gamma-ray burst. Phys. Rep. **325**, 83 (2000). doi:10.1016/S0370-1573(99)00084-8
75. Granot, J., Königl, A., Piran, T.: Implications of the early X-ray afterglow light curves of Swift gamma-ray bursts. Mon. Not. R. Astron. Soc. **370**, 1946 (2006). doi:10.1111/j.1365-2966.2006.10621.x

76. Abdo, A.A., Ackermann, M., Asano, K., et al.: Fermi observations of high-energy gamma-ray emission from GRB 080825C. Astrophys. J. **707**, 580 (2009). doi:10.1088/0004-637X/707/1/580
77. Granot, J., Cohen-Tanugi, J., do Couto e Silva, E.: Opacity buildup in impulsive relativistic sources. Astrophys. J. **677**, 92 (2008). doi:10.1086/526414
78. De Pasquale, M., Schady, P., Kuin, N.P.M., et al.: Swift and Fermi observations of the early afterglow of the short gamma-ray burst 090510. Astrophys. J. **709**, L146 (2010). doi:10.1088/2041-8205/709/2/L146
79. Rossi, E., Lazzati, D., Rees, M.J.: Afterglow light curves, viewing angle and the jet structure of $\gamma$-ray bursts. Mon. Not. R. Astron. Soc. **332**, 945 (2002). doi:10.1046/j.1365-8711.2002.05363.x
80. Peng, F., Königl, A., Granot, J.: Two-component jet models of gamma-ray burst sources. Astrophys. J. **626**, 966 (2005). doi:10.1086/430045
81. Wald, R.M.: General Relativity. The University of Chicago Press, Chicago (1984)
82. Thorne, K.: Nonspherical gravitational collapse: a short review. In: Klauder, J. (ed.) Magic Without Magic: John Archibald Wheeler, p. 231. Freeman, San Francisco (1972)
83. Senovilla, J.M.M.: A reformulation of the hoop conjecture. Europhys. Lett. **81**, 20004 (2008). doi:10.1209/0295-5075/81/20004
84. Argyres, P.C., Dimopoulos, S., March-Russell, J.: Black holes and sub-millimeter dimensions. Phys. Lett. B **441**, 96 (1998). doi:10.1016/S0370-2693(98)01184-8
85. Yoshino, H., Nambu, Y.: Black hole formation in the grazing collision of high-energy particles. Phys. Rev. D **67**(2), 024009 (2003). doi:10.1103/PhysRevD.67.024009
86. Yoo, C.M., Ishihara, H., Kimura, M., Tanzawa, S.: Hoop conjecture and the horizon formation cross section in Kaluza-Klein spacetimes. Phys. Rev. D **81**(2), 024020 (2010). doi:10.1103/PhysRevD.81.024020
87. Dimopoulos, S., Landsberg, G.: Black holes at the Large Hadron Collider. Phys. Rev. Lett. **87**(16), 161602 (2001). doi:10.1103/PhysRevLett.87.161602
88. Sperhake, U., Cardoso, V., Pretorius, F., Berti, E., Gonzalez, J.A.: The high-energy collision of two black holes. Phys. Rev. Lett. **101**, 161101 (2008). doi:10.1103/PhysRevLett.101.161101
89. Shibata, M., Okawa, H., Yamamoto, T.: High-velocity collision of two black holes. Phys. Rev. D **78**(10), 101501 (2008). doi:10.1103/PhysRevD.78.101501
90. Sperhake, U., Cardoso, V., Pretorius, F., et al.: Cross section, final spin, and zoom-whirl behavior in high-energy black-hole collisions. Phys. Rev. Lett. **103**(13), 131102 (2009). doi:10.1103/PhysRevLett.103.131102
91. Eardley, D.M., Giddings, S.B.: Classical black hole production in high-energy collisions. Phys. Rev. D **66**(4), 044011 (2002). doi:10.1103/PhysRevD.66.044011
92. D'Eath, P.D., Payne, P.N.: Gravitational radiation in black-hole collisions at the speed of light. I. Perturbation treatment of the axisymmetric collision. Phys. Rev. D **46**, 658 (1992). doi:10.1103/PhysRevD.46.658
93. D'Eath, P.D., Payne, P.N.: Gravitational radiation in black-hole collisions at the speed of light. II. Reduction to two independent variables and calculation of the second-order news function. Phys. Rev. D **46**, 675 (1992). doi:10.1103/PhysRevD.46.675
94. D'Eath, P.D., Payne, P.N.: Gravitational radiation in black-hole collisions at the speed of light. III. Results and conclusions. Phys. Rev. D **46**, 694 (1992). doi:10.1103/PhysRevD.46.694
95. Choptuik, M.W., Pretorius, F.: Ultrarelativistic particle collisions. Phys. Rev. Lett. **104**(11), 111101 (2010). doi:10.1103/PhysRevLett.104.111101
96. Rezzolla, L., Takami, K.: Black-hole production from ultrarelativistic collisions. Class. Quant. Grav. **30**(1), 012001 (2013). doi:10.1088/0264-9381/30/1/012001
97. East, W.E., Pretorius, F.: Ultrarelativistic black hole formation. ArXiv e-prints 1210.0443 [gr-qc] (2012)
98. Rischke, D.H., Bernard, S., Maruhn, J.A.: Relativistic hydrodynamics for heavy-ion collisions. I. General aspects and expansion into vacuum. Nucl. Phys. A **595**, 346 (1995). doi:10.1016/0375-9474(95)00355-1

99. Kellerman, T., Rezzolla, L., Radice, D.: Critical phenomena in neutron stars: II. Head-on collisions. Class. Quant. Grav. **27**(23), 235016 (2010). doi:10.1088/0264-9381/27/23/235016
100. Kellerman, T., Baiotti, L., Giacomazzo, B., Rezzolla, L.: An improved formulation of the relativistic hydrodynamics equations in 2D Cartesian coordinates. Class. Quant. Grav. **25**(22), 225007 (2008). doi:10.1088/0264-9381/25/22/225007
101. Radice, D., Rezzolla, L., Kellerman, T.: Critical phenomena in neutron stars: I. Linearly unstable nonrotating models. Class. Quant. Grav. **27**(23), 235015 (2010). doi:10.1088/0264-9381/27/23/235015
102. Bjorken, J.D.: Highly relativistic nucleus-nucleus collisions: the central rapidity region. Phys. Rev. D **27**, 140 (1983). doi:10.1103/PhysRevD.27.140
103. Huovinen, P., Ruuskanen, P.V.: Hydrodynamic models for heavy ion collisions. Annu. Rev. Nucl. Part. Sci. **56**, 163 (2006). doi:10.1146/annurev.nucl.54.070103.181236
104. Jin, K.J., Suen, W.M., et al.: Critical phenomena in head-on collision of neutron stars. Phys. Rev. Lett. **98**, 131101 (2007)
105. Wan, M.B.: Universality and properties of neutron star type I critical collapses. Class. Quant. Grav. **28**(15), 155002 (2011). doi:10.1088/0264-9381/28/15/155002
106. Heuer, R.D.: News from CERN, LHC status and strategy for linear colliders. ArXiv e-prints [arXiv:1202.5860 physics.acc-ph] (2012)
107. Blasi, P.: On the origin of very high energy cosmic rays. Mod. Phys. Lett. **A20**, 3055 (2005)
108. Peres, A.: Classical radiation recoil. Phys. Rev. **128**, 2471 (1962). doi:10.1103/PhysRev.128.2471
109. Fitchett, M.J., Detweiler, S.: Linear momentum and gravitational waves - circular orbits around a Schwarzschild black hole. Mon. Not. R. Astron. Soc. **211**, 933 (1984)
110. Wiseman, A.G.: Coalescing binary systems of compact objects to (post)5/2 Newtonian order. 2. Higher order wave forms and radiation recoil. Phys. Rev. D **46**, 1517 (1992). doi:10.1103/PhysRevD.46.1517
111. Andrade, Z., Price, R.H.: Head-on collisions of unequal mass black holes: close-limit predictions. Phys. Rev. D **56**, 6336 (1997). doi:10.1103/PhysRevD.56.6336
112. Baker, J.G., Centrella, J., Choi, D.I., et al.: Getting a kick out of numerical relativity. Astrophys. J. **653**, L93 (2006). doi:10.1086/510448
113. Gonzalez, J.A., Sperhake, U., Bruegmann, B., Hannam, M., Husa, S.: Total recoil: the maximum kick from nonspinning black-hole binary inspiral. Phys. Rev. Lett. **98**, 091101 (2007)
114. Campanelli, M., Lousto, C.O., Zlochower, Y., Merritt, D.: Maximum gravitational recoil. Phys. Rev. Lett. **98**, 231102 (2007). doi:10.1103/PhysRevLett.98.231102
115. Herrmann, F., Hinder, I., Shoemaker, D., Laguna, P., Matzner, R.A.: Gravitational recoil from spinning binary black hole mergers. Astrophys. J. **661**, 430 (2007). doi:10.1086/513603
116. Koppitz, M., et al.: Recoil velocities from equal-mass binary-black-hole mergers. Phys. Rev. Lett. **99**, 041102 (2007). doi:10.1103/PhysRevLett.99.041102
117. Lousto, C.O., Zlochower, Y.: Further insight into gravitational recoil. Phys. Rev. D **77**, 044028 (2008). doi:10.1103/PhysRevD.77.044028
118. Pollney, D., et al.: Recoil velocities from equal-mass binary black-hole mergers: a systematic investigation of spin-orbit aligned configurations. Phys. Rev. D **76**, 124002 (2007). doi:10.1103/PhysRevD.76.124002
119. Healy, J., et al.: Superkicks in hyperbolic encounters of binary black holes. Phys. Rev. Lett. **102**, 041101 (2009). doi:10.1103/PhysRevLett.102.041101
120. Lousto, C.O., Zlochower, Y.: Hangup kicks: still larger recoils by partial spin-orbit alignment of black-hole binaries. Phys. Rev. Lett. **107**(23), 231102 (2011). doi:10.1103/PhysRevLett.107.231102
121. Campanelli, M., Lousto, C.O., Zlochower, Y., Merritt, D.: Large merger recoils and spin flips from generic black-hole binaries. Astrophys. J. **659**, L5 (2007)
122. Gonzalez, J.A., Hannam, M.D., Sperhake, U., Bruegmann, B., Husa, S.: Supermassive kicks for spinning black holes. Phys. Rev. Lett. **98**, 231101 (2007). doi:10.1103/PhysRevLett.98.231101

123. Rezzolla, L.: Modelling the final state from binary black-hole coalescences. Class. Quant. Grav. **26**, 094023 (2009). doi:10.1088/0264-9381/26/9/094023
124. Lousto, C.O., Zlochower, Y.: Nonlinear gravitational recoil from the mergers of precessing black-hole binaries. ArXiv e-prints 1211.7099 [gr-qc] (2012)
125. Schnittman, J.D., Buonanno, A., van Meter, J.R., et al.: Anatomy of the binary black hole recoil: a multipolar analysis. Phys. Rev. D **77**(4), 044031 (2008). doi:10.1103/PhysRevD.77.044031
126. Mino, Y., Brink, J.: Gravitational radiation from plunging orbits—perturbative study. Phys. Rev. D **78**, 124015 (2008). doi:10.1103/PhysRevD.78.124015
127. Le Tiec, A., Blanchet, L., Will, C.M.: Gravitational-wave recoil from the ringdown phase of coalescing black hole binaries. Class. Quant. Grav. **27**, 012001 (2010). doi:10.1088/0264-9381/27/1/012001
128. Rezzolla, L., Macedo, R.P., Jaramillo, J.L.: Understanding the 'anti-kick' in the merger of binary black holes. Phys. Rev. Lett. **104**, 221101 (2010). doi:10.1103/PhysRevLett.104.221101
129. Jaramillo, J.L., Macedo, R.P., Moesta, P., Rezzolla, L.: Black-hole horizons as probes of black-hole dynamics. I. Post-merger recoil in head-on collisions. Phys. Rev. D **85**, 084030 (2012). doi:10.1103/PhysRevD.85.084030
130. Jaramillo, J.L., Macedo, R.P., Moesta, P., Rezzolla, L.: Black-hole horizons as probes of black-hole dynamics. II. Geometrical insights. Phys. Rev. D **85**, 084031 (2012). doi:10.1103/PhysRevD.85.084031
131. Lovelace, G., et al.: Momentum flow in black-hole binaries: II. Numerical simulations of equal-mass, head-on mergers with antiparallel spins. Phys. Rev. D **82**, 064031 (2010). doi:10.1103/PhysRevD.82.064031
132. Price, R.H., Khanna, G., Hughes, S.A.: Systematics of black hole binary inspiral kicks and the slowness approximation. Phys. Rev. D **83**(12), 124002 (2011). doi:10.1103/PhysRevD.83.124002
133. Robinson, I., Trautman, A.: Some spherical gravitational waves in general relativity. Proc. Roy. Soc. Lond. **A265**, 463 (1962). doi:10.1098/rspa.1962.0036
134. Penrose, R.: Naked singularities. Ann. N.Y. Acad. Sci. **224**, 125 (1973). doi:10.1111/j.1749-6632.1973.tb41447.x
135. Tod, K.P.: Analogue of the past horizon in the Robinson-Trautman metrics. Class. Quant. Grav. **6**, 1159 (1989). doi:10.1088/0264-9381/6/8/015
136. Chow, E.W.M., Lun, A.W.C.: Apparent horizons in vacuum Robinson-Trautman space-times. J. Austral. Math. Soc. Serv. B **41**, 217 (1999)
137. Natorf, W., Tafel, J.: Horizons in Robinson-Trautman space-times. Class. Quant. Grav. **25**, 195012 (2008). doi:10.1088/0264-9381/25/19/195012
138. Podolsky, J., Svitek, O.: Past horizons in Robinson-Trautman spacetimes with a cosmological constant. Phys. Rev. D **80**, 124042 (2009). doi:10.1103/PhysRevD.80.124042
139. Macedo, R.P., Saa, A.: Gravitational wave recoil in Robinson-Trautman spacetimes. Phys. Rev. D **78**, 104025 (2008). doi:10.1103/PhysRevD.78.104025
140. Chrusciel, P.T.: Semi-global existence and convergence of solutions of the Robinson-Trautman (2-dimensional Calabi) equation. Commun. Math. Phys. **137**, 289 (1991). doi:10.1007/BF02431882
141. de Oliveira, H.P., Damiao, I.: Soares, gravitational wave emission from a bounded source: the nonlinear regime. Phys. Rev. D **70**, 084041 (2004). doi:10.1103/PhysRevD.70.084041
142. Aranha, R.F., de Oliveira, H.P., Soares, I.D., Tonini, E.V.: The efficiency of gravitational bremsstrahlung production in the collision of two Schwarzschild black holes. Int. J. Mod. Phys. D **17**, 2049 (2008). doi:10.1142/S0218271808013704
143. Aranha, R.F., Soares, I.D., Tonini, E.V.: Mass-energy and momentum extraction by gravitational wave emission in the merger of two colliding black holes: the non-head-on case. Phys. Rev. D **85**(2), 024003 (2012). doi:10.1103/PhysRevD.85.024003
144. Ashtekar, A., Engle, J., Pawlowski, T., Van Den Broeck, C.: Multipole moments of isolated horizons. Class. Quant. Grav. **21**, 2549 (2004). doi:10.1088/0264-9381/21/11/003

145. Owen, R., Brink, J., Chen, Y., et al.: Frame-dragging vortexes and tidal tendexes attached to colliding black holes: visualizing the curvature of spacetime. Phys. Rev. Lett. **106**(15), 151101 (2011). doi:10.1103/PhysRevLett.106.151101
146. Nichols, D.A., Owen, R., Zhang, F., et al.: Visualizing spacetime curvature via frame-drag vortexes and tidal tendexes: general theory and weak-gravity applications. Phys. Rev. D **84**, 124014 (2011). doi:10.1103/PhysRevD.84.124014
147. Hayward, S.A.: General laws of black hole dynamics. Phys. Rev. D **49**(12), 6467 (1994). doi:10.1103/PhysRevD.49.6467
148. Ashtekar, A., Krishnan, B.: Dynamical horizons and their properties. Phys. Rev. D **68**, 104030 (2003). doi:10.1103/PhysRevD.68.104030
149. Ashtekar, A., Krishnan, B.: Isolated and dynamical horizons and their applications. Living Rev. Relativ. **7**, 10 (2004)
150. Sperhake, U., Berti, E., Cardoso, V., Pretorius, F., Yunes, N.: Superkicks in ultrarelativistic encounters of spinning black holes. Phys. Rev. D **83**, 024037 (2011). doi:10.1103/PhysRevD.83.024037

# Instabilities of Relativistic Stars

**John L. Friedman and Nikolaos Stergioulas**

**Abstract** Stable relativistic stars in uniform rotation form a two-parameter family, parametrized by mass and angular velocity. Limits on each of these quantities are associated with relativistic instabilities. A radial instability to gravitational collapse or explosion sets upper and lower limits on their mass, and an instability driven by gravitational waves may set an upper limit on their spin. Our summary of relativistic stability theory given here is based on and includes excerpts from the book *Rotating Relativistic Stars*, by the present authors.

## 1 Introduction

A neutron star in equilibrium is accurately approximated by a stationary self-gravitating perfect fluid.[1] The character of its oscillations and their stability, however, depend on bulk and shear viscosity, on the superfluid nature of its interior, and—for modes near the surface—on the properties of the crust and the strength of its magnetic field.

[1] Departures from the local isotropy of a perfect fluid are associated with the crust; with magnetic fields that are thought to be confined to flux tubes in the superfluid interior; and with a velocity field whose vorticity is similarly confined to vortex tubes. Departures from perfect fluid equilibrium due to a solid crust are expected to be smaller than one part in $\sim 10^{-3}$, corresponding to the maximum strain that an electromagnetic lattice can support. The vortex tubes are closely spaced; but the velocity field averaged over meter scales is that of a uniformly rotating configuration. Finally, the magnetic field contributes negligibly to the pressure support of the star, even in magnetars with fields of $10^{15}$ G.

J. L. Friedman (✉)
Department of Physics, University of Wisconsin-Milwaukee, Milwaukee, WI 53211, USA
e-mail: friedman@uwm.edu

N. Stergioulas
Department of Physics, Aristotle University of Thessaloniki, 56124Thessaloniki, Greece

J. Bičák and T. Ledvinka (eds.), *General Relativity, Cosmology and Astrophysics*,
Fundamental Theories of Physics 177, DOI: 10.1007/978-3-319-06349-2_20,

© Springer International Publishing Switzerland 2014

The stability of a rotating star is governed by the sign of the energy of its perturbations; and the amplitude of an oscillation that is damped or driven by gravitational radiation is governed by the rate at which its energy and angular momentum are radiated. Noether's theorem relates the stationarity and axisymmetry of the equilibrium star to conserved currents constructed from the perturbed metric and fluid variables. Their integrals, the canonical energy and angular momentum on a hypersurface can each be written as a functional quadratic in the perturbation, and the conservation laws express their change in terms of the flux of gravitational waves radiated to null infinity.

We begin with an action for perturbations of a rotating star from which these conserved quantities are obtained [1–6]. We next review local stability to convection and to differential rotation. A spherical star that is stable against convection is stable to all nonradial perturbations: Only the radial instability to collapse (or explosion) can remain. Instability to collapse sets upper and lower limits on the masses of stable relativistic stars, the analog for neutron stars of the Chandrasekhar limit. A turning-point criterion governs this axisymmetric instability of spherical stars against collapse and provides a sufficient condition for instability of rotating stars. Finally, we consider the additional instabilities of rotating stars. These are nonaxisymmetric instabilities that radiate gravitational waves. They may set an upper limit on the spin of old neutron stars spun up by accretion and on nascent stars that form with rapid enough rotation.

## 2 Action and Canonical Energy

The equations governing a perfect fluid are the Einstein equation and the equation of motion of the fluid,

$$G_{\alpha\beta} = 8\pi T_{\alpha\beta}, \quad \nabla_\beta T^{\alpha\beta} = 0, \tag{1}$$

together with an equation of state. We denote by $p$, $\varepsilon$, $\rho$ and $u^\alpha$ the fluid's pressure, energy density, rest-mass density and 4-velocity, respectively, and define a tensor

$$q^\alpha{}_\beta = \delta^\alpha_\beta + u^\alpha u_\beta \tag{2}$$

that is the spatial projection operator, the projection orthogonal to $u^\alpha$. The stress-energy tensor then has the form

$$T^{\alpha\beta} = \varepsilon u^\alpha u^\beta + p q^{\alpha\beta}.$$

Because the spatial projection $q^\alpha{}_\gamma \nabla_\beta T^{\beta\gamma} = 0$ is the relativistic Euler equation,

$$u^\beta \nabla_\beta u^\alpha = -\frac{q^{\alpha\beta}\nabla_\beta p}{\varepsilon + p}, \tag{3}$$

we call Eq. (1) the Einstein-Euler equations.

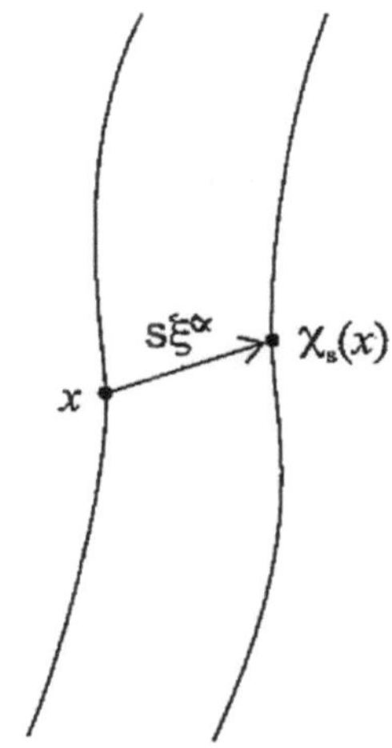


**Fig. 1** For small $s$, a Lagrangian displacement $\xi^\alpha$ can be regarded as a vector for which $s\xi^\alpha$ joins the position $x$ of a fluid element in an initial fluid flow to its position $\chi_s(x)$ in the perturbed fluid flow

One can obtain an action for stellar perturbations by introducing a Lagrangian displacement $\xi^\alpha$ joining each unperturbed fluid trajectory (the unperturbed worldline of a fluid element) to the corresponding trajectory of the perturbed fluid, as shown in Fig. 1.

The perturbative description is made precise by introducing a family of (time dependent) solutions

$$\mathcal{Q}(\lambda) = \left\{g_{\alpha\beta}(\lambda), u^\alpha(\lambda), \rho(\lambda), s(\lambda)\right\}, \tag{4}$$

and comparing to first order in $\lambda$ the perturbed variables $Q(\lambda)$ with their equilibrium values $Q(0)$. Eulerian and Lagrangian changes in the fluid variables are defined by

$$\delta\mathcal{Q} := \frac{d}{d\lambda}Q(\lambda)\bigg|_{\lambda=0}, \quad \Delta\mathcal{Q} = (\delta + \mathcal{L}_\xi)\mathcal{Q}, \tag{5}$$

with $\mathcal{L}_\xi$ the Lie derivative along $\xi^\alpha$.

Because oscillations of a neutron star proceed on a dynamical timescale, a timescale faster than that of heat flow, one requires that the Lagrangian change $\Delta s$ in the entropy per unit rest mass vanishes. With this condition, $\xi^\alpha$ and $h_{\alpha\beta} := \delta g_{\alpha\beta}$ completely specify a perturbation of a perfect-fluid spacetime with an equation of state of the form $\varepsilon = \varepsilon(\rho, s)$, $p = p(\rho, s)$. Perturbations of $u^\alpha$, $\rho$ and $\varepsilon$ are given by

$$\Delta u^\alpha = \frac{1}{2}u^\alpha u^\beta u^\gamma \Delta g_{\beta\gamma}, \quad \Delta\rho = -\frac{1}{2}\rho q^{\alpha\beta}\Delta g_{\alpha\beta}, \quad \Delta\varepsilon = -\frac{1}{2}(\varepsilon + p)q^{\alpha\beta}\Delta g_{\alpha\beta}, \tag{6}$$

with $\Delta g_{\alpha\beta} = h_{\alpha\beta} + \nabla_\alpha\xi_\beta + \nabla_\beta\xi_\alpha$. Our restriction to adiabatic perturbations means that the Lagrangian change in the pressure is given by

$$\frac{\Delta p}{p} = \Gamma_1 \frac{\Delta \rho}{\rho} = -\frac{1}{2} \Gamma_1 q^{\alpha\beta} \Delta g_{\alpha\beta}, \tag{7}$$

where the adiabatic index $\Gamma_1$ is defined by

$$\Gamma_1 = \frac{\partial \log p(\rho, s)}{\partial \log \rho} = \frac{\varepsilon + p}{p} \frac{\partial\, p(\varepsilon, s)}{\partial \varepsilon}. \tag{8}$$

The perturbed Einstein-Euler equations,

$$\delta(G^{\alpha\beta} - 8\pi\, T^{\alpha\beta}) = 0, \qquad \delta(\nabla_\beta T^{\alpha\beta}) = 0, \tag{9}$$

are self-adjoint in a weak and 4-dimensional sense that they are a symmetric system up to a total divergence: For any pairs $(\xi^\alpha, h_{\alpha\beta})$ and $(\widehat{\xi}^\alpha, \widehat{h}_{\alpha\beta})$, the symmetry relation has the form

$$\widehat{\xi}_\beta \delta(\nabla_\gamma T^{\beta\gamma} \sqrt{|g|}) + \frac{1}{16\pi} \widehat{h}_{\beta\gamma} \delta\left[(G^{\beta\gamma} - 8\pi T^{\beta\gamma})\sqrt{|g|}\right] = -\,2\mathscr{L}(\widehat{\xi}, \widehat{h}; \xi, h) + \nabla_\beta \Theta^\beta, \tag{10}$$

where $\mathscr{L}$ is symmetric under the interchange of $(\xi, h)$ and $(\widehat{\xi}, \hat{h})$. A symmetry relation of the form (10) implies that $\mathscr{L}^{(2)}(\xi, h) := \frac{1}{2}\mathscr{L}(\xi, h; \xi, h)$ is a Lagrangian density and

$$I^{(2)} = \int d^4x \mathscr{L}^{(2)} \tag{11}$$

is an action for the perturbed system.

The conserved canonical energy associated with the timelike Killing vector is the Hamiltonian of the perturbation, expressed in terms of configuration space variables,

$$E_c = \int_S d^3x\, (\Pi^\alpha \mathscr{L}_{\mathbf{t}} \xi_\alpha + \pi^{\alpha\beta} \mathscr{L}_{\mathbf{t}} h_{\alpha\beta} - \mathscr{L}^{(2)}), \tag{12}$$

where $\Pi^\alpha$ and $\pi^{\alpha\beta}$ are the momenta conjugate to $\xi^\alpha$ and $h_{\alpha\beta}$. On a spacelike hypersurface with future pointing unit normal $n^\alpha = -\alpha \nabla^\alpha t$ (where $\alpha$ is the lapse), the canonical momenta conjugate to $\xi^\alpha$ and $h_{\alpha\beta}$ are given by

$$\Pi^\alpha = \Pi^{\gamma\alpha} \nabla_\gamma t, \qquad \pi^{\alpha\beta} = \pi^{\gamma\alpha\beta} \nabla_\gamma t, \tag{13}$$

with

$$\Pi^{\alpha\beta} = \frac{1}{2}\frac{\partial\mathscr{L}(\xi, h; \xi, h)}{\partial\nabla_\alpha \xi_\beta}, \tag{14}$$

$$\pi^{\alpha\beta\gamma} = \frac{1}{2}\frac{\partial\mathscr{L}(\xi, h; \xi, h)}{\partial\nabla_\alpha h_{\beta\gamma}}. \tag{15}$$

The corresponding canonical momentum has the form

$$J_c = -\int_S d^3x\,(\Pi^\alpha \mathscr{L}_\phi \xi_\alpha + \pi^{\alpha\beta}\mathscr{L}_\phi h_{\alpha\beta}). \tag{16}$$

If one foliates the background spacetime by a family of spacelike but asymptotically null hypersurfaces, the difference $E_2 - E_1$ in $E_c$ from one hypersurface to another to its future is the energy radiated in gravitational waves to future null infinity. Because this energy is positive definite, $E_c$ can only decrease. This suggests that a condition for stability is that $E_c$ be positive for all initial data.

This is, in fact, an appropriate stability criterion, but there is a subtlety, associated with a gauge freedom in choosing a Lagrangian displacement: There is a class of *trivial* displacements, for which the Eulerian changes in all fluid variables vanish. For a one (two) parameter equation of state, these correspond to rearranging fluid elements with the same value of $\rho$ (and $s$).[2] For a trivial displacement $\eta^\alpha$, the same physical perturbation is described by the pairs $h_{\alpha\beta}, \xi^\alpha$ and $h_{\alpha\beta}, \xi^\alpha + \eta^\alpha$, but, for nonaxisymmetric perturbations, the canonical energy is not invariant under addition of a trivial displacement, and its sign depends on this kind of gauge freedom. There is, however, a preferred class of *canonical* displacements, the displacements $\xi^\alpha$ that are orthogonal to all trivial displacements, with respect to the symplectic product of two perturbations,

$$W(\widehat{\xi}, \widehat{h}; \xi, h) := \int_\Sigma (\widehat{\Pi}_\alpha \xi^\alpha + \widehat{\pi}^{\alpha\beta} h_{\alpha\beta} - \Pi_\alpha \widehat{\xi}^a - \pi^{\alpha\beta}\widehat{h}_{\alpha\beta}) d^3x. \tag{17}$$

The criterion for stability can then be phrased as follows:

1. If $E_c < 0$ for some canonical data on $\Sigma$, then the configuration is unstable or marginally stable: There exist perturbations on a family of asymptotically null hypersurfaces $\Sigma_u$ that do not die away in time.
2. If $E_c > 0$ for all canonical data on $\Sigma$, the magnitude of $E_c$ is bounded in time and only finite energy can be radiated.

The trivial displacements are relabelings of fluid elements with the same baryon density and entropy per baryon. They are Noether-related to conservation of circulation in surfaces of constant entropy per baryon [7–9], and canonical displacements

---

[2] This is not the gauge freedom associated with infinitesimal diffeos of the metric and matter, but a redundancy in the Lagrangian- displacement description of perturbations that is already present in a Newtonian context.

are displacements that preserve the circulation of each fluid ring—for which the Lagrangian change in the circulation vanishes.

For perturbations that are not spherical, stable perturbations have positive energy and die away in time; unstable perturbations have negative canonical energy and radiate negative energy to infinity, implying that $E_c$ becomes increasingly negative. One would like to show that when $E_c < 0$ a perfect-fluid configuration is strictly unstable, that within the linearized theory the time-evolved data radiates infinite energy and that $|E_c|$ becomes infinite along a family $\Sigma_u$ of asymptotically null hypersurfaces. There is no proof of this conjecture, but it is easy to see that if $E_c < 0$, the time derivatives $\dot{\xi}^\alpha$ and $\dot{h}_{\alpha\beta}$ must remain finitely large. Thus a configuration with $E_c < 0$ will be strictly unstable unless it admits perturbations that are time dependent but *nonradiative*.

For spherical stars, radial perturbations have this property, but in that case, the relativistic Euler equation has the form of a Sturm-Liouville equation, and perturbations with $E_c < 0$ are in fact strictly unstable.

The symplectic form provides an alternate form of the canonical energy, used in Wald's article in this volume. Because of the quadratic structure of the second-order Lagrangian, when the field equations are satisfied, Eq. (12) is equivalent to the expression

$$\begin{aligned} E_c &= \frac{1}{2} W(\mathcal{L}_t \xi^\alpha, \mathcal{L}_t h_{\alpha\beta}, \xi^\alpha, h_{\alpha\beta}) \\ &= \frac{1}{2} \int_\Sigma (\Pi_\alpha \mathcal{L}_t \xi^\alpha + \pi^{\alpha\beta} \mathcal{L}_t h_{\alpha\beta} - \mathcal{L}_t \Pi_\alpha \xi^\alpha - \mathcal{L}_t \pi^{\alpha\beta} h_{\alpha\beta}) d^3x. \end{aligned} \tag{18}$$

From this relation and Eq. (16), one has an immediate relation between $J_c$ and $E_c$ for a real-frequency mode with behavior $e^{i(m\phi+\omega t)}$:

$$J_c = -\frac{m}{\omega} E_c. \tag{19}$$

# 3 Local Stability

## *3.1 Convective Instability*

The criterion for the stability of a spherical star against convection is easy to understand. When a fluid element is displaced upward, if its density decreases more rapidly than the density of the surrounding fluid, then the element will be buoyed upward and the star will be unstable. If, on the other hand, the fluid element expands less than its surroundings it will fall back, and the star will be stable to convection.

As this argument suggests, criteria for convective stability are *local*, involving perturbations restricted to an arbitrarily small region of the star or, for axisymmetric perturbations, to an arbitrarily thin ring. For local perturbations, the Cowling approximation is valid: The change in the gravitational field can be ignored. The argument is this: A perturbation in density of order $\delta\varepsilon/\varepsilon$ that is restricted to a region of volume $V \ll R^3$ can be regarded as adding or subtracting from the source a mass $\delta m$ of order $\delta\varepsilon V$. Then

$$\frac{\delta m}{M} \sim \frac{V}{R^3}\frac{\delta\varepsilon}{\varepsilon} \ll \frac{\delta\varepsilon}{\varepsilon}. \tag{20}$$

The change in the metric is then also smaller than $\delta\varepsilon/\varepsilon$ by a factor $V/R^3$, arbitrarily small when the support of the matter perturbation is arbitrarily small. Note that, because the metric perturbation is gauge-dependent, this statement about the smallness of the perturbed metric is also gauge-dependent. A more precise way of stating this property of a local perturbation is that a gauge can be chosen in which the metric perturbation is smaller than the density perturbation by a factor of order $V/R^3$.

Convective instability of spherical relativistic stars was discussed by Thorne [10] and subsequently, with greater rigor, by Kovetz [11] and Schutz [12]. An initial heuristic treatment by Bardeen [13] of convective instability of differentially rotating stars was made more precise and extended to models with heat flow and viscosity by Seguin [14].

Consider a fluid element displaced radially outward from an initial position with radial coordinate $r$ to $r + \xi$. The fluid element expands (or, if displaced inward, contracts), with its pressure adjusting immediately—in sound travel time across the fluid element—to the pressure outside:

$$\Delta p = \boldsymbol{\xi} \cdot \nabla p = \frac{dp}{dr}\xi. \tag{21}$$

Heat diffuses more slowly, and the analysis assumes that the motion is faster than the time for heat to flow into or out of the fluid element: The perturbation is *adiabatic*:

$$\begin{aligned}\Delta\varepsilon &= \left(\frac{\partial\varepsilon}{\partial p}\right)_s \Delta p \\ &= \left(\frac{\partial\varepsilon}{\partial p}\right)_s \frac{dp}{dr}\xi = \Gamma_1 \frac{\varepsilon + p}{p}\frac{dp}{dr}\xi,\end{aligned} \tag{22}$$

where we have used the adiabatic conditions (6) and (7).

The difference $\Delta_\star\varepsilon$ in the density of the surrounding star between $r$ and $r + \xi$ is given by

$$\Delta_\star\varepsilon = \xi\frac{d\varepsilon}{dr}. \tag{23}$$

The displaced fluid element falls back if $|\Delta\varepsilon| < |\Delta_\star\varepsilon|$—if, that is, the fluid element's density decreases more slowly than the star's density:

$$\left(\frac{\partial\varepsilon}{\partial p}\right)_s \left|\xi\frac{dp}{dr}\right| < \left|\xi\frac{d\varepsilon}{dr}\right|. \tag{24}$$

The star is then stable against convection if the inequality,

$$\left(\frac{dp}{d\varepsilon}\right)_\star := \frac{dp/dr}{d\varepsilon/dr} < \left(\frac{\partial p}{\partial\varepsilon}\right)_s \tag{25}$$

is satisfied, unstable if the inequality is in the opposite direction.

In particular, in a homentropic star with no composition gradient, the adiabatic value of $dp/d\varepsilon$ coincides with its value in the equilibrium star,

$$\left(\frac{\partial\varepsilon}{\partial p}\right)_s = \left(\frac{dp}{d\varepsilon}\right)_\star, \tag{26}$$

implying that the star is marginally unstable.

For spherical stars Detweiler and Ipser [15] (generalizing a Newtonian result due to Lebovitz [16]), argue that, apart from local instability to convection, one need only consider radial perturbations: *If a nonrotating star is stable to radial oscillations and stable against convection, the star is stable*. The Detweiler-Ipser argument, however, relies on completeness of normal modes and the assumption that all modes are continuously joined to modes of a nearly Newtonian star, for which the Lebovitz result should imply that all modes are stable. Although the result is almost certainly true, the assumptions are not: There are outgoing modes—the $w$-modes—analogous to the outgoing modes of black holes, that have no Newtonian counterparts.

### 3.1.1 Research Problem

Prove that perturbations of spherical stars are stable if they are stable against convection and against radial perturbations.

This can be done by showing that, with reasonable assumptions about the EOS, the canonical energy of a nonradial perturbation is negative only if the Schwarzschild criterion is violated. The result may follow from an integral inequality (associated with Eq. (42) of [15]), that is central to the Detweiler-Ipser argument.

Within minutes after their birth, neutron stars cool to a temperature below the Fermi energy per nucleon, below $10^{12}$ K. Their neutrons are then degenerate, with a nearly isentropic equation of state: Convectively stable, but with convective modes having frequencies below 100 Hz, much lower than the kHz frequencies of the $f$- and $p$-modes.

### 3.2 Convective Instability due to Differential Rotation: The Solberg Criterion

Differentially rotating stars have one additional kind of convective (local) instability [17]. If the angular momentum per unit rest mass, $j = hu_\alpha \phi^\alpha$, decreases outward from the axis of symmetry, the star is unstable to perturbations that change the differential rotation law.

The criterion is easy to understand in a Newtonian context. Consider a ring of fluid in the star's equatorial plane that is displaced outward from $r$ to $r + \xi$, conserving angular momentum and mass. Again the displaced ring immediately adjusts its pressure to that of the surrounding star. If the ring's centripetal acceleration is larger that the net restoring force from gravity and the surrounding pressure gradient, it will continue to move outward. Now in the unperturbed star, the centripetal acceleration is equal to the restoring force. As in the discussion of convective instability, the displaced fluid element encounters the pressure gradient and gravitational field of the unperturbed star at its new position, and the restoring force is the restoring force on a fluid element at $r + \xi$ in the unperturbed star. Thus, if the displaced fluid ring has the same value of $v^2/r$ as the surrounding fluid it will be in equilibrium, and the star will be marginally stable. If a displaced fluid ring has larger $v^2/r$ than its surrounding fluid the star will be unstable.

The difference in acceleration for the background star is $\Delta_\star(v^2/r) = \xi^r \dfrac{d}{dr}(v^2/r)$, and stability then requires

$$\xi^r \frac{d}{dr}\left(\frac{v^2}{r}\right) - \Delta\frac{v^2}{r} > 0, \tag{27}$$

for $\xi^r > 0$.

Because $\Delta j = 0$ and $v(j, r) = j(r)/r$, we have

$$\Delta\frac{v^2}{r} = \Delta\frac{j^2}{r^3} = j^2\xi^r \frac{d}{dr}\frac{1}{r^3}, \tag{28}$$

while

$$\Delta_\star\frac{v^2}{r} = \xi^r \frac{d}{dr}\frac{j^2}{r^3}, \tag{29}$$

implying

$$\Delta_\star\frac{v^2}{r} - \Delta\frac{v^2}{r} = \xi^r \frac{1}{r^3}\frac{dj^2}{dr}; \tag{30}$$

and the star is stable only if $\dfrac{dj}{dr} > 0$ in the equatorial plane (for $j > 0$), or, equivalently, only if $\partial_\varpi(\varpi^2\Omega) > 0$.

Bardeen [13] gives a heuristic argument for a restricted version of this criterion, and a subsequent comprehensive and more precise treatment, including heat flow and viscosity, is due to Seguin [14]. Abramowicz [18] gives the relativistic version of the Newtonian argument summarized above; a presentation in [19] corrects some misprints and also relates the criterion to the sign of the canonical energy.

Here, for the relativistic case, we present Bardeen's simple argument. The relativistic angular momentum per unit rest mass is $j = \frac{\varepsilon + p}{\rho} u_\alpha \phi^\alpha$, with $\phi^\alpha$ the rotational Killing vector. The first law of thermodynamics for relativistic stars has the form

$$\delta M = \int_\Sigma \left( \frac{T}{u^t} \Delta dS + \frac{g}{u^t} \Delta dM_0 + \Omega \Delta dJ \right), \tag{31}$$

where $dM_0 = \rho dV$, $dJ = j dM_0$ and $g$ is the specific Gibbs free energy. If, in a homentropic, differentially rotating star, $j$ has an extremum as a function of radius in the the equatorial plane, then there are perturbations that conserve baryon number and that lower the energy of the system—for which $\delta M < 0$. The argument, for a homentropic star, is this: On opposite sides of the extremum, there are two rings, 1 and 2, with the same value of $j$ and with $\Omega_2 > \Omega_1$. A perturbation that transfers matter with baryon mass $\delta M_0$ from ring 2 to ring 1 then gives $\delta M = (\Omega_1 - \Omega_2) j\, \delta M_0 < 0$. That is, unless $j$ is a monotonic function, one can always find a perturbation with negative energy.

This is a simplest example of the turning-point criterion governing axisymmetric stability: A point of marginal stability along a sequence of circular orbits of a particle is a point at which $j$ is an extremum. The turning-point condition can be rephrased in terms of the particle's energy. For a particle of fixed rest mass, the difference in energy of adjacent orbits is related to the difference in its angular momentum by

$$\delta E = \Omega \delta J.$$

Then a point of marginal stability along a sequence of circular orbits of a particle of fixed baryon mass is a point at which its energy is an extremum.

## 4 Axisymmetric Instability and Turning Points

For spherical stars in Newtonian gravity, instability sets in when the matter becomes relativistic, when the adiabatic index $\Gamma_1$ (more precisely, its pressure-weighted average $\bar{\Gamma}_1$) reaches the value 4/3 characteristic of zero rest-mass particles. This can be seen from the Newtonian limit of the canonical energy,

$$E_c = \int \left\{ \frac{1}{2} \rho \dot{\xi}^2 + \frac{2}{r} p' \xi^2 + \frac{\Gamma_1 p}{2r^4} \left[ (r^2 \xi)' \right]^2 \right\} dV. \tag{32}$$

Choosing as initial data $\xi = r, \dot{\xi} = 0$, gives

$$E_c = \int \left(2rp' + \frac{9}{2}\Gamma_1 pr^2\right) dV = \frac{9}{2}\int \left(\Gamma_1 - \frac{4}{3}\right) p\, dV, \tag{33}$$

implying instability for $\bar{\Gamma}_1 < 4/3$. This shows only that $\Gamma_1 < 4/3$ is a sufficient condition for instability, but spherical Newtonian polytropes with $\Gamma_1 > 4/3$ are stable.

By, in effect, deriving the relativistic canonical energy,

$$E_c = \int_0^R \frac{1}{2} e^{\lambda+\nu} \left\{ \left[ \frac{4}{r} p' - \frac{p'^2}{\varepsilon + p} + 8\pi p(\varepsilon + p) \right] \xi^2 + \frac{e^{3\lambda-\nu}}{r^4} \Gamma_1 p \left[ (e^{-\nu} r^2 \xi)' \right]^2 \right\} r^2 dr, \tag{34}$$

Chandrasekhar [20, 21] showed that the stronger gravity of general relativity implies an earlier onset of instability: Even models with the stiffest equation of state must be unstable to collapse for some value of compactness $M/R < 9/8$, the value for the most compact uniform density model. The more stringent relativistic constraint on $\Gamma_1$ for a star to be stable against radial perturbations has the form

$$\Gamma_1 < \frac{4}{3} + K\frac{M}{R}, \tag{35}$$

where $K$ is positive and of order unity [20]. Because a gas of photons has $\Gamma_1 = 4/3$ and massive stars are radiation dominated, the instability can be important for stars with $M/R \ll 1$.

## 4.1 Turning Point Instability

The best-known instability result in general relativity is the statement that instability to collapse sets in at a point of maximum mass, along a sequence of spherical barotropic models. The configuration with maximum mass is called a turning point along the sequence, and it is also the configuration with maximum baryon mass. A similar result holds for uniformly rotating stars [22]: Instability to collapse is implied by a point of maximum mass and maximum baryon mass, along a sequence of uniformly rotating barotropic models with fixed angular momentum. As in the spherical case, stars with higher central density than that of the maximum-mass configuration are unstable. For rotating stars, however, the turning point is a sufficient but not a necessary condition for instability: The onset of instability is at a configuration with slightly lower central density (for fixed angular momentum) than that of the maximum-mass star. A formal symmetry in the way baryon mass and angular momentum occur in the first law implies that the line of turning points is also the line of extrema of angular momentum along sequences of fixed baryon mass.

For dynamical oscillations of neutron stars the adiabatic index does not coincide with the polytropic index, $\Gamma_1 \neq \Gamma := \dfrac{d\log p(r)/dr}{d\log \rho(r)/dr}$, and the turning point criterion implies *secular* instability—an instability whose growth time is long compared to the typical dynamical time of stellar oscillations. For spherical stars, the turning-point instability proceeds on a time scale slow enough to accommodate the nuclear reactions and energy transfer that accompany the change to a nearby equilibrium. For rotating stars, the time scale must also be long enough to accommodate a transfer of angular momentum from one fluid ring to another. That is, the growth rate of the instability is limited by the time required for viscosity to redistribute the star's angular momentum. For neutron stars, this is expected to be short, probably comparable to the spin-up time following a glitch, and certainly short compared to the lifetime of a pulsar or an accreting neutron star. For this reason, it is the secular instability, that sets the upper and lower limits on the mass of spherical and uniformly rotating neutron stars.

One can easily understand why the instability sets in at an extremum of the mass by looking at a radial mode of oscillation of a nonrotating star with an equation of state $p = p(\rho), \varepsilon = \varepsilon(\rho)$. Along the sequence of spherical equilibria, a radial mode changes from stability to instability when its frequency $\sigma$ changes from real to imaginary, with $\sigma = 0$ at the point of marginal stability. Now a zero-frequency mode is just a time-independent solution to the linearized Einstein-Euler equations - a perturbation from one equilibrium configuration to a nearby equilibrium with the same baryon number. From the first law of thermodynamics (31), a perturbation that keeps the star in equilibrium satisfies

$$\delta M = \frac{g}{u^t}\delta M_0. \tag{36}$$

The relation implies that, for a zero frequency perturbation involving no change in baryon number, the change $\delta M$ in mass must vanish. This is the requirement that the mass is an extremum along the sequence of equilibria. Models on the *high-density* side of the maximum-mass instability point are unstable: Because the turning point is a star with maximum baryon number as well as maximum mass, there are models on opposite sides of the turning point with the same baryon number. Because $g/u^t$ is a decreasing function of central density, the model on the high-density side of the turning point has greater mass than the corresponding model with smaller central density.

At the minimum mass, it is the *low-density* side that is unstable: Because the mass is a minimum, the model on the low-density side of the turning point has greater mass than the corresponding model with the same baryon number on the high-density side.

The precise statement of the turning-point criterion is the following result:

**Theorem 1** (Friedman et al. [22]): *Consider a continuous sequence of uniformly rotating stellar models based on an equation of state of the form* $p = p(\varepsilon)$. *Let* $\lambda$ *be the sequence parameter and denote the derivative* $d/d\lambda$ *along the sequence by* $(\dot{\ })$.

(i) Suppose that the total angular momentum is constant along the sequence and that there is a point $\lambda_0$ where $\dot{M} = 0$ and where $\mathscr{E} > 0$, $(\dot{\mathscr{E}}\dot{M}) \neq 0$. Then the part of the sequence for which $\dot{\mathscr{E}}\dot{M} > 0$ is unstable for $\lambda$ near $\lambda_0$.
(ii) Suppose that the total baryon mass $M_0$ is constant along the sequence and that there is a point $\lambda_0$ where $\dot{M} = 0$ and where $\Omega > 0$, $(\dot{\Omega}\dot{M}) \neq 0$. Then the part of the sequence for which $\dot{\Omega}\dot{M} > 0$ is unstable for $\lambda$ near $\lambda_0$.

In form (ii) of the theorem, the first law implies that the turning point is an extremum of angular momentum $J$ along a sequence of constant rest mass. Friedman et al. [22] point out the symmetry between $M_0$ and $J$ that implies this maximum-$J$ form of the theorem, and Cook et al. [23] first use the theorem in this form. For rotating stars, the turning point criterion is a sufficient condition for secular instability to collapse. In general, however, collapse can be expected to involve differential rotation, and the turning point identifies only nearby uniformly rotating configurations with lower energy. Rotating stars are therefore likely to be secularly unstable to collapse at densities slightly lower than the turning point density. The onset of secular instability to collapse is at or before the onset of dynamical instability along a sequence of uniformly rotating stars of fixed angular momentum, and recent work by Takami et al. [24] appears to show that rapidly rotating stars can also be dynamically unstable to collapse just prior to the turning point.

As illustrated in Fig. 2, they find a dynamical instability line that coincides with the turning-point line for spherical stars and that, for rapid uniform rotation, has a central density about 5 % below that of the turning point. This result somewhat overstates the difference between the two lines, because it ignores the difference $\Gamma - \Gamma_1$ between the indices governing dynamical oscillations and the equilibrium equation of state. The actual dynamical instability line begins at a spherical star with higher density than the marginally unstable turning point star and probably crosses the turning-point line to lower density at some angular velocity less than $\Omega_K$.

The greater significance of the Takami et al. result, however, is that stars along the line determined by using the equilibrium equation of state are guaranteed to be *secularly* unstable, because the diagnosed instability guarantees that the configurations have lower energy than equilibria with the same baryon mass and angular momentum. This means that the line of secular instability runs through rapidly rotating configurations with central densities more than 5 % below those along the line of turning points.

## 5 Nonaxisymmetric Instabilities

Rapidly rotating stars and drops of water are unstable to a bar mode that leads to fission in the water droplets; and a similar nonaxisymmetric instability is likely

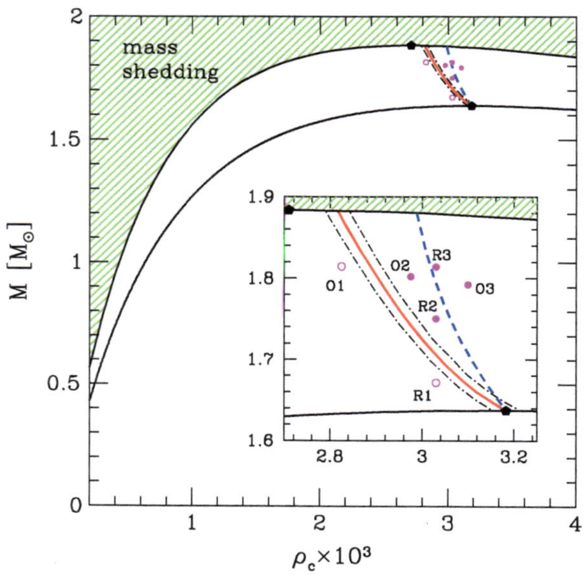


**Fig. 2** Stability lines in a $(\rho_c, M)$ diagram. The two *solid black lines* mark sequences with either zero (*lower line*) or mass-shedding angular momentum (*upper line*), with the filled symbols marking the corresponding maximum masses. The *solid grey line* is the neutral-stability line, "thickened" by the error bar (*dot-dashed lines*). The *grey dashed line* is instead the turning-point criterion for secular stability. Marked with empty or filled *circles* are representative models with constant angular velocity O1, O2, O3, or constant initial central rest-mass density R1, R2, R3. (Figure from Takami et al. [24]. Reproduced by permission of John Wiley and Sons.)

to be the reason most stars in the universe are in close binary systems. Galactic disks are unstable to nonaxisymmetric perturbations that lead to bars and to spiral structure. And a related instability of a variety of nonaxisymmetric modes, driven by gravitational waves, the Chandrasekhar-Friedman-Schutz (CFS) instability [8, 25]), may limit the rotation of neutron stars. The existence of the CFS-instability in rotating stars was first found by Chandrasekhar [25] in the case of the $l = 2$ mode in uniformly rotating, uniform density Maclaurin spheroids. Subsequently, Friedman and Schutz [8, 26] showed that this instability also appears in compressible stars and that all rotating, self-gravitating perfect fluid configurations are generically unstable to the emission of gravitational waves. We have seen that, along a sequence of stellar models, a mode changes from stable to unstable when its frequency vanishes. The generic-instability result means that zero-frequency nonaxisymmetric modes of rotating perfect-fluid stellar models are marginally stable.

Whereas axisymmetric instability to collapse sets in at points that are nearly independent of the magnitude of viscosity or the strength of gravitational waves, the opposite is true for the nonaxisymmetric case. Gravitational radiation drives a nonaxisymmetric instability that, if no other dissipation is present, makes *every* rotating star unstable. Viscosity can drive a nonaxisymmetric instability in rapidly rotating stars for which gravitational radiation is negligible. For slowly rotating stars, however (and nearly all neutron stars rotate slowly compared to the Kepler limit), viscosity simply damps out the gravitational-wave driven instability. That is, for slow rotation, we will see that the timescale of the CFS instability is longer than the timescale for viscous damping. On the other hand, for rapidly rotating neutron stars, the instability's timescale may be short enough that it limits the rotation of young neutron stars and of old neutron stars spun up by accretion.

This review begins with a discussion of the CFS instability for perfect-fluid models and then outlines the work that has been done to decide whether the instability is present in young neutron stars and in old neutron stars spun up by accretion. For very rapid rotation and for slower but highly differential rotation, nonaxisymmetric modes can be *dynamically unstable*, with growth times comparable to the period of a star's fundamental modes, and the review ends with a brief discussion of these related dynamical instabilities.

To understand the way the CFS instability arises, consider first a stable spherical star. All its modes have positive energy, and the sign of a mode's angular momentum $J_c$ about an axis depends on whether the mode moves clockwise or counterclockwise around the star. That is, a mode with angular and time dependence of the form $\cos(m\phi + \omega_0 t)e^{-\alpha_0 t}$, has positive angular momentum $J_c$ about the $z$-axis if and only if the mode moves in a positive direction: The pattern speed, $-\dfrac{\omega_0}{m}$, is positive. Because the wave moves in a positive direction relative to an observer at infinity, the star radiates positive angular momentum to infinity, and the mode is damped. Similarly, a mode with negative angular momentum has negative pattern speed, $-\dfrac{\omega_0}{m} < 0$, and radiates negative angular momentum to infinity; and the mode is again damped.

Now consider a slowly rotating star with a backward-moving mode, a mode that moves in a direction opposite to the star's rotation. Because a short-wavelength fluid mode (a mode with a Newtonian counterpart, not a $w$-mode) is essentially a wave in the fluid, the wave moves with nearly the same speed relative to a rotating observer that it had in the spherical star. That means that an observer at infinity sees the mode dragged forward by the fluid. The real part $\omega_r$ of the frequency seen in a rotating frame is the frequency associated with the $\phi$ coordinate $\phi_r = \phi - \Omega t$ of a rotating observer. Then

$$m\phi + \omega t = m\phi_r + (\omega + m\Omega)t = m\phi_r + \omega_r t,$$

implying that the frequency seen by the rotating observer is

$$\omega_r = \omega + m\Omega. \tag{37}$$

For a slowly rotating star, $\omega_r \approx \omega_0$. When the star rotates with an angular velocity greater than $|\omega_r/m|$, the backward-going mode is dragged *forward* relative to an observer at infinity, and $\omega_r$ and $\omega$ have opposite signs:

$$\omega_r \omega < 0. \tag{38}$$

Because the pattern speed is now positive, the mode radiates positive angular momentum to infinity. But the canonical angular momentum is still negative, because the mode is moving backward relative to the fluid: The angular momentum of the perturbed star is smaller than the angular momentum of the star without the backward-going mode. As the star radiates positive angular momentum to infinity, $J_c$ becomes increasingly negative, implying that the amplitude of the mode grows in time: *Gravitational radiation now drives the mode instead of damping it.*

For large $m$ or small $\omega_0$, the pattern speed will be positive when $\Omega \approx |\omega_0/m|$. This relation suggests two classes of modes that are unstable for arbitrarily slow rotation: Backward-moving modes with large values of $m$ and modes with any $m$ whose frequency is zero in a spherical star. Both classes of perturbations exist. The usual $p$-modes and $g$-modes have finite frequencies for a spherical star and are unstable for $\Omega \gtrsim |\omega_0|/m$; and $r$-modes, which have zero frequency for a non-rotating barotropic star, are unstable for all values of $m$ and $\Omega$ (that is, those $r$-modes are unstable that are backward-moving in the rotating frame of a slowly rotating star).

We have so far not mentioned the canonical energy, but our key criterion for the onset of instability is a negative $E_c$. If we ignore the imaginary part of the frequency, the change in the sign of $E_c$ follows immediately from the relation (19), $J_c = -\frac{m}{\omega} E_c$. To take the imaginary part $\mathrm{Im}\sigma = \alpha \neq 0$ of the frequency into account, we need to use the fact that energy is lost at a rate $\dot{E}_c \propto \dddot{Q}^2 \propto \sigma^6$ for quadrupole radiation, with $\dot{E}_c$ proportional to higher powers of $\sigma$ for radiation into higher multipoles. Because $E_c$ is quadratic in the perturbation, it is proportional to $e^{-2\alpha t}$, implying $\alpha \propto \sigma^6$. Thus $\alpha/\sigma \to 0$ as $\sigma \to 0$, implying that for a normal mode $E_c$ changes sign when $\omega$ changes sign.

Although the argument we have given so far is heuristic, there is a precise form of the statement that a stable, backward-moving mode becomes unstable when it is dragged forward relative to an inertial observer (see Friedman and Schultz [26] and Friedman and Stergioulas [19]).

**Theorem 2** *Consider an outgoing mode* $(h_{\alpha\beta}(\lambda), \xi^\alpha(\lambda))$, *that varies smoothly along a family of uniformly rotating perfect-fluid equilibria, labeled by* $\lambda$. *Assume that it has* $t$ *and* $\phi$ *dependence of the form* $e^{i(m\phi+\sigma t)}$, *that* $\omega = \mathrm{Re}\{\sigma\}$ *satisfies* $\omega/m + \Omega > 0$ *for all* $\lambda$, *and that the sign of* $\omega/m$ *is positive for* $\lambda < \lambda_0$ *and negative for* $\lambda > \lambda_0$. *Then in a neighborhood of* $\lambda_0$, $\alpha := \mathrm{Im}\{\sigma\} \leq 0$; *and if the mode has at least one nonzero asymptotic multipole moment with* $l \geq 2$ *at future null infinity, the mode is unstable* $(\alpha < 0)$ *for* $\lambda > \lambda_0$.

A corresponding result that does not rely on existence or completeness of normal modes is the statement that one can always choose canonical initial data to make $E_c < 0$ [8, 19].

The growth time $\tau_{GR}$ of the instability of a perfect fluid star is governed by the rate $\left.\dfrac{dE}{dt}\right|_{\mathrm{GR}}$ at which energy is radiated in gravitational waves:

$$\frac{1}{\tau_{\rm GR}} = -\frac{1}{2E_c}\left.\frac{dE}{dt}\right|_{\rm GR}, \tag{39}$$

where (Thorne [27])

$$\frac{dE}{dt} = -\sum_{l\geq m} \omega_i^{2l+2} N_l \left(|\delta D_{lm}|^2 + |\delta J_{lm}|^2\right). \tag{40}$$

Here $D_{lm}$ and $J_{lm}$ are the asymptotically defined mass and current multipole moments of the perturbation and $N_l = \dfrac{4\pi(l+1)(l+2)}{l(l-1)[(2l+1)!!]^2}$ is, for low $l$, a constant of order unity. In the Newtonian limit,

$$\delta D_{lm} = \int \delta\rho\, r^l Y_{lm} d^3x. \tag{41}$$

For a star to be unstable, the growth time $\tau_{GR}$ must be shorter than the viscous damping time $\tau_{\rm viscosity}$ of the mode, and the implications of this are discussed below. In particular because the growth time is longer for larger $l$, only low multipoles can be unstable in neutron stars.

**Modes with Polar and Axial Parity** The spherical symmetry of a nonrotating star and its spacetime implies that perturbations can be labeled by fixed values $l, m$ labeling an angular harmonic: The quantities $h_{\alpha\beta}$, $\xi^\alpha$, $\delta\rho$, $\delta\varepsilon$, $\delta p$, $\delta s$ that describe a perturbation are all proportional to scalar, vector and tensor spherical harmonics constructed from $Y_{lm}$, and perturbations with different $l, m$ values decouple. Similarly, because spherical stars are invariant under parity (a map of each point $P$ of spacetime to the diametrically opposite point on the symmetry sphere through $P$), perturbations with different parity decouple, the parity of a perturbation is conserved, and normal modes have definite parity. Perturbations associated with an $l, m$ angular harmonic are said to have *polar* parity if they have the same parity as the function $Y_{lm}$, $(-1)^l$. Perturbations having parity $(-1)^{l+1}$, opposite to that of $Y_{lm}$ have axial parity. In the Newtonian literature, modes of a rotating star that are continuously related to polar modes of a spherical star are commonly called *spheroidal*; while modes whose spherical limit is axial are called *toroidal*.

Every rotational scalar—$\varepsilon$, $p$, $\rho$, and the components in the $t$-$r$ subspace of the perturbed metric $h_{\alpha\beta}$ and the perturbed fluid velocity $\delta u^\alpha$—can be expressed as a superposition of scalar spherical harmonics $Y_{\ell m}$. As a result, modes of spherical stars that involve changes in any scalar are polar. On the other hand, the angular components of velocity perturbations can have either polar parity, with

$$\delta v = f(r)\nabla Y_{lm} \tag{42}$$

or axial parity, with Newtonian form

$$\delta v = f(r)\mathbf{r} \times \nabla Y_{lm}, \tag{43}$$

and the relativistic form $\delta u^\alpha \propto \varepsilon^{\alpha\beta\gamma\delta}\nabla_\beta t \nabla_\gamma r \nabla_\delta Y_{lm}$.

There are two families of polar modes of perfect-fluid Newtonian stars, $p$-modes (pressure modes) and $g$-modes (gravity modes). For short wavelengths, the $p$-modes are sound waves, with pressure providing the restoring force and frequencies

$$\sigma = c_s k, \tag{44}$$

where $k$ is the wavenumber and $c_s$ is the speed of sound. The short-wavelength $g$-modes are modes whose restoring force is buoyancy, and their frequencies are proportional to the Brunt-Väisälä frequency, related to the difference between $dp/d\varepsilon$ in the star and $c_s^2 = \partial p(\varepsilon, s)/\partial \varepsilon$. The fundamental modes of oscillation of a star ($f$-modes), with no radial nodes, can be regarded as a bridge between $g$-modes and $p$-modes.

Because axial perturbations of a spherical star involve no change in density or pressure, there is no restoring force in the linearized Euler equation, and the linear perturbation is a time-independent velocity field—a zero-frequency mode.[3] In a rotating star, the axial modes acquire a nonzero frequency proportional to the star's angular velocity $\Omega$, a frequency whose Newtonian limit has the simple form

$$\sigma = -\frac{(l-1)(l+2)}{l(l+1)} m\Omega, \tag{45}$$

where the harmonic time and angular dependence of the mode is $e^{i(m\phi - \sigma t)}$. These modes are called $r$-modes, their name derived from the Rossby waves of oceans and planetary atmospheres. The term $r$-mode can be usefully regarded as a mnemonic for a *rotationally restored* mode. Eq. (37) implies that the $r$-mode associated with every nonaxisymmetric multipole obeys the instability condition for every value of $\Omega$: It is forward moving in an inertial frame and backwards moving relative to a rotating observer:

$$\sigma_r = \frac{2m}{l(l+1)} \Omega, \tag{46}$$

with sign opposite to that of $\sigma$. Because the rate at which energy is radiated is greatest for the $r$-mode with $l = m = 2$, that is the mode whose instability grows most quickly and which determines whether an axial-parity instability can outpace viscous damping.

The instability of low-multipole $r$-modes for arbitrarily slow rotation is strikingly different from the behavior of the low-multipole $f$- and $p$-modes, which are unstable only for large values of $\Omega$. The reason is that the frequencies of $f$- and $p$-modes are high, and, from Eq. (38), a correspondingly high angular velocity is needed before a mode that moves backward relative to the star is dragged forward relative to an inertial observer at infinity. Of the polar modes, $f$-modes with $l = m$ have the fastest

[3] Axial perturbations of the spacetime of a spherical star include both axial perturbations of the fluid and gravitational waves with axial parity. The axial-parity waves do not couple to the fluid perturbation, which is stationary in the sense that $\partial_t \delta u_\alpha = 0$.

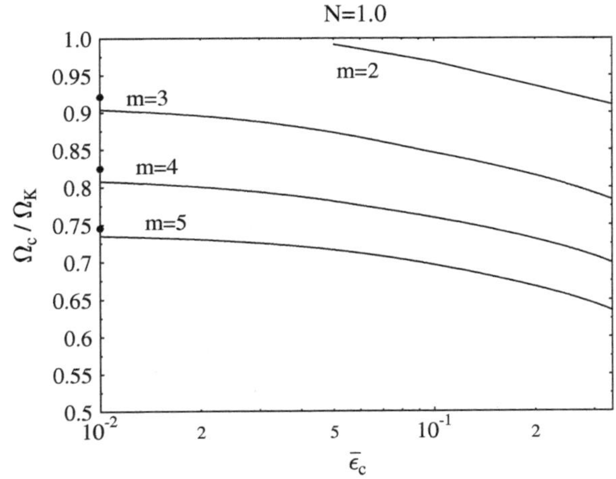


**Fig. 3** Critical angular velocity $\Omega/\Omega_K$ vs. the dimensionless central energy density $\bar{\varepsilon}_c$ for the $m = 2, 3, 4$ and 5 neutral modes of $N = 1.0$ polytropes. The filled *circles* on the vertical axis are the Newtonian values of the neutral points for each mode. (Reproduced from [28].)

growth rates; their instability points for uniformly rotating relativistic stars, found by Stergioulas and Friedman [28], are shown in Fig. 3. (Work on these stability points of relativistic stars is also reported in [29–32].)

The figure shows that, for uniform rotation, the $l = m = 2$ $f$-mode is unstable only for stars with relatively high central density or high mass. For tabulated EOSs, this practically applies to all neutron stars with masses greater than 1.3 $M_\odot$ and $T/|W| > 0.06$ [33]. Because neutron stars rotate differentially at birth, the $l = 2$ mode, as well as higher modes, could be initially unstable for a larger range of parameters.

**Implications of the Instability** The nonaxisymmetric instability may limit the rotation of nascent neutron stars and of old neutron stars spun up by accretion; and the gravitational waves emitted by unstable modes may be observable by gravitational wave detectors. Whether a limit on spin is in fact enforced depends on whether the instability of perfect-fluid models implies an instability of neutron stars; and the observability of gravitational waves also requires a minimum amplitude and persistence of an unstable mode. We briefly review observational support for an instability-enforced upper limit on spin and then turn to the open theoretical issues.

Evidence for an upper limit on neutron-star spin smaller than the Keplerian frequency $\Omega_K$ comes from nearly 30 years of observations of neutron stars with millisecond periods, seen as pulsars and as X-ray binaries. The observations reveal rotational frequencies ranging upward to 716 Hz and densely populating a range of frequencies below that. Selection biases against detection of the fastest millisecond radio pulsars have made conclusions about an upper limit on spin uncertain, but Chakrabarty argues that the class of sources whose pulses are seen in nuclear bursts

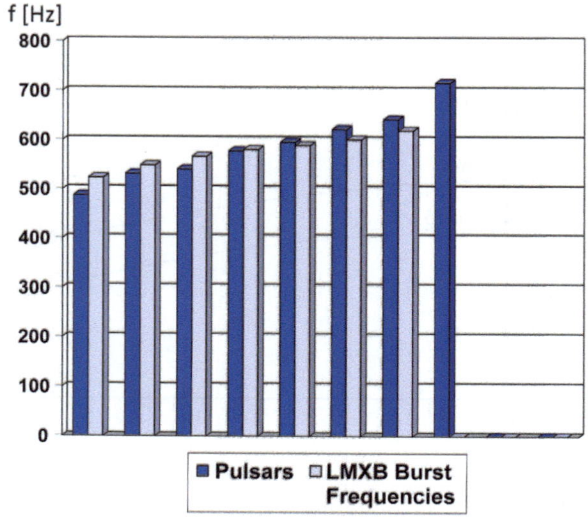


**Fig. 4** Highest observed neutron-star spin frequencies

(nuclear powered accreting millisecond X-ray pulsars) constitute a sample without significant bias [34]. Their distribution of spins, together with the spins of other millisecond pulsars, is shown in Fig. 4.

A magnetic field of order $10^8$ G can limit the spin of an accreting millisecond pulsar. Because matter within the magnetosphere corotates with the star, only matter that accretes from outside the magnetosphere can spin up the star, leading to an equilibrium period given approximately by Ghosh and Lamb [35][4]

$$P_{\rm eq} \sim 2 \times 10^{-3}{\rm s} \left(\frac{B}{10^8{\rm G}}\right)^{6/7} \left(\frac{\dot{M}}{10^{-10} M_\odot {\rm yr}^{-1}}\right)^{-3/7} . \tag{47}$$

Because this period depends on the magnetic field, a sharp cutoff in the frequency of accreting stars is not an obvious prediction of magnetically limited spins. For a magnetically set maximum rotation rate of order 700–800 Hz the range of magnetic fields would need to have a corresponding minimum cutoff value of about $10^8$ G; and the highest observed spin rates should correspond to the lowest magnetic fields. The required cutoff and a fairly narrow range of observed frequencies has made gravitational-wave limited spin a competitive possibility for accreting neutron stars. Arguments for and against this based on available observations are given by White and Zhang [37] and by Patruno et al. [38], respectively.

[4] Shapiro and Teukolsky [36] give a clear, simplified version, and Eq. (47) is their Eq. (15.2.22), with $M = 1.4\, M_\odot$, $R = 10$ km, and a ratio $\omega_s$ of the angular velocity to $\Omega_K$ at the inner edge of the disk set to 1.

Under what circumstances the CFS instability could limit the spin of recycled pulsars has now been studied in a large number of papers. References to this work can be found in the treatment in FS, on which the present review is based and in comprehensive earlier discussions by Stergioulas [39], by Andersson and Kokkotas [40], and by Kokkotas and Ruoff [41], while briefer reviews of more recent work are given in [42, 43]. References in the present review are generally limited to initial work and to a late paper that contains intervening references.

Whether the instability survives the complex physics of a real neutron star has been the focus of most recent work, but it remains an open question. Studies have focused on:

- Dissipation from bulk and shear viscosity and mutual friction in a superfluid interior;
- magnetic field wind-up;
- nonlinear evolution and the saturation amplitude; and
- the possibility that a continuous spectrum replaces $r$-modes in relativistic stars.

We discuss these in turn and then summarize their implications for nascent, rapidly rotating stars and for old stars spun up by accretions.

**Viscosity** When viscosity is included, the growth-time or damping time $\tau$ of an oscillation has the form

$$\frac{1}{\tau} = \frac{1}{\tau_{GR}} + \frac{1}{\tau_b} + \frac{1}{\tau_s}, \tag{48}$$

with $\tau_b$ and $\tau_s$ the damping times due to bulk and shear viscosity. Bulk viscosity is large at high temperatures, shear viscosity at low temperatures. This leaves a window of opportunity in which a star with large enough angular velocity can be unstable. The window for the $l = m = 2$ $r$-mode is shown in Fig. 5, for a representative computation of viscosity. The highest solid curves on left and right mark the critical angular velocity $\Omega_c$ above which the $l = m = 2$ $r$-mode is unstable. The curves on the left show the effect of shear viscosity at low temperature, allowing instability when $\Omega < \Omega_K$ only for $T > 10^6$ K; the curve on the right shows the corresponding effect of bulk viscosity, cutting off the instability at temperatures above about $4 \times 10^{10}$ K. There is substantial uncertainty in the positions of both of these curves.

Bulk viscosity arises from nuclear reactions driven by the changing density of an oscillating fluid element, with neutrons decaying, $n \to p + e + \bar{\nu}_e$, as the fluid element expands and protons capturing electrons, $p + e \to n + \nu_e$, as it contracts. The neutrinos leave the star, draining energy from the mode. The rates of these *URCA* reactions increase rapidly with temperature and are fast enough to be important above about $10^9$ K, with an expected damping time $\tau_b$ given by

$$\frac{1}{\tau_{\rm b}} = \frac{1}{2E_c} \int \zeta(\delta\theta)^2 d^3x, \tag{49}$$

where $\theta = \nabla_\alpha u^\alpha$ is the divergence of the fluid velocity and the coefficient of bulk viscosity $\zeta$ is given by [45]

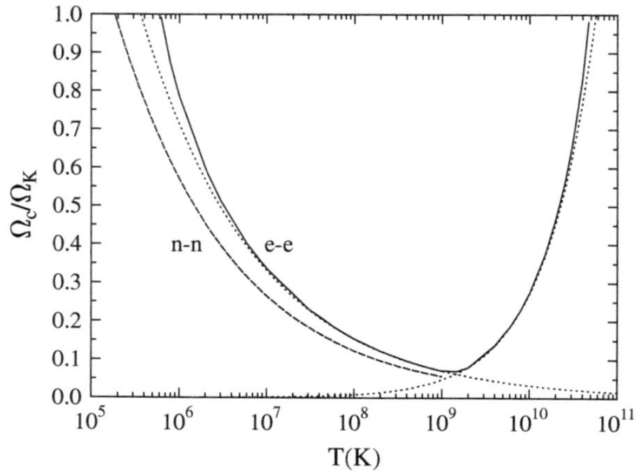


**Fig. 5** Critical angular velocity for the onset of the $r$-mode instability as a function of temperature (for a 1.5 $M_\odot$ neutron star model). The *solid line* corresponds to the $O(\Omega^2)$ result using electron-electron shear viscosity, and modified URCA bulk viscosity. The *dashed line* corresponds to the case of neutron-neutron shear viscosity. *Dotted lines* are $O(\Omega)$ approximations. (Reproduced from [44].)

$$\zeta = 6 \times 10^{25} \rho_{15}^2 T_9^6 \left(\frac{\omega_r}{1\text{Hz}}\right)^{-2} \ \text{g cm}^{-1}\ \text{s}^{-1}, \tag{50}$$

where $T_9 = T/(10^9\,\text{K})$. With these values, bulk viscosity suppresses the instability in all modes above a few times $10^{10}$ K (see also Ipser and Lindblom [46, 47] and Yoshida and Eriguchi [48]).[5]

These equations and Fig. 5 assume that only *modified URCA* reactions can occur, that the URCA reactions require a collision to conserve four-momentum, and this will be true when the proton fraction $x_p$ is less than about 1/9. Should the equation of state be unexpectedly soft (and if the mass is large enough), direct URCA reactions would be allowed, suppressing the instability for uniformly rotating stars at roughly $10^9$ K [52]. A soft equation of state would also more likely lead to stars with hyperons in their core with an additional set of nuclear reactions that dissipate energy and increase the bulk viscosity [53–57] or quarks [58–62]. However, the observation of a 2 $M_\odot$ pulsar makes the existence of hyperons or quarks in the core of 1.4 $M_\odot$ neutron stars less probable.

In contrast to bulk viscosity, shear viscosity increases as the temperature drops. In terms of the shear tensor $\sigma_{\alpha\beta} = (\delta_\alpha^\gamma + u_\alpha u^\gamma)(\delta_\beta^\delta + u_\beta u^\delta)(\nabla_\gamma u_\delta + \nabla_\delta u_\gamma - \frac{2}{3} g_{\gamma\delta} \nabla_\varepsilon u^\varepsilon)$, the damping time is given by

[5] At temperatures above roughly $10^{10}$ K, another complication appears: neutrino absorption increases with increasing temperature [49, 50], and the modified URCA bulk viscosity no longer rises, but is reduced by an order of magnitude between $10^{10}$ K and $10^{11}$ K, allowing the instability to operate in very hot proto-neutron stars [51].

$$\frac{1}{\tau_s} = \frac{1}{E_c} \int \eta \delta\sigma^{\alpha\beta} \delta\sigma_{\alpha\beta}\, d^3x, \tag{51}$$

where $\eta$ is the coefficient of shear viscosity. For nascent neutron stars hotter than the superfluid transition temperature (about $10^9$ K), a first estimate of the neutron-neutron shear viscosity coefficient is [63]

$$\eta_n = 2 \times 10^{18} \rho_{15}^{9/4} T_9^{-2} \text{ g cm}^{-1} \text{ s}^{-1}, \tag{52}$$

where $\rho_{15} = \rho/(10^{15}\text{ g cm}^{-3})$. Below the superfluid transition temperature, electron-electron scattering determines the shear viscosity in the superfluid core, giving [64]

$$\eta_e = 2.5 \times 10^{18} \left(\frac{x_p}{0.1}\rho_{15}\right)^2 T_9^{-5/3} \text{ g cm}^{-1} \text{ s}^{-1}. \tag{53}$$

Shear viscosity may be greatly enhanced after formation of the crust in a boundary layer (Ekman layer) between crust and core [65–69]. The enhancement depends on the extent to which the core participates in the oscillation, parametrized by the slippage at the boundary. The uncertainty in this slippage appears to be the greatest current uncertainty in dissipation of the mode by shear viscosity, and it significantly affects the critical angular velocity of the $r$-mode instability in accreting neutron stars.

For $f$-modes, the part of the instability window in Fig. 5 to the left of $10^9$ K is thought to be removed by another dissipative mechanism that comes into play below the superfluid transition temperature. Called mutual friction, it arises from the scattering of electrons off magnetized neutron vortices. Work by Lindblom and Mendell [70] shows that mutual friction in the superfluid core completely suppresses $f$- and $p$-mode instabilities below the transition temperature. For the $r$-mode instability, subsequent work by the same authors [71] finds that the mutual friction is much smaller, with a damping time of order $10^4$ s, too long to be important.

In a recent paper, Gaertig et al. [32] point out the possibility of an interaction between vortices and quantized flux tubes that would result in a much smaller value for the mutual friction. They argue that the resulting uncertainty is great enough that shear viscosity could be the dominant dissipative mechanism for $f$-modes as well as $r$-modes.

**Magnetic Field Windup** At second order in the perturbation, the nonlinear evolution of an unstable mode includes an axisymmetric part that describes a growing differential rotation. Because differential rotation will wind up magnetic field lines, the mode's energy could be transferred to the star's magnetic field [72–76]. Again there is large uncertainty about the strength of a toroidal magnetic field that will be generated by the differential rotation, what magnetic instabilities will arise, and what the effective dissipation will be. Apart from the studies cited here (all of which deal with $r$-modes) nearly all the remaining work on the evolution of unstable modes ignores magnetic fields.

**Relativistic *r*-Modes and a Possible Continuous Spectrum** Relativistic $r$-modes have been computed by a number of authors [41, 77–88]. Where the Newtonian approximation has purely axial $l = m$ $r$-modes for barotropic stars at lowest order in $\Omega$, in the full theory all rotationally restored modes include a polar part. The change in the structure of the computed $r$-modes are small, but that may not be the end of the story.

For non-barotropic stars Kojima found a single second-order eigenvalue equation for the frequency, to lowest nonvanishing order in $\Omega$. The coefficient of the highest derivative term in that equation vanishes at some value of the radial coordinate $r$, for typical candidate neutron-star equations of state, and that singular behavior gives a continuous spectrum. Lockitch et al. [83] consider the question of the continuous spectrum and the existence of $r$-modes in some detail. They argue that the singularity in the Kojima equation is an artifact of the slow-rotation approximation and is not present if one includes terms of order $\Omega^2$. Their work is a strong argument for the existence of $r$-modes in non-barotropic models.

Showing the existence of the mode, however, does not decide the question of whether a continuous spectrum is also present or whether the existence of a continuous or nearly continuous spectrum significantly alters the evolution of an initial perturbation.

**Nonlinear Evolution** Linear perturbation theory is valid only for small-amplitude oscillations; as the amplitude of an unstable mode grows, couplings to other modes become increasingly important, and the mode ultimately reaches a saturation amplitude or is disrupted, losing coherence. The first nonlinear studies of the $r$-mode instability involved fully nonlinear 3+1 evolutions by Stergioulas and Font [89], in which the $r$-mode was set at a large initial amplitude or Newtonian evolutions by Lindblom et al. [90, 91] in which the $r$-mode was driven to large amplitude by an artificially large gravitational-radiation reaction term. On a few tens of dynamical timescales, saturation was seen only at an amplitude of order unity. Subsequently, simulations on longer timescales showed a coupling to daughter modes [92, 93], suggesting that the actual saturation amplitude of the $r$-mode is smaller than the amplitude at which gravitational radiation reaction was switched off in the short-timescale simulations.

The grid resolution of 3+1 simulations, however, is currently too low to see couplings to short-wavelength modes, and they cannot run for a time long enough to see the growth from a realistic radiation reaction term. The alternative is to examine the nonlinear evolution in the context of higher-order perturbation theory. To do this, the Cornell group (initially with S. Morsink) [94–96] constructed a second-order perturbation theory for rotating Newtonian stars, and then used the formalism to study the nonlinear evolution of an unstable $r$-mode. Their series of papers leaves little doubt that nonlinear couplings sharply limit the amplitude of an unstable $r$-mode, with a possible range of $10^{-1}$–$10^{-5}$ (see Bondarescu et al. [97, 98] and references therein).

The nonlinear development of the $f$-mode instability has been modeled in three-dimensional, hydrodynamical simulations (in a Newtonian framework) by Ou et al. [99] and by Shibata and Karino [100], essentially confirming previous approximate results obtained by Lai and Shapiro [49]. Kastaun et al. [101] report an initial nonlinear study of $f$-modes in general relativity. In the framework of a 3+1 simulation in a Cowling approximation (a fixed background metric of the unperturbed rotating star), they find limits on the amplitude of less than 0.1, set by wave-breaking and by coupling to inertial modes. This can be regarded as an upper limit on the amplitude, with second-order perturbative computations still to be done.

**Instability Scenarios in Nascent Neutron Stars and in Old Accreting Stars** Both $r$-modes and $f$-modes may be unstable in nascent neutron stars that are rapidly rotating at birth. Recent work on $f$-modes in relativistic models [32, 102] finds growth times substantially shorter than previously computed Newtonian values. In a particular model (where the $l = m = 2$ mode becomes unstable only very near the mass-shedding limit), the $l = m = 3$ and $l = m = 4$ $f$-modes have growth times of $10^3$–$10^5$ s for $\Omega$ near $\Omega_K$. In a typical scenario, a star with rotation near the Kepler limit becomes unstable within a minute of formation, when the temperature has dropped below $10^{11}$ K. As the temperature drops further, the instability grows to saturation amplitude in days or weeks. Loss of angular momentum to gravitational waves spins down the star until the critical angular velocity is reached below which the star is stable, at or before the time at which the core becomes a superfluid. The $l = m = 3$ mode (or the $l = m = 2$ mode in models with different masses or equations of state than the one studied above) could be a source of observable gravitational waves for supernovae in or near the Galaxy (but with an uncertain event rate).

The time over which the instability is active depends on the saturation amplitude, the cooling rate, and the superfluid transition temperature, and all of these have large uncertainties. The time at which a superfluid transition occurs could be shorter than a year, but recent analyses of the cooling of a neutron star in Cassiopeia [103, 104] suggest a superfluid transition time for that star of order 100 years.

The scenario for the $l = m = 2$ $r$-mode instability of a nascent star is similar. The $r$-mode instability itself was pointed out by Andersson [78], with a mode-independent proof for relativistic stars given by Friedman and Morsink [105]. First computations of the growth and evolution were reported by Lindblom et al. [106] and by Andersson et al. [107], with effects of a crust discussed by Lindblom et al. [66]. Intervening work is referred to by Bondarescu et al. [98]; the simulations reported by Bondarescu et al. include nonlinear couplings that saturate the amplitude and the alternative possibilities for viscosity that we have discussed above. The $r$-mode's saturation amplitude is likely to be lower than that of the $f$-modes, and it is likely to persist longer because of its low mutual friction.

As mentioned above, the $r$-mode instability of neutron stars spun up by accretion has been more intensively studied in connection with the observed spins of LMXBs. Papaloizou and Pringle [108] suggested the possibility of accretion spinning up a star until it becomes unstable to the emission of gravitational waves and reaches a steady state, with the angular momentum gained by accretion equal to the angular momentum lost to gravitational waves. Following the discovery of the first millisecond pulsar, Wagoner examined the mechanism in detail for CFS unstable $f$-modes [109]. Although mutual friction appears to rule out the steady-state picture for $f$-modes, it remains a possibility for $r$-modes [67, 110–112]. Levin [113] and (independently) Spruit [72], however, pointed out that viscous heating of the neutron star by its unstable oscillations will lower the shear viscosity and so increase the mode's growth rate, leading to a runaway instability. The resulting scenario is a cycle in which a cold, stable neutron star is spun up over a few million years until it becomes unstable; the star then heats up, the instability grows, and the star spins down until it is again stable, all within a few months; the star then cools, and the cycle repeats.

This scenario would rule out $r$-modes in LMXBs as a source of detectable gravitational waves because the stars would radiate for only a small fraction of the cycle. A small saturation amplitude, however, lengthens the time spent in the cycle, possibly allowing observability [114]. The steady state itself remains a possible alternative in stars whose core contains hyperons or free quarks (or if the "neutron stars" are really strange quark stars) [54, 56, 57, 60, 112, 115]. Heating the core increases the bulk viscosity, and with an exotic core, this growth in the bulk viscosity is large enough to prevent the thermal runaway and allow a steady state. In Bondarescu et al. [97] the nonlinear evolutions (restricted to 3 coupled modes) include neutrino cooling, shear viscosity, hyperon bulk viscosity and dissipation at the core-crust boundary layer, with parameters to span a range of uncertainty in these various quantities. They display the regions of parameter space associated with the alternative scenarios just outlined—steady state, cycle, and fast and slow runaways. In all cases, the $r$-mode amplitude remains very small ($\sim 10^{-5}$), but because of the long duration of the instability, such systems are still good candidates for gravitational wave detection by advanced LIGO class interferometers [43, 97, 116].

**Dynamical Nonaxisymmetric Instability** Work on dynamical nonaxisymmetric instabilities is largely outside the scope of this review. They are most likely to be relevant to proto-neutron stars and to the short-lived hypermassive neutron stars that form in the merger of a double neutron star system. Unless the star has unusually high differential rotation, instability requires a large value of the ratio $T/|W|$ of rotational kinetic energy to gravitational binding energy: comparable to the value $T/|W| = 0.27$ that marks the dynamical instability of the $l = m = 2$ mode of uniformly rotating uniform density Newtonian models (the Maclauring spheroids). This bar instability, if present, will emit strong gravitational waves with frequencies in the kHz regime. The development of the instability and the resulting waveform have been computed numerically in the context of both Newtonian gravity and in full general relativity (see [117–120] for representative studies).

Uniformly rotating neutron stars have maximum values of $T/|W|$ smaller than 0.14, apparently precluding dynamical nonaxisymmetric instability. For highly differential rotation, however, Centrella et al. [121] found a one-armed ($m = 1$) instability for smaller rotation, for $T/|W| \sim 0.14$, but for a polytropic index of $N = 3$ which is not representative for neutron stars. Remarkably, Shibata et al. [122, 123] then found found an $m = 2$ instability for $T/|W|$ as low as 0.01, for models with polytropic index $N = 1$, representing a stiffness appropriate to neutron stars. These instabilities appear to be related to the existence of corotation points, where the pattern speed of the mode matches the star's angular velocity [124, 125]; Ou and Tohline [126] tie the growth of the instability to a resonant cavity associated with a minimum in the vorticity to density ratio (the so-called vortensity) . Collapsing cores in supernovae are differentially rotating, and these instabilities of proto-neutron stars arise in simulations of rotating core collapse [127, 128]. Because they can radiate more energy in gravitational waves than the post-bounce burst signal itself, interest in these dynamical instabilities is strong.

**Acknowledgments** J.F.'s work is supported in part by NSF Grant PHY 1001515. N.S.'s is supported in part by an Excellence Grant for Basic Research (Research Committee of the Aristotle University of Thessaloniki—87896) and by an IKY–DAAD exchange grant (IKYDA 2012).

## References

1. Taub, A.H.: General relativistic variational principal for perfect fluids. Phys. Rev. **94**, 1468 (1954). doi:10.1103/PhysRev.94.1468
2. Taub, A.H.: Stability of general relativistic gaseous masses and variational principles. Commun. Math. Phys. **15**, 235 (1969). doi:10.1007/BF01645677
3. Schutz, B.F.: Perfect fluids in general relativity: velocity potentials and a variational principle. Phys. Rev. D **2**, 2762 (1970). doi:10.1103/PhysRevD.2.2762
4. Schutz Jr, B.F.: Linear pulsations and stability of differentially rotating stellar models. II. General-relativistic analysis. Astrophys. J. **24**, 343 (1972). doi:10.1086/190258
5. Carter, B.: Elastic perturbation theory in general relativity and variational principle for a rotating solid star. Commun. Math. Phys. **30**, 261 (1973). doi:10.1007/BF01645505
6. Schutz, B.F., Sorkin, R.D.: Variational aspects of relativistic field theories with applications to perfect fluids. Ann. Phys. **107**, 1 (1977). doi:10.1016/0003-4916(77)90200-7
7. Calkin, M.G.: An action principle for magnetohydrodynamics. Can. J. Phys. **41**, 2241 (1963). doi:10.1139/p63-216
8. Friedman, J.L.: Generic instability of rotating relativistic stars. Commun. Math. Phys. **62**, 247 (1978)
9. Friedman, J.L., Schutz, B.F.: Lagrangian perturbation theory of nonrelativistic fluids. Astrophys. J. **221**, 937 (1978). doi:10.1086/156098
10. Thorne, K.S.: Validity in general relativity of the Schwarzschild criterion for convection. Astrophys. J. **144**, 201 (1966). doi:10.1086/148595
11. Kovetz, A.: Schwarzschild's criterion for convective instability in general relativity. Z. Astrophys. **66**, 446 (1967)
12. Schutz Jr, B.F.: Taylor instabilities in relativistic stars. Astrophys. J. **161**, 1173 (1970). doi:10.1086/150620
13. Bardeen, J.M.: A variational principle for rotating stars in general relativity. Astrophys. J. **162**, 71 (1970). doi:10.1086/150635
14. Seguin, F.H.: The stability of nonuniform rotation in relativistic stars. Astrophys. J. **197**, 745 (1975). doi:10.1086/153563
15. Detweiler, S.L., Ipser, J.R.: A variational principle and a stability criterion for the non-radial modes of pulsation of stellar models in general relativity. Astrophys. J. **185**, 685 (1973). doi:10.1086/152447
16. Lebovitz, N.R.: On Schwarzschild's criterion for the stability of gaseous masses. Astrophys. J. **142**, 229 (1965). doi:10.1086/148279
17. Solberg, H.: Le mouvement d'inertie de l'atmosphere stable et son role dans la theorie des cyclones, in Proces Verbaux de l' Association de Météorologie. International Union of Geodesy and Geophysics. 6th General Assembly (Edinburgh), vol. 2, pp. 66–82. International Union of Geodesy and Geophysics, Edinburgh (1936)
18. Abramowicz, M.A.: Rayleigh and Solberg criteria reversal near black holes: the optical geometry explanation. ArXiv:astro-ph/0411718. (2004)
19. Friedman, J.L., Stergioulas, N.: Rotating Relativistic Stars. Cambridge University Press (2013)

20. Chandrasekhar, S.: The dynamical instability of gaseous masses approaching the Schwarzschild limit in general relativity. Astrophys. J. **140**, 417 (1964). doi:10.1086/147938
21. Fowler, W.A.: The stability of supermassive stars. Astrophys. J. **144**, 180 (1966). doi:10.1086/148594
22. Friedman, J.L., Ipser, J.R., Sorkin, R.D.: Turning-point method for axisymmetric stability of rotating relativistic stars. Astrophys. J. **325**, 722 (1988). doi:10.1086/166043
23. Cook, G.B., Shapiro, S.L., Teukolsky, S.A.: Spin-up of a rapidly rotating star by angular momentum loss—Effects of general relativity. Astrophys. J. **398**, 203 (1992). doi:10.1086/171849
24. Takami, K., Rezzolla, L., Yoshida, S.: A quasi-radial stability criterion for rotating relativistic stars. Mon. Not. R. Astron. Soc. **416**, L1 (2011). doi:10.1111/j.1745-3933.2011.01085.x
25. Chandrasekhar, S.: Solutions of two problems in the theory of gravitational radiation. Phys. Rev. Lett. **24**, 611 (1970). doi:10.1103/PhysRevLett.24.611
26. Friedman, J.L., Schutz, B.F.: Secular instability of rotating Newtonian stars. Astrophys. J. **222**, 281 (1978). doi:10.1086/156143
27. Thorne, K.: Multipole expansions of gravitational radiation. Rev. Mod. Phys. **52**, 299 (1980). doi:10.1103/RevModPhys.52.299
28. Stergioulas, N., Friedman, J.L.: Nonaxisymmetric neutral modes in rotating relativistic stars. Astrophys. J. **492**, 301 (1998). doi:10.1086/305030
29. Yoshida, S., Eriguchi, Y.: Neutral points of oscillation modes along equilibrium sequences of rapidly rotating polytropes in general relativity: application of the Cowling approximation. Astrophys. J. **490**, 779 (1997)
30. Yoshida, S., Eriguchi, Y.: A numerical study of normal modes of rotating neutron star models by the Cowling approximation. Astrophys. J. **515**, 414 (1999). doi:10.1086/307012
31. Zink, B., Korobkin, O., Schnetter, E., Stergioulas, N.: Frequency band of the f-mode Chandrasekhar-Friedman-Schutz instability. Phys. Rev. D **81**(8), 084055 (2010). doi:10.1103/PhysRevD.81.084055
32. Gaertig, E., Glampedakis, K., Kokkotas, K.D., Zink, B.: The f-mode instability in relativistic neutron stars. Phys. Rev. Lett. **107**, 101102 (2011). doi:10.1007/s10714-010-1059-4
33. Morsink, S., Stergioulas, N., Blattnig, S.: Quasi-normal modes of rotating relativistic stars - neutral modes for realistic equations of state. Astrophys. J. **510**, 854 (1999). doi:10.1086/306630
34. Chakrabarty, D.: The spin distribution of millisecond X-ray pulsars. In: Wijnands, R. (ed.) A Decade of Accreting Millisecond X-Ray Pulsars, AIP Conference Proceedings, vol. 1068, pp. 67–74. (2008)
35. Ghosh, P., Lamb, F.K.: Accretion by rotating magnetic neutron stars. II–Radial and vertical structure of the transition zone in disk accretion. Astrophys. J. **232**, 259 (1979). doi:10.1086/157285
36. Shapiro, S.L., Teukolsky, S.A.: Black Holes, White Dwarfs and Neutron Stars. Wiley, New York (1983)
37. White, N.E., Zhang, W.: Millisecond X-ray pulsars in low-mass X-ray binaries. Astrophys. J. **490**, L87 (1997). doi:10.1086/311018
38. Patruno, A., Haskell, B., D'Angelo, C.: Gravitational waves and the maximum spin frequency of neutron stars. Astrophys. J. **746**, 9 (2012). doi:10.1088/0004-637X/746/1/9
39. Stergioulas, N.: Rotating stars in relativity. Living Rev. Relativ. **6**, 3 (2003)
40. Andersson, N., Kokkotas, K.D.: The r-mode instability in rotating neutron stars. Int. J. Mod. Phys. D **10**, 381 (2001). doi:10.1142/S0218271801001062
41. Kokkotas, K.D., Ruoff, J.: Instabilities of relativistic stars. In: Proceedings of 2001: A Relativistic Spacetime Odyssey 2002, Firenze (2001)
42. Andersson, N., Ferrari, V., Jones, D.I., et al.: Gravitational waves from neutron stars: promises and challenges. Gen. Relativ. Grav. **43**, 409 (2011). doi:10.1007/s10714-010-1059-4
43. Owen, B.J.: How to adapt broad-band gravitational-wave searches for r-modes. Phys. Rev. D **82**, 104002 (2010). doi:10.1103/PhysRevD.82.104002

44. Kokkotas, K.D., Stergioulas, N.: Analytic desctription of the r-mode instability in uniform density stars. Astron. Astrophys. **341**, 110 (1999)
45. Cutler, C., Lindblom, L., Splinter, R.J.: Damping times for neutron star oscillations. Astrophys. J. **363**, 603 (1990). doi:10.1086/169370
46. Ipser, J.R., Lindblom, L.: On the adiabatic pulsations of accretion disks and rotating stars. Astrophys. J. **379**, 285 (1991)
47. Ipser, J.R., Lindblom, L.: The oscillations of rapidly rotating Newtonian stellar models. II—dissipative effects. Astrophys. J. **373**, 213 (1991). doi:10.1086/170039
48. Yoshida, S., Eriguchi, Y.: Gravitational radiation driven secular instability of rotating polytropes. Astrophys. J. **438**, 830 (1995). doi:10.1086/175126
49. Lai, D., Shapiro, S.L.: Gravitational radiation from rapidly rotating nascent neutron stars. Astrophys. J. **442**, 259 (1995). doi:10.1086/175438
50. Bonazzola, S., Frieben, J., Gourgoulhon, E.: Spontaneous symmetry breaking of rapidly rotating stars in general relativity. Astrophys. J. **460**, 379 (1996). doi:10.1086/176977
51. Lai, D.: Secular bar-mode evolution and gravitational waves from neutron stars. Astrophysical sources for ground-based gravitational wave detectors. In: Proceedings of AIP Conference, vol. 575, pp. 246–257. (2001)
52. Zdunik, J.L.: Damping of GRR instability by direct URCA reactions. Astron. Astrophys. **308**, 828 (1996)
53. Jones, P.B.: Comment on Gravitational radiation instability in hot young neutron stars. Phys. Rev. Lett. **86**, 1384 (2001). doi:10.1103/PhysRevLett.86.1384
54. Lindblom, L., Owen, B.J.: Effect of hyperon bulk viscosity on neutron-star r-modes. Phys. Rev. D **65**, 063006 (2002). doi:10.1103/PhysRevD.65.063006
55. Haensel, P., Levenfish, K.P., Yakovlev, D.G.: Bulk viscosity in superfluid neutron star cores. III. Effects of $\Sigma^-$ hyperons. Astron. Astrophys. **381**, 1080 (2002). doi:10.1051/0004-6361:20011532
56. Nayyar, M., Owen, B.J.: R-modes of accreting hyperon stars as persistent sources of gravitational waves. Phys. Rev. D **73**, 084001 (2006). doi:10.1103/PhysRevD.73.084001
57. Haskell, B., Andersson, N.: Superfluid hyperon bulk viscosity and the r-mode instability of rotating neutron stars. Mon. Not. R. Astron. Soc. **408**, 1897 (2010)
58. Madsen, J.: How to identify a strange star. Phys. Rev. Lett. **81**, 3311 (1998). doi:10.1103/PhysRevLett.81.3311
59. Madsen, J.: Probing strange stars and color superconductivity by r-mode instabilities in millisecond pulsars. Phys. Rev. Lett. **85**, 10 (2000). doi:10.1103/PhysRevLett.85.10
60. Andersson, N., Jones, D.I., Kokkotas, K.D.: Strange stars as persistent sources of gravitational waves. Mon. Not. R. Astron. Soc. **337**, 1224 (2002). doi:10.1046/j.1365-8711.2002.05837.x
61. Jaikumar, P., Rupak, G., Steiner, A.W.: Viscous damping of r-mode oscillations in compact stars with quark matter. Phys. Rev. D **78**, 123007 (2008). doi:10.1103/PhysRevD.78.123007
62. Rupak, G., Jaikumar, P.: Constraining phases of quark matter with studies of r-mode damping in compact stars. Phys. Rev. C **82**, 055806 (2010). doi:10.1103/PhysRevC.82.055806
63. Flowers, E., Itoh, N.: Transport properties of dense matter. Astrophys. J. **206**, 218 (1976). doi:10.1086/154375
64. Shternin, P.S., Yakovlev, D.G.: Shear viscosity in neutron star cores. Phys. Rev. D **78**, 063006 (2008). doi:10.1103/PhysRevD.78.063006
65. Bildsten, L., Ushomirsky, G.: Viscous boundary-layer damping of r-modes in neutron stars. Astrophys. J. **529**, L33 (2000). doi:10.1086/312454
66. Lindblom, L., Owen, B.J., Ushomirsky, G.: Effect of a neutron-star crust on the r-mode instability. Phys. Rev. D **62**, 084030 (2000). doi:10.1103/PhysRevD.62.084030
67. Andersson, N., Jones, D.I., Kokkotas, K.D., Stergioulas, N.: R-mode runaway and rapidly rotating neutron stars. Astrophys. J. **534**, L75 (2000). doi:10.1086/312643
68. Glampedakis, K., Andersson, N.: Crust-core coupling in rotating neutron stars. Phys. Rev. D **74**, 044040 (2006). doi:10.1103/PhysRevD.74.044040
69. Glampedakis, K., Andersson, N.: Ekman layer damping of r modes revisited. Mon. Not. R. Astron. Soc. **371**, 1311 (2006). doi:10.1111/j.1365-2966.2006.10749.x

70. Lindblom, L., Mendell, G.: Does gravitational radiation limit the angular velocities of superfluid neutron stars. Astrophys. J. **444**, 804 (1995). doi:10.1086/175653
71. Lindblom, L., Mendell, G.: R-modes in superfluid neutron stars. Phys. Rev. D **61**, 104003 (2000). doi:10.1103/PhysRevD.61.104003
72. Spruit, H.C.: Gamma-ray bursts from X-ray binaries. Astron. Astrophys. **341**, L1 (1999)
73. Rezzolla, L., Lamb, F.K., Shapiro, S.L.: R-mode oscillations in rotating magnetic neutron stars. Astrophys. J. **531**, L139 (2000). doi:10.1086/312539
74. Rezzolla, L., Lamb, F.K., Marković, D., Shapiro, S.L.: Properties of r modes in rotating magnetic neutron stars. I. Kinematic secular effects and magnetic evolution equations. Phys. Rev. D **64**, 104013 (2001). doi:10.1103/PhysRevD.64.104013
75. Rezzolla, L., Lamb, F.K., Marković, D., Shapiro, S.L.: Properties of r modes in rotating magnetic neutron stars. II. Evolution of the r modes and stellar magnetic field. Phys. Rev. D **64**, 104014 (2001). doi:10.1103/PhysRevD.64.104014
76. Cuofano, C., Drago, A.: Magnetic fields generated by r-modes in accreting millisecond pulsars. Phys. Rev. D **82**, 084027 (2010). doi:10.1103/PhysRevD.67.084027
77. Kojima, Y.: Quasi-toroidal oscillations in rotating relativistic stars. Mon. Not. R. Astron. Soc. **293**, 49 (1998). doi:10.1046/j.1365-8711.1998.01119.x
78. Andersson, N.: A new class of unstable modes of rotating relativistic stars. Astrophys. J. **502**, 708 (1998). doi:10.1086/305919
79. Kojima, Y., Hosonuma, M.: The r-mode oscillations in relativistic rotating stars. Astrophys. J. **520**, 788 (1999). doi:10.1086/307481
80. Kojima, Y., Hosonuma, M.: Approximate equation relevant to axial oscillations on slowly rotating relativistic stars. Phys. Rev. D **62**, 044006 (2000). doi:10.1103/PhysRevD.62.044006
81. Lockitch, K.H., Andersson, N., Friedman, J.L.: Rotational modes of relativistic stars: Analytic results. Phys. Rev. D **63**, 024019 (2001). doi:10.1103/PhysRevD.63.024019
82. Lockitch, K.H., Friedman, J.L., Andersson, N.: Rotational modes of relativistic stars: Numerical results. Phys. Rev. D **68**, 124010 (2003). doi:10.1103/PhysRevD.68.124010
83. Lockitch, K.H., Andersson, N., Watts, A.L.: Regularizing the r-mode problem for nonbarotropic relativistic stars. Class. Quant. Grav. **21**, 4661 (2004). doi:10.1088/0264-9381/21/19/012
84. Ruoff, J., Kokkotas, K.D.: On the r-mode spectrum of relativistic stars in the low-frequency approximation. Mon. Not. R. Astron. Soc. **328**, 678 (2001). doi:10.1046/j.1365-8711.2001.04909.x
85. Ruoff, J., Kokkotas, K.D.: On the r-mode spectrum of relativistic stars: the inclusion of the radiation reaction. Mon. Not. R. Astron. Soc. **330**, 1027 (2002). doi:10.1046/j.1365-8711.2002.05169.x
86. Ruoff, J., Stavridis, A., Kokkotas, K.D.: Inertial modes of slowly rotating relativistic stars in the Cowling approximation. Mon. Not. R. Astron. Soc. **339**, 1170 (2003). doi:10.1046/j.1365-8711.2003.06267.x
87. Yoshida, S., Lee, U.: Relativistic r-modes in slowly rotating neutron stars: numerical analysis in the Cowling approximation. Astrophys. J. **567**, 1112 (2002). doi:10.1086/338663
88. Kastaun, W.: Inertial modes of rigidly rotating neutron stars in Cowling approximation. Phys. Rev. D **77**, 124019 (2008). doi:10.1103/PhysRevD.77.124019
89. Stergioulas, N., Font, J.A.: Nonlinear r-modes in rapidly rotating relativistic stars. Phys. Rev. Lett. **86**, 1148 (2001). doi:10.1103/PhysRevLett.86.1148
90. Lindblom, L., Tohline, J.E., Vallisneri, M.: Non-linear evolution of the r-modes in neutron stars. Phys. Rev. Lett. **86**, 1152 (2001). doi:10.1103/PhysRevLett.86.1152
91. Lindblom, L., Tohline, J.E., Vallisneri, M.: Numerical evolutions of nonlinear r-modes in neutron stars. Phys. Rev. D **65**, 084039 (2002). doi:10.1103/PhysRevD.65.084039
92. Gressman, P., Lin, L.M., Suen, W.M., Stergioulas, N., Friedman, J.L.: Nonlinear r-modes in neutron stars: instability of an unstable mode. Phys. Rev. D **66**, 041303 (2002). doi:10.1103/PhysRevD.66.041303
93. Lin, L.M., Suen, W.M.: Non-linear r-modes in neutron stars: a hydrodynamical limitation on r-mode amplitudes. Mon. Not. R. Astron. Soc. **370**, 1295 (2006). doi:10.1111/j.1365-2966.2006.10536.x

94. Schenk, A.K., Arras, P., Flanagan, E.E., Teukolsky, S.A., Wasserman, I.: Nonlinear mode coupling in rotating stars and the r-mode instability in neutron stars. Phys. Rev. D **65**, 024001 (2002). doi:10.1103/PhysRevD.65.024001
95. Arras, P., Flanagan, E.E., Morsink, S.M., et al.: Saturation of the r-mode instability. Astrophys. J. **591**, 1129 (2003). doi:10.1086/374657
96. Morsink, S.M.: Relativistic precession around rotating neutron stars: effects due to frame-dragging and stellar oblateness. Astrophys. J. **571**, 435 (2002)
97. Bondarescu, R., Teukolsky, S.A., Wasserman, I.: Spin evolution of accreting neutron stars: nonlinear development of the r-mode instability. Phys. Rev. D **76**, 064019 (2007). doi:10.1103/PhysRevD.76.064019
98. Bondarescu, R., Teukolsky, S.A., Wasserman, I.: Spinning down newborn neutron stars: non-linear development of the r-mode instability. Phys. Rev. D **79**, 104003 (2009). doi:10.1103/PhysRevD.79.104003
99. Ou, S., Tohline, J.E., Lindblom, L.: Nonlinear development of the secular bar-mode instability in rotating neutron stars. Astrophys. J. **617**, 490 (2004). doi:10.1086/425296
100. Shibata, M., Karino, S.: Numerical evolution of secular bar-mode instability induced by the gravitational radiation reaction in rapidly rotating neutron stars. Phys. Rev. D **70**, 084022 (2004). doi:10.1103/PhysRevD.70.084022
101. Kastaun, W., Willburger, B., Kokkotas, K.D.: Saturation amplitude of the f-mode instability. Phys. Rev. D **82**, 104036 (2010). doi:10.1103/PhysRevD.82.104036
102. Gaertig, E., Kokkotas, K.D.: Gravitational wave asteroseismology with fast rotating neutron stars. Phys. Rev. D **83**, 064031 (2011). doi:10.1103/PhysRevD.83.064031
103. Page, D., Prakash, M., Lattimer, J.M., Steiner, A.W.: Rapid cooling of the neutron star in Cassiopeia a triggered by neutron superfluidity in dense matter. Phys. Rev. Lett. **106**, 081101 (2011). doi:10.1103/PhysRevLett.106.081101
104. Shternin, P.S., Yakovlev, D.G., Heinke, C.O., Ho, W.C.G., Patnaude, D.J.: Cooling neutron star in the Cassiopeia a supernova remnant: evidence for superfluidity in the core. Mon. Not. R. Astron. Soc. **412**, L108 (2011). doi:10.1111/j.1745-3933.2011.01015.x
105. Friedman, J.L., Morsink, S.M.: Axial instability of rotating relativistic stars. Astrophys. J. **502**, 714 (1998). doi:10.1086/305920
106. Lindblom, L., Owen, B.J., Morsink, S.M.: Gravitational radiation instability in hot young neutron stars. Phys. Rev. Lett. **80**, 4843 (1998). doi:10.1103/PhysRevLett.80.4843
107. Andersson, N., Kokkotas, K.D., Schutz, B.F.: Gravitational radiation limit on the spin of young neutron stars. Astrophys. J. **510**, 846 (1999). doi:10.1086/306625
108. Papaloizou, J., Pringle, J.E.: Gravitational radiation and the stability of rotating stars. Mon. Not. R. Astron. Soc. **184**, 501 (1978)
109. Wagoner, R.V.: Gravitational radiation from accreting neutron stars. Astrophys. J. **278**, 345 (1984). doi:10.1086/161798
110. Bildsten, L.: Gravitational radiation and rotation of accreting neutron stars. Astrophys. J. **501**, L89 (1998). doi:10.1086/311440
111. Andersson, N., Kokkotas, K.D., Stergioulas, N.: On the relevance of the r-mode instability for accreting neutron stars and white dwarfs. Astrophys. J. **516**, 307 (1999). doi:10.1086/307082
112. Wagoner, R.V.: Conditions for steady gravitational radiation from accreting neutron stars. Astrophys. J. **578**, L63 (2002). doi:10.1086/344502
113. Levin, Y.: Runaway heating by r-modes of neutron stars in low-mass X-ray binaries. Astrophys. J. **517**, 328 (1999). doi:10.1086/307196
114. Heyl, J.: Low-mass X-ray binaries may be important laser interferometer gravitational-wave observatory sources after all. Astrophys. J. **574**, L57 (2002). doi:10.1086/342263
115. Reisenegger, A., Bonacić, A.: Millisecond pulsars with r-modes as steady gravitational radiators. Phys. Rev. Lett. **91**, 201103 (2003). doi:10.1103/PhysRevLett.91.201103
116. Watts, A.L., Krishnan, B.: Detecting gravitational waves from accreting neutron stars. Adv. Space Res. **43**, 1049 (2009). doi:10.1016/j.asr.2009.01.006
117. Houser, J.L., Centrella, J.M., Smith, S.C.: Gravitational radiation from nonaxisymmetric instability in a rotating star. Phys. Rev. Lett. **72**, 1314 (1994). doi:10.1103/PhysRevLett.72.1314

118. Tohline, J.E., Durisen, R.H., McCollough, M.: The linear and nonlinear dynamic stability of rotating $n = 3/2$ polytropes. Astrophys. J. **298**, 220 (1985). doi:10.1086/163600
119. Shibata, M.: Axisymmetric simulations of rotating stellar collapse in full general relativity—criteria for prompt collapse to black holes. Progress Theoret. Phys. **104**, 325 (2000). doi:10.1143/PTP.104.325
120. Manca, G.M., Baiotti, L., De Pietri, R., Rezzolla, L.: Dynamical non-axisymmetric instabilities in rotating relativistic stars. Class. Quant. Grav. **24**, S171 (2007). doi:10.1088/0264-9381/24/12/S12
121. Centrella, J.M., New, K.C.B., Lowe, L.L., Brown, J.D.: Dynamical rotational instability at low T/W. Astrophys. J. **550**, L193 (2001). doi:10.1086/319634
122. Shibata, M., Karino, S., Eriguchi, Y.: Dynamical instability of differentially rotating stars. Mon. Not. R. Astron. Soc. **334**, L27 (2002). doi:10.1046/j.1365-8711.2002.05724.x
123. Shibata, M., Karino, S., Eriguchi, Y.: Dynamical bar-mode instability of differentially rotating stars: effects of equations of state and velocity profiles. Mon. Not. R. Astron. Soc. **343**, 619 (2003). doi:10.1046/j.1365-8711.2003.06699.x
124. Watts, A.L., Andersson, N., Jones, D.I.: The nature of low $T/|W|$ dynamical instabilities in differentially rotating stars. Astrophys. J. **618**, L37 (2005). doi:10.1086/427653
125. Saijo, M., Yoshida, S.: Low $T/|W|$ dynamical instability in differentially rotating stars: diagnosis with canonical angular momentum. Mon. Not. R. Astron. Soc. **368**, 1429 (2006). doi:10.1111/j.1365-2966.2006.10229.x
126. Ou, S., Tohline, J.E.: Unexpected dynamical instabilities in differentially rotating neutron stars. Astrophys. J. **651**, 1068 (2006). doi:10.1086/507597
127. Ott, C.D., Ou, S., Tohline, J.E., Burrows, A.: One-armed spiral instability in a low-$T/|W|$ postbounce supernova core. Astrophys. J. **625**, L119 (2005). doi:10.1086/431305
128. Ott, C.D.: The gravitational-wave signature of core-collapse supernovae. Class. Quant. Grav. **26**, 063001 (2009). doi:10.1088/0264-9381/26/6/063001

# Gravity Talks: Observing the Universe with Gravitational Waves

**Bernard F. Schutz**

**Abstract** When the current upgrade of the large ground-based gravitational wave detectors LIGO and VIRGO is completed, the new science of gravitational wave astronomy will begin. In this overview I review the current status of the detector projects on the ground and in space (LISA), the kinds of signals and sources they expect to observe, and the science returns that are anticipated.

## 1 Introduction

The effort to detect gravitational waves has been one of the most remarkable examples of sustained technology development in the history of physics and astronomy. Like most kinds of instrumentation, gravitational wave detectors have been developed through many steps, each one bringing a significant improvement in performance. Unlike all other kinds of instrumentation that I am aware of, improvements in the sensitivity of gravitational wave detectors have not yet—after over 50 years of improvements—led to a single detection. In all other areas of physics and astronomy, early instruments have made at least some measurements or detections of interest, and these have stimulated and justified the next improvements in performance. This cycle of observation/improvement/observation/improvement is clearly evident in the histories of optical, radio, X-ray, and cosmic-ray astronomy during the last 50 years. For gravitational wave detectors, evolving over the same period of time, the cycle was simply improvement/improvement/improvement. The effort has been sustained, not by the success of serendipitous observation, but by the dedication of those working in the field and by the deep conviction (shared, of course, by the science-funding bodies) not only that gravitational waves of a certain amplitude are certainly passing through the Earth regularly, but that detecting these waves will dramatically change our view of the universe.

B. F. Schutz(✉)
Max Planck Institute for Gravitational Physics (Albert Einstein Institute), Potsdam, Germany
e-mail: bernard.schutz@aei.mpg.de

J. Bičák and T. Ledvinka (eds.), *General Relativity, Cosmology and Astrophysics*, Fundamental Theories of Physics 177, DOI: 10.1007/978-3-319-06349-2_21,

© Springer International Publishing Switzerland 2014

The 50-year effort has not been without other rewards for those involved. On the experiment side, many fascinating new technologies have been developed to allow us to make the most sensitive measurements of distance changes ever accomplished. Detector sensitivity after the current upgrades will be something like $10^7$ times better than the original instruments built in the 1960s [1]. This maps directly into range: our detectors will be able to detect signals from sources $10^7$ times further away than the early detectors could. On the theory side, the possibility of observing gravitational waves has stimulated many important developments, not least the ability to solve Einstein's equations on computers and to simulate systems that cannot be studied analytically. But the big goal of direct detections has not yet been reached.

This will not be the case for much longer. The large ground-based detectors LIGO, VIRGO, and GEO600 have already taken data with such high sensitivity that detections could have happened if we had been a bit lucky with how close the nearest event would happen; that no detections were registered was not a surprise, even if a slight disappointment. But the next phase of data-taking with LIGO and VIRGO will be different: if the instruments now being upgraded perform to specification, and if the astrophysical event rate is close to the estimates made by the collaboration [2], then the first detection is likely by around 2017, and it is very unlikely that nothing will be registered before, say, 2020.

Unlike in the other areas of astronomy that have had the normal observation/improvement cycle, we believe we actually have a pretty good idea of what the first detected gravitational wave signals will be: they will be from binary systems in which compact objects (neutron stars and/or black holes) spiral together and merge. This expectation is based on theory—extensive studies of the motion of compact objects in fully general-relativistic binaries, and of the astrophysics of these objects—which was needed during the last 50 years in order to justify the effort being put into the technology development, but which has even more importantly become an essential part of the detection chain. The accurate predictions we now have of what the gravitational waveforms from binaries and other systems (for example, gravitational wave pulsars) should look like permit us to dig deep into instrumental noise and detect with confidence signals that might otherwise have to wait for yet another cycle of instrument improvement. Good data analysis based on theoretical waveform predictions probably improves the sensitivity (equivalently, the distance reach) of our detectors by factors of 10–30 for binaries, and up to several thousand for pulsars.

Theoretical studies have led also to the development during the last 10 years of a new paradigm for the kind of observing that will be done by gravitational wave detectors. We now think of it as "listening" to the universe, rather than "watching" it, as one does with normal telescopes. Our detectors are not pointed, but rather omni-directional, like microphones. The data stream is one-dimensional, like an audio stream. The waves are detected coherently, like audio waves, rather than just bolometrically, like photons; in fact, the phase evolution of the waves contains more information about the sources than the amplitude does. All these analogies help to understand how we will extract information from our arrays of detectors. And in the same way that hearing complements vision in animals, gravitational wave observing will add a qualitatively new "sense" to our ability to record the universe around us: relativistic systems we don't know about, or are hidden, or are unable to radiate light can be discovered, located, and studied using our network of microphone-like detectors.

With first observations not far in the future, this meeting is a good time to look forward toward the coming gravitational wave astronomy: what the instruments are, what the sources are that are most likely to be detected, and what the kinds of information are that we are likely to be able to infer when we are finally able to hear the universe speaking to us through the medium of gravity.

## 2 Light Deflection and Gravitational Wave Detection

Since this meeting marks Einstein's work in Prague, it seems appropriate to draw a link between his work and the detection of gravitational waves. One of the subjects that drew Einstein's attention in Prague in 1911-12 was the deflection of light by gravity. It is perhaps amusing that our present method of detecting gravitational waves by interferometry also relies on the action of gravity on light.

Einstein's work in Prague [3] was his famous demonstration that light, on passing the Sun or another body, will be deflected. Because he relied essentially only on the equivalence principle (the curvature of time) and did not yet have a theory that included the curvature of space, he got only half of the right value; nevertheless it was an important advance in his own thinking.

In principle he could have predicted also that the propagation time of light would change due to the deflection, something we now call the Shapiro Effect. But as there was no way to measure this with light from a distant star, he would not have given the idea much thought. Nevertheless it is precisely these propagation-time changes that we use in detectors today. The gravitational wave makes a time-dependent alteration in the time it takes light to move up and back along the arms of a detector, and by interferometry we can compare this to the propagation time in the perpendicular arm. The signature of a gravitational wave is a *difference* in the propagation time alterations between the two arms.

## 3 The Global Interferometer Network

The worldwide network of gravitational wave detectors, illustrated in Fig. 1, consists of three large instruments (two LIGO and VIRGO), one medium-sized detector (GEO600), and two projects: KAGRA in Japan and LIGO-India. The existing detectors have successfully reached their first-stage sensitivity goals and demonstrated that they could operate reliably and produce data. The data have been extensively analyzed, leading to over 70 papers reporting methods and results [4], but no gravitational wave signals were found in the data. This was not a surprise, since the first sensitivity goals were modest relative to the expected event rates, so now the three large instruments are upgrading their sensitivities by a factor of about 10, at which level it would be very surprising if no detections were made. During this upgrade, the

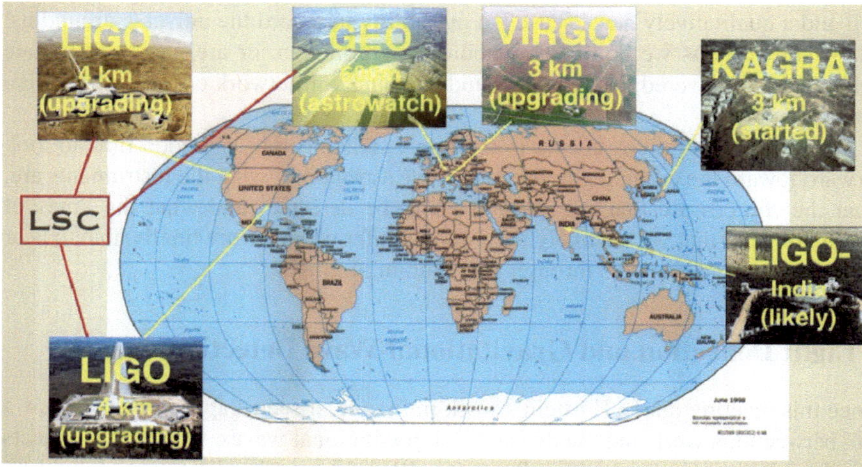


**Fig. 1** The six large interferometers currently operating, upgrading, or planned

GEO600 detector is spending about 70 % of its time in "Astrowatch" mode, meaning it is taking data just in case an interesting astronomical event occurs nearby, such as a supernova in our galactic neighborhood. The rest of the time it is also upgrading its sensitivity. The LIGO and GEO600 detectors are jointly managed and developed within the LIGO Scientific Collaboration (LSC).

GEO600 has functioned as a technology development platform as well as a detector, and a number of the important technologies that are being used in the upgrades of the larger instruments were first developed and/or tested in GEO600: high laser powers, monolithic suspensions for controlling thermal noise, signal recycling, and squeezed light. The basic installation that will turn LIGO into Advanced LIGO will be completed in 2015, and there will follow a series of commissioning periods alternating with observing runs, as the sensitivity is improved to the final goal. By 2017 it seems reasonable to expect the first detections to have occurred. Advanced VIRGO is on the same trajectory, perhaps 6 months to a year later.

KAGRA (previously LCGT) is a 3-km scale instrument with very ambitious technology: underground and cryogenic. It is now digging its tunnels in the Kamiokande mine in Japan, and should be producing data with a sensitivity comparable to that of Advanced LIGO and VIRGO before the end of the decade. The newest development is LIGO-India, in which LIGO will install a detector into a vacuum system built by India. The project has been endorsed in the US by the National Science Board, and is awaiting final funding approval at cabinet level in India.

In the longer-term future, scientists have studied so-called third-generation detectors and their technology. The most complete study was for the European Einstein Telescope [5, 6], which will probably bid for first funding toward the end of the current decade.

All of these instruments share data (or in the case of KAGRA and LIGO-India, plan to do so) and publish results jointly, with more than 600 authors on each paper. When one thinks about it, this degree of cooperation is unusual among physicists. Large collaborations are not uncommon, but normally there are two or more collaborations that compete with one another. In gravitational waves, all the detectors cooperate.

Such harmony is driven by the science: there are good reasons why the science benefits from pooling data. First, one could not reliably claim a detection of a short burst of gravitational radiation on the basis of the data of a single detector, because the signals are so weak that it would be easy for some un-modeled noise in a detector to masquerade as signal. Second, detectors are almost omni-directional, so to get direction information and be able to solve for the polarization of a short burst signal one needs three or more detectors, triangulating the position on the sky from time-delays among the detectors. That means that the global network itself forms an interferometer in the same sense that radio telescopes around the globe join in VLBI networks. And with four or more detectors the science gets even better: since instruments have duty cycles (for first-stage LIGO it was around 80 %), more detectors means that there is better coverage of the sky in both direction and time. And with more observed signals it is possible to extract better waveform information by averaging over the (independent) noise in the various detectors.

These advantages can be made quantitative in terms of some figures of merit for different networks, introduced in [7]. The Triple Detection Rate is a number that reflects the sensitivity of a network (as measured by the spatial volume inside its antenna pattern) and the time-coverage for an assumed duty cycle. For networks of more than three detectors it weights the product of time and volume for all three-detector sub-networks, on the grounds that one needs at least three detectors to get a reasonable amount of science from a detection. The Sky Resolution figure of merit measures the inverse of the typical area of the error box on the sky for locating an observed event. Both of these measures are given for three detector networks in Table 1. The networks are: the originally planned LIGO-VIRGO network, with two detectors at LIGO Hanford in Washington, one at LIGO Livingston in Louisiana, and the VIRGO detector (HHLV); VIRGO with the extended LIGO network with the second Hanford detector moved to its new home in India (HILV); and the ultimate worldwide network including Japan (HIJLV). The numbers are used only to compare networks, so we do not give scalings that allow one to go from a figure of merit to a measurable quantity like the area of the error box. But the relative values are significant: HIJLV produces sky location error boxes that are a factor of 7.6 smaller in area than HHLV, for example. This and the improvement in location going from HHLV to HILV are the main motivation for building the two detectors in India and Japan. However, one also sees that the added sensitivity and time-coverage if one assumes 80 % duty cycle for the new detectors will practically double the number of events that are detected with three or more detectors, comparing HHLV to HIJLV.

Anyone interested in the sociology of the international gravitational wave collaboration, about how scientists manage to cooperate and reach decisions without having a strong hierarchical organization, should read the study [8]. The author,

**Table 1** Figures of merit for three networks (see text for explanation). Data from [7]

| Network | Triple rate | Resolution |
|---|---|---|
| HHLV | 4.86 | 0.65 |
| HILV | 5.94 | 2.96 |
| HIJLV | 8.62 | 4.60 |

H Collins, has privileged access to the inner discussions of the collaboration, which he uses for his sociological study.

# 4 Data Analysis

Data analysis is an important part of any physics experiment or astronomical observation, but it plays a particularly important role in gravitational waves detection. In fact, detection with the ground-based instruments relies on three key components: the detectors' sensitivity, the accuracy of the source modeling (e.g. predicting waveforms from black hole mergers), and the sophistication of the data analysis.

The signals we expect will not be easily visible in the data streams: they will emerge only after processing the data through intelligent filters that remove noise and enhance signal. This so-called matched filtering relies, in turn, on good predicted waveforms. So, unlike in most branches of astronomy, where the modeling of the source comes after the data have been analyzed, here we need the modeling first as an input to the data analysis. And because noise can always masquerade as signal, any detection statement will be a statistical one. The LSC and VIRGO collaborations have agreed that their first claim to detection will need to have a "$5\sigma$" level of confidence, that there is less than one chance in three million that noise could have created the claimed signal at any time during the entire data run in which the detection was made. This will need to take into account the fact that searches are done over large parameter spaces of possible signals (different masses, spins, sky locations, etc), so that there is a large "trials factor" for all the different independent filters that have been employed, which additionally discounts the significance of a detection.

Because the noise is not an ideal Gaussian distribution, but rather has a population of random instrumental "glitches" that have to be discriminated from signals, and because the number of filters is so large that the computational demands sometimes exceed the available resources, it is generally not possible to do fully optimum signal analysis. Instead, there is a premium on clever data analysis algorithms that get closer to the theoretical optimum sensitivity. This means that if it is possible to develop an algorithm that digs deeper into the noise by a factor of 2, then that algorithm effectively improves the sensitivity of the detectors to that kind of signal by a factor of 2. Developing such algorithms is usually a lot less expensive than upgrading the hardware to achieve a sensitivity change by a factor of 2. That is why, alongside the ongoing hardware upgrades, there is a large amount of work going into algorithm development.

But besides algorithms, accurate input from the wider research community on source and waveform models can be similarly effective in improving sensitivity. The NINJA [9] and NRAR [10] collaborations are working hard to bring numerical relativity results into the data analysis systems.

To cope with the unusually large demand for computing power of some of the searches (particularly for low-amplitude continuous signals from pulsars) the gravitational wave community has developed the Einstein@Home platform [11]. This is a screen saver that performs data analysis when a computer is otherwise idle. So many people have downloaded and participate in the search that the platform currently delivers almost 700 Tflops of performance continuously! It has also been used recently to search through radio and gamma-ray data for pulsars, with great success, discovering systems that had been missed by previous data analysis. This validates the methods and approach of the gravitational wave community in the search for long-duration signals.

Because the data analysis involves such a large amount of preparation and resource, there has been considerable discussion about the value of public data releases. Current policy by the LSC is to release small segments of data that contain confirmed signals as soon as possible. Full data releases will occur after a period of perhaps 2 years, once the full data analysis apparatus of the collaboration has been applied and the results understood. The collaboration is also preparing software interfaces to allow scientists not experienced in the projects to perform their own analyses. One unknown is how much support the collaboration will be funded to provide for outside analysis: much of the complication of the analysis has to do with characteristics of the detectors (un-modeled noise, etc) that may prove difficult for non-members to deal with.

## 5 Observables

The data analysis ultimately determines the values of the parameters that are the unknowns in the family of signals that are used as the template of the search, within the uncertainties of the observations. For the most commonly expected signals, from binary coalescence, the principal parameters are:

- *Location on the sky.* Fundamentally this comes from the differences in the arrival times of the signals at the various detectors. Actually the arrival time is not well-defined for a long-duration signal like that from a binary, so some fiducial time has to be defined. One such time would be the *expected* arrival time of the final coalescence signal of the system, if it consisted of two point particles. This can be predicted from the inspiral waveform. The actual merger will be more complicated, but the (fictional) coalescence arrival time provides a well defined time-parameter determined by fitting the theoretical waveform to the data. Three such times of arrival are enough to define two possible locations on the sky. With four detectors this ambiguity is resolved.

- *Polarization.* Once the direction to the source is known, two detector signals suffice to determine the two polarizations. The polarization tells us the inclination of the plane of the binary orbit to the line of sight. If the polarization changes (say through orbit precession) then only one of the two possible directions determined by three detectors will remain fixed in the sky. If there are three or more detectors and the location of the signal is known, then the two polarization signals are over-determined, at least if general relativity is valid, and that redundancy is useful: there is a linear combination of outputs that contains no gravitational wave signal: this is called a *null stream*, and it can be used, for example, to veto spurious signals of instrumental origin. For $N > 2$ detectors there are $N-2$ null streams. With three or more detectors it is possible to test the general-relativity model of polarization; this could reveal new gravitational fields, which are a feature of many unified field theories for quantum gravity.
- *Masses and spins of the component masses.* For signals that can be well modeled, the phase/frequency information tells us about the masses. For a binary, fitting a post-Newtonian waveform description to the orbit is enough to determine the individual masses and spins, provided the signal is strong enough. The parameter that is most reliably determined is the so-called *chirp mass* $\mathcal{M} = m_1^{3/5} m_2^{3/5}/M^{1/5} = \mu^{3/5}M$, where $m_k$ is the mass of the $k$th star, $M = m_1 + m_2$ the total mass, and $\mu = m_1 m_2/M^2$ the symmetric mass ratio [12].
- *Distance to the source.* Remarkably, binary signals from systems whose orbital frequency changes during the observation due to gravitational radiation reaction contain enough information to determine the distance to the binary system: they are standard candles, or (more appropriately) *standard sirens* [12]. To understand how this is possible, consider first the three observables that one can measure for a circular binary with equal masses: the frequency $f$ of the signal, its time-scale for changing $\tau_c = f/\dot{f}$ (called the chirp time), and the intrinsic amplitude of the wave $h$ (which is only known if the location of the signal on the sky and its polarization have also been measured). These three depend on the mass $m$ of both components, their orbital radius $a$, and the distance $d$ to the source. With three measured values, each of these three unknowns can be determined, and in fact to within factors of order unity
$$d \sim c/(f^2 \tau_c h).$$

If the system has unequal mass components, then this formula still applies, because only the chirp mass $\mathcal{M}$ is needed. If the system is eccentric, then there is enough information in the phase of the signal to determine the eccentricity and come back to the same formula. And if the system is at cosmological distances, then the distance measured is the luminosity distance $D_L$. We expect this to be a powerful added tool in astronomy, by checking the standard astronomical distance ladder (which this method is completely independent of), by helping to identify binary sources and their host galaxies, and even by providing a local value of the Hubble constant [12, 13].

It is of course important to understand that not all parameters are determined equally well; in fact the errors in some of them can have strong covariances with one another. This issue is studied in detail by the search teams as they do their data analysis. The location on the sky can have strong covariances with polarization, for example. Determining the position accurately is a key for doing Multimessenger Astronomy: finding correlated signals in the gravitational wave and electromagnetic spectra. Conversely, an independently determined position (from an electromagnetic observation) will reduce the position errors and thereby improve the determination of other parameters.

# 6 Gravitational Waves from Neutron Stars and Black Holes

## *6.1 Neutron Star Binary Coalescence*

The in-spiral and coalescence of two neutron stars is the top candidate for the first-ever detection by the ground-based network. This is because the rate of such events in the universe is fairly well understood [2] and the signal template is very accurately modeled. The best estimate is that a three-detector network of interferometers at the advanced level of sensitivity (which we expect to have from 2019 onwards) should detect something like 40 events per year. The first event could come in 2017, earlier if we are fortunate. With an enlarged network that includes LIGO-India and KAGRA, the event rate could go up to 100–150 events per year. The maximum distance for such events will be in the range 400–600 Mpc. The enlarged network does not have a greatly increased reach. Instead, it covers the sky more isotropically, so it does not miss very many events out to this distance [7].

These events will bring a great deal of exciting science. We will have a much better sampling of the binary pulsar population, leading to mass distributions that could give clues to the prior evolution of such systems. The strongest (closest) event in any year should have a high enough signal-to-noise ratio (SNR) to provide a strong constraint on the neutron-star equation of state. These mergers are likely to produce gamma-ray bursts [14]; see Fig. 2. Because such bursts are strongly beamed, they are not likely to be seen with each gravitational wave detection, but there should be a few coincidences per year. The combined gamma-ray and gravitational-wave data could also provide further insight into the physics of neutron stars [15].

By comparing the time of arrival of gravitational waves and gamma rays, one can constrain the difference in the speed of electromagnetic and gravitational waves; again, this would be a fundamental test of general relativity [16]. And by comparing the arrival times of the left- and right-handed circular polarization components of the waves (which will be possible by detailed fitting to the signal template) one can look for birefringence in the propagation of gravitational waves, something again that is possible in modified theories of gravity [17].

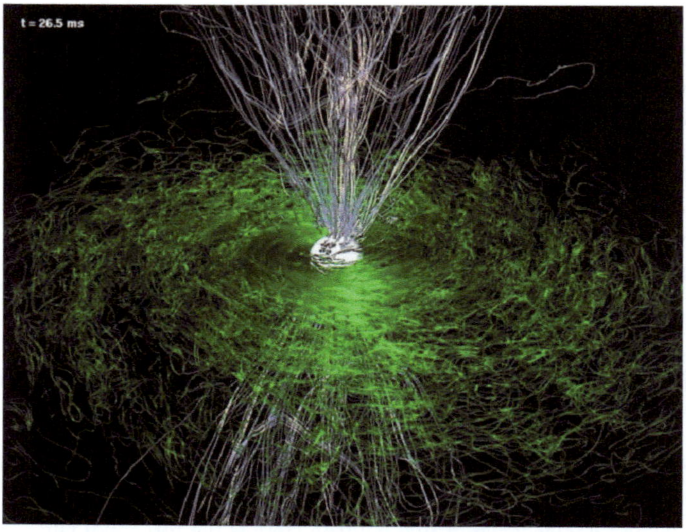


**Fig. 2** A poloidal magnetic field structure that arises naturally from binary neutron-star coalescence, as computed using fully general-relativistic MHD [14]

The distance measures to binary systems described above, when combined with galaxy surveys and other information, could determine the local value of the Hubble constant to uncertainties less than 1 % within the first 3 years [13]. When combined with accurate measurements of cosmic-microwave background measurements (as expected from the Planck mission), these coincidences might even determine the dark energy parameter $w$ to accuracies of 1 % (D Holz, private communication).

## *6.2 Neutron Star Interiors*

Neutron stars may be the most complex and extreme physical systems we know of, and because of that there is still considerable uncertainty about the physics that goes on inside them. The nuclear physics that underlies their equation of state is not accessible experimentally, and there are consequently considerable differences among the various proposed theoretical model equations. Pulsar observations tell us about strong magnetic fields, but their origin and interior distribution and strength are basically still a mystery. Therefore any information that gravitational wave observations can shed on these objects will be welcome.

Modeling binary coalescence will certainly be one way of learning about them, as mentioned in the previous section. Another avenue will be to detect gravitational radiation from neutron star pulsation modes that may be excited by outbursts on their surfaces, which occur when newly accreted matter undergoes a thermonuclear explosion. The amount of energy that might be converted into gravitational waves is not known, but searches in coincidence with X-ray and gamma-ray observations are planned.

## 6.3 Black Hole Binary Coalescence

Like neutron stars, black holes are expected to be found in binary systems, some of which will decay and lead to the merger of the black holes. However, we cannot observe a population of binary black holes so we have less reliable statistics on the rates we can expect. The best estimates suggest that the observed rate of coalescence will be similar to that of neutron stars: a sparser population will be detected to greater distances due to the larger masses [2]. Within the uncertainties it is perfectly possible that the first detection will be a binary black hole merger.

These observations will enormously increase our understanding of the black hole population: masses, spins, binary mass ratios. By comparing with numerical-relativity simulations, which are now very accurate, it may be possible to test general relativity, particularly cosmic censorship, which is the hypothesis (still unproven) that a merger will always lead to a black hole and not a naked singularity. The properties of the final black hole can be inferred from its ringdown pulsation spectrum, if observed.

Black holes probably also form binaries with neutron stars, although as yet no pulsar has been detected in orbit around a black hole. Observations of these mergers with gravitational waves would be very interesting from the point of view of binary evolution theory and also to constrain the mechanisms for producing gamma-ray bursts.

## 6.4 Gravitational Wave Pulsars

Spinning isolated neutron stars will emit gravitational waves if they are asymmetric. The asymmetries must be small, of course, or known spinning pulsars would by now have lost their spin to the emission of gravitational waves. This consideration sets an upper limit on the expected amplitude (the spindown limit). For two pulsars, the Crab and Vela, current gravitational wave observations have constrained the amplitudes more tightly than the spindown limit [18, 19]. Advanced detectors may detect such radiation or constrain many more systems.

The search for such pulsars is very demanding computationally, and blind searches (for systems not already known as radio pulsars) are only possible with the volunteer-computing platform Einstein@Home [20]. In fact, the power of this system is already being used to find weak pulsar signals in radio telescope data [21], as mentioned earlier.

# 7 Other Gravitational Wave Sources

Supernova explosions motivated the first bar gravitational wave detectors [1]; the expected kHz frequencies matched what bars could detect, and it seemed reasonable to think that these explosions were asymmetric enough to produce strong radiation. Recent studies, however, suggest that the amplitudes to be expected are rather small. This has to do with the difficulty that stars have in producing supernovae in the first place: the collapsed interior hangs up at large densities for a relatively long time before finally acquiring enough energy to explode. During this time, any asymmetries from the initial collapse are reduced, and the final explosion is dominated by outflow rather than oscillation, although it is possibly very turbulent (which can lead to stochastic gravitational wave emission). See, e.g. [22], for a recent study in this rapidly developing field.

Another signal that is certain to be present, but is expected to be very weak, is a stochastic gravitational wave background. There are many sources of such backgrounds, including all the binary systems in the Galaxy. At nanohertz frequencies, where pulsar timing arrays operate, this is expected to be the dominant signal and to be detectable within the next 10 years [23]. But in the ground-based frequency band these signals are likely to be weaker. A big prize would be to detect a background generated by the Big Bang, but current estimates suggest it is rather weak.

# 8 Detecting Gravitational Waves from Space

## *8.1 LISA and eLISA*

So far this review has focussed on the imminent detection of gravitational waves by ground-based interferometers. But many of the people involved in the ground-based activity are also helping to develop the project to put a detector into space. The proposal has long been known as LISA, although events have recently led to a descoped version called eLISA, described below. I shall use the term LISA as a generic descriptor of a concept that is at least the long-term science goal of the field, of which eLISA will be perhaps its first realization.

Gravitational waves at frequencies below about 1 Hz will be very difficult or impossible to detect from the Earth because the Earth's gravity is noisy at these frequencies, and no detector could be made to respond to cosmic gravitational waves and not also to terrestrial Newtonian gravity fluctuations. Seismic density disturbances, air pressure/density changes, and a host of other phenomena with timescales between 1 s and 1 h would produce responses in detectors that are larger than those expected from cosmic sources. The only way to observe in this frequency band is to get away from the Earth, into space. The LISA proposal has been developed since 1995 within ESA, and then since 1998 jointly with NASA. Along with this proposal, ESA has been developing the LISA Pathfinder (LPF) mission, whose purpose is to demonstrate and prove the novel aspects of the LISA measurement technology that cannot be tested on the ground. LPF must be flown before the final go-ahead can be given to build a LISA mission. NASA is a minor partner in LPF.

Unfortunately, in 2011 NASA, beset with cost overruns on other missions, withdrew from its partnership with ESA on LISA, and also on two other proposed large missions. ESA then asked the three proposals to descope and enter a competition for an ESA-only large mission launch around 2021. This competition was won in mid-2012 by JUICE, a mission to the moons of Jupiter. Interestingly, eLISA/NGO, the descoped gravitational wave mission, seems to have received the highest scientific rating by the ESA committees, but was nevertheless not selected, primarily because LPF has not yet launched. Since LISA Pathfinder is expected to be launched by 2015, it is not unreasonable to think that the chances of success for eLISA in the selection for the next large mission will be very good indeed. The eLISA team is therefore working hard to understand better the science capabilities of the instrument and at the same time to explore the possibility of a partnership with one or more other agencies (including NASA) that might bring in enough resources to restore some of the features of the original LISA proposal. The science document of the 2012 eLISA/NGO proposal is on the arXiv [24] and a summary appeared as a conference proceeding [25].

The LISA concept is to put a long-baseline interferometer into space, with arm-lengths not much shorter than the wavelengths of the gravitational waves being detected. This requires three spacecraft in an L-shaped configuration with separations of order $10^6$ km. In the original LISA concept these spacecraft form an equilateral triangle, and laser light is used along all three arms to monitor the small changes in light travel-time along the arms produced by gravitational waves, as illustrated in Fig. 3. In this way one can construct three different interferometers by using, in turn, each spacecraft as the central hub of a two-arm interferometer. These three interferometers measure different polarizations of the incoming radiation. If general relativity is correct then the output signals will be linearly dependent; checking this is a good test of the general relativity model for gravitational waves.

If the array is placed with its center on the Earth's orbit, some 20° or so away from the Earth, and if the plane of the array is tilted by 60° to the plane of the Earth's orbit, then when the spacecraft are given suitable initial velocities they will remain in an equilateral triangle, which rotates counter to the direction of the orbit, without any need to maintain the positions of the spacecraft: they simply follow their Newtonian orbits around the Sun. The spacecraft do need to carry thrusters, but only to resist external perturbations, such as fluctuations in solar radiation pressure. The ends of the interferometers' arms are defined by small cubes (called test masses) that fall freely in cavities inside the spacecraft (not attached to the spacecraft). The spacecraft effectively act as shields protecting the cubes from external perturbations. In this way the cubes follow geodesics of the gravitational field closely enough to track the disturbances produced by gravitational waves.

The LPF mission consists of a single spacecraft within which are two experiment packages like those that mark the ends of the LISA arms. Therefore, LPF contains a single LISA arm, only very short: half a meter instead of several million meters.

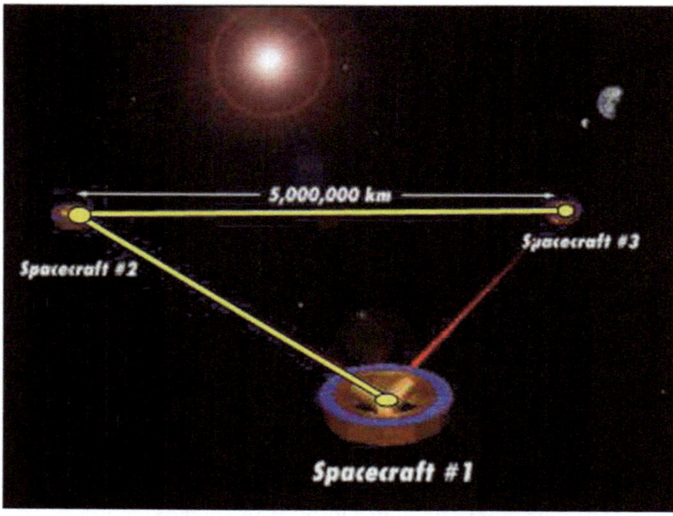


**Fig. 3** A schematic illustration of the LISA concept of three equidistant spacecraft in orbit around the Sun, forming three interferometers. With an appropriately chosen orbit, the array remains an equilateral triangle even as the spacecraft orbit the Sun freely

This means that LPF can prove the measurement technology of LISA (by measuring changes $\delta L$ in the proof-mass separation to the accuracy needed by LISA) but cannot measure gravitational waves because the baseline $L$ is so short that the smallest strain that can be measured, $\delta L/L$, is far larger than any expected from astronomical systems.

The eLISA proposal was descoped to save launch weight and some component costs. It must still have three spacecraft, but only two of the arms are illuminated with laser light, so there is only one interferometer signal. Although this reduces the information available from the mission, eLISA would still do spectacular science. In the following I describe this science without distinguishing too much between LISA and eLISA; this is because events in the near future might lead to further changes in the design of the mission that eventually is flown by ESA.

## 8.2 LISA Science

By virtue of its long arm lengths, any LISA-like mission will detect some events with extraordinarily high signal-to-noise ratios, as compared with our expectations for ground-based observing. Mergers of comparably sized massive black holes ($10^6 M_\odot$) can register SNR of up to 1000 in eLISA, and 5 or ten times higher in LISA. EMRI events, in which a stellar-mass black hole is captured from a nearly circular orbit by a massive black hole, can typically have SNR around 50–100. With such strong signals, it is possible to do stringent tests of gravity theory. EMRI events will probe whether the metric around central black holes in galaxies really is the Kerr metric. Merger events will give a detailed picture of how horizons merge (which can be compared with accurate numerical simulations) and will measure the spectrum of the ring-down radiation emitted as the final black hole settles down. From this we expect to measure the mass and spin of the hole, and thereby test cosmic censorship: is it a "clothed" Kerr metric or is it a naked singularity with $a > M$?

The astrophysics return will be similarly impressive and important. Any LISA-like mission will be able to detect comparable-mass black hole mergers at redshifts of 15 and beyond, if there are any that happen in the frequency band of the instrument (equivalently in the mass range $10^4$–$10^5 M_\odot$). These would be the earliest individual astronomical systems ever detected and studied by any observing method. They would tell us much about how galaxies formed, since black holes are believed to be tracers, embedded in proto-galactic clouds, merging with one another when their host clouds merge. A three-arm LISA would have enough information to measure distances to individual events, which when combined with known cosmological parameters would give the redshifts directly. With a two-arm eLISA mission, distance measurements have large but finite errors, and when these are averaged over an ensemble of detected events it will still be possible to discriminate among different models of early galaxy and black hole evolution [24, 25].

The mergers of relatively nearby pairs of massive black holes (out to z = 2) are likely to provide a link between mergers and their host galaxies. With positions provided by the three-arm LISA configuration, it should be possible to identify the galaxies since their morphology should be disturbed by having experienced the gradual spiralling together of the holes over hundreds of millions of years. This will be harder with eLISA data, but with both missions there will also be an intensive search for time-linked counterparts to the merger: enhanced X-ray, optical, or radio emission compared to observations of the same area of the sky in previous epochs. Such studies should provide clues to the mechanisms of jet emission and other phenomena associated with active galaxies.

The EMRI population will for the first time give us a relatively unbiased sample of massive black holes in the centres of galaxies. This will determine the mass distribution of these black holes and also help understand the nature of the star clusters in their immediate neighbourhoods.

White dwarf binary systems in our Galaxy will actually dominate the LISA data stream at low frequencies: there are enough systems in tight orbits to provide a confusion background for LISA below 1 mHz, and a large number of these systems will be directly detectable by both LISA and eLISA. This will again provide, for the first time, an unbiased sampling of this important population, which represents the endpoint of evolution of most binary star systems. The soon-to-be-launched Gaia [26] mission will identify thousands of such systems, which can then be matched with their gravitational wave counterparts by identifying systems with the same orbital periods. Distances provided the gravitational wave measurements (for systems which change their orbital period because of the emission of gravitational waves) will then calibrate the absolute brightness of the components.

## 9 Conclusion

The development of detectors has taken place over many decades, essentially because most of the key technologies had to be developed to the required sensitivity level within the field itself and could not be taken over from other areas of astronomy of physics. But within a very few years, ground-based detectors should be making observations with regularity, and as the network grows there could be events every couple of days. Space-based detectors will take longer to realise, again because the technology has to be developed and proved to the required sensitivity and degree of reliability demanded by space missions.

Driving these proposals is the completely new science that detectors can return, the new way of listening to the universe that provides unique information essentially orthogonal to that provided by telescopes. We will see these science returns affecting astronomy very soon, in the areas of neutron star physics, gamma-ray bursts, stellar evolution, black hole studies, and the search for gravitational wave pulsars. The science that can be delivered by a LISA-like space-based detector is recognised as having a high priority, and I am confident that we will see the launch of such a detector approved within the next few years, allowing us finally to listen to the the low-frequency whispers of our universe.

## References

1. Weber, J.: Gravitational radiation. Phys. Rev. Lett. **18**, 498 (1967)
2. Abadie, J., Belcznski, K., et al. (LIGO Scientific Collaboration, V. Collaboration): Predictions for the rates of compact binary coalescences observable by ground-based gravitational-wave detectore. Class. Quantum Gravity **27**(17), 173001 (2010). doi:10.1088/0264-9381/27/17/173001
3. Einstein, A.: Über den Einfluß der Schwerkraft auf die Ausbreitung des Lichtes. Annalen der Physik **35**, 898 (1911)
4. LSC Publications. http://www.lsc-group.phys.uwm.edu/ppcomm/Papers.html
5. Punturo, M., Abernathy, M., Acernese, F., et al.: The third generation of gravitational wave observatories and their science reach. Class. Quantum Gravity **27**(8), 084007 (2010)
6. Sathyaprakash, B., Abernathy, M., Acernese, F., et al.: Scientific objectives of Einstein telescope. Class. Quantum Gravity **29**(12), 124013 (2012). doi:10.1088/0264-9381/29/12/124013
7. Schutz, B.F.: Networks of gravitational wave detectors and three figures of merit. Class. Quantum Gravity **28**(12), 125023 (2011). doi:10.1088/0264-9381/28/12/125023
8. Collins, H.: Gravity's Shadow: The Search for Gravitational Waves. University of Chicago Press, Chicago (2004)
9. Ajith, P., Boyle, M., Brown, D.A., et al.: The NINJA-2 catalog of hybrid post-Newtonian/numerical-relativity waveforms for non-precessing black-hole binaries. Classical and Quantum Gravity **29**(12), 124001 (2012). doi:10.1088/0264-9381/29/12/124001
10. Home Page. https://www.ninja-project.org/doku.php?id=nrar:home
11. Einstein@Home Project Home Page. http://einstein.phys.uwm.edu/
12. Schutz, B.F.: Determining the Hubble constant from gravitational wave observations. Nature **323**, 310 (1986)
13. Del Pozzo, W.: Inference of cosmological parameters from gravitational waves: applications to second generation interferometers. Phys. Rev. D **86**(4), 043011 (2012). doi:10.1103/PhysRevD.86.043011

14. Rezzolla, L., Giacomazzo, B., Baiotti, L., et al.: The missing link: merging neutron stars naturally produce jet-like structures and can power short gamma-ray bursts. Astrophys. J. Lett. **732**, L6 (2011). doi:10.1088/2041-8205/732/1/L6
15. Nissanke, S., Holz, D.E., Hughes, S.A., Dalal, N., Sievers, J.L.: Exploring short gamma-ray bursts as gravitational-wave standard sirens. Astrophys. J. **725**, 496 (2010). doi:10.1088/0004-637X/725/1/496
16. Will, C.M.: Bounding the mass of the graviton using gravitional-wave observations of inspiralling compact binaries. Phys. Rev. D **57**, 2061 (1998)
17. Alexander, S., Yunes, N.: Chern-Simons modified general relativity. Phys. Rep. **480**, 1 (2009). doi:10.1016/j.physrep.2009.07.002
18. Abbott, B.: Beating the spin-down limit on gravitational wave emission from the Crab pulsar. Astrophys. J. Lett. **683**, L45 (2008). doi:10.1086/591526
19. Abadie, J., Abbott, B.P., Abbott, R., et al.: Beating the spin-down limit on gravitational wave emission from the Vela pulsar. Astrophys. J. **737**, 93 (2011). doi:10.1088/0004-637X/737/2/93
20. Aasi, J., Abadie, J., Abbott, B.P., et al.: Einstein@Home all-sky search for periodic gravitational waves in LIGO S5 data, ArXiv e-prints arXiv:1207.7176 [gr-qc] (2012)
21. Knispel, B., Lazarus, P., Allen, B., et al.: Arecibo PALFA survey and Einstein@Home: binary pulsar discovery by volunteer computing. Astrophys. J. Lett. **732**, L1 (2011). doi:10.1088/2041-8205/732/1/L1
22. Ott, C.D., Abdikamalov, E., Moesta, P., et al.: General-relativistic simulations of three-dimensional core-collapse supernovae. ArXiv e-prints arXiv:1210.6674 [astro-ph.HE] (2012)
23. Sesana, A., Vecchio, A.: Gravitational waves and pulsar timing: stochastic background, individual sources and parameter estimation. Class. Quant. Grav. **27**, 084016 (2010). doi:10.1088/0264-9381/27/8/084016
24. Amaro-Seoane, P., Aoudia, S., Babak, S., et al.: eLISA: astrophysics and cosmology in the millihertz regime. ArXiv e-prints arXiv:1201.3621 [astro-ph.CO] (2012)
25. Amaro-Seoane, P., Aoudia, S., Babak, S., et al.: Low-frequency gravitational-wave science with eLISA/NGO. Class. Quantum Gravity **29**(12), 124016 (2012)
26. Turon, C., O'Flaherty, K.S., Perryman, M.A.C.: The three-dimensional universe with Gaia. ESA Special Publication, SP-576 (2005). http://sci.esa.int/gaia/37100

# LISA in 2012 and Beyond: 20 Years After the First Proposal

**Gerhard Heinzel and Karsten Danzmann**

**Abstract** After 20 years of study as a joint ESA-NASA mission, LISA had to be redesigned as an ESA-only mission in 2011/2012 to meet programmatic and budgetary constraints of the space agencies. The result is a mission concept called "eLISA" or "NGO" with two arms instead of three and one million km armlengths instead of 5, which results in smaller launch mass but still provides revolutionary science. Nevertheless, even the reduced science is expected to be revolutionary for the study of black holes and other astrophysical and cosmological questions. "eLISA"/"NGO" was not selected in ESA's call for the first ("L1") large mission in the Cosmic Vision program, but is a strong candidate for the L2 call, with possible international contributions from the US and/or China.

## 1 Introduction

Gravitational Waves have been predicted in 1916 by Albert Einstein as a consequence of GR. So far they have not yet been directly observed, despite the enormous efforts invested in resonant detectors and ground-based interferometers since the 1960s, culminating in the present LIGO, VIRGO, GEO600 and similar detectors. These are ultra-sensitive Michelson-type laser interferometers with km-scale armlengths and sophisticated vibration-isolation systems. Nevertheless, they are insensitive at frequencies below about 10 Hz because of inevitable disturbances on Earth such as seismic and gravity gradient noise. A first direct detection with these instruments is expected in a few years when the next generation with yet another huge step in sensitivity improvement goes online.

G. Heinzel (✉) · K. Danzmann
Max-Planck-Institut für Gravitationsphysik (Albert-Einstein-Institut), Callinstrasse 38, 30167 Hannover, Germany
e-mail: gerhard.heinzel@aei.mpg.de

K. Danzmann
e-mail: karsten.danzmann@aei.mpg.de

J. Bičák and T. Ledvinka (eds.), *General Relativity, Cosmology and Astrophysics*, Fundamental Theories of Physics 177, DOI: 10.1007/978-3-319-06349-2_22,
© Springer International Publishing Switzerland 2014

The majority of expected sources, however, have frequencies well below 10 Hz and are therefore only accessible from space. For their detection, space-based laser interferometers have been studied since the 1980s. LISA was selected as an ESA cornerstone mission in 1995, and the concept of a collaborative ESA/NASA mission with 3 spacecraft in heliocentric orbits trailing the Earth and inclined by 60° against the ecliptic first appeared in 1997. This basic mission concept is basically unchanged until today.

The mission design was refined in great detail between 2004 and 2010 in an joint ESA/NASA Mission Formulation study that included an ESA-sponsored industrial study by EADS Astrium. Both the feasibility and scientific case were scrutinized in numerous reviews both in Europe and the US, such as the NRC's Beyond Einstein Program Assessment Committee (BEPAC) in 2006–7 and the Astro2010 Astronomy and Astrophysics Decadal Survey. All these reviews attested LISA a compelling and convincing science case as well as technical feasibility. Literally thousands of papers have been published on LISA sources, data analysis and instrumentation.

## 2 Planned ESA Cosmic Vision L1 Selection in 2011

In preparation for the call for proposals for the first "Large" mission (L1) in the ESA Cosmic Vision program, an Assessment Study Report ("Yellow Book") was prepared in February 2011 [1] that summarizes the scientific objectives, mission design and most important literature.

At that time three missions were in competition: LISA, the International X-ray Observatory (IXO), and the Europa Jupiter System Mission (EJSM-Laplace), all of which were conceived as ESA-NASA partnerships with about equal contributions.

In March 2011, however, ESA announced that due to budgetary constraints in NASA related to (among other reasons) the James Webb Space Telescope, none of these three missions could rely on the required NASA contribution [2]. Thus, the L1 downselection was postponed by one year, and the three missions were given the homework to modify their mission concepts such that they fit into an ESA-only envelope.

The LISA team studied different options, supported by industry and a mission concept study in ESA's concurrent design facility (CDF) in June/July 2011, and concluded that the "eLISA"/"NGO" concept described below would be the optimal mission within the given constraints. A new version of the "Yellow Book" was prepared and published in January 2012 [3].

## 3 The eLISA/NGO Mission Concept

Since the name LISA was considered to refer to the specific ESA/NASA mission described in [1], with 6 arms of 5 million km length etc., it was required to find a new name for the rescoped mission concept. The two names eLISA ("evolved LISA") and NGO ("New Gravitational wave Observatory") were chosen and used interchangeably for this purpose. See, however, Sect. 6 below regarding current naming conventions. The given constraints were a total cost for ESA of 850 M€, and member state contributions of about 200 M€. It quickly turned out that the most efficient way to reduce the predicted cost while maintaining as much of the science as possible is to reduce the launch mass and volume and the mission lifetime.

This was achieved by

- shrinking the telescope from 38 cm diameter to 20 cm, with subsequent reductions also in length and height of the payload,
- shrinking the armlength from 5 to 1 million km, which requires less fuel for plane-change maneuvers to leave the ecliptic plane and helps to restore the reduced light power levels due to the smaller telescope,
- omitting the last maneuver of the cruise phase, resulting in a "drift-away" orbit with ever increasing distance to the Earth,
- omitting the third arm, resulting in a 'V' shaped configuration instead of a full triangle.

Especially the omission of the third arm obviously leads to further cost reductions beyond the reduced launch mass, since less payload hardware is required. Other factors that contributed to the cost saving are:

- reducing the required mission lifetime from 5 to 2 years, with associated savings in on-ground testing and operations cost,
- providing the science instrument by an ESA memberstate consortium.

Apart from the above changes, the NGO payload is basically unchanged from LISA, and would be mounted into 3 identical spacecraft buses based on the LISA Pathfinder design, one 'mother' at the vertex of the 'V' and two simpler 'daughter' spacecrafts at the ends. The two lighter daughter would together fit into one Soyuz launcher, while a second launcher is used for the mother spacecraft.

These reductions come, of course, at a cost in instrument performance. Figure 1 shows the strain sensitivity in comparison.

A significant qualitative reduction in science output results from the omission of the third arm: The NGO standard Michelson interferometer cannot instantaneously disentangle two possible mechanisms for a reduction in signal amplitude: larger distance to the source or polarization mismatch to the antenna orientation. The LISA triangle with 6 links, however, can simultaneously measure both polarizations and thereby provide absolute distance measurements to sources whose brightness is known, e.g., through the time evolution of the waveform. This shortcoming of NGO is somewhat mitigated, however, for sources that are observed over a significant fraction of a year, since the rotation of the constellation restores some of the polarization sensitivity. NGO would observe thousands of Compact White Dwarf Binaries (CWDB), hundreds of black hole binary inspirals and dozens of Extreme Mass Ratio Inspirals (EMRI).

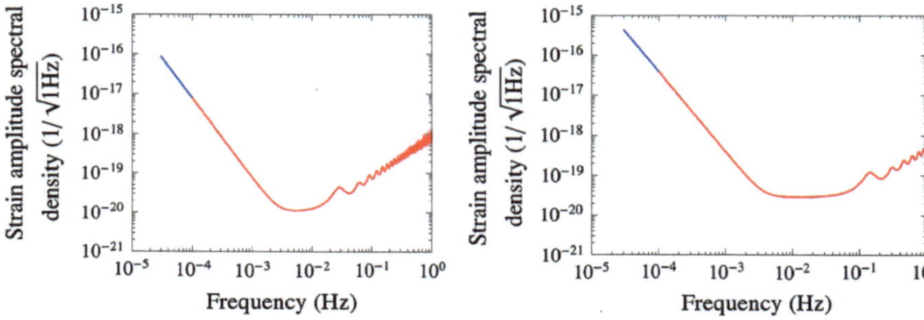


**Fig. 1** Strain sensitivity of LISA (*left*) and NGO (*right*) in comparison (*source* [1, 3])

# 4 The eLISA Consortium

A significant programmatic change between LISA and NGO was the definition of an 'instrument' that would be delivered by the member states. That instrument consists of most of the scientific payload, with the exception of the lasers and telescopes. Hence, the instrument contains:

- the test masses in their Gravitational Reference Sensors, including front-end electronics and discharging equipment,
- the optical benches, including mechanisms and photoreceivers,
- the phasemeters and
- integration and testing tasks.

An important difference to the LISA planning is the phasemeter, which used to be a NASA contribution, but for eLISA/NGO needs to come from Europe. An ESA technology development contract granted to a Danish/German team will be completed in early 2013 and will produce a European phasemeter with all required functionality and performance. A consortium consisting of institutes and space agencies in Denmark, France, Germany, Italy, Spain, Switzerland and UK was formed and committed to develop and deliver the instrument.

# 5 The ESA L1 Decision and Its Aftermath

In early May 2012, the ESA Science Programme Committee (SPC) chose the Jupiter moon mission JUICE (evolved from EJSM-Laplace) as L1 mission to be launched in 2022. Reasons for not selecting eLISA/NGO in spite of its repeatedly praised scientific value and technological readiness, as judged by ESA itself, were not clear; possibly having to do with remaining concerns about the risk of such a revolutionary new instrument and the desire to wait for the results of LISA Pathfinder, now planned to be launched in 2015. At the same time ESA announced to continue the technology development for eLISA/NGO.

The 9th LISA Symposium was held in Paris from May 21–25, 2012, with wide participation of scientists from Europe, the US and China. The community resolved to push forward with the LISA concept and submit an unbeatable entry for the L2 call for proposals in ESA's Cosmic Vision program, expected to occur in 2013 or 2014. The 'instrument' definition and consortium were maintained and are in function now. The first meeting of the new eLISA consortium took place October 22–23, 2012, also in Paris, and it was decided to form several working groups for science and technology. Both the consortium and ESA agree that third-party contributions of up to 20 % of the mission cost are welcome, e.g. from the US or China, provided they are non-essential, i.e. European alternative suppliers are available in case of necessity.

## 6 Naming the Mission Concepts

At the 9th LISA Symposium in Paris it was decided to retain the name LISA for the general concept of a gravitational wave detector in space, consisting of a triangle in inclined heliocentric orbits, with drag-free operation, armlengths of some million km and heterodyne laser interferometers along the arms. "eLISA"/"NGO" is one specific incarnation of that concept under study in 2012.

## 7 Conclusion

Despite the drawbacks in 2011 and 2012, the scientific interest in LISA is stronger than ever before, the technology is well developed, the team is strong and convinced that LISA must fly in the early 2020s, and is committed to work hard to make that happen.

## References

1. LISA: Unveiling a hidden universe. Technical report ESA/SRE(2011) 3, Noordwijk, Paris (2011). http://sci.esa.int/lisa/48364
2. Reich, E.S.: Europe makes do without NASA. Nature **471**, 421 (2011). doi:10.1038/471421a
3. NGO: Revealing a hidden universe: opening a new chapter of discovery. Technical report ESA/SRE(2011) 19, Noordwijk, Paris (2011). http://sci.esa.int/ngo/49839

# Einstein's Gravity as Seen by a Cosmic Lighthouse Keeper

**Michael Kramer**

**Abstract** The last years have seen continuing activities in the exploration of our understanding of gravity, motivated by results from precision cosmology and new precision astrophysical experiments. At the centre of attention lies the question as to whether general relativity is the correct theory of gravity. In answering this question, we work not only towards correctly interpreting the phenomenon of "dark energy" but also towards the goal of achieving a quantum theory of gravity. In these efforts, the observations of pulsars, especially those in binary systems, play an important role. Pulsars do not only provide the only evidence for the existence of gravitational waves so far, but they also provide precision tests of general relativity and alternative theories of gravity. This talk summarizes the current state-of-art in these experiments and looks into the future.

## 1 Introduction

This conference celebrated Einstein's time in Prague 100 years ago. As detailed during the conference and in these proceedings, important groundwork for the later theory of general relativity (GR) was achieved by Einstein during this time. Now, at the time of writing, we are less than 3 years away from celebrating the centenary of Einstein's greatest achievement. And yet, we have also seen a lot of presentations at this conference that addressed the properties and experimental consequences of alternative theories of gravity. As a matter of fact, nearly a 100 years later, efforts in testing GR and its concepts are still being made by many colleagues around the world, using many different approaches. To date GR has passed all these experimental and

M. Kramer (✉)
Max-Planck-Institut für Radioastronomie, Auf dem Hügel 69, 53121 Bonn, Germany
e-mail: mkramer@mpifr.de

M. Kramer
Jodrell Bank Centre for Astrophysics, University of Manchester, Oxford Road, Manchester M13 9PL, UK

J. Bičák and T. Ledvinka (eds.), *General Relativity, Cosmology and Astrophysics*, Fundamental Theories of Physics 177, DOI: 10.1007/978-3-319-06349-2_23,

© Springer International Publishing Switzerland 2014

observational tests with flying colours, but in light of recent progress in observational cosmology in particular, the question of as to whether alternative theories of gravity need to be considered is as topical as ever.

Many experiments are designed to achieve ever more stringent tests by either increasing the precision of the tests or by testing different, new aspects. Some of the most stringent tests are obtained by satellite experiments in the solar system, providing exciting limits on the validity of GR and alternative theories of gravity. However, solar-system experiments are made in the gravitational weak-field regime, while deviations from GR may appear only near the Planck scale or in strong gravitational fields.

We are all very much looking forward to the first direct detection of gravitational waves with ground-based (and hopefully, eventually, space-based) detectors, which will not only open a completely new window to the Universe but which will also provide superb tests of GR under strong-field conditions. Meanwhile, it happens that nature provides us with an almost perfect laboratory to test the strong-field regime—in the form of binary radio pulsars.

## *1.1 Pulsars*

Pulsars are rotating neutron stars that emit a radio beam that is eventually powered by the pulsars' rotational energy and that is centred on the magnetic axis of the neutron star. As the magnetic axis and hence the beam are inclined to the rotation axis, the pulsar acts as a cosmic lighthouse, and a pulsar appears as a pulsating radio source. The moment of inertia and the stored rotational energy of pulsars are large, so that in particular the fast rotating millisecond pulsars deliver a radio "tick" per rotation with a precision that rivals the best atomic clocks on Earth. Corresponding pulse (or spin) periods range from 1.4 ms to 8.5 s. As they concentrate an average of 1.4 solar masses on a diameter of only about 20 km, pulsars are exceedingly dense and compact, representing the densest matter in the observable universe. The resulting gravitational field near the surface is large, enabling strong-field tests of gravity.

For these tests, it is irrelevant how this emission is created, as long the lighthouse effects sends us a regular beacon. That is useful, because after more than 40 years of pulsar research, the details of the actual emission process still elude us. However, we have a basic understanding that is sufficient to perform the experiments described later. In our straw-man model, the high magnetic field of the rotating neutron star ($B_{\rm surf} \sim 10^8$ to $10^{14}$ Gauss) induces a huge electric quadrupole field and an electromagnetic force that exceeds gravity by 10–12 orders of magnitudes. Charges are pulled out easily from the surface, and the result is a dense, magnetized plasma that surrounds the pulsar. The strong magnetic field forces the plasma to co-rotate with the pulsar like a rigid body. This co-rotating *magnetosphere* can only extend up to a distance where the co-rotation velocity reaches the speed of light.[1] This distance defines the so-called light cylinder which separates the magnetic field lines into two

[1] Strictly speaking, the Alfvén velocity will determine the co-rotational properties of the magnetosphere.

distinct groups, i.e. *open and closed field lines*. Closed field lines are those which close within the light cylinder, while open field lines would close outside. The plasma on the closed field lines is trapped and will co-rotate with the pulsar forever. In contrast, plasma on the open field lines can reach highly relativistic velocities and can leave the magnetosphere, creating the observed well-confined radio beam at a distance of a few tens to hundreds of km above the pulsar surface. It is this beam which creates the pulsating signals once per rotation when it points towards Earth.

## 1.2 Pulsars and Their Companions

The idea behind the usage of pulsars for testing GR and alternative theories of gravity is straightforward: if the pulsar is in orbit with a binary companion, we use the measured variation in the arrival times of the received signal to determine and trace the orbit of the pulsar about the common centre of mass as it moves in the curved space-time of the companion.

This "pulsar timing" experiment is simultaneously clean, conceptually simple and very precise. The latter is true since when measuring the exact arrival time of pulses at our telescope on Earth, we do a ranging experiment that is vastly superior in precision than a simple measurement of Doppler-shifts in the pulse period. This is possible, since the pulsed nature of our signal links tightly and directly to the rotation of the neutron star, allowing us to count every single rotation of a neutron star. Furthermore, in this experiment we can consider the pulsar as a test mass that has a precision clock attached to it.

While, strictly speaking, binary pulsars move in the weak gravitational field of a companion, they do provide precision tests of the strong-field regime. This becomes clear when considering that the majority of alternative theories predicts strong self-field effects which would clearly affect the pulsars' orbital motion. Hence, tracing their fall in a gravitational potential, we can search for tiny deviations from GR, providing us with unique precision strong-field tests of gravity.

As a result, a wide range of relativistic effects can be observed, identified and studied. These include so far:

- Precession of periastron,
- Gravitational redshift,
- Shapiro delay due to curved space-time,
- Gravitational wave emission,
- Geodetic precession, relativistic spin-orbit coupling,
- Speed of gravity.

But we can also convert our observations in tests of concepts and principles deeply embedded in theoretical frameworks, such as

- Strong Equivalence Principle (grav. Stark effect),
- Lorentz invariance of gravitational interaction,

- Non-existence of preferred frames,
- Conservation of total momentum,
- Non-variation of gravitational constant,

which also leads to stringent limits on alternatives theories of gravity, e.g. tensor–scalar theories, Tensor-Vector-Scalar (TeVeS) theories.

## 1.3 Nature is Kind...

The various effects or concepts to be tested require sometimes a rather different type of laboratory. For instance, to test the violation of the Strong-Equivalence-Principle, one would like to use a binary system that consists of different types of masses (i.e. with different gravitational self-energy), rather than a system made of very similar bodies. Fortunately, nature has been kind.

At the moment, we know about 2,000 pulsars, with about 10 % of these in binary systems. The shortest orbital period is about 90 min while the longest period is 5.3 yr (e.g. [1]). We find different types of components, i.e. main-sequence stars, white dwarfs (WD), neutron stars (NS) and even planets. Unfortunately, despite past and on-going efforts, we have not yet found a pulsar about a stellar black hole companion or about the supermassive black hole in the centre of our Galaxy [2]. Double neutron star systems are rare but usually produce the largest observable relativistic effects in their orbital motion and, as we will see, produce the best tests of general relativity for strongly self-gravitating bodies. In comparison, pulsar—white dwarf systems are much more common. Indeed, most pulsar companions are white dwarfs, with a wide range of orbital periods, ranging from hours to days, weeks and months. Still, many of them can be used for tests of gravitational theories where we utilize the fact that white dwarfs and neutron stars differ very significantly in their structure and, consequently, self-energies.

## 1.4 Precision Experiments

Using pulsar timing techniques, we make extremely precise measurements that allow us to probe gravitation with exquisite accuracy. Table 1 gives an idea about the precision that we already achieve today. As discussed later in this contribution, with future telescopes like the "Square Kilometre Array", the precision will even be enhanced by at least two orders of magnitudes and should, for instance, allow us to find a pulsar orbiting SGR A*, which would provide the mass of the central black hole to a precision of an amazing $1 M_{\odot}$! It would also allow us to measure the spin of the black hole with a precision of $10^{-4}$ to $10^{-3}$ (enabling tests of the "cosmic censorship conjecture") and the quadrupole moment with a precision to $10^{-4}$ to $10^{-3}$ (thus enabling tests of the "no-hair theorem"). See later and [2] for more details.

**Table 1** Examples of precision measurements using pulsar timing as a variation demonstration what is possible today

| | | |
|---|---|---|
| *Masses* | | |
| Masses of neutron stars | $m_1 = 1.4398(2)\ M_\odot$ | [3] |
| | $m_2 = 1.3886(2)\ M_\odot$ | [3] |
| Mass of WD companion | $0.207(2)\ M_\odot$ | [4] |
| Mass of millisecond pulsar | $1.67(2)\ M_\odot$ | [5] |
| Main sequence star companion | $1.029(8)\ M_\odot$ | [5] |
| Mass of Jupiter and moons | $9.547921(2) \times 10^{-4}\ M_\odot$ | [6] |
| *Spin parameters* | | |
| Period | 5.757451924362137(2) ms | [7] |
| *Orbital parameters* | | |
| Period | 0.102251562479(8) day | [8] |
| Eccentricity | $3.5(1.1) \times 10^{-7}$ | [9] |
| *Astrometry* | | |
| Distance | 157(1) pc | [7] |
| Proper motion | 140.915(1) mas $\text{yr}^{-1}$ | [7] |
| *Tests of general relativity* | | |
| Periastron advance | 4.226598(5) deg $\text{yr}^{-1}$ | [3] |
| Shrinkage due to GW emission | 7.152(8) mm/day | [8] |
| GR validity (obs/exp) | 1.0000(5) | [8] |
| Constancy of grav. constant, $\dot{G}/G$ | $-0.6(1.6) \times 10^{-12}\ \text{yr}^{-1}$ | [9] |

The digit in bracket indicates the uncertainty in the last digit of each value. References are cited

## 2 The Hulse-Taylor Pulsar: Gravitational Wave Damping

The first binary pulsar to ever be discovered happened to be a rare double neutron star system. It was discovered by Hulse and Taylor in 1974 [10]. The pulsar, B1913+16, has a period of 59 ms and is in eccentric ($e = 0.61$) orbit around a unseen companion with an orbital period of less than 8 h. It became soon clear that the pulsar does not follow the movement expected from a simple Keplerian description of the binary orbit, but that it shows the impact of relativistic effects. In order to describe the relativistic effects in a theory-independent fashion, one introduces so-called "Post-Keplerian" (PK) parameters that are included in a timing model to accurately describe the measured pulse times-of-arrival (see e.g. [1] for more details).

For the Hulse-Taylor pulsar, it was soon measured that the system showed a relativistic advance of its periastron, comparable to what is seen in the solar system for Mercury, albeit with a much larger amplitude of $\dot{\omega} = 4.226598 \pm 0.000005$ deg/yr [3]. GR predicts a value for the periastron advance that depends on the Keplerian parameters and the masses of the pulsar and its companion:

$$\dot{\omega} = 3T_\odot^{2/3} \left(\frac{P_b}{2\pi}\right)^{-5/3} \frac{1}{1-e^2} (m_p + m_c)^{2/3}. \tag{1}$$

Here, $T_\odot$ is a constant, $P_b$ the orbital period, $e$ the eccentricity, and $m_p$ and $m_c$ the masses of the pulsar and its companion. See [1] for further details.

The Hulse-Taylor pulsar also shows the effects of gravitational redshift (including a contribution from a second-order Doppler effect) as the pulsar moves in its elliptical orbit at varying distances from the companion and with varying speeds. The result is a variation in the clock rate with an amplitude of $\gamma = 4.2992 \pm 0.0008$ ms [3]. In GR, the observed value is related to the Keplerian parameters and the masses as

$$\gamma = T_\odot^{2/3} \left(\frac{P_b}{2\pi}\right)^{1/3} e \frac{m_c(m_p + 2m_c)}{(m_p + m_c)^{4/3}}. \tag{2}$$

We can now combine these measurements. We have two equations with a measured left-hand side. On the right-hand side, we measured everything apart from two unknown masses. We solve for those and obtain, $m_p = 1.4398 \pm 0.0002\ M_\odot$ and $m_c = 1.3886 \pm 0.0002\ M_\odot$ [3]. These masses are correct if GR is the right theory of gravity. If that is indeed the case, we can make use of the fact that (for point masses with negligible spin contributions), the PK parameters in each theory should only be functions of the a priori unknown masses of pulsar and companion, $m_p$ and $m_c$, and the easily measurable Keplerian parameters [11].[2] With the two masses now being determined using GR, we can compare any observed value of a third PK parameter with the predicted value. A third such parameter is the observed decay of the orbit which can be explained fully by the emission of gravitational waves. And indeed, using the derived masses, and the prediction of general relativity, i.e.

$$\dot{P}_b = -\frac{192\pi}{5} T_\odot^{5/3} \left(\frac{P_b}{2\pi}\right)^{-5/3} \frac{\left(1 + \frac{73}{24}e^2 + \frac{37}{96}e^4\right)}{(1 - e^2)^{7/2}} \frac{m_p m_c}{(m_p + m_c)^{1/3}}, \tag{3}$$

one finds an agreement with the observed value of $\dot{P}_b^{obs} = (-2.423 \pm 0.001) \times 10^{-12}$ [3]—however, only if a correction for a relative acceleration between the pulsar and the solar system barycentre is taken into account. As the pulsar is located about 7 kpc away from Earth, it experiences a different acceleration in the Galactic gravitational potential than the solar system (see e.g. [1]). The precision of our knowledge to correct for this effect eventually limits our ability to compare the GR prediction to the observed value. Nevertheless, the agreement of observations and prediction, today within a 0.2 % (systematic) uncertainty [3], represented the first evidence for the existence of gravitational waves. Today we know many more binary pulsars where we can detect gravitational wave emission. In one particular case, the measurement uncertainties are not only more precise, but also the systematic uncertainties are much smaller, as the system is much more nearby. This system is the Double Pulsar.

---

[2] For alternative theories of gravity this statement may only be true for a given equation of state.

# 3 The Double Pulsar

The Double Pulsar was discovered in 2003 [12, 13]. It does not only show larger relativistic effects and is much closer to Earth (about 1 kpc) than the Hulse-Taylor pulsar, allowing us to largely neglect the relative acceleration effects, but the defining unique property of the system is that it does not consist of one active pulsar and its *unseen* companion, but that it harbours two *active* radio pulsars.

One pulsar is mildly recycled with a period of 22 ms (named "A"), while the other pulsar is young with a period of 2.8 s (named "B"). Both orbit the common centre of mass in only 147 min with orbital velocities of 1 Million km per hour. Being also mildly eccentric ($e = 0.09$), the system is an ideal laboratory to study gravitational physics and fundamental physics in general. A detailed account of the exploitation for gravitational physics has been given, for instance, by [14–16]. An update on those results is in preparation [8], with the largest improvement undoubtedly given by a large increase in precision when measuring the orbital decay. Not even 10 years after the discovery of the system, the Double Pulsar provides the best test for the accuracy of the gravitational quadrupole emission prediction by GR far below the 0.1 % level.

In order to perform this test, we first determine the mass ratio of pulsar A and B from their relative sizes of the orbit, i.e. $R = x_B/x_A = m_A/m_B = 1.0714 \pm 0.0011$ [14]. Note that this value is theory-independent to the 1PN level [17]. The most precise PK parameter that can be measured is a large orbital precession, i.e. $\dot{\omega} = 16.8991 \pm 0.0001$ deg/yr. Using Eq. (1), this measured value and the mass ratio, we can determine the masses of the pulsars, assuming GR is correct, to be $m_A = (1.3381 \pm 0.0007)\, M_\odot$ and $m_B = (1.2489 \pm 0.0007)\, M_\odot$.

We can use these masses to compute the expected amplitude for the gravitational redshift, $\gamma$, if GR is correct. Comparing the result with the observed value of $\gamma = 383.9 \pm 0.6\ \mu$s, we find that theory (GR) agrees with the observed value to a ratio of $1.000 \pm 0.002$, as a first of five tests of GR in the Double Pulsar.

The Double Pulsar also has the interesting feature that the orbit is seen nearly exactly edge-on. This leads to a 30 s long eclipse of pulsar A due to the blocking magnetosphere of B that we discuss further below, but it also leads to a "Shapiro delay": whenever the pulse needs to propagate through curved space-time, it takes a little longer than travelling through flat space-time. At superior conjunction, when the signal of pulsar A passes the surface of B in only 20,000 km distance, the extra path length due to the curvature of space-time around B leads to an extra time delay of about 100 $\mu$s. The shape and amplitude of the corresponding Shapiro delay curve yield two PK parameters, $s$ and $r$, known as *shape* and *range*, allowing two further tests of GR. $s$ is measured to $s = \sin(i) = 0.99975 \pm 0.00009$ and is in agreement with the GR prediction of

$$s = T_\odot^{-1/3} \left(\frac{P_\mathrm{b}}{2\pi}\right)^{-2/3} x \, \frac{(m_A + m_B)^{2/3}}{m_B}, \tag{4}$$

(where $x$ is the projected size of the semi-major axis measured in lt-s) within a ratio of $1.0000 \pm 0.0005$. It corresponds to an orbital inclination angle of $88.7 \pm 0.2°$, which is indeed very close to 90° as suggested by the eclipses. $r$ can be measured with much less precision and yields an agreement with GR's value given by

$$r = T_{\odot} m_B, \tag{5}$$

to within a factor of $0.98 \pm 0.02$.

A fourth test is given by comparing an observed orbital decay of $107.79 \pm 0.11$ ns/day to the GR prediction. Unlike the Hulse-Taylor pulsar, extrinsic effects are negligible and the values agree with each other without correction to within a ratio of $1.000 \pm 0.001$. This is already a better test for the existence of GW than possible with the Hulse-Taylor pulsar and will continue to improve with time. Indeed, at the time of writing the agreement has already surpassed the 0.1 % level significantly [8].

## 4 Relativistic Spin-Orbit Coupling

Apart from the Shapiro delay, the impact of curved space time is also immediately measurable by its effect on the orientation of the pulsar spin in a gyroscope experiment. This effect, known as geodetic precession or de Sitter precession, represents the effect on a vector carried along with an orbiting body such that the vector points in a different direction from its starting point (relative to a distant observer) after a full orbit around the central object. Experimental verification has been achieved by precision tests in the solar system, e.g. by Lunar Laser Ranging (LLR) measurements, or recently by measurements with the Gravity Probe-B satellite mission (see [18] for a review of experimental tests). However, these tests are done in the weak field conditions of the solar system, so that pulsars provide the only access to the strong-field regime.

In binary systems one can interpret the observations, depending on the reference frame, as a mixture of different contributions to relativistic spin-orbit interaction. One contribution comes from the motion of the first body around the centre of mass of the system (de Sitter-Fokker precession), while the other comes from the dragging of the inertial frame at the first body due to the translational motion of the companion [19]. Hence, even though we loosely talk about geodetic precession, the result of the spin-orbit coupling for binary pulsar is more general, and hence we will call it *relativistic spin-precession*. The consequence of relativistic spin precession is a precession of the pulsar spin about the total angular moment vector, changing the orientation of the pulsar relative to Earth.

Since the orbital angular momentum is much larger than the angular momentum of the pulsar, the orbital spin practically represents a fixed direction in space, defined by the orbital plane of the binary system. Therefore, if the spin vector of the pulsar is misaligned with the orbital spin, relativistic spin-precession leads to a change in viewing geometry, as the pulsar spin precesses about the total angular momentum

**Table 2** DNSs sorted according to the expected relativistic spin precession rate

| PSR | $P$ (ms) | $P_b$ (d) | $x$ (lt-s) | $e$ | $\Omega_p$ (deg yr$^{-1}$) |
|---|---|---|---|---|---|
| J0737 − 3039A/B* | 22.7/2770 | 0.10 | 1.42/1.51 | 0.09 | 4.8/5.1 |
| J1906 + 0746* | 144.1 | 0.17 | 1.42 | 0.09 | 2.2 |
| B2127 + 11C* | 30.5 | 0.34 | 2.52 | 0.68 | 1.9 |
| B1913 + 16* | 59.0 | 0.33 | 2.34 | 0.62 | 1.2 |
| J1756 − 2251 | 28.5 | 0.32 | 2.76 | 0.18 | 0.8 |
| B1534 + 12* | 37.9 | 0.42 | 3.73 | 0.27 | 0.5 |
| J1829 + 2456 | 41.0 | 1.18 | 7.24 | 0.14 | 0.08 |
| J1518 + 4904 | 40.9 | 8.64 | 20.0 | 0.25 | – |
| J1753 − 2240 | 95.1 | 13.63 | 18.1 | 0.30 | – |
| J1811 − 1736 | 104.2 | 18.8 | 34.8 | 0.83 | – |
| J1141 − 6545* | 394.0 | 0.20 | 1.89 | 0.17 | 1.4 |

Also included is PSR J1141−6545 which is in a relativistic orbit about a white dwarf companion. Pulsars marked with an asterisk have been identified of showing spin precession. For sources where no precession rate is listed, the companion mass could not be accurately measured yet, indicating however, that the precession rate is low

vector. Consequently, as many of the observed pulsar properties are determined by the relative orientation of the pulsar axes towards the distant observer on Earth, we should expect a modulation in the measured pulse profile properties, namely its shape and polarisation characteristics [20]. The precession rate is another PK parameter and given in GR by (e.g. [1])

$$\Omega_p = T_\odot^{2/3} \times \left(\frac{2\pi}{P_b}\right)^{5/3} \times \frac{m_c(4m_p + 3m_c)}{2(m_p + m_c)^{4/3}} \times \frac{1}{1 - e^2}. \tag{6}$$

In order to see a measurable effect in any binary pulsar, (a) the spin axis of the pulsar needs to be misaligned with the total angular momentum vector and (b) the precession rate must be sufficiently large compared to the available observing time to detect a change in the emission properties. Table 2 lists the known Double Neutron Star Systems which typically show the largest degree of relativistic effects due to the often short eccentric binary orbits. However, the last entry in the table is PSR J1141−6545 which is a relativistic system with a white dwarf companion. Those pulsars that are marked with an asterisk have been identified as pulsars showing relativistic spin precession. Note that the top 5 out of 8 sources (with a known expected precession rate) indeed show the effect.

As the most relativistic binary system known to date, we expect a large amount of spin precession in the Double Pulsar system. Despite careful studies, profile changes for A have not been detected, suggesting that A's misalignment angle is rather small (e.g. Ferdman et al. in prep.). In contrast, changes in the light curve and pulse shape on secular timescales [21] reveal that this is not the case for B. In fact, B had been becoming progressively weaker and disappeared from our view in 2009 [22]. Making the valid assumption that this disappearance is solely caused by relativistic spin precession, it will only be out of sight temporarily until it reappears later. Modelling suggests that, depending on the beam shape, this will occur in about 2035 but an earlier time cannot be excluded. The geometry that is derived from this modelling is consistent with the results from complementary observations of spin precession, visible via a rather unexpected effect described in the following.

The change in the orientation of B also changes the observed eclipse pattern in the Double Pulsar, where we can see periodic bursts of emission of A during the dark eclipse phases, with the period being the full- or half-period of B. As this pattern is caused by the rotation of B's blocking magnetospheric torus that allows light to pass B when the torus rotates to be seen from the side, the resulting pattern is determined by the three-dimensional orientation of the torus, which is centred on the precessing pulsar spin. Eclipse monitoring over the course of several years shows exactly the expected changes, allowing to determine the precession rate to $\Omega_{\mathrm{p,B}} = 4.77^{+0.66}_{-0.65}$ deg/yr. This value is fully consistent with the value expected GR, providing a fifth test [23]. This measurement also allows to test alternative theories of gravity and their prediction for relativistic spin-precession in strongly self-gravitating bodies for the first time (see [16] for details).

## 5 Alternative Theories

Despite the successes of GR, a range of observational data have fuelled the continuous development of alternative theories of gravity. Such data include the apparent observation of "dark matter" or the cosmological results interpreted in the form of "inflation" and "dark energy". Confronting alternative theories with data also in other areas of the parameter space (away from the CMB or Galactic scales), requires that these theories are developed sufficiently in order to make predictions. A particular sensitive criterion is if the theory is able to make a statement about the existence and type of gravitational waves. Most theories cannot do this (yet), but a class of theories where this has been achieved is the class of tensor-scalar theories as discussed and demonstrated by Damour and Esposito-Farèse in a series of works (e.g. [24]). For corresponding tests, the choice of a double neutron star system is not ideal, as the difference in scalar change (that would be relevant, for instance, for the emission of gravitational *dipole* radiation) is small. The ideal laboratory would be a pulsar orbiting a black hole, as the black hole would have zero scalar charge and the difference would be maximised. The next best laboratory is a pulsar-white dwarf system. Indeed, such binary systems are able to provide constraints for alternative theories of gravity that are equally good or even better than solar system limits [9].

A recently studied pulsar-white dwarf system turns out to be very exciting: PSR J0348+0432 harbours a white dwarf whose composition and orbital motion can be precisely derived from optical observations. The results allow us to measure the mass of the neutron star, showing that it has a record-braking value of $2.01 \pm 0.04 M_{\odot}$! This is not only the most massive neutron star known (at least with reliable precision), but the 39 ms pulsar and the white dwarf orbit each other in only 2.46 h, i.e. the orbit is only tens of seconds longer than that of the Double Pulsar. Even though the orbital motion is nearly circular, the effect of gravitational wave damping is clearly measured. Thereby, the high pulsar mass and the compact orbit make this system a sensitive laboratory of a previously untested strong-field gravity regime. Thus far, the observed orbital decay agrees with GR, supporting its validity even for the extreme conditions present in the system (Antoniadis et al. submitted).

## 6 Detecting Gravitational Waves with Pulsars

The observed orbital decay in binary pulsars detected via precision timing experiments so far offers the only evidence for the existence of gravitational wave (GW) emission. Intensive efforts are therefore on-going world-wide to make a direct detection of gravitational waves that pass over the Earth. Ground-based detectors like GEO600, VIRGO or LIGO use massive mirrors, the relative distance of which are measured by a laser interferometer set-up, while the future space-based LISA detector uses formation flying of three test-masses that are housed in satellites. The change of the space-time metric around the Earth also influences the arrival times of pulsar signals measured at the telescope, so that high-precision MSP timing can also potentially directly detect GWs. Because pulsar timing requires the observations of a pulsar for a full Earth orbit before the relative position between pulsar, Solar System Barycentre and Earth can be precisely determined, only GWs with periods of more than a year can usually be detected. In order to determine possible uncertainties in the used atomic clocks, planetary ephemerides used, and also since GWs are expected to produce a characteristic quadrupole signature on the sky, several pulsars are needed to make a detection. The sensitivity of such a "Pulsar Timing Array" (PTA) increases with the number of pulsars and should be able to detect pulsars in the nHz regime, hence below the frequencies of LIGO ($\sim$kHz and higher) and LISA ($\sim\mu$Hz) (see Fig. 1).

A number of PTA experiments are ongoing, namely in Australia, Europe and North America (see [25] for a summary). The currently derived upper limits on a stochastic GW background (e.g. [26, 27]) are very close to the theoretical expectation for a signal that originates from binary super-massive black holes expected from the hierarchical galaxy evolution model [28, 29]. The best limit has recently been published by the European Pulsar Timing Array (EPTA) that uses the telescopes at Effelsberg, Jodrell Bank, Nancay and Westerbork [30]. Soon the Sardinia Radio Telescope will be added to the EPTA once it is completed.

While the limits are close, it seems that "simply a bit of extra sensitivity is needed" to make a first detection. This is the motivation for the Large European Array for Pulsars (LEAP) project in Europe. It aims to phase-coherently connect Europe's largest radio telescopes to form an Arecibo-sized dish that can observe a large number of millisecond pulsars with high sensitivity enabling high precision pulsar timing. LEAP is part of the EPTA and also acts as a test-bed for technology development for the SKA [27].

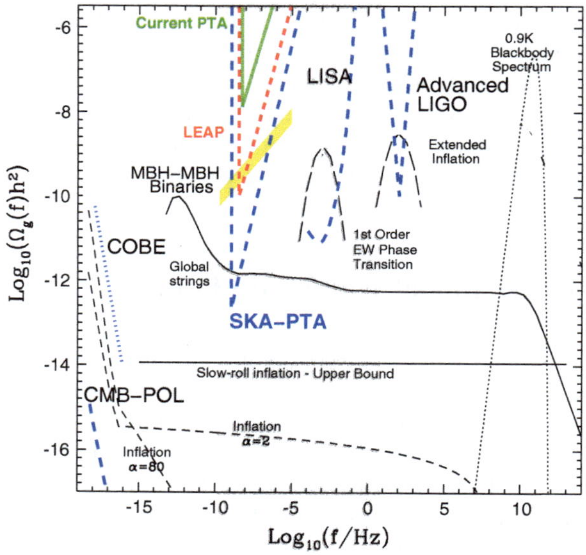


**Fig. 1** Summary of the potential cosmological sources of a stochastic gravitational wave (GW) background overlaid with bounds from COBE, current pulsar timing array (PTA) experiments and the goals of CMB polarization experiments, LISA and Advanced LIGO. LEAP will improve on the current best PTA limits by more than two orders of magnitude, enabling the detection of a GW background caused by the merger of massive *black holes* (MBHs) in early galaxy formation. The amplitude depends on the MBH mass function and merger rate, so that uncertainty is indicated by the size of the *shaded area*. LEAP is the next logical step towards a PTA realized with the SKA which will improve on the current sensitivity by about four orders of magnitude

Demonstrating the power of PTA experiments, Champion et al. [6] recently used data of PTA observations to determine the mass of the Jovian system independently of the spacecraft data obtained by fly-bys. Here, the idea is that an incorrectly known planet mass will result in an incorrect model of the location of the Solar System Barycentre relative to the Earth. However, as discussed the SSB is the reference point for pulsar arrival time measurements, so that a mismatch between assumed and actual position would lead to a periodic signal in the pulsar data with the period being that of the planet with the ill-measured mass. This measurement technique is sensitive to a mass difference of only 0.003 % of the mass of the Earth, and $10^{-7}$-th of Jupiter's mass.

If LEAP or other experiments do not detect GWs in the next few years, a first detection is virtually guaranteed with the more sensitive Phase I of the SKA. But the science that can eventually be done with the full SKA goes far beyond simple GW detection—a whole realm of astronomy and fundamental physics studies will become possible. For instance, it will be possible to study the properties of gravitational waves, such as their polarisation properties or the mass of the graviton [32, 33]. This is achieved by measuring the degree of correlation in the arrival time variation of pairs of pulsars separated by a certain angle on the sky. A positive correlation is expected for pulsars in the same direction or 180° apart on the sky, while pulsars separated by 90° should be anti-correlated. The exact shape of this correlation curve obviously depends on the GW polarisation properties [32] but also on the mass of the graviton [33]. The latter becomes clear when we consider that a non-zero mass leads to a dispersion relation and a cut-off frequency $\omega_{\rm cut} = m_{\rm g}c^2/\hbar$, below which a propagation is not possible anymore, affecting the degree of correlation possible between two pulsars. With a 90 % probability, massless gravitons can be distinguished from gravitons heavier than $3 \times 10^{-22}$ eV (Compton wavelength $\lambda_{\rm g} = 4.1 \times 10^{12}$ km), if bi-weekly observation of 60 pulsars are performed for 5 years with pulsar RMS timing accuracy of 100 ns. If 60 pulsars are observed for 10 years with the same accuracy, the detectable graviton mass is reduced to $5 \times 10^{-23}$ eV ($\lambda_{\rm g} = 2.5 \times 10^{13}$ km) [33].

In addition to detecting a *background* of GW emission, the probability of detecting a *single* GW source increases from a few percent now to well above 95 % with the full SKA. We can, for instance, expect to find the signal of a single supermassive black hole binary. Considering the case when the orbit is effectively not evolving over the observing span, we can show that, by using information provided by the "pulsar term" (i.e. the retarded effect of the GW acting on the pulsar's surrounding spacetime), we can achieve a rather astounding source localization. For a GW with an amplitude exceeding $10^{-16}$ and PTA observations of 40 pulsars with weekly timing to 30 ns precision, one can measure the GW source position to an accuracy of better than ~1 arcmin (Fig. 2, [31]). With such an error circle, an identification of the GW source in the electromagnetic spectrum should be easily feasible. We note that in order to achieve such a result, a precise distance measurement to the pulsars is needed, which in turn can then be improved further during the fitting process that determines the orbital parameters of the GW source. Fortunately, the SKA will be a superb telescope to do astrometry with pulsars [34].

## 7 The Future and the Ultimate Laboratory

Essentially all upcoming and future telescopes will contribute in one way or the other to advances in the field of pulsar astrophysics. LOFAR will find a large number of neutron stars that are potential sources for tests of relativity or pulsar timing array experiments that attempt to directly detect gravitational waves [35]. Even though these experiments are performed at high frequencies, LOFAR can find appropriate sources and also monitor the interstellar weather that needs to be corrected for in high precision timing. The Chinese FAST telescope will have a collecting area that will allow us to find and time pulsars that will be significantly better than currently achieved with Arecibo [36]. However, the real big advance for pulsars and their applications will be achieved with the Square Kilometre Array (SKA).

It is clear that the SKA will have a huge impact on the study of pulsars and their applications, in particular in using them for our understanding gravity. Some of the questions directly addressed with SKA pulsar studies are:

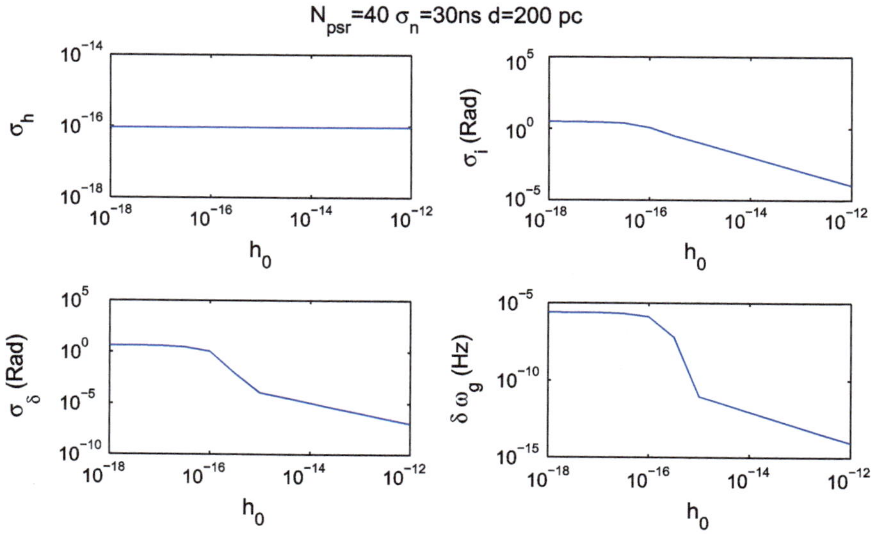


**Fig. 2** Results of computations for the detection of a single GW source with the help of a PTA consisting of 40 pulsars with an average timing precision of 300 ns at a typical distance of 200 pc [31]. *Top left* Error in measuring the characteristic strain amplitude of a single GW source for a variety of signal strengths. *Top right* Corresponding error in orbital inclination measurement, *bottom left* positional error on the sky and *bottom right* error in determining the gravitational wave frequency. See [31] for details

- What is the nature of gravity? Was Einstein right? Is gravity described by a tensor field or are there additional scalar fields, as it is sometimes proposed to explain Dark Energy?
- What are the properties of gravitational waves? Do gravitons have spin 2? What is the mass of gravitons and hence the propagation speed of gravitational waves?
- What happens in strong gravitational fields, in conditions of extreme curvature and near singularities? What are the properties of Black Holes? Do the no-hair and cosmic censorships theorems hold?

Answering these questions requires a survey for pulsars and the high-precision timing of a selected sample of those. In these searches, we can in particular expect to find the first pulsar-black hole systems.

What makes a binary pulsar with a black-hole companion so interesting is that it has the potential of providing a superb new probe of relativistic gravity. As pointed out by Damour and Esposito-Farèse [37], the discriminating power of this probe might supersede all its present and foreseeable competitors. The reason lies in the fact that such a system would be very sensitive to strong gravitational self-field effects, making it for instance an excellent probe for tensor-scalar theories.

But also for testing the black hole properties predicted by GR, a pulsar-BH system will be superb laboratory. Wex and Kopeikin [38] showed that the measurement

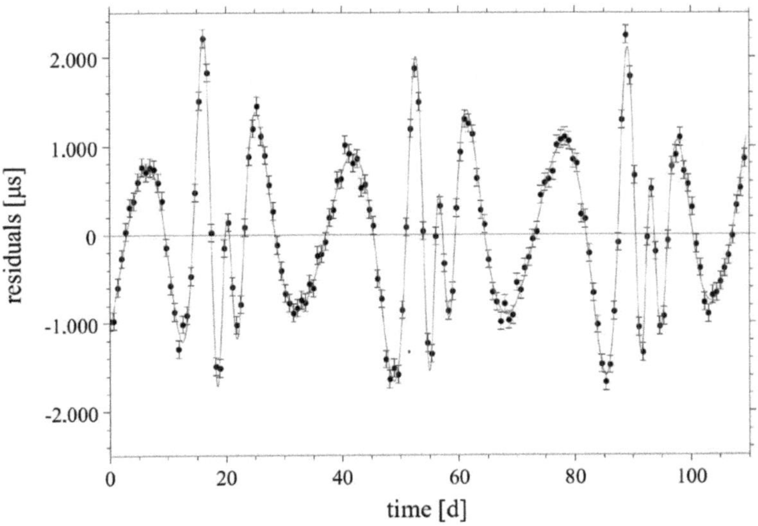


**Fig. 3** Timing residuals of a pulsar orbiting the supermassive BH in the Galactic Centre in a 0.1 year orbit. Shown are three orbits with an eccentricity of $e = 0.4$ for an extreme Kerr BH. Even with a timing precision of only 100 $\mu$s, the characteristic periodic residuals caused by the BH's quadrupole moment are clearly visible [2]

of classical and relativistic spin-orbit coupling in a pulsar-BH binary, in principle, allows us to determine the spin and the quadrupole moment of the black hole. This would test the "cosmic censorship conjecture" and the "no-hair theorem". While [38] showed that with current telescopes such an experiment would be almost impossible to perform (with the possible exception of pulsars about the Galactic centre black hole), Kramer et al. [39] pointed out that the SKA sensitivity should be sufficient. Indeed, this experiment benefits from the SKA sensitivity in multiple ways. On one hand, it provides the required timing precision but it also allows to perform the Galactic Census which should eventually deliver the sample of pulsars with a BH companion. As shown recently by Liu et al. [2] it should be "fairly easy" to measure the spin of the GC black hole with a precision of $10^{-4}$ to $10^{-3}$. Even for a pulsar with a timing precision of only 100 $\mu$s, characteristic periodic residuals would enable tests of the no-hair theorem with a precision of one percent or better (see Fig. 3)!

## 8 Summary and Conclusions

A variety of experiments and observational data exist that allow us to test our understanding of gravity with increased precision. So far, general relativity has passed all tests with flying colours but the apparent existence of "Dark Energy" challenges this simple picture. It is clear that the observations of pulsars will continue to play an important part in testing general relativity and its alternatives. We expect to detect gravitational waves not only indirectly but also directly using pulsar observations, and we have all reasons to believe that future searches will yield pulsars that can probe the space-time around black holes. Combined with the results of other experiments, namely the detection of gravitational waves with ground based detectors, we can expect a bright future for our understanding of gravity.

I want to conclude with a quote from Einstein, which he made in a letter to Arnold Sommerfeld on December 9th, 1915, a few days after he presented his final field equations. Delighted by the fact that his theory could predict correctly the Mercury perihelion advance, he wrote:

> Wie kommt uns da die pedantische Genauigkeit der Astronomie zu Hilfe, über die ich mich im Stillen früher oft lustig machte!

Unfortunately, Einstein died more than 10 years before pulsars were discovered. But he once also said that if he were not a physicist, he would like to be a lighthouse keeper. I dare to think that he would also have liked, perhaps, to be a pulsar astronomer, i.e. a cosmic lighthouse keeper who can enjoy precision astronomy and the science possible thanks to it.

## References


1. Lorimer, D.R., KramerM.: Handbook of Pulsar Astronomy, Cambridge Observing Handbooks for Research Astronomers, vol. 4, 1st edn. Cambridge University Press, New York (2005)
2. Liu, K., Wex, N., Kramer, M., Cordes, J.M., Lazio, T.J.W.: Prospects for probing the spacetime of Sgr A* with pulsars. Astrophys. J. **747**, 1 (2012). doi:10.1088/0004-637X/747/1/1
3. Weisberg, J.M., Nice, D.J., Taylor, J.H.: Timing measurements of the relativistic binary pulsar PSR B1913+16. Astrophys. J. **722**, 1030 (2010). doi:10.1088/0004-637X/722/2/1030
4. Hotan, A.W., Bailes, M., Ord, S.M.: High-precision baseband timing of 15 ms pulsars. Mon. Not. R. Astron. Soc. **369**, 1502 (2006). doi:10.1111/j.1365-2966.2006.10394.x
5. Freire, P.C.C., Bassa, C.G., Wex, N., et al.: On the nature and evolution of the unique binary pulsar J1903+0327. Mon. Not. R. Astron. Soc. **412**, 2763 (2011). doi:10.1111/j.1365-2966.2010.18109.x
6. Champion, D.J., Hobbs, G.B., Manchester, R.N., et al.: Measuring the mass of solar system planets using pulsar timing. Astrophys. J. **720**, L201 (2010). doi:10.1088/2041-8205/720/2/L201
7. Verbiest, J.P.W., Bailes, M., van Straten, W., et al.: Precision timing of PSR J0437–4715: an accurate pulsar distance, a high pulsar mass, and a limit on the variation of Newton's gravitational constant. Astrophys. J. **679**, 675 (2008). doi:10.1086/529576
8. Kramer, M., et al.: Tests of gravity from timing the Double Pulsar (2013). In preparation.
9. Freire, P.C.C., Wex, N., Esposito-Farèse, G., et al.: The relativistic pulsar-white dwarf binary PSR J1738+0333—II. The most stringent test of scalar-tensor gravity. Mon. Not. R. Astron. Soc. **423**, 3328 (2012). doi:10.1111/j.1365-2966.2012.21253.x
10. Hulse, R.A., Taylor, J.H.: Discovery of a pulsar in a binary system. Astrophys. J. **195**, L51 (1975). doi:10.1086/181708
11. Damour, T., Taylor, J.H.: Strong-field tests of relativistic gravity and binary pulsars. Phys. Rev. D **45**, 1840 (1992). doi:10.1103/PhysRevD.45.1840
12. Burgay, M., D'Amico, N., Possenti, A., et al.: An increased estimate of the merger rate of double neutron stars from observations of a highly relativistic system. Nature **426**, 531 (2003). doi:10.1038/nature02124

13. Lyne, A.G., Burgay, M., Kramer, M., et al.: A double-pulsar system: a rare laboratory for relativistic gravity and plasma physics. Science **303**, 1153 (2004). doi:10.1126/science.1094645
14. Kramer, M., Stairs, I.H., Manchester, R.N., et al.: Tests of general relativity from timing the double pulsar. Science **314**, 97 (2006). doi:10.1126/science.1132305
15. Kramer, M., Stairs, I.H.: The double pulsar. Annu. Rev. Astron. Astrophys. **46**, 541 (2008). doi:10.1146/annurev.astro.46.060407.145247
16. Kramer, M., Wex, N.: The double pulsar system: a unique laboratory for gravity. Class Quantum Gravity **26**, 073001 (2009). doi:10.1088/0264-9381/26/7/073001
17. Damour, T., Deruelle N.: General relativistic celestial mechanics of binary systems. II. The Post-Newtonian timing formula, Ann. Inst. Henri Poincare (A) 44, 263 (1986)
18. Will, C.M.: The confrontation between general relativity and experiment. Living Rev. Relativity 9, lrr-2006-3 (2006). http://www.livingreviews.org/lrr-2006-3
19. Börner, G., Ehlers, J., Rudolph, E.: Relativistic spin precession in two-body systems. Astron. Astrophys. **44**, 417 (1975)
20. Damour, T., Ruffini, R.: Certain new verifications of general relativity made possible by the discovery of a pulsar belonging to a binary system. C. R. Acad. Sci. Ser. A **279**, 971 (1974)
21. Burgay, M., Possenti, A., Manchester, R.N., et al.: Long-term variations in the pulse emission from PSR J0737–3039b. Astrophys. J. **624**, L113 (2005). doi:10.1086/430668
22. Perera, B.B.P., McLaughlin, M.A., Kramer, M., et al.: The evolution of PSR J0737–3039b and a model for relativistic spin precession. Astrophys. J. **721**, 1193 (2010). doi:10.1088/0004-637X/721/2/1193
23. Breton, R.P., Kaspi, V.M., Kramer, M., et al.: Relativistic spin precession in the double pulsar. Science **321**, 104 (2008). doi:10.1126/science.1159295
24. Damour, T., Esposito-Farèse, G.: Tensor-scalar gravity and binary-pulsar experiments. Phys. Rev. D **54**, 1474 (1996). doi:10.1103/PhysRevD.54.1474
25. Hobbs, G., Archibald, A., Arzoumanian, Z., et al.: The international pulsar timing array project: using pulsars as a gravitational wave detector. Class. Quant. Grav. **27**, 084013 (2010). doi:10.1088/0264-9381/27/8/084013
26. Jenet, F.A., Hobbs, G.B., van Straten, W., et al.: Upper bounds on the low-frequency stochastic gravitational wave background from pulsar timing observations: current limits and future prospects. Astrophys. J. **653**, 1571 (2006). doi:10.1086/508702
27. Ferdman, R.D., van Haasteren, R., Bassa, C.G., et al.: The European pulsar timing array: current efforts and a leap toward the future. Class. Quant. Grav. **27**, 084014 (2010). doi:10.1088/0264-9381/27/8/084014
28. Sesana, A., Vecchio, A., Colacino, C.N.: The stochastic gravitational-wave background from massive black hole binary systems: implications for observations with pulsar timing arrays. Mon. Not. R. Astron. Soc. **390**, 192 (2008). doi:10.1111/j.1365-2966.2008.13682.x
29. Sesana, A., Vecchio, A.: Gravitational waves and pulsar timing: stochastic background, individual sources and parameter estimation. Class. Quant. Grav. **27**, 084016 (2010). doi:10.1088/0264-9381/27/8/084016
30. van Haasteren, R., Levin, Y., Janssen, G.H., et al.: Placing limits on the stochastic gravitational-wave background using European pulsar timing array data. Mon. Not. R. Astron. Soc. **414**, 3117 (2011). doi:10.1111/j.1365-2966.2011.18613.x
31. Lee, K.J., Wex, N., Kramer, M., et al.: Gravitational wave astronomy of single sources with a pulsar timing array. Mon. Not. R. Astron. Soc. **414**, 3251 (2011). doi:10.1111/j.1365-2966.2011.18622.x
32. Lee, K.J., Jenet, F.A., Price, R.H.: Pulsar timing as a probe of Non-Einsteinian polarizations of gravitational waves. Astrophys. J. **685**, 1304 (2008). doi:10.1086/591080
33. Lee, K., Jenet, F.A., Price, R.H., Wex, N., Kramer, M.: Detecting massive gravitons using pulsar timing arrays. Astrophys. J. **722**, 1589 (2010). doi:10.1088/0004-637X/722/2/1589
34. Smits, R., Tingay, S.J., Wex, N., Kramer, M., Stappers, B.: Prospects for accurate distance measurements of pulsars with the square kilometre array: enabling fundamental physics. Astron. Astrophys. **528**, A108 (2011). doi:10.1051/0004-6361/201016141

35. Stappers, B.W., Hessels, J.W.T., Alexov, A., et al.: Observing pulsars and fast transients with LOFAR. Astron. Astrophys. **530**, A80 (2011). doi:10.1051/0004-6361/201116681
36. Smits, R., Lorimer, D.R., Kramer, M., et al.: Pulsar science with the five hundred metre aperture spherical telescope. Astron. Astrophys. **505**, 919 (2009). doi:10.1051/0004-6361/200911939
37. Damour, T., Esposito-Farèse, G.: Gravitational-wave versus binary-pulsar tests of strong-field gravity. Phys. Rev. D **58**, 042001 (1998). doi:10.1103/PhysRevD.58.042001
38. Wex, N., Kopeikin, S.: Frame dragging and other precessional effects in black hole-pulsar binaries. Astrophys. J. **513**, 388 (1999). doi:10.1086/306933
39. Kramer, M., Backer, D.C., Cordes, J.M., et al.: Strong-field tests of gravity using pulsars and black holes. New Astron. Rev. **48**, 993 (2004). doi:10.1016/j.newar.2004.09.020

# The Astrophysical Signatures of Black Holes: The Horizon, The ISCO, The Ergosphere and The Light Circle

**Marek A. Abramowicz**

**Abstract** Three advanced instruments planned for a near future (LOFT, GRAVITY, THE EVENT HORIZON TELESCOPE) provide unprecedented angular and time resolutions, which allow to probe regions in the immediate vicinity of black holes. We may soon be able to search for the signatures of the super-strong gravity that is characteristic to black holes: the event horizon, the ergosphere, the innermost stable circular orbit (ISCO), and the photon circle. This review discusses a few fundamental problems concerning these theoretical concepts.

## 1 Introduction

Undoubtedly, the existence of black holes is one of a few most bizarre predictions ever made in the whole acts of physics. The existence of massive objects with gravity strong enough to prevent light from escaping their surfaces was anticipated already in the XVIII century by John Mitchel, and later but independently by Pierre–Simon Laplace. However, the fundamental properties of black holes have been discovered, understood and described in terms of brilliant mathematical developments of

M. A. Abramowicz (✉)
Nicolaus Copernicus Astronomical Center, ul. Bartycka 18, 00-716 Warszawa, Poland
e-mail: marek.abramowicz@physics.gu.se

M. A. Abramowicz
Physics Department, Institute of Astronomy (Prague), Silesian University, Opava, The Czech Republic

M. A. Abramowicz
Department of Physics, University of Gothenburg, Gothenburg, Sweden

J. Bičák and T. Ledvinka (eds.), *General Relativity, Cosmology and Astrophysics*, Fundamental Theories of Physics 177, DOI: 10.1007/978-3-319-06349-2_24,
© Springer International Publishing Switzerland 2014

Albert Einstein's general theory of relativity, only much later—indeed in our times.[1] Astrophysically, the most important of these fundamental properties are[2]:

1. **Event horizon**: This may be imagined as a sphere (but *not* as a rigid surface) with the "gravitational radius" $r_G \sim GM/c^2$ surrounding the black hole of the mass $M$, from within which nothing may emerge. This is a unique signature of black holes.
2. **Ergosphere**: This is a region around a rotating black hole where spacetime itself is dragged along in the direction of rotation at a speed greater than the local speed of light in relation to the rest of the universe. In this region, negative energy states are possible, which means that the rotational energy of the black hole can be tapped through the "Penrose process".
3. **Innermost stable circular orbit (ISCO)**: This is the smallest circle ($r = r_{ms}$) along which free particles may stably orbit around a black hole. No stable circular motion is possible for $r < r_{ms}$. The presence of ISCO in the black hole case is one of the most important features of the black hole accretion [4].
4. **Circular photon orbit**: At a specific radius, often called "the light circle", photons may circle freely around a black hole.

In the last three decades, robust detections were made of several astrophysical black hole candidates within our Galaxy and in many others galaxies. However, no direct and unambiguous observational signatures of the horizon, the ergosphere, the ISCO and the light circle have been found. The obvious difficulty here is the high angular resolution that is needed to observe the black hole signatures—the "angular size" in the sky of a black hole at a distance $D$ is $\delta = r_G/D$. For SgrA*, the black hole at the center of our Galaxy, this implies that the smallest observed images of accretion structures around it have a size of a few $\mu$as. At the moment, they cannot be observed, they are too small, but the angular resolution in this range will be reached by advanced new detectors that are planned for the near future.

1. `GRAVITY` is planned for 2014 by the Max Planck Institute for the Extraterrestrial Physics in Garching, Germany.[3] It is the second-generation VLTI instrument for precision narrow-angle astrometry and interferometric imaging, consisting of four 8m telescopes and a total collecting area of 200 m$^2$. It is the only interferometer to allow direct imaging at high sensitivity and high image quality in a large ($\sim 2''$) field of view. It provides precision astrometry with resolution $\sim 10$ $\mu$as (microarcsecond), and imaging with resolution $\sim 4$ mas (milliarcsecond). Note

[1] The modern history of black holes started thanks to great discoveries of Schwarzschild [1], Chandrasekhar [2] and Oppenheimer [3], and the follow-up work in the 1960s and the 1970s, done mostly by collaborators and students of Dennis W. Sciama in Cambridge (Carter, Ellis, Gibbons, Hawking, Penrose), John A. Wheeler in Princeton (Bekenstein, Ruffini, Thorne) and Yakov B. Zel'dovich in Moscow (Novikov, Starobinsky) and other researchers (e.g. Israel, Damour, Kerr, Kruskal, Wald).

[2] Here, and in a few other places, I directly quote a `Living Review` devoted to the subject: Abramowicz and Fragile, *Foundations of Black Hole Accretion Disk Theory*.

[3] Homepage http://www.mpe.mpg.de/ir/gravity.

that the apparent size of the gravitational radius of SgrA*, the Galactic center black hole, is 10 $\mu$ as.

2. THE EVENT HORIZON TELESCOPE (EHT) The EHT uses the technique of Very Long Baseline Interferometry (VLBI) to synthesize an Earth-sized telescope in order to achieve the highest resolution possible using ground-based instrumentation. Several existing (or planned) radio telescopes are the part of the baseline: ARO/SMT 10 m, APEX 12 m, ASTE 10 m CARMA array of six 10 m and nine of 6 m antennas, CSO 10 m, IRAM 30 m, JCMT 15 m, SMA array eight 6m antennas. A few more telescopes and arrays will join. The target source is observed simultaneously at all telescopes. The data are recorded at each of the sites and later brought back to a processing facility where they are passed through a special purpose supercomputer. It will be fully operational in 2015.
3. LOFT is a newly proposed space mission selected by ESA as one of the four missions that will compete for a launch opportunity in the 2020s. The Large Area Detector (LAD) on board of LOFT achieves an effective area of $\sim$10 m$^2$ i.e. more than an order of magnitude larger than current spaceborne X-ray detectors in the 2-30 keV range (up to 80 keV in expanded mode). LOFT will have improved energy resolution (better than 260 eV) and will be able to investigate range from submillisecond quasi-periodic oscillations (QPOs) to years long transient outbursts. LOFT will provide access (for the first time) to types of information in these signals that are qualitatively new due to the capability to measure dynamical timescale phenomena within their coherence time, where so far only statistical averages of signals were accessible.

It is hoped that data from these new detectors will provide information sufficient to answer several of the questions concerning horizon, ergosphere, ISCO and the light circle that I describe in this review.

## 2 The Kerr Metric and Its Symmetries

The famous Kerr solution [5] describes the spacetime metric of a rotating, uncharged black hole with the total mass $M$ and the spin $a = J/Mc$ (here $J$ is the angular momentum). In the Boyer–Lindquist coordinates the Kerr metric takes the form (in the usual $c = 1 = G$ units and $+ - - -$ signature),

$$\begin{aligned} g_{\mu\nu}dx^{\mu}dx^{\nu} = & \left(1 - \frac{2Mr}{\varrho^2}\right) dt^2 + 2\,\frac{2Mar\sin^2\theta}{\varrho^2}\,dt d\phi \\ & - \left(r^2 + a^2 + \frac{2Ma^2 r\sin^2\theta}{\varrho^2}\right)\sin^2\theta\, d\phi^2 \\ & - \frac{\varrho^2}{\Delta}\,dr^2 - \varrho^2\, d\theta^2, \end{aligned} \tag{1}$$

where $\Delta = r^2 - 2\,M\,r + a^2$ and $\varrho^2 = r^2 + a^2\,\cos^2\theta$.

The Boyer-Lindquist coordinates are most widely used by astrophysicists. They are singular at the horizon, which in these coordinates is given by $\Delta = 0$. The Kerr metric is itself not singular at the horizon, which is apparent in the Kerr-Schild coordinates. The Kerr-Schild version of the Kerr metric takes the form,

$$
\begin{aligned}
g_{\mu\nu} dx^\mu dx^\nu = & \left(1 - \frac{2Mr_*}{\varrho_*^2}\right) dt_*^2 - \frac{4Mr_*}{\varrho_*^2}\, dt_* dr_* + \frac{4Mr_*}{\varrho_*^2} \sin^2\theta_* dt_* d\phi_* \\
& - \left(r_*^2 + a^2 + \frac{2Ma^2 r_* \sin^2\theta_*}{\varrho_*^2}\right) \sin^2\theta_*\, d\phi_*^2 - \varrho_*^2\, d\theta_*^2 \\
& - \left(1 + \frac{2Mr_*}{\varrho_*^2}\right) dr_*^2 + 2a \sin^2\theta_* \left(1 + \frac{2Mr_*}{\varrho_*^2}\right) dr_* d\phi_*, \qquad (2)
\end{aligned}
$$

where $\varrho_*^2 = r_*^2 + a^2 \cos^2\theta_*$.

The Kerr metric in the Boyer–Lindquist coordinates does not depend on $t$, $\phi$. In the Kerr–Schild coordinates it does not depend on $t_*$, $\phi_*$. This is a sign of the Killing symmetries. The Killing vectors defined by (in the Boyer–Lindquist coordinates)

$$
\eta^\mu = \delta^\mu_{\ t}, \quad \xi^\mu = \delta^\mu_{\ \phi}, \qquad (3)
$$

obey the Killing equation and commute,

$$
\begin{aligned}
& \nabla_\nu \eta_\mu + \nabla_\mu \eta_\nu = 0, \quad \nabla_\nu \xi_\mu + \nabla_\mu \xi_\nu = 0, \qquad (4) \\
& \xi^\nu \nabla_\nu \eta_\mu = \eta^\nu \nabla_\nu \xi_\mu. \qquad (5)
\end{aligned}
$$

For commuting Killing vectors one may write convenient relations,

$$
\begin{aligned}
\xi^\nu \nabla_\nu \eta_\mu = \eta^\nu \nabla_\nu \xi_\mu &= -\frac{1}{2} \nabla_\nu (\xi^\mu \eta_\mu), \\
\eta^\nu \nabla_\nu \eta_\mu &= -\frac{1}{2} \nabla_\nu (\eta^\mu \eta_\mu), \\
\xi^\nu \nabla_\nu \xi_\mu &= -\frac{1}{2} \nabla_\nu (\xi^\mu \xi_\mu), \qquad (6)
\end{aligned}
$$

which often allow to keep Christoffel's symbols out of calculations.[4]

The existence of the Killing vectors $\eta^\mu$, $\xi^\mu$, implies conservation of energy $\mathscr{E}$ and angular momentum $\mathscr{L}$ in geodesic motion,

$$
\{u^\mu \nabla_\mu u_\nu = 0\} \Rightarrow \left\{ \begin{array}{c} u^\mu \nabla_\mu \mathscr{E} = 0 \text{ for } \mathscr{E} \equiv \eta^\mu u_\mu \\ \\ u^\mu \nabla_\mu \mathscr{L} = 0 \text{ for } \mathscr{L} \equiv -\xi^\mu u_\mu \end{array} \right\} \qquad (7)
$$

---

[4] At http://www.physics.uci.edu/~etolleru/KerrOrbitProject.pdf the Christoffel symbols are given (by Tollerud 2007) in the form of a `Mathematica` package. Unfortunately, there is an error in the Kerr metric: the $g_{t\phi}$ metric component is (consistently everywhere) factor of 2 too big. See also [6].

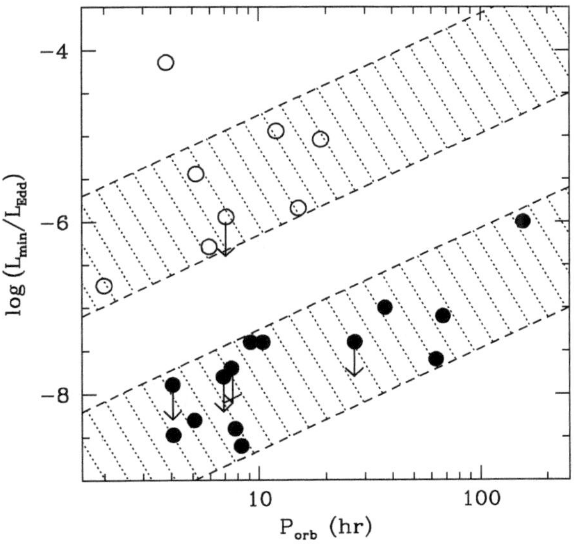


**Fig. 1** Eddington-scaled luminosities (0.5–10 keV) of BH transients (*filled circles*) and NS transients (*open circles*) versus the orbital period (see [7, 8]). Only the lowest quiescent detections or Chandra/XMM *upper limits* are shown. The plot shows all systems with known orbital periods, which have optical counterparts and good distance estimates. The *diagonal* hatched areas delineate the regions occupied by the two classes of sources and indicate the observed dependence of luminosity on orbital period. Note that the BH systems are on average nearly 3 orders of magnitude fainter than the NS systems with similar orbital periods. Figure and caption from [9]

The specific angular momentum $\ell \equiv \mathcal{L}/\mathcal{E}$ is obviously also a constant of geodesic motion. In a general, not necessarily geodesic case, the *circular* orbits are given by $u^\mu = A(\eta^\mu + \Omega\xi^\mu)$, where $\Omega$ is the angular velocity, connected to the specific angular momentum by

$$\ell = (\Omega - \omega)\,\frac{\tilde{r}^2}{1 + \Omega\,\omega\,\tilde{r}^2}, \tag{8}$$

$$\omega = -\frac{\xi^\mu \eta_\mu}{\xi^\nu \xi_\nu} = \text{(frame dragging)}, \tag{9}$$

$$\tilde{r}^2 = -\frac{\xi^\nu \xi_\nu}{\eta^\mu \eta_\mu} = \text{(gyration radius)}^2. \tag{10}$$

From (3) it follows that in the Boyer-Lindquist coordinates these quantities are expressed by

$$\mathcal{E} = u_t, \quad \mathcal{L} = -u_\phi, \quad \ell = -\frac{u_\phi}{u_t}, \quad \Omega = \frac{u^\phi}{u^t} = \frac{d\phi}{dt}. \tag{11}$$

## 3 The Horizon

A coordinate independent condition for the Kerr black hole horizon could be expressed by

$$(\eta^{\nu}\xi_{\nu})^2 - (\eta^{\nu}\eta_{\nu})(\xi^{\mu}\xi_{\mu}) = 0. \tag{12}$$

The presence of a horizon is most often argued from observational estimates of the "compactness parameter" $\xi \sim R_G/R$. Here $R$ is the measured size of the black hole candidate, and the gravitational radius $R_G$ is known from a mass measurement. Mass measurements are *very* accurate in a few cases (based on Kepler's laws and precise orbit measurements), but in most cases they are not accurate (orbits are unobserved directly). Size measurements are not yet accurate, and in addition, the measured size is only an upper limit. In most of the considered cases, one estimates not the size of the horizon but the size of ISCO or the size of the light circle radius.

Narayan [9–11] presented arguments that point to the horizon presence more directly. When a central object accretes matter through an accretion disk, a part of its radiation, $L_{disk}$, originates at the disk, and a part $L_{surf}$ at the surface of the object. The surface may radiate itself, or reflect (re-radiate) a part of the disk radiation. The total is therefore $L = L_{disk} + L_{surf}$. The horizon of the black hole cannot radiate or reflect $L_{surf} = 0$, and therefore in the black hole case, $L = L_{disk}$. Thus, black holes should look dimmer than non-black-hole objects with the similar accretion disks.

This indeed is observed in the case of accreting neutron stars vs. accreting black holes, in the galactic X-ray binaries, as explained in Fig. 1. In the figure luminosities of the binary black hole sources are compared with luminosities of the binary neutron star sources with the same orbital binary period. In these binaries, the compact object (a black hole or a neutron star) has a close companion, which is a low mass "normal" star. The companion loses mass which is accreted onto the compact object via the accretion disk. The luminosity of the accretion disk depends on the accretion rate, which is determined by the mass loss rate from the companion. The accretion rate cannot be directly measured, but the precisely and directly observed orbital period is a rather accurate indicator of the accretion rate—sources with the same orbital periods should have the same accretion rates. In Fig. 1, luminosities are given in terms of the Eddington luminosity, $L_{\rm Edd} = 10^{38} M/M_{\odot}$ [erg/s] which is proportional to the mass of the compact object $M$. One should note that the neutron star sources are order of magnitudes more luminous than the black hole sources in binaries with the same binary periods. This difference may be indeed due to the (already mentioned) fact that in the neutron star sources, matter that arrives at the surface may emit radiation, while in the black hole sources matter is lost without a trace inside the black hole. However, it was pointed out by several authors, e.g. [12, 13], for various reasons, rigid surfaces of *strange stars* may also absorb all matter without a trace.

Narayan's argument based on the evidence of some radiation lost inside the black hole works, in a different version, also for the super-massive black holes, in particular for SgrA*, the accreting black hole in the center of our Galaxy, for which the mass, $M = 4 \times 10^6 M_{\odot}$, was accurately measured from the analysis of stellar orbits in

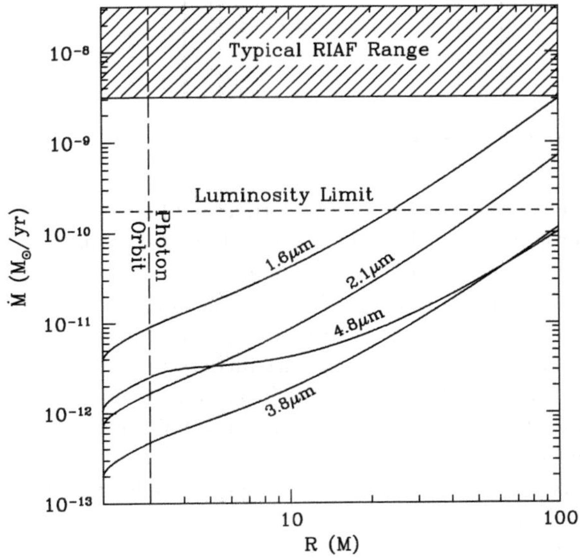


**Fig. 2** The four *solid lines* show independent *upper limits* on the mass accretion rate at the surface of Sgr A* (assuming the source has a surface) as a function of the surface radius R. Each *limiting curve* is derived from a limit on the quiescent flux of Sgr A* in an infrared band. The hatched area at the *top* labeled "Typical RIAF Range" corresponds to the mass accretion rate in typical ADAF models of Sgr A* (e.g. [14]). The *horizontal dashed line* represents the minimum accretion rate needed to power the bolometric luminosity of Sgr A*. Figure and caption from [9], after [15]

its vicinity. The accreting black hole has a very low luminosity. As Fig. 2 explains, a hypothesis that SgrA* is a dark object with a rigid surface is inconsistent with observations and the standard accretion theory of radiatively inefficient accretion flows (RIAF).

# 4 The Ergosphere

The surface of ergosphere in Kerr geometry is given by the covariant condition,[5]

$$\eta^{\mu}\eta_{\mu} = 0. \tag{13}$$

In the Boyer–Lindquist coordinates this is equivalent to $g_{tt} = 0$. Thus, inside the ergosphere, the Killing vector $\eta^{\mu}$ is spacelike, and therefore it is possible that, for a timelike four-velocity $u^{\mu}$ of a free particle, the conserved energy is negative,

[5] In this Section I quote extensively from a paper in preparation: Abramowicz, Gourgoulhon, Lasota, Narayan and Tchekhovskoy (2013), *Blandford-Znajek mechanism as the Penrose process. Application to Magnetically Arrested Disks*.

$$u^\mu \eta_\mu = \mathscr{E} < 0. \tag{14}$$

### 4.1 The Penrose Process

Penrose [16] considered[6] a freely falling particle with energy $\mathscr{E}_{\text{in}}$, which enters the ergosphere of a rotating black hole. He imagined that the particle disintegrates there into two particles, with energies $\mathscr{E}_{(-)} < 0$ and $\mathscr{E}_{(+)} > 0$. Then, the particle with negative energy $\mathscr{E}_{(-)}$ falls into the black hole, and the other one escapes to infinity with positive energy, $\mathscr{E}_{\text{out}} = \mathscr{E}_{(+)}$. Since energy conservation implies that $\mathscr{E}_{(-)} + \mathscr{E}_{(+)} = \mathscr{E}_{\text{in}}$, one deduces that $\mathscr{E}_{\text{out}} > \mathscr{E}_{\text{in}}$, so there is a gain of energy "at infinity". The negative energy particle absorbed by the black hole has also a negative angular momentum, $\mathscr{L}_{(-)} < 0$. The source for the gain of energy at infinity is therefore the rotational energy of the black hole.

Soon after Penrose's discovery that rotating black holes may be energy sources, several authors suggested that the Penrose process may power relativistic jets observed in quasars (and later in microquasars). However, a careful analysis by [18–22] and others shown that it is unlikely that negative energy states, necessary for the Penrose process to work, may be achieved through the particles disintegration inside ergosphere. The same conclusion was reached more recently by [23] for high-energy collision of particles. The reason is that in the case of disintegration, for a negative energy state to occur, velocities of fragments measured in their center of mass frame should be very relativistic, $v > c/2$. In the case of collisions, the particles with positive energies cannot escape, as they must have large and negative radial momenta. Thus, they are captured (together with the negative energy particles) by the black hole.

In the context of the super-energetic collisions there is a disagreement between two opinions, based on recently obtained results. Firstly, Silk and his collaborators [24] claimed that in the center-of-mass frame, the energy of particular types of collisions may be arbitrary large and that this may lead to astrophysically important and interesting consequences. Secondly, Bejger and collaborators [23] claimed that consequences of such collisions are unobservable. Both results have been published and attracted a considerable attention. There is a vigorous follow-up going on—see [25] for the most recent significant result, which confirms and expands results obtained by [23]: *black holes are neither particle accelerators nor dark matter probes*. This leaves magnetic processes as the only astrophysically realistic way to extract rotational energy from a rotating black hole.

---
[6] See also [17].

### 4.2 The Blandford-Znajek Mechanism

The presence of ergosphere is today discussed mostly in two contexts: the origin of relativistic jets and super-energetic collisions of particles deep inside the ergosphere.

In the jet context, Tchekhovskoy, Narayan and their collaborators [26–28] have found that the famous Blandford–Znajek mechanism [29] works for “magnetically arrested disks”, i.e. a special type of magnetized black hole accretion [30]. There are several issues here that need to be studied and explained.

## 5 The ISCO

The existence of ISCO is probably the single most important strong-gravity effect in the whole black hole accretion disk theory. There is, however, a controversy concerning the ISCO. According to one view [31, 32], for small accretion rates, location of ISCO determines the “inner edge” of the disk which separates the part of the disk where matter rotates on almost Keplerian, almost circular orbits, from the plunging-in region, where matter falls in into the black hole almost freely. There is no (significant) radiation coming from the plunging-in region, and stresses there are negligible. Numerous well-known and widely used results in accretion theory depend on the assumption that ISCO is the sharp boundary between the two different accretion regimes. In particular, works on the black hole spin estimate based on spectral fitting adopt this assumption [33, 34]. According to the opposite view [35], most recently eloquently summarized by Balbus [36], ISCO is not an important feature of black hole accretion even for small accretion rates, because the magnetohydrodynamical MRI instability makes the flow unstable and turbulent on both sides of ISCO.

It is obvious that because the ISCO appears as an important ingredient in numerous specific (and important) results of the black hole accretion disk theory, it is one of its few pillars on which the theory rests. For all these reasons, the ISCO receives more attention in my review than the other three black hole accretion disk signatures (horizon, ergosphere, circular photon orbit).

### 5.1 ISCO for Keplerian, Circular Orbits (Circular Geodesics)

In this section we use the Boyer–Lindquist coordinates (1). Let us consider a test particle with mass $m$ which moves on the equatorial plane, $\theta = \pi/2$. For such a motion, the polar components of the four velocity vanish, $u^\theta = 0 = u_\theta$, and from $u_\mu u_\nu g^{\mu\nu} = 1$ it follows that,

$$(u_t)^2 g^{tt} + 2u_t u_\phi g^{t\phi} + (u_\phi)^2 g^{\phi\phi} + (u^r)^2 g_{rr} = 1. \tag{15}$$

Let us now assume that the motion is almost circular in the sense that $r(s) = r_0 + \delta r(s)$ and $|\delta r(s)| \ll r_0$ for some constant positive $r_0$, which corresponds to the radius of the circle in the Boyer–Lindquist coordinates ($s$ is the spacetime length along the trajectory). Using the Boyer–Lindquist expressions (11) for the conserved energy $\mathcal{E}$ and the conserved specific angular momentum $\ell$, one may write equation (15) as

$$\frac{1}{2}|g_{rr}|(\delta\dot{r})^2 = E - \mathcal{U}_{\rm eff}(r,\ell), \quad (\delta\dot{r}) \equiv \frac{d(\delta r)}{ds}, \tag{16}$$

$$\mathcal{U}_{\rm eff} = -\frac{1}{2}\ln\left[g^{tt} - 2\ell g^{t\phi} + \ell^2 g^{\phi\phi}\right], \tag{17}$$

where $E = \ln\mathcal{E}$ may also be rightly called the energy because in the Newtonian limit $E \to E_{\rm Newton}$, and $\mathcal{E} \to E_{\rm Newton} + 1$. The effective potential $\mathcal{U}_{\rm eff}(r,\ell)$ equals its Newtonian counterpart in the Newtonian limit. Except of $|g_{rr}| \to 1$, Eq. (16) has its form identical with the corresponding Newtonian one, very familiar from the classical mechanics. Thus, using the same procedure as in the Newtonian case, one may derive equations that govern the strictly circular geodesic motion $\delta\dot{r} = 0$, and a slightly perturbed $\delta\dot{r} \neq 0$ one,

$$\left(\frac{\partial\mathcal{U}_{\rm eff}}{\partial r}\right)_\ell = 0 \Rightarrow \partial_r g^{tt} - 2\ell\,\partial_r g^{t\phi} + \ell^2\,\partial_r g^{\phi\phi} = 0, \tag{18}$$

$$\frac{1}{|g_{rr}|}\left(\frac{\partial^2\mathcal{U}_{\rm eff}}{\partial r^2}\right)_\ell = \kappa^2 \quad \text{and} \quad \boxed{\delta\ddot{r} + \kappa^2\delta r = 0}. \tag{19}$$

As the Kerr metric components $g^{\mu\nu} = g^{\mu\nu}(r)$ are known functions of the Boyer-Lindquist coordinate $r$ on the equatorial plane, the quadratic equation in (18) may be easily solved. Its solution $\ell = \ell_K(r)$ represents the Keplerian angular momentum distribution,

$$\ell_K = \frac{M^{1/2}(r^2 - 2aM^{1/2}r^{1/2} + a^2)}{r^{3/2} - 2Mr^{1/2} + aM^{1/2}}, \tag{20}$$

$$(\Omega_K)^2 = \frac{GM}{\left(r^{3/2} + aM^{1/2}\right)^2}, \tag{21}$$

$$\omega_r = \frac{\kappa}{A} = \Omega_K\left(1 - 6x^{-1} + 8a_* x^{-3/2} - 3a_*^2 x^{-2}\right)^{1/2}. \tag{22}$$

Here $a_* = a/M$. In the second line (21) we give the Keplerian angular velocity $\Omega_K$, and in the third line (22) the radial epicyclic frequency, which is the redshifted eigenvalue $\kappa$. In the Boyer–Lindquist coordinates the redshift factor is $A = u^t$, and covariantly,

$$A = \frac{1}{\left[(\eta^\mu\eta_\mu) + 2\Omega(\xi^\mu\eta_\mu) + \Omega^2(\xi^\mu\xi_\mu)\right]^{1/2}}. \tag{23}$$

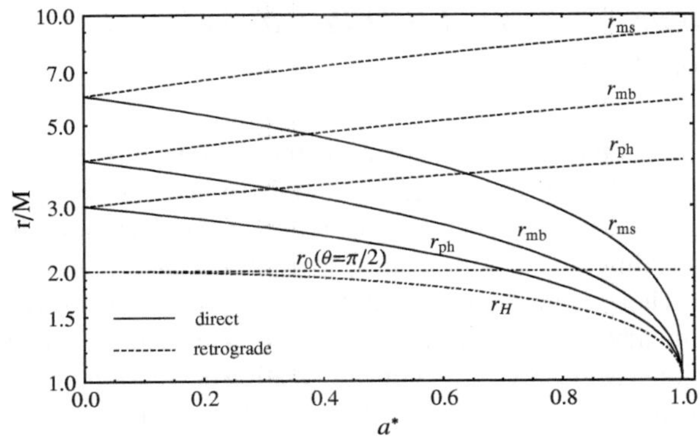


**Fig. 3** The location of ISCO for different Kerr black hole spins, $r_{ms}(a)$, together with locations of other characteristic radii (in terms of the Boyer–Lindquist coordinate $r$). Adapted from [18]

From the simple harmonic oscillator equation in (19) it follows that for $\kappa^2 < 0$ the circular Keplerian orbits are unstable. The Innermost Stable Circular Orbit (ISCO) is located therefore at the radius $r_{ms}$ defined by equation $\kappa^2(r_{ms}) = 0$. Its solution yields

$$\begin{aligned} &\text{ISCO location}: \; r_{ms} = M\left\{3 + Z_2 - [(3 - Z_1)(3 + Z_1 + 2Z_2)]^{1/2}\right\}, \\ &Z_1 = 1 + (1 - a_*^2)^{1/3}[(1 + a_*)^{1/3} + (1 - a_*)^{1/3}], \\ &Z_2 = (3a_*^2 + Z_1^2)^{1/2}. \end{aligned} \tag{24}$$

The location of ISCO for different Kerr black hole spins, $r_{ms}(a)$, is shown in Fig. 3, together with locations of other characteristic radii (in terms of the Boyer–Lindquist coordinate $r$),

$$\text{photon } r_{\text{ph}} = 2M\left\{1 + \cos\left[\frac{2}{3}\cos^{-1}(a_*)\right]\right\}, \tag{25}$$

$$\text{bound } r_{\text{mb}} = 2M\left(1 - \frac{a_*}{2} + \sqrt{1 - a_*}\right), \tag{26}$$

$$\text{horizon } r_{\text{H}} = M\left(1 + \sqrt{1 - a_*^2}\right), \tag{27}$$

$$\text{ergosphere } r_0 = M\left(1 + \sqrt{1 - a_*^2\cos^2\theta}\right). \tag{28}$$

## 5.2 The Shakura–Sunyaev ISCO Paradigm for Thin Accretion Disks

Many basic results of the black hole accretion theory are a direct consequence of the existence of the ISCO.[7] They are reviewed and discussed e.g. in [4]. The "standard" black hole stationary, non-magnetized, accretion disks disks are characterized by small (very sub-Eddington) accretion rates, $\dot{M} \ll \dot{M}_{\rm Edd}$, and small vertical thickness, $h \equiv H/r \ll 1$. Their mathematical model was developed (in Newton's theory) by Shakura and Sunyaev in the most influential and important paper in the accretion disk theory [37], hereafter SS73. Its general relativistic version, proper for the Kerr black holes, was found by Novikov and Thorne [38], hereafter NT73.

For such disks, the ISCO behaves like a physical boundary, separating the *disk proper* from the *plunging* region. In the disk proper (we will call it simply "disk"), i.e. in the region $r > r_{\rm ms}$, matter moves on nearly circular, nearly Keplerian orbits. Consequently, the angular momentum of the disk is nearly Keplerian, $\ell(r) \approx \ell_K(r)$. Dissipation of orbital energy and angular momentum is significant, and most (~98 %) of the accretion flow radiation comes from there. In the plunging region, $r < r_{\rm ms}$, matters nearly falls freely. Stresses are ineffective in dissipating energy and transporting angular momentum. Consequently, very little radiation (~2 %) comes from this region, and the angular momentum distribution there is almost constant, $\ell(r) \approx$ const.

The SS73 and NT73 models were developed two decades before Balbus and Hawley [39] made the seminal discovery that the torque needed for accretion discs to operate originates from the turbulence caused by the Magneto Rotational Instability (MRI) which weakly magnetized, differentially rotating fluids suffer. The view that the MRI is crucial in providing that necessary torque, was supported by numerous follow-up works, mostly based on magnetohydrodynamical numerical simulations. Today it is generally accepted by a vast majority of the black hole accretion disk researchers.[8] For the review of successes of the hypothesis of the MRI induced torque, see [42] and [43].

With no knowledge of the true nature of viscus torques in accretion disks, the SS73 and NT73 models adopted a heuristic viscosity prescription based on the assumption that the torque is proportional to the total pressure,

$$\mathscr{T} = \alpha p, \tag{29}$$

with $0 < \alpha < 1$ being a parameter. The SS73 and NT73 models have also adopted other simplifying assumptions, among them that at the ISCO the torque is zero,

[7] In this Section I quote *in extenso* a few paragraphs from an unfinished draft of an unpublished paper by Abramowicz, Horák and Kluźniak, *The MRI in the plunge-in region: the Shakura–Sunyaev ISCO paradigm confirmed*, 2013, in preparation.

[8] However, [40, 41] and others pointed to some difficulties with the MRI concept and also with its description in the shearing box simulations.

$$\mathscr{T}(r_{\rm ms}) = 0, \tag{30}$$

and that there is no radial (advective) heat transport in the disk. These assumptions introduced an artificial singularity at the ISCO, not relevant for calculating spectra, but obviously calling for a more accurate treatment of the flow near ISCO. This more accurate treatment was initiated in works [44] and [45]. It eventually maturated as the *slim disk* model [46].

Slim disks assume the alpha-viscosity prescription (29), but do *not* assume the zero-torque condition at ISCO (30). Instead, they consistently solve (numerically) the full set of the (relativistic, in the Kerr geometry) Navier–Stokes equations, $\nabla_\nu T^\nu_{\ \mu} = 0$, expanded in terms of the small vertical disk thickness $h$, up to the quadratic order $\mathscr{O}(h^2)$. This includes solving the vertical radiative transfer in the diffusion approximation [47]. Mathematically, the slim disk equations form an eigen-value problem, with the eigen-value being the angular momentum of matter crossing the black hole horizon. The physical reason for the eigen-value nature of the problem is that the black hole accretion must necessarily be transonic, with the sonic point $r_S$ being the critical (saddle) point of the slim disk differential equations. The regularity conditions at $r_S$ assure that its location is very close to ISCO, $r_S = r_{\rm ms} - A^2h^2$, where $A$ depends on the equation of state, with the the sound speed $C_S \ll c$, which is always true for very thin disks, $h \ll 1$ [48].

The slim disk calculations fully confirmed that, for $\dot{M} \ll \dot{M}_{edd}$ and $h \ll 1$, the ISCO was a (quite sharp) boundary between disk and plunging region, and that the stress at ISCO was indeed small. Despite that, the ISCO paradigm was challenged mostly by Krolik [35], but also by others, on the ground that the alpha viscosity prescription (29) used in the slim disk calculations was not adequate, because (according to their view) it cannot properly describe the MRI torque across the ISCO. Paczynski [31] and Afshordi and Paczynski [32] argued that *independently of the physical nature of the torque*, for the thin ($h \ll 1$) disks the small stress at ISCO is a consequence of the angular momentum conservation. However, Paczyński's clear arguments have not been accepted (or understood) by a few opponents of the standard ISCO paradigm. It was obvious that the opponents could be convinced only by calculations or arguments that specifically include the MRI induced turbulence.

Fortunately for the validity of the standard ISCO paradigm, such calculations and arguments are now available. Firstly, Hirose et al. [49] convincingly demonstrated in terms of MHD shear box simulations (which included vertical stratification and radiative cooling) that the standard Shakura-Sunyaev viscosity prescription (29) adequately describes the MRI torque (see also [50]). Secondly, Shafee et al. [51], Reynolds and Fabian [52] and Penna et al. [53] performed recently MHD global simulations of the black hole accretion disks, concluding that for geometrically thin accretion disks used in their simulations (i.e. with $h \sim 0.05 - 0.1$):

1. The angular momentum distribution in the flow is characterized by $\ell(r) \sim \ell_K(r)$ outside ISCO (i.e. for $r > r_{ms}$) and $\ell(r) \sim$ const inside ISCO (i.e. for $r < r_{ms}$), see Fig. 4. Correspondingly to this (as $\delta E \sim \Omega \delta \ell$), there is very small energy dissipation inside ISCO, and therefore very little radiation may originate there.

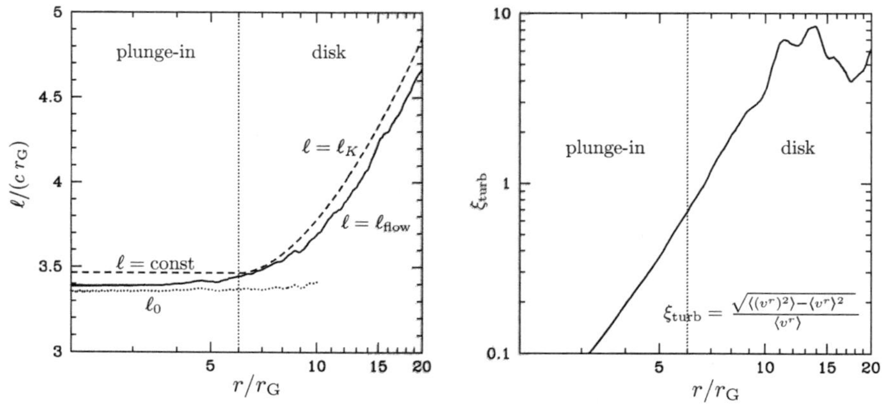


**Fig. 4** *Left:* Recent three-dimensional MHD simulations by Shafee et al. [51], Reynolds and Fabian [52] and Penna etal. [53] proved that the angular momentum distribution, $\ell_{flow}$, in the *vertically thin* accretion flow ($h < 0.1$) agrees within ~2 % with that predicted by the standard ISCO paradigm in SS73, NT73 and the *slim* disk models, i.e. that $\ell = \ell_K$ in the "*disk*" *region outside* ISCO, and $\ell =$ const in *the "plunge-in" region inside* ISCO. The *vertical dotted line* indicates the ISCO. The *horizontal dotted line* corresponds to the *total* angular momentum transport, divided by accretion rate, $\ell_0 = \dot{J}/\dot{M}$. Its constancy over the large range of radii proves that the steady state has been achieved in the simulations. Figure adopted from [51]. *Right:* Time averaged profile of $\xi_{turb}$, which measures the relative magnitude of turbulent fluctuations in the accreting gas. The fluid becomes mostly laminar inside the ISCO. Figure taken from [51]

2. Turbulent activity is pronounced at radii beyond about $10r_G$. Flow is nearly laminar inside the ISCO, see Fig. 4. This agrees with $\ell(r) \sim$ const there, and also suggests that the bulk of the dissipation occurs in the disk outside the ISCO, but *not* in the plunging region.

The above mentioned new results could be taken as the final proof of the validity of the standard ISCO paradigm if not one remaining subtle point. Most of the present understanding of the properties of MIR is based on the MHD shear box simulations, that are usually done with two particular assumptions—that the radial component of the flow is negligible, and that the radial profile of angular momentum corresponds to the Newtonian-Keplerian one,

$$V_r = 0, \tag{31}$$

$$\ell(r) \sim r^s, \quad \text{with } s = 1/2. \tag{32}$$

In the disk proper these conditions are satisfied. However, in the plunging region they are *not* fulfilled. Instead of $V_r = 0$ one has $V_r \gg C_S$ (indeed $V_r \sim c$), and instead of $s = 1/2$, one has $s = 0$. These differences could *very* significantly affect the strength of the MRI, as argued by [54].

However, works on slim disks and MHD simulations that support the standard ISCO paradigm have adopted assumption (directly or not) that the MRI induced torque is not qualitatively different on both sides of ISCO. For example, in terms of the alpha prescription this corresponds to $\alpha(r) \approx$ const.[9]

If one could argue that the MRI properties on both sides of ISCO are indeed qualitatively the same (despite the differences in the behavior of $V_r$ and $s$), then the validity of the ISCO paradigm would be finally proven.

Most recently such arguments have been put forward by [36], who stated in the Abstract of his paper that: *the defining properties of the MRI—its maximum growth rate and the direction of the associated eigenvector displacement—remain unchanged as the Rayleigh discriminant passes from positive to negative values*. In other words, because the "Rayleigh discriminant" is simply $\kappa^2$, i.e. the square of the radial epicyclic frequency which changes its sign at the ISCO, *the properties of MRI are the same on both sides of the ISCO*. Thus, if the arguments in [36] are correct, they provide the last missing point to complete the proof of the validity of the standard ISCO paradigm.

## *5.3 Leaving the ISCO*

Abramowicz et al. [4] revisited Krolik and Hawley's [56] discussion of the location of the "inner edge" $r_{in}$ of accretion disks around black holes, and expanded it to include disks with around Eddington accretion rates. The concept of the inner edge may be introduced by several empirical definitions of the accretion disk inner edge, each serving a different practical purpose:

1. **The potential spout edge** $r_{in} = r_{pot}$, where the effective potential $\mathscr{U}_{\rm eff}$, see equation (17) forms a self-crossing Roche lobe, and accretion is governed by the Roche lobe overflow.
2. **The sonic edge** $r_{in} = r_{son}$, where the transition from subsonic to transonic accretion occurs. Hydrodynamical disturbances do not propagate upstream a supersonic flow, and therefore the subsonic part of the flow is "causally" disconnected from the supersonic part.
3. **The variability edge** $r_{in} = r_{var}$, the smallest radius where orbital motion of coherent spots may produce quasi periodic variability.
4. **The stress edge** $r_{in} = r_{str}$, the outermost radius where the Reynolds stress is small, and plunging matter has no dynamical contact with the outer accretion flow.
5. **The radiation edge** $r_{in} = r_{rad}$, the innermost place from which significant luminosity emerges.
6. **The reflection edge** $r_{in} = r_{ref}$, the smallest radius capable of producing significant fluorescent iron line.

They found [4] that: *for black hole accretion disks with very sub-Eddington luminosities all these inner edges locate at ISCO. Thus, in this case, one may rightly*

[9] But see [55].

*consider ISCO as the unique inner edge of the black hole accretion disk. However, even for moderately higher luminosities, there is no such unique inner edge as differently defined edges locate at different places. Several of them are significantly closer to the black hole than ISCO. The differences grow with the increasing luminosity. For nearly Eddington luminosities, they are so huge that the notion of the inner edge loses all practical significance.*

## 5.4 Evidence for ISCO from the Observed Variability

Quasi periodic oscillations (called QPOs) are high frequency (about kHz) oscillations in the X-ray fluxes from neutron star and black hole sources in the Galactic X-ray sources (they have been also observerd in a few extra-Galactic sources and in SgrA*). In many sources they appear as a pair of oscillations, and are called the "twin-peak" QPOs. Before 2000, twin peak kHz QPOs have been observed only in the neutron star sources. It was believed that they must be connected to the neutron star rigid surfaces, and cannot occur in the black hole sources. Kluźniak and Abramowicz suggested (see [57]) that the twin peak kHz QPOs are due to a non-linear resonance in accretion disks oscillations, and for this reason their frequencies should have ratios close to those of small natural numbers, for example 3:2. Their prediction that the twin peak QPOs should also appear in the BH sources was soon confirmed by Strohmayer [58], who observed a twin peak QPOs in the BH candidate GRO J1655-40. The 3:2 non-linear resonance explanation is now generally accepted, but despite its successes in finding general signatures of the resonance in the observational data, several questions remain unanswered, in particular the behavior of the quality factor Q of the QPOs. Barret et al. (see [59]) found that Q of the lower-frequency QPO in the neutron-star sources increase with increasing QPO frequency up to $Q \gg 200$ and then it drops. The high-frequency QPO has $Q \sim 10$ and does not follow the same pattern. There is a consensus that this rules out any kinematical model of QPOs as orbiting clumps or spots. The drop in Q in the lower-frequency QPO is attributed to the existence of the ISCO (see Fig. 5), but this interpretation is not generally accepted.

# 6 The Light Circle

The light circle issue is an emerging topic [61] that is connected to improving observational potentials in the sub-milliarcsecond radio imaging of the black hole sources, in particular SgrA*, i.e. the black hole in the center of our Galaxy.[10]

The shortest timescale that may be in principle observed in the accreting black hole sources is not connected to the accretion flow, but to the strong gravity itself

[10] Based on a lecture by M. Bursa given at the 9 RAGtime Workshop in Opava, 19–21 September, 2007 [61].

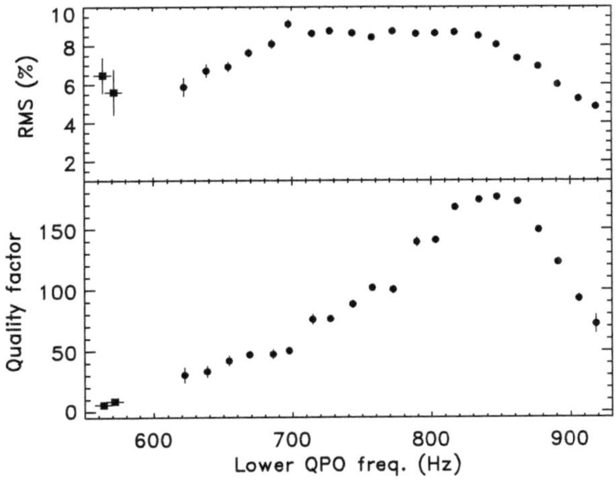


**Fig. 5** A sudden drop in the quality factor $Q$ of the observed twin peak QPO (lower frequency) was suggested as an evidence for the existence of the ISCO. The data presented here is from the Rossi X-ray Timing Explorer satellite observations of the neutron star binary source 4U 1636-536. The source shows quasi-periodic oscillations with *varying* frequencies in the range 650 Hz–900 Hz. The sharp drop in the quality factor (*bottom panel*) was recorded at the frequency ~S870 Hz. Figure from [60]

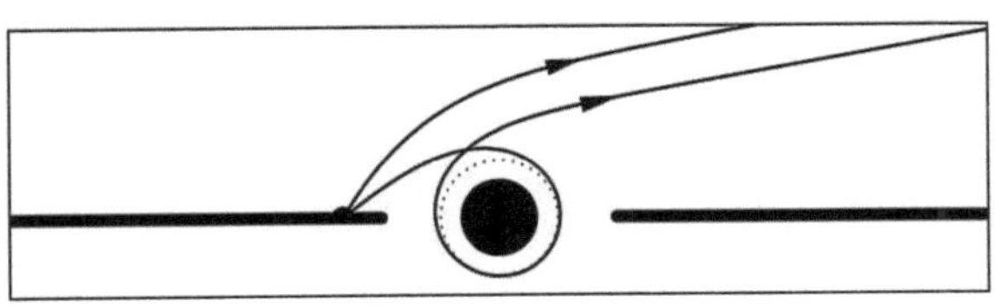

**Fig. 6** Trajectories of direct and looped photons emerging from a flare on the surface of an accretion disk. Figure reproduced from [61]

which close to the horizon loops photon trajectories around the black hole (see Fig. 6). Signals from some transient events in accretion disk, e.g., random short-lived flares, may therefore reach the observer repeatedly with delays corresponding to the travel time around the circular photon orbit. The looped signals will introduce a correlation in the variability data, with the timescale $T_{photon}$ shorter than the timescale at ICSO, $T_{ISCO}$. For a non-rotating black hole it is, (Fig. 7)

$$T_{photon} = 32.6M, \quad T_{ISCO} = 92.3M. \tag{33}$$

If the conditions are right, signal from some transient events in accretion disk, e.g., random short-lived flares, may reach the observer repeatedly with delays corre-

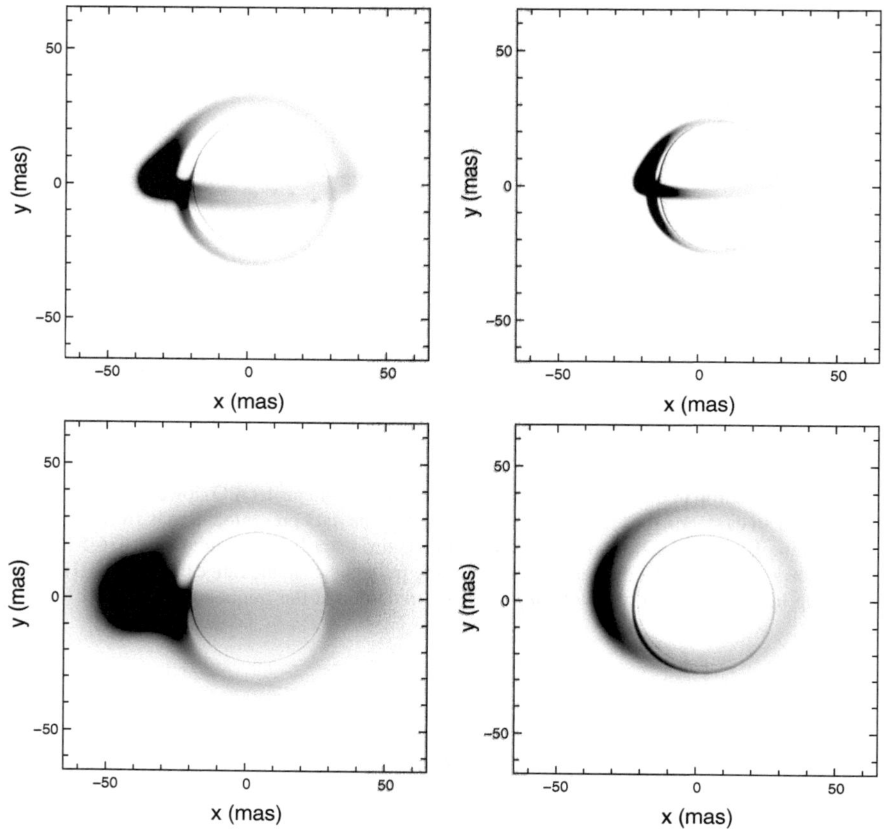


**Fig. 7** The *circular* light trajectory shows up in the theoretically calculated images of SgrA*. Images taken from [62]

sponding to the travel time around the photon orbit and still with a sufficient intensity to be practically detected. If the delay in arrival time from the vicinity of a black hole between "direct" and "looped" photons could indeed be found in the light curves of AGNs or microquasars, it would not only provide an excellent tool to measure the mass and spin of the black hole, but it would also provide direct evidence for the existence of (nearly) circular photon orbits. In this way one would demonstrate the validity of an important prediction of general relativity in the regime of extremely strong gravitational field, see Table 1.

**Table 1** Light circle timescales for different black hole sources: AGN, i.e. supermassive black holes in galactic centers, ULX i.e. hypothetical "intermediate mass" black holes postulated as an explanation of the Ultra Luminous X-ray sources, and GBH, i.e. microquasars in the Galactic black holes (X-ray binaries)

| | Luminosity [erg/s] | Distance [kpc] | Timescale $T_{photon}$ [s] |
|---|---|---|---|
| AGN | $10^{41} - 10^{43}$ | $10^{3} - 10^{4}$ | $10^{2} - 10^{4}$ |
| ULX | $10^{39} - 10^{41}$ | $10^{3} - 10^{4}$ | $10^{-2} - 10^{-1}$ |
| GBH | $10^{36} - 10^{38}$ | $10^{0} - 10^{1}$ | $10^{-4} - 10^{-3}$ |

## 7 Conclusions

Today we have at hand strong observational arguments that, for all practical purposes, prove that the compact objects detected in the Galactic X-ray binaries and at the centers of our Galaxy, and other galaxies, are indeed black holes. We do not have yet similarly strong arguments, based on observational data, to prove that these are the Kerr black holes. Advanced instruments, planned for a near future, will provide opportunities to probe the spacetime metric around the black hole candidates with a sufficient space (i.e. angular) and time resolution to obtain constrains.

**Acknowledgments** My work at the Silesian University in Opava was supported by the Czech CZ.1.07/2.3.00/20.0071 "Synergy" grant for international collaboration, and at the Institute of Astronomy in Prague by the Czech ASCR M100031242 grant. I also acknowledge support from the Polish NCN grant UMO-2011/01/B/ST9/05439.

## References

1. Schwarzschild, K.: Über das Gravitationsfeld eines Massenpunktes nach der Einsteinschen Theorie. Sitzungsberichte der Königlich Preussischen Akademie der Wissenschaften **1**, 189 (1916)
2. Chandrasekhar, S.: The maximum mass of ideal white dwarfs. Astrophys. J. **74**, 81 (1931). doi:10.1086/143324
3. Oppenheimer, J.R., Snyder, H.: On continued gravitational contraction. Phys. Rev. **56**, 455 (1939). doi:10.1103/PhysRev.56.455
4. Abramowicz, M.A., Jaroszyński, M., Kato, S., et al.: Leaving the innermost stable circular orbit: the inner edge of a black-hole accretion disk at various luminosities. A&A **521**, A15 (2010). doi:10.1051/0004-6361/201014467
5. Kerr, R.P.: Gravitational field of a spinning mass as an example of algebraically special metrics. Phys. Rev. Lett. **11**, 237 (1963). doi:10.1103/PhysRevLett.11.237
6. Dovciak, M.: Radiation of accretion discs in strong gravity. ArXiv e-prints arXiv:astro-ph/0411605 (2004)
7. Garcia, M.R., McClintock, J.E., Narayan, R., et al.: New evidence for black hole event horizons from Chandra. Astrophys. J. Letters **553**, L47 (2001). doi:10.1086/320494
8. McClintock, J.E., Narayan, R., Rybicki, G.B.: On the lack of thermal emission from the quiescent black hole XTE J1118+480: Evidence for the event horizon. Astrophys. J. **615**, 402 (2004). doi:10.1086/424474

9. Narayan, R., McClintock, J.E.: Advection-dominated accretion and the black hole event horizon. New Astr. Rev. **51**, 733 (2008). doi:10.1016/j.newar.2008.03.002
10. Menou, K., Esin, A.A., Narayan, R., et al.: Black hole and neutron star transients in quiescence. Astrophys. J. **520**, 276 (1999). doi:10.1086/307443
11. Narayan, R.: George Darwin lecture: Evidence for the black hole event horizon. Astron. Geophys. **44**(6), 060000 (2003). doi:10.1046/j.1468-4004.2003.44622.x
12. Abramowicz, M.A., Kluźniak, W., Lasota, J.P.: No observational proof of the black-hole event-horizon. A&A **396**, L31 (2002). doi:10.1051/0004-6361:20021645
13. Lemos, J.P.S., Zaslavskii, O.B.: Black hole mimickers: regular versus singular behavior. Phys. Rev. D **78**(2), 024040 (2008). doi:10.1103/PhysRevD.78.024040
14. Yuan, F., Quataert, E., Narayan, R.: Nonthermal electrons in radiatively inefficient accretion flow models of sagittarius a*. Astrophys. J. **598**, 301 (2003). doi:10.1086/378716
15. Broderick, A.E., Narayan, R.: On the nature of the compact dark mass at the galactic center. Astrophys. J. Letters **638**, L21 (2006). doi:10.1086/500930
16. Penrose, R.: Gravitational collapse: the role of general relativity. Nuovo Cimento Rivista Serie **1**, 252 (1969)
17. Penrose, R., Floyd, R.M.: Extraction of rotational energy from a black hole. Nat. Phys. Sci. **229**, 177 (1971). doi:10.1038/physci229177a0
18. Bardeen, J.M., Press, W.H., Teukolsky, S.A.: Rotating black holes: locally nonrotating frames, energy extraction, and scalar synchrotron radiation. Astrophys. J. **178**, 347 (1972). doi:10.1086/151796
19. Wald, R.M.: Energy limits on the Penrose process. Astrophys. J. **191**, 231 (1974). doi:10.1086/152959
20. Kovetz, A., Piran, T.: The efficiency of the Penrose process. Nuovo Cimento Lettere **12**(2), 39–42 (1975)
21. Piran, T., Shaham, J.: Upper bounds on collisional Penrose processes near rotating black-hole horizons. Phys. Rev. D **16**, 1615 (1977). doi:10.1103/PhysRevD.16.1615
22. Wald, R.M.: General Relativity. University of Chicago Press, Chicago (1984)
23. Bejger, M., Piran, T., Abramowicz, M., Håkanson, F.: Collisional Penrose process near the horizon of extreme Kerr black holes. Phys. Rev. Lett. **109**(12), 121101 (2012). doi:10.1103/PhysRevLett.109.121101
24. Bañados, M., Silk, J., West, S.M.: Kerr black holes as particle accelerators to arbitrarily high energy. Phys. Rev. Lett. **103**(11), 111102 (2009). doi:10.1103/PhysRevLett.103.111102
25. McWilliams, S.T.: Black holes are neither particle accelerators nor dark matter probes. Phys. Rev. Lett. **110**(1), 011102 (2013). doi:10.1103/PhysRevLett.110.011102
26. Tchekhovskoy, A., Narayan, R., McKinney, J.C.: Black hole spin and the radio loud/quiet dichotomy of active galactic nuclei. Astrophys. J. **711**, 50 (2010). doi:10.1088/0004-637X/711/1/50
27. Tchekhovskoy, A., Narayan, R., McKinney, J.C.: Efficient generation of jets from magnetically arrested accretion on a rapidly spinning black hole. Mon. Not. Roy. astr. Soc. **418**, L79 (2011). doi:10.1111/j.1745-3933.2011.01147.x
28. McKinney, J.C., Tchekhovskoy, A., Blandford, R.D.: General relativistic magnetohydrodynamic simulations of magnetically choked accretion flows around black holes. Mon. Not. Roy. astr. Soc. **423**, 3083 (2012). doi:10.1111/j.1365-2966.2012.21074.x
29. Blandford, R.D., Znajek, R.L.: Electromagnetic extraction of energy from Kerr black holes. Mon. Not. Roy. astr. Soc. **179**, 433 (1977)
30. Narayan, R., Igumenshchev, I.V., Abramowicz, M.A.: Magnetically arrested disk: an energetically efficient accretion flow. Pub. Astr. Soc. Jap. **55**, L69 (2003)
31. Paczyński, B.: The inner boundary condition for a thin disk accreting into a black hole. ArXiv e-prints arXiv:astro-ph/0004129 (2000)
32. Afshordi, N., Paczyński, B.: Geometrically thin disk accreting into a black hole. Astrophys. J. **592**, 354 (2003). doi:10.1086/375559
33. Shafee, R., McClintock, J.E., Narayan, R., et al.: Estimating the spin of stellar-mass black holes by spectral fitting of the X-ray continuum. Astrophys. J. **636**, L113 (2006). doi:10.1086/498938

34. Straub, O., Bursa, M., Sadowski, A., et al.: Testing slim-disk models on the thermal spectra of LMC X-3. A&A 533, A67 (2011). doi:10.1051/0004-6361/201117385
35. Krolik, J.H.: Magnetized accretion inside the marginally stable orbit around a black hole. Astrophys. J. **515**, L73 (1999). doi:10.1086/311979
36. Balbus, S.A.: On the behaviour of the magnetorotational instability when the rayleigh criterion is violated. Mon. Not. Roy. astr. Soc. **423**, L50 (2012). doi:10.1111/j.1745-3933.2012.01255.x
37. Shakura, N.I., Sunyaev, R.A.: Black holes in binary systems. Observational appearance. A&A **24**, 337 (1973)
38. Novikov, I.D., Thorne, K.S.: Astrophysics of black holes. In: Dewitt, C., Dewitt, B.S. (eds.) Black Holes (Les Astres Occlus), pp. 343–450 (1973)
39. Balbus, S.A., Hawley, J.F.: A powerful local shear instability in weakly magnetized disks. I—Linear analysis. II—Nonlinear evolution. Astrophys. J. **376**, 214 (1991). doi:10.1086/170270
40. Umurhan, O.M., Menou, K., Regev, O.: Weakly nonlinear analysis of the magnetorotational instability in a model channel flow. Phys. Rev. Lett. **98**(3), 034501 (2007). doi:10.1103/PhysRevLett.98.034501
41. Regev, O., Umurhan, O.M.: On the viability of the shearing box approximation for numerical studies of mhd turbulence in accretion disks. A&A **481**, 21 (2008). doi:10.1051/0004-6361:20078413
42. Balbus, S.A., Hawley, J.F.: Instability, turbulence, and enhanced transport in accretion disks. Rev. Mod. Phys. **70**, 1 (1998). doi:10.1103/RevModPhys.70.1
43. Balbus, S.A.: Enhanced angular momentum transport in accretion disks. Ann. Rev. Astr. Ap. **41**, 555 (2003). doi:10.1146/annurev.astro.41.081401.155207
44. Paczynski, B., Bisnovatyi-Kogan, G.: A model of a thin accretion disk around a black hole. Acta Astr. **31**, 283 (1981)
45. Muchotrzeb, B., Paczynski, B.: Transonic accretion flow in a thin disk around a black hole. Acta Astr. **32**, 1 (1982)
46. Abramowicz, M.A., Czerny, B., Lasota, J.P., Szuszkiewicz, E.: Slim accretion disks. Astrophys. J. **332**, 646 (1988). doi:10.1086/166683
47. Sadowski, A., Abramowicz, M., Bursa, M., et al.: Relativistic slim disks with vertical structure. A&A **527**, A17 (2011). doi:10.1051/0004-6361/201015256
48. Abramowicz, M.A., Zurek, W.H.: Rotation-induced bistability of transonic accretion onto a black hole. Astrophys. J. **246**, 314 (1981). doi:10.1086/158924
49. Hirose, S., Blaes, O., Krolik, J.H.: Turbulent stresses in local simulations of radiation-dominated accretion disks, and the possibility of the lightman-eardley instability. Astrophys. J. **704**, 781 (2009). doi:10.1088/0004-637X/704/1/781
50. Hirose, S., Krolik, J.H., Blaes, O.: Radiation-dominated disks are thermally stable. Astrophys. J. **691**, 16 (2009). doi:10.1088/0004-637X/691/1/16
51. Shafee, R., McKinney, J.C., Narayan, R., et al.: Three-dimensional simulations of magnetized thin accretion disks around black holes: stress in the plunging region. Astrophys. J. **687**, L25 (2008). doi:10.1086/593148
52. Reynolds, C.S., Fabian, A.C.: Broad iron-K$\alpha$ emission lines as a diagnostic of black hole spin. Astrophys. J. **675**, 1048 (2008). doi:10.1086/527344
53. Penna, R.F., McKinney, J.C., Narayan, R., et al.: Simulations of magnetized discs around black holes: effects of black hole spin, disc thickness and magnetic field geometry. Mon. Not. Roy. astr. Soc. **408**, 752 (2010). doi:10.1111/j.1365-2966.2010.17170.x
54. Abramowicz, M., Brandenburg, A., Lasota, J.P.: The dependence of the viscosity in accretion discs on the shear/vorticity ratio. Mon. Not. Roy. astr. Soc. **281**, L21 (1996)
55. R.F. Penna, A. Sadowski, A.K. Kulkarni, R. Narayan, The Shakura-Sunyaev viscosity prescription with variable $\alpha(r)$, ArXiv e-prints arXiv:1211.0526 [astro-ph.HE] (2012)
56. Krolik, J.H., Hawley, J.F.: Where is the inner edge of an accretion disk around a black hole? Astrophys. J. **573**, 754 (2002). doi:10.1086/340760
57. Abramowicz, M.A., Kluźniak, W.: A precise determination of black hole spin in GRO J1655–40. A&A **374**, L19 (2001). doi:10.1051/0004-6361:20010791

58. Strohmayer, T.E.: Discovery of a 450 Hz quasi-periodic oscillation from the microquasar GRO J1655-40 with the Rossi X-ray timing explorer. Aastrophys. J. Letters **552**, L49 (2001). doi:10.1086/320258
59. Barret, D., Kluźniak, W., Olive, J.F., Paltani, S., Skinner, G.K.: On the high coherence of kHz quasi-periodic oscillations. Mon. Not. Roy. astr. Soc. **357**, 1288 (2005). doi:10.1111/j.1365-2966.2005.08734.x
60. Barret, D., Olive, J.F., Miller, M.C.: An abrupt drop in the coherence of the lower kHz quasi-periodic oscillations in 4U 1636–536. Mon. Not. Roy. astr. Soc. **361**, 855 (2005). doi:10.1111/j.1365-2966.2005.09214.x
61. Bursa, M., Abramowicz, M.A., Karas, V., Kluźniak, W., Schwarzenberg-Czerny, A.: The timescale of encircling light. In: Hledík, S., Stuchlík, Z. (eds.) Proceedings of RAGtime 8/9: Workshops on Black Holes and Neutron Stars, pp. 21–25 (2007)
62. Straub, O., Vincent, F.H., Abramowicz, M.A., Gourgoulhon, E., Paumard, T.: Modelling the black hole silhouette in Sagittarius A* with ion tori. A&A **543**, A83 (2012). doi:10.1051/0004-6361/201219209

# Energy Extraction from Spinning Black Holes Via Relativistic Jets

**Ramesh Narayan, Jeffrey E. McClintock and Alexander Tchekhovskoy**

**Abstract** It has for long been an article of faith among astrophysicists that black hole spin energy is responsible for powering the relativistic jets seen in accreting black holes. Two recent advances have strengthened the case. First, numerical general relativistic magnetohydrodynamic simulations of accreting spinning black holes show that relativistic jets form spontaneously. In at least some cases, there is unambiguous evidence that much of the jet energy comes from the black hole, not the disk. Second, spin parameters of a number of accreting stellar-mass black holes have been measured. For ballistic jets from these systems, it is found that the radio luminosity of the jet correlates with the spin of the black hole. This suggests a causal relationship between black hole spin and jet power, presumably due to a generalized Penrose process.

## 1 Introduction

Relativistic jets are a common feature of accreting black holes (BHs). They are found in both stellar-mass BHs and supermassive BHs, and are often very powerful. Understanding how jets form and where they obtain their enormous power is an active area of research in astrophysics.

R. Narayan (✉) · J. E. McClintock
Harvard-Smithsonian Center for Astrophysics, Harvard University,
60 Garden St, Cambridge, MA 02138, USA
e-mail: rnarayan@cfa.harvard.edu

J. E. McClintock
e-mail: jmcclintock@cfa.harvard.edu

A. Tchekhovskoy
Department of Astrophysical Sciences, Peyton Hall, Princeton University,
Princeton, NJ 08544, USA
e-mail: atchekho@princeton.edu

J. Bičák and T. Ledvinka (eds.), *General Relativity, Cosmology and Astrophysics*,
Fundamental Theories of Physics 177, DOI: 10.1007/978-3-319-06349-2_25,

© Springer International Publishing Switzerland 2014

In seminal work, Penrose [1] showed that a spinning BH has free energy that is, in principle, available to be tapped. This led to the popular idea that the energy source behind relativistic jets might be the rotational energy of the accreting BH. A number of astrophysical scenarios have been described in which magnetic fields enable this process [2–11]. Field lines are kept confined around the BH by an accretion disk, and the rotation of space-time near the BH twists these lines into helical magnetic springs which expand under their own pressure, accelerating any attached plasma. Energy is thereby extracted from the spinning BH and is transported out along the magnetic field, making a relativistic jet. Although this mechanism requires accretion of magnetized fluid and is thus not the same as Penrose's original proposal,[1] we will still refer to it as the "generalized Penrose process" since ultimately the energy comes from the spin of the BH.

It is not easy to prove that the generalized Penrose process is necessarily in operation in a given jet. The reason is that jets are always associated with accretion disks, and the accretion process itself releases gravitational energy, some of which might flow into the jet. Let us define a jet efficiency factor $\eta_{\rm jet}$,

$$\eta_{\rm jet} = \frac{\langle P_{\rm jet}\rangle}{\langle \dot{M}(r_{\rm H})\rangle c^2}, \tag{1}$$

where $\langle P_{\rm jet}\rangle$ is the time-average power flowing out through the jet and $\langle \dot{M}(r_{\rm H})\rangle c^2$ is the time-average rate at which rest-mass energy flows in through the BH horizon. Many jets, both those observed and those seen in computer simulations, have values of $\eta_{\rm jet}$ quite a bit less than unity. With such a modest efficiency, the jet power could easily come from the accretion disk [13–15].

The situation has improved considerably in the last couple of years. As we show in Sect. 2, numerical simulations have now been carried out where it can be demonstrated beyond reasonable doubt that the simulated jet obtains power directly from the BH spin energy. Furthermore, as we discuss in Sect. 3, the first observational evidence for a correlation between jet power and BH spin has finally been obtained. The correlation appears to favor a Penrose-like process being the energy source of jets.

## 2 Computer Simulations of Black Hole Accretion and Jets

For the last decade or so, it has been possible to simulate numerically the dynamics of MHD accretion flows in the fixed Kerr metric of a spinning BH. The dynamics

---

[1] Penrose considered a simple model in which particles on negative energy orbits fall into a spinning BH. Wagh and Dadhich [12] extended the analysis to discrete particle accretion in the presence of a magnetic field, which introduces additional interesting effects. We do not discuss these particle-based mechanisms, but focus purely on fluid dynamical processes within the magnetohydrodynamic (MHD) approximation. We also do not discuss an ongoing controversy on whether or not different mechanisms based on magnetized fluids differ from one another [6].

of the magnetized fluid are described using the general relativistic MHD (GRMHD) equations in a fixed space-time, and the simulations are carried out in 3D in order to capture the magnetorotational instability (MRI), the agency that drives accretion [16]. Radiation is usually ignored, but this is not considered a problem since jets are usually found in systems with geometrically thick accretion disks, which are radiatively inefficient. We describe here one set of numerical experiments [10, 17] which have been run using the GRMHD code HARM [18] and which are particularly relevant for understanding the connection between the generalized Penrose process and jets.

As is standard, the numerical simulations are initialized with an equilibrium gas torus orbiting in the equatorial plane of a spinning BH. The torus is initially embedded with a weak seed magnetic field, as shown in panel (a) of Fig. 1. Once the simulation begins, the magnetic field grows as a result of the MRI [16]. This leads to MHD turbulence, which in turn drives accretion of mass and magnetic field into the BH. We define the mass accretion rate,

$$\dot{M}(r) = -\iint_{\theta,\varphi} \rho u^r dA_{\theta,\varphi}, \tag{2}$$

where the integration is over all angles on a sphere of radius $r$ (Boyer-Lindquist or Kerr-Schild coordinates), $dA_{\theta,\varphi} = \sqrt{-g}d\theta d\varphi$ is the surface area element, $\rho$ is the density, $u^r$ is the contravariant radial component of 4-velocity, and $g$ is the determinant of the metric. The sign in Eq. (2) is chosen such that $\dot{M} > 0$ corresponds to mass inflow. One is usually interested in the mass accretion rate at the horizon, $\dot{M}(r = r_H)$. In computing $\dot{M}(r_H)$ from the simulations, one waits until the system has reached approximate steady state. One then computes $\dot{M}(r_H)$ over a sequence of many snapshots in time and then averages to eliminate turbulent fluctuations. This gives the time-average mass accretion rate $\langle \dot{M}(r_H) \rangle$.

Panels (b)–(d) in Fig. 1 show the time evolution of the accretion flow and jet in a simulation with BH spin $a_* \equiv a/M = 0.99$, where $M$ is the BH mass [10]. The steady accretion of magnetized fluid causes magnetic field to accumulate in the inner regions near the BH. After a while, the field becomes so strong that it compresses the inner part of the otherwise geometrically thick accretion flow into a thin sheet (panel b). The effect is to obstruct the accretion flow (panels c and d), leading to what is known as a magnetically-arrested disk [19, 20] or a magnetically choked accretion flow [21]. The strong field extracts BH spin energy and forms a powerful outflow. To understand the energetics, consider the rate of flow of energy,

$$\dot{E}(r) = \iint_{\theta,\varphi} T^r_t \, dA_{\theta,\varphi}, \tag{3}$$

where the stress-energy tensor of the magnetized fluid is

$$T^\mu_\nu = \left(\rho + u_g + p_g + \frac{b^2}{4\pi}\right) u^\mu u_\nu + \left(p + \frac{b^2}{8\pi}\right)\delta^\mu_\nu - \frac{b^\mu b_\nu}{4\pi}, \tag{4}$$

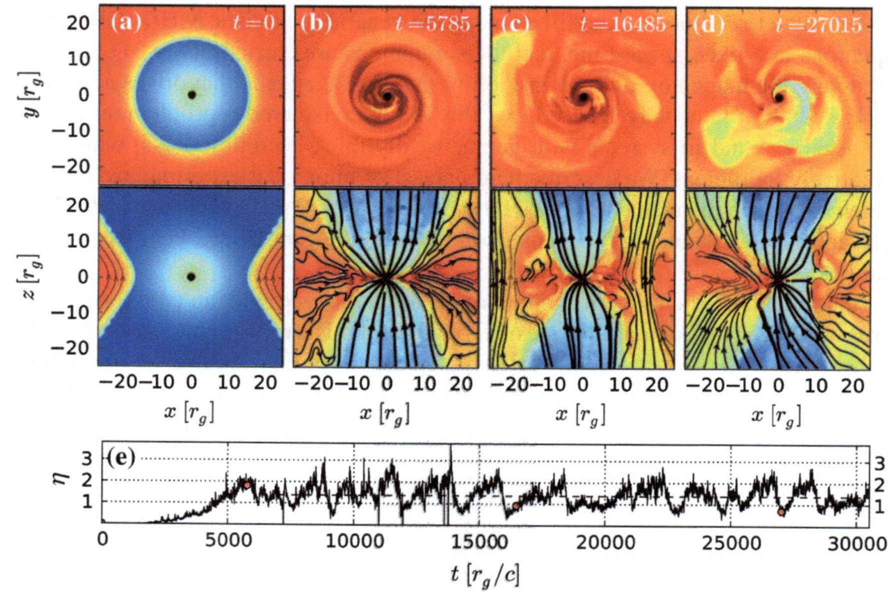


**Fig. 1** Formation of a magnetically-arrested disk and ejection of powerful jets in a GRMHD simulation of magnetized accretion on to a rapidly spinning BH with $a_* = 0.99$ [10]. The *top* and *bottom* rows in panels (**a**-**d**) show a time sequence of equatorial and meridional slices through the accretion flow. *Solid lines* show magnetic field lines in the image plane, and color shows $\log \rho$ (*red* high, *blue* low). The simulation starts with an equilibrium torus embedded with a weak magnetic field (panel a). The weakly magnetized orbiting gas is unstable to the MRI, which causes gas and field to accrete. As large-scale magnetic flux accumulates at the center, a coherent bundle of field lines forms at the center, which threads the BH and has the configuration of bipolar funnels along the (vertical) BH rotation axis. These funnels contain strong field and low mass density (lower panels b, c, d). Helical twisting of the field lines as a result of dragging of frames causes a powerful outflow of energy through the funnels in the form of twin jets. The outflow efficiency $\eta$ (panel e), calculated as in Eq. (5), becomes greater than unity once the flow achieves quasi-steady state at time $t \gtrsim 5000 r_g/c$. This is the key result of the simulation. Having a time-average $\eta > 1$ means that there is a net energy flow *out* of the BH, i.e., spin energy is extracted from the BH by the magnetized accretion flow. This constitutes a demonstration of the generalized Penrose process in the astrophysically relevant context of a magnetized accretion flow

$u_g$ and $p_g$ are the internal energy and pressure of the gas, $b^\mu$ is the fluid-frame magnetic field 4-vector (see Gammie and McKinney [18] for the definition), and $b^2 = b^\mu b_\mu$ is the square of the fluid-frame magnetic field strength. The sign of Eq. (3) is chosen such that $\dot{E}(r) > 0$ corresponds to energy inflow. Note that $T^r_t$ includes the inflow of rest mass energy via the term $\rho u^r u_t$.

Let us define the efficiency with which the accreting BH produces outflowing energy as

$$\eta = \frac{\dot{M}(r_{\rm H})c^2 - \dot{E}(r_{\rm H})}{\langle \dot{M}(r_{\rm H}) \rangle c^2}, \tag{5}$$

where we have made $\eta$ dimensionless by normalizing the right-hand side by the time-average mass energy accretion rate. To understand the meaning of Eq. (5), consider the simple example of gas falling in radially from infinity, with no radiative or other energy losses along the way. In this case, we have $\dot{E}(r_{\rm H}) = \dot{M}(r_{\rm H})c^2$, i.e., the gas carries an energy equal to its rest mass energy into the BH. Hence $\eta = 0$, as appropriate for this example. For a more realistic accretion flow, some energy is lost by the gas via radiation, winds and jets, and one generally expects the energy flowing into the BH to be less than the rest mass energy: $\dot{E}(r_{\rm H}) < \dot{M}(r_{\rm H})c^2$. This will result in an efficiency $\eta > 0$, where $\eta$ measures the ratio of the energy returned to infinity, $\dot{M}(r_{\rm H})c^2 - \dot{E}(r_{\rm H})$, to the energetic price paid in the form of rest mass energy flowing into the BH, $\dot{M}(r_{\rm H})c^2$.

Usually, $\dot{E}(r_{\rm H})$ is positive, i.e., there is a net flow of energy into the BH through the horizon, and $\eta < 1$. However, there is no theorem that requires this. Penrose's [1] great insight was to realize that it is possible to have $\dot{E}(r_{\rm H}) < 0$ (net *outward* energy flow as measured at the horizon), and thus $\eta > 1$. In the context of an accretion flow, $\dot{E}(r_{\rm H}) < 0$ means that, even though rest mass flows steadily into the BH, there is a net energy flow out of the BH. As a result, the gravitational mass of the BH decreases with time. It is the energy associated with this decreasing mass that enables $\eta$ to exceed unity. Of course, as the BH loses gravitational mass, it also loses angular momentum and spins down. This can be verified by considering the angular momentum flux at the horizon, $\dot{J}(r_{\rm H})$, which may be computed as in Eq. (3) but with $T_t^r$ replaced by $T_\phi^r$ (e.g., Tchekhovskoy and McKinney [22]).

Returning to the simulation under consideration, Fig. 1e shows the outflow efficiency $\eta$ as a function of time. It is seen that the average efficiency exceeds unity once the flow achieves steady state at time $t \gtrsim 5000 r_g/c$, where $r_g = GM/c^2$. The outflow thus carries away more energy than the entire rest mass energy brought in by the accretion flow. This is an unambiguous demonstration of the generalized Penrose process in the astrophysically plausible setting of a magnetized accretion flow on to a spinning BH. Of course, it is not obvious that the energy necessarily flows out in a collimated relativistic jet. The quantity $\eta$ is defined via global integrals (Eqs. 2, 5) and it does not specify exactly where the outflowing energy ends up. A more detailed analysis reveals that the bulk of the energy does indeed go into a relativistic jet, while about 10 % goes into a quasi-relativistic wind [17].

Figure 1 corresponds to an extreme example, viz., a very rapidly spinning BH with $a_* = 0.99$. Figure 2 shows results from a parameter study that investigated the effect of varying $a_*$. It is seen that the time-average $\eta$ increases steeply with increasing $a_*$. For an accretion flow that corotates with the BH, the power going into the jet can be well-fit with a power-law dependence,

$$\eta_{\rm jet} \approx 0.65 a_*^2 (1 + 0.85 a_*^2). \qquad (6)$$

This approximation remains accurate to within 15 % for $0.3 \le a_* \le 1$. For low spins, the net efficiency derived from the simulations is greater than that predicted by Eq. (6). For example, as Fig. 2 shows, the simulation gives a non-zero value of $\eta$ for $a_* = 0$, which is inconsistent with Eq. (6). This is because, for $a_* = 0$, all the

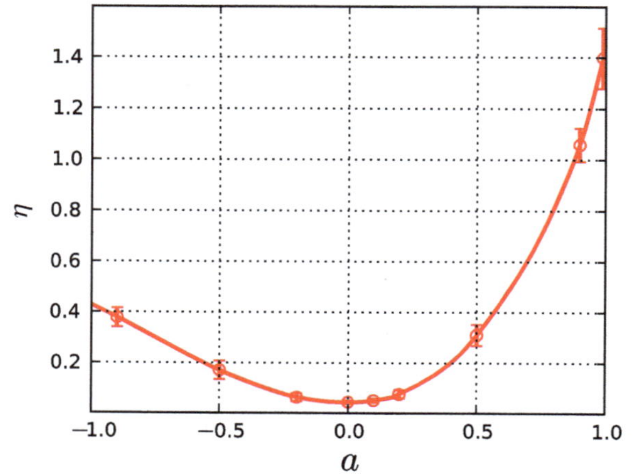


**Fig. 2** Time-average outflow efficiency $\eta$ versus BH spin parameter $a_*$ for a sequence of GRMHD simulations of non-radiative BH accretion flows [17]. The efficiency exceeds unity for $a_* \gtrsim 0.9$. Negative values of $a_*$ correspond to the accretion flow counter-rotating with respect to the BH

outflow energy comes directly from the accretion flow, most of which goes into a wind. Nothing comes from the BH, whereas Eq. (6) refers specifically to the efficiency $\eta_{\rm jet}$ associated with jet power from the BH. With increasing BH spin, both the disk and the hole contribute to energy outflow, with the latter becoming more and more dominant. For spin values $a_* > 0.9$, the BH's contribution is so large that the net efficiency exceeds unity.

Before leaving this topic, we note that other numerical simulations have used geometrically thicker accretion configurations than the one shown in Fig. 1 and find even larger values of $\eta$ [21, 23].

## 3 Empirical Evidence for the Generalized Penrose Process

As discussed in Sect. 2, there is definite evidence from computer simulations that the generalized Penrose process is feasible, and even quite plausible, with magnetized accretion flows. We discuss here recent progress on the observational front. In Sect. 3.1 we briefly summarize efforts to measure spin parameters of astrophysical BHs. Then in Sect. 3.2 we discuss a correlation that has been found between jet power and BH spin. Finally in Sect. 3.3 we explain why we think the observational evidence favors a Penrose-like process rather than disk power.

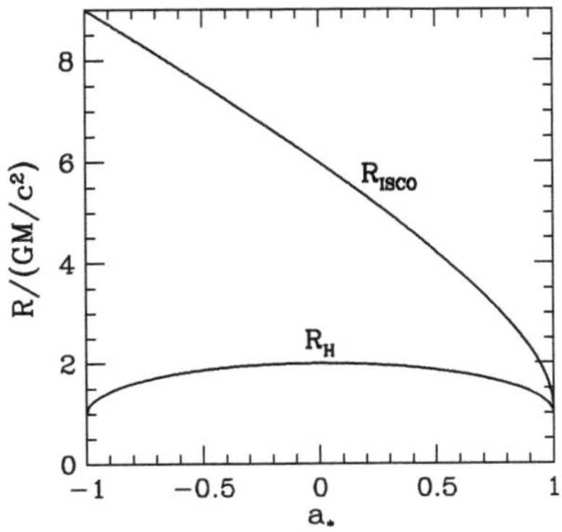


**Fig. 3** Radius of the ISCO $R_{\rm ISCO}$ and of the horizon $R_{\rm H}$ in units of $GM/c^2$ plotted as a function of the black hole spin parameter $a_*$. Negative values of $a_*$ correspond to retrograde orbits. Note that $R_{\rm ISCO}$ decreases monotonically from $9GM/c^2$ for a retrograde orbit around a maximally spinning black hole, to $6GM/c^2$ for a non-spinning black hole, to $GM/c^2$ for a prograde orbit around a maximally spinning black hole

## *3.1 Spin Parameters of Stellar-Mass Black Holes*

In 1989, the first practical approach to measuring black hole spin was suggested [24], viz., modeling the relativistically-broadened Fe K emission line emitted from the inner regions of an accretion disk. The first compelling observation of such a line was reported 6 years later [25]. Presently, the spins of more than a dozen black holes have been estimated by modeling the Fe K line (see Reynolds et al. [26] for a recent review).

In 1997, a second approach to measuring black hole spin, the "continuum-fitting method," was proposed [27]. In this method, one fits the thermal continuum spectrum of a black hole's accretion disk to the relativistic model of Novikov and Thorne [28]. One then identifies the inner edge of the modeled disk with the radius $R_{\rm ISCO}$ of the innermost stable circular orbit (ISCO) in the space-time metric. Since $R_{\rm ISCO}$ varies monotonically with respect to the dimensionless BH spin parameter $a_*$ (see Fig. 3), a measurement of the former immediately provides an estimate of the latter.

In 2006, the continuum-fitting method was employed to estimate the spins of three stellar-mass BHs [30, 31]. Seven additional spins have since been measured. Table 1 lists the masses and spins of these ten BHs. Readers are referred to a recent review by the authors [29] for details of the continuum-fitting method and uncertainties in the derived spin estimates.

The continuum-fitting method is simple and demonstrably robust. It does not make many assumptions; those few it makes have nearly all been tested and shown to be valid (see Steiner et al. [29, 32] for details). A significant limitation of the method is that it is only readily applicable to stellar-mass BHs. For such BHs, however, we would argue that it is the method of choice. The Fe K method can be applied to both

**Table 1** The spins and masses of ten stellar-mass black holes [29]

| System | $a_*$ | $M/M_\odot$ |
|---|---|---|
| *Persistent* | | |
| Cygnus X-1 | >0.95 | $15.8 \pm 1.0$ |
| LMC X-1 | $0.92^{+0.05}_{-0.07}$ | $10.9 \pm 1.4$ |
| M33 X-7 | $0.84 \pm 0.05$ | $15.65 \pm 1.45$ |
| *Transient* | | |
| GRS 1915+105 | >0.95 | $10.1 \pm 0.6$ |
| 4U 1543–47 | $0.80 \pm 0.10$ | $9.4 \pm 1.0$ |
| GRO J1655–40 | $0.70 \pm 0.10$ | $6.3 \pm 0.5$ |
| XTE J1550-564 | $0.34 \pm 0.24$ | $9.1 \pm 0.6$ |
| H1743-322 | $0.2 \pm 0.3$ | ~8 |
| LMC X-3 | <0.3 | $7.6 \pm 1.6$ |
| A0620-00 | $0.12 \pm 0.19$ | $6.6 \pm 0.25$ |

stellar-mass and supermassive BHs. For the latter, it is the only method currently available.

## 3.2 Correlation Between Black Hole Spin and Jet Radio Power

The 10 stellar-mass BHs in Table 1 are divided into two classes: "persistent" sources, which are perennially bright in X-rays at a relatively constant level, and "transient" sources, which have extremely large amplitude outbursts. During outburst, the transient sources generally reach close to the Eddington luminosity limit (see [33] for a quantitative discussion of this point). Close to the peak, these systems eject blobs of plasma that move ballistically outward at relativistic speeds (Lorentz factor $\Gamma > 2$). These ballistic jets are often visible in radio and sometimes in X-rays out to distances of order a parsec from the BH, i.e., to distances $> 10^{10}GM/c^2$. Because ballistic jets resemble the kiloparsec-scale jets seen in quasars, stellar-mass BHs that produce them are called microquasars [34].

On general principles, one expects jet power to depend on the BH mass $M$, its spin $a_*$, and the mass accretion rate $\dot{M}$ (plus perhaps other qualitative factors such as the topology of the magnetic field [35, 36]). If one wishes to investigate the dependence of jet power on $a_*$, one needs first to eliminate the other two variables. Ballistic jets from transient stellar-mass BHs are very well-suited for this purpose. First, the BH masses are similar to better than a factor of two (see Table 1). Second, all these sources have similar accretion rates, close to the Eddington limit, at the time they eject their ballistic jets [33]. This leaves $a_*$ as the only remaining variable.

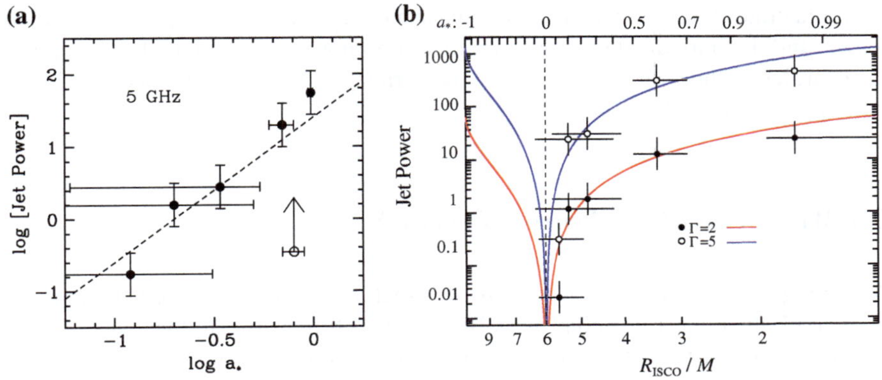


**Fig. 4** **a** Plot of jet power, estimated from 5 GHz radio flux at light curve maximum, versus black hole spin, measured via the continuum-fitting method, for five transient stellar-mass BHs [33, 37]. The *dashed line* has slope fixed to 2 (see Eq. 6) and is not a fit. **b** Plot of jet power versus $R_{\rm ISCO}/(GM/c^2)$. Here jet power has been corrected for beaming assuming jet Lorentz factor $\Gamma = 2$ (*filled circles*) or $\Gamma = 5$ (*open circles*). The two *solid lines* correspond to fits of a relation of the form, "Jet Power" $\propto \Omega_{\rm H}^2$, where $\Omega_{\rm H}$ is the angular frequency of the black hole horizon [33]. Note that jet power varies by a factor $\gtrsim 1000$ among the five objects shown

Narayan and McClintock [37] considered the peak radio luminosities of ballistic jet blobs in four transient BHs, A0620-00, XTE J1550-564, GRO J1655-40, GRS 1915+105, and showed that they correlate well with the corresponding black hole spins measured via the continuum-fitting method.[2] Later, Steiner et al. [33] included a fifth BH, H1743-322, whose spin had been just measured. Figure 4a shows the inferred ballistic jet powers of these five objects plotted versus black hole spin. The quantity "Jet Power" along the vertical axis refers to $(\nu S_\nu)D^2/M$, where $\nu$ (= 5 GHz) is the radio frequency at which the measurements are made, $S_\nu$ is the flux density in janskys at this frequency at the peak of the ballistic jet radio light curve, $D$ is the distance in kiloparsecs, and $M$ is the black hole mass in solar units. There is unmistakeable evidence for a correlation between jet power and $a_*$. Although there are only five data points, note that jet power varies by nearly three orders of magnitude as the spin parameter varies from $\approx 0.1$ to 1.

The very unequal horizontal errorbars in Fig. 4a are a feature of the continuum-fitting method of measuring $a_*$. Recall that the method in effect measures $R_{\rm ISCO}$ and then deduces the value of $a_*$ using the mapping shown in Fig. 3. Since the mapping is highly non-linear, especially as $a_* \to 1$, comparable errors in $R_{\rm ISCO}$ correspond to vastly different uncertainties in $a_*$. In addition, the use of $\log a_*$ along the horizontal axis tends to stretch errorbars excessively for low spin values. Figure 4b, based on

[2] In the case of a fifth transient BH, 4U1543-47, radio observations did not include the peak of the light curve, so one could only deduce a lower limit to the jet power, which is shown as an open circle in Fig. 4a.

[33], illustrates these effects. Here the horizontal axis tracks $\log R_{\rm ISCO}$ rather than $\log a_*$, and the horizontal errorbars are therefore more nearly equal. The key point is, regardless of how one plots the data, the correlation between jet power and black hole spin appears to be strong.

### 3.3 Why Generalized Penrose Process?

Assuming the correlation shown in Fig. 4 is real, there are two immediate implications: (i) Ballistic jets in stellar-mass BHs are highly sensitive to the spins of their underlying BHs. (ii) Spin estimates of stellar-mass BHs obtained via the continuum-fitting method are sufficiently reliable to reveal this long-sought connection between relativistic jets and BH spin.

With respect to (i), the mere existence of a correlation does not necessarily imply that the generalized Penrose process is at work. We know that the accretion disk itself is capable of producing a jet-like outflow [13–15]. Furthermore, the gravitational potential well into which an accretion disk falls becomes deeper with increasing BH spin, since the inner radius of the disk $R_{\rm ISCO}$ becomes smaller (Fig. 3). Therefore, a disk-driven jet is likely to become more powerful with increasing spin. Could this be the reason for the correlation between jet power and spin seen in Fig. 4? We consider it unlikely. The radiative efficiency $\eta_{\rm disk}$ of a Novikov-Thorne thin accretion disk increases only modestly with spin; for the spins of the five objects shown in Fig. 4, $\eta_{\rm disk} = 0.061, 0.069, 0.072, 0.10$ and $0.19$, respectively, varying by only a factor of three. Of course, there is no reason why the power of a disk-driven jet should necessarily scale like $\eta_{\rm disk}$. Nevertheless, the fact that $\eta_{\rm disk}$ shows only a factor of three variation makes it implausible that a disk-powered jet could vary in power by three orders of magnitude.

In contrast, any mechanism that taps directly into the BH spin energy via some kind of generalized Penrose process can easily account for the observed variation in jet power. Analytical models of magnetized accretion predict that the jet efficiency factor should vary as $\eta_{\rm jet} \propto a_*^2$ [2, 3] or $\eta_{\rm jet} \propto \Omega_{\rm H}^2$ [38], where $\Omega_{\rm H}$ is the angular frequency of the BH horizon.[3] The dashed line in Fig. 4a corresponds to the former scaling, and the solid lines in Fig. 4b to the latter scaling; Equation (6), which is obtained by fitting simulation results, is intermediate between the two. The observational data agree remarkably well with the predicted scalings, strongly suggesting that the generalized Penrose process is in operation.

We cannot tell whether the energy extraction in the observed systems is mediated specifically by magnetic fields (as in the simulations), since there is no way to observe what is going on near the BH (where the jet is initially launched). Where the ballistic jet blobs are finally observed they are clearly magnetized—it is what enables the charged particles to produce radiation via the synchrotron mechanism—but this is at distances $\sim 10^{10} GM/c^2$.

[3] The two scalings agree for small values of $a_*$, but differ as $a_* \to 1$.

## 4 Summary

In summary, the case for a generalized version of the Penrose process being the power source behind astrophysical jets has become significantly stronger in the last few years. Computer simulations have been very helpful in this regard since they enable one to study semi-realistic configurations of magnetized accretion flows and to explore quantitatively how mass, energy and angular momentum flow through the system. Recent computer experiments find unambiguous indications for energy extraction from spinning BHs via magnetic fields. Whether these simulated models describe real BHs in nature is not yet certain. However, completely independent observational data suggest a link between the spins of transient stellar-mass BHs and the energy output in ballistic jets ejected from these systems. The jet power increases steeply with BH spin (Fig. 4), and the dependence is quite similar to that found both in simple analytical models [2, 3] and in simulations (Fig. 2). Taking all the evidence into account, the authors believe that Penrose's seminal ideas on energy extraction from spinning BHs are relevant for the production of at least some categories of relativistic astrophysical jets.

RN's work was supported in part by NASA grant NNX11AE16G. AT was supported by a Princeton Center for Theoretical Science fellowship and an XSEDE allocation TG-AST100040 on NICS Kraken and Nautilus and TACC Ranch.

## References

1. Penrose, R.: Gravitational collapse: the role of general relativity. Riv. Nuovo Cimento **1**, 252 (1969)
2. Ruffini, R., Wilson, J.R.: Relativistic magnetohydrodynamical effects of plasma accreting into a black hole. Phys. Rev. D **12**, 2959 (1975). doi:10.1103/PhysRevD.12.2959
3. Blandford, R.D., Znajek, R.L.: Electromagnetic extraction of energy from Kerr black holes. Mon. Not. R. Astron. Soc. **179**, 433 (1977)
4. Damour, T., Ruffini, R., Hanni, R.S., Wilson, J.R.: Regions of magnetic support of a plasma around a black hole. Phys. Rev. D **17**, 1518 (1978). doi:10.1103/PhysRevD.17.1518
5. Koide, S., Shibata, K., Kudoh, T., Meier, D.L.: Extraction of black hole rotational energy by a magnetic field and the formation of relativistic jets. Science **295**, 1688 (2002). doi:10.1126/science.1068240
6. Komissarov, S.S.: Blandford-Znajek mechanism versus Penrose process. J. Korean Phys. Soc. **54**, 2503 (2009). doi:10.3938/jkps.54.2503
7. McKinney, J.C., Gammie, C.F.: A measurement of the electromagnetic luminosity of a Kerr black hole. Astrophys. J. **611**, 977 (2004). doi:10.1086/422244
8. McKinney, J.C.: General relativistic magnetohydrodynamic simulations of the jet formation and large-scale propagation from black hole accretion systems. Mon. Not. R. Astron. Soc. **368**, 1561 (2006). doi:10.1111/j.1365-2966.2006.10256.x
9. Beskin, V.S.: MHD Flows in Compact Astrophysical Objects. Astronomy and astrophysics library. Springer, (2010). doi:10.1007/978-3-642-01290-7
10. Tchekhovskoy, A., Narayan, R., McKinney, J.C.: Efficient generation of jets from magnetically arrested accretion on a rapidly spinning black hole. Mon. Not. R. Astron. Soc. **418**, L79 (2011). doi:10.1111/j.1745-3933.2011.01147.x

11. Meier, D.L.: Black Hole Astrophysics: The Engine Paradigm. Praxis Books in Astronomy and Planetary Sciences. Springer, Heidelberg; New York (2012)
12. Wagh, S.M., Dadhich, N.: The energetics of black holes in electromagnetic fields by the Penrose process. Phys. Rep. **183**, 137 (1989). doi:10.1016/0370-1573(89)90156-7
13. Blandford, R.D., Payne, D.G.: Hydromagnetic flows from accretion discs and the production of radio jets. Mon. Not. R. Astron. Soc. **199**, 883 (1982)
14. Ghosh, P., Abramowicz, M.A.: Electromagnetic extraction of rotational energy from disc-fed black holes: the strength of the Blandford-Znajek process. Mon. Not. R. Astron. Soc. **292**, 887 (1997)
15. Livio, M., Ogilvie, G.I., Pringle, J.E.: Extracting energy from black holes: the relative importance of the Blandford-Znajek mechanism. Astrophys. J. **512**, 100 (1999). doi:10.1086/306777
16. Balbus, S.A., Hawley, J.F.: Instability, turbulence, and enhanced transport in accretion disks. Rev. Mod. Phys. **70**, 1 (1998). doi:10.1103/RevModPhys.70.1
17. Tchekhovskoy, A., McKinney, J.C., Narayan, R.: General relativistic modeling of magnetized jets from accreting black holes. J. Phys. Conf. Ser. **372**, 012040 (2012). doi:10.1088/1742-6596/372/1/012040
18. Gammie, C.F., McKinney, J.C., Tóth, G.: HARM: a numerical scheme for general relativistic magnetohydrodynamics. Astrophys. J. **589**, 444 (2003). doi:10.1086/374594
19. Narayan, R., Igumenshchev, I.V., Abramowicz, M.A.: Magnetically arrested disk: an energetically efficient accretion flow. Publ. Astron. Soc. Jpn **55**, L69 (2003)
20. Igumenshchev, I.V.: Magnetically arrested disks and the origin of Poynting jets: a numerical study. Astrophys. J. **677**, 317 (2008). doi:10.1086/529025
21. McKinney, J.C., Tchekhovskoy, A., Blandford, R.D.: General relativistic magnetohydrodynamic simulations of magnetically choked accretion flows around black holes. Mon. Not. R. Astron. Soc. **423**, 3083 (2012). doi:10.1111/j.1365-2966.2012.21074.x
22. Penna, R.F., McKinney, J.C., Narayan, R., et al.: Simulations of magnetized discs around black holes: effects of black hole spin, disc thickness and magnetic field geometry. Mon. Not. R. Astron. Soc. **408**, 752 (2010). doi:10.1111/j.1365-2966.2010.17170.x
23. Tchekhovskoy, A., McKinney, J.C.: Prograde and retrograde black holes: whose jet is more powerful? Mon. Not. R. Astron. Soc. **423**, L55 (2012). doi:10.1111/j.1745-3933.2012.01256.x
24. Fabian, A.C., Rees, M.J., Stella, L., White, N.E.: X-ray fluorescence from the inner disc in Cygnus X-1. Mon. Not. R. Astron. Soc. **238**, 729 (1989)
25. Tanaka, Y., Nandra, K., Fabian, A.C., et al.: Gravitationally redshifted emission implying an accretion disk and massive black hole in the active galaxy MCG-6-30-15. Nature **375**, 659 (1995). doi:10.1038/375659a0
26. Reynolds, C.S., Brenneman, L.W., Lohfink, A.M., et al.: Probing relativistic astrophysics around SMBHs: The Suzaku AGN spin survey. In: Petre, R., Mitsuda, K., Angelini, L. (eds.) SUZAKU 2011: Exploring the X-ray Universe: Suzaku and Beyond, AIP Conference Proceedings, vol. 1427, pp. 157–164. American Institute of Physics, Melville, NY (2012). doi:10.1063/1.3696170
27. Zhang, S.N., Cui, W., Chen, W.: Black hole spin in X-ray binaries: observational consequences. Astrophys. J. Lett. **482**, L155 (1997). doi:10.1086/310705
28. Novikov, I.D., Thorne, K.S.: Astrophysics of Black Holes. In: DeWitt, C., DeWitt, B.S. (eds.) Black Holes. Les Astres Occlus, pp. 343–450. Gordon and Breach, New York; London (1973)
29. McClintock, J.E., Narayan, R.: Black Hole Spin Via Continuum Fitting and the Role of Spin in Powering Transient Jets. Space Sci Rev. (2013). doi:10.1007/s11214-013-0003-9
30. Shafee, R., McClintock, J.E., Narayan, R., et al.: Estimating the spin of stellar-mass black holes by spectral fitting of the X-ray continuum. Astrophys. J. Lett. **636**, L113 (2006). doi:10.1086/498938
31. McClintock, J.E., Shafee, R., Narayan, R., et al.: The spin of the near-extreme Kerr black hole GRS 1915+105. Astrophys. J. **652**, 518 (2006). doi:10.1086/508457
32. McClintock, J.E., Narayan, R., Davis, S.W., et al.: Measuring the spins of accreting black holes. Class. Quantum Grav. **28**, 114009 (2011). doi:10.1088/0264-9381/28/11/114009

33. Steiner, J.F., McClintock, J.E., Narayan, R.: Jet power and black hole spin: testing an empirical relationship and using it to predict the spins of six black holes. Astrophys. J. **762**, 104 (2013). doi:10.1088/0004-637X/762/2/104
34. Mirabel, I.F., Rodríguez, L.F.: Sources of relativistic jets in the Galaxy. Annu. Rev. Astron. Astrophys. **37**, 409 (1999). doi:10.1146/annurev.astro.37.1.409
35. Beckwith, K., Hawley, J.F., Krolik, J.H.: The influence of magnetic field geometry on the evolution of black hole accretion flows: similar disks, drastically different jets. Astrophys. J. **678**, 1180 (2008). doi:10.1086/533492
36. McKinney, J.C., Blandford, R.D.: Stability of relativistic jets from rotating, accreting black holes via fully three-dimensional magnetohydrodynamic simulations. Mon. Not. R. Astron. Soc. **394**, L126 (2009). doi:10.1111/j.1745-3933.2009.00625.x
37. Narayan, R., McClintock, J.E.: Observational evidence for a correlation between jet power and black hole spin. Mon. Not. R. Astron. Soc. **419**, L69 (2012). doi:10.1111/j.1745-3933.2011.01181.x
38. Tchekhovskoy, A., Narayan, R., McKinney, J.C.: Black hole spin and the radio loud/quiet dichotomy of active galactic nuclei. Astrophys. J. **711**, 50 (2010). doi:10.1088/0004-637X/711/1/50

MIX
Papier aus verantwortungsvollen Quellen
Paper from responsible sources
FSC® C105338

If you have any concerns about our products,
you can contact us on
**ProductSafety@springernature.com**

In case Publisher is established outside the EU,
the EU authorized representative is:
**Springer Nature Customer Service Center GmbH**
**Europaplatz 3, 69115 Heidelberg, Germany**

Printed by Libri Plureos GmbH
in Hamburg, Germany